EDA for IC System Design, Verification, and Testing

Electronic Design Automation for Integrated Circuits Handbook

Edited by
Louis Scheffer, Luciano Lavagno, and Grant Martin

EDA for IC System Design, Verification, and Testing

EDA for IC Implementation, Circuit Design, and Process Technology

EDA for IC System Design, Verification, and Testing

Edited by

Louis Scheffer
Cadence Design Systems
San Jose, California, U.S.A.

Luciano Lavagno
Cadence Berkeley Laboratories
Berkeley, California, U.S.A.

Grant Martin
Tensilica Inc.
Santa Clara, California, U.S.A.

CRC Taylor & Francis
Taylor & Francis Group
Boca Raton London New York

A CRC title, part of the Taylor & Francis imprint, a member of the Taylor & Francis Group, the academic division of T&F Informa plc.

Published in 2006 by
CRC Press
Taylor & Francis Group
6000 Broken Sound Parkway NW, Suite 300
Boca Raton, FL 33487-2742

Printed in the United States of America on acid-free paper
10 9 8 7 6 5 4 3 2 1

International Standard Book Number-10: 0-8493-7923-7 (Hardcover)
International Standard Book Number-13: 978-0-8493-7923-9 (Hardcover)
Library of Congress Card Number 2005052924

Library of Congress Cataloging-in-Publication Data

EDA for IC system design, verification, and testing / editors, Louis Scheffer, Luciano Lavagno, Grant Martin.
p. cm. -- (Electronic design and automation for integrated circuits handbook)
Includes bibliographical references and index.
ISBN 0-8493-7923-7
1. Integrated circuits--Computer-aided design. 2. Integrated circuits--Verification--Data processing. I. Title: Electronic design automation for integrated circuit system design, verification, and testing. II. Scheffer, Louis. III. Lavagno, Luciano, 1959- IV. Martin, Grant (Grant Edmund) V. Series.

TK7874.E26 2005
621.3815--dc22 2005052924

Visit the Taylor & Francis Web site at
http://www.taylorandfrancis.com
and the CRC Press Web site at
http://www.crcpress.com

Acknowledgments and Dedication for the EDA Handbook

The editors would like to acknowledge the unsung heroes of EDA, those who have worked to advance the field, in addition to advancing their own personal, corporate, or academic agendas. These are the men and women who have played a variety of key roles — they run the smaller conferences, they edit technical journals, and they serve on standards committees, just to name a few. These largely volunteer jobs do not make anyone rich or famous despite the time and effort that goes into them, but they do contribute mightily to the remarkable and sustained advancement of EDA. Our kudos to these folks, who do not get the credit they deserve.

On a more personal note, Louis Scheffer would like to acknowledge the love, support, encouragement, and help of his wife Lynde, his daughter Lucynda, and his son Loukos. Without them this project would not have been possible.

Luciano Lavagno would like to thank his wife Paola and his daughter Alessandra Chiara for making his life so wonderful.

Grant Martin would like to acknowledge, as always, the love and support of his wife, Margaret Steele, and his two daughters, Jennifer and Fiona.

Preface

Preface for Volume 1

Electronic Design Automation (EDA) is a spectacular success in the art of engineering. Over the last quarter of a century, improved tools have raised designers' productivity by a factor of more than a thousand. Without EDA, Moore's law would remain a useless curiosity. Not a single billion-transistor chip could be designed or debugged without these sophisticated tools — without EDA we would have no laptops, cell phones, video games, or any of the other electronic devices we take for granted.

Spurred on by the ability to build bigger chips, EDA developers have largely kept pace, and these enormous chips can still be designed, debugged, and tested, even with decreasing time-to-market.

The story of EDA is much more complex than the progression of integrated circuit (IC) manufacturing, which is based on simple physical scaling of critical dimensions. EDA, on the other hand, evolves by a series of paradigm shifts. Every chapter in this book, all 49 of them, was just a gleam in some expert's eye just a few decades ago. Then it became a research topic, then an academic tool, and then the focus of a start-up or two. Within a few years, it was supported by large commercial EDA vendors, and is now part of the conventional wisdom. Although users always complain that today's tools are not quite adequate for today's designs, the overall improvements in productivity have been remarkable. After all, in which other field do people complain of *only* a 21% compound annual growth in productivity, sustained over three decades, as did the International Technology Roadmap for Semiconductors in 1999?

And what is the future of EDA tools? As we look at the state of electronics and IC design in 2005–2006, we see that we may soon enter a major period of change in the discipline. The classical scaling approach to ICs, spanning multiple orders of magnitude in the size of devices over the last 40+ years, looks set to last only a few more generations or process nodes (though this has been argued many times in the past, and has invariably been proved to be too pessimistic a projection). Conventional transistors and wiring may well be replaced by new nano- and biologically based technologies that we are currently only beginning to experiment with. This profound change will surely have a considerable impact on the tools and methodologies used to design ICs. Should we be spending our efforts looking at Computer Aided Design (CAD) for these future technologies, or continue to improve the tools we currently use?

Upon further consideration, it is clear that the current EDA approaches have a lot of life left in them. With at least a decade remaining in the evolution of current design approaches, and hundreds of thousands or millions of designs left that must either craft new ICs or use programmable versions of them, it is far too soon to forget about today's EDA approaches. And even if the technology changes to radically new forms and structures, many of today's EDA concepts will be reused and built upon for design of technologies well beyond the current scope and thinking.

Preface

The field of EDA for ICs has grown well beyond the point where any single individual can master it all, or even be aware of the progress on all fronts. Therefore, there is a pressing need to create a snapshot of this extremely broad and diverse subject. Students need a way of learning about the many disciplines and topics involved in the design tools in widespread use today. As design grows multi-disciplinary, electronics designers and EDA tool developers need to broaden their scope. The methods used in one subtopic may well have applicability to new topics as they arise. All of electronics design can utilize a comprehensive reference work in this field.

With this in mind, we invited many experts from across all the disciplines involved in EDA to contribute chapters summarizing and giving a comprehensive overview of their particular topic or field. As might be appreciated, such chapters represent a snapshot of the state of the art in 2004–2005. However, as surveys and overviews, they retain a lasting educational and reference value that will be useful to students and practitioners for many years to come.

With a large number of topics to cover, we decided to split the Handbook into two volumes. Volume One covers system-level design, micro-architectural design, and verification and test. Volume Two covers the classical "RTL to GDS II" design flow, incorporating synthesis, placement and routing, along with related topics; analog and mixed-signal design, physical verification, analysis and extraction, and technology CAD topics for IC design. These roughly correspond to the classical "front-end/back-end" split in IC design, where the front-end (or logical design) focuses on making sure that the design does the right thing, assuming it can be implemented, and the back-end (or physical design) concentrates on generating the detailed tooling required, while taking the logical function as given. Despite limitations, this split has persisted through the years — a complete and correct logical design, independent of implementation, remains an excellent handoff point between the two major portions of an IC design flow. Since IC designers and EDA developers often concentrate on one side of this logical/physical split, this seemed to be a good place to divide the book as well.

In particular, Volume One starts with a general introduction to the topic, and an overview of IC design and EDA. System-level design incorporates many aspects — application-specific tools and methods, special specification and modeling languages, integration concepts including the use of Intellectual Property (IP), and performance evaluation methods; the modeling and choice of embedded processors and ways to model software running on those processors; and high-level synthesis approaches. ICs that start at the system level need to be refined into micro-architectural specifications, incorporating cycle-accurate modeling, power estimation methods, and design planning. As designs are specified and refined, verification plays a key role — and the handbook covers languages, simulation essentials, and special verification topics such as transaction-level modeling, assertion-based verification, and the use of hardware acceleration and emulation as well as emerging formal methods. Finally, making IC designs testable and thus cost-effective to manufacture and package relies on a host of test methods and tools, both for digital and analog and mixed-signal designs.

This handbook with its two constituent volumes is a valuable learning and reference work for everyone involved and interested in learning about electronic design and its associated tools and methods. We hope that all readers will find it of interest and that it will become a well-thumbed resource.

Louis Scheffer
Luciano Lavagno
Grant Martin

Editors

Louis Scheffer

Louis Scheffer received the B.S. and M.S. degrees from Caltech in 1974 and 1975, and a Ph.D. from Stanford in 1984. He worked at Hewlett Packard from 1975 to 1981 as a chip designer and CAD tool developer. In 1981, he joined Valid Logic Systems, where he did hardware design, developed a schematic editor, and built an IC layout, routing, and verification system. In 1991, Valid merged with Cadence, and since then he has been working on place and route, floorplanning systems, and signal integrity issues.

His main interests are floorplanning and deep submicron effects. He has written many technical papers, tutorials, invited talks, and panels, and has served the DAC, ICCAD, ISPD, SLIP, and TAU conferences as a technical committee member. He is currently the general chair of TAU and ISPD, on the steering committee of SLIP, and an associate editor of IEEE Transactions on CAD. He holds five patents in the field of EDA, and has taught courses on CAD for electronics at Berkeley and Stanford. He is also interested in SETI, and serves on the technical advisory board for the Allen Telescope Array at the SETI institute, and is a co-author of the book SETI-2020, in addition to several technical articles in the field.

Luciano Lavagno

Luciano Lavagno received his Ph.D. in EECS from U.C. Berkeley in 1992 and from Politecnico di Torino in 1993. He is a co-author of two books on asynchronous circuit design, of a book on hardware/software co-design of embedded systems, and of over 160 scientific papers.

Between 1993 and 2000, he was the architect of the POLIS project, a cooperation between U.C. Berkeley, Cadence Design Systems, Magneti Marelli and Politecnico di Torino, which developed a complete hardware/software co-design environment for control-dominated embedded systems.

He is currently an Associate Professor with Politecnico di Torino, Italy and a research scientist with Cadence Berkeley Laboratories. He serves on the technical committees of several international conferences in his field (e.g., DAC, DATE, ICCAD, ICCD) and of various workshops and symposia. He has been the technical program and tutorial chair of DAC, and the technical program and general chair of CODES. He has been associate and guest editor of IEEE Transactions on CAD, IEEE Transactions on VLSI and ACM Transactions on Embedded Computing Systems.

His research interests include the synthesis of asynchronous and low-power circuits, the concurrent design of mixed hardware and software embedded systems, as well as compilation tools and architectural design of dynamically reconfigurable processors.

Grant Martin

Grant Martin is a Chief Scientist at Tensilica, Inc. in Santa Clara, California. Before that, Grant worked for Burroughs in Scotland for 6 years; Nortel/BNR in Canada for 10 years; and Cadence Design Systems for 9 years, eventually becoming a Cadence Fellow in their Labs. He received his Bachelors and Masters degrees in Mathematics (Combinatorics and Optimization) from the University of Waterloo, Canada, in 1977 and 1978.

Grant is a co-author of *Surviving the SOC Revolution: A Guide to Platform-Based Design*, 1999, and *System Design with SystemC*, 2002, and a co-editor of the books *Winning the SoC Revolution: Experiences in Real Design*, and *UML for Real: Design of Embedded Real-Time Systems*, June 2003, all published by Springer (originally by Kluwer). In 2004, he co-wrote with Vladimir Nemudrov the first book on SoC design published in Russian by Technosphera, Moscow. Recently, he co-edited *Taxonomies for the Development and Verification of Digital Systems* (Springer, 2005), and *UML for SoC Design* (Springer, 2005).

He has also presented many papers, talks and tutorials, and participated in panels, at a number of major conferences. He co-chaired the VSI Alliance Embedded Systems study group in the summer of 2001, and is currently co-chair of the DAC Technical Programme Committee for Methods for 2005 and 2006. His particular areas of interest include system-level design, IP-based design of system-on-chip, platform-based design, and embedded software. He is a senior member of the IEEE.

Contributors

Iuliana Bacivarov
SLS Group, TIMA Laboratory
Grenoble, France

Mike Bershteyn
Cadence Design Systems, Inc.
Cupertino, California

Shuvra Bhattacharyya
University of Maryland
College Park, Maryland

Joseph T. Buck
Synopsys, Inc.
Mountain View, California

Raul Camposano
Synopsys Inc.
Mountain View, California

Naehyuck Chang
Seoul National University
Seoul, South Korea

Kwang-Ting (Tim) Cheng
University of California
Santa Barbara, California

Alain Clouard
STMicroelectronics
Crolles, France

Marcello Coppola
STMicroelectronics
Grenoble, France

Robert Damiano
Synopsys Inc.
Hillsboro, Oregon

Marco Di Natale
Scuola Superiore S. Anna
Pisa, Italy

Nikil Dutt
Donald Bren School of Information and
Computer Sciences,
University of California, Irvine
Irvine, California

Stephen A. Edwards
Columbia University
New York, New York

Limor Fix
Design Technology, Intel
Pittsburgh, Pennsylvania

Harry Foster
Jasper Design Automation
Mountain View, California

Frank Ghenassia
STMicroelectronics
Crolles, France

Miltos D. Grammatikakis
ISD S.A.
Athens, Greece

Rajesh Gupta
University of California, San Diego
San Diego, California

Sumit Gupta
Tensilica Inc.
Santa Clara, California

Ahmed Jerraya
SLS Group, TIMA Laboratory, INPG
Grenoble, France

Bozena Kaminska
Simon Fraser University and
Pultronics Incorporated
Burnaby, British Colombia,
Canada

Bernd Koenemann
Mentor Graphics, Inc.
San Jose, California

Luciano Lavagno
Cadence Berkeley Laboratories
Berkeley, California

Steve Leibson
Tensilica, Inc.
Santa Clara, California

Enrico Macii
Politecnico di Torino
Torino, Italy

Laurent Maillet-Contoz
STMicroelectronics
Crolles, France

Erich Marschner
Cadence Design Systems
Berkeley, California

Grant Martin
Tensilica Inc.
Santa Clara, California

Ken McMillan
Cadence Berkeley Laboratories
Berkeley, California

Renu Mehra
Synopsys, Inc.
Mountain View, California

Prabhat Mishra
University of Florida
Gainesville, Florida

Ralph H.J.M. Otten
Eindhoven University of Technology
Eindhoven, Netherlands

Massimo Poncino
Politecnico di Torino
Torino, Italy

John Sanguinetti
Forte Design Systems, Inc.
San Jose, California

Louis Scheffer
Cadence Design Systems
San Jose, California

Sandeep Shukla
Virginia Tech
Blacksburg, Virginia

Gaurav Singh
Virginia Tech
Blacksburg, Virginia

Jean-Philippe Strassen
STMicroelectronics
Crolles, France

Vivek Tiwari
Intel Corp.
Santa Clara, California

Ray Turner
Cadence Design Systems
San Jose, California

Li-C. Wang
University of California
Santa Barbara, California

John Wilson
Mentor Graphics
Berkshire, United Kingdom

Wayne Wolf
Princeton University
Princeton, New Jersey

Contents

SECTION IV Logical Verification

1 Overview

Luciano Lavagno
Cadence Berkeley Laboratories
Berkeley, California

Grant Martin
Tensilica Inc.
Santa Clara, California

Louis Scheffer
Cadence Design Systems
San Jose, California

Introduction to Electronic Design Automation for Integrated Circuits

Modern integrated circuits (ICs) are enormously complicated, often containing many millions of devices. Design of these ICs would not be humanly possible without software (SW) assistance at every stage of the process. The tools used for this task are collectively called electronic design automation (EDA).

EDA tools span a very wide range, from purely logical tools that implement and verify functionality, to purely physical tools that create the manufacturing data and verify that the design can be manufactured. The next chapter, *The IC Design Process and EDA*, by Robert Damiano and Raul Camposano, discusses the IC design process, its major stages and design flow, and how EDA tools fit into these processes and flows. It particularly looks at interfaces between the major IC design stages and the kind of information — abstractions upwards, and detailed design and verification information downwards — that must flow between these stages.

A Brief History of Electronic Design Automation

This section contains a *very* brief summary of the origin and history of EDA for ICs. For each topic, the title of the relevant chapter(s) is mentioned in *italics*.

The need for tools became clear very soon after ICs were invented. Unlike a breadboard, ICs cannot be modified easily after fabrication, so testing even a simple change involves weeks of delay (for new masks and a new fabrication run) and considerable expense. Furthermore, the internal nodes of an IC are difficult to probe because they are physically small and may be covered by other layers of the IC. Even if these problems can be worked around, the internal nodes often have very high impedances and hence are difficult to measure without dramatically changing the performance. Therefore circuit simulators were crucial to IC design almost as soon as ICs came into existence. These programs are covered in the chapter *Analog Simulation: Circuit Level and Behavioral Level*, and appeared in the 1960s.

Next, as the circuits grew bigger, clerical help was required in producing the masks. At first there were digitizing programs, where the designer still drew with colored pencils but the coordinates were transferred to the computer, written to magnetic tape, and then transferred to the mask making machines. Soon, these early programs were enhanced into full-fledged layout editors. These programs were first developed in the late 1960s and early 1970s. Analog designs in the modern era are still largely laid out manually, with some tool assistance, as *Layout Tools for Analog ICs and Mixed-Signal SoCs: A Survey* will attest, although some developments in more automated optimization have been occurring, along with many experiments in more automated layout techniques.

As the circuits grew larger, getting the logic design correct became difficult, and *Digital Simulation* (i.e., logic simulation) was introduced into the IC design flow. Also, testing of the completed chip proved to be difficult, since unlike circuit boards, internal nodes could not be observed or controlled through a "bed of nails" fixture. Therefore automatic test pattern generation (ATPG) programs were developed that generate test vectors that only refer to the visible pins. Other programs that modified designs to make them more controllable, observable, and testable were not far behind. These programs, covered in *Design-for-Test* and *Automatic Test Pattern Generation*, were first available in the mid-1970s. Specialized *Analog and Mixed-Signal Test* needs were met by special testers and tools.

As the number of design rules, number of layers, and chip sizes all continued to increase, it became increasingly difficult to verify by hand that a layout met all the manufacturing rules, and to estimate the parasitics of the circuit. Therefore *Design Rule Checking*, and *Layout Extraction* programs were developed, starting in the mid-1970s. As the processes became more complex, with more layers of interconnect, the original analytic approximations to R, C, and L values became inadequate, and *High-Accuracy Parasitic Extraction* programs were required to determine more accurate values, or at least calibrate the parameter extractors.

The next bottleneck was doing the detailed designing of each polygon. Placement and routing programs allowed the user to specify only the gate-level netlist — the computer would then decide on the location of the gates and the wires connecting them. Although some silicon efficiency was lost, productivity was greatly improved, and IC design opened up to a wider audience of logic designers. The chapters *Digital Layout — Placement* and *Routing* cover these programs, which became popular in the mid-1980s.

Even just the gate-level netlist soon proved to be of too much detail, and synthesis tools were developed to create such a netlist from a higher level specification, usually expressed in a hardware description language (HDL). This is called *Logic Synthesis* and became available in the mid-1980s. In the last decade, *Power Analysis and Optimization from Circuit to Register Transfer Levels* has become a major area of concern and is becoming the number one optimization criterion for many designs, especially portable and battery powered ones.

Around this time, the large collection of tools that need to be used to complete a single design became a serious problem. Electronic design automation *Design Databases* were introduced to cope with this problem. In addition, *Design Flows* began to become more and more elaborate in order to hook tools together, as well as to develop and support both methodologies and use models for specific design groups, companies, and application areas.

In the late 1990s, as the circuits continued to shrink, noise became a serious problem. Programs that analyzed power and ground networks, cross-talk, and substrate noise in systematic ways became commercially available. The chapters *Design and Analysis of Power Supply Networks*, *Mixed-Signal Noise Coupling in System-on-Chip Design: Modeling, Analysis and Validation*, and *Noise Considerations in Digital ICs* cover these topics.

Gradually through the 1990s and early 2000s, chips and processes became sufficiently complex that the designs that optimize yield were no longer only a minimization of size. *Design for Manufacturability in the Nanometer Era*, otherwise known as "Design for Yield", became a field of its own. Also in this time frame, the size of the features on the chip became comparable to, or less than, the wavelength of the light used to create them. To compensate for this as much as possible, the masks were no longer a direct copy of what the designer intended. The creation of these more complex masks is covered in *Resolution Enhancement Techniques and Mask Data Preparation.*

On a parallel track, developing the process itself was also a difficult problem. *Process Simulation* tools were developed to predict the effects of changing various process parameters. The output from these programs, such as doping profiles, was useful to process engineers but too detailed for electrical analysis. Another suite of tools (see *Device Modeling — from Physics to Electrical Parameter Extraction*) that predict device performance from a physical description of devices was needed and developed. These models were particularly useful when developing a new process.

One of the areas that developed very early in the design of electronic systems, at least in part, but which is the least industrialized as a standard process, is that of system-level design. As the chapter on *Using Performance Metrics to Select Microprocessor Cores for IC Designs* points out, one of the first instruction set simulators appeared soon after the first digital computers did. However, until the present day, system-level design has consisted mainly of a varying collection of tricks, techniques, and *ad hoc* modeling tools.

The logic simulation and synthesis processes introduced in the 1970s and 1980s, respectively, are, as was discussed earlier, much more standardized. The front–end IC design flow would not have been possible to standardize without the introduction of standard HDLs. Out of a huge variety of HDLs introduced from the 1960s to the 1980s, Verilog and VHDL have become the major *Design and Verification Languages.* For a long time — till the mid to late 1990s, verification of digital design seemed stuck at standard digital simulation — although at least since the 1980s, a variety of *Hardware Acceleration and Emulation* solutions have been available to designers. However, advances in verification languages and the growth in design complexity have triggered interest in more advanced verification methods, and the last decade has seen considerable interest in *Using Transactional Level Models in a SoC Design Flow, Assertion-based Verification,* and *Formal Property Verification. Equivalence Checking* has been the formal technique most tightly integrated into design flows, since it allows designs to be compared before and after various optimizations and back-end-related modifications, such as scan insertion.

For many years, specific systems design domains have fostered their own application-specific *Tools and Methodologies for System-Level Design* — especially in the areas of algorithm design from the late 1980s through to this day. The late 1990s saw the emergence of and competition between a number of C/C++-based *System-Level Specification and Modeling Languages.* With the possibility of now incorporating the major functional units of a design (processors, memories, digital and mixed-signal HW blocks, peripheral interfaces, and complex hierarchical buses) all onto a single silicon substrate, the mid-1990s to the present day have also seen the rise of the System-on-chip (SoC). It is thus that the area of *SoC Block-Based Design and IP Assembly* has grown, in which the complexity possible with advanced semiconductor processes is ameliorated to some extent via reuse of blocks of design. Concomitant with the SoC approach has been the development, during the last decade, of *Performance Evaluation Methods for MPSoC Design,* development of embedded processors through specialized *Processor Modelling and Design Tools,* and gradual and still-forming links to *Embedded Software Modelling and Design.* The desire to raise HW design productivity to higher levels has spawned considerable interest in (*Parallelizing*) *High Level Synthesis* over the years. It is now seeing something of a resurgence driven by C/C++/SystemC as opposed to the first-generation high-level synthesis (HLS) tools driven by HDLs in the mid-1990s.

After the system level of design, architects need to descend one level of abstraction to the micro-architectural level. Here, a variety of tools allow one to look at the three main performance criteria: timing or delay (*Cycle-accurate System-Level Modeling and Performance Evaluation*), power (*Micro-Architectural Power Estimation and Optimization*), and physical *Design Planning.* Micro-architects need to make trade-offs between the timing, power, and cost/area attributes of complex ICs at this level.

The last several years have seen a considerable infilling of the design flow with a variety of complementary tools and methods. Formal verification of function is only possible if one is assured that the timing is correct, and by keeping a lid on the amount of dynamic simulation required, especially at the postsynthesis and postlayout gate levels, good *Static Timing Analysis* tools provide the assurance that timing constraints are being met. It is also an underpinning to timing optimization of circuits and for the design of newer mechanisms for manufacturing and yield. Standard cell-based placement and routing are not appropriate for *Structured Digital Design* of elements such as memories and register files, leading to specialized tools. As design groups began to rely on foundries and application specific integrated circuit (ASIC) vendors and as the IC design and manufacturing industry began to "de-verticalize", design libraries, covered in *Exploring Challenges of Libraries for Electronic Design*, became a domain for special design flows and tools. It ensured the availability of a variety of high performance and low power libraries for optimal design choices and allowed some portability of design across processes and foundries. *Tools for Chip-Package Codesign* began to link more closely the design of IOs on chip, the packages they fit into, and the boards on which they would be placed. For implementation "fabrics" such as field-programmable gate arrays (FPGAs), specialized *FPGA Synthesis and Physical Design Tools* are necessary to ensure good results. And a renewed emphasis on *Design Closure* allows a more holistic focus on the simultaneous optimization of design timing, power, cost, reliability, and yield in the design process.

Another area of growing but specialized interest in the analog design domain is the use of new and higher level modeling methods and languages, which are covered in *Simulation and Modeling for Analog and Mixed-Signal Integrated Circuits.*

A much more detailed overview of the history of EDA can be found in [1]. A historical survey of many of the important papers from the International Conference on Computer-Aided Design (ICCAD) can be found in [2].

Major Industry Conferences and Publications

The EDA community, formed in the early 1960s from tool developers working for major electronics design companies such as IBM, AT&T Bell Labs, Burroughs, Honeywell, and others, has long valued workshops, conferences, and symposia, in which practitioners, designers, and later, academic researchers, could exchange ideas and practically demonstrate the techniques. The Design Automation Conference (DAC) grew out of workshops, which started in the early 1960s, and although held in a number of U.S. locations, has in recent years tended to stay on the west coast of the United States or a bit inland. It is the largest combined EDA trade show and technical conference held annually anywhere in the world. In Europe, a number of country-specific conferences held sporadically through the 1980s, and two competing ones held in the early 1990s, led to the creation of the consolidated Design Automation and Test in Europe (DATE) conference, which started in the mid-1990s and has grown consistently in strength ever since. Finally, the Asia-South Pacific DAC (ASP-DAC) started in the mid to late 1990s and completes the trio of major EDA conferences spanning the most important electronics design communities in the world.

Complementing the larger trade show/technical conferences has been ICCAD, which for over 20 years has been held in San Jose, and has provided a more technical conference setting for the latest algorithmic advances in EDA to be presented, attracting several hundred attendees. Various domain areas of EDA knowledge have sparked a number of other workshops, symposia, and smaller conferences over the last 15 years, including the International Symposium on Physical Design (ISPD), International Symposium on Quality in Electronic Design (ISQED), Forum on Design Languages in Europe (FDL), HDL and Design and Verification conferences (HDLCon, DVCon), High-level Design, Verification and Test (HLDVT), International Conference on Hardware–Software Codesign and System Synthesis (CODES+ISSS), and many other gatherings. Of course, the area of Test has its own long-standing International Test Conference (ITC); similarly, there are specialized conferences for FPGA design (e.g., Forum on Programmable Logic [FPL]) and a variety of conferences focusing on the most advanced IC designs such as the International Solid–State Circuits Conference (ISSCC) and its European counterpart (ESSCC).

There are several technical societies with strong representation of design automation: one is the Institute of Electrical and Electronics Engineers (IEEE, pronounced "eye-triple-ee"), and the other is the Association for Computing Machinery (ACM).

Various IEEE and ACM transactions contain major work on algorithms and design techniques in print — a more archival-oriented format than conference proceedings. Among these, the IEEE Transactions on computer-aided design (CAD), the IEEE Transactions on VLSI systems, and the ACM Transactions on Design Automation of Electronic Systems are notable. A more general readership magazine devoted to Design and Test and EDA topics is IEEE Design and Test.

As might be expected, the EDA community has a strong online presence. All the conferences mentioned above have web pages describing locations, dates, manuscript submission and registration procedures, and often detailed descriptions of previous conferences. The journals above offer online submission, refereeing, and publication. Online, the IEEE (http://ieee.org), ACM (http://acm.org), and *CiteSeer* (http://citeseer.ist.psu.edu) offer extensive digital libraries, which allow searches through titles, abstracts, and full text. Both conference proceedings and journals are available. Most of the references found in this volume, at least those published after 1988, can be found in at least one of these libraries.

Structure of the Book

In the simplest case of digital design, EDA can be divided into system-level design, micro-architecture design, logical verification, test, synthesis-place-and-route, and physical verification. System-level design is the task of determining which components (bought and built, HW and SW) should comprise a system that can do what one wants. Micro-architecture design fills out the descriptions of each of the blocks, and sets the main parameters for their implementation. Logical verification verifies that the design does what is intended. Test ensures that functional and nonfunctional chips can be told apart reliably, and inserts testing circuitry if needed to ensure that this is the case. Synthesis, place, and route take the logical description, and map it into increasingly detailed physical descriptions, until the design is in a form that can be built with a given process. Physical verification checks that the design is manufacturable and will be reliable. In general, each of these stages works with an increasingly detailed description of the design, and may fail due to problems unforeseen at earlier stages. This makes the flow, or sequence of steps that the users follow to finish their design, a crucial part of any EDA methodology.

Of course not all, or even most chips, are fully digital. Analog chips and chips with a mixture of analog and digital signals (commonly called mixed-signal chips) require their own specialized tool sets.

All these tools must work on circuits and designs that are quite large, and do so in a reasonable amount of time. In general, this cannot be done without models, or simplified descriptions of the behavior of various chip elements. Creating these models is the province of Technology CAD (TCAD), which in general treats relatively small problems in great physical detail, starting from very basic physics and building the more efficient models needed by the tools that must handle higher data volumes.

The division of EDA into these sections is somewhat arbitrary, and below a brief description of each of the chapters of the book is given.

System Level Design

Tools and Methodologies for System-Level Design by Bhattacharyya and Wolf

This chapter covers very high level system-level design approaches and associated tools such as Ptolemy, the Mathworks tools, and many others, and uses video applications as a specific example illustrating how these can be used.

System-Level Specification and Modeling Languages by Buck

This chapter discusses the major approaches to specify and model systems, and the languages and tools in this domain. It includes issues of heterogeneous specifications, models of computation and linking multidomain models, requirements on languages, and specialized tools and flows in this area.

SoC Block-Based Design and IP Assembly by Wilson

This chapter approaches system design with particular emphasis on SoC design via IP-based reuse and block-based design. Methods of assembly and compositional design of systems are covered. Issues of IP reuse as they are reflected in system-level design tools are also discussed.

Performance Evaluation Methods for Multiprocessor Systems-on-Chip Design by Bacivarov and Jerraya

This chapter surveys the broad field of performance evaluation and sets it in the context of multi-processor systems-on-chip (MPSoC). Techniques for various types of blocks — HW, CPU, SW, and interconnect — are included. A taxonomy of performance evaluation approaches is used to assess various tools and methodologies.

System-Level Power Management by Chang, Macii, Poncino and Tiwari

This chapter discusses dynamic power management approaches, aimed at selectively stopping or slowing down resources, whenever this is possible while still achieving the required level of system performance. The techniques can be applied both to reduce power consumption, which has an impact on power dissipation and power supply, and energy consumption, which improves battery life. They are generally driven by the software layer, since it has the most precise picture about both the required quality of service and the global state of the system.

Processor Modeling and Design Tools by Mishra and Dutt

This chapter covers state-of-the-art specification languages, tools, and methodologies for processor development used in academia and industry. It includes specialized architecture description languages and the tools that use them, with a number of examples.

Embedded Software Modeling and Design by di Natale

This chapter covers models and tools for embedded SW, including the relevant models of computation. Practical approaches with languages such as unified modeling language (UML) and specification and description language (SDL) are introduced and how these might link into design flows is discussed.

Using Performance Metrics to Select Microprocessor Cores for IC Designs by Leibson

This chapter discusses the use of standard benchmarks, and instruction set simulators, to evaluate processor cores. These might be useful in nonembedded applications, but are especially relevant to the design of embedded SoC devices where the processor cores may not yet be available in HW, or be based on user-specified processor configuration and extension. Benchmarks drawn from relevant application domains have become essential to core evaluation and their advantages greatly exceed that of the general-purpose ones used in the past.

Parallelizing High-Level Synthesis: A Code Transformational Approach to High-Level Synthesis by Singh, Gupta, Shukla, and Gupta

This chapter surveys a number of approaches, algorithms, and tools for HLS from algorithmic or behavioral descriptions, and focuses on some of the most recent developments in HLS. These include the use of techniques drawn from the parallel compiler community.

Micro-Architecture Design

Cycle-Accurate System-Level Modeling and Performance Evaluation by Coppola and Grammatikakis

This chapter discusses how to use system-level modeling approaches at the cycle-accurate micro-architectural level to do final design architecture iterations and ensure conformance to timing and performance specifications.

Micro-Architectural Power Estimation and Optimization by Macii, Mehra, and Poncino

This chapter discusses the state of the art in estimating power at the micro-architectural level, consisting of major design blocks such as data paths, memories, and interconnect. *Ad hoc* solutions for optimizing both specific components and the whole design are surveyed.

Design Planning by Otten

This chapter discusses the topics of physical floor planning and its evolution over the years, from dealing with rectangular blocks in slicing structures to more general mathematical techniques for optimizing physical layout while meeting a variety of criteria, especially timing and other constraints.

Logical Verification

Design and Verification Languages by Edwards

This chapter discusses the two main HDLs in use — VHDL and Verilog, and how they meet the requirements for design and verification flows. More recent evolutions in languages, such as SystemC, System Verilog, and verification languages such as OpenVera, e, and PSL are also described.

Digital Simulation by Sanguinetti

This chapter discusses logic simulation algorithms and tools, as these are still the primary tools used to verify the logical or functional correctness of a design.

Using Transactional Level Models in a SoC Design Flow by Clouard, Ghenassia, Maillet-Contoz, and Strassen

This chapter discusses a real design flow at a real IC design company to illustrate the building, deployment, and use of transactional-level models to simulate systems at a higher level of abstraction, with much greater performance than at register transfer level (RTL), and to verify functional correctness and validate system performance characteristics.

Assertion-Based Verification by Foster and Marschner

This chapter introduces the relatively new topic of assertion-based verification, which is useful for capturing design intent and reusing it in both dynamic and static verification methods. Assertion libraries such as OVL and languages such as PSL and System Verilog assertions are used for illustrating the concepts.

Hardware Acceleration and Emulation by Bershteyn and Turner

This chapter discusses HW-based systems including FPGA, processor based accelerators/emulators, and FPGA prototypes for accelerated verification. It compares the characteristics of each type of system and typical use models.

Formal Property Verification by Fix and McMillan

This chapter discusses the concepts and theory behind formal property checking, including an overview of property specification and a discussion of formal verification technologies and engines.

Test

Design-for-Test by Koenemann

This chapter discusses the wide variety of methods, techniques, and tools available to solve design-for-test (DFT) problems. This is a huge area with a huge variety of techniques, many of which are implemented in tools that dovetail with the capabilities of the physical test equipment. The chapter surveys the specialized techniques required for effective DFT with special blocks such as memories as well as general logic cores.

Automatic Test Pattern Generation by Wang and Cheng

This chapter starts with the fundamentals of fault modeling and combinational ATPG concepts. It moves on to gate-level sequential ATPG, and discusses satisfiability (SAT) methods for circuits. Moving on beyond traditional fault modeling, it covers ATPG for cross talk faults, power supply noise, and applications beyond manufacturing test.

Analog and Mixed-Signal Test by Kaminska

This chapter first overviews the concepts behind analog testing, which include many characteristics of circuits that must be examined. The nature of analog faults is discussed and a variety of analog test equipment and measurement techniques surveyed. The concepts behind analog built-in-self-test (BIST) are reviewed and compared with the digital test.

RTL to GDS-II, or Synthesis, Place, and Route

Design Flows by Hathaway, Stok, Chinnery, and Keutzer

The RTL to GDSII flow has evolved considerably over the years, from point tools hooked loosely together, to a more integrated set of tools for design closure. This chapter addresses the design flow challenges based on the rising interconnect delays and new challenges to achieve closure.

Logic Synthesis by Khatri and Shenoy

This chapter provides an overview and survey of logic synthesis, which has since the early 1980s, grown to be the vital center of the RTL to GDSII design flow for digital design.

Power Analysis and Optimization from Circuit to Register Transfer Levels by Monteiro, Patel, and Tiwari

Power has become one of the major challenges in modern IC design. This chapter provides an overview of the most significant CAD techniques for low power, at several levels of abstraction.

Equivalence Checking by Kuehlmann and Somenzi

Equivalence checking can formally verify whether two design specifications are functionally equivalent. The chapter defines the equivalence-checking problem, discusses the foundation for the technology, and then discusses the algorithms for combinational and sequential equivalence checking.

Digital Layout — Placement by Reda and Kahng

Placement is one of the fundamental problems in automating digital IC layout. This chapter reviews the history of placement algorithms, the criteria used to evaluate quality of results, many of the detailed algorithms and approaches, and recent advances in the field.

Static Timing Analysis by Sapatnekar

This chapter overviews the most prominent techniques for static timing analysis. It then outlines issues relating to statistical timing analysis, which is becoming increasingly important to handle process variations in advanced IC technologies.

Structured Digital Design by Mo and Brayton

This chapter covers the techniques for designing regular structures, including data paths, programable logic arrays, and memories. It extends the discussion to include regular chip architectures such as gate arrays and structured ASICs.

Routing by Scheffer

Routing continues from automatic placement as a key step in IC design. Routing creates all the wires necessary to connect all the placed components while obeying the process design rules. This chapter discusses various types of routers and the key algorithms.

Exploring Challenges of Libraries for Electronic Design by Hogan and Becker

This chapter discusses the factors that are most important and relevant for the design of libraries and IP, including standard cell libraries, cores, both hard and soft, and the design and user requirements for the same. It also places these factors in the overall design chain context.

Design Closure by Cohn and Osler

This chapter describes the common constraints in VLSI design, and how they are enforced through the steps of a design flow that emphasizes design closure. A reference flow for ASIC is used and illustrated. Future design closure issues are also discussed.

Tools for Chip-Package Codesign by Franzon

Chip-package co-design refers to design scenarios, in which the design of the chip impacts the package design or vice versa. This chapter discusses the drivers for new tools, the major issues, including mixed-signal needs, and the major design and modeling approaches.

Design Databases by Bales

The design database is at the core of any EDA system. While it is possible to build a bad EDA tool or flow on *any* database, it is impossible to build a good EDA tool or flow on a bad database. This chapter describes the place of a design database in an integrated design system. It discusses databases used in the past, those currently in use as well as emerging future databases.

FPGA Synthesis and Physical Design by Betz and Hutton

Programmable logic devices, both complex programmable logic devices (CPLDs) and FPGAs, have evolved from implementing small glue-logic designs to large complete systems. The increased use of such devices — they now are the majority of design starts — has resulted in significant research in CAD algorithms and tools targeting programmable logic. This chapter gives an overview of relevant architectures, CAD flows, and research.

Analog and Mixed-Signal Design

Analog Simulation: Circuit Level and Behavioral Level by Mantooth and Roychowdhury

Circuit simulation has always been a crucial component of analog system design and is becoming even more so today. In this chapter, we provide a quick tour of modern circuit simulation. This includes starting on the ground floor with circuit equations, device models, circuit analysis, more advanced analysis techniques motivated by RF circuits, new advances in circuit simulation using multitime techniques, and statistical noise analysis.

Simulation and Modeling for Analog and Mixed-Signal Integrated Circuits by Gielen and Philips

This chapter provides an overview of the modeling and simulation methods that are needed to design and embed analog and RF blocks in mixed-signal integrated systems (ASICs, SoCs, and SiPs). The role of behavioral models and mixed-signal methods involving models at multiple hierarchical levels is covered. The generation of performance models for analog circuit synthesis is also discussed.

Layout Tools for Analog ICs and Mixed-Signal SoCs: A Survey by Rutenbar and Cohn

Layout for analog circuits has historically been a time-consuming, manual, trial-and-error task. In this chapter, we cover the basic problems faced by those who need to create analog and mixed-signal layout, and survey the evolution of design tools and geometric/electrical optimization algorithms that have been directed at these problems.

Physical Verification

Design Rule Checking by Todd, Grodd, and Fetty

After the physical mask layout is created for a circuit for a specific design process, the layout is measured by a set of geometric constraints or rules for that process. The main objective of design rule checking is to achieve high overall yield and reliability. This chapter gives an overview of design rule checking (DRC) concepts and then discusses the basic verification algorithms and approaches.

SECTION I

INTRODUCTION

Resolution Enhancement Techniques and Mask Data Preparation by Schellenberg

With more advanced IC fabrication processes, new physical effects, which could be ignored in the past, are being found to have a strong impact on the formation of features on the actual silicon wafer. It is now essential to transform the final layout via new tools in order to allow the manufacturing equipment to deliver the new devices with sufficient yield and reliability to be cost-effective. This chapter discusses the compensation schemes and mask data conversion technologies now available to accomplish the new design for manufacturability (DFM) goals.

Design for Manufacturability in the Nanometer Era by Dragone, Guardiani, and Strojwas

Achieving high yielding designs in state-of-the-art IC process technology has become an extremely challenging task. Design for manufacturability includes many techniques to modify the design of ICs in order to improve functional and parametric yield and reliability. This chapter discusses yield loss mechanisms and fundamental yield modeling approaches. It then discusses techniques for functional yield maximization and parametric yield optimization. Finally, DFM-aware design flows and the outlook for future DFM techniques are discussed.

Design and Analysis of Power Supply Networks by Blaauw, Pant, Chaudhry, and Panda

This chapter covers design methods, algorithms, tools for designing on-chip power grids, and networks. It includes the analysis and optimization of effects such as voltage drop and electro-migration.

Noise Considerations in Digital ICs by Kariat

On-chip noise issues and impact on signal integrity and reliability are becoming a major source of problems for deep submicron ICs. Thus the methods and tools for analyzing and coping with them, which are discussed in this chapter, have been gaining importance in recent years.

Layout Extraction by Kao, Lo, Basel, Singh, Spink, and Scheffer

Layout extraction is the translation of the topological layout back into the electrical circuit it is intended to represent. This chapter discusses the distinction between designed and parasitic devices, and discusses the three major parts of extraction: designed device extraction, interconnect extraction, and parasitic device extraction.

Mixed-Signal Noise Coupling in System-on-Chip Design: Modeling, Analysis, and Validation by Vergese and Nagata

This chapter describes the impact of noise coupling in mixed-signal ICs, and reviews techniques to model, analyze, and validate it. Different modeling approaches and computer simulation methods are presented, along with measurement techniques. Finally, the chapter reviews the application of substrate noise analysis to placement and power distribution synthesis.

Technology Computer-Aided Design

Process Simulation by Johnson

Process simulation is the modeling of the fabrication of semiconductor devices such as transistors. The ultimate goal is an accurate prediction of the active dopant distribution, the stress distribution, and the device geometry. This chapter discusses the history, requirements, and development of process simulators.

Device Modeling — from Physics to Electrical Parameter Extraction by Dutton, Choi, and Kan

Technology files and design rules are essential building blocks of the IC design process. Development of these files and rules involves an iterative process that crosses the boundaries of technology and device development, product design, and quality assurance. This chapter starts with the physical description of IC devices and describes the evolution of TCAD tools.

High-Accuracy Parasitic Extraction by Kamon and Iverson

This chapter describes high-accuracy parasitic extraction methods using fast integral equation and random walk-based approaches.

References

[1] A. Sangiovanni-Vincentelli, The tides of EDA, *IEEE Des. Test Comput.*, 20, 59–75, 2003.
[2] A. Kuelhmann, Ed., *20 Years of ICCAD*, Kluwer Academic Publishers (now Springer), Dordrecht, 2002.

2

The Integrated Circuit Design Process and Electronic Design Automation

Robert Damiano
Synopsys Inc.
Hillsboro, Oregon

Raul Camposano
Synopsys Inc.
Mountain View, California

2.1 Introduction

In this chapter, we describe the design process, its major stages, and how electronic design automation (EDA) tools fit into these processes. We also examine the interfaces between the major integrated circuit (IC) design stages as well as the kind of information — both abstractions upwards, and detailed design and verification information downward — that must flow between stages. We assume Complementary Metal Oxide Semiconductor (CMOS) is the basis for all technologies.

We will illustrate with a continuing example. A company wishes to create a new system on chip (SoC). The company assembles a product team, consisting of a project director, system architects, system verification engineers, circuit designers (both digital and analog), circuit verification engineers, layout engineers, and manufacturing process engineers. The product team determines the target technology and geometry as well as the fabrication facility or foundry. The system architects initially describe the system-level design (SLD) through a transaction-level specification in a language such as C++, SystemC, or Esterel. The system verification engineers determine the functional correctness of the SLD through simulation. The engineers validate the transaction processing through simulation vectors. They monitor the results for errors. Eventually, these same engineers would simulate the process with an identical set of vectors through the system implementation to see if the specification and the implementation match. There is some ongoing research to check this equivalence formally.

The product team partitions the SLD into functional units and hands these units to the circuit design teams. The circuit designers describe the functional intent through a high-level design language (HDL). The most popular HDLs are Verilog and VHDL. SystemVerilog is a new language, adopted by the IEEE, which contains design, testbench, and assertion syntax. These languages allow the circuit designers to express the behavior of their design using high-level functions such as addition and multiplication. These languages allow expression of the logic at the register transfer level (RTL), in the sense that an assignment of registers expresses functionality. For the analog and analog mixed signal (AMS) parts of the design, there are also high-level design languages such as Verilog-AMS and VHDL-AMS. Most commonly, circuit designers use Simulation Program with Integrated Circuit Emphasis (SPICE) transistor models and netlists to describe analog components. However, high-level languages provide an easier interface between analog and digital segments of the design and they allow writing higher-level behavior of the analog parts. Although the high-level approaches are useful as simulation model interfaces, there remains no clear method of synthesizing transistors from them. Therefore, transistor circuit designers usually depend on schematic capture tools to enter their data.

The design team must consider functional correctness, implementation closure (reaching the prioritized goals of the design), design cost, and manufacturability of a design. The product team takes into account risks and time to market as well as choosing the methodology. Anticipated sales volume can reflect directly on methodology; whether it is better to create a fully custom design, semicustom design, use standard cells , gate arrays, or a field programmable gate array (FPGA). Higher volume mitigates the higher cost of fully custom or semicustom design, while time to market might suggest using an FPGA methodology. If implementation closure for power and speed is tantamount, then an FPGA methodology might be a poor choice. Semicustom designs, depending on the required volume, can range from microprocessor central processor units (CPUs), digital signal processors (DSPs), application-specific standard parts (ASSP) or application-specific integrated circuits (ASIC). In addition, for semicustom designs, the company needs to decide whether to allow the foundry to implement the layout, or whether the design team should use customer owned tools (COT). We will assume that our product team chooses semicustom COT designs. We will mention FPGA and fully custom methodologies only in comparison.

In order to reduce cost, the product team may decide that the design warrants reuse of intellectual property (IP). Intellectual property reuse directly addresses the increasing complexity of design as opposed to feature geometry size. Reuse also focuses on attaining the goals of functional correctness. One analysis estimates that it takes 2000 engineering years and 1 trillion simulation vectors to verify 25 million lines of RTL code. Therefore, verified IP reuse reduces cost and time to market. Moreover, IP blocks themselves have become larger and more complex. For example, the 1176JZ-S ARM core is 24 times larger than the older 7TDI-S ARM core. The USB 2.0 Host is 23 times larger than the Universal Serial Bus (USB) 1.1 Device. PCI Express is 7.5 times larger than PCI v 1.1.

Another important trend is that SoC-embedded memories are an increasingly large part of the SoC real estate. While in 1999, 20% of a 180-nm SoC was embedded memory, roadmaps project that by 2005, embedded memory will consume 71% of a 90-nm SoC. These same roadmaps indicate that by 2014, embedded memory will grow to 94% of a 35-nm SoC.

Systems on chips typically contain one or more CPUs or DSPs (or both), cache, a large amount of embedded memory and many off-the-shelf components such as USB, Universal Asynchronous Receiver-Transmitter (UART), Serial Advanced Technology Attachment (SATA), and Ethernet (cf. Figure 2.1). The differentiating part of the SoC contains the new designed circuits in the product.

The traditional semicustom IC design flow typically comprises up to 50 steps. On the digital side of design, the main steps are functional verification, logical synthesis, design planning, physical implementation which includes clock-tree synthesis, placement and routing, extraction, design rules checking (DRC) and layout versus schematic checking (LVS), static timing analysis, insertion of test structures, and test pattern generation. For analog designs, the major steps are as follows: schematic entry, SPICE simulation, layout, layout extraction, DRC, and LVS. SPICE simulations can include DC, AC, and transient analysis, as well as noise, sensitivity, and distortion analysis. Analysis and implementation of corrective procedures for the manufacturing process such as mask synthesis and yield analysis, are critical at smaller geometries. In order to verify an SoC system where many components reuse IP, the IP provider may supply verification IP, monitors, and checkers needed by system verification.

There are three basic areas where EDA tools assist the design team. Given a design, the first is verification of functional correctness. The second deals with implementation of the design. The last area deals with analysis and corrective procedures so that the design meets all manufacturability specifications. Verification, layout, and process engineers on the circuit design team essentially own these three steps.

SPICE reportedly is an acronym for Simulation Program with Integrated Circuit Emphasis

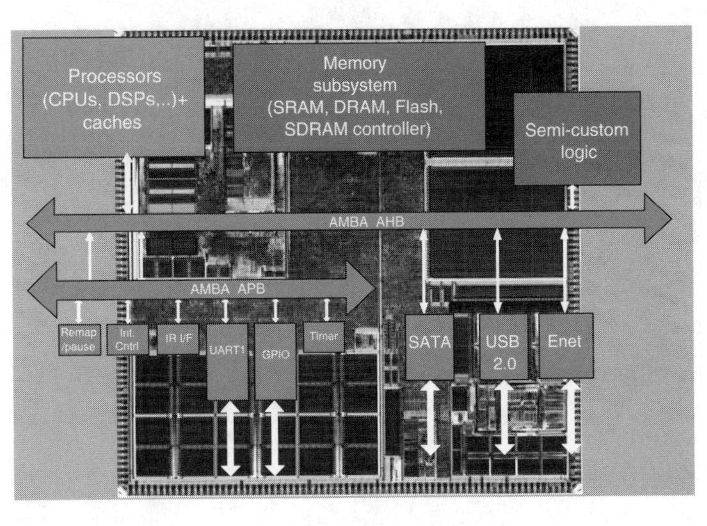

FIGURE 2.1. SoC with IP.

2.2 Verification

The design team attempts to verify that the design under test (DUT) functions correctly. For RTL designs, verification engineers rely highly on simulation at the cycle level. After layout, EDA tools, such as equivalence checking, can determine whether the implementation matches the RTL functionality. After layout, the design team must check that there are no problem delay paths. A static timing analysis tool can facilitate this. The team also needs to examine the circuit for noise and delay due to parasitics. In addition, the design must obey physical rules for wire spacing, width, and enclosure as well as various electrical rules. Finally, the design team needs to simulate and check the average and transient power. For transistor circuits, the design team uses SPICE circuit simulation or fast SPICE to determine correct functionality, noise, and power.

We first look at digital verification (cf. Figure 2.2). RTL simulation verifies that the DUT behavior meets the design intent. The verification engineers apply a set of vectors, called a testbench, to the design through an event-driven simulator, and compare the results to a set of expected outputs. The quality of the verification depends on the quality of the testbench. Many design teams create their testbench by supplying a list of the vectors, a technique called directed test. For a directed test to be effective, the design team must know beforehand what vectors might uncover bugs. This is extremely difficult since complex sequences of vectors are necessary to find some corner case errors. Therefore, many verification engineers create testbenches that supply stimulus through random vectors with biased inputs, such as the clock or reset signal. The biasing increases or decreases the probability of a signal going high or low. While a purely random testbench is easy to create, it suffers from the fact that vectors may be illegal as stimulus. For better precision and wider coverage, the verification engineer may choose to write a constrained random testbench. Here, the design team supplies random input vectors that obey a set of constraints.

The verification engineer checks that the simulated behavior does not have any discrepancies from the expected behavior. If the engineer discovers a discrepancy, then the circuit designer modifies the HDL and the verification engineer resimulates the DUT. Since exhaustive simulation is usually impossible, the design team needs a metric to determine quality. One such metric is coverage. Coverage analysis considers how well the test cases stimulate the design. The design team might measure coverage in terms of number of lines of RTL code exercised, whether the test cases take each leg of each decision, or how many "reachable" states encountered.

Another important technique is for the circuit designer to add assertions within the HDL. These assertions monitor whether internal behavior of the circuit is acting properly. Some designers embed tens of thousands of assertions into their HDL. Languages like SystemVerilog have extensive assertion syntax based

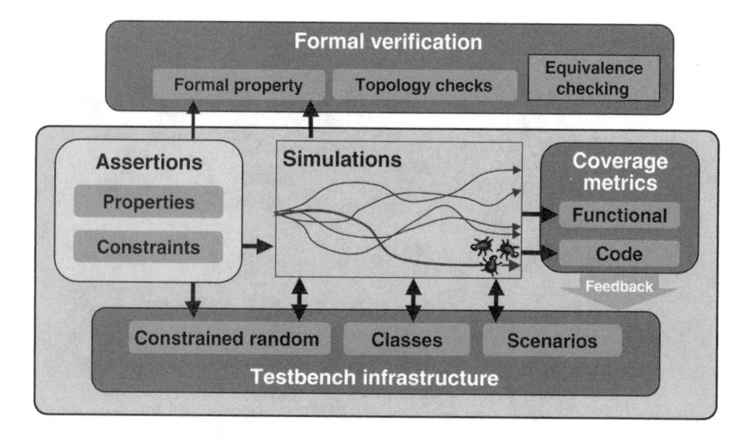

FIGURE 2.2. Digital Simulation/Formal Verification.

on linear temporal logic. Even for languages without the benefit of assertion syntax, tool-providers supply an application program interface (API), which allows the design team to build and attach its own monitors.

The verification engineer needs to run a large amount of simulation, which would be impractical if not for compute farms. Here, the company may deploy thousands of machines, 24/7, to enable the designer to get billions of cycles a day; sometimes the machines may run as many as 200 billion cycles a day. Best design practices typically create a highly productive computing environment. One way to increase throughput is to run a cycle simulation by taking a subset of the chosen verification language which is both synchronous and has a set of registers with clear clock cycles. This type of simulation assumes a uniformity of events and typically uses a time wheel with gates scheduled in a breadth first manner.

Another way to tackle the large number of simulation vectors during system verification is through emulation or hardware acceleration. These techniques use specially configured hardware to run the simulation. In the case of hardware acceleration, the company can purchase special-purpose hardware, while in the case of emulation the verification engineer uses specially configured FPGA technology. In both cases, the system verification engineer must synthesize the design and testbench down to a gate-level model. Tools are available to synthesize and schedule gates for the hardware accelerator. In the case of an FPGA emulation system, tools can map and partition the gates for the hardware.

Of course, since simulation uses vectors, it is usually a less than exhaustive approach. The verification engineer can make the process complete by using assertions and formal property checking. Here, the engineer tries to prove that an assertion is true or to produce a counterexample. The trade-off is simple. Simulation is fast but by definition incomplete, while formal property checking is complete but may be very slow. Usually, the verification engineer runs constrained random simulation to unearth errors early in the verification process. The engineer applies property checking to corner case situations that can be extremely hard for the testbench to find. The combination of simulation and formal property checking is very powerful. The two can even be intermixed, by allowing simulation to proceed for a set number of cycles and then exhaustively looking for an error for a different number of cycles. In a recent design, by using this hybrid approach , a verification engineer found an error 21,000 clock cycles from an initial state. Typically, formal verification works well on specific functional units of the design. Between the units, the system engineers use an “assume/guarantee” methodology to establish block pre- and postconditions for system correctness.

During the implementation flow, the verification engineer applies equivalence checking to determine whether the DUT preserves functional behavior. Note that functional behavior is different from functional intent. The verification engineer needs RTL verification to compare functional behavior with functional intent. Equivalence checking is usually very fast and is a formal verification technology, which is exhaustive in its analysis. Formal methods do not use vectors.

For transistor-level circuits, such as analog, memory, and radio frequency (RF), the event-driven verification techniques suggested above do not suffice (cf. Figure 2.3). The design team needs to compute signals

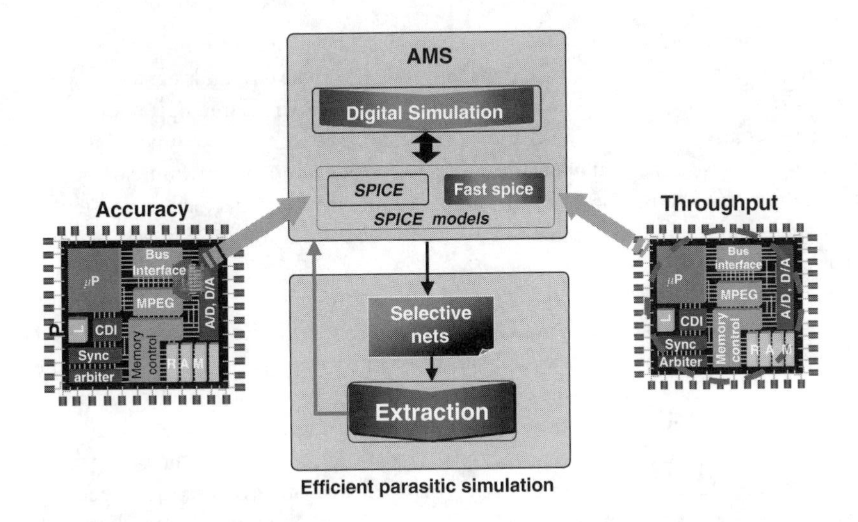

FIGURE 2.3. Transistor simulation with parasitics.

accurately through SPICE circuit simulation. SPICE simulation is very time consuming because the algorithm solves a system of differential equations. One way to get around this cost is to select only a subset of transistors, perform an extraction of the parasitics, and then simulate the subset with SPICE. This reduction gives very accurate results for the subset, but even so, the throughput is still rather low. Another approach is to perform a fast SPICE simulation. This last SPICE approach trades some accuracy for a significant increase in throughput. The design team can also perform design space exploration by simulating various constraint values on key goals such as gain or phase margin to find relatively optimal design parameters. The team analyzes the multiple-circuit solutions and considers the cost trade-offs. A new generation of tools performs this "design exploration" in an automatic manner. Mixed-level simulation typically combines RTL, gate and transistor parts of the design and uses a communication back-plane to run the various simulations and share input and output values.

Finally, for many SoCs, both hardware and software comprise the real system. System verification engineers may run a hardware–software co-simulation before handing the design to a foundry. All simulation system components mentioned can be part of this co-simulation. In early design stages, when the hardware is not ready, the software can simulate ("execute") an instruction set model (ISM), a virtual prototype (model), or an early hardware prototype typically implemented in FPGAs.

2.3 Implementation

This brings us to the next stage of the design process, the implementation and layout of the digital design. Circuit designers implement analog designs by hand. Field programmable gate array technologies usually have a single basic combinational cell, which can form a variety of functions by constraining inputs. Layout and process tools are usually proprietary to the FPGA family and manufacturer. For semicustom design, the manufacturer supplies a precharacterized cell library, either standard cell or gate array. In fact, for a given technology, the foundry may supply several libraries, differing in power, timing, or yield. The company decides on one or more of these as the target technology. One twist on the semicustom methodology is structured ASIC. Here, a foundry supplies preplaced memories, pad-rings and power grids as well as sometimes preplaced gate array logic, similar to the methodology employed by FPGA families. The company can use semicustom techniques for the remaining combinational and sequential logic. The goal is to reduce nonrecurring expenses by limiting the number of mask-sets needed and by simplifying physical design.

By way of contrast, in a fully custom methodology, one tries to gain performance and limit power consumption by designing much of the circuit as transistors. The circuit designers keep a corresponding RTL design. The verification engineer simulates the RTL and extracts a netlist from the transistor description. Equivalence checking compares the extracted netlist to the RTL. The circuit designer manually places and routes the transistor-level designs. Complex high-speed designs, such as microprocessors, sometimes use full custom methodology, but the design costs are very high. The company assumes that the high volume will amortize the increased cost. Fully custom designs consider implementation closure for power and speed as most important. At the other end of the spectrum, FPGA designs focus on design cost and time to market. Semicustom methodology tries to balance the goals of timing and power closure with design cost (cf. Figure 2.4).

In the semicustom implementation flow, one first attempts to synthesize the RTL design into a mapped netlist. The circuit designers supply their RTL circuit along with timing constraints. The timing constraints consist of signal arrival and slew (transition) times at the inputs, and required times and loads (capacitances) at the outputs. The circuit designer identifies clocks as well as any false or multiple-cycle paths. The technology library is usually a file that contains a description of the function of each cell along with delay, power, and area information. Either the cell description contains the pin-to-pin delay represented as look-up table functions of input slew, output load, and other physical parameters such as voltage and temperature, or as polynomial functions that best fit the parameter data. For example, foundries provide cell libraries in Liberty or OLA (Open Library Application Programming Interface) formats. The foundry also provides a wire delay model, derived statistically from previous designs. The wire delay model correlates the number of sinks of a net to capacitance and delay.

Several substages comprise the operation of a synthesis tool. First, the synthesis tool compiles the RTL into technology-independent cells and then optimizes the netlist for area, power, and delay. The tool maps the netlist into a technology. Sometimes, synthesis finds complex functions such as multipliers and adders in parameterized (area/timing) reuse libraries. For example, the tool might select a Booth multiplier from the reuse library to improve timing. For semicustom designs, the foundry provides a standard cell or gate array library, which describes each functional member. In contrast, the FPGA supplier describes a basic combinational cell from which the technology mapping matches functional behavior of subsections of the design. To provide correct functionality, the tool may set several pins on the complex gates to constants. A post-process might combine these functions for timing, power, or area.

A final substage tries to analyze the circuit and performs local optimizations that help the design meet its timing, area and power goals. Note that due to finite number of power levels of any one cell, there are limits to the amount of capacitance that functional cell types can drive without the use of buffers. Similar restrictions apply to input slew (transition delay). The layout engineer can direct the synthesis tool by enhancing or omitting any of these stages through scripted commands. Of course, the output must be a mapped netlist.

To get better timing results, foundries continue to increase the number of power variations for some cell types. One limitation to timing analysis early in the flow is that the wire delay models are statistical estimates of the real design. Frequently, these wire delays can differ significantly from those found after routing. One interesting approach to synthesis is to extend each cell of the technology library so

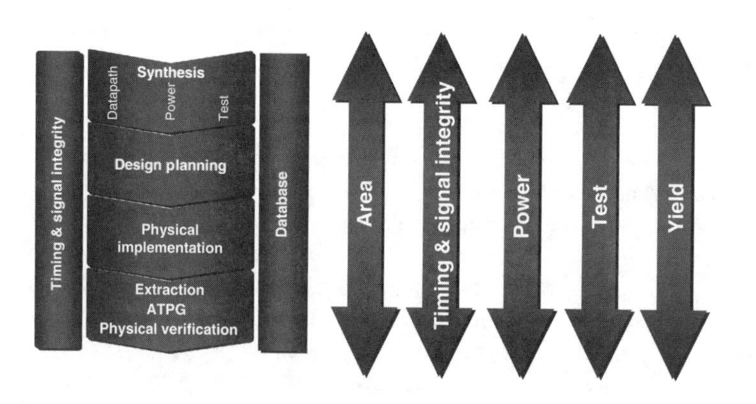

FIGURE 2.4. Multi-objective implementation convergence.

that it has an infinite or continuous variation of power. This approach, called gain-based synthesis, attempts to minimize the issue of inaccurate wire delay by assuming cells can drive any wire capacitance through appropriate power level selection. In theory, there is minimal perturbation to the natural delay (or gain) of the cell. This technique makes assumptions such as that the delay of a signal is a function of capacitance. This is not true for long wires where resistance of the signal becomes a factor. In addition, the basic approach needs to include modifications for slew (transition delay).

To allow detection of manufacturing faults, the design team may add extra test generation circuitry. Design for test (DFT) is the name given to the process of adding this extra logic (cf. Figure 2.5). Sometimes, the foundry supplies special registers, called logic-sensitive scan devices. At other times, the test tool adds extra logic called Joint Test Action Group (JTAG) boundary scan logic that feeds the registers. Later in the implementation process, the design team will generate data called scan vectors that test equipment uses to detect manufacturing faults. Subsequently, tools will transfer these data to automatic test equipment (ATE), which perform the chip tests.

As designs have become larger, so has the amount of test data. The economics of the scan vector production with minimal cost and design impact leads to data compression techniques. One of most widely used techniques is deterministic logic built in self-test(BIST). Here, a test tool adds extra logic on top of the DFT to generate scan vectors dynamically.

Before continuing the layout, the engineer needs new sets of rules, dealing with the legal placement and routing of the netlist. These libraries, in various exchange formats, e.g., LEF for logic, DEF for design and PDEF for physical design, provide the layout engineer physical directions and constraints. Unlike the technology rules for synthesis, these rules are typically model-dependent. For example, there may be information supplied by the circuit designer about the placement of macros such as memories. The routing tool views these macros as blockages. The rules also contain information from the foundry.

Even if the synthesis tool preserved the original hierarchy of the design, the next stages of implementation need to view the design as flat. The design-planning step first flattens the logic and then partitions the flat netlist as to assist placement and routing;—in fact, in the past, design planning was sometimes known as floor planning. A commonly used technique is for the design team to provide a utilization ratio to the design planner. The utilization ratio is the percentage of chip area used by the cells as opposed to the nets. If the estimate is too high, then routing congestion may become a problem. If the estimate is too low, then the layout could waste area. The design-planning tool takes the locations of hard macros into account. These macros are hard in the sense that they are rectangular with a fixed length, fixed width, and sometimes a fixed location on the chip. The design-planning tool also tries to use the logical hierarchy of the design as a guide to the partitioning. The tool creates, places and routes a set of macros that have fixed lengths, widths, and locations. The tool calculates timing constraints for each macro and routes the power

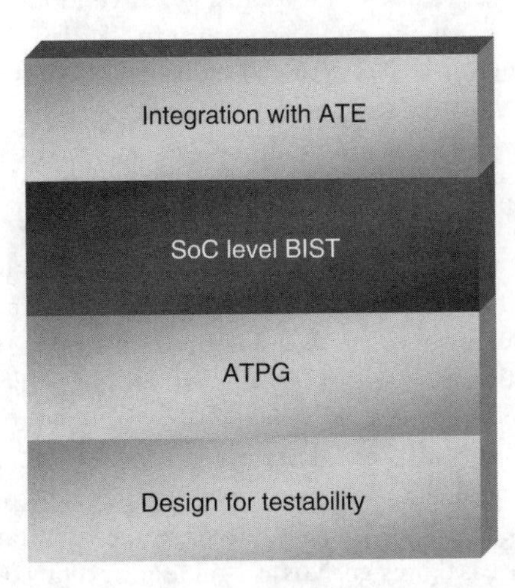

FIGURE 2.5. Design for test.

and ground grids. The power and ground grids are usually on the chip's top levels of metal and then distributed to the lower levels. The design team can override these defaults and indicate which metal layers should contain these grids. Sometimes design planning precedes synthesis. In these cases, the tool partitions the RTL design and automatically characterizes each of the macros with timing constraints.

After design planning, the layout engineer runs the physical implementation tools on each macro. First, the placer assigns physical locations to each gate of the macro. The placer typically moves gates while minimizing some cost, e.g., wire length or timing. Legalization follows the coarse placement to make sure the placed objects fit physical design rules. At the end of placement, the layout engineer may run some more synthesis, like re-sizing of gates. One of the major improvements to placement over the last decade is the emergence of physical synthesis. In physical synthesis, the tool interleaves synthesis and placement. Recall that previously, logic synthesis used statistical wire capacitance. Once the tool places the gates, it can perform a global route and get capacitances that are more accurate for the wires, based on actual placed locations. The physical synthesis tool iterates this step and provides better timing and power estimates.

Next, the layout engineer runs a tool that buffers and routes the clock tree. Clock-tree synthesis attempts to minimize the delay while assuring that skew, that is the variation in signal transport time from the clock to its corresponding registers, is close to zero.

Routing the remaining nets comes after clock-tree synthesis. Routing starts with a global analysis called global route. Global route creates coarse routes for each signal and its outputs. Using the global routes as a guide, a detailed routing scheme, such as a maze channel or switchbox, performs the actual routing. As with the placement, the tool performs a final legalization to assure that the design obeys physical rules. One of the major obstacles to routing is signal congestion. Congestion occurs when there are too many wires competing for a limited amount of chip wire resource. Remember that the design team gave the design planner a utilization ratio in the hope of avoiding this problem.

Both global routing and detailed routing take the multilayers of the chip into consideration. For example, the router assumes that the gates are on the polysilicon layer, while the wires connect the gates through vias on 3–8 layers of metal. Horizontal or vertical line segments comprise the routes, but some recent work allows 45° lines for some foundries. As with placement, there may be some resynthesis, such as gate resizing, at the end of the detailed routing stage.

Once the router finishes, an extraction tool derives the capacitances, resistances, and inductances. In a two-dimensional (2-D) parasitic extraction, the extraction tool ignores 3-D details and assumes that each chip level is uniform in one direction. This produces only approximate results. In the case of the much slower 3-D parasitic extraction, the tool uses 3-D field solvers to derive very accurate results. A 2½-D extraction tool compromises between speed and accuracy. By using multiple passes, it can access some of the 3-D features. The extraction tool places its results in a standard parasitic exchange format file (SPEF).

During the implementation process, the verification engineer continues to monitor behavioral consistency through equivalence checking and using LVS comparison. The layout engineer analyzes timing and signal integrity issues through timing analysis tools, and uses their results to drive implementation decisions. At the end of the layout, the design team has accurate resistances, capacitances, and inductances for the layout. The system engineer uses a sign-off timing analysis tool to determine if the layout meets timing goals. The layout engineer needs to run a DRC on the layout to check for violations.

Both the Graphic Data System II (GDSII) and the Open Artwork System Interchange Standard (OASIS) are databases for shape information to store a layout. While the older GDSII was the database of choice for shape information, there is a clear movement to replace it by the newer, more efficient OASIS database. The LVS tool checks for any inconsistencies in this translation.

What makes the implementation process so difficult is that multiple objectives need consideration. For example, area, timing, power, reliability, test, and yield goals might and usually cause conflict with each other. The product team must prioritize these objectives and check for implementation closure.

Timing closure—that is meeting all timing requirements—by itself is becoming increasingly difficult and offers some profound challenges. As process geometry decrease, the significant delay shifts from the cells to the wires. Since a synthesis tool needs timing analysis as a guide and routing of the wires does not occur until after synthesis, we have a chicken and egg problem. In addition, the thresholds for noise sensitivity also shrink with smaller geometries. This along with increased coupling capacitances, increased current densities and sensitivity to inductance, make problems like crosstalk and voltage (IR) drop increasingly familiar.

Since most timing analysis deals with worst-case behavior, statistical variation and its effect on yield add to the puzzle. Typically timing analysis computes its cell delay as function of input slew (transition delay) and output load (output capacitance or RC). If we add the effects of voltage and temperature variations as well as circuit metal densities, timing analysis gets to be very complex. Moreover, worst-case behavior may not correlate well with what occurs empirically when the foundry produces the chips. To get a better predictor of parametric yield, some layout engineers use statistical timing analysis. Here, rather than use single numbers (worst case, best case, corner case, nominal) for the delay-equation inputs, the timing analysis tool selects probability distributions representing input slew, output load, temperature, and voltage among others. The delay itself becomes a probability distribution. The goal is to compute the timing more accurately in order to create circuits with smaller area and lower power but with similar timing yield.

Reliability is also an important issue with smaller geometries. Signal integrity deals with analyzing what were secondary effects in larger geometries. These effects can produce erratic behavior for chips manufactured in smaller geometries. Issues such as crosstalk, IR drop, and electromigration are factors that the design team must consider in order to produce circuits that perform correctly.

Crosstalk noise can occur when two wires are close to each other (cf. Figure 2.6). One wire, the aggressor, switches while the victim signal is in a quiet state or making an opposite transition. In this case, the aggressor can force the victim to glitch. This can cause a functional failure or can simply consume additional power. Gate switching draws current from the power and ground grids. That current, together with the wire resistance in the grids, can cause significant fluctuations in the power and ground voltages supplied to gates. This problem, called IR drop, can lead to unpredictable functional errors. Very high frequencies can produce high current densities in signals and power lines, which can lead to the migration of metal ions. This power electromigration can lead to open or shorted circuits and subsequent signal failure.

Power considerations are equally complex. As the size of designs grow and geometries shrink, power increases. This can cause problems for batteries in wireless and hand-held devices, and thermal management in microprocessor, graphic and networking applications. Power consumption falls into two areas: dynamic power (cf. Figure 2.7), the power consumed when devices switch value; and leakage power (cf. Figure 2.8), the power leaked through the transistor. Dynamic power consumption grows directly with increased capacitance and voltage. Therefore, as designs become larger, dynamic power increases. One easy way to reduce dynamic power is to decrease voltage. However, decreased voltage leads to smaller noise margins and less speed.

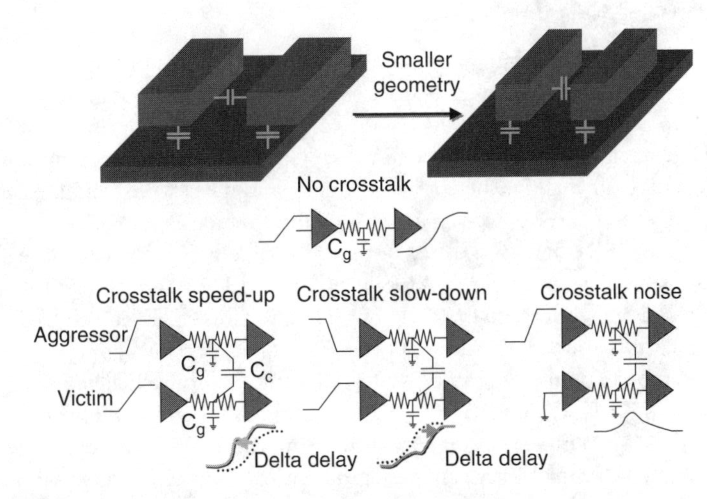

FIGURE 2.6. Crosstalk.

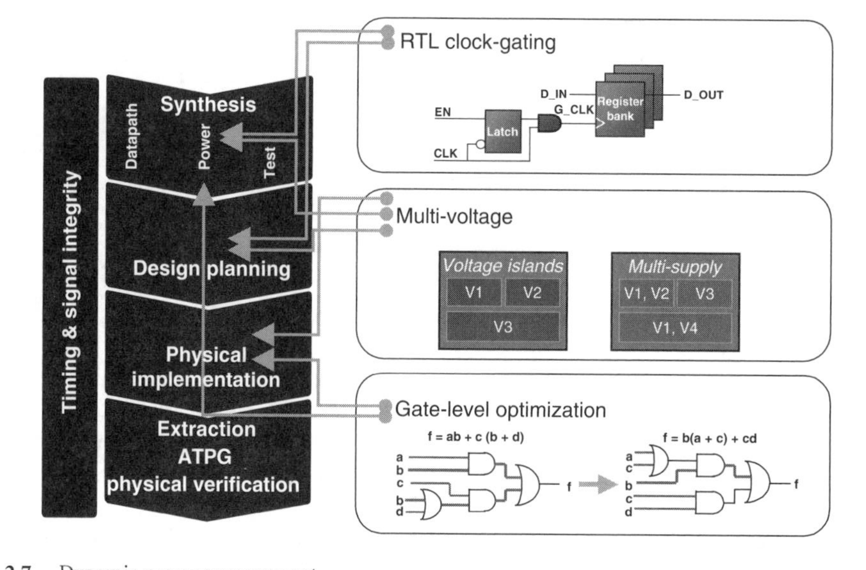

FIGURE 2.7. Dynamic power management.

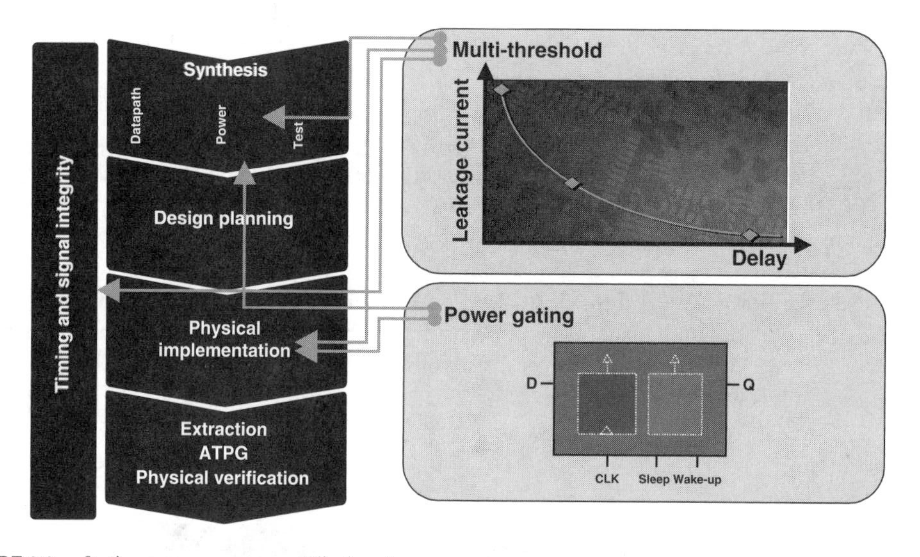

FIGURE 2.8. Static power management (leakage).

A series of novel design and transistor innovations can reduce the power consumption. These include operand isolation, clock gating, and voltage-islands. Timing and power considerations are very often in conflict with each other, so the design team must employ these remedies carefully.

A design can have part of its logic clock-gated by using logic to enable the bank of registers. The logic driven by the registers is quiescent until the clock-gated logic enables the registers. Latches at the input can isolate parts of a design that implement operations (e.g. an arithmetic logic unit (ALU)), when results are unnecessary for correct functionality, thus preventing unnecessary switching. Voltage-islands help resolve the timing vs. power conflicts. If part of a design is timing critical, a higher voltage can reduce the delay. By partitioning the design into voltage-islands, one can use lower voltage in all but the most timing-critical parts of the design. An interesting further development is dynamic voltage/frequency scaling, which consists of scaling the supply voltage and the speed during operation to save power or increase performance temporarily.

The automatic generation of manufacturing fault detection tests was one of the first EDA tools. When a chip fails, the foundry wants to know why. Test tools produce scan vectors that can identify various manufacturing faults within the hardware. The design team translates the test vectors to standard test data format and the foundry can inject these inputs into the failed chip through automated test equipment (ATE). Remember that the design team added extra logic to the netlist before design planning, so that test equipment could quickly insert the scan vectors, including set values for registers, into the chip. The most common check is for stuck at 0 or stuck at 1 faults where the circuit has an open or short at a particular cell. It is not surprising that smaller geometries call for more fault detection tests. An integration of static timing analysis with transition/path delay fault automatic test pattern generation (ATPG) can help, for example, to detect contact defects; while extraction information and bridging fault ATPG can detect metal defects.

Finally, the design team should consider yield goals. Manufacturing becomes more difficult as geometries shrink. For example, thermal stress may create voids in vias. One technique to get around this problem is to minimize the vias inserted during routing, and for those inserted, to create redundant vias. Via doubling, which converts a single via into multiple vias, can reduce resistance and produce better yield. Yield analysis can also suggest wire spreading during routing to reduce cross talk and increase yield. Manufacturers also add a variety of manufacturing process rules needed to guarantee good yield. These rules involve antenna checking and repair through diode insertion as well as metal fill needed to produce uniform metal densities necessary for copper wiring chemical–mechanical polishing (CMP). Antenna repair has little to do with what we typically view as antennas. During the ion-etching process, charge collects on the wires connected to the polysilicon gates. These charges can damage the gates. The layout tool can connect small diodes to the interconnect wires as a discharge path.

Even with all the available commercial tools, there are times when layout engineers want to create their own tool for analysis or small implementation changes. This is analogous to the need for an API in verification. Scripting language and C-language-based APIs for design databases such as MilkyWay and OpenAccess are available. These databases supply the user with an avenue to both the design and rules. The engineer can directly change and analyze the layout.

2.4 Design for Manufacturing

One of the newest areas for EDA tools is design for manufacturing. As in other areas, the driving force of the complexity is the shrinking of geometries. After the design team translates their design to shapes, the foundry must transfer those shapes to a set of masks. Electron beam (laser) equipment then creates the physical masks for each layer of the chip from the mask information. For each layer of the chip, the foundry applies photoresistive material, and then transfers the mask structures by the stepper optical equipment onto the chip. Finally, the foundry etches the correct shapes by removing the excess photoresist material.

Since the stepper uses light for printing, it is important that the wavelength is small enough to transcribe the features accurately. When the chip's feature size was 250 nm, we could use lithography equipment that produced light at a wavelength of 248 nm. New lithography equipment that produces light of lower wavelength needs significant innovation and can be very expensive. When the feature geometry gets significantly smaller than the wavelength, the detail of the reticles (fine lines and wires), transferred to the chip from the mask can be lost. Electronic design automation tools can analyze and correct this transfer operation without new equipment, by modifying the shapes data— a process known as mask, synthesis (cf. Figure 2.9). This process uses resolution enhancement techniques and methods to provide dimensional accuracy.

One mask synthesis technique is optimal proximity correction (OPC). This process takes the reticles in the GDSII or OASIS databases and modifies them by adding new lines and wires, so that even if the geometry is smaller than the wavelength, optical equipment adequately preserves the details. This technique successfully transfers geometric features of down to one-half of the wavelength of the light used. Of course given a fixed wavelength, there are limits beyond which the geometric feature size is too small for even these tricks.

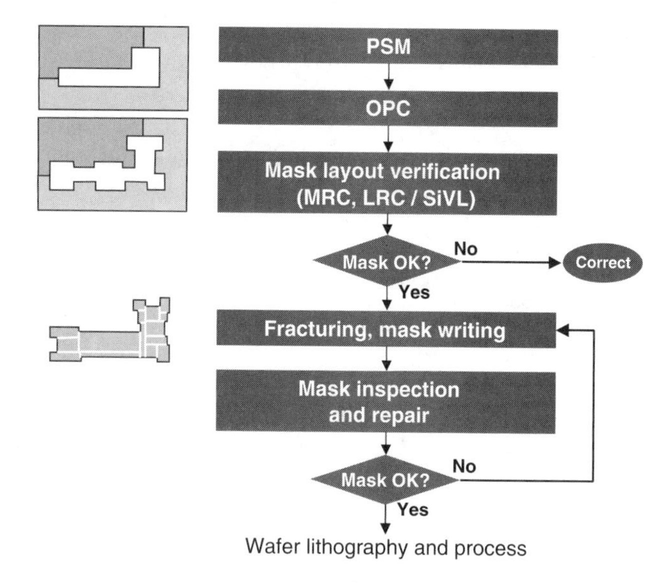

FIGURE 2.9. Subwavelength: from layout to masks.

For geometries of 90 nm and below, the lithography EDA tools combine OPC with other mask synthesis approaches such as phase shift mask (PSM), off-axis illumination and assist features (AF). For example, PSM is a technique where the optical equipment images dark features at critical dimensions with 0° illumination on one side and 180° illumination on the other side. There are additional manufacturing process rules needed such as minimal spacing and cyclic conflict avoidance, to avoid situations where the tool cannot map the phase.

In summary, lithography tools proceed through PSM, OPC, and AF to enhance resolution and make the mask more resistive to process variations. The process engineer can perform a verification of silicon vs. layout and a check of lithography rule compliance. If either fails, the engineer must investigate and correct, sometimes manually. If both succeed, another EDA tool "fractures" the design, subdividing the shapes into rectangles (trapezoids), which can be fed to the mask writing equipment. The engineer can then transfer the final shapes file to a database, such as the manufacturing-electron-beam-exposure system (MEBES). Foundry equipment uses the MEBES database (or other proprietary formats) to create the physical masks. The process engineer can also run a "virtual" stepper tool to pre-analyze the various stages of the stepper operation. After the foundry manufactures the masks, a mask inspection and repair step ensures that they conform to manufacturing standards.

Another area of design for manufacturing analysis is prediction of yield (cf. Figure 2.10). The design team would like to correlate some of the activities during route with actual yield. Problems with CMP, via voids and cross talk can cause chips to unexpectedly fail. EDA routing tools offer some solutions in the form of metal fill, via doubling and wire spacing. Library providers are starting to develop libraries for higher yields that take into account several yield failure mechanisms. There are tools that attempt to correlate these solutions with yield. Statistical timing analysis can correlate timing constraints to parametric circuit yield.

Finally, the process engineer can use tools to predict the behavior of transistor devices or processes. Technology computer aided design (TCAD) deals with the modeling and simulation of physical manufacturing process and devices. Engineers can model and simulate individual steps in the fabrication process. Likewise, the engineer can model and simulate devices, parasitics or electrical/thermal properties, therefore providing insights into their electrical, magnetic or optical properties.

For example, because of packing density, foundries may switch isolation technology for an IC from the local oxidation of silicon model toward the shallow trench isolation (STI) model. Under this model, the

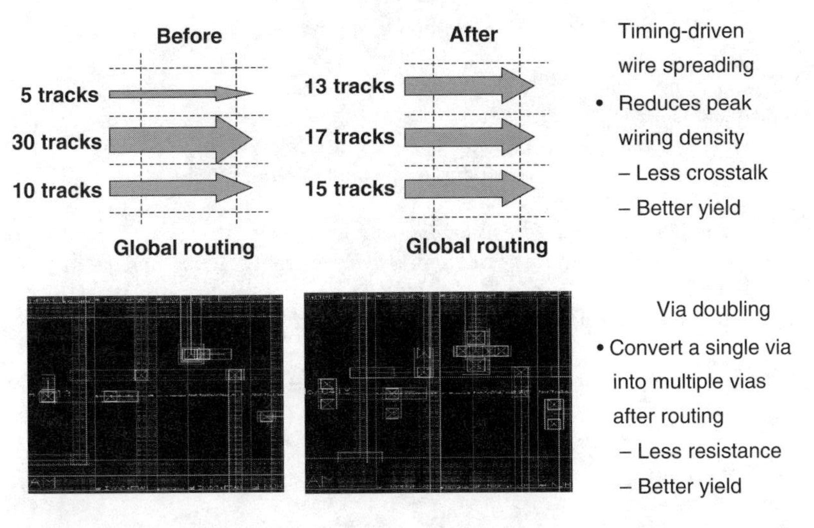

FIGURE 2.10. Yield enhancement features in routing.

process engineer can analyze breakdown stress, electrical behavior such as leakage, or material vs. process dependencies. Technology computer aided design tools can simulate STI effects, extract interconnect parasitics, such as diffusion distance, and determine SPICE parameters.

References

[1] M. Smith, *Application Specific Integrated Circuits*, Addison-Wesley, Reading, MA, 1997.
[2] A. Kuehlmann, *The Best of ICCAD*, Kluwer Academic Publishers, Dordrecht, 2003.
[3] D. Thomas and P. Moorby, *The Verilog Hardware Description Language*, Kluwer Academic Publishers, Dordrecht, 1996.
[4] D. Pellerin and D. Taylor, *VHDL Made Easy*, Pearson Education, Upper Saddle River, N.J., 1996.
[5] S. Sutherland, S. Davidson, and P. Flake, *SystemVerilog For Design: A Guide to Using SystemVerilog for Hardware Design and Modeling*, Kluwer Academic Publishers, Dordrecht, 2004.
[6] T. Groetker, S. Liao, G. Martin, and S. Swan, *System Design with SystemC*, Kluwer Academic Publishers, Dordrecht, 2002.
[7] G. Peterson, P. Ashenden, and D. Teegarden, *The System Designer's Guide to VHDL-AMS*, Morgan Kaufman Publishers, San Francisco, CA, 2002.
[8] K. Kundert and O. Zinke, *The Designer's Guide to Verilog-AMS*, Kluwer Academic Publishers, Dordrecht, 2004.
[9] M. Keating and P. Bricaud, *Reuse Methodology Manual for System-on-a-Chip Designs*, Kluwer Academic Publishers, Dordrecht, 1998.
[10] J. Bergeron, *Writing Testbenches*, Kluwer Academic Publishers, Dordrecht, 2003.
[11] E. Clarke, O. Grumberg, and D. Peled, *Model Checking*, The MIT Press, Cambridge, MA, 1999.
[12] S. Huang and K. Cheng, *Formal Equivalence Checking and Design Debugging*, Kluwer Academic Publishers, Dordrecht, 1998.
[13] R. Baker, H. Li, and D Boyce, *CMOS Circuit Design, Layout, and Simulation*, Series on Microelectronic Systems, IEEE Press, New York, 1998.
[14] L. Pillage, R. Rohrer, and C. Visweswariah, *Electronic Circuit and System Simulation Methods*, McGraw-Hill, New York, 1995.
[15] J. Elliott, *Understanding Behavioral Synthesis: A Practical Guide to High-Level Design*, Kluwer Academic Publishers, Dordrecht, 2000.
[16] S. Devadas, A. Ghosh, and K. Keutzer, *Logic Synthesis*, McGraw-Hill, New York, 1994.
[17] G. DeMicheli, *Synthesis and Optimization of Digital Circuits*, McGraw-Hill, New York, 1994.
[18] I. Sutherland, R. Sproull, and D. Harris, *Logical Effort: Defining Fast CMOS Circuits*, Academic Press, New York, 1999.

[19] N. Sherwani, *Algorithms for VLSI Physical Design Automation*, Kluwer Academic Publishers, Dordrecht, 1999.

[20] F. Nekoogar, *Timing Verification of Application-Specific Integrated Circuits (ASICs)*, Prentice-Hall PTR, Englewood Cliffs, NJ, 1999.

[21] K. Roy and S. Prasad, *Low Power CMOS VLSI: Circuit Design*, Wiley, New York, 2000.

[22] C-K.Cheng, J. Lillis, S. Lin, and N. Chang, *Interconnect Analysis and Synthesis*, Wiley, New York, 2000.

[23] W. Dally and J. Poulton, *Digital Systems Engineering*, Cambridge University Press, Cambridge, 1998.

[24] M. Abramovici, M. Breuer, and A. Friedman, *Digital Systems Testing and Testable Design*, Wiley, New York, 1995.

[25] A. Wong, *Resolution Enhancement Techniques in Optical Lithography*, SPIE Press, Bellingham, WA, 2001.

[26] International Technology Roadmap for Semiconductors (ITRS), 2004, URL: http://public.itrs.net/.

SECTION II

SYSTEM LEVEL DESIGN

3

Tools and Methodologies for System-Level Design

Shuvra Bhattacharyya
University of Maryland
College Park, Maryland

Wayne Wolf
Princeton University
Princeton, New Jersey

3.1 Introduction

System-level design has long been the province of board designers, but levels of integration have increased to the point that chip designers must concern themselves about system-level design issues. Because chip design is a less forgiving design medium — design cycles are longer and mistakes are harder to correct — system-on-chip (SoC) designers need a more extensive tool suite than may be used by board designers.

System-level design is less amenable to synthesis than are logic or physical design. As a result, system-level tools concentrate on modeling, simulation, design space exploration, and design verification. The goal of modeling is to correctly capture the system's operational semantics, which helps with both implementation and verification. The study of models of computation provides a framework for the description of digital systems. Not only do we need to understand a particular style of computation such as dataflow, but we also need to understand how different models of communication can reliably communicate with each other. Design space exploration tools, such as hardware/software codesign, develop candidate designs to understand trade-offs. Simulation can be used not only to verify functional correctness but also to supply performance and power/energy information for design analysis.

We will use video applications as examples in this chapter. Video is a leading-edge application that illustrates many important aspects of system-level design. Although some of this information is clearly specific to video, many of the lessons translate to other domains.

The next two sections briefly introduce video applications and some SoC architecture that may be the targets of system-level design tools. We will then study models of computation and languages for system-level modeling. Following this, we will survey simulation technique. We will close with a discussion of hardware/software codesign.

3.2 Characteristics of Video Applications

The primary use of SoCs for multimedia today is for video encoding — both compression and decompression. In this section, we review the basic characteristics of video compression algorithms and the implications for video SoC design.

Video compression standards enable video devices to inter-operate. The two major lines of video compression standards are MPEG and H.26x. The MPEG standards concentrate on broadcast applications, which allow for a more expensive compressor on the transmitter side in exchange for a simpler receiver. The H.26x standards were developed with videoconferencing in mind, in which both sides must encode and decode. The advanced video codec (AVC) standard, also known as H.264, was formed by the confluence of the H.26x and MPEG efforts.

Modern video compression systems combine lossy and lossless encoding methods to reduce the size of a video stream. Lossy methods throw away information as a result of which the uncompressed video stream is not a perfect reconstruction of the original; lossless methods do allow the information provided to them to be perfectly reconstructed. Most modern standards use three major mechanisms:

- The discrete cosine transform (DCT) together with quantization
- Motion estimation and compensation
- Huffman-style encoding

The first two are lossy while the third is lossless. These three methods leverage different aspects of the video stream's characteristics to encode it more efficiently.

The combination of DCT and quantization was originally developed for still images and is used in video to compress single frames. The DCT is a frequency transform that turns a set of pixels into a set of coefficients for the spatial frequencies that form the components of the image represented by the pixels. The DCT is preferred over other transforms because a two-dimensional (2D) DCT can be computed using two one-dimemsional (1D) DCTs, making it more efficient. In most standards, the DCT is performed on an 8×8 block of pixels. The DCT does not by itself lossily compress the image; rather, the quantization phase can more easily pick out information to acknowledge the structure of the DCT. Quantization throws out fine details in the block of pixels, which correspond to the high-frequency coefficients in the DCT. The number of coefficients set to zero is determined by the level of compression desired.

Motion estimation and compensation exploit the relationships between frames provided by moving objects. A reference frame is used to encode later frames through a motion vector, which describes the motion of a macroblock of pixels (16×16 in many standards). The block is copied from the reference frame into the new position described by the motion vector. The motion vector is much smaller than the block it represents. Two-dimensional correlation is used to determine the position of the macroblock's position in the new frame; several positions in a search area are tested using 2D correlation. An error signal encodes the difference between the predicted and the actual frames; the receiver uses that signal to improve the predicted picture.

MPEG distinguishes several types of frames: I (inter) frames, which are not motion-compensated; P (predicted) frames, which have been predicted from earlier frames; and B (bidirectional) frames, which have been predicted from both earlier and later frames.

The results of these lossy compression phases are assembled into a bit stream and compressed by using lossless compression such as Huffman encoding. This process reduces the size of the representation without further compromising image quality.

It should be clear that video compression systems are actually heterogeneous collections of algorithms. We believe that this is true of other applications of SoCs as well. A video platform must run several algorithms; those algorithms perform very different types of operations, imposing very different requirements on the architecture.

This has two implications for tools: first, we need a wide variety of tools to support the design of these applications; second, the various models of computation and algorithmic styles used in different parts of an application must at some point be made to communicate to create the complete system.

Several studies of multimedia performance on programmable processors have remarked on the significant number of branches in multimedia code. These observations contradict the popular notion of video as regular operations on streaming data. Fritts and Wolf [1] measured the characteristics of the MediaBench benchmarks.

They used path ratio to measure the percentage of instructions in a loop body that were actually executed. They found that the average path ratio of the MediaBench suite was 78%, which indicates that a significant number of loops exercise data-dependent behavior. Talla et al. [2] found that most of the available parallelism in multimedia benchmarks came from inter-iteration parallelism.

3.3 Other Application Domains

Video and multimedia are not the only application domains for SoCs. Communications and networking are the other areas in which SoCs provide cost/performance benefits. In all these domains, the SoC must be able to handle multiple simultaneous processes. However, the characteristics of those processes do vary. Networking, for example, requires a lot of packet-independent operations. While some networking tasks do require correlating multiple packets, the basic work is packet independent. The large extent of parallelism in packet-level processing can be exploited in the micro-architecture. In the communications world, SoCs are used today primarily for baseband processing, but we should expect SoCs to take over more traditional high-frequency radio functions over time. Since radio functions can operate at very high frequencies, the platform must be carefully designed to support these high rates while providing adequate programmability of radio functions. We should expect highly heterogeneous architectures for high-frequency radio operations.

3.4 Platform Characteristics

Many SoCs are heterogeneous multiprocessors and the architectures designed for multimedia applications are no exceptions. In this section, we review several SoCs, including some general-purpose SoC architectures as well as several designed specifically for multimedia applications.

Two very different types of hardware platforms have emerged for large-scale applications. On the one hand, many custom SoCs have been designed for various applications. These custom SoCs are customized by loading software onto them for execution. On the other hand, platform field-programmable gate arrays (FPGAs) provide FPGA fabrics along with CPUs and other components; the design can be customized by programming the FPGA as well as the processor(s). These two styles of architecture represent different approaches for SoC architecture and they require very different sorts of tools: custom SoCs require large-scale software support, while platform FPGAs are well suited to hardware/software codesign.

3.4.1 Custom System-on-Chip Architectures

The Viper chip [3], shown in Figure 3.1, was designed for high-definition video decoding and set-top box support. Viper is an instance of the Philips Nexperia™ architecture, which is a platform for multimedia applications.

The Viper includes two CPUs: a MIPS32 processor and a Trimedia processor. The MIPS32 is an RISC architecture. The Trimedia is a 5-issue VLIW designed to support video applications. The Trimedia handles all video tasks. It also schedules operations since it is faster than the MIPS32. The MIPS32 runs the operating system and the application stack provided by the service provider (i.e., non-Philips-provided applications). It also performs graphics operations and handles access requests (Figure 3.1).

Each processor has its own bus. A third large, point-to-point bus connects to the off-chip bulk memory used for video memory. Several bridges connect the three high-speed busses. The Viper includes many on-chip devices. The peripheral system provides various general-purpose types of I/Os such as GPIO and IEEE 1394. The audio subsystem includes I/O units that support the SPDIF standard. The video system includes a number of units that video perform processing as well as I/O: picture input processing, a video scaler, an advanced image composition processor, a processor for the MPEG system layer, etc. The infrastructure system includes a memory management unit and the busses.

The Texas Instruments OMAP processor (http://www.omap.com) is designed for mobile multimedia. As shown in Figure 3.2, it includes two processors, an ARM9 CPU and a TI C5x-series digital signal, image and video processing (DSP), connected by a shared memory. OMAP implementations include a wide variety of peripherals. Some peripherals are shared between the two processors, including UARTs and mailboxes for interprocessor communication. Many peripherals, however, are owned by one processor. The DSP owns serial ports, DMA, timers, etc. The ARM9 owns the majority of the I/O devices, including USB, LCD controller, camera interface, etc.

The ST Microelectronics Nomadik is also designed for mobile multimedia applications. (Both the TI OMAP and Nomadik implement the OMAPI software standard.) As shown in Figure 3.3, it can be viewed as a hierarchical multiprocessor. The overall architecture is a heterogeneous multiprocessor organized around a bus, with an ARM processor attached to audio and video processors. But the audio and video processors are each heterogeneous multiprocessors in their own right. The video processor provides an extensive set of hardwired accelerators for video operations organized around two busses; a small processor known as an MMDSP+ provides programmability. The audio processor makes heavier use of an MMDSP+ processor to implement audio-rate functions; it also includes two busses.

All the above SoCs have been architected for particular markets. The ARM multiprocessor core takes a more general-purpose approach. Its MPCore™ is a core designed to be used in SoCs. It can implement

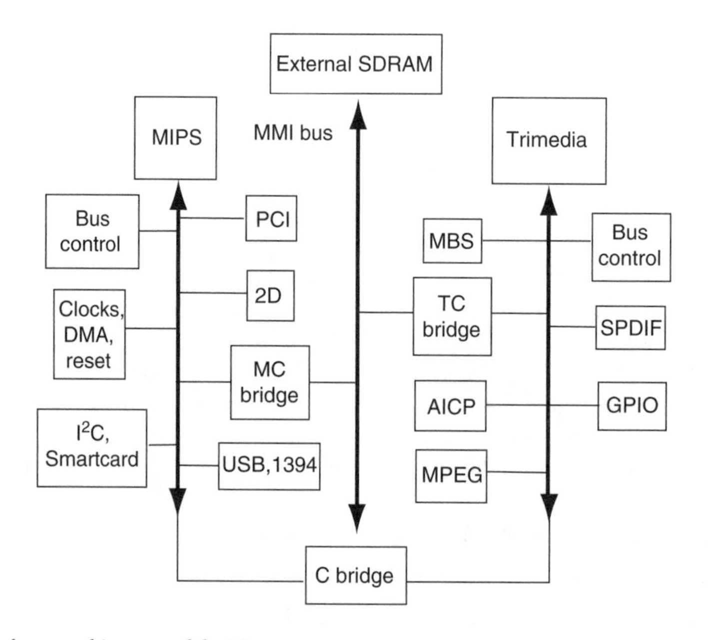

FIGURE 3.1 Hardware architecture of the Viper.

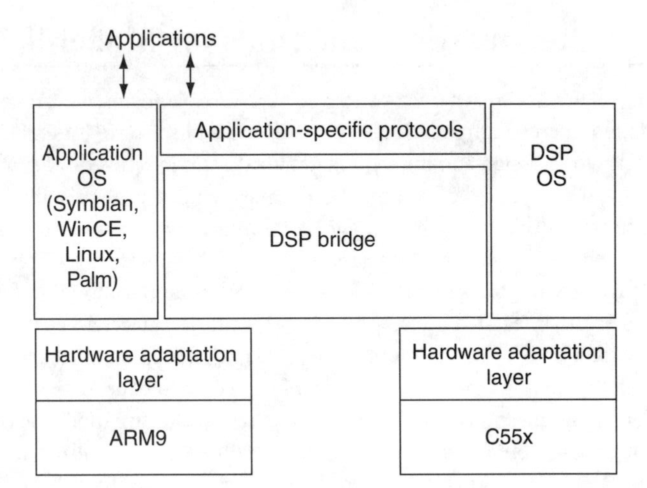

FIGURE 3.2 Hardware and software architecture of the Texas Instruments OMAP.

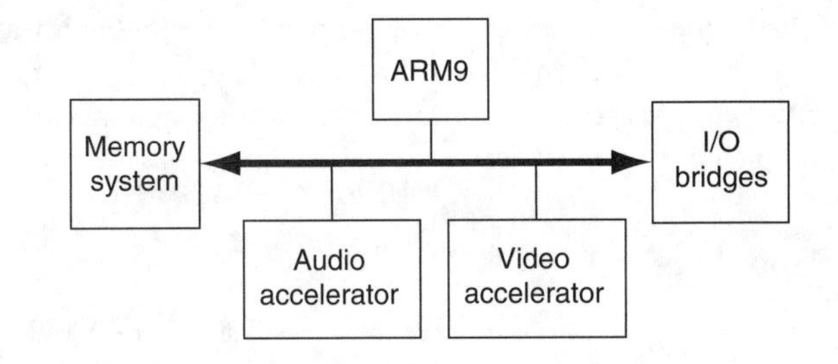

FIGURE 3.3 Architecture of the ST Micro Nomadik.

up to four ARM cores. Those cores can be arranged in any combination of symmetric and asymmetric multiprocessing. Symmetric multiprocessing requires cache coherence mechanisms, which may slow down execution to some extent and cost additional energy, but simplify software design. Asymmetric multiprocessing does not rely on shared memory.

3.4.2 Platform Field-Programmable Gate Arrays

Field-programmable gate arrays [4] have been used for many years to implement logic designs. The FPGA provides a more general structure than programmable logic devices, allowing denser designs. They are less energy-efficient than custom ASICs but do not require the long ASIC design cycle.

Field-programmable gate array logic has now become dense enough that manufacturers provide platform FPGAs. While this term has been used in more than one way, most platform FPGAs provide one or more CPUs in addition to a programmable FPGA fabric. Platform FPGAs provide a very different sort of heterogeneous platform than custom SoCs. The FPGA fabric allows the system designer to implement new hardware functions. While they generally do not allow the CPU itself to be modified, the FPGA logic is a closely coupled device with high throughput to the CPU and to memory. The CPU can also be programmed using standard tools to provide functions that are not well suited to FPGA implementation.

The ATMEL AT94 family (http://www.atmel.com) provides an RISC CPU along with an FPGA core. The AVR CPU core is relatively simple, without a memory management unit. The FPGA core is reprogrammable and configured at power-up.

The Xilinx Virtex II Pro (http://www.xilinx.com) is a high-performance platform FPGA. Various configurations include one to four PowerPCs. The CPUs are connected to the FPGA fabric through the PowerPC bus. The large FPGA fabric can be configured to provide combinations of logic and memory functions.

3.5 Models of Computation and Tools for Model-Based Design

Based on our discussion of application characteristics and hardware platforms, we can now consider tools for multimedia system design. Increasingly, developers of hardware and software for embedded computer systems are viewing aspects of the design and implementation processes in terms of domain-specific models of computation. A domain-specific model of computation is designed to represent applications in a particular functional domain such as DSP; control system design; communication protocols or more general classes of discrete, control-flow intensive decision-making processes; graphics; and device drivers. For discussions of some representative languages and tools that are specialized for these application domains, see [5–13] .

Chapter 4 includes a broad, integrated review of domain-specific languages and programming models for embedded systems. This section discusses in more detail particular modeling concepts and tools that are particularly relevant for multimedia systems. Since DSP techniques constitute the "computational core" of many multimedia applications, a significant part of this section focuses on effective modeling of DSP functionality.

As described in Chapter 4, domain-specific computational models specialize the types of functional building blocks (components) that are available to the programmer and the interactions between components to characterize more intuitively or explicitly relevant characteristics of the targeted class of applications compared to general-purpose programming models. In doing so, domain-specific models of computation often exchange some of the *Turing-complete* expressive power (ability to represent arbitrary computations) of general-purpose models and often also a large body of legacy infrastructure for advantages such as increased intuitive appeal, support for formal verification, and more thorough optimization with respect to relevant implementation metrics.

3.5.1 Dataflow Models

For most DSP applications, a significant part of the computational structure is well suited to modeling in a dataflow model of computation. In the context of programming models, dataflow refers to a modeling methodology where computations are represented as directed graphs in which vertices (actors) represent functional components and edges between actors represent first-in-first-out (FIFO) channels that buffer data values (tokens) as they pass from an output of one actor to an input of another. Dataflow actors can represent computations of arbitrary complexity; typically in DSP design environments, they are specified using conventional languages such as C or assembly language, and their associated tasks range from simple, "fine-grained" functions such as addition and multiplication to "coarse-grain" DSP kernels or subsystems such as FFT units and adaptive filters. The development of application modeling and analysis techniques based on dataflow graphs was inspired significantly by the computation graphs of Karp and Miller [14], and the process networks of Kahn [15]. A unified formulation of dataflow modeling principles, as they apply to DSP design environment, is provided by the dataflow process networks model of computation of Lee and Parks [16].

A dataflow actor is enabled for execution whenever it has sufficient data on its incoming edges (i.e., in the associated FIFO channels) to perform its specified computation. An actor can execute whenever it is enabled (data-driven execution). In general, the execution of an actor results in some number of tokens being removed (consumed) from each incoming edge, and some number being placed (produced) on each outgoing edge. This production activity in general leads to the enabling of other actors.

The order in which actors execute, called the schedule, is not part of a dataflow specification, and is constrained only by the simple principle of data-driven execution defined above. This is in contrast to many alternative computational models, such as those that underlie procedural languages, in which execution order is overspecified by the programmer [17]. The schedule for a dataflow specification may be determined at compile time (if sufficient static information is available), at run-time, or using a mixture of compile- and run-time techniques. A particularly powerful class of scheduling techniques, referred to as quasi-static scheduling (see, e.g., [18]), involves most, but not all of the scheduling decisions being made at compile time.

Figure 3.4 shows an illustration of a video processing subsystem that is modeled using dataflow semantics. This is a design, developed using the Ptolemy II tool for model-based embedded system design [19]

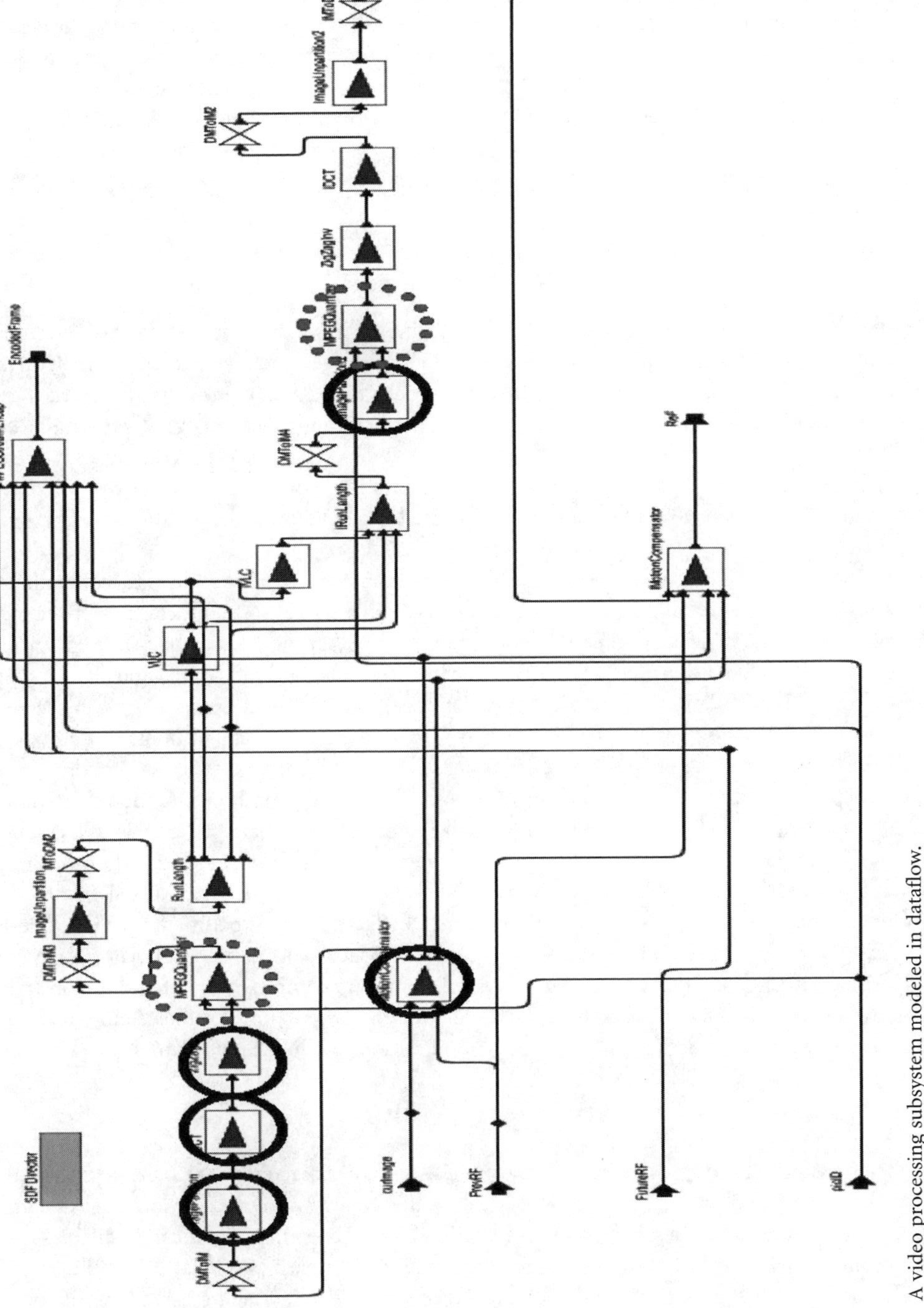

FIGURE 3.4 A video processing subsystem modeled in dataflow.

of an MPEG2 subsystem for encoding the P frames that are processed by an enclosing MPEG2 encoder system. A thorough discussion of this MPEG2 system and its comparison to a variety of other modeling representations is presented in [20]. The components in the design of Figure 3.4 include actors for the DCT, zig-zag scanning, quantization, motion compensation and run length coding. The arrows in the illustration correspond to the edges in the underlying dataflow graph. The actors and their interactions all conform to the semantics of *synchronous dataflow* (SDF), which is a restricted form of dataflow that is efficient for describing a broad class of DSP applications and has particularly strong formal properties and optimization advantages [21,22]. Specifically, SDF imposes the restriction that the number of tokens produced and consumed by each actor on each incident edge is constant. Many commercial DSP design tools have been developed that employ semantics that are equivalent to or closely related to SDF. Examples of such tools include Agilent's ADS, Cadence's SPW (now developed by CoWare), and MCCI's Autocoding Toolset.

3.5.2 Dataflow Modeling for Video Processing

In the context of video processing, SDF permits accurate representation of many useful subsystems, such as the P-frame encoder shown in Figure 3.4. However, such modeling is often restricted to a highly coarse level of granularity, where actors process individual frames or groups of successive frames on each execution. Modeling at such a coarse granularity can provide compact, top-level design representations, but greatly limits the benefits offered by the dataflow representation since most of the computation is subsumed by the general-purpose, intra-actor program representation. For example, the degree of parallel processing and memory management optimizations exposed to a dataflow-based synthesis tool becomes highly limited at such coarse levels of actor granularity.

A number of alternative dataflow modeling methods have been introduced to address this limitation of SDF modeling for video processing and more generally, multidimensional signal processing applications. For example, the multidimensional synchronous dataflow (MD-SDF) model extends SDF semantics to allow constant-sized, *n*-dimensional vectors of data to be transferred across graph edges, and provides support for arbitrary sampling lattices and lattice-changing operations [23]. The computer vision synchronous dataflow (CV-SDF) model is designed specifically for computer vision applications, and provides a notion of *structured buffers* for decomposing video frames along graph edges; accessing neighborhoods of image data from within actors, in addition to the conventional production and consumption semantics of dataflow, and allowing actors to access efficiently previous frames of image data [24,25]. Blocked dataflow (BLDF) is a metamodeling technique for efficiently incorporating hierarchical, block-based processing of multidimensional data into a variety of dataflow modeling styles, including SDF and MD-SDF [20].

Multidimensional synchronous dataflow, CV-SDF, and BLDF are still at experimental stages of development and to our knowledge, none of them has yet been incorporated into a commercial design tool, although practical image and video processing systems of significant complexity have been demonstrated using these techniques. Integrating effective dataflow models of computation for image and video processing into commercial DSP design tools is an important area for further exploration.

3.5.3 Control Flow

As described previously, modern video processing applications are characterized by some degree of control flow processing for carrying out data-dependent configuration of application tasks and changes across multiple application modes. For example, in MPEG2 video encoding, significantly different processing is required for I, P, and B frames. Although the processing for each particular type of frame (I, P, or B) conforms to the SDF model, as illustrated for P-frame processing in Figure 3.4, a layer of control flow processing is needed to integrate efficiently these three types of processing methods into a complete MPEG2 encoder design. The SDF model is not well suited for performing this type of control flow processing and, more generally, for any functionality that requires dynamic communication patterns or activation/deactivation across actors.

A variety of alternative models of computation have been developed to address this limitation, and integrate flexible control flow capability with the advantages of dataflow modeling. In Buck's Boolean dataflow model [26], and the subsequent generalization as integer-controlled dataflow [27], provisions for such flexible processing were incorporated without departing from the framework of dataflow, and in a manner that facilitates construction of efficient hybrid compile-/run-time schedules. In Boolean dataflow, the number of tokens produced or consumed on an edge is either fixed or is a two-valued function of a control token present on a control terminal of the same actor. It is possible to extend important SDF analysis techniques to Boolean dataflow graphs by employing symbolic variables. In particular, in constructing a schedule for Boolean dataflow actors, Buck's techniques attempt to derive a quasi-static schedule, where each conditional actor execution is annotated with the run-time condition under which the execution should occur. Boolean dataflow is a powerful modeling technique that can express arbitrary control flow structures. However, as a result, key formal verification properties of SDF, such as bounded memory and deadlock detection, are lost in the context of general Boolean dataflow specifications.

In recent years, several modeling techniques have also been proposed that enhance expressive power by providing precise semantics for integrating dataflow or dataflow-like representations with finite-state machine (FSM) models. These include El Greco [28], which has evolved into the Synopsys Cocentric System Studio, and that provides facilities for "control models" to configure dynamically specification parameters; *charts (pronounced starcharts) with heterochronous dataflow as the concurrency model [29]; the FunState intermediate representation [30]; the DF* framework developed at K. U. Leuven [31]; and the control flow provisions in bounded dynamic dataflow [32].

Figure 3.5 shows an illustration of a specification of a complete MPEG2 video encoder system that builds on the P-frame-processing subsystem of Figure 3.4, and employs multiple dataflow graphs nested within an FSM representation. Details on this specification can be found in [20].

Cocentric System Studio is a commercial tool for system design that employs integrated representations using FSMs and dataflow graphs. We will discuss the Cocentric tool in more detail later in this chapter.

3.5.4 Ptolemy

The Ptolemy project at U.C. Berkeley has had considerable influence on models of computation for DSP and multimedia, and also on the general trend toward viewing embedded systems design in terms of models of computation. The first origins of this project are in the BLOSIM [33] tool, which developed block diagram simulation capability for signal processing systems. Work on BLOSIM led to the development of the SDF formalism [21] and the Gabriel design environment [10], which provided simulation, static scheduling, and single- and multiprocessor software synthesis capability for SDF-based design.

Ptolemy (now known as Ptolemy Classic) is the third-generation tool that succeeded Gabriel [34]. The design of Ptolemy Classic emphasized efficient modeling and simulation of embedded systems based on the interaction of heterogeneous models of computation. A key motivation was to allow designers to represent each subsystem of a design in the most natural model of computation associated with that subsystem, and allow subsystems expressed in different models of computation to be integrated seamlessly into an overall system design. A key constraint imposed by the Ptolemy Classic approach to heterogeneous modeling is the concept of *hierarchical heterogeneity*. It is widely understood that in hierarchical modeling, a system specification is decomposed into a set C of subsystems in which each subsystem can contain one or more hierarchical components, each of which represents another subsystem in C. Under hierarchical heterogeneity, each subsystem in C must be described using a uniform model of computation, but the nested subsystem associated with a hierarchical component H can be expressed in a model of computation that is different from the model of computation that expresses the subsystem containing H.

Thus, under hierarchical heterogeneity, the integration of different models of computation must be achieved entirely through the hierarchical embedding of heterogeneous models. A key consequence is that whenever a subsystem S_1 is embedded in a subsystem S_2 that is expressed in a different model of computation, the subsystem S_1 must be abstracted by a hierarchical component in S_2 that conforms to the model of computation associated with S_2. This provides precise constraints for interfacing different models

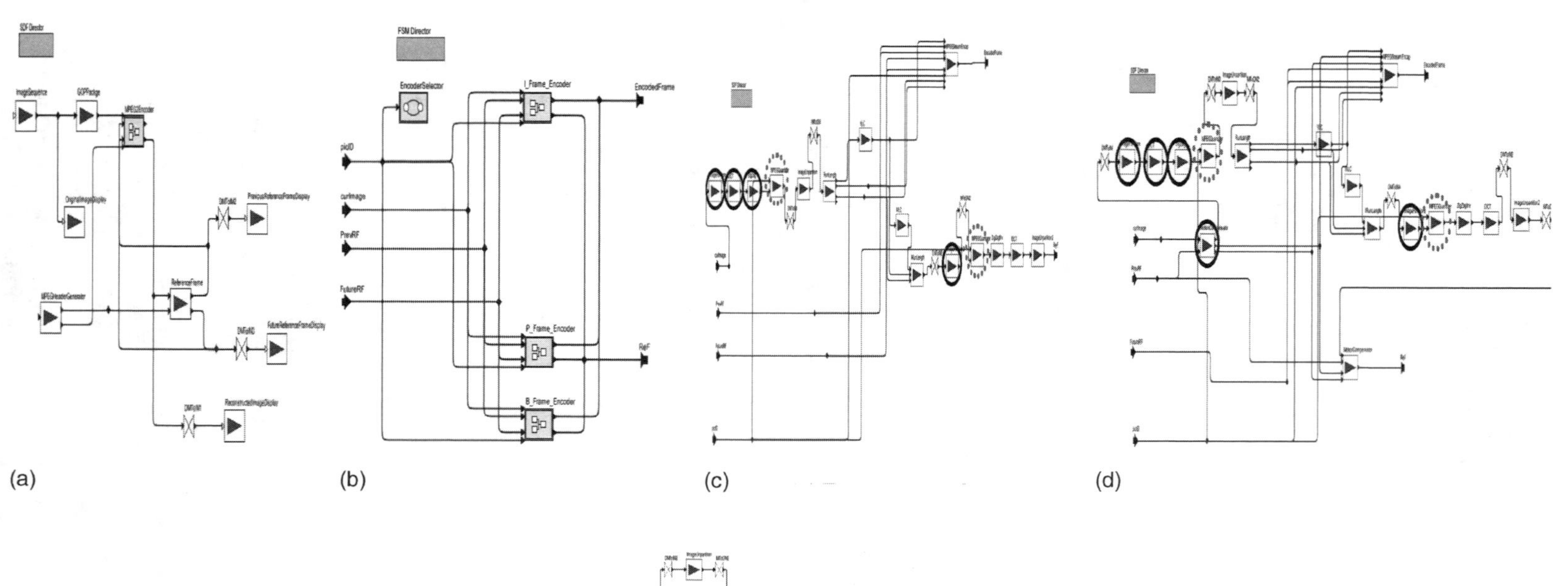

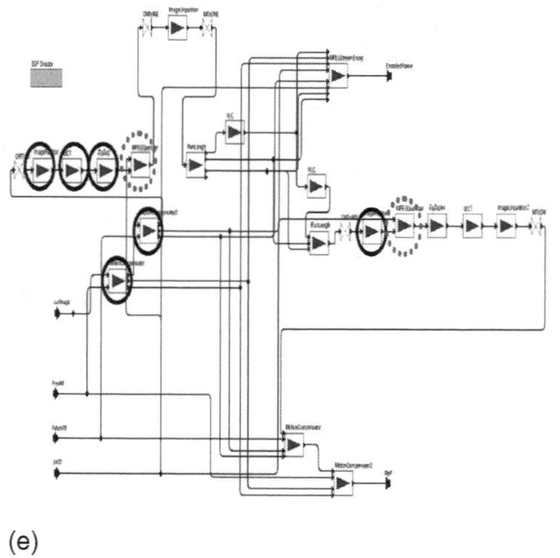

FIGURE 3.5 An MPEG2 video encoder specification. (a) MPEG2 encoder (top); (b) inside the FSM; (c) I-frame encoder; (d) P-frame encoder; and (e) B-frame encoder.

of computation. Although these constraints may not always be easy to conform to, they provide a general and unambiguous convention for heterogeneous integration and perhaps even more importantly, the associated interfacing methodology allows each subsystem to be analyzed using the techniques and tools available for the associated model of computation.

Ptolemy Classic was developed through a highly flexible, extensible, and robust software design, and this has facilitated experimentation with the underlying modeling capabilities in various aspects of embedded systems design by many research groups. Other major areas of contribution associated with development of Ptolemy Classic include hardware/software codesign, as well as further contributions in dataflow-based modeling and synthesis (see, e.g., [23,35,36]).

The current incarnation of the Ptolemy project, called Ptolemy II, is a Java-based tool that furthers the application of model-based design and hierarchical heterogeneity [19], and provides an even more malleable software infrastructure for experimentation with new techniques involving models of computation.

An important theme in Ptolemy II is the reuse of actors across multiple models and computational models. Through an emphasis in Ptolemy II on support for *domain polymorphism*, the same actor definition can in general be applicable across a variety of models of computation. In practice, domain polymorphism greatly increases reuse of actor code. Techniques based on interface automata [37] have been developed to characterize systematically the interactions between actors and models of computation, and reason about their compatibility (i.e., whether or not it makes sense to instantiate the actor in specifications that are based on the model) [38].

3.5.5 Compaan

MATLAB® is one of the most popular programming languages for algorithm development and high-level functional simulation for DSP applications. In the Compaan project developed at Leiden University, systematic techniques have been developed for synthesizing embedded software and FPGA-based hardware implementations from a restricted class of MATLAB programs known as parameterized, static nested loop programs [39]. In Compaan, an input MATLAB specification is first translated into an intermediate representation based on the Kahn process network model of computation [15]. The Kahn process network model is a general model of data-driven computation that subsumes as a special case as in the dataflow process networks mentioned earlier in this chapter. Like dataflow process networks, Kahn process networks also consist of concurrent functional modules that are connected by FIFO buffers with non-blocking writes and blocking reads; however, unlike in the dataflow process network model, modules in Kahn process networks do not necessarily have their execution decomposed *a priori* into well-defined, discrete units of execution [16].

Through its aggressive dependence analysis capabilities, Compaan combines the widespread appeal of MATLAB at the algorithm development level, with the guaranteed determinacy, compact representation, simple synchronization, and distributed control features of Kahn process networks for efficient hardware/software implementation.

Technically, the Kahn process networks derived in Compaan can be described as equivalent cyclo-static dataflow graphs [40], which we discuss in more detail later in this chapter, and therefore fall under the category of dataflow process networks. However, these equivalent cyclo-static dataflow graphs can be very large and unwieldy to work with and, therefore, analysis in terms of the Kahn process network model is often more efficient and intuitive.

Development of the capability for translation of MATLAB to Kahn process networks was originally developed by Kienhuis et al. [41], and this capability has since then evolved into an elaborate suite of tools for mapping Kahn process networks into optimized implementations on heterogeneous hardware/software platforms consisting of embedded processors and FPGAs [39]. Among the most interesting optimizations in the Compaan tool suite are dependence analysis mechanisms that determine the most specialized form of buffer implementation, with respect to reordering and multiplicity of buffered values, for implementing interprocess communication in Kahn process networks [42].

3.5.6 CoWare

CoWare, developed originally by IMEC and now by CoWare, Inc., is a tool for system-level design, simulation, and hardware/software synthesis of DSP systems [43]. Integration of dataflow and control flow modeling is achieved in CoWare through networks of processes that communicate through remote procedure call semantics. Processes can have one or more flows of control, which are called threads, and through a remote procedure call, a process can invoke a thread in another process that is connected to it. A thread in a process can also execute autonomously, independent of any remote procedure call, in what is effectively an infinite loop that is initiated at system start-up. During run time, the overhead of switching between threads can be significant, and furthermore, such points of switching present barriers for important compiler optimizations. Therefore, remote procedure calls are inlined during synthesis, thereby merging groups of compatible, communicating processes into monolithic processes.

Buffered communication of data can also be carried out between processes. A variety of protocols can be used for this purpose, including support for the FIFO buffering associated with dataflow-style communication. Thus CoWare provides distinct mechanisms for interprocess control flow and dataflow that can be flexibly combined within the same system or subsystem specification. This is in contrast to control flow/dataflow integration techniques associated with methods such as Boolean dataflow and general dataflow process networks, where all intermodule control flow is mapped into a more general dataflow framework, and also with techniques based on hierarchical heterogeneity, where control flow and dataflow styles are combined only through hierarchical embedding.

3.5.7 Cocentric System Studio

At Synopsys, a design tool for hardware/software cosimulation and synthesis has been developed based on principles of hierarchical heterogeneity, in particular, hierarchical combinations of FSMs and dataflow graphs. This tool, originally called El Greco [28], along with its predecessor COSSAP [44], which was based on a close variant of SDF semantics [45], evolved into Synopsys's Cocentric System Studio.

Cocentric incorporates two forms of dataflow modeling — cyclo-static dataflow modeling for purely deterministic dataflow relationships, and a general form of dataflow modeling for expressing dynamic dataflow constructs. Cyclo-static dataflow developed at Katholieke Universiteit Leuven, is a useful extension of SDF, in which the numbers of tokens produced and consumed by an actor are allowed to vary at run time as long as the variation takes the form of a fixed, periodic pattern [40]. At the expense of some simplicity in defining actors, cyclo-static dataflow has been shown to have many advantages over SDF, including more economical buffering of data, more flexible support for hierarchical specifications, improved ability to expose opportunities for behavioral optimization, and facilitation of more compact design representations (see, e.g., [40,46,47]).

Control flow models in Cocentric can be integrated hierarchically with dataflow models. Three types of hierarchical control flow models are available: an *or-model* represents conditional transitions, and associated actions, among a collection of mutually exclusive states; an *and-model* represents a group of subsystems that execute in parallel with broadcast communication across the subsystems; and a *gated model* is used to conditionally switch models between states of activity and suspension. To increase the potential for compile-time analysis and efficient hardware implementation, a precise correspondence is maintained between the control semantics of Cocentric and the synchronous language Esterel [48].

3.5.8 Handel-C

One important direction in which innovative models of computation are being applied to embedded systems is in developing variants of conventional procedural languages, especially variants of the C language, for the design of hardware and for hardware/software codesign. Such languages are attractive because they are based on a familiar syntax, and are thus easier to learn and to adapt to existing designs. A prominent example of one such language is Handel-C, which is based on embedding concepts of the abstract Handel language [49] into a subset of ANSI C. Handel in turn is based on the communicating sequential processes (CSPs) model of computation [50]. CSP provides a simple and intuitive form of support to coordinate parallel execution of multiple hardware subsystems through its mechanism of synchronized, point-to-point communication channels for interprocess communication. Handel was developed at Oxford University and prototyped originally by embedding its semantics in the SML language [51]; subsequently, the concepts of Handel were ported to ANSI C and commercialized through Celoxica Limited as Handel-C (see, e.g., [52]).

In Handel-C, parallelism is specified explicitly by the programmer using a *par* statement, and communication across parallel subsystems through CSP semantics is specified using an associated *chan* (short for "channel") construct.

Handel-C has been targeted to platform FPGAs, such as Altera Excalibur and Xilinx Virtex II Pro.

3.5.9 Simulink

At present, Simulink [53], developed by the The MathWorks, is perhaps the most widely used commercial tool for model-based design of DSP hardware and software. Simulink provides a block-diagram interface and extensive libraries of predefined blocks; highly expressive modeling semantics with support for continuous time, discrete time, and mixed-signal modeling; support for fixed-point data types; and capabilities for incorporating code in a variety of procedural languages, including Matlab and C. Simulink can be augmented with various capabilities to enhance greatly its utility in DSP and multimedia system implementation. For example, various add-on libraries provide rich collections of blocks geared toward signal processing, communications, and image/video processing. The Real-time Workshop provides code generation capabilities to translate automatically Simulink models into ANSI C. Stateflow provides the capability to augment Simulink with sophisticated control flow modeling features. The Xilinx System Generator is a plug-in tool for Simulink, which generates synthesizable hardware description language code targeted to Xilinx devices. Supported devices include the Virtex-II Pro platform FPGA, which was described earlier in this chapter. The Texas Instruments Embedded Target links Simulink and the Real-time Workshop with the Texas Instruments Code Composer Studio to provide a model-based environment for design and code generation that is targeted to fixed- and floating-point programmable digital signal processors in the Texas Instruments C6000 series.

3.5.10 Prospects for Future Development of Tools

The long list (over 300 at present) of other third-party products and services that work with Simulink demonstrates the significance of this tool in the embedded systems industry as well as the growing adoption of model-based design techniques. Although model-based design is used widely for signal processing, which is an important building block for multimedia applications, model-based techniques are not yet so extensively used for the overall multimedia systems design process. This section has discussed key challenges in developing effective design tools that are centered around domain-specific computational models for multimedia. These include efficient, integrated representation of multidimensional data streams, high-level control flow, and reconfigurable application behavior and closer interaction, such as that facilitated by the relationship between Cocentric and Esterel, of models of computation with design representations that are used in the back end of the hardware/software synthesis process. As these challenges are studied further, and more experience is gained in the application of models of computation to multimedia hardware and software implementation, we expect increasing deployment of model-based tools in the multimedia domain.

3.6 Simulation

Simulation is very important in SoC design. Simulation is not limited to functional verification, as with logic design. System-on-chip designers use simulation to measure the performance and power consumption of their SoC designs. This is due in part to the fact that much of the functionality is implemented in software, which must be measured relative to the processors on which it runs. It is also due to the fact that the complex input patterns inherent in many SoC applications do not lend themselves to closed-form analysis.

SystemC (http://www.systemc.org) is a simulation language that is widely used to model SoCs. SystemC leverages the C++ programming language to build a simulation environment. SystemC classes allow designers to describe a digital system using a combination of structural and functional techniques. SystemC supports simulation at several levels of abstraction. Register-transfer level simulations, for example, can be performed with the appropriate SystemC model. SystemC is most often used for more abstract models. A common type of model built in SystemC is a transaction-level model. This style of modeling describes the SoC as a network of communicating machines, with explicit connections between the models and functional descriptions for each model. The transaction-level model describes how data are moved between the models.

Hardware/software cosimulators are multimode simulators that simultaneously simulate different parts of the system at different levels of detail. For example, some modules may be simulated in register-transfer mode while software running on a CPU is simulated functionally. Cosimulation is particularly useful for debugging the hardware/software interface, such as debugging driver software. The Seamless cosimulator from Mentor Graphics is a well-known example of a hardware/software cosimulator. The VaST Systems CoMET simulator is designed to simulate networks of processors and hardware devices.

Functional validation, performance analysis, and power analysis of SoCs require simulating large number of vectors. Video and other SoC applications allow complex input sequences. Even relatively compact tests can take up tens of millions of bytes. These long input sequences are necessary to run the SoC through a reasonable number of the states implemented in the system. The large amounts of memory that can be integrated into today's systems, whether they are on-chip or off-chip, allow the creation of SoCs with huge numbers of states that require long simulation runs.

Simulators for software running on processors have been developed over the past several decades. Both computer architects and SoC designers need fast simulators to run the large benchmarks required to evaluate architectures. As a result, a number of simulation techniques covering a broad range of accuracy and performance have been developed.

A simple method of analyzing CPU is to sample the program counter (PC) during program execution. The Unix *prof* command is an example of a PC-sampling analysis tool. Program counter sampling is subject to the same limitations on sampling rate as any other sampling process, but sampling rate is usually not a major concern in this case. A more serious limitation is that PC sampling gives us relative performance but not absolute performance. A sampled trace of the PC tells us where the program spent its time during execution, which gives us valuable information about the relative execution time of program modules that can be used to optimize the program. But it does not give us the execution time on a particular platform — especially if the target platform is different from the platform on which the trace is taken — and so we must use other methods to determine the real-time performance of programs.

Some simulators concentrate on the behavior of the cache, given the major role of the cache in determining overall system performance. The *dinero* simulator is a well-known example of a cache simulator. These simulators generally work from a trace generated from the execution of a program. The program is to be analyzed with additional code that records the execution behavior of the program. The *dinero* simulator then reconstructs the cache behavior from the program trace. The architect can view the cache in various states or calculate cache statistics.

Some simulation systems model the behavior of the processor itself. A functional CPU simulator models instruction execution and maintains the state of the programming model, that is, the set of registers visible to the programmer. The functional simulator does not, however, model the performance or energy consumption of the program's execution.

A cycle-accurate simulator of a CPU is designed to predict accurately the number of clock cycles required to execute every instruction, taking into account pipeline and memory system effects. The CPU model must therefore represent the internal structure of the CPU accurately enough to show how resources in the processor are used. The SimpleScalar simulation tool [54] is a well-known toolkit for building cycle-accurate simulators. SimpleScalar allows a variety of processor models to be built by a

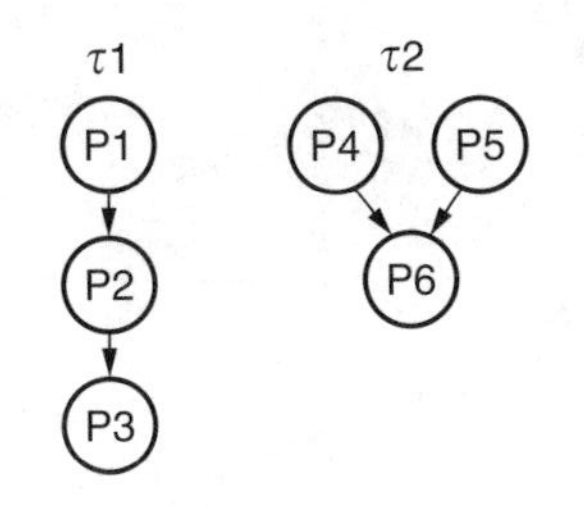

FIGURE 3.6 A task graph.

combination of parameterization of existing models and linking new simulation modules into the framework.

Power simulators are related to cycle-accurate simulators. Accurate power estimation requires models of the CPU micro-architecture, which are at least as detailed as those used for performance evaluation. A power simulator must model all the important wires in the architecture since capacitance is a major source of power consumption. Wattch [55] and SimplePower [56] are the two best-known CPU power simulators.

3.7 Hardware/Software Cosynthesis

Hardware/software cosynthesis tools allow system designers to explore architectural trade-offs. These tools take a description of a desired behavior that is relatively undifferentiated between hardware and software. They produce a heterogeneous hardware architecture and the architecture for the software to run on that platform. The software architecture includes the allocation of software tasks to the processing elements of the platform, and the scheduling of computation and communication.

The functional description of an application may take several forms. The most basic is a task graph, as shown in Figure 3.6. A task graph is a simplified form of dataflow graph in which each graph component runs at its own rate; however, the rates of the tasks need not be integrally related. The graph describes data dependencies between a set of processes. Each component of the graph (i.e., each set of connected nodes) forms a task. Each task runs periodically and every task can run at a different rate. The task graph model generally does not concern itself with the details of operations within a process. The process is characterized by its execution time. Several variations of task graphs that include control information have been developed. In these models, the output of a process may enable one of several different processes.

An alternative representation for behavior is a programming language. Several different codesign languages have been developed and languages such as SystemC have been used for cosynthesis as well. These languages may make use of constructs to describe parallelism that were originally developed for parallel programming languages. Such constructs are often used to capture operator-level concurrency. The subroutine structure of the program can be used to describe task-level parallelism.

The most basic form of hardware/software cosynthesis is hardware/software partitioning. As shown in Figure 3.7, this method maps the design into an architectural template. The basic system architecture is bus-based, with a CPU and one or more custom hardware processing elements attached to the bus. The type of CPU is determined in advance, which allows the tool to estimate accurately software performance. The tool must decide what functions go into the custom processing elements; it must also schedule all the operations, whether implemented in hardware or software. This approach is known as hardware/software partitioning because the bus divides the architecture into two partitions, and partitioning algorithms can be used to explore the design space.

Two important approaches to searching the design space during partitioning were introduced by early tools. The Vulcan system [57] starts with all processes in the custom processing elements and iteratively moves selected processes to the CPU to reduce the system cost. The COSYMA system [58] starts with all operations running on the CPU and moves selected operations from loop nests into the custom processing element to increase performance.

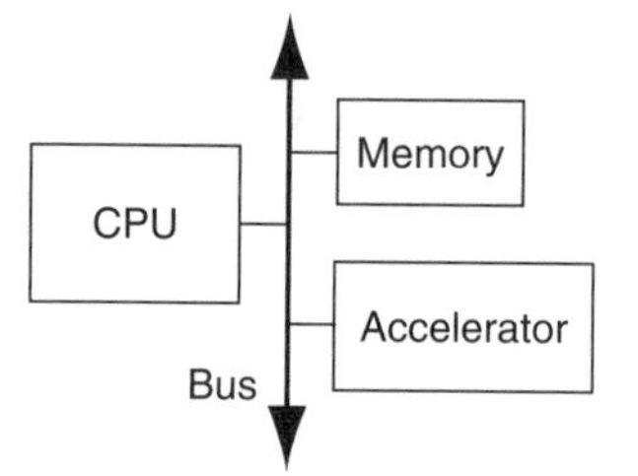

FIGURE 3.7 A template for hardware/software partitioning.

Hardware/software partitioning is ideally suited to platform FPGAs, which implement the bus-partitioned structure and use FPGA fabrics for the custom processing elements. However, the cost metric is somewhat different than in custom designs. Because the FPGA fabric is of a fixed size, using more or less of the fabric may not be important so long as the design fits into the amount of logic available.

Other cosynthesis algorithms have been developed that do not rely on an architectural template. Kalavade and Lee [59] alternately optimize for performance and cost to generate a heterogeneous architecture. Wolf [60] alternated cost reduction and load balancing while maintaining a performance-feasible design. Dick and Jha [61] used genetic algorithms to search the design space.

Scheduling is an important task during cosynthesis. A complete system schedule must ultimately be constructed; an important aspect of scheduling is the scheduling of multiple processes on a single CPU. The study of real-time scheduling for uniprocessors was initiated by Liu and Layland [62], who developed rate-monotonic scheduling (RMS) and earliest-deadline-first (EDF) scheduling. Rate-monotonic scheduling and EDF are priority-based schedulers, which use priorities to determine which process to run next. Many cosynthesis systems use custom, state-based schedulers that determine the process to be executed based upon the state of the system.

Design estimation is an important aspect of cosynthesis. While some software characteristics may be determined by simulation, hardware characteristics are often estimated using high-level synthesis. Henkel and Ernst [63] used forms of high-level synthesis algorithms to synthesize quickly a hardware accelerator unit and estimate its performance and size. Fornaciari et al. [64] used high-level synthesis and libraries to estimate power consumption.

Software properties may be estimated in a variety of ways, depending on the level of abstraction. For instance, Li and Wolf [65] built a process-level model of multiple processes interacting in the cache to provide an estimate of the performance penalty due to caches in a multitasking system. Tiwari et al. [66] used measurements to build models of the power consumption of instructions executing on processors.

3.8 Summary

System-level design is challenging because it is heterogeneous. The applications that we want to implement are heterogeneous — we may mix data and control, synchronous and asynchronous, etc. The architectures on which we implement these applications are also heterogeneous combinations of custom hardware, processors and memory. As a result, system-level tools help designers manage and understand complex, heterogeneous systems. Models of computation help designers cast their problem in a way that can be clearly understood by both humans and tools. Simulation helps designers gather important design characteristics. Hardware/software cosynthesis helps explore design spaces. As applications become more complex, we should expect to see tools continue to reach into the application space to aid with the transition from algorithm to architecture.

Over the next few years, we should expect to see simulation tools improve. Commercial simulation tools are well suited to networks of embedded processors or to medium-size SoCs, but new tools may be needed for large heterogeneous single-chip multiprocessors. Most multiprocessor simulators are designed for symmetric multiprocessors, but the large number of heterogeneous multiprocessor SoCs being designed assures that more general simulators will be developed.

Design modeling languages continue to evolve and proliferate. As CAD companies introduce new languages, it becomes harder to support consistent design flows. We hope that the design community will settle on a small number of languages that can be supported by a flexible set of interoperable tools.

References

[1] J. Fritts and W. Wolf, Evaluation of static and dynamic scheduling for media processors, *Proceedings, MICRO-33 MP-DSP2 Workshop*, ACM, Monterey, CA, December 2000.

[2] D. Talla, L. John, V. Lapinskii, and B.L. Evans, Evaluating signal processing and multimedia applications on SIMD, VLIW and superscalar architectures, *Proceedings of the IEEE International Conference on Computer Design*, Austin, TX, 2000.

[3] S. Dutta et al., Viper: a multiprocessor SOC for advanced set-top box and digital TV systems, *IEEE Design Test Comput.*, 18, 21–31, 2001.

[4] W. Wolf, *FPGA-Based System Design*, PTR Prentice Hall, New York, 2004.

[5] A. Basu, M. Hayden, G. Morrisett, and T. von Eicken, A language-based approach to protocol construction, *ACM SIGPLAN Workshop on Domain-Specific Languages*, Paris, France, January 1997.

[6] C.L. Conway and S.A. Edwards, NDL: a domain-specific language for device drivers, *Proceedings of the Workshop on Languages Compilers and Tools for Embedded Systems*, Washington, D.C., June 2004.

[7] S.A. Edwards, *Languages for Digital Embedded Systems*, Kluwer Academic Publishers, Dordrecht, 2000.

[8] K. Konstantinides and J.R. Rasure, The Khoros software-development environment for image-processing and signal-processing, *IEEE Trans. Image Process.*, 3, 243–252, 1994.

[9] R. Lauwereins, M. Engels, M. Ade, and J.A. Peperstraete, Grape-II: a system-level prototyping environment for DSP applications, *IEEE Comput. Mag.*, 28, 35–43, 1995.

[10] E.A. Lee, W.H. Ho, E. Goei, J. Bier, and S.S. Bhattacharyya, Gabriel: a design environment for DSP, *IEEE Trans. Acoust., Speech, Signal Process.*, 37, 1751–1762, 1989.

[11] V. Manikonda, P.S. Krishnaprasad, and J. Hendler, Languages, behaviors, hybrid architectures and motion control, in *Essays in Mathematical Control Theory (in Honor of the 60th Birthday of Roger Brockett)*, J. Baillieul and J.C. Willems, Eds., Springer, Heidelberg, 1998, pp. 199–226.

[12] K. Proudfoot, W.R. Mark, S. Tzvetkov, and P. Hanrahan, A real-time procedural shading system for programmable graphics hardware, *Proceedings of SIGGRAPH*, Los Angeles, CA, 2001.

[13] S.A. Thibault, R. Marlet, and C. Consel, Domain-specific languages: from design to implementation application to video device drivers generation, *IEEE Trans. Software Eng.*, 25, 363–377, 1999.

[14] R.M. Karp and R.E. Miller, Properties of a model for parallel computations: determinacy, termination, queuing, *SIAM J. Appl. Math.*, 14, 1966.

[15] G. Kahn, The semantics of a simple language for parallel programming, *Proceedings of the IFIP Congress*, Stockholm, Sweden, 1974.

[16] E.A. Lee and T.M. Parks, Dataflow process networks, *Proceedings of the IEEE*, May 1995, pp. 773–799.

[17] A.L. Ambler, M.M. Burnett, and B.A. Zimmerman, Operational versus definitional: a perspective on programming paradigms, *IEEE Comput. Mag.*, 25, 28–43, 1992.

[18] S. Ha and E.A. Lee, Compile-time scheduling and assignment of data-flow program graphs with data-dependent iteration, *IEEE Trans. Comput.*, 40, 1225–1238, 1991.

[19] J. Eker, J.W. Janneck, E.A. Lee, J. Liu, X. Liu, J. Ludvig, S. Neuendorffer, S. Sachs, and Y. Xiong, Taming heterogeneity — the Ptolemy approach, *Proceedings of the IEEE*, January 2003.

[20] D. Ko and S.S. Bhattacharyya, Modeling of block-based DSP systems, *J. VLSI Signal Process. Syst. Signal, Image, and Video Technol.*, 40, 289–299, 2005.

[21] E.A. Lee and D.G. Messerschmitt, Synchronous dataflow, *Proc. IEEE*, 75, 1235–1245, 1987.

[22] S.S. Bhattacharyya, P.K. Murthy, and E.A. Lee, *Software Synthesis from Dataflow Graphs*, Kluwer Academic Publishers, Dordrecht, 1996.

[23] P.K. Murthy and E.A. Lee, Multidimensional synchronous dataflow, *IEEE Trans. Signal Process.*, 50, 2064–2079, 2002.

[24] D. Stichling and B. Kleinjohann, CV-SDF — a model for real-time computer vision applications, *Proceedings of the IEEE Workshop on Application of Computer Vision*, Orlando, FL, December 2002.

[25] D. Stichling and B. Kleinjohann, CV-SDF — a synchronous data flow model for real-time computer vision applications, *Proceedings of the International Workshop on Systems, Signals and Image Processing*, Manchester, UK, November 2002.

[26] J.T. Buck and E.A. Lee, Scheduling dynamic dataflow graphs using the token flow model, *Proceedings of the International Conference on Acoustics, Speech, and Signal Processing*, Minneapolis, MN, April 1993.

[27] J.T. Buck, Static scheduling and code generation from dynamic dataflow graphs with integer-valued control systems, *Proceedings of the IEEE Asilomar Conference on Signals, Systems, and Computers*, Pacific Grove, CA, October 1994, pp. 508–513.

[28] J. Buck and R. Vaidyanathan, Heterogeneous modeling and simulation of embedded systems in El Greco, *Proceedings of the International Workshop on Hardware/Software Co-Design*, San Diego, CA, May 2000.

[29] A. Girault, B. Lee, and E.A. Lee, Hierarchical finite state machines with multiple concurrency models, *IEEE Trans. Comput.-Aid. Design Integrated Circuits Syst.*, 18, 742–760, 1999.

[30] L. Thiele, K. Strehl, D. Ziegenbein, R. Ernst, and J. Teich, FunState — an internal representation for codesign, *Proceedings of the International Conference on Computer-Aided Design*, San Jose, CA, November 1999.

[31] N. Cossement, R. Lauwereins, and F. Catthoor, DF*: an extension of synchronous dataflow with data dependency and non-determinism, *Proceedings of the Forum on Design Languages*, Tubingen, Germany, September 2000.

[32] M. Pankert, O. Mauss, S. Ritz, and H. Meyr, Dynamic data flow and control flow in high level DSP code synthesis, *Proceedings of the International Conference on Acoustics, Speech, and Signal Processing*, Adeliade, Australia, 1994.

[33] D.G. Messerschmitt, Structured interconnection of signal processing programs, *Proceedings of the IEEE Global Telecommunications Conference*, Atlanta, GA, 1984.

[34] J.T. Buck, S. Ha, E.A. Lee, and D.G. Messerschmitt, Ptolemy: a framework for simulating and prototyping heterogeneous systems, *Int. J. Comput. Simul.*, 4, 155–182, 1994.

[35] S.S. Bhattacharyya, J.T. Buck, S. Ha, and E.A. Lee, Generating compact code from dataflow specifications of multirate signal processing algorithms, *IEEE Trans. Circuits Syst. — I: Fundam. Theory Appl.*, 42, 138–150, 1995.

[36] A. Kalavade and E.A. Lee, A hardware/software codesign methodology for DSP applications, *IEEE Design Test Comput. Mag.*, 10, 16–28, 1993.

[37] L. de Alfaro and T. Henzinger, Interface automata, *Proceedings of the Joint European Software Engineering Conference and ACM SIGSOFT International Symposium on the Foundations of Software Engineering*, Vienna, Austria, 2001.

[38] E.A. Lee and Y. Xiong, System-level types for component-based design, *Proceedings of the International Workshop on Embedded Software*, Tahoe City, CA, October 2001, pp. 148–165.

[39] T. Stefanov, C. Zissulescu, A. Turjan, B. Kienhuis, and E. Deprettere, System design using Kahn process networks: the Compaan/Laura approach, *Proceedings of the Design, Automation and Test in Europe Conference and Exhibition*, Paris, France, February 2004.

[40] G. Bilsen, M. Engels, R. Lauwereins, and J.A. Peperstraete, Cyclo-static dataflow, *IEEE Trans. Signal Process.*, 44, 397–408, 1996.

[41] B. Kienhuis, E. Rijpkema, and E. Deprettere, Compaan: deriving process networks from Matlab for embedded signal processing architectures, *Proceedings of the International Workshop on Hardware/Software Co-Design*, San Diego, CA, May 2000.

[42] A. Turjan, B. Kienhuis, and E. Deprettere, Approach to classify inter-process communication in process networks at compile time, *Proceedings of the International Workshop on Software and Compilers for Embedded Processors*, Amsterdam, The Netherlands, September 2004.

[43] D. Verkest, K. Van Rompaey, I. Bolsens, and H. De Man, CoWare — a design environment for heterogeneous hardware/software systems, *Readings in Hardware/Software Co-design*, Kluwer Academic Publishers, Dordrecht, 2001.

[44] S. Ritz, M. Pankert, and H. Meyr, High level software synthesis for signal processing systems, *Proceedings of the International Conference on Application Specific Array Processors*, Berkeley, CA, August 1992.

[45] S. Ritz, M. Pankert, and H. Meyr, Optimum vectorization of scalable synchronous dataflow graphs, *Proceedings of the International Conference on Application Specific Array Processors*, Venice, Italy, October 1993.

[46] T.M. Parks, J.L. Pino, and E.A. Lee, A comparison of synchronous and cyclo-static dataflow, *Proceedings of the IEEE Asilomar Conference on Signals, Systems, and Computers*, Pacific Grove, CA, November 1995.

[47] S.S. Bhattacharyya, Hardware/software co-synthesis of DSP systems, in *Programmable Digital Signal Processors: Architecture, Programming, and Applications*, Y.H. Hu, Ed., Marcel Dekker, New York, 2002, pp. 333–378.

[48] G. Berry and G. Gonthier, The Esterel synchronous programming language: design, semantics, implementation, *Sci. Comput. Programming*, 19, 87–152, 1992.

[49] I. Page, Constructing hardware/software systems from a single description, *J. VLSI Signal Process.*, 12, 87–107, 1996.

[50] C.A.R. Hoare, *Communicating Sequential Processes*, Prentice-Hall, New York, 1985.

[51] L.C. Paulson, *ML for the Working Programmer*, Cambridge University Press, London, 1996.

[52] S. Chappell and C. Sullivan, Handel-C for Co-processing & Co-design of Field Programmable System on Chip, Technical report, Celoxica Limited, September 2002.

[53] J.B. Dabney and T.L. Harman, *Mastering Simulink*, Prentice Hall, New York, 2003.

[54] D.C. Burger and T.M. Austin, The SimpleScalar Tool Set, Version 2.0, U W Madison Computer Sciences Technical Report #1342, June, 1997.

[55] D. Brooks, V. Tiwari, and M. Martonosi, Wattch: a framework for architectural-level power analysis and optimizations, *Proceedings of the 27th International Symposium on Computer Architecture*, Vancouver, Canada, June 2000.

[56] N. Vijaykrishnan, M. Kandemir, M.J. Irwin, H.Y. Kim, and W. Ye, Energy-driven integrated hardware-software optimizations using SimplePower, *Proceedings of the International Symposium on Computer Architecture*, Vancouver, Canada, June 2000.

[57] R.K. Gupta and G. De Micheli, Hardware-software cosynthesis for digital systems, *IEEE Design Test Comput.*, 10, 29–41, 1993.

[58] R. Ernst, J. Henkel, and T. Benner, Hardware-software cosynthesis for microcontrollers, *IEEE Design Test Comput.*, 10, 64–75, 1993.

[59] A. Kalavade and E.A. Lee, The extended partitioning problem: hardware/software mapping, scheduling, and implementation-bin selection, *Design Autom. Embedded Syst.*, 2, 125–163, 1997.

[60] W. Wolf, An architectural co-synthesis algorithm for distributed, embedded computing systems, *IEEE Trans. VLSI Syst.*, 5, 218–229, 1997.

[61] R.P. Dick and N.K. Jha, MOGAC: a multiobjective genetic algorithm for the hardware-software co-synthesis of distributed embedded systems, *IEEE Trans. Comput.-Aid. Design*, 17, 920–935, 1998.

[62] C.L. Liu and J.W. Layland, Scheduling algorithms for multiprogramming in a hard- real-time environment, *J. ACM*, 20, 46–61, 1973.

[63] J. Henkel and R. Ernst, A path-based estimation technique for estimating hardware runtime in HW/SW-cosynthesis, *Proceedings of the 8th IEEE International Symposium on System Level Synthesis*, Cannes, France, 1995, pp. 116–121.

[64] W. Fornaciari, P. Gubian, D. Sciuto, and C. Silvano, Power estimation of embedded systems: a hardware/software codesign approach, *IEEE Trans. VLSI Syst.*, 6, 266–275, 1998.

[65] Y. Li and W. Wolf, Hardware/software co-synthesis with memory hierarchies, *IEEE Trans. CAD*, 18, 1405–1417, 1999.

[66] V. Tiwari, S. Malik, and A. Wolfe, Power analysis of embedded software: a first step towards software power minimization, *IEEE Trans. VLSI Syst.*, 2, 437–445, 1994.

4

System-Level Specification and Modeling Languages

Joseph T. Buck
Synopsys, Inc.
Mountain View, California

4.1 Introduction

This chapter is an overview of high-level, abstract approaches to the specification and modeling of systems, and the languages and tools in this domain. Many of the most effective approaches are domain-specific, and derive their power from the effective exploitation of a *model of computation* that is effective for expressing the problem at hand. The chapter first surveys some of the most used models of computation, and introduces the specialized tools and techniques used for each. Heterogeneous approaches that allow for the combination of more than one model of computation are then discussed.

It is hardly necessary to explain the reader that because of Moore's law, the size and complexity of electronic circuit design is increasing exponentially. For a number of years, most advanced chip designs have included at least one programmable processor, meaning that what is being designed truly qualifies for the name "system-on-chip" (SoC), and success requires that both hardware and embedded software be designed, integrated, and verified. It was common in the late 1990s to see designs that included a general-purpose microprocessor core, a programmable digital signal processor, as well as application-specific synthesized digital logic circuitry. Leading-edge designs now include many processors as well as on-chip networks. As a consequence, the task of the SoC architect has become not merely quantitatively but also qualitatively more complex.

To keep up with the increasing design complexity, there are two principal means of attacking the problem. The first tactic is to increase design reuse, i.e., assemble new designs from previously designed and verified components. The second tactic is to raise the level of abstraction, so that larger designs can be kept manageable. The purpose of the languages, approaches and tools described in this chapter is to raise the level of abstraction. Given suitable higher-level models for previously designed components, these approaches can also help with design reuse.

However, there is a price to be paid; some of the most powerful system-level language approaches are also domain-specific. Data flow-oriented approaches (Kahn process networks [7]) have great utility for digital signal processing intensive subsystems (including digital audio, image and video processing as well as digital communication), but are cumbersome to the point of unsuitability if complex control must be implemented. As a consequence, there has been great interest in hybrid approaches.

It has long been common practice for designers to prototype components of systems in a traditional higher level language, most commonly C or C++. A high level language raises the level of abstraction to some extent, but for problems involving signal processing, many designers find Matlab [1] preferable. However, since these are sequential languages, they do not provide systematic facilities for modeling concurrency, reactivity, interprocess communication, or synchronization.

When the system designer begins to think about concurrency and communication, it quickly becomes apparent that suitable models for explaining/understanding how this will work are problem-specific. A designer of synchronous logic can consider a decomposition of a problem into a network of extended finite-state machines (FSMs), operating off a common clock. A designer of a video processing application, which will run on a network of communicating processors and sharing a common bus, will require a very different model.

The organizing concept of this chapter is the notion of a *model of computation*. Edward Lee, who is largely responsible for drawing attention to the concept, describes a model of computation as "the laws of physics that govern component interactions" [2]. In [3], Berry takes the analogy further, comparing the instantaneous communication in the synchronous-reactive model to Newtonian gravitation and the bounded delays of the discrete-event (DE) model to the theory of vibration.

Models of computation are related to languages, but the relationship is not exact. For example, while Esterel [4] is specifically designed to support the synchronous-reactive model of computation, it is also common to implement support for a model of computation with a C++ class library (e.g., SystemC [5]) or by using a subset of a language. Using the synthesizable subset of VHDL or Verilog, and working with synthesis semantics, is to use a very different model of computation than to use the full languages and to work with the simulation behavior.

4.2 A Survey of Domain-Specific Languages and Methods

Any designer manages complexity by subdividing a design into manageable components or modules, and block diagrams are also common in many disciplines. To a hardware designer, the components or blocks have a direct physical significance, as logic blocks to be instantiated or synthesized, and they communicate by means of wires. To a designer of a digital communication system, the blocks might be processes communicating by means of streams of data and connections between components might be first-in first-out (FIFO) queues. Alternatively, the blocks might be tasks or system modes, and the connections might be precedence constraints or even state transitions. The model of computation in the sense we use it here, forms the "rules of the game" and describes how the components execute and communicate.

This chapter will first discuss those models of computation that have proved effective in hardware and embedded software design and modeling, and will then discuss hybrid approaches, where more than one model of computation is used together. The Ptolemy project [6] was, to the best of the author's knowledge, the first systematic attempt to allow designers to combine many models of computation in a systematic way, but there are also many other important hybrid approaches.

4.2.1 Kahn Process Networks and Dataflow

A Kahn process network (KPN) [7] is a network of processes that communicate by means of unbounded FIFO queues. Each process has zero or more input FIFOs, and zero or more output FIFOs. Each FIFO is connected to one input process and one output process. When a process writes data to an output FIFO, the write always succeeds; the FIFO grows as needed to accommodate the data written. Processes may read their inputs only by means of blocking reads; if there are insufficient data to satisfy a request, the reading process blocks until the writing process for the FIFO provides more. The data written by a process depend only on the data read by that process and the initial state of the process. Under these conditions, Kahn [7] showed that the trace of data written on each FIFO is independent of process scheduling order. As a result, KPNs are a popular paradigm for the description and implementation of systems for the parallel processing of streaming data.

Because any implementation that preserves the semantics will compute the same data streams, KPN representations or special cases of them, are a useful starting point for the system architect, and are often used as executable specifications.

Dataflow process networks are a special case of KPNs, in which the behavior of each process (often called an *actor* in the literature) can be divided into a sequence of execution steps called *firings* by Lee and Parks [8]. A firing consists of zero or more read operations from input FIFOs, followed by a computation, and then by zero or more write operations to output queues (and possibly a modification to the process's internal state). This model is widely used in both commercial and academic software tools such as SPW [9], COSSAP [10], System Studio [11], and Ptolemy. The subdivision into firings, which are treated as indivisible quanta of computation, can greatly reduce the context switching overhead in simulation, and can enable synthesis of software and hardware. In some cases (e.g., Yapi [12]), the tools permit processes to be written as if they were separate threads, and then split the threads into individual firings by means of analysis. The thread representation allows read and write directives to occur anywhere, while the firing representation can make it easier to understand the data rates involved, which is important for producing consistent designs.

In an important special case of dataflow process networks, the number of values read and written by each firing of each process is fixed, and does not depend on the data. This model of computation was originally called synchronous dataflow (SDF) [13], which is now widely considered an unfortunate choice of terminology because of confusion with synchronous languages, and because SDF is an untimed model. In fact, the term was originally used for the very different LUSTRE language [14]; LUSTRE is synchronous but not dataflow; SDF is dataflow but not synchronous. The term "static dataflow" is now considered preferable; Lee himself uses it in [8]. Fortunately, the widely used acronym "SDF" still applies.

Figure 4.1 shows a simple SDF graph. In the diagram, the numbers adjacent to the inputs and outputs of the actors indicate how many data values are written to, or read from, the attached FIFO queue on each actor execution. While not shown in this example, it is possible for an edge of an SDF graph to have initial logical delays which can be thought of as initial values in the queues. If the graph contains cycles, initial values must be present to avoid a deadlock.

The first task in the analysis of an SDF graph is to derive an equation from each FIFO that constrains the number of actor firings the schedule must contain: the number of values written to the FIFO must equal the number of values read from the FIFO. Let n_{ij} be the number of values written by actor i to the edge (FIFO queue) j. If actor i reads, rather than writes from edge j, then a negative number is stored in n_{ij} (the negative of the number of values read). Now, let r_i be the number of times actor i is "fired" in one iteration of the schedule. It is required that the number of values in each queue be the same after a completion of the schedule, so we obtain a series of *balance equations:*

$$\sum_{i=1}^{K} n_{ij} r_i, \; j=1,\ldots,L$$

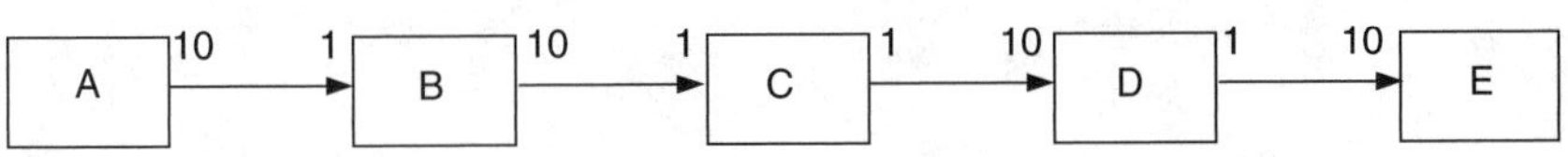

FIGURE 4.1 A simple static dataflow graph. Numbers indicate the number of values read or written on each firing of the adjacent dataflow actor.

where K is the number of actors and L is the number of edges. Alternatively, treating N as the matrix formed by the n_{ij} terms, and $\hat{r}$ as the vector of rates, we have

$$N\hat{r}=\hat{0}$$

If the matrix N is nonsingular, only the trivial solution that sets all rates to zero exists, and we say that the graph is inconsistent. In Figure 4.2, graph (a) is an example of an inconsistent graph. Such a graph can be executed by a dynamic dataflow simulator, but the number of values in some queues must increase without bound. In [13] it is shown that, for a connected graph, if the matrix is nonsingular the null space has rank 1, so that all solutions for $\hat{r}$ are proportional. We are interested in the least integral solution; as shown in [13], there is a simple algorithm to find the minimal integer solution in $O(K + L)$ time. Note that even if a graph has consistent rates, if there are cycles with insufficient delay, we have a deadlock condition and a schedule is not possible. Figure 4.2(b) is an example of such a graph.

In the case of Figure 4.1, a minimal schedule must execute A and E once, B and D 10 times, and C 100 times. Linear schedules containing 100 invocations of C are clearly not efficient, so looped schedules are of great interest for implementations, especially for sequential software implementations. Looped schedules, in which the invocation of each actor appears exactly once — the so-called *single appearance schedules* — are treated in detail in [15]. One possible single appearance schedule for our example graph can be written as A,10(B,10(C),D),E.

For static dataflow networks, efficient static schedules are easily produced, and bounds can be determined for all of the FIFO buffers, whether for a single programmable processor, multiple processors, or hardware implementations. An excellent overview of the analysis of SDF designs as well as the synthesis of software for a single processor from such designs, can be found in [16].

For multiple processors, one obvious alternative is to form a task dependence graph from the actor firings that make up one iteration of the SDF system, and apply standard task scheduling techniques such as list scheduling to the result. Even for uniform-rate graphs where there is no looping, the scheduling problem is NP-hard. However, because it is likely that implicit loops are present, a linear schedule is likely to be too large to be handled successfully. Efficient multiprocessor solutions usually require preservation of the hierarchy introduced by looping — see [17] for one approach to do just that.

Engels et al. [18] proposed an extension of static dataflow that allows the input–output pattern of an actor to vary in a periodic manner; this model is called cyclo-static dataflow. Complete static schedules can still be obtained, but in most cases the interconnecting FIFO queues can be made much shorter, which is particularly advantageous for hardware implementation.

The Gabriel system [19] was one of the earliest examples of a design environment that supported the SDF model of computation for both simulation and code generation for DSPs. Gabriel's successor, Ptolemy, extended and improved Gabriel's dataflow simulation and implementation capabilities [20]. Another successful early SDF-based code generation system was Descartes [21], which was later commercialized as part of COSSAP by Cadis (later acquired by Synopsys). The GRAPE-II system [22] supported implementation using the cyclo-static dataflow model.

The term "dynamic dataflow" is often used to describe dataflow systems that include data-dependent firing and therefore are not static. COSSAP [10] was apparently the first true dynamic dataflow simulator;

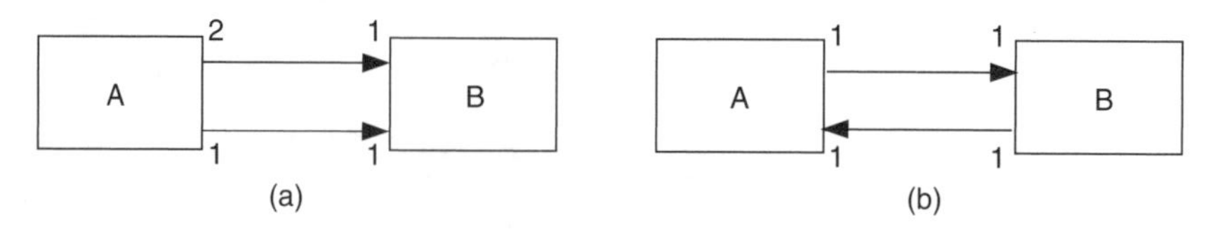

Figure 4.2 Inconsistent graphs. (a) has inconsistent rates; (b) is rate-consistent but deadlocked.

Messerschmitt's Blosim [23], while older, required the user to specify sizes of all FIFO buffers and writes to full FIFOs blocked (while COSSAP buffers grow as needed), so it was not a true dataflow (or KPN) simulator. However, the method used by COSSAP for dynamic dataflow execution is suitable for simulation only, and not for embedded systems implementation.

While there have been many graphical, block diagram-based systems supporting simulation as well as software and hardware implementation, there are also textual languages whose model of computation corresponds to SDF. The first of these was Silage [24]. Silage is a declarative language, in which all loops are bounded; it was designed to allow DSP algorithms to be efficiently implemented in software or hardware. The DFL language [25] was derived from Silage and a set of DFL-based implementation tools was commercialized by Mentor Graphics as DSPstation.

While there are many algorithmic problems or subproblems that can be modeled as SDF, at least some dynamic behavior is required in most cases. Hence there has long been a interest in providing for at least some data-dependent execution of actors in tools, without paying for the cost of full dynamic dataflow.

The original SPW tool from Comdisco (later Cadence, now CoWare) [9] used a dataflow-like model of computation that was restricted in a different way: each actor had a hold signal. If connected, the actor reads a value from the hold signal. If a "true" value is read, the actor does not execute; otherwise the actor reads one value from each input, does a computation, and writes one value to each output. This model is more cumbersome than SDF for static multirate operation, but can express dynamic behaviors that SDF cannot express, and the one place buffers simplified the generation of hardware implementations. It is a special case of dynamic dataflow (although limited to one place buffers). Later versions of SPW added full SDF and dynamic dataflow support.

Boolean-controlled dataflow (BDF) [26], later extended to allow for integer control streams (IDF) [27], was an attempt to extend SDF analysis and scheduling techniques to a subset of dynamic dataflow. While in SDF, the number of values read or written by each I/O port is fixed, in BDF the number of values read by any port can depend on the value of a Boolean data value read by, or written by some other port, called a control port. In the BDF model, as in SDF, each port of each actor is annotated with the number of values transferred (read or written) during one firing. However, in the case of BDF, instead of a compile-time constant, the number of values transferred can be an expression containing Boolean-valued variables. These variables are the data values that are arrived at or are written by a *control port*, a port of the BDF actor that must transfer one value per firing. The SPW model, then, is a special case of BDF where there is one control port, which controls all other ports, and the number of values transferred must be 1 or 0. Integer control streams allow control streams to be integers.

While BDF is still restricted compared to general dataflow, it is sufficiently expressive to be Turing-equivalent. Unfortunately, this means that a number of analysis problems, including the important question of whether buffer sizes can be bounded, are undecidable in general (as shown in [26]). Nevertheless, clustering techniques can be used to convert an IDF graph into a reduced graph consisting of clusters; each individual cluster has a static or quasi-static schedule, and only a subset of the buffers connecting clusters can potentially grow to an unbounded size. This approach was taken in the dataflow portion of Synopsys's System Studio [11], for example.

The BDF model has also been used for hardware synthesis by Zepter [28]. Zepter's approach can be thought of as a form of interface synthesis: given a set of predesigned components that read and write data periodically, perhaps with different periods and perhaps controlled by enable signals, together with a behavioral description of each component as a BDF model, Zepter's Aden tool synthesized the required control logic and registers to correctly interface the components.

There is a close correspondence between the cases that can be handled by Aden and in the cases handled by the LUSTRE language (see Section 4.2.5), because the Boolean conditions that control the dataflow become gated clocks in the implementation.

Later work based on Zepter's concept [29] relaxed the requirement of periodic component behavior and provided more efficient solutions, but handled only the SDF case.

4.2.2 Dataflow with Bounded Queues

Many implementers that start from a general dataflow description of the design choose an implementation with fixed bounds on the size of the queues. Given a fixed queue size, a write to a full queue can get blocked, but if analysis has already shown that the sizes of the queues suffice, this does not matter.

One of the most commonly used fixed-sized queue implementations at the prototyping level is the sc_fifo<T> channels in SystemC [5]. An example of implementation-oriented approach that uses fixed-size queues is the task transfer level (TTL) approach [30], in which a design implemented in terms of fixed-sized queues is obtained by refinement from an executable specification written in KPN form in an environment called Yapi [12].

4.2.3 Matlab

The most widely used domain-specific tool for digital signal processing intensive applications is Matlab [1], from the MathWorks, Inc. Matlab is both a high-level language and an interactive environment. The language (sometimes called M-code, to contrast it with C-code) treats vectors and arrays as first-class objects, and includes a large set of built-in functions as well as libraries for the manipulation of arrays, signal processing, linear algebra, and numerical integration; the integration features can be used effectively in continuous-time modeling. The environment provides extensive data visualization capabilities. Matlab is an interpreted language, but for code that manipulates vectors and matrices, the interpreter overhead is small.

An algorithm written in Matlab is sequential. Other tools build on Matlab by allowing functional blocks to be written in Matlab or C. The MathWorks provides Simulink [31], a block diagram dataflow tool that allows functional blocks to be written in C or in M-code. A number of other tools in this category (including SPW, COSSAP, System Studio, and others) all provide some form of interface to Matlab.

4.2.4 Statecharts and its Variants

A very influential methodology for representing hierarchical control structures is Statecharts, invented by Harel [32], and commercialized by the company i-Logix, which Harel cofounded, with the name Statemate.

Statecharts representations are hierarchical extended finite state machines (FSMs). Two types of models are supported, called *or-states* and *and-states*. An or-state Statecharts model looks like an FSM diagram, with guard conditions and actions associated with each transition between states. The constituent states of a Statecharts model can be atomic states, or can be Statecharts models themselves. The other form of Statecharts model is called an and-state, and consists of two or more Statecharts models that execute in parallel. In Harel's original version of Statecharts, transition arrows can cross levels, in either

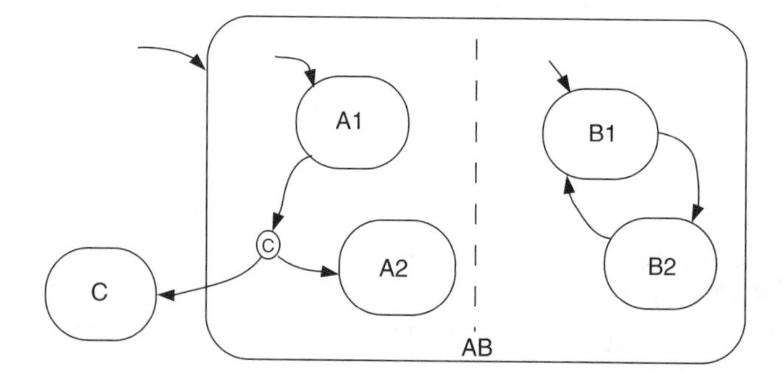

FIGURE 4.3 A simple Statechart. The graph as a whole is an or-state (either in AB or in C); AB is an and-state (we enter both A1 and B1 at the same instant), which is composed of two or-states (A1 and A2; B1 and B2). Transition arcs are annotated with conditions and actions (not shown here).

direction, which provides for great expressive power, but at the price of modularity. Figure 4.3 is an example Statechart.

Statecharts can be used for generation of C or hardware description language (HDL) code for implementation purposes; however the representation seems to be used most commonly today as a specification language rather than as an implementation tool [33]; Harel's Statecharts were combined with a variety of other specification methodologies to form Unified Modeling Language (UML) [34].

The form of Statecharts has been widely accepted, but there has been a great deal of controversy about the semantics, and as a result many variants of Statecharts have been created. In [35], von der Beek identifies 20 Statecharts variants in the literature and proposes a 21st !

The main points of controversy are:

- *Modularity.* Many researchers, troubled by Harel's level-crossing transitions, eliminated them and came up with alternative approaches to make them unnecessary. In many cases, signals are used to communicate between levels.
- *Microsteps.* Harel's formulation handles cascading transitions, which occur based on one transition of the primary inputs by using delta cycles, as is done in VHDL and Verilog. Others, such as Argos [36] and SyncCharts [37], have favored synchronous-reactive semantics and find a fixpoint, so that what requires a series of microsteps in Harel's statecharts becomes a single atomic transition. If a unique fixpoint does not exist, then formulations of Statecharts that require a fixpoint reject the specification as ill formed.
- *Strong preemption vs. weak preemption.* When a hierarchical state and an interior state both have an outgoing transition that is triggered by the same event, in a sense we have a race. With strong preemption, the outer transition "wins"; the inner transition is completely preempted. With weak preemption, both transitions take place (meaning that the action associated with the inner transition is performed), with the inner action taking place first (this order is required because the outer transition normally causes the inner state to terminate). Strong preemption can create causality violations, since the action on an inner transition can cause an outer transition that would preempt the inner transition. Many Statecharts variants reject specifications with this kind of causality violation as ill formed. Some variants permit strong or weak preemption to be specified separately for each transition.
- *History and suspension.* When a hierarchical state is exited and reentered, does it "remember" its previous state? If it does, is the current state remembered at all levels of hierarchy (deep history), or only at the top level (shallow history)? In Harel's Statecharts, shallow history or deep history can be specified as an attribute of a hierarchical state. In some other formulations, a suspension mechanism is used instead, providing the equivalent of deep history by "freezing" a state (which might correspond to the gating of a clock in a hardware implementation). Figure 4.4 shows an oversimplified traffic light state machine with a "pause mode", implemented first with a Harel-style history mechanism, and then with a SyncCharts-like suspension mechanism.

Most of these issues are also relevant for control-oriented synchronous languages, and will be discussed in the next section.

4.2.5 Synchronous/Reactive Languages

In the synchronous-reactive model of computation, we consider reactive systems, meaning systems that respond to events from the environment. A reactive system described with a synchronous model instantaneously computes its reaction, modifies its state, and produces outputs. Of course, physical realizations cannot react instantaneously but as for synchronous circuits, what is really required is that the reaction can be computed before the deadline arrives for the processing of the next event (typically the next clock cycle boundary, minus any setup and hold times for a synchronous circuit). The concepts used in synchronous languages originate in the synchronous process calculi of Milner [38]. The best known textual synchronous languages are Esterel [4], LUSTRE [14], and SIGNAL [39]. There are also several variants of

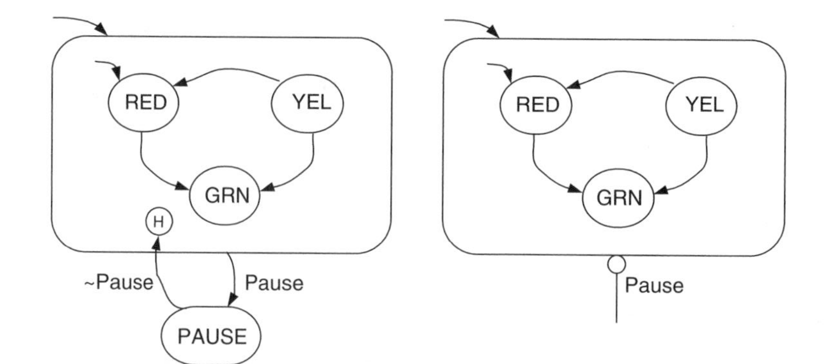

FIGURE 4.4 History vs. suspension. The traffic light controller freezes when the pause signal is true, and resumes when it is false. On the left, this is done using a history mechanism; on the right, it is done with suspension, which works like gating a clock.

Statecharts that have synchronous-reactive semantics:* Argos [36] and SyncCharts [37]. An excellent survey of the synchronous model and principal languages can be found in [40].

The synchronous languages can be subdivided into two groups: those that support hierarchical control, and those that define time series by means of constraints. These two groups of languages appear very different in form, but it turns out that they share an underlying structure.

Of the languages that we have mentioned, Esterel, Argos, and SyncCharts fall into the first category; in fact, SyncCharts can be thought of as a successful effort to extend Argos into a graphical form of the Esterel language.

Here is a simple example of an Esterel program from [3]. In this case, a controller has two Boolean input signals, A and B, as well as a reset signal, R. The task is to wait for an event on both A and B; the events can be received in either order or simultaneously. When both events are seen, an output event is to be produced on O and then the controller should restart. However, if a reset event is received on R, the controller should restart without producing an event on O. The code looks like

```
module ABO:
input A, B, R;
output O; loop
    [ await A | await B ];
    emit O
each R end
module
```

In this example, we have synchronous parallelism (for the two awaited statements) as well as strong preemption (the loop ... each construct aborts and restarts the loop on each receipt of the reset event).

The corresponding representation of this example in SyncCharts is shown in Figure 4.5; the graphical form has the same semantics as the textual form.

Esterel's well-behaved hierarchical control structure makes it a good language to write model checkers for; XEVE [41] is one such checker. Esterel and XEVE have been successfully used to achieve verifiably high state coverage for the control-dominated parts of a commercial DSP [42], by writing the reference model in a mixture of C (for datapath) and Esterel (for control).

The first implementation of Esterel required that the entire program be flattened into a state transition graph (essentially a flat FSM). While this approach yielded fast software implementations, the code size tended to explode for large program sizes. For many years, the implementation of choice was to translate the Esterel program into a circuit representation that preserves the control hierarchy of the original

*Harel's original Statecharts has discrete event semantics, including microsteps that resemble those of VHDL and Verilog.

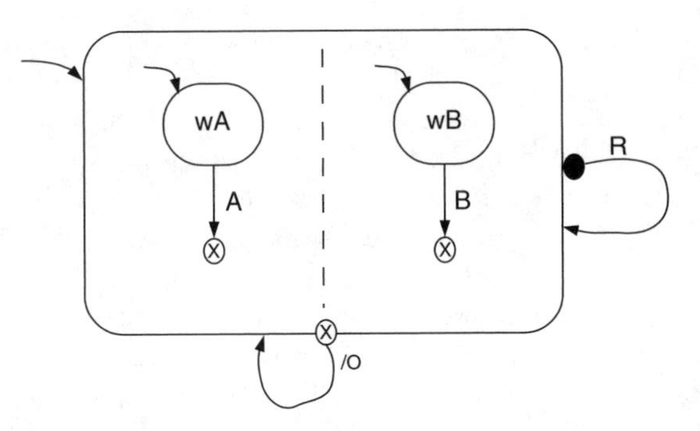

FIGURE 4.5 Syncharts version of "ABO". The x-in a-circle denotes normal termination; if on the edge of a subsystem, it indicates the action to be taken on termination. The circle on the edge triggered by R indicates strong preemption; strong preemption edges take priority over termination edges.

program; the two implementation approaches are described and compared in [43]. The software implementation produced by the popular "version 5" Esterel software release was in effect a simulator for this hierarchical control circuit, which had the disadvantage of being slower than necessary due to the frequent update of circuit signals that had no effect. There has been recent interest in more efficient software implementations of Esterel, particularly due to the work of Edwards [44].

An approach with close connections to Esterel is Seawright's production-based specification system, Clairvoyant [45]. The "productions" correspond to regular expressions, and actions (written in VHDL or Verilog) are associated with these expressions. The specification style closely resembles that of the Unix tool *lex* [46]. Seawright et al. [47] later extended Clairvoyant, designed a graphical interface, and produced a commercial implementation called Protocol Compiler. While not a commercial success, the tool was effectively used to synthesize and verify complex controllers for SDH/Sonet applications [48]. The circuit synthesis techniques used were quite similar to those used for Esterel in [43].

LUSTRE and SIGNAL are a very different style of synchronous language, in that their variables are streams; specifically, they are time series. With each time series there is a clock associated; two streams that are defined at the same points in time are said to have the same clock. While clocks impose a global ordering, there is neither a mechanism nor a need to define time scales in more detail.

It is possible to sub-sample a stream, based on a Boolean-valued stream, to produce a stream with a slower clock; this is analogous to a down-sampling operation in digital signal processing.

In LUSTRE, variables correspond to streams, and unary and binary arithmetic and logical operations are extended to operate pointwise on the elements of those streams. LUSTRE is a declarative language, meaning that streams are defined once in terms of other streams. The following is a simplified example from one in [14]: here

```
node COUNTER(reset: bool) returns (n: int)
let
      n = 0 -> if reset then 0 else pre(n) + 1;
tel.
```

defines a counter with reset, that increments with every basic clock cycle and that resets when the reset input is true. Every node (corresponding to what might be called a module or an entity in other languages) has a basic clock, corresponding to the fastest clock that is input to the node. Note that both n and reset are streams.

The keyword produces a sub-sampled stream, and the current keyword interpolates a stream (using a sample-and-hold mechanism) to match the current basic clock. Definitions can be cyclic as long as each cycle contains at least one preoperator.

The LUSTRE language does not allow basic clock time intervals to be split into smaller ones; we can sub-sample a stream or interpolate it to match the clock of a faster stream. The SIGNAL language (which we will not discuss further) allows this kind of subdivision. This distinction makes it much easier to map

LUSTRE programs into digital circuits than SIGNAL; pre makes a register, when gates the clock, current is effectively a latch.

There is a strong connection between the stream-based synchronous languages and certain dataflow languages such as Silage and SDF-based systems. Even the name "synchronous dataflow" suggests the connection. The important distinction between the two approaches is in the concept of time. In LUSTRE, every pair of events in the system has a defined global ordering, and the location of each event can be correlated to a particular tick of the highest-rate clock in the system. In Silage or SDF, there is only data, and the specification gives only a partial ordering of data production events (all required inputs must be computed before an output is produced). Edward Lee has long argued (e.g., in [8]) that the specification of global ordering where it is not warranted can over-constrain the solution space.

However, this distinction is perhaps less important that it appears as a practical matter, in that an implementer could start with a LUSTRE specification and ignore timing issues. Similarly, in the Ptolemy system [6], it is necessary to assign time values to stream elements produced by an SDF "master" system to interface it to a DE subsystem, and in DSP applications assigning equally spaced times to sample points is often the most useful choice, which makes the SDF system look much like Esterel.

4.2.6 Communicating Sequential Processes

Communicating sequential processes (CSP) with rendezvous is a model of computation first proposed by Hoare [49]. Like dataflow, CSP is an untimed model of computation, in which processes synchronize based on the availability of data. Unlike KPNs, however, there is no storage element at all between the connected processes (so for example, a hardware implementation might have only wires, and no registers, at the point of connection).

Given a pair of communicating processes, the read request and matching write request form a rendezvous point; whichever process performs the I/O request first, waits for its peer. The Occam language [50] was designed around CSP.

The use of remote procedure calls (RPC) as a form of inter-process communication is a form of CSP. This is because the communication between the client and server is completely synchronized; there are no buffers to store outstanding requests (though protocols built on top of RPC often implement buffering at a higher level).

4.2.7 Polis and Related Models

The Polis system [51] uses a locally synchronous and globally asynchronous model of computation. Each process in a design is a "codesign finite state machine" (CFSM), a form of extended FSM. The interaction between the CFSM tasks differs from the standard model of interacting concurrent FSMs in that an unbounded (and nonzero) delay is added to each interaction between CFSMs. The intent of this design style is to avoid introducing a bias toward hardware or software implementation: a synchronous hardware design might introduce one cycle of delay for communication, while a software implementation or a partition of the design that requires tasks to communicate over a network, might require a much larger time for a communication. The processes communicate with each other asynchronously using one place buffers; this means that data loss is a possibility.

The unconstrained model, where there are no time deadlines or protection against overwritten buffers, is only intended as a starting point. The designer is expected to add constraints sufficient to assure that deadlines are met. It is not strictly necessary to guard against overwritten communication buffers, as there are some cases where the data loss is not a problem or is even desirable. For example, in an automotive application, a speedometer task might repeatedly compute a current speed estimate, while a digital display of current speed only needs to sample these estimates occasionally.

Cadence produced a commercial tool, VCC, that was based on the Polis model.

Polis permits the CFSM tasks to be written in any of several forms that can be translated into FSMs, including Esterel, graphical FSMs, or HDL subsets. Many of the analysis tools in the Polis framework require that the FSM tasks be completely flattened and that individual states be enumerated, and as a result Polis has been most effective on designs with many relatively simple communicating tasks. Automotive applications have been a principal driver for Polis and VCC and seem most suitable for this design style; for example, Ref. [52], which describes the design of an engine management system in VCC.

Specification and description language (SDL) [53] is a standard specification language used widely in telephony, particularly in Europe. Like Polis, SDL represents tasks as communicating FSMs, and communication between tasks can introduce an unbounded delay. There is more detail about SDL in Chapter 9.

4.2.8 Discrete Events and Transaction-Level Modeling

In the discrete event (DE) model of computation, no doubt most familiar to the reader as the underlying model for hardware description languages (HDLs) such as Verilog and VHDL, each event has associated with it a specific time. However, unlike the synchronous-reactive model, reactions to events are not instantaneous, instead simulated time elapses. Even for reactions that conceptually take zero time, there is still an event ordering that takes place, often in the form of microsteps or delta cycles. SystemC also has discrete event semantics.

We will not deal further with Verilog or VHDL in this chapter; the languages are discussed in detail in Chapter 15. In this chapter, we are interested in DE modeling at higher levels of abstraction.

Transaction-level modeling (TLM) has become an extremely popular term in recent years. Grötker et al., [5] define TLM as "a high-level approach to modeling digital systems where details of communication among modules are separated from the details of the implementation of the functional units or of the communication architecture." The term "transaction-level modeling" was coined by the original SystemC development team; an alternative, "transaction-based modeling", was also considered and might have been a preferable choice, as TLM does not correspond to a particular level of abstraction in the same sense that, for example, register transfer level (RTL) does [54]. However, distinctions between TLM approaches and register-transfer level can clearly be made: while in an RTL model of a digital system, the detailed operation of the protocol, address, and data signals on a bus are represented in the model, with TLM a client of a bus-based interface might simply issue a call to high-level read() or write () functions. Chapter 8 of Grötker et al.[5] gives a simple but detailed example of a TLM approach to the modeling of a bus-based system with multiple masters, slaves, and arbiters.

While the SystemC project coined the TLM term, it clearly did not invent the concept. The SpecC language [55], through its channel feature, permits the details of communication to be abstracted away in much the same manner that SystemC supports. Furthermore, the SystemVerilog language [56] supports TLM as well, through a new type of port that is declared with the interface keyword. It is possible to use the separation of functional units and communication simply as a cleaner way to organize one's SystemVerilog code, and still represent the full RTL representation of the design in all of its detail, and Sutherland et al. [56] recommend just this.

With TLM as it is commonly used, the model of computation is still DE simulation, and it is common for a design that is in the process of being refined to mix levels, with detailed RTL representation of some components and TLM representations of others. It is possible, however, to disregard time in a very high-level TLM representation, resulting in a model of computation that more closely resembles communicating sequential processes with rendezvous, or even dataflow with bounded queues.

At present, methodology for TLM in system design is not particularly mature, and there is disagreement about the number of abstraction levels that need to be represented in the design flow. One example of an attempt to make TLM and refinement more rigorous is given by Cai and Gajski [57]. In their model, the designer starts with an untimed specification model, and separately refines the model of communication and of computation. Many intermediate points are identified, for example, their bus-functional models have accurate timing for communication and approximate timing for computation.

4.3 Heterogeneous Platforms and Methodologies

The first systematic approach that accepted the value of domain-specific tools and models of computation, yet sought to allow designers to combine more than one model in the same design, was Ptolemy [6]. The approach to heterogeneity followed in the original Ptolemy was to consistently use block diagrams for designs, but to assign different semantics to a design based on its "domain", roughly corresponding to a model of computation. Primitive actors were designed to function only in particular domains, and hierarchical designs were simulated based on the rules of the current domain. To achieve heterogeneity, Ptolemy allowed a hierarchical design belonging to one domain (e.g., SDF) to appear as an atomic actor following the rules of another domain (e.g., DE simulation). For this to work correctly, mechanisms had to be developed to synchronize schedulers operating in different domains.

The original Ptolemy, now called Ptolemy Classic, was written in C++ as an extensible class library; its successor, Ptolemy II [58] was thoroughly rearchitected and written in Java.

Ptolemy Classic classed domains as either untimed (e.g., dataflow) or timed (e.g., DE). From the perspective of a timed domain, actions in an untimed domain will appear to be instantaneous. In the case of a mixture of SDF and DE for example, one might represent a computation with SDF components and the associated delay in the DE domain, thus separating the computation from the delay involved in a particular implementation. When two distinct timed domains are interfaced, a global time is maintained, and the schedulers for the two domains are kept synchronized. The details of implementation of scheduler synchronization across domain boundaries are described in [6].

Ptolemy Classic was successful as a heterogeneous simulation tool, but it possessed a path to implementation (in the form of generated software or HDL code) only for dataflow domains (SDF and BDF). Furthermore, all of its domains shared the characteristic that atomic blocks represented processes and connections represented data signals.

One of the more interesting features added by Ptolemy II was its approach to hierarchical control [59]. The concept, flowing logically out of the Ptolemy idea, was to extend Statecharts to allow for complete nesting of data-oriented domains (e.g., SDF and DE) as well as synchronous-reactive domains, working out the semantic details as required. Ptolemy II calls designs that are represented as state diagrams, modal designs. If the state symbols represent atomic states, we simply have an extended FSM (extended because, as in Statecharts, the conditions and actions on the transition arcs are not restricted to Boolean signals). However, the states can also represent arbitrary Ptolemy subsystems. When a state is entered, the subsystem contained in the state begins execution. When an outgoing transition occurs, the subsystem halts its execution. The so-called time-based signals that are presented to the modal model propagate downward to the subsystems that are "inside" the states.

Girault et al. [59] claim that the semantic issues with Statecharts variants identified by von der Beek [35] can be solved by orthogonality: nesting FSMs together with domains providing the required semantics, thereby obtaining, for example, either synchronous-reactive or DE behavior. In the author's view, this only partially solves the problem, because there are choices to be made about the semantics of FSMs nested inside of other FSMs.

A similar project to allow for full nesting of hierarchical FSMs and dataflow with a somewhat different design, was part of Synopsys's System Studio [11] (originally code-named "El Greco"). System Studio's functional modeling combines dataflow models with Statecharts-like control models that have semantics very close to those of SyncCharts [37], as both approaches started with Esterel semantics. Like Ptolemy II, System Studio permits any kind of model to be nested inside of any other, and state transitions cause interior subsystems to start and stop. One unique feature of System Studio is that parameter values, which in Ptolemy are set at the start of a simulation run or are compiled in when code generation is performed, can be reset to different values each time a subsystem inside a state transition diagram is started.

There have been several efforts to make the comparison and the combination of models of computation more theoretically rigorous. Lee and Sangiovanni-Vincentelli [60] introduced a meta-model that represents signals as sets of events. Each event is a pair consisting of a value and a tag, where the tags can come from either a totally ordered or a partially ordered set. Synchronous events share the same tag. The model can represent important features of a wide variety of models of computation, including most of those discussed in this chapter; however, in itself it does not lead to any new results. It can be thought of, perhaps, as an educational tool.

Metropolis [61] is a new design environment for heterogeneous systems. Like Ptolemy, it supports more than one model of computation. Unlike Ptolemy, Metropolis attempts to treat functionality and architecture as orthogonal, to make it easier to create mappings between functional and architectural representations, and to make it possible to represent explicitly constraints as well as to verify properties with model checkers [62].

4.4 Conclusions

This chapter has described a wide variety of approaches, all of them deserving of more depth than could be presented here. Some approaches have been more successful than others.

It should not be surprising that there is resistance to learning new languages. It has long been argued that system-level languages are most successful when the user does not recognize that what is being proposed is, in fact, a new language. Accordingly there has sometimes been less resistance to graphical approaches (especially when the text used in such an approach is from a familiar programming language such as C++ or an HDL), and to class libraries that extend C++ or Java. Lee [2] makes a strong case that such approaches are really languages, but acceptance is sometimes improved if users are not told about this. A strong case can be made that approaches based on KPNs, dataflow, and Matlab have been highly successful in a variety of application areas that require digital signal processing. These include wireless; audio, image and video processing; radar; 3-D graphics, and many others. Hierarchical control tools, such as those based on Statecharts and Esterel, have also been successful, though their use is not quite as widespread. Most of the remaining tools described here have been found useful in smaller niches, though some of these are important. However, it is the author's belief that higher-level tools are underused, and that as the complexity of systems to be implemented continues to increase, designers who exploit domain-specific system-level approaches will benefit by doing so.

References

[1] The MathWorks Inc., *MATLAB Reference Guide*, Natick, MA, USA, 1993.

[2] E.A. Lee, Embedded software, in M. Zelowitz, Ed., *Advances in Computers*, Vol. 56, Academic Press, New York, 2002.

[3] G. Berry, The foundations of Esterel, in G. Plotkin, C. Stirling, and M. Tofte, Eds., *Proof, Language and Interaction: Essays in Honor of Robin Milner*, MIT Press, Cambridge, MA, 2000, pp. 425–454.

[4] G. Berry and G. Gonthier, The Esterel synchronous programming language: design, semantics, implementation, *Sci. Comput. Progr.*, 19, 87–152, 1992.

[5] T. Grotker, S. Liao, G. Martin, and S. Swan. *System Design with SystemC*, Kluwer Academic Publishers, Dordrecht, 2002.

[6] J.T. Buck, S. Ha, E. A. Lee, and D.G. Messerschmitt, Ptolemy: a framework for simulating and prototyping heterogeneous systems, *Int. J. Comput. Simul.*, 4, 155–182, 1994.

[7] G. Kahn, The semantics of a simple language for parallel programming, *Information Processing 74:Proceedings of IFIP Congress 74*, Stockholm, Sweden,1974, pp. 471–475.

[8] E.A. Lee and T.M. Parks, Dataflow process networks, *Proc. IEEE*, 83, 773–801, 1995.

[9] Cadence Design Systems, *Signal Processing WorkSystem (SPW)*, 1994.

[10] J. Kunkel, COSSAP: a stream driven simulator, *IEEE Int. Workshop Microelectrion. Commun.*, 1991.

[11] J.T. Buck and R. Vadyanathan, Heterogeneous modeling and simulation of embedded systems in El greco, *Proceedings of the Eighth International Workshop on Hardware/Software Codesign (CODES)*, 2000.

[12] E.A. de Kock, G. Essink, W.J.M. Smits, P. van der Wolf, J.Y. Brunel, W.M. Kruijtzer, P. Lieverse, and K.A. Vissers, Yapi: application modeling for signal processing systems, *Proceedings of the 37th Design Automation Conference*, Los Angeles, CA, 2000.

[13] E.A. Lee and D.G. Messerschmitt, Synchronous data flow, *Proc. IEEE*, 75(9), 1235–1245, 1987.

[14] N. Halbwachs, P. Caspi, P. Raymond, and D. Pilaud, The synchronous data flow programming language LUSTRE, *Proc. IEEE*, 79, 1305–1320, 1991.

[15] S.S. Bhattacharyya and E.A. Lee, Looped schedules for dataflow descriptions of multirate signal processing algorithms, *Form. Method. Sys. Design*, 5, 183–205, 1994.

[16] S.S. Bhattacharyya, P.K. Murthy, and E.A. Lee, *Software Synthesis from Dataflow Graphs*, Kluwer Academic Publishers, Dordrecht, 1996.

[17] J.L. Pino, S.S. Bhattacharyya, and E.A. Lee, A hierarchical multiprocessor scheduling system for DSP applications, *Proceedings of IEEE Asilomar Conference on Signals, Systems, and Computers*, Pacific Grove, CA, 1995.

[18] M. Engels, G. Bilsen, R. Lauwereins, and J. Peperstraete, Cyclo-static data flow: model and implementation, *Proceedings of the 28th Asilomar Conference on Signals, Systems, and Computers*, Pacific Grove, CA, 1994, pp. 503–507.

[19] J. Bier, E. Goei, W. Ho, P. Lapsley, M. O'Reilly, G. Sih, and E.A. Lee, Gabriel: a design environment for DSP, *IEEE Micro Mag.*, 10, 28–45, 1990.

[20] J.L. Pino, S. Ha, E.A. Lee, and J.T. Buck. Software synthesis for DSP using Ptolemy, *J. VLSI Signal Process.*, 9, 7–21, 1995.

[21] S. Ritz, M. Pankert, and H. Meyr, High level software synthesis for signal processing systems, *International Conference on Application Specific Array Processors*, 1992, pp. 679–693.

[22] R. Lauwereins, M. Engels, M. Ade, and J.A. Perperstraete, GRAPE-II: a tool for the rapid prototyping of multi-rate asynchronous DSP applications on heterogeneous multiprocessors, *IEEE Comput.*, 28, 35–43, 1995.

[23] D.G. Messerschmitt, A tool for structured functional simulation, *IEEE J. Selec. Area. Comm.*, SAC-2, 137–147, 1984.

[24] P.N. Hilfinger, J. Rabaey, D. Genin, C. Scheers, and H. De Man, DSP specification using the Silage language, *Proceedings of the International Conference on Acoustics, Speech, and Signal Processing*, Albuquerque, NM, USA, 1990, pp. 1057–1060.

[25] P. Willekens, D. Devisch, M. Ven Canneyt, P. Conflitti, and D. Genin, Algorithm specification in DSP station using data flow language, *DSP Appl.*, 3, 8–16, 1994.

[26] J.T. Buck, Scheduling Dynamic Dataflow Graphs with Bounded Memory Using the Token Flow Mode,. Ph.D. thesis, University of California at Berkeley, 1993.

[27] J.T. Buck, Static scheduling and code generation from dynamic dataflow graphs with integer-valued control systems, *Proceedings of the IEEE Asilomar Conference on Signals, Systems, and Computers*, Pacific Grove, CA, 1994.

[28] P. Zepter and T. Grotker, Generating synchronous timed descriptions of digital receivers from dynamic data flow system level configurations, *Proceedings of European Design And Test Conference*, Paris, France, 1994.

[29] J. Horstmannshoff and H. Meyr, Efficient building block based RTL/HDL code generation from synchronous data flow graphs, *Proceedings of the 37th Design Automation Conference*, Los Angeles, CA, 2000.

[30] P. van der Wolf, E. de Kock, T. Henriksson, W.M. Kruijtzer, and G. Essink, Design and programming of embedded multiprocessors: an interface-centric approach, *Proceedings of CODES+ISSS 2004*, 2004, pp. 206–216.

[31] The MathWorks Inc., *Simulink: dynamic system simulation for MATLAB*, Natick, MA, USA, 1997.

[32] D. Harel, Statecharts: a visual approach to complex systems, *Sci. Comput. Progr.*, 8, 231–275, 1987.

[33] D. Harel and E. Gery, Executable object modeling with statecharts, *Computer*, 30, 31–42, 1997.

[34] J. Rumbaugh, I. Jacobson, and G. Booch, *Unified Modelling Language Reference Manual*, Addison Wesley Longman, Inc., Reading, MA, USA, 1998.

[35] M. von der Beek, A comparison of statechart variants, in L. de Roever and J. Vytopil, Eds., *Formal Techniques in Real-Time and Fault Tolerant Systems*, Springer, Berlin, 1994, pp. 128–148.

[36] F. Maraninchi, The Argos language: graphical representation of automata and description of reactive systems, *IEEE International Conference on Visual Languages*, Kobe, Japan, 1991.

[37] C. Andre, Representation and analysis of reactive behaviors: a synchronous approach, *Proceedings of CESA 1996*, 1996.

[38] R. Milner, Calculi for synchrony and asynchrony, *Theor. Comput. Sci.*, 25 ,267–310, 1983.

[39] P. Le Guernic and T. Gautier, Data-flow to von Neumann: the SIGNAL approach, in L. Bic and J.-L. Gaudiot, Eds., *Advanced Topics in Dataflow Computing*, Prentice-Hall, New York, 1991.

[40] N. Halbwachs, *Synchronous Programming of Reactive Systems*, Kluwer Academic Publishers, Dordrecht, 1993.

[41] A. Bouali, XEVE, an Esterel verification environment, *Proceedings of the International Conference on Computer-Aided Verification (CAV)*, 1998, pp. 500–504.

[42] L. Arditi, A. Bouali, H. Boufaied, G. Clave, M. Hadj-Chaib, and R. de Simone, Using Esterel and formal methods to increase the confidence in the functional validation of a commercial DSP, *Proceedings of ERCIM Workshop on Formal Methods for Industrial Critical Systems*, 1999.

[43] G. Berry, A hardware implementation of pure Esterel, *Sadhana-Acad. P. Eng. S.*, 17, 95–139, 1992. (Special issue on real time systems.)

[44] S.A. Edwards, Compiling Esterel into sequential code, *Proceedings of the 37th Design Automation Conference*, 2000.

[45] A. Seawright and F. Brewer, Clairvoyant: a synthesis system for production-based specification, *Proc. IEEE Trans. VLSI Syst.*, 2, 172–185, 1994.

[46] M.E. Lesk, LEX: A Lexical Analyzer Generator, Technical Report Computing Science Technical Report 39, AT&T Bell Laboratories, 1975.

[47] A. Seawright, U. Holtmann, W. Meyer, B. Pangrle, R. Verbrugghe, and J.T. Buck, A system for compiling and debugging structured data processing controllers, *Proceedings of EuroDAC 96*, 1996.

[48] W. Meyer, A. Seawright, and F. Tada, Design and synthesis of array structured telecommunication processing applications, *Proceedings of the 34th Design Automation Conference*, Anaheim, CA, 1997, pp. 486–491.

[49] C.A.R. Hoare, CSP — Communicating Sequential Processes, *International Series in Computer Science*, Prentice-Hall, New York, 1985.

[50] INMOS Limited, *Occam Programming Manual*, Prentice-Hall, New York, 1984.

[51] F. Balarin, E. Sentovich, M. Chiodo, P. Giusto, H. Hsieh, B. Tabbara, A. Jurecska, L. Lavagno, C. Passerone, K. Suzuki, and A. Sangiovanni-Vincentelli, *Hardware-Software Co-design of Embedded Systems — The POLIS Approach*, Kluwer Academic Publishers, Dordrecht, 1997.

[52] M. Baleani, A. Ferrari, A. Sangiovanni-Vincentelli, and C. Turchetti, Hw/sw codesign of an engine management system, *Proceedings of DATE 2000*, 2000.

[53] CCITT, *Specification and Description Language, CCITT Z.100*, International Consultative Committee on Telegraphy and Telephony, 1992.

[54] T. Grotker, private communication, 2004.

[55] D. Gajski, J. Zhu, R. Dömer, A. Gerstlauer, and S. Zhao, *SpecC: Specification Language and Methodology*, Kluwer Academic Publishers, Dordrecht, 2000.

[56] S. Sutherland, S. Davidmann, and P. Flake, *SystemVerilog For Design: A Guide to Using SystemVerilog for Hardware Design and Modeling*, Kluwer Academic Publishers, Dordrecht, 2004.

[57] L. Cai and D. Gajski, Transaction level modeling: an overview, *First International Conference on HW/SW Codesign and System Synthesis (CODES+ISSS 2003)*, Newport Beach, CA, 2003.

[58] J. Davis, R. Galicia, M. Goel, C. Hylands, E.A. Lee, J. Liu, X. Liu, L. Muliadi, S. Neuendorffer, J. Reekie, N. Smyth, J. Tsay, and Y. Xiong, Heterogeneous Concurrent Modeling and Design in Java, Technical Report UCB/ERL M98/72, EECS, University of California, Berkeley, 1998.

[59] A. Girault, B. Lee, and E.A. Lee, Hierarchical finite state machines with multiple concurrency models, *IEEE T. Comput. Aid. D.*, 18, 742–760, 1999.

[60] E.A. Lee and A. Sangiovanni-Vincentelli, Comparing models of computation, *Proceedings of the IEEE/ACM International Conference on Computer-aided Design*, 1996, pp. 234–241.

[61] F. Balarin, H. Hsieh, L. Lavagno, C. Passerone, A. Sangiovanni-Vincentelli, and Y. Watanabe, Metropolis: an integrated environment for electronic system design, *IEEE Comput.*, 36(4), 45–52, 2003.

[62] X. Chen, F. Chen, H. Hsieh, F. Balarin, and Y. Watanabe, Formal verification of embedded system designs at multiple levels of abstraction, *Proceedings of the International Workshop on High Level Design, Validation, and Test*, Cannes, France, 2002, pp. 125–130.

5 SoC Block-Based Design and IP Assembly

John Wilson
Mentor Graphics
Berkshire, United Kingdom

Block-based design strategies, where existing IP is reused in new designs, are reaching prominence as designers start to examine how best to take advantage of the capacity now available on silicon.

Traditional design processes start with an initial concept, and are followed by a process of refinement and detailing that eventually results in an implementation of the original idea. This craftsman's approach, where each module in the implementation is created specifically for the design in which it ultimately exists, results in a cohesive design, but at some cost. If an IP module that implements a required function already exists, it may be more costly to train a designer to acquire enough knowledge to reproduce that functionality in-house [1]. With first time deployment of a design module, there is always an increased risk of encountering design bugs.

A block-based approach offers the possibility of reusing existing design modules in a new design by trading off the design optimization benefits of custom design work against the time and cost savings offered by reusing existing components. In making the "design" or "reuse" decision, designers have to weigh a large number of factors, and many of these factors are economic rather than design driven.

Processor vendors such as ARM and MIPS are prominent amongst the top IP revenue generators [2]. Processors are relatively small IP modules (in terms of gate count) and are not especially hard to design. So there must be some characteristics that make processor IP compelling to buy rather than build.

The first factor is the development effort required to create the software tools like assemblers, compilers, linkers, debuggers, and OS ports. Without this support infrastructure, the processor IP may be easy to deploy in designs, but impossible to utilize effectively.

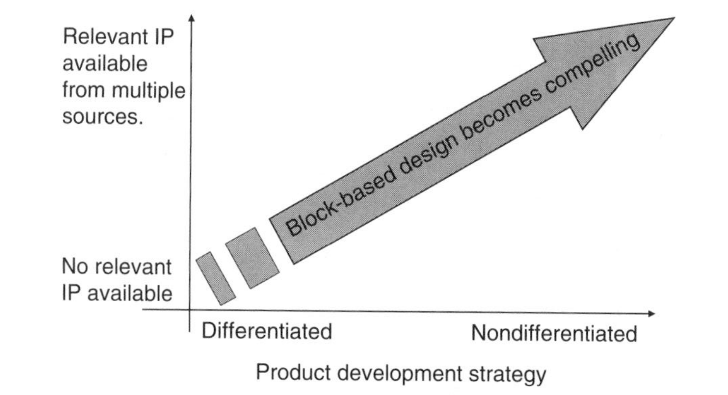

FIGURE 5.1 When block-based design becomes compelling.

The second factor is the design maturity. In creating home-grown IP, designers are deploying prototype modules with the inherent risk of initial deployment. There may be significant cost and reliability advantages in deploying IP that is already mature and improved through iteration [3].

As more IP begins to exhibit the characteristics of processors (extended infrastructure, maturity acquired over multiple projects), deploying reusable IP becomes a viable design choice.

In reality, designers should not be choosing between the block-based and traditional approaches, but examining where best to deploy those strategies on each project. Design modules that embody the design group's key knowledge and competencies — the differentiated part of a design — are the key areas on which to focus project resources and effort. Nondifferentiated parts of a design are ideal candidates for buying and configuring external IP, instead of using up valuable design team resources on designing something that already exists and offers no competitive advantage to the product (Figure 5.1).

There are some significant challenges in reusing IP in a new design, and these should not be dismissed lightly. However, a number of different industry trends are converging to ease these problems.

Design teams need to pay careful attention to these industry trends because they have the potential to offer significant competitive advantage and, increasingly teams will be expected to deliver their designs as reusable modules for use by other design teams within and outside their organizations.

5.1 The Economics of Reusable IP and Block-Based Design

For both IP creators and designers, IP becomes compellingly reusable as it increases in complexity, and standards emerge to enable that IP to be deployed rapidly into designs. A rough rule of thumb is that IP complexity (as measured by gate count and functionality) increases by five times over 3 years. Figure 5.2 illustrates the complexity growth by comparing the gate count of the highest performance members of two commonly used IP families over a 4-year period. It is unlikely that designer productivity can match this complexity growth without aggressive adoption of reuse strategies.

Standards are important for IP providers because that increases the potential market for their products. Lack of standards means that requests to customize IP for each customer are very likely. Customizing IP increases design, verification, testing, and support costs. Requirements to customize can increase design costs by such an extent that the design project is more appropriately priced as a consulting engagement. At this point, all the advantages of reusing IPs are lost. To make reuse a reality, it is incumbent on the IP providers to build enough flexibility and customization options into their designs to allow the IP to be used in an SoC, and for designers to create their designs according to standards to maximize the chances of being able to deploy reusable IP without requiring customization.

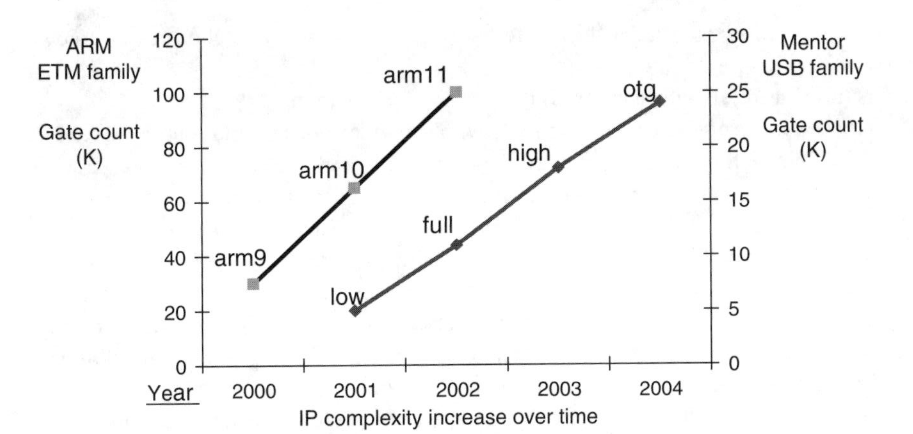

FIGURE 5.2 Complexity (as measured by gate count) increase in two IP product families over successive generations.

Many IP providers were caught up in the HDL language wars where customers were demanding Verilog or VHDL models. Many IP providers had to deliver the same IP in multiple model formats, doubling the modeling and verification effort and the costs.

The emergence of multilanguage HDL simulators and synthesis tools has made the choice of language almost totally irrelevant. It is straightforward to mix and match IP modules written in different languages. Most IP companies can be confident that supplying IP in one language is not a barrier to adoption by potential customers using the other.

In fact, some commercial IP is now supplied as mixed VHDL/Verilog design modules. An example of this is the Mentor Graphics Ethernet controller, where one of the major functional blocks was designed as a Verilog module by an external organization, and was subsequently integrated with other blocks written in VHDL, to create the overall IP module.

Up to now, IP reuse has focused on individual IP modules and how they should be structured [4] but much less emphasis was placed on how the IP modules could be deployed in a design.

Many large organizations use proprietary IP formats and demand that external IP supplied to the organization meet those format requirements [5],[6]. Reformatting and validating IP packages to meet those proprietary standards is a significant burden for IP providers.

Under the auspices of VSIA (www.vsia.org), an IP infrastructure (IPI) standard is being developed to enable a standard IP packaging format to be followed, and to allow the creation of IP readers that can reformat the IP package to meet each organization's IP structure requirements. This offers the possibility that IP deliverables can be packaged in a generic format by the IP creators, and unpackaged according to the format required by the recipient organization. Many large organizations with internal IP standards spend considerable amounts of time qualifying bought-in IP packages. In many cases, qualifying means rearranging the contents of the IP package to match the internal company standards, and that is just another customization cost.

Each standard that emerges makes block-based design more compelling for both IP creators and designers who want to buy and deploy that IP. Creating IP that conforms to recognized standards gives IP providers a bigger, more viable market for their products, and the power to resist costly requests for customization [7].

5.2 Standard Bus Interfaces

The first point of contact for most designers integrating IP into systems is the bus interfaces and protocols. Nonstandard and custom interfaces make integration a time-consuming task. On-chip bus standards such as AMBA [8], OCP [9], and CoreConnect [10] have become established, and most IP will be expect to conform to one of those interfaces unless there is a very good reason to deviate from that standard.

The benefits of using IP with standard interfaces are overwhelming.

Designers who have integrated an IP module with a standard bus interface into a system are able to integrate a different IP module using the same interface with ease.

Because this task has moved from a custom to a repeatable operation, specialist designers who understand how to craft optimally standard interconnect between IP modules have emerged. In turn, their understanding has resulted in tools that can automatically generate the required interconnect infrastructure for each bus [11]. Some of these tools are extremely sophisticated, taking into account bandwidth and power requirements to generate bus infrastructure IP that may be significantly more complex than the IP to which it is connected. A typical On-The-Go USB IP configuration might use 30K gates, while a fully configured adapter for the Sonics [11]On Chip bus can exceed 100K gates.

Verification also becomes easier because the same test modules used to validate a bus interface on one IP module can be reused on another. Again, the emergence of standard interfaces has stimulated new markets — there are many companies now supplying verification IP [12] to validate standard interface protocols. Using verification IP from one source to check IP modules from another is a very robust check.

So far, most of the focus has been on system bus for interconnecting CPUs, memory, and peripheral devices. But standard interfaces are emerging for other IP functions: for instance, many USB providers now have a standard interface to the PHY, enabling the designer to choose the best mix of USB and PHY for their design [13].

Designers may think interfaces only refer to hardware but consistent software interfaces are equally important. There have been attempts to create standards around hardware-dependent software [14], but these have not gained widespread acceptance, partly because of concerns over debug capabilities and performance (driver functions are called frequently and can have a disproportionate effect on system performance). However, some proprietary formats (for example, Xilinx's MLD) have been interfaced successfully with multiple commercial operating systems without any significant loss of performance. So it seems that some level of software driver standardization can be expected to gain acceptance in the near future.

The use of standard interfaces, wherever possible, is a key strategy for IP reuse and deployment, and enables the automation of block-based design.

5.3 Use of Assertion-Based Verification

IP creators spend significant time and effort creating testbenches to exercise and validate their "standalone" IP over a wide range of design scenarios. Testbenches are delivered to customers as part of the IP package to help designers build up confidence that the IP is functional.

There is one fundamental limitation in this approach: testbenches cannot be easily reused when the IP is embedded in a design. The decision to buy in the IP indicates the lack of a certain level of design knowledge and sophistication in the design team, so most designers do not have the specialized knowledge to develop their own design-dependent functional tests for that IP module. The testbenches supplied with the IP are not modular, so when designers integrate the IP into their designs, they have to start from scratch to develop testbenches that are appropriate for the new IP design context.

This has caused a "chicken-and-egg" problem for the IP industry. It is difficult to convince designers to use IP if it cannot be demonstrated to be successful, but how can the IP be shown to be successful if it does not get deployed in designs? Currently, the gold standard indicator that an IP is reusable is the length of the list of working designs in which that IP has been successfully used.

One technology that has emerged to ease this problem is assertion-based verification (ABV). When a design is created, designers can also write down a series of assertions about how the IP module should react to various events. When the IP is simulated as part of a design, the assertions can be used to check that the IP is not encountering previously unknown scenarios that are untested or cause malfunctions.

```
-- Generate parity serially, clearing down after every
character TX
  Process (CLOCK, RESETN)
  begin
    if (RESETN = '0') then
      DP <= '0';
    elsif (CLOCK' event and CLOCK = '1') then
      if (TxState = C_TX_START) and (TxClkEnab = '1') then
         DP <= '0' --initialise parity = 0
      else
         if ((HStep and TxClkEnab) = '1') then
           DP <= (TXMD xor DP); --EXOR DP and current data bit
         end if;
      end if;
    end if;
  end process;

-- Generate parity, if required of the right polarity
process (EPS, SP, DP)
begin
PTY <= not (EPS xor (not (SP or not DP)));
end process;


-- Assert 0: check illegal Transmit State Machine states
-- psl    property TXE_FSM_illegal is never
--       ((TxState = "1101") or (TxState = "1110") or
--        (TxState = "1111") @ rose(CLOCK));
-- psl    assert TXE_FSM_illegal;
```

In the same way that code-coverage tools can be used to indicate areas of HDL that have not been adequately tested, designers can also use code coverage to check that the assertions associated with the IP module have been activated. Unexecuted assertions would indicate that the original IP designer's intent has not been checked for the design in which the IP is included. More tests are needed to prove that the IP is executing correctly (or not, as the case may be) in the design.

There are a number of different assertion languages, some proprietary and some standards-based. PSL [15] and System Verilog Assertions (SVA)[16] are emerging as the languages most likely to be supported in standard HDL simulators.

5.4 Use of IP Configurators and Generators

IP reuse becomes compelling for the IP creator if there are opportunities to reuse the IP in a wide range of designs. A common strategy is to make the IP highly configurable, giving designers the best chance of setting up an optimal configuration that will work in the target design.

The first generation of configurable IP was statically configurable. That is, parameterization features of the underlying HDL were used to allow designers to select and set particular options.

However, many of the IP configurations are so complex and interdependent that this is well outside the scope of HDL configuration capabilities. A single configuration choice may result in complete sets of interfaces being added or deleted from the IP. Increasingly, IP is provided with their own configuration scripts and programs enabling designers to customize the IP and the supporting files delivered with that IP. This is dynamically configurable IP. A simple example of dynamically configurable IP is the ARM MultiLayer AMBA switched bus infrastructure IP. The version included in the Arm Design Kit v1.1 did not include the IP directly, but did include a program to be run by the designer to generate the HDL for the IP that matched the designer's specific requirements.

For some highly complex IP, the delivery package contains no IP at all, but only the generator program. Running the generator program creates the IP and support files, customized according to each designer's requirements.

The effort required to create generator programs is far greater than creating just one instance of the IP, but the economics can be compelling if the potential market for the IP increases commensurately, and will become even more compelling as design environments that support heterogeneous IP integration emerge.

One interesting aspect of IP configuration is how processor vendors choose to handle extensions to the core processor architecture. The traditional approach is to create a standard co-processor interface onto which designers can add additional functions that are tightly bound into the processor operation to augment the standard processor capabilities. This approach generally lacks support from the software tools associated with the processor, which can make it difficult for programmers to take full advantage of the co-processor.

An alternative approach pioneered by Tensilica and ARC is to create generators that create custom instances of a processor, including the ability to add custom instructions. The software tools designed for these processors also follow the configuration, so that the software tools match the processor capabilities.

Both approaches have merit. Configurable processors come more into their own if significant additions are required to processor capabilities. The co-processor approach has been boosted by the emergence of tools that can synthesize software system functionality into hardware co-processors using standard interfaces to accelerate overall system performance. It is still too early to say how much utility these tools will have: software underperformance is often not diagnosed until the hardware prototype is complete, at which stage it may be too late to change the hardware design.

Some companies offer IP already integrated into larger processor-based design modules. The IP in the design is often configurable but cannot be deleted, and standard bus interfaces are exposed for designers to connect on additional IP [17].

There are significant advantages providing highly integrated modules. Additional levels of high-value IP integration can be offered — the module test structures, and OSs can be pre-configured to run on such systems. This is platform-based design (PBD).

The first generation of PBD IP was very generic in nature. Often consisting of a processor, essential core IP such as interrupt and memory controllers, and a set of standard peripherals such as UARTs, timers and GPIO modules, it was also supplied with parameterizable boot code to enable the user to start executing the code quickly. The push behind the creation of this PBD IP was processor companies who were finding that it was very difficult for designers to understand how to create the support infrastructure required to make the processors execute the code efficiently.

Second-generation PBD IP tends to be strongly themed for particular classes of target applications. Alongside the standard CPU subsystems, specialist IP and software are integrated for applications such as mobile phones [18] or video applications [19], and significant effort is applied to provide customized configuration tools for the second-generation PBD IP [20].

Second-generation PBD IP providers have taken nondifferentiated processor subsystems and added their specialist knowledge and IP to create a value-added differentiated design.

Platform-based design will dominate nondifferentiated design strategies. Many designers will start by searching for the platform IP that is closest to their requirements and build, from that point, IP connected using standard buses and interfaces.

5.5 The Design Assembly and Verification Challenge

Designers creating designs using block-based design techniques face an interesting dichotomy. The design choice to use external IP is almost certainly driven by the lack of resources or expertise to create that IP on the project.

However, when the IP is integrated into the design, the designer is still faced with the task of understanding the IP enough to create appropriate system-level verification tests, and this is a significant and often underestimated task.

Where IP with standard interfaces are deployed, there are often standard protocol tests available as verification IP.

ABV-enabled IP, in addition to enabling in-system checks, provide some new possibilities when used in conjunction with code coverage tools. Assertions state designer intent. Measuring the extent of activation of these assertions gives designers a good indication of how well the IP functionality has been covered.

Building system-level verification environments involves many of the same characteristics as automating the design build. The same tools that automate the design-build processes would be expected to generate appropriate system-level tests.

As an example, it is interesting to think about a simple processor-based SoC design that incorporates some external IP such as a USB and Bluetooth module connected on a standard system bus.

If the system bus infrastructure is generated ("stitched"), then it is reasonable to expect monitors to be automatically enabled in the design environment to identify any deviations from the bus standard. The monitor may be used to check specific protocols at recognized interfaces on the bus, or examine the overall block operation to help gather and evaluate performance and other data.

Providing stimulus to exercise automatically the design is more challenging. It is quite easy to build automatically software programs to run on the processor to provide "ping" tests for the IP, which can be used to verify that the processor-to-peripheral functionality is correct.

It is also reasonable to provide "traffic generator" IP to stimulate external standard interfaces. If a design exports (makes visible externally) an interface, it is not too difficult to automatically construct a testbench that attaches a traffic generator module to that interface. However, when more than one traffic generator is used in a design, there are major problems in automating the intent and synchronization of the activities of different traffic generators to create desired test scenarios.

For processor-based designs, one common technique is to construct a "loopback" style of testbench where a component in the design with a processor and communications interface (for instance, a UART, USB, or Bluetooth controller) is connected to a matching component in the testbench. The processor interfaces on the testbench components are connected back to the processor in the design (Figure 5.3).

This allows test programs to be created that can be run on the processor, enabling both the design under test to be set up and activated, and cause events via the testbench components which can be used to verify if the design components are operational.

This methodology meshes neatly with the techniques used by software programmers to test their programs outside the host-system environment.

While this methodology is not generally applicable, associating test modules with IP modules is the first step to partially automating testbench creation in parallel with design creation.

There are limits to what can be automated. It would be very difficult for a designer to create a system with Bluetooth and USB modules under the control of an RTOS running on a processor, and expect automatic generation of tests to insert a message into the Bluetooth module and see the message emerge from the USB port. Automation tools cannot reasonably be expected to understand design intent for complex test scenarios.

Ideally, a designer would get involved in verification test development for higher level system functionality, and leave the tools to automate the mechanical parts of the testbench construction, analogous to the way that the designer selects the IP for inclusion in a design but uses automated tools for the detailed integration.

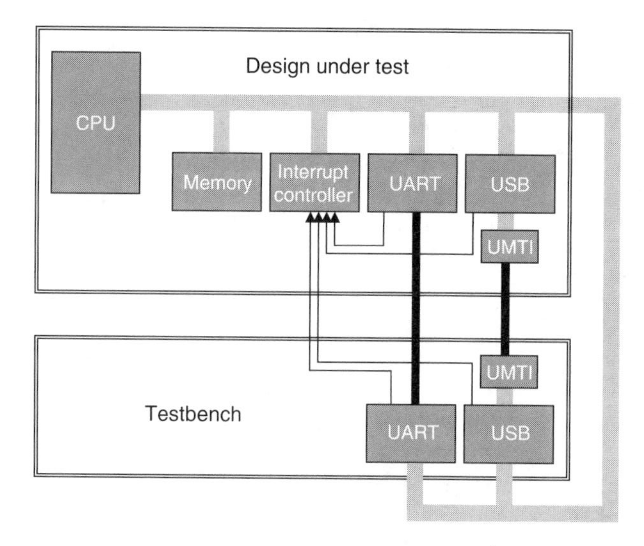

FIGURE 5.3 "Loopback" style of testbench.

However, for the moment, it seems that the lack of agreed standards in this area restricts even this level of automation to proprietary systems where organizations have full control over the test interfaces and methodology.

5.6 The SPIRIT XML Databook Initiative

Abstraction level paradigm shifts are common when designers need an order of magnitude increase in design productivity. Working with logic gates instead of transistors, and with MSI modules instead of gates were consistent trends in previous decades.

Each of these abstractions is consistent within its own reference framework: there is no logic gate that could not be connected to another logic gate, for instance. But block-based design is different.

Links to many disparate sources of information are required to make sensible design choices. For instance, if an IP module is supplied with an OS driver stack, then it may be easy to connect the module to the processor from an HW viewpoint, but this does not mean that the IP is functional. The driver software may be supplied in a form that is incompatible with the processor or the operating system running on that processor.

There have been many credible attempts to create a unified design flow [21] but most put a significant burden on IP creators to model IP in a specific way, and do not deal with enough of the myriad disparate design aspects to be immediately compelling. It is very difficult for the IP creators to justify the effort to support such tools within a nascent market environment and this, rather than technical considerations, have tended to restrict promising initiatives.

The SPIRIT initiative (www.spiritconsortium.com) changes the emphasis from requiring IP to be supplied in a fixed format, to one of documenting whatever aspects of the IP are available in XML [22]. Typically, this would include, but not be restricted to, information about the configurable IP parameters, bus interface information, and pointers to the various resources — like ABV files and software driver modules — associated with that IP module [23].

The power of SPIRIT [24] is that the information enables a consistent HW, SW, and system view of a design to be maintained. The same information can be used to create HW views and software programmer's views for a design.

Design environments capable of reading this XML information can make intelligent design configuration and integration choices [25][24]. From an HW designer's point of view, that could be an IP "stitcher" creating detailed HDL descriptions. The same information can be used to build SW programs to run on the design, design documentation, and setup verification environments, etc.

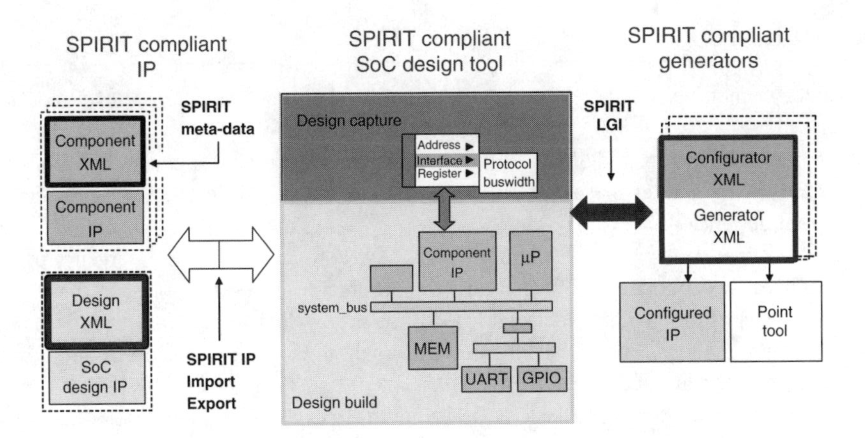

FIGURE 5.4 The SPIRIT design environment architecture. (From Mentor Graphics Platform Express, www.-mentor.com/products/embedded_software/platform_baseddesign/index.cfm.)

SPIRIT enables the deployment of disparate IP generators within design environments so many of the IP configurators and generators already in existence can be reused. In fact, their use is enhanced because instead of designers being required to guess at the parameters to be supplied to configure the IP appropriately, the XML-based design database can be used to extract and calculate the required settings automatically (Figure 5.4).

In turn, this enables the creation of new classes of "design-wide" generators, which assess the information available to individual IP modules and build aspects of the design, e.g., HDL, software programs to run on the design or design documentation. Nobody expects all IP to be delivered with every conceivable view: but where information is missing (from the design generator's point of view), the generators are expected to have a strategy to handle that missing information.

An example of a tool that can process SPIRIT formatted information is Mentor Graphic's Platform Express [25]. SPIRIT XML component and design information is read into the tool, along with other XML information, to build the design database. Generators extract information from this design database, sometimes creating design modules such as HDL netlists, software programs, and documentation; sometimes creating tool modules such as configuration files for external tools used in the design process; or creating new information to add to the design database for use by other generators later in the design creation process.

There is also a realization in SPIRIT that it is impossible to create a single global unifying standard that can encompass all aspects of design. Instead, the approach has been to agree on core information, and provide mechanisms for extending and customizing the data. This allows arbitrary file types to be associated with IP. It enables individual organizations to add their own custom data where specialist data processing is required (this is a common requirement for highly configurable IP).

Most importantly, it enables *de facto* data standards for areas not directly covered by SPIRIT to be developed by expert interested groups and used in conjunction with SPIRIT.

An organic, evolutionary approach to evolve data standards may seem a strange path to follow, but where the information domain is characterized by high complexity and unstructured relationships, enabling different data models to be supported and watching for winners to emerge is a powerful strategy. The moderating factor that will prevent an explosion of incompatible or overlapping standards is the availability of generators and features inherent in XML. IP providers will provide data in the format that can be utilized by the most functionally useful generators, and as more IP support the data, more generators will emerge to use that data. For data provided in different formats, XSL [26] (a standard within the XML standards family) can be used to extract and reformat that data for use with existing generators.

Ultimately, the goal of SPIRIT is to enable designers to mix and match IP and design generators from many different sources, and leverage the emerging block-based design standards to build and verify high-quality designs rapidly and reliably.

5.7 Conclusions

No single strategy enables block-based design strategies, but a number of important initiatives are converging to make block-based design compelling. For example, IP supplied with standard SoC bus interfaces ready for integration into a design; configurators to help designers to choose valid options; ABV modules to catch deployment errors; and XML databook and catalog information will be easier to deploy in many different designs.

SPIRIT does not mandate specific IP standards, but is likely to accelerate the emergence of various *de facto* standards because generators will search out compatible information supplied in the IP package to build various aspects of designs. IP blocks which do not supply that information can be handled, but designers are likely to choose IP that the generators can help deploy most effectively, and ignore the IP that requires additional manual design effort.

The real challenge for design teams will be deciding on their core design competencies on which to focus their expertise and resources, and being ruthless in deciding that compromising on bought-in IP modules is a more efficient design choice than requiring custom IP design modifications for noncritical system components.

References

[1] K. Pavitt, What we know about strategic management of technology, in *Implementing New Technologies: Innovation and the Management of Technology*, 2nd ed., E. Rhodes and D. Wield, Eds., NCC Blackwell, Oxford, UK,1994, pp. 176–178.

[2] G. Moretti, Third Party IP: A Shaky Foundation for SoC Design, *EDN*, February 3, magazine (online), 2005.

[3] C.M. Christensen, *The Innovator's Dilemma*, Harvard Business School Press, Cambridge, Mass, USA, 1997, p. 9.

[4] M. Keating and P. Bricaud, *Reuse Methodology Manual for System-on-a-Chip Designs,* Kluwer Academic Publishers, Boston, 1998, Chap. 9.

[5] Repositories are Better for Reuse, *Electronics Times* (U.K.), March 27, 2000.

[6] Philips Execs Find IP Unusable, but Necessary, *Electronics Times* (U.K.), March 22, 2000.

[7] M. E. Porter, How Competitive Forces Shape Strategy, *Harvard Business Review*, March 1, 1979.

[8] AMBA Specification, www.arm.com/products/solutions/AMBAHomePage.html

[9] OCP Specification, www.ocpip.org/getspec

[10] CoreConnect Specification, www-03.ibm.com/chips/products/coreconnect

[11] Sonics, www.sonicsinc.com

[12] For example, Synopsys (www.synopsys.com/products/designware/dwverificationlibrary.html), Cadence (www.cadence.com/products/ip) are just two of many companies. The Design and Reuse catalogue gives many examples — see www.us.design-reuse.com/verificationip.

[13] S. McGowan, *USB 2.0 Transceiver Macrocell Interface (UTMI)*, Intel, March 29, 2001. www.intel.com/technology/usb/download/2_0_Xcvr_Macrocell_1_05.pdf

[14] Hardware-dependent software, in *Taxonomies for the Development and Verification of Digital Systems*, B. Bailey, G. Martin and T. Anderson, Eds., Springer, Heidelberg, 2005, pp. 135–168.

[15] Accellera, Property Specification Language Reference Manual, June 9, 2004, www.eda.org/vfv/docs/PSL-v1.1.pdf

[16] System Verilog Assertions — see the System Verilog LRM, www.eda.org/sv/SystemVerilog_3.1a.pdf

[17] H. Chang, L. Cooke, M. Hunt, G. Martin, A. McNelly, and L. Todd, *Surviving the SoC Revolution,* Kluwer Academic Publishers, Dordrecht, 1999, pp. 155–182.

[18] P. Cumming, The TI OMAP™ platform approach to SoC, in *Winning the SoC Revolution*, G. Martin and H. Chang, Eds., Kluwer Academic Publishers, Boston, 2003, Chap. 5, pp. 97–118.

[19] J. Augusto de Oliveira and H. Van Antwerpen, The Philips Nexperia digital video platform, in *Winning the SoC Revolution*, G. Martin and H. Chang, Eds., Kluwer Academic Publishers, Boston, 2003, Chap. 4, pp. 67–96.

[20] G. Mole, Philips Semiconductors Next Generation Architectural IP ReUse Developments for SoC Integration, *IP/SOC 2004*, Grenoble, France, December 2004.

[21] S. Krolikoski, F. Schirrmeister, B. Salefski, J. Rowson, and G. Martin, Methodology and Technology for Virtual Component Driven Hardware/Software Co-Design on the System Level, paper 94.1, *ISCAS 1999*, Orlando, FL, 1999.
[22] W3C Consortium, XML Standard, www.w3.org/XML
[23] SPIRIT Consortium, SPIRIT — Examples [presentation], www.spiritconsortium.com, December 8, 2004.
[24] SPIRIT Consortium, "SPIRIT — The Dream" [presentation], www.spiritconsortium.com, December 8, 2004.
[25] Mentor Graphics Platform Express, www.mentor.com/products/embedded_software/platform_baseddesign/index.cfm
[26] W3C Consortium, XSL Standard, www.w3.org/Style/XSL

6

Performance Evaluation Methods for Multiprocessor System-on-Chip Design

Ahmed Jerraya
SLS Group, TIMA Laboratory, INPG
Grenoble, France

Iuliana Bacivarov
SLS Group, TIMA Laboratory
Grenoble, France

6.1 Introduction

Multi-processor systems-on-chip (MPSoCs) require the integration of heterogeneous components (e.g., micro-processors, DSP, ASIC, memories, buses, etc.) on a single chip. The design of MPSoC architectures requires the exploration of a huge space of architectural parameters for each component. The challenge of building high-performance MPSoCs is closely related to the availability of fast and accurate performance evaluation methodologies. This chapter provides an overview of the performance evaluation methods developed for specific subsystems. It then proposes to combine subsystem performance evaluation methods to deal with MPSoC.

Performance evaluation is the process that analyzes the capabilities of a system in a particular context, i.e., a given behavior, a specific load, or a specific set of inputs. Generally, performance evaluation is used to validate design choices before implementation or to enable architecture exploration and optimization from very early design phases.

A plethora of performance evaluation tools have been reported in the literature for various subsystems. Research groups have approached various types of subsystems i.e., software (SW), hardware (HW), or interconnect, differently, by employing different description models, abstraction levels, performance metrics, or technology parameters. Consequently, there is currently a broad range of methods and tools for performance evaluation, addressing virtually any kind of design and level of hierarchy, from very specific subsystems to generic, global systems.

Multi-processor system-on-chip (MPSoC) is a concept that aims at integrating multiple subsystems on a single chip. Systems that put together complex HW and SW subsystems are difficult to analyze. Additionally, in this case, the design space exploration and the parameter optimization can quickly become intractable. Therefore, the challenge of building high-performance MPSoCs is closely related to the availability of fast and accurate performance evaluation methodologies.

Existing performance evaluation methods have been developed for specific subsystems. However, MPSoCs require new methods for evaluating their performance. Therefore the purpose of this study is to explore different methodologies used for different subsystems in order to propose a general framework that tackles the problem of performance evaluation for heterogeneous MPSoC. The long-term goal of this work is to build a global MPSoC performance evaluation by composing different tools. This kind of evaluation will be referred to as holistic performance evaluation.

The chapter is structured as follows: Section 6.2 defines the key characteristics of performance evaluation environments. It details the analyzed subsystems, their description models and environments, and the associated performance evaluation tools and methods. Section 6.3 is dedicated to the study of MPSoC performance evaluation. Section 6.4 proposes several trends that could guide future research toward building efficient MPSoC performance evaluation environments.

6.2 Overview of Performance Evaluation in the Context of System Design Flow

This section defines typical terms and concepts used for performance evaluation. First, the performance-evaluation process is positioned within a generic design flow. Three major axes define existing performance evaluation tools: the subsystem under analysis, the performance model, and the performance evaluation methodology. They are detailed in this section. An overview of different types of subsystems is provided, focusing on their abstraction levels, performance metrics, and technology parameters. In the end, the main performance evaluation approaches are introduced.

6.2.1 Major Steps in Performance Evaluation

This section analyzes the application domain of the performance evaluation process within the systems design flow. A designed system is evaluated by a suitable performance evaluation tool where it is represented by a performance model. Next, this section presents how evaluation results may influence decisions during system design.

A design flow may include one or more performance evaluation tools. These evaluation tools could be used for different purposes, e.g., to verify if a system meets the constraints imposed or runs properly and to help making design choices.

Figure 6.1 presents a generic design flow. The initial specification is split into a functional and a nonfunctional part of the subsystem to be analyzed. The functional part contains the behavior of the subsystem under analysis, described it as an executable program or as a formal model (e.g., equation). However, the set of evaluation constraints or quality criteria selected from the initial specification constitute the nonfunctional part.

The performance model cannot be separated from the evaluation methodology, because it provides the technology characteristics used to compute performance results prior to the real implementation. Moreover, it selects the critical characteristics to be analyzed, such as the model performance metrics: timing, power, and area. Eventually it chooses the measurement strategy (e.g., an aggregation approach). Both performance metrics and technology parameters may be built into the evaluation tool or given as external libraries.

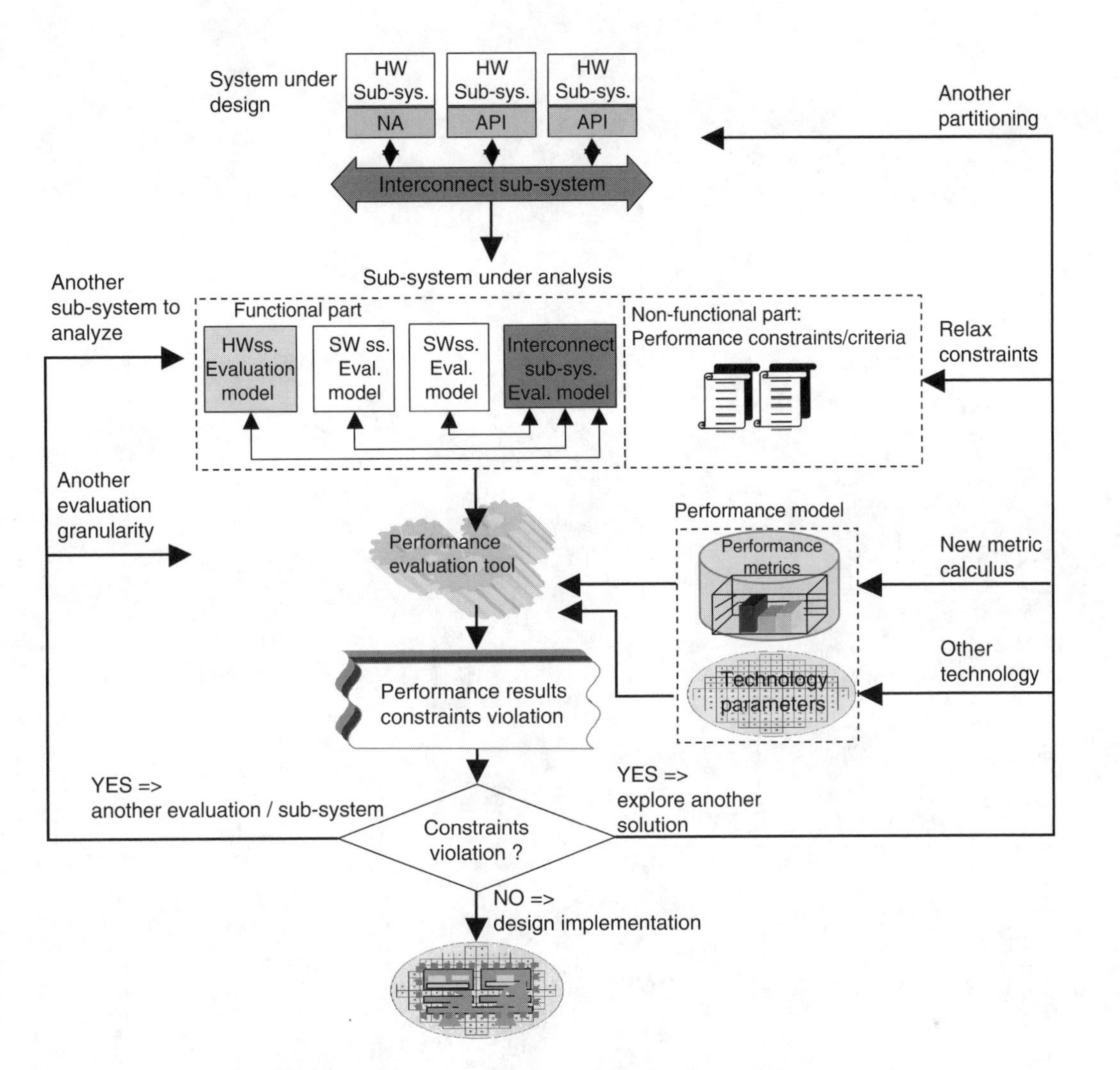

FIGURE 6.1 General performance evaluation and design optimization flow.

The design process may be made of several iterations when the system needs to be tuned or partially redesigned. Several design loops may modify the performance model. Each time a new calculation of the metrics (e.g., the chip area can be enlarged in order to meet timing deadlines) or technological parameters (e.g., a new underlying technology or increased clock frequency) is initiated.

6.2.2 The Key Characteristics of Performance Evaluation

Figure 6.2 details on three axes, the three main characteristics of the performance evaluation process: the subsystem under analysis, its abstraction level, and the performance evaluation methodology. Analysis of five kinds of subsystems will be considered: HW, SW, CPUs, interconnect subsystems, and MPSoCs. Each basic subsystem has specific design methods and evaluation tools. They may be designed at different abstraction levels, of which we will consider only three. Also, three performance evaluation methodologies will be considered, with metrics and technology parameters specific to different subsystems.

A subsystem under analysis is characterized by its type (e.g., HW, SW, etc.), and its abstraction level. The performance evaluation may be applied to different kinds of subsystems varying from simple devices to

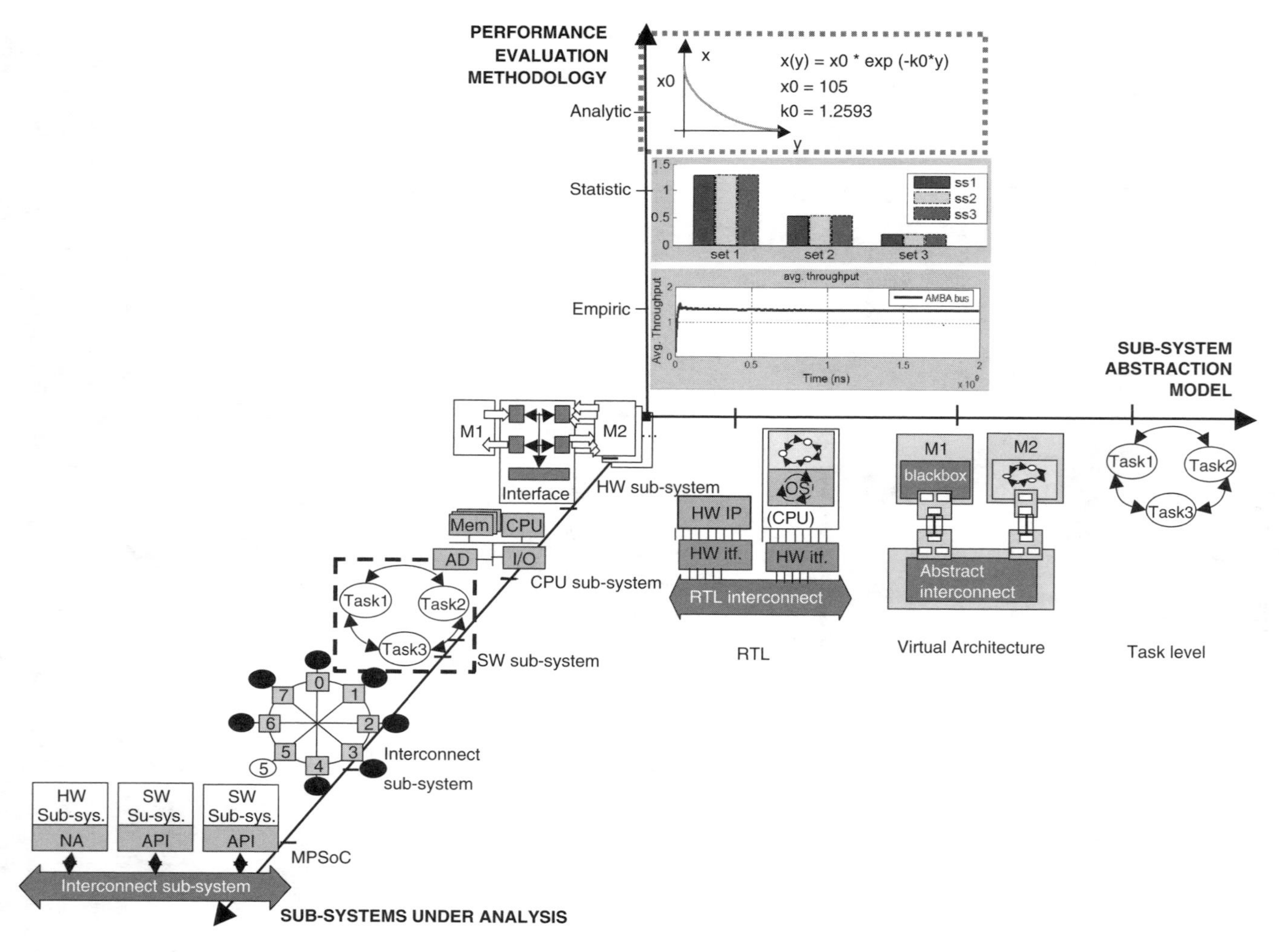

FIGURE 6.2 Performance evaluation environment characteristics.

sophisticated modules. We will consider five main kinds of subsystems: HW subsystems, CPU subsystems, SW subsystems, interconnect subsystems, and constituent parts of MPSoC. Traditionally they are studied by five different research communities.

Classically, the HW community [1] considers HW subsystems as HDL models. They are designed with electronic design automation (EDA) tools that include specific performance analysis [2–5]. The computer architecture community e.g., [6] considers CPU subsystems as complex microarchitectures. Consequently, specialized methodologies for CPU design and performance evaluation have been developed [7]. The SW community [8,9] considers SW subsystems as programs running parallel tasks. They are designed with computer-aided SW engineering (CASE) tools and evaluated with specific methods [10–22]. The networking community [23–30] considers interconnect subsystems as a way to connect diverse HW or SW components. The network performance determines the overall system efficiency, and consequently it is an intensively explored domain.

Each of these communities uses different *abstraction levels* to represent their subsystem. Without any loss of generality, in this study only three levels will be considered: register transfer level (RTL), virtual-architecture level, and task level. These may be adapted for different kinds of subsystems.

Performance evaluation uses a specific methodology and a system performance model. The methodology may be simulation-based, analytic (i.e., using a mathematical description), or statistical. The system performance model takes into account the performance metrics and technology parameters.

The performance metrics are used for assessing the system under analysis. They may be physical metrics related to real system functioning or implementation (e.g., execution timings, occupied area, or consumed power), or quality metrics that are related to nonfunctional properties (e.g., latency, bandwidth, throughput, jitter, or errors).

The technology parameters are required to fit the performance model to an appropriate analysis domain or to customize given design constraints. The technology parameters may include architectural features of higher level (e.g., the implemented parallelism, the network topology), or lower level (e.g., the silicon technology, the voltage supply).

6.2.3 Performance Evaluation Approaches

The two main classes of performance evaluation reported in literature are: statistical approaches and deterministic approaches. For statistical approaches, the performance is a random variable characterized by several parameters such as a probability distribution function, average, standard deviation, and other statistical properties. Deterministic approaches are divided into empirical and analytical approaches. In this case, the performance cost function is defined as a deterministic variable, a function of critical parameters. Each of these approaches is defined as follows:

The *statistical approach* [17,19] proceeds in two phases. The first phase finds the most suitable model to express the system performance. Usually parameters are calibrated by running random benchmarks. The second phase makes use of the statistical model previously found to predict the performance of new applications. In most cases, this second phase provides a feedback for updating the initial model.

The *empirical* approach can be accomplished either by measurement or simulation. *Measurement* is based on the real measurement of an already built or prototyped system. It generally provides extremely accurate results. Because this approach can be applied only late in the design cycle when a prototype can be made available, we do not include it in this study. The *simulation approach* [3,16,21,24–28,31] relies on the execution of the complete system using input scenarios or representative benchmarks. It may provide very good accuracy. Its accuracy and speed depend on the abstraction level used to describe the simulated system.

The *analytical approach* [2,10–12,14,15,32] formally investigates system capabilities. The subsystem under analysis is generally described at a high level of abstraction by means of algebraic equations. Mathematical theories applied to performance evaluation make possible a complete analysis of the full system performance at an early design stage. Moreover, such approaches provide fast evaluation because they replace time-consuming system compilation and execution. Building an analytical model could be very complex. The dynamic behavior (e.g., program context switch and wait times due to contentions or collisions) and refinement steps (e.g., compiler optimizations) are hard to model. However, this approach may be useful for worst-case analysis or to find corner cases that are hard to cover with simulation.

6.2.4 Hardware Subsystems

6.2.4.1 Definition

An HW subsystem is a cluster of functional units with a low programmability level like FPGA or ASIC devices. It can be specified by finite state machines (FSMs) or logic functions. In this chapter, the HW concept excludes any modules that are either CPUs or interconnection like. We also restrain the study to digital HW.

6.2.4.2 Abstraction Levels

HW abstraction is related to system timing, of which we consider three levels: high-level language (HLL), bus cycle-accurate (BCA), and RTL. At HLL, the behavior and communication may hide clock cycles, by using abstract channels and high-level communication of primitives, e.g., a system described by untimed computation and transaction-level communication. At BCA level, only the communication of the subsystem is detailed to the clock cycle level, while the computation may still be untimed. At RTL, both the computation and communication of the system are detailed to clock cycle level. A typical example is a set of registers and some combinatorial functions.

6.2.4.3 Performance Metrics

Typical performance metrics are power, execution time, or size, which could accurately be extracted during low-level estimations and used in higher abstraction models.

6.2.4.4 Technology Parameters

Technology parameters abstract implementation details of the real HW platform, depending on the abstraction level. At HLL, as physical signals and behavior are abstracted, the technology parameters denote high-level partitioning of processes with granularity of functions (e.g., C function) and with reference to the amount of exchanged transactions. At BCA level, the technology parameters concern data formats (e.g., size, coding, etc.), or behavioral data processing (e.g., number of bytes transferred, throughputs, occupancies, and latencies). At RTL, the HW subsystem model is complete. It requires parameters denoting structural and timing properties (e.g., for the memory or communication subsystems) and implementation details (e.g., the FPGA mapping or ASIC implementation).

There are different performance evaluation tools for HW subsystems, which make use of different performance evaluation approaches: simulation-based approaches [3], analytical approaches [2], mixed analytical and statistical approaches [18], mixed simulation and statistical approaches [5], and mixed analytical, simulation, and statistical approaches [4].

6.2.5 CPU Modules

6.2.5.1 Definition

A CPU module is a hardware module executing a specific instruction set. It is defined by an instruction set architecture (ISA) detailing the implementation and interconnection of the various functional units, the set of instructions, register utilization, and memory addressing.

6.2.5.2 Abstraction Levels

For CPU modules, three abstraction levels can be considered: RTL, also known as the micro-architecture level, the cycle-accurate ISA level, and the assembler (ASM) ISA level. The RTL (or micro-architecture level) offers the most detailed view of a CPU. It contains the complete detailed description of each module, taking into account the internal data, control, and memory hierarchy. The cycle-accurate ISA level details the execution of instructions with clock accuracy. It exploits the real instruction set model and internal resources, but in an abstract CPU representation. The ASM ISA level increases the abstraction, executing programs on a virtual CPU representation, with abstract interconnections and parameters, e.g., an instruction set simulator.

6.2.5.3 Performance Metrics

The main performance metrics for CPU subsystems are related to timing behavior. We can mention among these the throughput that expresses the number of instructions executed by CPU per time unit, the utilization that represents the time ratio spent on executing tasks, and the time dedicated to the execution of a program or to respond to a peripheral. Other performance evaluation metrics are power consumption and memory size.

6.2.5.4 Technology Parameters

Technology parameters abstract the CPU implementation details, depending on the abstraction level. At RTL, only the technology for the CPU physical implementation is abstract. The ISA level abstracts the control and data path implementation, but it still details the execution with clock-cycle accuracy using the real instruction set (load/store, floating point, or memory management instructions), the internal register set, and internal resources. And finally, the ASM level abstracts the micro-architecture (e.g., pipeline and cache memory), providing only the instruction set to program it.

Different performance evaluation tools for CPU subsystems exist, making use of different performance evaluation approaches: simulation-based approaches [31,33] analytical approaches [32,10], statistical approaches [19], mixed analytical and statistical approaches [34], and mixed analytical, simulation, and statistical approaches [7].

Chapter 9 of this book ("Using Performance Metrics to Select Microprocessor Cores for IC Designs") has as its main objective, the measurement of CPU performance. For a thorough discussion on CPU performance evaluation utilizing benchmarking techniques, we refer the reader to Chapter 10.

6.2.6 Software Modules

6.2.6.1 Definition

A software module is defined by the set of programs to be executed on a CPU. They may have different representations (procedural, functional, object-oriented, etc.), different execution models (single- or multi-threaded), different degrees of responsiveness (real time, nonreal time), or different abstraction levels (from HLL down to ISA level).

6.2.6.2 Abstraction Levels

We will consider three abstraction levels for SW modules. At HLL, parallel programs run independently on the underlying architecture, interacting by means of abstract communication models. At transaction-level modeling (TLM) level, parallel programs are mapped and executed on generic CPU subsystems. They communicate explicitly but their synchronization remains implicit. And finally, at ISA level, the code is targeted at a specific CPU and it targets explicit interconnects.

6.2.6.3 Performance Metrics

The metrics most used for SW performance evaluation are run time, power consumption, and occupied memory (footprint). Additionally, SW performance may consider concurrency, heterogeneity, and abstraction at different levels [35].

6.2.6.4 Technology Parameters

For SW performance evaluation, technology parameters abstract the execution platform. At HLL, technology parameters abstract the way different programs communicate using, for example, coarse-grain send()/receive() primitives. At TLM level, technology parameters hide the technique or resources used for synchronization, such as using a specific Operating System (OS), application program interfaces (APIs) and mutex_lock()/unlock()-like primitives. At ISA level, technology parameters abstract the data transfer scheme, the memory mapping, and the addressing mode.

Different performance evaluation tools for SW models exist, making use of different performance evaluation approaches: simulation-based [21], analytical [12], and statistical [17].

6.2.6.5 Software Subsystems

6.2.6.5.1 Definition.

When dealing with system-on-chip design, the SW is executed on a CPU subsystem, made of a CPU and a set of peripherals. In this way, the CPU and the executed SW program are generally combined in an SW subsystem.

6.2.6.5.2 Abstraction Levels.

The literature presents several classifications for the abstraction of SW subsystems, among which we will consider three abstraction levels: The HLL, the OS level, and the hardware abstraction layer (HAL) level. At the HLL, the application is composed of a set of tasks communicating through abstract HLL primitives provided by the programming languages (e.g., send()/receive()). The architecture, the interconnections, and the synchronization are abstract. At the OS level, the SW model relies on specific OS APIs, while the interconnections and the architecture still remain abstract. Finally, at the HAL level, the SW is bound to use a specific CPU insturtion set and may run on an RTL model of the CPU. In this case, the interconnections are described as an HW model, and the synchronization is explicit.

6.2.6.5.3 Performance Metrics.

The main performance metrics are the timing behavior, the power consumption, and the occupied area. They are computed by varying the parameters related to the SW program and to the underlying CPU architecture.

6.2.6.5.4 Technology Parameters.

In SW subsystem performance evaluation, technology parameters abstract the execution platform, and characterize the SW program. At HLL, technology parameters mostly refer to application behavioral features, abstracting the communication details. At the OS level, technology parameters include OS features (e.g., interrupts, scheduling, and context switching delays), but their implementation remains abstract. At HLL, technology parameters abstract only the technology of implementation, while all the other details, such as the data transfer scheme, the memory mapping or the addressing mode are explicitly referred to.

Different performance evaluation tools for SW subsystems exist, making use of different performance evaluation approaches: simulation-based approaches [16,20], analytical approaches [11,13], statistical approaches [19], and mixed analytical, simulation, and statistical approaches [22].

6.2.7 Interconnect Subsystems

6.2.7.1 Definition

The interconnect subsystem provides the media and the necessary protocols for commusnication between different subsystems.

6.2.7.2 Abstraction Levels

In this study, we will consider RTL, transactional, and service or HLL models. At the HLL, different modules communicate by requiring services using an abstract protocol, via abstract network topologies. The transactional level still uses abstract communication protocols (e.g., send/receive) but it fixes the communication topology. The RTL communication is achieved by means of explicit interconnects like physical wires or buses, driving explicit data.

6.2.7.3 Performance Metrics

The performance evaluation of interconnect subsystem focuses on the traffic, interconnection topology (e.g., network topology, path routing, and packet loss within switches), interconnection technology (e.g., total wire length and the amount of packet switch logic), and application demands (e.g., delay, throughput, and bandwidth).

6.2.7.4 Technology Parameters

A large variety of technology parameters emerge at different levels. Thus, at HLL, parameters are the throughput or latency. At TLM level, the parameters are the transaction times and the arbitration strategy. At the RTL, the wires and pin-level protocols allow system delays to be measured accurately.

Simulation is the performance evaluation strategy most used for interconnect subsystems at different abstraction levels: behavioral [24], cycle-accurate level [25], and TLM level [27]. Interconnect simulation models can be combined with HW/SW co-simulation at different abstraction levels for the evaluation of full MPSoC [26,28].

6.2.8 Multiprocessor Systems-on-Chip Models

6.2.8.1 Definition

An MPSoC is a heterogeneous system built of several different subsystems like HW, SW, and interconnect, and it takes advantage of their synergetic collaboration.

6.2.8.2 Abstraction levels

MPSoCs are made of subsystems that may have different abstraction levels. For example, in the same system, RTL HW components can be coupled with HLL SW components, and they can communicate at the RTL or by using transactions. In this study, we will consider the interconnections, synchronization, and interfaces between the different subsystems. The abstraction levels considered are the functional level, the virtual architecture model level, and the level that combines RTL models of the hardware with instruition set architecture models of the CPU. At the functional level (like message passing interface (MPI)) [36], the HW/SW interfaces, the interconnections, and synchronization are abstracted and the subsystems interact through high-level primitives (send, receive). For the virtual architecture level, the interconnections and synchronization are explicit but the HW/SW interfaces are still abstract. The lowest level considered deals with an RTL architecture for the HW-related sections coupled with the ISA level for the SW. This kind of architecture explicitly presents the interconnections, synchronization, and interfaces. In order to master the complexity, most existing methods used to assess heterogeneous MPSoC systems are applied at a high abstraction level.

6.2.8.3 Performance Metrics

MPSoC performance metrics can be viewed as the union of SW, HW and interconnect subsystems performance metrics, for instance, execution time, memory size, and power consumption.

6.2.8.4 Technology Parameters

A large variety of possible technology parameters emerge for each of the subsystems involved, mostly at different levels and describing multiple implementation alternatives. They are considered during system analysis, and exploited for subsystem performance optimization.

Different performance-evaluation tools for MPSoC exist. They are developed for specific subsystems [37–41], or for the complete MPSoC [42–44]. Section 6.3 deals with performance-evaluation environments for MPSoCs.

6.3 MPSoC Performance Evaluation

As has been shown in previous sections, many performance evaluation methods and tools are available for different subsystems: HW, SW, interconnect and even for MPSoCs. They include a large variety of measurement strategies, abstraction levels, evaluated metrics, and techniques. However, there is still a considerable gap between particular evaluation tools that consider only isolated components and performance evaluation for a full MPSoC.

The evaluation of a full MPSoC design containing a mixture of HW, SW and interconnect subsystems, needs to cover the evaluation of all the subsystems, at different abstraction levels.

Few MPSoC evaluation tools are reported in the literature [37–39,41,42,44,45]. The key restriction with existing approaches is the use of a homogeneous model to represent the overall system, or the use of slow evaluation methods that cannot allow the exploration of the architecture by evaluating a massive number of solutions.

For example, the SymTA/S approach [45] makes use of a standard event model to represent communication and computation for complex heterogeneous MPSoC. The model allows taking into account complex behavior such as interrupt control and data dependant computation. The approach allows accurate performance analysis; however, it requires a specific model of the MPSoC to operate.

The ChronoSym approach [41] makes use of a time-annotated native execution model to evaluate SW execution times. The timed simulation model is integrated into an HW/SW co-simulation framework to consider complex behaviors such as interactions with the HW resources and OS performance. The approach allows fast and accurate performances analysis of the SW subsystem. However, for the entire MPSoC evaluation, it needs to be combined with other approaches for the evaluation of interconnect and HW subsystems.

Co-simulation approaches are also well suited for the performance evaluation of heterogeneous systems. The co-simulation offers flexibility and modularity to couple various subsystem executions at different abstraction levels and even specified in different languages. The accuracy of performance evaluation by co-simulation depends on the chosen subsystem model and on their global synchronization.

A complete approach aiming at analyzing the power consumption for the entire MPSoC by using several performance models is presented in [42]. It is based on interconnecting different simulations (e.g., SystemC simulation and ISS execution) and different power models for different components, in a complete system simulation platform named MPARM. A related work is [43], where the focus is on MPSoC communication-performance analysis.

The co-estimation approach in [44] is based on the concurrent and synchronized execution of multiple power estimators for HW/SW system-on-chip-power analysis. Various power estimators can be plugged into the co-estimation framework, possibly operating at different levels of abstraction. The approach is integrated in the POLIS system design environment and PTOLEMY simulation platform. The power co-estimation framework drives system design trade-offs, e.g., HW/SW partitioning, component, or parameter selection. A similar approach is represented in [46,47]. The tool named ROSES allows different subsystems that may be described at different abstraction levels and in different languages to be co-simulated.

However, when applied to low abstraction levels, evaluation approaches based on co-simulation [43,44,46] appear to be slow. They cannot explore large solution spaces. An alternative would be to combine co-simulation with analytical methods in order to achieve faster evaluation. This is similar to methods used for CPU architecture exploration [31,33].

Figure 6.3 shows such a scheme for MPSoC. The key idea is to use co-simulation for computing extensive data for one architectural solution. The results of co-simulation will be further used to parameterize an analytical model. A massive number of new design solutions can be evaluated faster using the newly designed analytic model.

The left branch of Figure 6.3 represents the performance evaluation of the full system by co-simulation. This can be made by using any existing co-simulation approach [43,44,46]. The input of the evaluation process is the specification of the MPSoC architecture to be analyzed. The specification defines each subsystem, the communication model, and the interconnection interfaces.

The right branch of Figure 6.3 describes the evaluation of the full system using an analytical model. The figure represents the analytical model construction with dotted lines. This is done through component characterizations and parameter extraction from the base co-simulation model.

After the construction phase, the analytical branch can be decoupled from the co-simulation. The stand-alone analytical performance evaluation provides quick and still accurate evaluations for new designs. The two branches of Figure 6.3, i.e., the co-simulation and the analytical approach, lead to similar performance results, but they are different in terms of evaluation speed and accuracy.

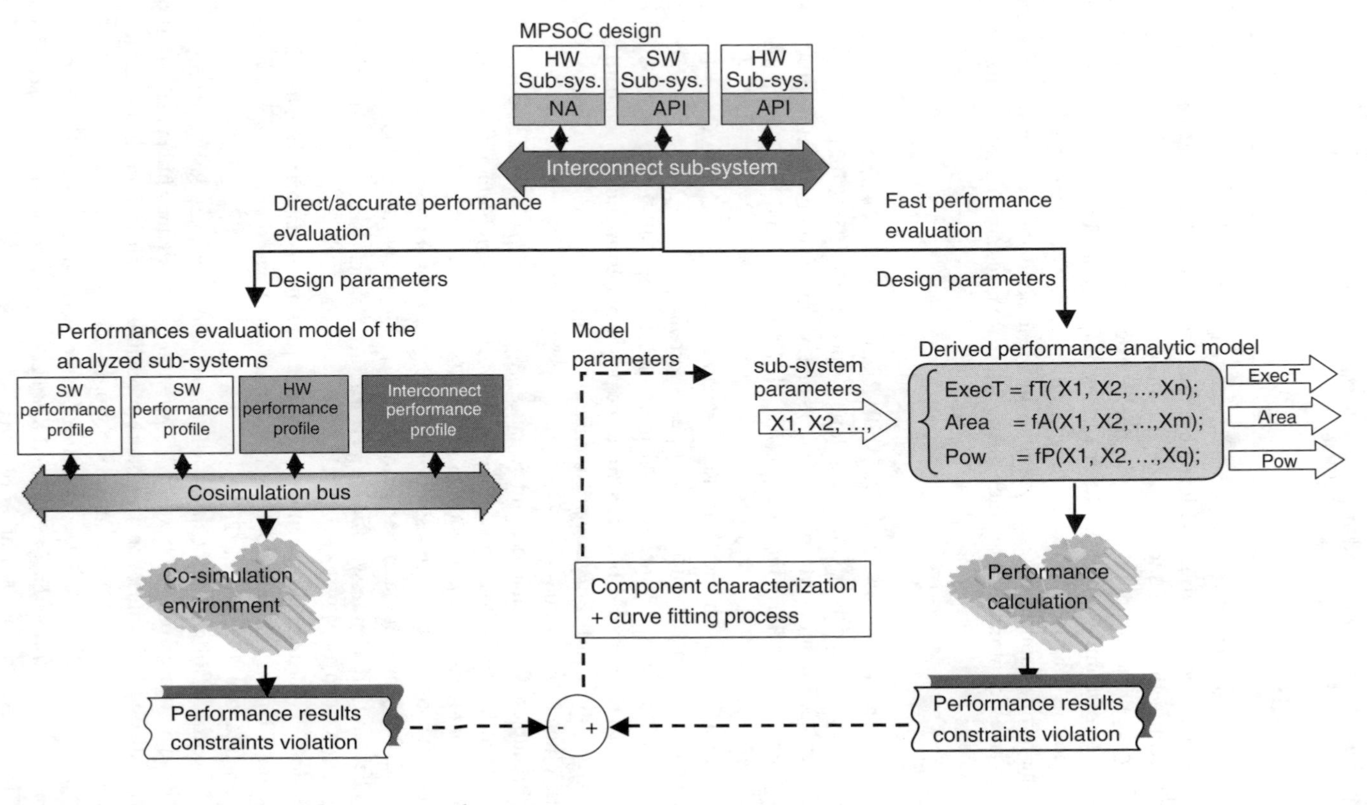

FIGURE 6.3 Global heterogeneous MPSoC evaluation approach.

The proposed strategy is based on the composition of different evaluation models, for different MPSoC subsystems. It combines simulation and analytical models for fast and accurate evaluation of novel MPSoC designs. The further objective is to develop a generic framework for design space exploration and optimization, where different simulation based or analytical evaluation methods could be applied to different subsystems and at different levels of abstraction.

6.4 Conclusion

MPSoC is an emerging community trying to integrate multiple subsystems on a single chip and consequently requiring new methods for performance evaluation [48]. Therefore, the aim of this study was to explore different methodologies for the different subsystems that may compose the MPSoC.

We defined a general taxonomy to handle the heterogeneity and diversity of performance-evaluation solutions. This taxonomy introduced the attributes of an evaluation tool: abstraction levels, modeling techniques, measured metrics, and technology parameters. Finally, we proposed an evaluation framework based on the composition of different methods in which simulation and analytical methods could be combined in an efficient manner, to guide the design space exploration and optimization.

References

[1] G. de Micheli, R. Ernst, and W. Wolf, *Readings in Hardware/Software Co-Design*, Morgan Kaufmann, San Francisco, CA, 1st ed., June 1, 2001, ISBN 1558607021.

[2] S. Dey and S. Bommu, Performance analysis of a system of communicating processes, *International Conference on Computer-Aided Design (ICCAD 1997)*, ACM and IEEE Computer Society, San Jose, CA, 1997, pp. 590–597.

[3] V. Agarwal, M.S. Hrishikesh, S.W. Keckler, and D. Burger, Clock rate versus IPC: The end of the road for conventional microarchitectures, *Proceedings of the 27th International Symposium on Computer Architecture*, Vancouver, Canada, June 2000, pp. 248–259.

[4] H. Yoshida, K. De, and V. Boppana, Accurate pre-layout estimation of standard cell characteristics, *Proceedings of ACM/IEEE Design Automation Conference (DAC)*, San Diego, California, United States, June 2004, pp. 208–211.

[5] C. Brandolese, W. Fornaciari, and F. Salice, An area estimation methodology for FPGA based designs at SystemC-level, *Proceedings of ACM/IEEE Design Automation Conference (DAC)*, San Diego, California, USA, June 2004, pp. 129–132.

[6] A. D. Patterson and L. J. Hennessy, *Computer Organization and Design, the Hardware/SW Interface*, 2nd ed., Morgan-Kaufmann, San Francisco, CA, 1998, ISBN 155860 -491-X.

[7] D. Ofelt and J.L. Hennessy, Efficient performance prediction for modern microprocessors, *SIGMETRICS*, 2000, pp. 229–239.

[8] B. Selic, An efficient object-oriented variation of the statecharts formalism for distributed real-time systems, CHDL 1993, *IFIP Conference on Hardware Description Languages and their Applications*, Ottawa, Canada, 1993, pp. 28–28.

[9] B. Selic and J. Rumbaugh, *Using UML for Modeling Complex Real-Time Systems*, Whitepaper, Rational Software Corp., 1998, http://www.rational.com/media/whitepapers/umlrt.pdf

[10] J. Walrath and R. Vemuri, A performance modeling and analysis environment for reconfigurable computers, IPPS/SPDP Workshops, 1998, pp. 19–24.

[11] B. Spitznagel and D. Garlan, Architecture-based performance analysis, *Proceedings of the 1998 Conference on Software Engineering and Knowledge Engineering*, San Francisco, California, 1998.

[12] P. King and R. Pooley, Derivation of petri net performance models from UML specifications of communications SW, *Proceedings of Computer Performance Evaluation Modelling Techniques and Tools: 11th International Conference, TOOLS 2000*, Schaumburg, IL, 2000.

[13] F. Balarin, STARS of MPEG decoder: a case study in worst-case analysis of discrete event systems, *Proceedings of the International Workshop on HW/SW Codesign*, Copenhagen, Denmark, April 2001, pp. 104–108.

[14] T. Schuele and K. Schneider, Abstraction of assembler programs for symbolic worst case execution time analysis, *Proceedings of ACM/IEEE Design Automation Conference (DAC)*, San Diego, California, USA, June 2004, pp. 107–112.

[15] C Lu, J.A. Stankovic, T.F. Abdelzaher, G. Tao, S.H. Son, and M. Marley, Performance specifications and metrics for adaptive real-time systems, *IEEE Real-Time Systems Symposium (RTSS 2000)*, Orlando, FL, 2000.

[16] M. Lajolo, M. Lazarescu, and A. Sangiovanni-Vincentelli, A compilation-based SW estimation scheme for hardware/SW cosimulation, *Proceedings of the 7th IEEE International Workshop on Hardware/SW Codesign*, Rome, Italy, 3–5, 1999, pp. 85–89.

[17] E.J. Weyuker and A. Avritzer, A metric for predicting the performance of an application under a growing workload, *IBM Syst. J.*, 41, 45–54, 2002.

[18] V.D. Agrawal and S.T. Chakradhar, Performance estimation in a massively parallel system, *SC*, 1990, pp. 306–313.

[19] E.M. Eskenazi, A.V. Fioukov, and D.K. Hammer, Performance prediction for industrial software with the APPEAR method, *Proceedings of STW PROGRESS Workshop*, Utrecht, The Netherlands, October 2003.

[20] K. Suzuki, and A.L. Sangiovanni-Vincentelli, Efficient software performance estimation methods for hardware/software codesign, *Proceedings ACM/IEEE Design Automation Conference (DAC)*, Los Vegas, Nevada, United States, 1996, ISBN:0-89791-779-0, pp.605–610.

[21] Y. Liu, I. Gorton, A. Liu, N. Jiang, and S.P. Chen, Designing a test suite for empirically-based middleware performance prediction, *The Proceedings of TOOLS Pacific 2002*, Sydney, Australia, 2002.

[22] A. Muttreja, A. Raghunathan, S. Ravi, and N.K. Jha, Automated energy/performance macromodeling of embedded software, *Proceedings ACM/IEEE Design Automation Conference (DAC)*, San Diego, CA, USA, June 2004, ISBN:1-58113-828-8, pp. 99–102.

[23] K. Lahiri, A. Raghunathan, and S. Dey, Fast performance analysis of bus-based system-on-chip communication architectures, *Proceedings ACM/IEEE Design Automation Conference (DAC)*, San Jose, CA, United States, June 1999, ISBN:0-7803-5862-5, pp. 566–572.

[24] M. Lajolo, A. Raghunathan, S. Dey, L. Lavagno, and A. Sangiovanni-Vincentelli, A case study on modeling shared memory access effects during performance analysis of HW/SW systems, *Proceedings of the 6th IEEE International Workshop on Hardware/SW Codesign*, Seattle, WA, 15–18, 1998, pp. 117–121.

[25] J.A. Rowson and A.L. Sangiovanni-Vincentelli, Interface-based design, *Proceedings of the 34th Conference on Design Automation*, Anaheim Convention Center, ACM Press, Anaheim, CA, ISBN 0-89791-920-3, 9–13, 1997, pp. 178–183.

[26] K. Hines and G. Borriello, Optimizing communication in embedded system co-simulation, *Proceedings of the 5th International Workshop on Hardware/Software Co-Design*, Braunschweig, Germany, March 1997, ISBN:0-8186-7895-X, p. 121.

[27] S. G. Pestana, E. Rijpkema, A. Radulescu, K.G.W. Goossens, and O.P. Gangwal, Cost-performance trade-offs in networks on chip: a simulation-based approach, *DATE*, 2004, 764–769.

[28] F. Poletti, D. Bertozzi, L. Benini, and A. Bogliolo, Performance analysis of arbitration policies for SoC communication architectures, *Kluwer J. Des. Autom. Embed. Syst.*, 8, 189–210, 2003.

[29] L. Benini and G.D. Micheli, Networks on chips: a new SoC paradigm, *IEEE Comp.*, 35, 70–78, 2002.

[30] E. Bolotin, I. Cidon, R. Ginosar, and A. Kolodny, QNoC: QoS architecture and design process for network on chip, *J. Syst. Archit.*, 49, 2003. Special issue on Networks on Chip.

[31] K. Chen, S. Malik, and D.I. August, Retargetable static timing analysis for embedded SW, *International Symposium on System Synthesis ISSS*, 2001, 39–44.

[32] A. Hergenhan and W. Rosenstiel, Static timing analysis of embedded SW on advanced processor architectures, *Proceedings of Design, Automation and Test in Europe*, Paris, 2000, pp. 552–559.

[33] V. Tiwari, S. Malik, and A. Wolfe, Power analysis of embedded SW: a first step towards SW power minimization, *IEEE T. VLSI Syst.*, 2, 437–445, 1994.

[34] P.E. McKenney, Practical performance estimation on shared-memory multiprocessors, *Parall. Distr. Comput. Syst.*, 1999, pp. 125–134.

[35] M.K. Nethi and J.H. Aylor, Mixed level modelling and simulation of large scale HW/SW systems, *High Performance Scientific and Engineering computing*: Hardware/Software Support, Kluwer Academic Publishers, Norwell, MA, USA, 2004, ISBN:1-4020-7580-4. pp. 157–166.

[36] The MPI Standard, http://www-unix.mcs.anl.gov/mpi/standard.html.

[37] V. Gupta, P. Sharma, M. Balakrishnan, and S. Malik, Processor evaluation in an embedded systems design environment, *13th International Conference on VLSI Design (VLSI-2000)*, Calcutta, India, 2000, pp. 98–103.

[38] A. Maxiaguine, S. Künzli, S. Chakraborty, and L. Thiele, Rate analysis for streaming applications with on-chip buffer constraints, *ASP-DAC*, Yokohama, Japan, 2004.

[39] R. Marculescu, A. Nandi, L. Lavagno, and A. Sangiovanni-Vincentelli, System-level power/performance analysis of portable multimedia systems communicating over wireless channels, *Proceedings of IEEE/ACM International Conference on Computer Aided Design*, San Jose, CA, 2001.

[40] Y. Li and W. Wolf, A task-level hierarchical memory model for system synthesis of multiprocessors, *Proceedings of the ACM/IEEE Design Automation Conference (DAC)*, Anaheim, CA, United States, June 1997, pp. 153–156.

[41] I. Bacivarov, A. Bouchhima, S. Yoo, and A.A. Jerraya, ChronoSym — a new approach for fast and accurate SoC cosimulation, *Int. J. Embed. Syst.*, Interscience Publishers, ISSN (Print): 1741-1068, in press.

[42] M. Loghi, M. Poncino, and L. Benini, Cycle-accurate power analysis for multiprocessor systems-on-a-chip, *ACM Great Lakes Symposium on VLSI*, 2004, 410–406.

[43] M. Loghi, F. Angiolini, D. Bertozzi, L. Benini, and R. Zafalon, Analyzing on-chip communication in a MPSoC environment, *Proceedings of the Design, Automation and Test in Europe (DATE)*, Vol. 2, Paris, France, February 2004, pp. 752–757.

[44] M. Lajolo, A. Raghunathan, S. Dey, and L. Lavagno, Efficient power co-estimation techniques for systems-on-chip design, *Proceedings of Design Automation and Test in Europe*, Paris, 2000.

[45] R. Henia, A. Hamann, M. Jersak, R. Racu, K. Richter, and R. Ernst, System level performance analysis — the SymTA/S approach, *IEE Proceedings Comput. Dig. Tech.*, 152, 148–166, 2005.

[46] W. Cesario, A. Baghdadi, L. Gauthier, D. Lyonnard, G. Nicolescu, Y. Paviot, S. Yoo, A.A. Jerraya, and M. Diaz-Nava, Component-based design approach for multicore SoCs, *Proceedings of the ACM/IEEE Design Automation Conference (DAC)*, New Orleans, LA, USA, June 2002, ISBN~ISBN:0738-100X, 1-58113-461-4, pp. 789–794.

[47] A. Baghdadi, D. Lyonnard, N-E. Zergainoh, and A.A. Jerraya, An efficient architecture model for systematic design of application-specific multiprocessor SoC, *Proceedings of the Conference on Design, Automation and Test in Europe (DATE)*, Munich, Germany, March 2001, ISBN:0-7695-0993-2, pp. 55–63.

[48] *4th International Seminar on Application-Specific Multi-Processor SoC Proceedings*, 2004, Saint-Maximin la Sainte Baume, France, available at http://tima.imag.fr/mpsoc/2004/index.html.

7

System-Level Power Management

Naehyuck Chang
Seoul National University
Seoul, South Korea

Enrico Macii
Politecnico di Torino
Torino, Italy

Massimo Poncino
Politecnico di Torino
Torino, Italy

Vivek Tiwari
Intel Corp.
Santa Clara, California

7.1 Introduction

Power consumption can be drastically reduced if it is considered from the very early stages of the design flow. Power-aware system design has thus become one of the most important areas of investigation in the recent past, although only a few techniques that have been proposed for addressing the problem have already undergone the automation process. One of the approaches that has received a lot of attention, both from the conceptual and the implementation sides is certainly the so-called dynamic power management (DPM). The idea behind this technique, which is very broad and thus may come in very many different flavors, is that of selectively stopping or under-utilizing for some time the system resources that for some periods are not executing useful computation or that do not have to provide results at maximum speed. The landscape of DPM techniques is wide, and exhaustively surveying it is a hard task, which goes beyond the scope of this handbook (the interested reader may refer to the excellent literature on the subject, for instance [1–4]). However, some of the solutions that have been proposed so far have shown to be particularly effective, and are thus finding their way in commercial products. This chapter will review such techniques in some detail.

In general terms, a system is a collection of components whose combined operations provide a service to a user. In the specific case of electronic systems, the components are processors, digital signal processors (DSPs), memories, buses, and macro-cells. Power efficiency in an electronic system can be achieved by optimizing: (1) the architecture and the implementation of the components, (2) the communication between the components, and (3) the usage of the components.

In this chapter, we will restrict our attention to point (3), that is, to the issues of reducing the power consumed by an electronic system by means of properly managing its resources during the execution of the tasks the system is designed for.

The underlying assumption for all the solutions we will discuss is that the activity of the system components is event-driven; for example, the activity of display servers, communication interfaces, and user-interface functions is triggered by external events and it is often interleaved with long periods of quiescence. An intuitive way of reducing the average power dissipated by the whole system consists of shutting down (or reducing the performance of) the components during their periods of inactivity (or under-utilization). In other words, one can adopt a dynamic power management (DPM) policy that dictates how and when the various components should be powered (or slowed) down in order to minimize the total system power budget under some performance/throughput constraints.

A component of an event-driven system can be modeled through a finite state machine that, in the simplest case, has two states: *active* and *idle*. When the component is idle, it is desirable to shut it down by lowering its power supply or by turning off its clock; in this way, its power dissipation can be drastically reduced. Modern components are able to operate in a much wider range of modes, making the issue of modeling the power behavior of an electronic system a much more complex task. We will discuss a modeling approach based on the concept of power and energy state machine (ESM) at the beginning of the next section.

The simplest dynamic power management policies that have been devised are time-out-based: a component is put in its power-down mode only T time units after its finite state machine model has entered the idle state. This is because it is assumed that there will be a very high chance for the component to be idle for a much longer time if it has been in the idle state for at least T time units.

Time-out policies can be inefficient for three reasons: first, the assumption that if the component is idle for more than T time units, then it will remain idle for much longer, may not be true in many cases. Second, whenever the component enters the *idle* state, it stays powered for at least T time units, wasting a considerable amount of power in that period. Third, speed and power degradations due to shut downs performed at inappropriate times are not taken into account; in fact, it should be kept in mind that the transition from power-down to fully functional mode has an overhead: it takes some time to bring the system up to speed, and it may also take more power than the average, steady-state power.

To overcome the limitations of time-out-based policies, more complex strategies have been developed. Predictive and stochastic techniques are at the basis of effective policies; they rely upon complex mathematical models and they all try to exploit the past history of the active and idle intervals to predict the length of the next idle period, and thus decide whether it may be convenient to turn off or power-down a system component.

Beside component idleness, also component under-utilization can be successfully used to reduce power consumption by a fair amount. For example, if the time a processing unit requires to complete a given task is shorter than the actual performance constraint, execution time can be increased by making the processor run slower; this can be achieved by either reducing the clock frequency, or by lowering the supply voltage, or both, provided that the hardware and the software enable this kind of operation. In all cases, substantial power savings are guaranteed basically with no performance penalty. Policies for dynamic clock and voltage control are of increasing popularity in electronic products that feature dynamic power management capabilities.

The core of the chapter focuses on DPM policies, including those targeting dynamic voltage scaling (DVS). Some attention, in this context, will be devoted to battery-driven DPM and, more in general, to the problem of optimizing the usage of batteries, a key issue for portable systems. Finally, the software perspective on the DPM problem, which entails the availability of some capabilities for software-level power estimation, is briefly touched upon in the last section of the chapter.

7.2 Dynamic Power Management

7.2.1 Power Modeling for DPM: Power and Energy State Machines

The formalism of finite state machines can be extended to model power consumption of an electronic system. A power state machine (PSM) illustrates power consumption variations as the state of the system changes. PSMs associate the workload values with a set of resources relevant to its power behavior [5]. At any given point of time, each resource is in a specific power state, which consumes a discrete amount of power.

Figure 7.1 shows the example of the PSM for the Intel StrongARM SA-1100 processor. Each state denotes the power value or performance of the system in a specific state. The transition of the state machine often denotes transition overhead or transition probability. Once we come up with a PSM of all the components in a system, we can easily track the current power values of all resources as the state of the system changes. This approach enables us to have the average or peak power dissipation by looking at the power state transition over time for a given set of environmental conditions. The power consumption in each state may be a designer's estimation, a value obtained from simulation, or a manufacturer's specification.

The PSM formalism can be enhanced to account for the power cost (in addition to the time cost) due to speed changes, such as clock frequency changes, if a device consumes a nontrivial amount of dynamic power. In fact, dynamic power in a state depends on frequency, and it increases as the clock frequency of the device decreases.

Energy state machines (ESMs) separately denote the dynamic energy and leakage power, while PSMs do not distinguish them [6]. To eliminate time dependency, the dynamic portion is represented by energy, while the leakage portion is represented by power. Figure 7.2 shows the concept of ESM. Dynamic energy consumption, ξ_i, is associated with a transition, and leakage power consumption, ϕ_i, is associated with a state. Each state change clearly requires a different amount of dynamic energy. If we slow down the clock frequency, only the tenure time of the states becomes longer.

The ESM represents the actual behavior of a system and its energy consumption. However, measuring leakage power and dynamic energy consumption separately is far from trivial, without a very elaborate power estimation framework. Special measurement and analysis tools that separately handle dynamic energy and leakage power in system level are often required to annotate the ESM [7,8].

7.2.2 Requirements and Implementation of Dynamic Power Management

In addition to models for power-managed systems, as those discussed in the previous section, several other issues must be considered when implementing DPM in practice. These include the choice of an implementation style for the power manager (PM) that must guarantee accuracy in measuring inter-arrival times and service times for the system components, flexibility in monitoring multiple types of components, low perturbation, and marginal impact on component usage. Also key is the choice of the style for monitoring component activity; options here are off-line (traces of events are dumped for later analysis) and online (traces of events are analyzed while the system is running, and statistics related to component usage are constantly updated) monitoring. Finally, of utmost importance is the definition of an appropriate power management policy, which is essential for achieving good power/performance

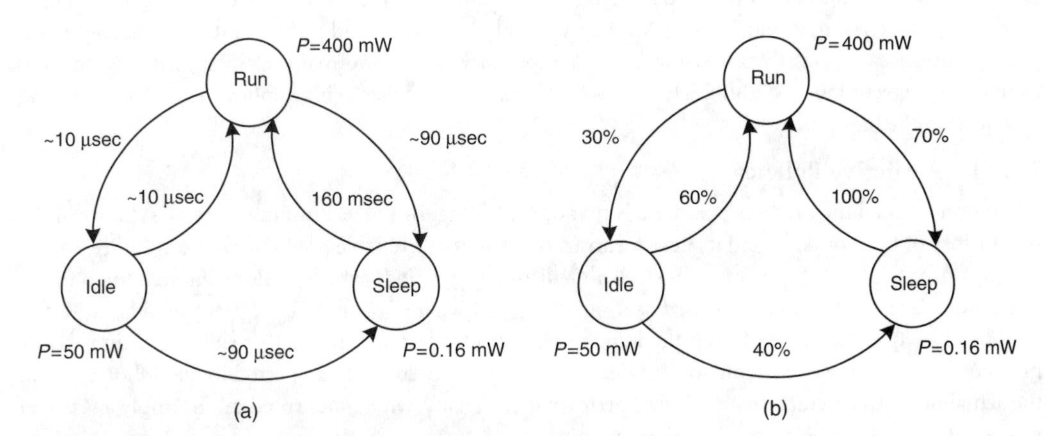

FIGURE 7.1 Power state machine for StrongARM SA-1100: (a) with transition overhead; (b) with transition probability.

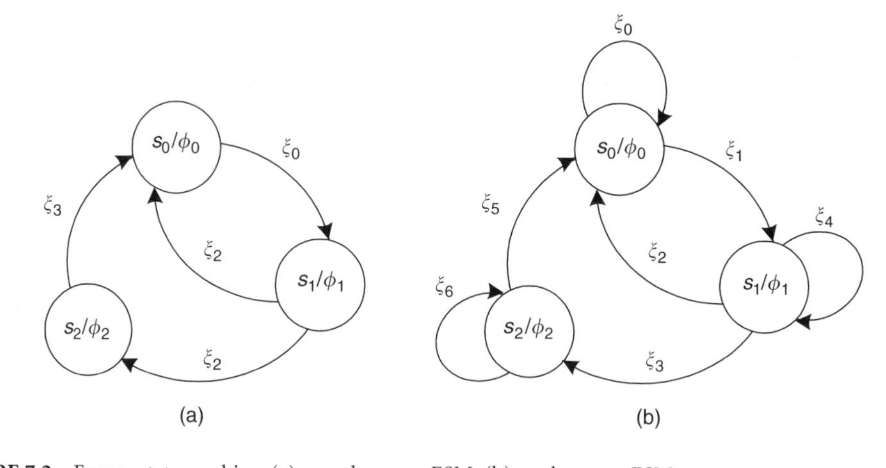

FIGURE 7.2 Energy state machine: (a) asynchronous ESM; (b)synchronous ESM.

trade-offs. The remainder of this section is devoted to the discussion of various options for DPM policies, including those regarding DVS.

7.2.3 Dynamic Power Management Policies

Dynamic power management is the process of dynamically reconfiguring the operating mode of the resources of a system. Its goal is to activate only a minimum number of components or a minimum load on such components, while providing the requested service and performance levels.

DPM encompasses a set of policies that achieve energy-efficient computation by selectively turning off or reducing the performance of system components when they are idle or partially unexploited. The fundamental premise for the applicability of DPM is that systems and their components experience nonuniform workloads during operation time. Such an assumption is generally valid for most battery-operated systems. The second assumption of DPM is that it is possible to predict, with a certain degree of confidence, the fluctuations of workloads. Obviously, workload observation and prediction should not consume significant energy in order to minimize the overhead.

A successful implementation of DPM relies on two key assumptions: (1) the availability of power manageable components that support multiple operational states, which express a trade-off between power and performance; and (2) the existence of a *power manager* that drives component transitions with the correct protocol and implements a power management policy (an algorithm that decides when to shut down idle components and to wake up sleeping components). This section aims at covering and relating different approaches to system-level DPM. We will highlight benefits and pitfalls of different power management policies. We classify power management approaches into two major classes: *predictive* and *stochastic* control techniques. Within each class, we introduce approaches being applied to system design and described in the literature.

7.2.3.1 Predictive Policies

The rationale in all the predictive policies is that of exploiting the correlation that may exist between the past history of the workload and its near future in order to make reliable predictions about future events. For the purpose of DPM, we are interested in *predicting idle periods*, which are long enough to go to the low-power state. This can be expressed as $T_{idle} > T_{breakeven}$, where T_{idle} and $T_{breakeven}$ are the idle time and the transition overhead to and from the idle mode, respectively. Good predictors should minimize mispredictions. We define overprediction (underprediction) as a predicted idle period longer (shorter) than the actual one. Overpredictions result in a performance penalty, while underpredictions imply a waste of power without incurring a performance penalty.

The quality of a predictor is measured by two figures: *safety* and *efficiency.* Safety is the complement of the risk of making overpredictions, and efficiency is the complement of the risk of making underpredictions. We call a predictor with maximum safety and efficiency an *ideal predictor.* Unfortunately, predictors of practical interest are neither safe nor efficient, thus resulting in suboptimum controls.

There are a lot of power management policies with predictive behavior, the simplest one being the *fixed time-out policy.* The fundamental assumption in the fixed time-out policy is that the probability of T_{idle} being longer than $T_{breakeven} + T_{timeout}$, given that $T_{idle} > T_{timeout}$, is close to 1:

$$P(T_{idle} > T_{breakeven} + T_{timeout} \mid T_{idle} \geq T_{timeout}) \approx 1 \tag{7.1}$$

Thus, the critical design parameter is the choice of the time-out value, $T_{timeout}$. The time-out policy has two main advantages: first, it is general; second, its safety can be improved by simply increasing the time-out value. Unfortunately, large timeout values compromise the efficiency, because of the trade-off between safety and efficiency. In addition, as mentioned in the introduction, timeout policies have two more disadvantages. First, they tend to waste a sizeable amount of power while waiting for the timeout to expire. Second, they suffer from a performance penalty upon wake-up. These two disadvantages can be addressed by using a predictive shut-down or wake-up policy, respectively.

The *predictive shut-down policy* forces the PM to make a decision about whether or not to transit the system to the low-power state as soon as a new idle period starts. The decision is based on the observation of past idle/busy periods, and it eliminates the waste of power due to the waiting for the time-out to expire.

Two kinds of predictive shut-down policies were proposed in [9]. In the first policy, a nonlinear regression equation is obtained from the previous "on" and "off" time, and the next "turn-on" time is estimated. If the predicted idle period (T_{pred}) is longer than $T_{breakeven}$, the system is immediately shut down as soon as it becomes idle. This policy, however, has two disadvantages. First, there is no automatic way to decide the type of regression equation. Second, it requires one to perform on-line data collection and analysis for constructing and fitting the regression model.

In the second predictive shut-down policy, the idle period is predicted based on a threshold. The duration of the busy period preceding the current idle period is observed. If the previous busy period is longer than the threshold, the current idle period is assumed to be longer than $T_{breakeven}$, and the system is shut down. The rationale of this policy is that short busy periods are often followed by long idle periods. Similar to the time-out policy, the critical design decision parameter is the choice of the threshold value $T_{threshold}$.

The predictive wake-up policy is proposed in [10], and it addresses the second limitation of the fixed time-out policy, namely the performance penalty that is always paid upon wake-up. To reduce this cost, the power manager performs predictive wake-up when the predicted idle time expires, even if no new request has arrived. This policy may increase power consumption if the idle time has been under-predicted, but it decreases the response time of the first incoming request after an idle period. As usual, in this context, power is traded for performance.

7.2.3.1.1 *Adaptive Predictive Policies.*

The aforementioned static predictive policies are all ineffective when the workload is either unknown *a priori* or it is nonstationary, since the optimality of DPM strategies depends on the workload statistics. Several adaptive techniques have been proposed to deal with nonstationary workloads.

One option is to maintain a set of time-out values, each of which is associated with an index indicating how successful it would have been [11]. The policy chooses, at each idle time, the time-out value that would have performed best among the set of available ones. In alternative, a weighted list of time-out values is kept, where the weight is determined by relative performance of past requests with respect to the optimal strategy [12], and the actual time-out is calculated as a weighted average of all the time-out values in the list [13]. The time-out value can be increased when it results in too many shut-downs, and it can otherwise be decreased when more shut-downs are desirable.

Another adaptive shut-down policy has been proposed in [10]. The idle time prediction is calculated as a weighted sum of the last idle period and the last prediction (exponential average):

$$T^{n}_{\text{pred}} = \alpha T^{\text{n-1}}_{\text{idle}} + (1 - \alpha) T^{\text{n-1}}_{\text{pred}} \tag{7.2}$$

The last underprediction is mitigated by reevaluating T_{pred} periodically, if the system is idle and it has not been shut down. The last overprediction is reduced by imposing a saturation constraint, C_{max}, on predictions:

$$T^{\text{n}}_{\text{pred}} \leq C_{\text{max}}\, T^{\text{n}-1}_{\text{pred}} \tag{7.3}$$

7.2.3.2 Stochastic Policies

Although all the predictive techniques address workload uncertainty, they assume deterministic response time and transition time of a system. However, the system model for policy optimization is very abstract, and abstraction introduces uncertainty. Hence, it is safer and more general to assume a stochastic model for the system as well as the workload. Moreover, real-life systems support multiple power states, which cannot be handled by simple predictive techniques. In fact, the latter are based on heuristic algorithms, and their optimality can be gauged only through comparative simulations. Finally, predictive techniques are only geared toward the power minimization, and cannot finely control performance penalty.

Stochastic control techniques formulate policy optimization as an optimization problem under the uncertainty of the system as well as of the workload. They assume that both the system and the workload can be modeled as Markov chains, and offer significant improvement over previous power management techniques in terms of theoretical foundations and of robustness of the system model. Using stochastic techniques allows one to (1) model the uncertainty in the system power consumption and the state-transition time; (2) model complex systems with multiple power states; and (3) compute globally optimum power management policies, which minimize the energy consumption under performance constraints or maximize the performance under power constraints.

A typical Markov model of a system consists of the following entities [14]:

- A service requester (SR) that models the arrival of service requests.
- A service provider (SP) that models the operation states of the system.
- A PM that observes the state of the system and the workload, makes a decision, and issues a command to control the next state of the system.
- Some cost metrics that associate power and performance values to each command.

7.2.3.2.1 Static Stochastic Policies.

Static stochastic control techniques [14] perform policy optimization based on the fixed Markov chains of the SR and of the SP. Finding a globally power-optimal policy that meets given performance constraints can be formulated as a linear program (LP), which can be solved in polynomial time in the number of variables. Thus, policy optimization for Markov processes is exact and computationally efficient. However, several important points must be understood: (1) The performance and power obtained by a policy are expected values, and there is no guarantee of the optimality for a specific workload instance. (2) We cannot assume that we always know the SR model beforehand. (3) The Markov model for the SR or SP is just an approximation of a much more complex stochastic process, and thus the power-optimal policy is also just an approximate solution.

The Markov model in [14] assumes a finite set of states, a finite set of commands, and discrete time. Hence, this approach has some shortcomings: (1) The discrete-time Markov model limits its applicability since the power-managed system should be modeled in the discrete-time domain. (2) The PM needs to send control signals to the components in every time-slice, which results in heavy signal traffic and heavy load on the system resources (therefore more power dissipation). Continuous-time Markov models [15] overcome these shortcomings by introducing the following characteristics: (1) A system model based on continuous-time Markov decision process is closer to the scenarios encountered in practice.

(2) The resulting power management policy is asynchronous, which is more suitable for implementation as a part of the operating system (OS).

7.2.3.2.2 Adaptive Stochastic Policies.

An adaptive extension of the stochastic control techniques is proposed to overcome a limitation of the static approach. As it is not possible to have the complete knowledge of the system (SP) and its workload (SR) *a priori*, even though it is possible to construct a model for the SP once and for all, the system workload is generally much harder to characterize in advance. Furthermore, workloads are often nonstationary. Adaptation consists of three main phases: policy precharacterization, parameter learning, and policy interpolation [16]. Policy precharacterization constructs an n-dimensional table addressed by n parameters for the Markov model of the workload. The table contains the optimal policy for the system under different workloads. Parameter learning is performed online during system operation by short-term averaging techniques. The parameter values are then used for addressing the table and for obtaining the power management policy. If the estimated parameter values are not in accordance with the exact values, which are used for addressing the table, interpolation may obtain an appropriate policy as a combination of the policies in the table. Experimental results reported in [16] indicate that the adaptive technique performs nearly as well as the ideal policy computed off-line, assuming perfect knowledge of the workload parameters over time.

7.2.4 Dynamic Voltage Scaling

Supply voltage scaling is one of the most effective techniques in power minimization of CMOS circuits because the dynamic energy consumption of CMOS devices is quadratically related to the supply voltage. Unfortunately, the supply voltage has a strong relation to the circuit delay: the lower the supply voltage, the larger the circuit delay and the smaller the maximum operating frequency, which may degrade the performance of the target system.

Dynamic voltage scaling is the power management technique that controls the supply voltage according to the current workload at run-time to minimize the energy consumption without having an adverse effect on system performance. Dynamic voltage scaling can be viewed as a variant of DPM in which DPM is applied not just to *idle* components but also to those resources that are *noncritical* in terms of performance, running the resource at different power/speed points. In other words, DVS introduces the notion of *multiple active states*, besides multiple idle states exploited by traditional DPM. Moreover, since in DVS power/speed trade-off points are defined by different supply voltage levels, DVS is traditionally applied to CPUs, rather than to other components, and it is thus exploited at the task granularity.

The DVS technique typically utilizes the slack time of tasks to avoid performance degradation. For example, when the current task has less remaining time than the expected execution time at the maximum frequency, the voltage scheduler lowers the supply voltage, and extends the execution time of this task up to the arrival time of the next task. To apply DVS to real systems, hardware support for voltage scaling is required [17], and the software support that monitors the task execution and gives the voltage control command to the DVS hardware is needed as well.

The issues related to the implementation of DVS have been investigated in many studies. A majority of them developed an energy-efficient scheduling method for a system with real-time requirements. Each work suggests different run-time slack estimation and distribution schemes [18] that are trying to achieve the theoretical lower bound of energy consumption that is calculated statically for a given workload [19].

7.2.4.1 Task Scheduling Schemes for Dynamic Voltage Scaling

The various voltage scheduling methods in DVS-enabled systems differ by when to adjust frequency and voltage, how to estimate slacks, and how to distribute these to waiting tasks. Several DVS scheduling schemes are summarized and evaluated in [18]. In this section, DVS scheduling algorithms for hard-real time systems are classified by the *granularity* of the voltage adjustments. Also, the slack estimation methods for some DVS scheduling algorithms are introduced.

7.2.4.1.1 *Voltage Scheduling Granularity.*

DVS scheduling schemes are classified by voltage scheduling granularity and fall into two categories: inter-task DVS algorithms and intra-task DVS algorithms. In the intra-task DVS algorithms, a task is partitioned into multiple pieces such as time slots [20] or basic blocks [21], and a frequency and consequent voltage assignment is applied during the task execution. The actual execution time variation is estimated at the boundary of each time slot or each basic block and used for the input of adaptive operation frequency and voltage assignment.

In the inter-task DVS algorithms, voltage assignment is executed at the task's boundaries. After a task is completed, a new frequency and consequent voltage setting are applied by static or dynamic slack time estimation. The slack time estimation method has to be aggressive for the system energy reduction. At the same time, it must be conservative so that every task is successfully scheduled within its deadline. These two rules are conflicting with each other, making it difficult to develop an effective slack estimation algorithm.

7.2.4.1.2 *Slack Time Estimation.*

DVS techniques for hard-real-time systems enhance the traditional earliest deadline first (EDF) or rate monotonic (RM) scheduling to exploit slack time, which is used to adjust voltage and frequency of voltage-scalable components. Therefore, the primary objective is to estimate the slack time accurately for more energy reduction. Various kinds of static and dynamic slack estimation methods have been proposed to exploit most of the slack time without violating the hard-real-time constraints.

Many approaches for inter-task DVS algorithms deal with static slack estimation methods [22–26]. These methods typically aim at finding the lowest possible operating frequency at which all the tasks meet their deadlines. These methods rely on the worst-case execution time (WCET) to guarantee hard-real-time demands. Therefore, the operating frequency can be lowered to the extent that each task's WCET does not exceed the deadline. The decision of each task's frequency can be done statically because it is a function of WCET and deadline, which are not changed during run-time.

In general, the actual execution time is quite shorter than WCET. Therefore, WCET-based static slack time estimation cannot fully exploit actual slacks. To overcome this limitation, various dynamic slack estimation methods have been proposed.

The cycle-conserving RT-DVS technique utilizes the extra slack time to run other remaining tasks at a lower clock frequency [24]. In this approach, operating frequency is scaled by the CPU utilization factor. The utilization factor is updated when any task is released or completed. When any task is released, the utilization factor is calculated according to the task's WCET. After a task is completed, the utilization factor is updated by using the actual execution time. The operation frequency may be lowered until the next arrival time of that task.

The next release time of a task can be used to calculate the slack budget [23]. This approach maintains two queues: the run queue and the delay queue. The former holds tasks that are waiting by their priority order, while the latter holds tasks that are waiting for next periods, ordered by their release schedule. When the active queue is empty and the required execution time of an active task is less than its allowable time frame, the operation frequency is lowered using that slack time. According to Shin et al. [23], the allowable time frame is defined by

$$\min(\text{active\_task.deadline}, \text{delay\_queue.head.release\_time}) - \text{current\_time} \tag{7.4}$$

A path-based slack estimation method for intra-task DVS algorithms is presented in [21]. The control flow graph (CFG) is used for slack time estimation during the execution of the given hard-real-time program. Each node of the CFG is a basic block of the target program, and each edge indicates the control dependency between basic blocks. When the thread of execution control branches to the next basic block, the expected execution time is updated. If the expected execution time is smaller than the tasks's WCET, the operating frequency can be lowered.

7.2.4.2 Practical Considerations in Dynamic Voltage Scaling

Many DVS studies, especially those focusing on task scheduling, have assumed a target system: (1) consisting of all voltage-scalable components whose supply voltage can be set to any value within a given range of supply voltage;(2) in which only dynamic energy is considered; and (3) in which the speed settings of the tasks do not affect other components of the system. Although these assumptions simplify the calculation of energy consumption and development of a scheduling scheme, the derived scheduling may not perform well because the assumption does not reflect a realistic setting. In fact, recent advancement in CMOS technology makes leakage power consumption significant. Recent studies have addressed the impact of these nonidealities on DVS.

7.2.4.2.1 *Discretely Variable Voltage Levels.*

Unlike assumption (1) above, most commercial microprocessors supporting supply voltage scaling (e.g., Intel XScale, Transmeta Crusoe) can select only a small number of predefined voltages as supply voltage. To get a more practical DVS schedule, some scheduling techniques are proposed for discretely variable supply voltage levels instead of continuously variable ones.

An optimal voltage allocation technique for a single task with discontinuously variable voltages is proposed in [27] using integer linear programming. In case only a small number of discrete voltages can be used, they show that the schedule with at most two voltages for each task minimizes the energy consumption under a timing constraint. Another work deals with the static voltage allocation problem for circuits with multiple supply voltages [28]. This scheduling problem for discretely variable voltage levels is extended to get an energy-optimal schedule of multiple tasks for dynamically voltage-variable microprocessors controlled by software [29].

7.2.4.2.2 *Leakage-Aware Dynamic Voltage Scaling.*

As the supply voltage of CMOS devices becomes lower, the threshold voltage should also be reduced, which results in dramatic increase of the subthreshold leakage current. Therefore, the static power including the leakage current as well as the dynamic power are major contributors to the total power dissipation.

If static energy consumption is considered, the total energy consumption is no longer a monotonically increasing function of the supply voltage. Since the performance degradation due to the reduction of the supply voltage may increase the execution (or wake-up) time, this may result an increase in the static energy consumption. Consequently, if the supply voltage is reduced below a certain limit, the energy consumption becomes larger again. Inspired by this convex energy curve, which is no longer monotonically increasing with the supply voltage, a leakage-aware DVS scheduling scheme is proposed in [30]. It finds the voltage that minimizes the total energy consumption including leakage current, and avoids supply voltage scaling below the limit, even though there is still some slack time available.

7.2.4.2.3 *Memory-Aware Dynamic Voltage Scaling.*

As the complexity of modern systems increases, other components beside microprocessors, e.g., memory devices, contribute more to system power consumption. Thus, their power consumption must be considered when applying a power management technique. Unfortunately, many offchip components do not allow supply voltage scaling. Since they are controlled by the microprocessor, their active periods may become longer when we slow down the microprocessor using DVS. The delay increase due to lower supply voltage of a microprocessor may increase the power consumption of devices that do not support DVS. In these cases, the energy gain achieved from DVS on the processor must be traded for the energy increase of other devices.

There are some related studies about DVS in systems including devices without scalable supply voltages, especially focusing on memories. In both [31] and [32], it is shown that aggressive reduction of supply voltage of a microprocessor results in an increase of total energy consumption, because the static energy consumption of memory devices becomes larger as the execution time gets longer, and thus chances of actuating some power-down decreases. In [32], an analytic method to obtain an energy-optimal frequency assignment of memories not supporting DVS and CPUs supporting DVS is proposed as a viable improvement in this context.

7.2.4.3 Nonuniform Switching Load Capacitances

Even if multiple tasks are scheduled, most studies characterize these tasks only by the timing constraints. This assumes that all tasks with the same voltage assignment and the same period will consume the same amount of energy irrespective of their operation. This means that a uniform switching load capacitance in the energy consumption equation is assumed for all the tasks.

However, in practice, different tasks may utilize different data-path units that produce different energy consumption even for the tasks with the same period. A scheduling method for multiple tasks with nonuniform switching load capacitances has been proposed in [29]. This approach modifies the algorithm of Yao et al. [19] so as to obtain an energy-optimal schedule for nonuniform switching load capacitances, which helps in better matching the real power behavior of the overall system.

7.3 Battery-Aware Dynamic Power Management

All the dynamic power management techniques described in some detail in the previous section do implicitly assume an ideal power supply. While this simplifying assumption may be considered very reasonable for electronic systems connected to a fixed power supply, it is simply not correct in the case of battery-operated systems.

7.3.1 Battery Properties

Batteries are nonideal charge storage units, as pointed out in any battery handbook [33]. From a designer's standpoint, there are two main nonidealities of real-life battery cells that need to be considered:

- *The capacity of a battery depends on the discharge current.* At higher currents, a battery is less efficient in converting its chemically stored energy into available electrical energy. This fact is shown on the top panel of Figure 7.3, where the capacity of the battery is plotted as a function of the average-current load. We observe that, for increasing load currents, the battery capacity progressively deviates from the nominal value (broken line). Moreover, battery capacity is also affected by the discharge rate: At higher rates, the cell is less efficient at converting its stored energy into available electrical energy. This fact implies that battery lifetime will be negatively correlated to the *variance* of the current load; for a given average current value, a constant load will result in the longest battery lifetime of all load profiles.
- *Batteries have some (limited) recovery capacity when they are discharged at high current loads.* A battery can recover some of its deliverable charge if the periods of discharge are interleaved with rest periods, i.e., periods in which no current is drawn. The bottom panel of Figure 7.3 shows how an intermittent current load (broken line) results in a longer battery lifetime than a constant current load (solid line), for identical discharge rate. The x-axis represents the actual elapsed time of discharge, that is, it does not include the time during which the battery has been disconnected from the load.

Accounting for the aforementioned nonidealities is essential, since it can be shown that power management techniques (both with and without DVS) that neglect these issues may actually result in an increase in energy [34,35].

7.3.2 Battery-Aware Dynamic Power Management

The most intuitive solution consists of incorporating *battery-driven* policies into the DPM framework, either implicitly (i.e., using a battery-driven metric for a conventional policy) [36] or explicitly (i.e., a truly battery-driven policy) [37,38]. A simple example of the latter type could be a policy whose decision rules used to control the system operation state are based on the observation of a battery's output voltage, which is (nonlinearly) related to the state of charge.

More generally, it is possible to directly address the above-mentioned nonidealities to shape the load current profile so as to increase as much as possible the effective capacity of the battery.

The issue of load-dependent capacity can be faced along two dimensions. The dependency on the average current can be tackled by shaping the current profile in such a way that high-current-demanding operations are executed first (i.e., with fully charged battery), and low-current-demanding ones are executed later (i.e., with a reduced-capacity battery) [39].

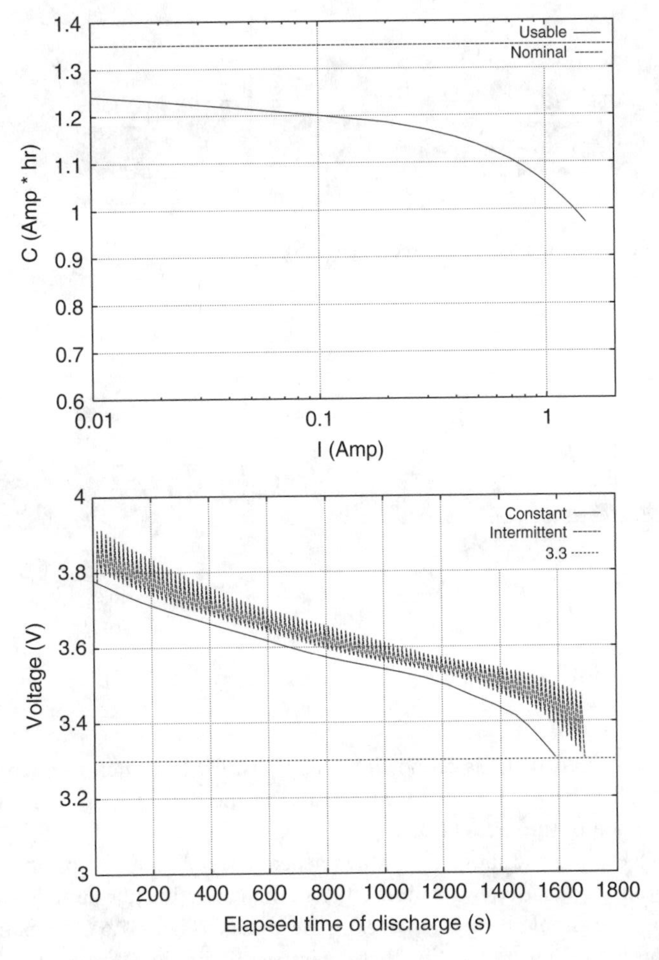

FIGURE 7.3 Capacity variation as a function of load and charge recovery in intermittent discharge.

This principle fits well at the task granularity, where the shape of the profile corresponds to task scheduling. Intuitively, the solution that maximizes battery efficiency and thus its lifetime is the one in which tasks are scheduled in nondecreasing order of their average current demand (Figure 7.4), compatibly with possible deadline or response time constraints [40].

The issue of charge recovery can be taken into account by properly arranging the idle times in the current profile. In particular, idle periods can be *inserted* between the execution of tasks. Notice that this is different from typical current profiles, where active and idle intervals alternate in relatively long bursts (Figure 7.5a). Inserting idle slots between task execution will allow the battery to recover some of the charge so that lifetime can be increased (Figure 7.5b). In the example, it may happen that after execution of T_2 the battery is almost exhausted and execution of T_3 will not complete; conversely, the insertion of an idle period will allow the battery to recover part of the charge so that execution of T_3 becomes possible. Idle time insertion can be combined with the ordering of tasks to exploit both properties and achieve longer lifetimes [41].

7.3.3 Battery-Aware Dynamic Voltage Scaling

The possibility of scaling supply voltages at the task level adds further degrees of freedom in choosing the best shaping of the current loads. Voltage scaling can be in fact viewed as another opportunity for reducing the current demand of a task, at the price of increased execution time. Under a simple, first-order

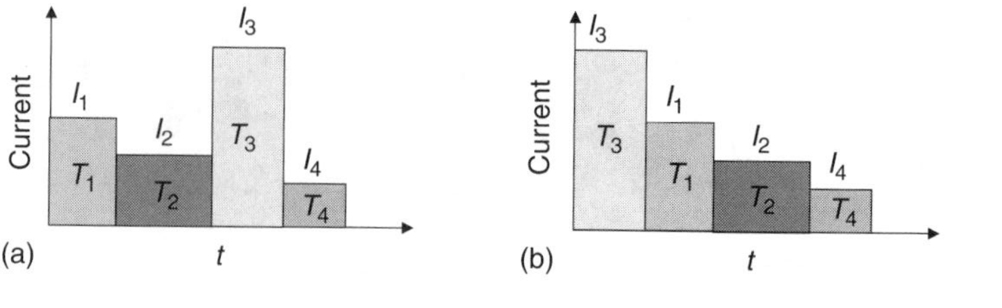

FIGURE 7.4 A set of tasks (a) and its optimal sequencing (b).

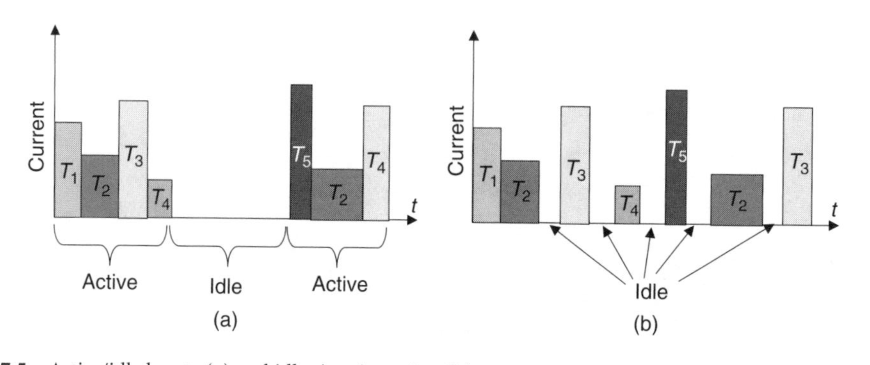

FIGURE 7.5 Active/idle bursts (a) and idle time insertion (b).

approximation, the drawn current I is proportional to V^3, while delay increases proportionally with V. Therefore, the trade-off is between a decrease (increase) in the the discharge current, and an increase (decrease) in the duration of the stress [41].

Battery-aware DVS exploits the above-mentioned dependency of battery capacity on the load, since it can be proved that scaling voltage is always more efficient than inserting idle periods [40]. Therefore, anytime a slack is available, it should be filled by scaling the voltage (Figure 7.6), compatibly with performance constraints. This is equivalent to stating that the impact on lifetime of the rate-dependent behavior of batteries dominates that due to the charge recovery effect.

Solutions proposed in the literature typically start from a nondecreasing current profile and reclaim the available slack from the tail of the schedule by slowing down tasks according to the specified constraints [40,42,43].

7.3.4 Battery Scheduling

In the case of supply systems consisting of multiple batteries, the load-dependent capacity of the batteries has deeper implications, which open additional opportunities for optimization. In fact, since at a given point in time the load is connected to only one battery, the other ones are idle. Dynamic Power Management with multiple batteries amounts thus to the problem of assigning (i.e., scheduling) battery usage over time; hence this problem is often called *battery scheduling.*

The default battery scheduling policy in use in most devices is a nonpreemptive one that sequentially discharges one battery after another, in some order. Because of the above-mentioned nonidealities, this is clearly a suboptimal policy [44].

Similar to task scheduling, battery scheduling can be either independent or dependent from the workload. In the former case, batteries are attached to the load in a round-robin fashion for a fixed amount of time. Unlike task scheduling, the choice of this quantity is dictated by the physical property of batteries. It can be shown, in fact, that the smaller this interval, the higher is the equivalent capacity of the battery sets [45]. This is because by rapidly switching between full load and no load, each battery perceives an

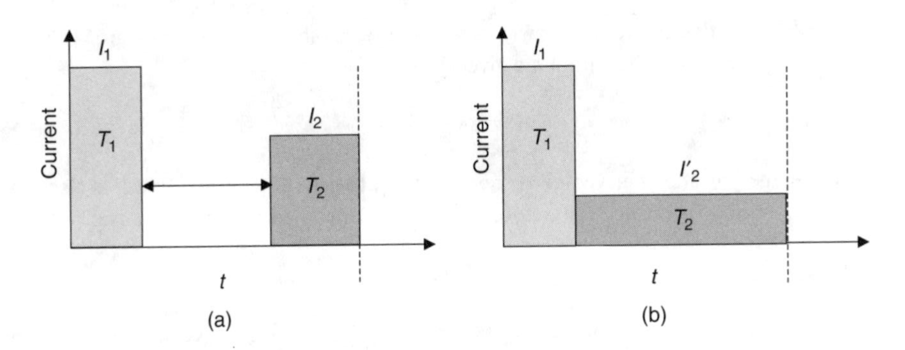

FIGURE 7.6 Idle period insertion (a) vs. voltage scaling (b).

effective averaged discharge current proportional to the fraction of time it is connected to the load. In other words, if a battery is connected to the load current I for a fraction $\alpha < 1$ of the switching period, it will perceive a load current αI. This is formally equivalent to connecting the two batteries in parallel, without incurring into the problem of mutually discharging the batteries.

In the latter case, we assign a battery to the load depending on its characteristics. More precisely, one should select which battery to connect to the load based on run-time measurement of the current drawn, in an effort to match a load current to the battery that better responds to it [37,44,45]. A more sophisticated, workload-dependent scheme consists of adapting the round-robin approach to heterogeneous multibattery supplies (i.e., having different nominal capacities and discharge curves). In these cases, the current load should be split nonuniformly over all the cells in the power supply. Therefore, the round-robin policy can be modified so that the time slice has different durations. For instance, in a two-battery system, this is equivalent to connecting the batteries to the load following a square wave with unbalanced duty cycle [46].

7.4 Software-Level Dynamic Power Management

7.4.1 Software Power Analysis

Software constitutes a major component of systems where power is a constraint. Its presence is very visible in a mobile computer, in the form of the system software and application programs running on the main processor. But software also plays an even greater role in general digital applications. An ever-growing fraction of these applications are now being implemented as embedded systems. In these systems the functionality is partitioned between a hardware and a software component.

The software component usually consists of application-specific software running on a dedicated processor. Relating the power consumption in the processor to the instructions that execute on it provides a direct way of analyzing the impact of the processor on the system power consumption. Software impacts system power at various levels. At the highest level, this is determined by the way functionality is partitioned between hardware and software. The choice of the algorithm and other higher level decisions about the design of the software component can affect system power consumption significantly. The design of the system software, the actual application source code, and the process of translation into machine instructions — all of these determine the power cost of the software component.

In order to systematically analyze and quantify this cost, however, it is important to start at the most fundamental level, that is, at the level of the individual instructions executing on the processor. Just as logic gates are the fundamental units of computation in digital hardware circuits, instructions can be thought of as the fundamental unit of software. Accurate modeling and analysis at this level is therefore essential. Instruction level models can then be used to quantify the power costs of the higher constructs of software (application programs, system software, algorithm, etc.).

It would be helpful to define the terms "power" and "energy," as they relate to software. The average power consumed by a processor while running a given program is given by

$$P = IV_{dd}$$

where P is the average power, I the average current, and V_{dd} the supply voltage. Power is also defined as the rate at which energy is consumed. Therefore, the energy consumed by a program is given by

$$E = P\,T$$

where T is the execution time of the program. This, in turn, is given by $T = N\tau$, where N is the number of clock cycles taken by the program and τ the clock period. Energy is thus given by

$$E = I\,V_{dd}\,N\,\tau$$

Note that if the processor supports dynamic voltage and frequency switching, then V_{dd} and τ can vary over the execution of the program. It is then best to consider the periods of code execution with different (V_{dd},τ) combinations as separate components of the power/energy cost. As it can be seen from the above discussion, the ability to obtain an estimate of the current drawn by the processor during the execution of the program is essential for evaluating the power/energy cost of software. These estimates can either be obtained through simulations or through direct measurements.

7.4.1.1 Software Power Estimation through Simulation

The most commonly used method for power analysis of circuits is through specialized power analysis tools that operate on abstract models of the given circuits. These tools can be used for software power evaluation too. A model of the given processor and a suitable power analysis tool are required. The idea is to simulate the execution of the given program on the model. During simulation the power analysis tool estimates the power (current) drawn by the circuit using predefined power estimation formulas, macro-models, heuristics, or algorithms.

However, this method has some drawbacks. It requires the availability of models that capture the internal implementation details of the processor. This is proprietary information, which most software designers do not have access to. Even if the models are available, there is an accuracy vs. efficiency trade-off. The most accurate power analysis tools work at the lower levels of the design — switch- or circuit-level. These tools are slow and impractical for analyzing the total power consumption of a processor as it executes entire programs. More efficient tools work at the higher levels — register transfer or architectural. However, these are limited in the accuracy of their estimates.

7.4.1.2 Measurement-Based Instruction-Level Power Modeling

The above problems can be overcome if the current being drawn by the processor during the execution of a program is physically measured. A practical approach to current measurement as applied to the problem of instruction-level power analysis has been proposed [47]. Using this approach, empirical instruction-level power models were developed for three commercial microprocessors. Other researchers have subsequently applied these concepts to other microprocessors. The basic idea is to assign power costs to individual instructions (or instruction pairs to account for inter-instruction switching effects) and to various inter-instruction effects like pipeline stalls and cache misses. These power costs are obtained through experiments that involve the creation of specific programs and measurement of the current drawn during their execution. These costs are the basic parameters that define the instruction-level power models. These models form the basis of estimating the energy cost of entire programs. For example, for the processors studied in [47], for a given program, the overall energy cost is given by

$$E = \sum_{i} (B_i \cdot N_i) + \sum_{i,j}(O_{i,j} \cdot N_{i,j}) + \sum_{k} E_k$$

The base power cost, B_i, of each instruction, i, weighted by the number of times it will be executed, N_i, is added up to give the base cost of the program. To this the circuit state switching overhead, $O_{i,j}$, for each pair of consecutive instructions, i,j , weighted by the number of times the pair is executed, $N_{i,j}$, is added. The energy contribution, E_k, of the other inter-instruction effects, k (stalls and cache misses) that would occur during the execution of the program, is finally added.

The base costs and overhead values are empirically derived through measurements. The other parameters in the above formula vary from program to program. The execution counts N_i and $N_{i,j}$ depend on the execution path of the program. This is dynamic, run-time information that has to be obtained using software performance analysis techniques. In certain cases, it can be determined statically, but in general it is best obtained from a program profiler. For estimating E_k, the number of times the pipeline stalls and cache misses occur have to be determined. This is again dynamic information that can be statically predicted only in certain cases. In general, it can be obtained from a program profiler and a cache simulator, which are able to track the dynamic nature of program execution.

The processors whose power models have been published using the ideas above, have so far been in-order machines with relatively simple pipelines. Out-of-order superscalar machines present a number of challenges to an instruction oriented modeling approach and provide good opportunities for future research.

7.4.1.3 Idle Time Evaluation

The above discussion is for the case when the processor is active and is constantly executing instructions. However, a processor may not always be performing useful work during program execution. For example, during the execution of a word processing program, the processor may simply be waiting for keyboard input from the user and may go into a low-power state during such idle periods. To account for these low-power periods, the average power cost of a program is thus given by

$$P = P_{active}\, T_{active} + P_{idle}\, T_{idle}$$

where P_{active} represents the average power consumption when the microprocessor is active, T_{active} the fraction of the time the microprocessor is active, and P_{idle} and T_{idle} the corresponding parameters for when the microprocessor is idle and has been put in a low-power state. T_{active} and T_{idle} need to be determined using dynamic performance analysis techniques. In modern microprocessors, a hierarchy of low-power states is typically available, and the average power and time spent for each state would need to be determined.

7.4.2 Software-Controlled Power Management

For systems in which part of the functionality is implemented in software, it is natural to expect that there is potential for power reduction through modification of the software. Software power analysis (whether achieved through physical current measurements or through simulation of models of the processors) as described in Section 7.4.1, helps in identifying the reasons for the variation in power from one program to another. These differences can then be exploited in order to search for best low-power alternatives for each program.

The information provided by the instruction-level analysis can guide higher-level design decisions like hardware–software partitioning and choice of algorithm. It can also be directly used by automated tools like compilers, code generators, and code schedulers for generating code targeted toward low power. Several ideas in this regard have been published, starting with the work summarized in [47]. Some of these ideas are based on specific architectural features of the subject processors and memory systems. The most important conclusion though is that in modern general-purpose CPUs, software energy and performance track each other, i.e., for a given task, a faster program implementation will also have lower energy. Specifically, it is observed that the difference in average current for instruction sequences that perform the same function is not large enough to compensate for any difference in the number of execution cycles.

Thus, given a function, the least energy implementation for it is the one with the faster running time. The reason for this is that CPU power consumption is dominated by a large common cost factor (power consumption due to clocks, caches, instruction fetch hardware, etc.) that for the most part is independent of instruction functionality and does not vary much from one cycle to the other. This implies that the large body of work devoted to software performance optimization provides direct benefits for energy as well. Power management techniques, such as increased use of clock gating and multiple on-chip voltages, indicate that future CPUs may show greater variation in power consumption from cycle to cycle. However, CPU design and power consumption trends do suggest that the relationship between software energy and power that was observed before will continue to hold. In any case, it is important to realize that software directly impacts energy/power consumption, and thus it should be designed to be efficient with respect to these metrics. A classic example of inefficient software is "busy wait loops." Consider an application such as a spreadsheet that requires frequent user input. During the times when the spreadsheet is recalculating values, high CPU activity is desired in order to complete the recalculation in a short time. In contrast, when the application is waiting for the user to type in values, the CPU should be inactive and in a low-power state. However, a busy wait loop will prevent this from happening, and will keep the CPU in a high-power state. The power wastage is significant. Converting the busy wait loops into an instruction or system call that puts the CPU into a low-power state from which it can be woken up on an I/O interrupt will eliminate this wasted power.

7.4.2.1 OS-Directed Dynamic Power Management

An important trend in modern CPUs is the ability for software to control the operating voltage and frequency of the processor. These different voltage/frequency operating points, which represent varying levels of power consumption, can be switched dynamically under software control. This opens up additional opportunities for energy-efficient software design, since the CPU can be made to run at the lowest-power state that still provides enough performance to meet the task at hand. In practice, the DPM and DVS techniques described in Section 7.2 from a hardware-centric perspective (because they managed multiple power states of *hardware components)* can be also viewed from the software perspective.

In this section, we describe implementation issues of power management techniques. In general, the OS is the best software layer where a DPM policy can be implemented. OS-directed power management (OSPM) has several advantages: (1) The power/performance dynamic control is performed by the software layer that manages computational, storage, and I/O tasks of the system. (2) Power management algorithms are unified in the OS, yielding much better integration between the OS and the hardware, (3) Moving the power management functionality into the OS makes it available on every machine on which the OS is installed. Implementation of OSPM is a hardware/software co-design problem because the hardware resources need to interface with OS-directed software PM, and because both hardware resources and the software applications need to be designed so that they could cooperate with OSPM.

The advanced configuration and power interface (ACPI) specification [48] was developed to establish industry common interfaces enabling a robust OSPM implementation for both devices and entire systems. Currently, it is standardized by Hewlett-Packard, Intel, Microsoft, Phoenix, and Toshiba. It is the key element in OSPM, since it facilitates and accelerates the co-design of OSPM by providing a standard interface to control system components. From a power management perspective, OSPM/ACPI promotes the concept that systems should conserve energy by transitioning unused devices into lower power states, including placing the entire system in a low-power state (sleep state) when possible. ACPI-defined interfaces are intended for wide adoption to encourage hardware and software designers to build ACPI-compatible (and, thus, OSPM-compatible) implementations. Therefore, ACPI promotes the availability of power-manageable components that support multiple operational states. It is important to notice that ACPI specifies neither how to implement hardware devices nor how to realize power management in the OS. No constraints are imposed on implementation styles for hardware and on power management policies. The implementation of ACPI-compliant hardware can leverage any technology or architectural optimization as long as the power-managed device is controllable by the standard interface specified by ACPI. The power management module of the OS can be implemented using any kind of DPM policy.

Experiments were carried out in [49,50] to measure the effectiveness of different DPM policies on ACPI-compliant computers.

7.5 Conclusions

The design of energy-efficient systems goes through the optimization of the architecture of the individual components, the communication between them, and their utilization. Out of these three dimensions, the latter offers a lot of opportunities from the system perspective, as it enables the exploitation of very abstract models of the components and it does not usually require detailed information about their implementation.

In this chapter, we have analyzed the problem of optimizing the usage of components of an electronic system, and we have discussed a number of solutions belonging to the wide class of DPM techniques. DPM aims at dynamically choosing the operating states of the components that best fit the required performance levels, thus reducing the power wasted by idle or under-utilized components.

In its simplest form, DPM entails the decision between keeping a given component active or turning it off, where the "off" state may consist of different power/performance trade-off values. When combined with the possibility of dynamically varying the supply voltage, DPM generalizes to the more general problem of DVS, in which multiple active states are possible. Different policies for actuating DPM/DVS solutions have been studied in the literature, accounting for different aspects of the problem, including, for instance, nonidealities such as the use of a discrete set of voltage levels, the impact of leakage power, and the presence of nonideal supply sources (i.e., batteries).

As mentioned earlier in this chapter, run-time power management is only one facet of the problem of optimizing power/energy consumption of electronic systems. Component and communication design are equally important in the global context, but the way they can be addressed differs substantially from the case of DPM/DVS, as the effectiveness of the latter is mainly dependent on the system workload.

References

[1] L. Benini and G. De Micheli, *Dynamic Power Management: Design Techniques and CAD Tools*, Kluwer Academic Publishers, Dordrecht, 1998.

[2] L. Benini and G. De Micheli, System level power optimization: techniques and tools, *ACM Trans. Des. Autom. Electron. Syst.*, 5, 115–192, 2000.

[3] L. Benini, A. Bogliolo, and G. De Micheli, A survey of design techniques for system-level dynamic power management, *IEEE Trans. VLSI Syst.*, 8, 299–316, 2000.

[4] E. Macii, Ed., *IEEE Design Test Comput.*, Special Issue on Dynamic Power Management, 18, 2001.

[5] L. Benini, R. Hodgson, and P. Siegel, System-level power estimation and optimization, *ISLPED-98: International Symposium on Low Power Electronics and Design*, August 1998, pp. 173–178.

[6] H. Shim, Y. Joo, Y. Choi, H. G. Lee, K. Kim, and N. Chang, Low-energy off-chip SDRAM memory systems for embedded applications, *ACM Trans. Embed. Comput. Syst.*, Special Issue on Memory Systems, 2, 98–130, 2003.

[7] N. Chang, K. Kim, and H. G. Lee, Cycle-accurate energy consumption measurement and analysis: case study of ARM7TDMI, *IEEE Trans. VLSI Syst.*, 10, 146–154, 2002.

[8] I. Lee, Y. Choi, Y. Cho, Y. Joo, H. Lim, H. G. Lee, H. Shim, and N. Chang, A web-based energy exploration tool for embedded systems, *IEEE Des. Test Comput.*, 21, 572–586, 2004.

[9] M. Srivastava, A. Chandrakasan, and R. Brodersen, Predictive system shutdown and other architectural techniques for energy efficient programmable computation, *IEEE Trans. VLSI Syst.*, 4, 42–55, 1996.

[10] C.-H. Hwang and A. Wu, A predictive system shutdown method for energy saving of event-driven computation, *ICCAD-97: International Conference on Computer-Aided Design*, November 1997, pp. 28–32.

[11] P. Krishnan, P. Long, and J. Vitter, Adaptive Disk Spin-Down via Optimal Rent-to-Buy in Probabilistic Environments, *International Conference on Machine Learning*, July 1995, pp. 322–330.

[12] D. Helmbold, D. Long, and E. Sherrod, Dynamic disk spin-down technique for mobile computing, *Proceedings of the International Conference on Mobile Computing,* November 1996, pp. 130–142.

[13] F. Douglis, P. Krishnan, and B. Bershad, Adaptive disk spin-down policies for mobile computers, *Proceedings of the 2nd USENIX Symposium on Mobile and Location-Independent Computing,* April 1995, pp. 121–137.

[14] L. Benini, G. Paleologo, A. Bogliolo, and G. D. Micheli, Policy optimization for dynamic power management, *IEEE Trans. Comput.-Aid. Des.,* 18, 813–833, 1999.

[15] Q. Qiu and M. Pedram, Dynamic power management based on continuous-time Markov decision processes, *DAC-36: Design Automation Conference,* June 1999, pp. 555–561.

[16] E. Chung, L. Benini, A. Bogliolo, and G. D. Micheli, Dynamic power management for non-stationary service requests, *DATE-99: Design Automation and Test in Europe,* March 1999, pp. 77–81.

[17] T. Burd and R. Brodersen, Design issues for dynamic voltage scaling, *ISLPED-00: International Symposium on Low Power Electronics and Design,* July 2000, pp. 9–14.

[18] W. Kim, D. Shin, H. S. Yun, J. Kim, and S. L. Min, Performance comparison of dynamic voltage scaling algorithms for hard real-time systems, *Real-Time and Embedded Technology and Applications Symposium,* September 2002, pp. 219–228.

[19] F. Yao, A. Demers, and A. Shenker, A scheduling model for reduced CPU energy, *FOCS-95: Foundations of Computer Science,* October 1995, pp. 374–382.

[20] S. Lee and T. Sakurai, Run-time voltage hopping for low-power real-time systems, *DAC-37: Design Automation Conference,* June 2000, pp. 806–809.

[21] D. Shin, J. Kim, and S. Lee, Low-energy intra-task voltage scheduling using static timing analysis, *DAC-38: Design Automation Conference,* June 2001, pp. 438–443.

[22] Y.-H. Lee and C. M. Krishna, Voltage-clock scaling for low energy consumption in real-time embedded systems, *6th International Workshop on Real-Time Computing Systems and Applications Symposium,* December 1999, pp. 272–279.

[23] Y. Shin, K. Choi, and T. Sakurai, Power optimization of real-time embedded systems on variable speed processors, *ICCAD-00: International Conference on Computer-Aided Design,* November 2000, pp. 365–368.

[24] P. Pillai and K. G. Shin, Real-time dynamic voltage scaling for low-power embedded operating systems, *Proceedings of the 18th ACM Symposium on Operating Systems Principles,* October 2001, pp. 89–102.

[25] F. Gruian, Hard real-rime scheduling using stochastic data and DVS processors, *ISLPED-01: International Symposium on Low Power Electronics and Design,* August 2001, pp. 46–51.

[26] G. Quan and X. S. Hu, Energy efficient fixed-priority scheduling for real-time systems on variable voltage processors, *DAC-38: Design Automation Conference,* June 2001, pp. 828–833.

[27] T. Ishihara and H. Yasuura, Voltage scheduling problem for dynamically variable voltage processors, *ISLPED-98: International Symposium on Low Power Electronics and Design,* August 1998, pp. 197–202.

[28] Y. Lin, C. Hwang, and A. Wu, Scheduling techniques for variable voltage low power designs, *ACM Trans. Des. Autom. Electron. Syst.,* 2, 81–97, 1997.

[29] W. Kwon and T. Kim, Optimal voltage allocation techniques for dynamically variable voltage processors, *DAC-40: Design Automation Conference,* June 2003, pp. 125–130.

[30] R. Jejurikar, C. Pereira, and R. Gupta, Leakage aware dynamic voltage scaling for real-time embedded systems, *DAC-41: Design Automation Conference,* June 2004, pp. 275–280.

[31] X. Fan, C. Ellis, and A. Lebeck, The synergy between power-aware memory systems and processor voltage, *Workshop on Power-Aware Computer Systems,* December 2003, pp. 130–140.

[32] Y. Cho and N. Chang, Memory-aware energy-optimal frequency assignment for dynamic supply voltage scaling, *ISLPED-04: International Symposium on Low Power Electronics and Design,* August 2004, pp. 387–392.

[33] D. Linden, *Handbook of Batteries,* 2nd ed., McGraw-Hill, Hightstown, NJ, 1995.

[34] T. Martin and D. Sewiorek, Non-ideal battery and main memory effects on CPU speed-setting for low power, *IEEE Trans. VLSI Syst.,* 9, 29–34, 2001.

[35] L. Benini, G. Castelli, A. Macii, E. Macii, M. Poncino, and R. Scarsi, Discrete-time battery models for system-level low-power design, *IEEE Trans. VLSI Syst.,* 9, 630–640, 2001.

[36] M. Pedram and Q. Wu. Design considerations for battery-powered electronics, *IEEE Trans. VLSI Syst.,* 10, 601–607, 2002.

[37] L. Benini, G. Castelli, A. Macii, and R. Scarsi, Battery-driven dynamic power management, *IEEE Des. Test Comput.*, 18, 53–60, 2001.

[38] P. Rong and M. Pedram, Extending the lifetime of a network of battery-powered mobile devices by remote processing: a Markovian decision-based approach, *DAC-40: Design Automation Conference*, June 2003, pp. 906–911.

[39] A. Macii, E. Macii, and M. Poncino, Current-controlled battery management policies for lifetime extension of portable systems, *ST J. Syst. Res.*, 3, 92–99, 2002.

[40] P. Chowdhury and C. Chakrabarti, Static task-scheduling algorithms for battery-powered DVS systems, *IEEE Trans. VLSI Syst.*, 13, 226–237, 2005.

[41] D. Rakhmatov and S. Vrudhula, Energy management for battery-powered embedded systems, *ACM Trans. Embed. Comput. Syst.*, 2, 277–324, 2003.

[42] J. Luo and N. Jha, Battery-aware static scheduling for distributed real-time embedded systems, *DAC-38: Design Automation Conference*, June 2001, pp. 444–449.

[43] D. Rakhmatov, S. Vrudhula, and C. Chakrabarti, Battery-conscious task sequencing for portable devices including voltage/clock scaling, *DAC-39: Design Automation Conference*, June 2002, pp. 211–217.

[44] Q. Wu, Q. Qiu, and M. Pedram, An interleaved dual-battery power supply for battery-operated electronics, *ASPDAC-00: Asia and South Pacific Design Automation Conference*, January 2000, pp. 387–390.

[45] L. Benini, G. Castelli, A. Macii, E. Macii, M. Poncino, and R. Scarsi, Scheduling battery usage in mobile systems, *IEEE Trans. VLSI Syst.*, 11, 1136–1143, 2003.

[46] L. Benini, D. Bruni, A. Macii, E. Macii, and M. Poncino, Extending lifetime of multi-battery mobile systems by discharge current steering, *IEEE Trans. Comput.*, 53, 985–995, 2003.

[47] V. Tiwari, S. Malik, A. Wolfe, and T. C. Lee, Instruction level power analysis and optimization software, *J. VLSI Signal Process.*, 13, 1–18, 1996.

[48] Hewlett-Packard Corporation, Intel Corporation, Microsoft Corporation, Phoenix Technologies Ltd., and Toshiba Corporation, *Advanced Configuration and Power Interface Specification Revision 3.0*, September 2004.

[49] Y. Lu, T. Simunic, and G. D. Micheli, Software controlled power management, *CODES-99: International Workshop on Hardware-Software Codesign*, May 1999, pp. 151–161.

[50] Y. Lu, E. Y. Chung, T. Simunic, L. Benini, and G. D. Micheli, Quantitative comparison of power management algorithms, *DATE-00: Design Automation and Test in Europe*, March 2000, pp. 20–26.

8

Processor Modeling and Design Tools

Prabhat Mishra
University of Florida
Gainesville, Florida

Nikil Dutt
Donald Bren School of Information and Computer Sciences,
University of California, Irvine
Irvine, California

8.1 Introduction

This chapter covers state-of-the-art specification languages, tools, and methodologies for processor development in academia as well as industry. Time-to-market pressure coupled with short product lifetimes creates a critical need for design automation in processor development. The processor is modeled using a specification language such as Architecture description language (ADL). The ADL specification is used to generate various tools (e.g., simulators, compilers, and debuggers) to enable exploration and validation of candidate architectures. The goal is to find the best possible processor architecture for the given set of application programs under various design constraints such as cost, area, power, and performance. The ADL specification is also used to perform various design automation tasks, including hardware generation and functional verification of processors.

Computing is an integral part of daily life. We encounter two types of computing devices everyday: desktop-based computing devices and embedded computer systems. Desktop-based computing systems encompass traditional "computers", including personal computers, notebook computers, workstations, and servers. Embedded computer systems are ubiquitous — they run the computing devices hidden inside a vast array of everyday products and appliances such as cell phones, toys, handheld PDAs, cameras, and microwave ovens. Both types of computing devices use programmable components such as processors, coprocessors, and memories to execute the application programs. These programmable components are also referred as *programmable architectures.* Figure 8.1 shows an example embedded system with programmable architectures. Depending on the application domain, the embedded system can have application-specific hardwares, interfaces, controllers, and peripherals. The complexity of the programmable architectures is increasing at an exponential rate due to technological advances as well as demand for realization of ever more complex applications in communication,

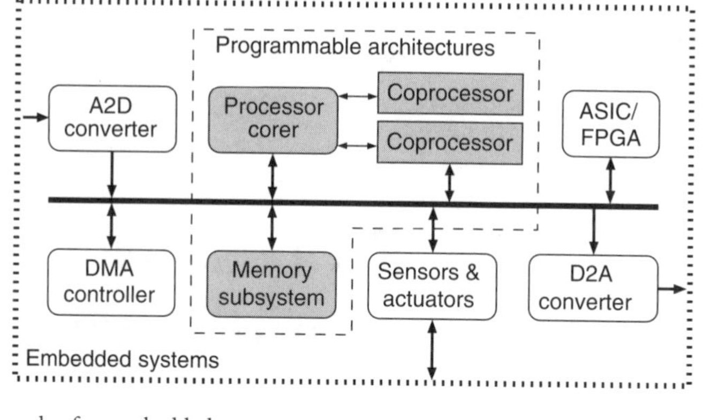

FIGURE 8.1 An example of an embedded system.

multimedia, networking, and entertainment. Shrinking time-to-market coupled with short product lifetimes create a critical need for design automation of increasingly sophisticated and complex programmable architectures.

Modeling plays a central role in design automation of processors. It is necessary to develop a specification language that can model complex processors at a higher level of abstraction and also enable automatic analysis and generation of efficient prototypes. The language should be powerful enough to capture high-level description of the programmable architectures. On the other hand, the language should be simple enough to allow correlation of the information between the specification and the architecture manual. Specifications widely in use today are still written informally in natural language like English. Since natural language specifications are not amenable to automated analysis, there are possibilities of ambiguity, incompleteness, and contradiction: all problems that can lead to different interpretations of the specification. Clearly, formal specification languages are suitable for analysis and verification. Some have become popular because they are input languages for powerful verification tools such as a model checker. Such specifications are popular among verification engineers with expertise in formal languages. However, these specifications are not acceptable by designers and other tool developers. Therefore, the ideal specification language should have formal (unambiguous) semantics as well as easy correlation with the architecture manual.

Architecture description languages (ADLs) have been successfully used as a specification language for processor development. The ADL specification is used to perform early exploration, synthesis, test generation, and validation of processor-based designs as shown in Figure 8.2. The ADL specification can also be used for generating hardware prototypes [1,2]. Several researches have shown the usefulness of ADL-driven generation of functional test programs [3] and JTAG interface [4]. The specification can also be used to generate device drivers for real-time operating systems (RTOS)[5]. The rest of the chapter is organized as follows: Section 2 describes processor modeling using ADLs. Section 3 presents ADL-driven methodologies for software toolkit generation, hardware synthesis, exploration, and validation of programmable architectures. Finally, Section 4 concludes the chapter.

8.2 Processor Modeling Using ADLs

The phrase ADL has been used in the context of designing both software and hardware architectures. Software ADLs are used for representing and analyzing software architectures [6,7]. They capture the behavioral specifications of the components and their interactions that comprise the software architecture. However, hardware ADLs capture the structure (hardware components and their connectivity), and the behavior (instruction-set) of processor architectures. The concept of using machine description languages for specification of architectures has been around for a long time. Early ADLs such as ISPS [8] were used for simulation, evaluation, and synthesis of computers and other digital systems. This section describes contemporary hardware ADLs. First, it tries to answer why ADLs (not other languages) are used

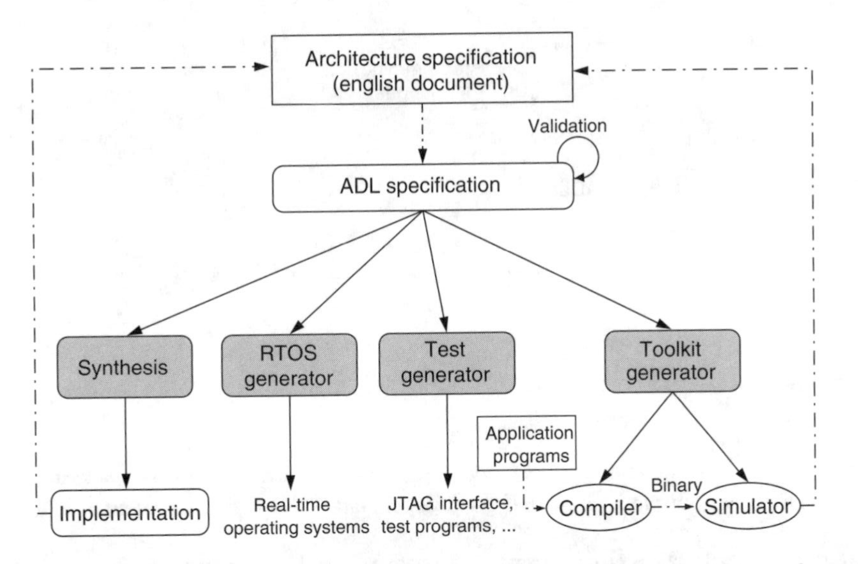

FIGURE 8.2 Architecture description languages-driven exploration, synthesis, and validation of programmable architectures.

for modeling and specification. Next, it surveys contemporary ADLs to compare their relative strengths and weaknesses in the context of processor modeling and ADL-driven design automation.

8.2.1 ADLs and other Languages

How do ADLs differ from programming languages, hardware description languages (HDLs), modeling languages, and the like? This section attempts to answer this question. However, it is not always possible to answer the following question: Given a language for describing an architecture, what are the criteria for deciding whether it is an ADL or not?

In principle, ADLs differ from programming languages because the latter bind all architectural abstractions to specific point solutions whereas ADLs intentionally suppress or vary such binding. In practice, architecture is embodied and recoverable from code by reverse engineering methods. For example, it might be possible to analyze a piece of code written in C language and figure out whether it corresponds to *Fetch* unit or not. Many languages provide architecture level views of the system. For example, C++ language offers the ability to describe the structure of a processor by instantiating objects for the components of the architecture. However, C++ offers little or no architecture-level analytical capabilities. Therefore, it is difficult to describe architecture at a level of abstraction suitable for early analysis and exploration. More importantly, traditional programming languages are not natural choice for describing architectures due to their inability in capturing hardware features such as parallelism and synchronization.

ADLs differ from modeling languages (such as UML) because the latter are more concerned with the behaviors of the whole rather than the parts, whereas ADLs concentrate on the representation of components. In practice, many modeling languages allow the representation of cooperating components and can represent architectures reasonably well. However, the lack of an abstraction would make it harder to describe the instruction set (IS) of the architecture. Traditional HDLs, such as VHDL and Verilog, do not have sufficient abstraction to describe architectures and explore them at the system level. It is possible to perform reverse-engineering to extract the structure of the architecture from the HDL description. However, it is hard to extract the instruction set behavior of the architecture. In practice, some variants of HDLs work reasonably well as ADLs for specific classes of programmable architectures.

There is no clear line between ADLs and non-ADLs. In principle, programming languages, modeling languages, and hardware description languages have aspects in common with ADLs, as shown in Figure 8.3. Languages can, however, be discriminated from one another according to how much architectural

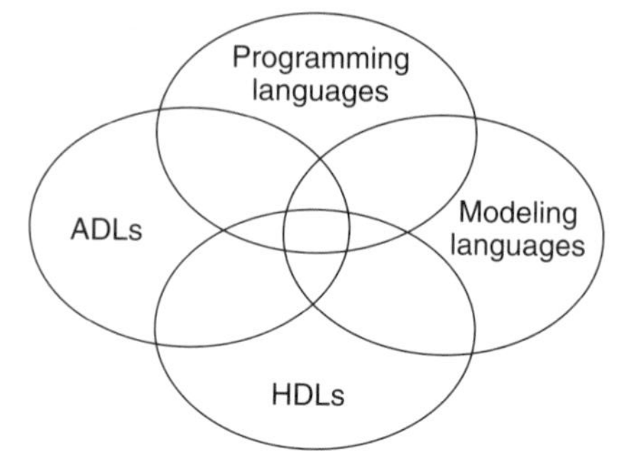

FIGURE 8.3 ADLs vs. non-ADLs.

information they can capture and analyze. Languages that were born as ADLs show a clear advantage in this area over languages built for some other purpose and later co-opted to represent architectures.

8.2.2 Contemporary Architecture Description Languages

This section briefly surveys some of the contemporary ADLs in the context of processor modeling and design automation. There are many comprehensive ADL surveys available in the literature including ADLs for retargetable compilation [9], programmable embedded systems [10], and system-on-chip (SOC) design [11]. Figure 8.4 shows the classification of ADLs based on two aspects: *content* and *objective.* The content-oriented classification is based on the nature of the information an ADL can capture, whereas the objective-oriented classification is based on the purpose of an ADL. Contemporary ADLs can be classified into six categories based on the following objectives: simulation, synthesis, test generation, compilation, validation, and operating system (OS) generation.

ADLs can be classified into four categories based on the nature of the information: structural, behavioral, mixed, and partial. The structural ADLs capture the structure in terms of architectural components and their connectivity. The behavioral ADLs capture the instruction-set behavior of the processor architecture. The mixed ADLs capture both structure and behavior of the architecture. These ADLs capture complete description of the structure or behavior or both. However, the partial ADLs capture specific information about the architecture for the intended task. For example, an ADL intended for interface synthesis does not require internal structure or behavior of the processor. Traditionally, structural ADLs are suitable for synthesis and test-generation. Similarly, behavioral ADLs are suitable for simulation and compilation. It is not always possible to establish a one-to-one correspondence between content- and objective-based classifications. For example, depending on the nature and amount of information captured, partial ADLs can represent any one or more classes of the objective-based ADLs. This section presents the survey using content-based classification of ADLs.

8.2.2.1 Structural ADLs

ADL designers consider two important aspects: the level of abstraction vs. generality. It is very difficult to find an abstraction to capture the features of different types of processors. A common way to obtain generality is to lower the abstraction level. Register transfer level (RTL) is a popular abstraction level — low enough for detailed behavior modeling of digital systems, and high enough to hide gate-level implementation details. Early ADLs are based on RTL descriptions. This section briefly describes a structural ADL: MIMOLA [12].

8.2.2.1.1 MIMOLA

MIMOLA [12] is a structure-centric ADL developed at the University of Dortmund, Germany. It was originally proposed for micro-architecture design. One of the major advantages of MIMOLA is that the same description can be used for synthesis, simulation, test generation, and compilation. A tool chain including the MSSH hardware synthesizer, the MSSQ code generator, the MSST self-test program compiler, the MSSB functional simulator, and the MSSU RTL simulator were developed based on the MIMOLA language [12].

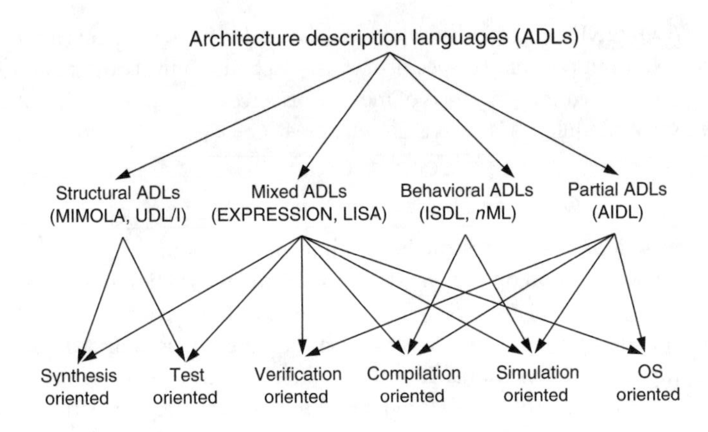

FIGURE 8.4 Taxonomy of ADLs.

MIMOLA has also been used by the RECORD [12] compiler. MIMOLA description contains three parts: the algorithm to be compiled, the target processor model, and additional linkage and transformation rules. The software part (algorithm description) describes application programs in a PASCAL-like syntax. The processor model describes micro-architecture in the form of a component netlist. The linkage information is used by the compiler in order to locate important modules such as program counter and instruction memory. The following code segment specifies the program counter and instruction memory locations [12]:

```
LOCATION_FOR_PROGRAMCOUNTER  PCReg;
LOCATION_FOR_INSTRUCTIONS  IM[0..1023];
```

The algorithmic part of MIMOLA is an extension of PASCAL. Unlike other high-level languages, it allows references to physical registers and memories. It also allows the use of hardware components using procedure calls. For example, if the processor description contains a component named MAC, programmers can write the following code segment to use the multiply-accumulate operation performed by MAC:

```
res  := MAC(x, y, z);
```

The processor is modeled as a netlist of component modules. MIMOLA permits modeling of arbitrary (programmable or nonprogrammable) hardware structures. Similar to VHDL, a number of predefined, primitive operators exist. The basic entities of MIMOLA hardware models are modules and connections. Each module is specified by its port interface and its behavior. The following example shows the description of a multi-functional ALU module [12]:

```
MODUL ALU
  (IN inpl inp2: 31:0);
  OUT outp (31:0) ;
  IN ctrl ;
  )
CONBEGIN
   outp <—    CASE ctrl OF
     0: inpl + inp2 ;
     1: inpl - inp2 ;
     END;
CONEND;
```

The CONBEGIN/CONEND construct includes a set of concurrent assignments. In the example a conditional assignment to output port *outp* is specified, which depends on the two-bit control input *ctrl*. The netlist structure is formed by connecting ports of module instances. For example, the following MIMOLA description connects two modules: *ALU* and accumulator *ACC*.

```
CONNECTIONS ALU.outp → ACC.inp
            ACC.outp → ALU.inp
```

The MSSQ code generator extracts instruction-set information from the module netlist. It uses two internal data structures: connection operation graph (COG) and instruction tree (I-tree). It is a very difficult task to extract the COG and I-trees even in the presence of linkage information due to the flexibility of an RTL structural description. Extra constraints need to be imposed in order for the MSSQ code generator to work properly. The constraints limit the architecture scope of MSSQ to micro-programmable controllers, in which all control signals originate directly from the instruction word. The lack of explicit description of processor pipelines or resource conflicts may result in poor code quality for some classes of VLIW or deeply pipelined processors.

8.2.2.2 Behavioral ADLs

The difficulty of instruction-set extraction can be avoided by abstracting behavioral information from the structural details. Behavioral ADLs explicitly specify the instruction semantics and ignore detailed hardware structures. Typically, there is a one-to-one correspondence between behavioral ADLs and instruction-set reference manual. This section briefly describes two behavioral ADLs: nML [13] and Instruction set description language [14].

8.2.2.2.1 nML

nML is an instruction-set-oriented ADL proposed at the Technical University of Berlin, Germany. nML has been used by code generators CBC [15] and CHESS [16], and instruction-set simulators (ISSs) Sigh/Sim [17] and CHECKERS. Currently, CHESS/CHECKERS environment is used for automatic and efficient software compilation and instruction-set simulation [18]. nML developers recognized the fact that several instructions share common properties. The final nML description would be compact and simple if the common properties are exploited. Consequently, nML designers used a hierarchical scheme to describe instruction-sets. The instructions are the topmost elements in the hierarchy. The intermediate elements of the hierarchy are partial instructions (PI). The relationship between elements can be established using two composition rules: AND- and OR-rules. The AND-rule groups several PIs into a larger PI and the OR-rule enumerates a set of alternatives for one PI. Therefore, instruction definitions in nML can be in the form of an and–or tree. Each possible derivation of the tree corresponds to an actual instruction.

To achieve the goal of sharing instruction descriptions, the instruction-set is enumerated by an attributed grammar [19]. Each element in the hierarchy has a few attributes. A nonleaf element's attribute values can be computed based on its children's attribute values. Attribute grammar is also adopted by other ADLs such as ISDL [14] and TDL [20]. The following nML description shows an example of instruction specification [13]:

```
op numeric_instruction(a:num_action, src:SRC, dst:DST)
action {
  temp_src = src;
  temp_dst = dst;
  a.action;
  dst = temp_dst;
}
op num_action = add|sub
op add()
action = {
  temp_dst = temp_dst + temp_src
}
```

The definition of *numeric_instruction* combines three PI with the AND-rule: *num_action,* SRC, and DST. The first PI, *num_action,* uses OR-rule to describe the valid options for actions: *add* or *sub.* The number of all possible derivations of *numeric_instruction* is the product of the size of *num_action, SRC* and *DST.* The common behavior of all these options is defined in the *action* attribute of *numeric_instruction.* Each option for *num_action* should have its own action attribute defined as its specific behavior, which is referred by the *a.action* line. For example, the above code segment has action description for *add* operation. Object-code image and assembly syntax can also be specified in the same hierarchical manner.

nML also captures the structural information used by instruction-set architecture. For example, storage units should be declared since they are visible to the instruction-set. nML supports three types of storages: RAM, register, and transitory storage. Transitory storage refers to machine states that are retained only for a limited number of cycles, e.g., values on buses and latches. Computations have no delay in nML timing model — only storage units have delay. Instruction delay slots are modeled by introducing storage units as pipeline registers. The result of the computation is propagated through the registers in the behavior specification. nML models constraints between operations by enumerating all valid combinations. The enumeration of valid cases can make nML descriptions lengthy. More complicated constraints, which often appear in DSPs with irregular instruction-level parallelism (ILP) constraints or VLIW processors with multiple issue slots, are hard to model with nML. For example, nML cannot model the constraint that operation *I1* cannot directly follow operation *I0.* nML explicitly supports several addressing modes. However, it implicitly assumes an architecture model, which restricts its generality. As a result it is hard to model multicycle or pipelined units and multiword instructions explicitly. A good critique of nML is given in [61].

8.2.2.2.2 ISDL

Instruction Set Description Language (ISDL) was developed at MIT and used by the Aviv compiler [21] and GENSIM simulator generator [22]. The problem of constraint modeling is avoided by ISDL with explicit specification. ISDL language is mainly targeted toward VLIW processors. Similar to nML, ISDL primarily describes the instruction-set of processor architectures. ISDL consists of mainly five sections: instruction word format, global definitions, storage resources, assembly syntax, and constraints. It also contains an optimization information section that can be used to provide certain architecture specific hints for the compiler to make better machine-dependent code optimizations. The section on instruction word format defines fields of the instruction word. The instruction word is separated into multiple fields each containing one or more subfields. The global definition section describes four main types: tokens, nonterminals, split functions, and macro definitions. Tokens are the primitive operands of instructions. For each token, assembly format and binary encoding information must be defined. An example token definition of a binary operand is

```
Token X[0..1]   X_R ival {yylval.ival = yytext[1] – '0';}
```

In this example, following the keyword *Token* is the assembly format of the operand. *X_R* is the symbolic name of the token used for reference. The *ival* is used to describe the value returned by the token. Finally, the last field describes the computation of the value. In this example, the assembly syntax allowed for the token *XR* is *X0* or *X1,* and the values returned are 0 or 1, respectively. The value (last) field is used for behavioral definition and binary encoding assignment by nonterminals or instructions. Nonterminal is a mechanism provided to exploit commonalities among operations. The following code segment describes a nonterminal named *XYSRC:*

```
Non_Terminal ival XYSRC:   X_D {$$ = 0;}        |
                           Y_D {$$ = Y_D + 1;};
```

The definition of *XYSRC* consists of the keyword *Nonterminal,* the type of the returned value, a symbolic name as it appears in the assembly, and an action that describes the possible token or nonterminal combinations, and the return value associated with each of them. In this example, *XYSRC* refers to tokens *X_D* and *Y_D* as its two options. The second field *(ival)* describes the returned value type. It returns 0 for *X_D* or incremented value for *Y_D.* Similar to nML, storage resources are the only structural information modeled by ISDL. The storage section lists all the storage resources visible to the programmer. It lists the names and sizes of the memory, register files, and special registers. This information is used by the compiler to determine the available resources and how they should be used.

The assembly syntax section is divided into fields corresponding to the separate operations that can be performed in parallel within a single instruction. For each field, a list of alternative operations can be described. Each operation description consists of a name, a list of tokens or nonterminals as parameters, a set of commands that manipulate the bitfields, RTL description, timing details, and costs. RTL description captures the effect of the operation on the storage resources. Multiple costs are allowed including operation execution time, code size, and costs due to resource conflicts. The timing model of ISDL describes when the various effects of the operation take place (e.g., because of pipelining). In contrast to nML, which enumerates all valid combinations, ISDL defines invalid combinations in the form of Boolean expressions. This often leads to a simple constraint specification. It also enables ISDL to capture irregular instruction sequencing constraints. The following example shows how to describe the constraint that instruction I1 cannot directly follow instruction I0. The "[1]" indicates a time shift of one instruction fetch for the I0 instruction. The ~ is a symbol for NOT and '&' is for logical AND.

```
~(I1 *) & ([1] I0 *, *)
```

ISDL provides the means for compact and hierarchical instruction-set specification. However, it may not be possible to describe instruction-sets with multiple encoding formats using the simple tree-like instruction structure of ISDL.

8.2.2.3 Mixed ADLs

Mixed ADLs capture both structural and behavioral details of the architecture. This section briefly describes three mixed ADLs: HMDES, EXPRESSION, and LISA.

8.2.2.3.1 HMDES

Machine description language HMDES was developed at the University of Illinois at Urbana-Champaign for the IMPACT research compiler [23]. C-like preprocessing capabilities such as file inclusion, macro expansion, and conditional inclusion are supported in HMDES. An HMDES description is the input to the MDES machine description system of the Trimaran compiler infrastructure, which contains IMPACT as well as the Elcor research compiler from HP Labs. The description is first pre-processed, then optimized and translated to a low-level representation file. A machine database reads the low-level files and supplies information for the compiler backend through a predefined query interface.

MDES captures both structure and behavior of target processors. Information is broken down into sections such as format, resource usage, latency, operation, and register. For example, the following code segment describes register and register file. It describes 64 registers. The register file describes the width of each register and other optional fields such as generic register type (virtual field), speculative, static, and rotating registers. The value '1' implies speculative and '0' implies nonspeculative.

```
SECTION Register {
       R0();R1(); ... R63 ();
  'R[0]'()... 'R[63]'();
...
}

SECTION Register_File {
RF_i(width(32) virtual(i) speculative (1)
  static(R0 ...R63 rotating ('R[0]' ... 'R[63]'));
  ...
}
```

MDES allows only a restricted retargetability of the cycle-accurate simulator to the HPL-PD processor family [24]. MDES permits description of memory systems, but limited to the traditional hierarchy, i.e., register files, caches, and main memory.

8.2.2.3.2 EXPRESSION

Traditional mixed ADLs require explicit description of Reservation Tables (RT). Processors that contain complex pipelines, large amounts of parallelism, and complex storage sub-systems, typically contain a large number of operations and resources (and hence RTs). Manual specification of RTs on a per-operation basis thus becomes cumbersome and error-prone. The manual specification of RTs (for each configuration) becomes impractical during rapid architectural exploration. The EXPRESSION ADL [25] describes a processor as a netlist of units and storages to generate automatically RTs based on the netlist [26]. Unlike MIMOLA, the netlist representation of EXPRESSION is coarse grain. It uses a higher level of abstraction similar to block-diagram level description in architecture manual.

EXPRESSION ADL was developed at the University of California, Irvine. The ADL has been used by the retargetable compiler (EXPRESS [27]) and simulator (SIMPRESS [28]) generation framework. The framework also supports a graphical user interface (GUI) and can be used for design space exploration of programmable architectures consisting of processor cores, coprocessors, and memories [29]. An EXPRESSION description is composed of two main sections: behavior and structure. The behavior section has three subsections: operations, instruction, and operation mappings. Similarly, the structure section consists of three subsections: components, pipeline/data-transfer paths, and memory subsystem.

The operation subsection describes the instruction-set of the processor. Each operation of the processor is described in terms of its opcode and operands. The types and possible destinations of each operand are also specified. A useful feature of EXPRESSION is operation group that groups similar operations together for the ease of later reference. For example, the following code segment shows an operation group *(alu_ops)* containing two ALU operations: *add* and *sub*.

```
(OP_GROUP alu_ops
  (OPCODE add
    (OPERANDS (SRC1 reg)(SRC2  reg/imm) (DEST  reg))
    (BEHAVIOR DEST = SRC1 + SRC2)
  ...)
(OPCODE sub
    (OPERANDS (SRC1 reg)  (SRC2 reg/imm) (DEST reg))
    (BEHAVIOR DEST = SRC1 − SRC2)
  ...)
)
```

The instruction subsection captures the parallelism available in the architecture. Each instruction contains a list of slots (to be filled with operations), with each slot corresponding to a functional unit. The operation mapping subsection is used to specify the information needed by instruction selection and architecture-specific optimizations of the compiler. For example, it contains mapping between generic and target instructions. The component subsection describes each RTL component in the architecture. The components can be pipeline units, functional units, storage elements, ports, and connections. For multicycle or pipelined units, the timing behavior is also specified. The subsection on pipeline/data-transfer path describes the netlist of the processor. The *pipeline path description* provides a mechanism to specify the units, which comprise the pipeline stages, while the *data-transfer path description* provides a mechanism for specifying the valid data transfers. This information is used to both retarget the simulator, and to generate reservation tables needed by the scheduler [26]. An example path declaration for the DLX architecture [30] (Figure 8.5) is shown below. It describes that the processor has five pipeline stages. It also describes that the *Execute* stage has four parallel paths. Finally, it describes each path, e.g., it describes that the *FADD* path has four pipeline stages.

```
(PIPELINE Fetch Decode Execute MEM WriteBack)
(Execute (ALTERNATE IALU MULT FADD DIV))
(MULT (PIPLINE MUL1 MUL2...MUL7))
(FADD(PIPELINE FADD1 FADD2 FADD3 FADD4))
```

The memory subsection describes the types and attributes of various storage components (such as register files, SRAMs, DRAMs, and caches). The memory netlist information can be used to generate memory aware compilers and simulators [31]. Memory aware compilers can exploit the detailed information to hide the latency of the lengthy memory operations [32]. EXPRESSION captures the data path information in the processor. The control path is not explicitly modeled. The instruction model requires extension to capture inter-operation constraints such as sharing of common fields. Such constraints can be modeled by ISDL through cross-field encoding assignment.

8.2.2.3.3 LISA

Language for Instruction Set Architecture (LISA) [33] was developed at Aachen University of Technology, Germany, with a simulator centric view. The language has been used to produce production quality simulators [34]. An important aspect of LISA is its ability to capture control path explicitly. Explicit modeling of both data path and control is necessary for cycle-accurate simulation. LISA has also been used to generate retargetable C compilers [35,36]. Its descriptions are composed of two types of declarations: resource and operation. The resource declarations cover hardware resources such as registers, pipelines, and memories. The pipeline model defines all possible pipeline paths that operations can go through. An example pipeline description for the architecture shown in Figure 8.5 is as follows:

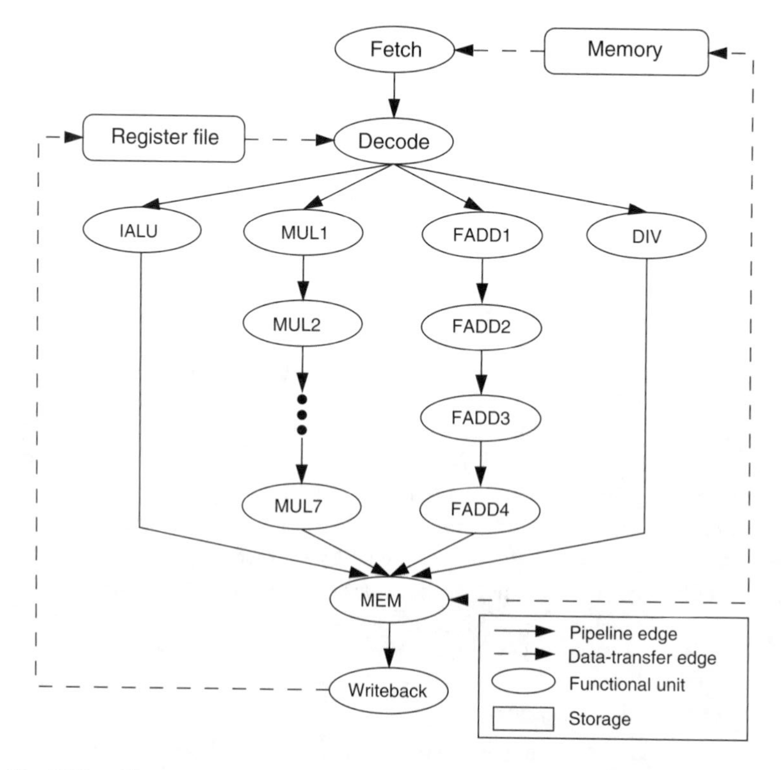

FIGURE 8.5 The DLX architecture.

```
PIPELINE int   = {Fetch; Decode; IALU; MEM; WriteBack}
PIPELINE flt   = {Fetch; Decode; FADD1; FADD2;
                  FADD3; FADD4; MEM; WriteBack}
PIPELINE mul   = {Fetch; Decode; MUL1; MUL2; MUL3; MUL4;
                  MUL5; MUL6; MUL7; MEM; WriteBack)
PIPELINE div   = {Fetch; Decode; DIV; MEM; WriteBack}
```

Operations are the basic objects in LISA. They represent the designer's view of the behavior, the structure, and the instruction-set of the programmable architecture. Operation definitions capture the description of different properties of the system such as operation behavior, instruction-set information, and timing. These operation attributes are defined in several sections. LISA exploits the commonality of similar operations by grouping them into one. The following code segment describes the decoding behavior of two immediate-type (i-type) operations (ADDI and SUBI) in the DLX *Decode* stage. The complete behavior of an operation can be obtained by combining its behavior definitions in all the pipeline stages.

```
OPERATION i-type IN pipe_int.Decode {
  DECLARE {
    GROUP opcode = {ADD|| SUBI}
    GROUP rsi, rd = {fix_register};
  }
  CODING {opcode rsi rd immediate}
  SYNTAX {opcode rd " , " rsi " , " immediate}
  BEHAVIOUR {reg_a = rsi; imm = immediate; cond = 0;
  }
  ACTIVATION {opcode, writeback}
}
```

A language similar to LISA is RADL. RADL [37] was developed at Rockwell, Inc. USA as an extension of the LISA approach that focuses on explicit support of detailed pipeline behavior to enable generation of production quality cycle- and phase-accurate simulators.

8.2.2.4 Partial ADLs

The ADLs discussed so far capture complete description of the processor's structure, behavior, or both. There are many description languages such as AIDL [38] that capture partial information of the architecture needed to perform a specific task. AIDL is an ADL developed at the University of Tsukuba for design of high-performance superscalar processors [38]. It seems that AIDL does not aim at data path optimization but aims at the validation of the pipeline behavior such as data-forwarding and out-of-order completion. In AIDL, the timing behavior of the pipeline is described using interval temporal logic. AIDL does not support software toolkit generation. However, AIDL descriptions can be simulated using the AIDL simulator.

8.3 ADL-Driven Methodologies

This section describes the ADL-driven methodologies used for processor development. It presents the following three methodologies that are used in academic research as well as industry:

- software toolkit generation and exploration;
- generation of hardware implementation;
- top–down validation.

8.3.1 Software Toolkit Generation and Exploration

Embedded systems present a tremendous opportunity to customize designs by exploiting the application behavior. Rapid exploration and evaluation of candidate architectures are necessary due to time-to-market pressure and short product lifetimes. ADLs are used to specify processor and memory architectures and generate software toolkit including compiler, simulator, assembler, profiler, and debugger. Figure 8.6 shows a traditional ADL-based design space exploration flow. The application programs are compiled and simulated, and the feedback is used to modify the ADL specification with the goal of finding the best possible architecture for the given set of application programs under various design constraints such as area, power, and performance. An extensive body of recent work addresses ADL-driven software toolkit generation and design space exploration of processor-based embedded systems, in both academia: ISDL [14], Valen-C [39], MIMOLA [12], LISA [33], nML [13], Sim-nML [40], EXPRESSION [25], and industry: ARC [41], Axys [42], RADL [37], Target [18], Tensilica [43], MDES [24].

One of the main purposes of an ADL is to support the automatic generation of a high-quality software toolkit including at least an ILP compiler and a cycle-accurate simulator. However, such tools require detailed information about the processor, typically in a form that is not concise and easily specifiable. Therefore, it becomes necessary to develop procedures to generate automatically such tool-specific information from the ADL specification. For example, RTs are used in many ILP compilers to describe resource conflicts. However, manual description of RTs on a per-instruction basis is cumbersome and error-prone. Instead, it is easier to specify the pipeline and datapath resources in an abstract manner, and generate RTs on a per-instruction basis [26]. This section describes some of the challenges in automatic generation of software tools (focusing on compilers and simulators) and survey some of the approaches adopted by current tools.

8.3.1.1 Compilers

Traditionally, the software for embedded systems was hand-tuned in assembly. With increasing complexity of embedded systems, it is no longer practical to develop software in assembly language or to optimize it manually except for the critical sections of the code. Compilers which produce optimized machine-specific code from a program specified in a high-level language (HLL) such as C/C++ and Java are necessary in order to produce efficient software within the time budget. Compilers for embedded systems have been the focus of several research efforts recently [44].

The compilation process can be broadly broken into two steps: analysis and synthesis. During analysis, the program (in HLL) is converted into an intermediate representation (IR) that contains all the desired information such as control and data dependences. During synthesis, the IR is transformed and

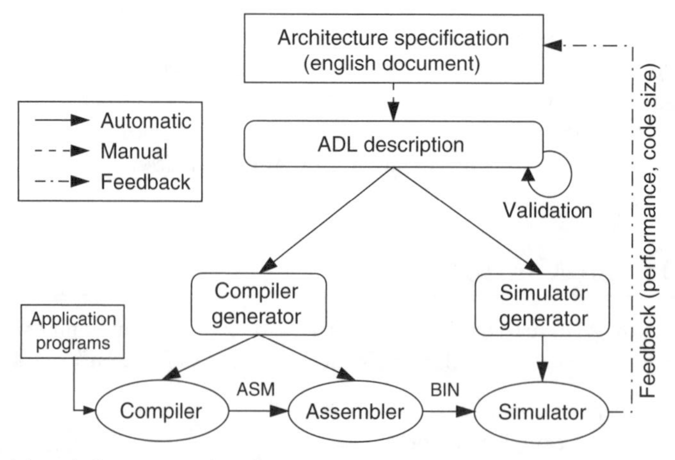

FIGURE 8.6 ADL-driven design space exploration.

optimized in order to generate efficient target-specific code. The synthesis step is more complex and typically includes the following phases: instruction selection, scheduling, resource allocation, code optimizations/transformations, and code generation. The effectiveness of each phase depends on the algorithms chosen and the target architecture. A further problem during the synthesis step is that the optimal ordering between these phases is highly dependent on the target architecture and the application program. As a result, traditionally, compilers have been painstakingly hand-tuned to a particular architecture (or architecture class) and application domain(s). However, stringent time-to-market constraints for SOC designs no longer make it feasible to generate manually compilers tuned to particular architectures. Automatic generation of an efficient compiler from an abstract description of the processor model becomes essential.

A promising approach to automatic compiler generation is the "retargetable compiler" approach. A compiler is classified as retargetable if it can be adapted to generate code for different target processors with significant reuse of the compiler source code. Retargetability is typically achieved by providing target machine information (in an ADL) as input to the compiler along with the program corresponding to the application. The complexity in retargeting the compiler depends on the range of target processors it supports and also on its optimizing capability. Due to the growing amount of ILP features in modern processor architectures, the difference in quality of code generated by a naive code conversion process and an optimizing ILP compiler can be enormous. Recent approaches on retargetable compilation have focused on developing optimizations/transformations that are "retargetable" and capturing the machine-specific information needed by such optimizations in the ADL. The retargetable compilers can be classified into three broad categories, based on the type of the machine model accepted as input.

8.3.1.1.1 Architecture Template Based

Such compilers assume a limited architecture template, which is parameterizable for customization. The most common parameters include operation latencies, number of functional units, number of registers, etc. Architecture template-based compilers have the advantage that both optimizations and the phase ordering between them can be manually tuned to produce highly efficient code for the limited architecture space. Examples of such compilers include the Valen-C compiler [39] and the GNU-based C/C++ compiler from Tensilica Inc. [43]. The Tensilica GNU-based C/C++ compiler is geared toward the Xtensa parameterizable processor architecture. One important feature of this system is the ability to add new instructions (described through an instruction extension language) and automatically generate software tools tuned to the new instruction-set.

8.3.1.1.2 Explicit Behavioral Information Based

Most compilers require a specification of the behavior in order to retarget their transformations (e.g., instruction selection requires a description of the semantics of each operation). Explicit behavioral information-based retargetable compilers require full information about the instruction-set as well as explicit resource conflict information. Examples include the AVIV [21] compiler using ISDL, CHESS [16] using nML, and Elcor [24] using MDES. The AVIV retargetable code generator produces machine code, optimized for minimal size, for target processors with different instruction-set. It solves the phase-ordering problem by performing a heuristic branch-and-bound step that performs resource allocation/assignment, operation grouping, and scheduling concurrently. CHESS is a retargetable code generation environment for fixed-point DSP processors. CHESS performs instruction selection, register allocation, and scheduling as separate phases (in that order). Elcor is a retargetable compilation environment for VLIW architectures with speculative execution. It implements a software pipelining algorithm (modulo scheduling) and register allocation for static and rotating register files.

8.3.1.1.3 Behavioral Information Generation Based

Recognizing that the architecture information needed by the compiler is not always in a form that may be well suited for other tools (such as synthesis) or does not permit concise specification, some research has focused on extraction of such information from a more amenable specification. Examples include the MSSQ and RECORD compiler using MIMOLA [12], retargetable C compiler based on LISA [35], and the

EXPRESS compiler using EXPRESSION [25]. MSSQ translates Pascal-like HLL into microcode for micro-programmable controllers, while RECORD translates code written in a DSP-specific programming language, called data flow language, into machine code for the target DSP. The retargetable C compiler generation using LISA is based on the reuse of a powerful C compiler platform with many built-in code optimizations and generation of mapping rules for code selection using the instruction semantics information [35]. The EXPRESS compiler tries to bridge the gap between explicit specification of all information (e.g., AVIV) and implicit specification requiring extraction of instruction-set (e.g., RECORD) by having a mixed behavioral/structural view of the processor.

8.3.1.2 Simulators

Simulators are critical components of the exploration and software design toolkit for the system designer. They can be used to perform diverse tasks such as verifying the functionality and/or timing behavior of the system (including hardware and software), and generating quantitative measurements (e.g., power consumption), which can be used to aid the design process. Simulation of the processor system can be performed at various abstraction levels. At the highest level of abstraction, a functional simulation of the processor can be performed by modeling only the instruction-set. Such simulators are termed instruction-set simulators (ISS) or instruction-level simulators (ILS). At lower levels of abstraction are the cycle- and phase-accurate simulation models that yield more detailed timing information. Simulators can be further classified based on whether they provide bit-accurate models, pin-accurate models, exact pipeline models, and structural models of the processor.

Typically, simulators at higher levels of abstraction (e.g., ISS and ILS) are faster but gather less information as compared to those at lower levels of abstraction (e.g., cycle- and phase-accurate). Retargetability (i.e., ability to simulate a wide variety of target processors) is especially important in the arena of embedded system design with emphasis on the exploration and co-development of hardware and software. Simulators with limited retargetability are very fast but may not be useful in all aspects of the design process. Such simulators typically incorporate a fixed architecture template and allow only limited retargetability in the form of parameters such as the number of registers and ALUs. Examples of such simulators are numerous in the industry and include the HPL-PD [24] simulator using the MDES ADL. The model of simulation adopted has significant impact on the simulation speed and flexibility of the simulator. On the basis of the simulation model, simulators can be classified into three types: interpretive, compiled, and mixed.

8.3.1.2.1 Interpretation-Based

Such simulators are based on an interpretive model of the processors instruction-set. Interpretive simulators store the state of the target processor in host memory. It then follows a fetch, decode, and execute model: instructions are fetched from memory, decoded and then executed in serial order. Advantages of this model include ease of implementation, flexibility, and the ability to collect varied processor state information. However, it suffers from significant performance degradation as compared to other approaches primarily due to the tremendous overhead in fetching, decoding, and dispatching instructions. Almost all the commercially available simulators are interpretive. Examples of research interpretive retargetable simulators include SIMPRESS [28] using EXPRESSION, and GENSIM/XSIM [22] using ISDL.

8.3.1.2.2 Compilation-Based

Compilation-based approaches reduce the runtime overhead by translating each target instruction into a series of host machine instructions, which manipulate the simulated machine state. Such translation can be done either at compile time (static compiled simulation) where the fetch-decode-dispatch overhead is completely eliminated, or at load time (dynamic compiled simulation), which amortizes the overhead over repeated execution of code. Simulators based on the static compilation model are presented by Zhu et al. [45] and Pees et al. [34]. Examples of dynamic compiled code simulators include the Shade simulator [46], and the Embra simulator [47].

8.3.1.2.3 Interpretive + Compiled

Traditional interpretive simulation is flexible but slow. Instruction decoding is a time-consuming process in a software simulation. Compiled simulation performs compile time decoding of application programs to improve the simulation performance. However, all compiled simulators rely on the assumption that the complete program code is known before the simulation starts and is furthermore run-time static. Due to the restrictiveness of the compiled technique, interpretive simulators are typically used in embedded systems design flow. Two recently proposed simulation techniques (JIT-CCS [48] and IS-CS [49]) combine the flexibility of interpretive simulation with the speed of the compiled simulation. The JIT-CCS technique compiles an instruction during runtime, *just-in-time* before the instruction is going to be executed. Subsequently, the extracted information is stored in a simulation cache for direct reuse in a repeated execution of the program address. The simulator recognizes if the program code of a previously executed address has changed and initiates a re-compilation. The IS-CS technique performs time-consuming instruction decoding during compile time. In case, an instruction is modified at run-time, the instruction is re-decoded prior to execution. The IS-CS technique also uses an *instruction abstraction* technique to generate aggressively optimized decoded instructions that further improves simulation performance [49,50].

8.3.2 Generation of Hardware Implementation

Recent approaches on ADL-based software toolkit generation enable performance-driven exploration. The simulator produces profiling data and thus may answer questions regarding the instruction-set, the performance of an algorithm and the required size of memory and registers. However, the required silicon area, clock frequency, and power consumption can only be determined in conjunction with a synthesizable HDL model. There are two major approaches in the literature for synthesizable HDL generation. The first one is a parameterized processor core-based approach. These cores are bound to a single processor template, whose architecture and tools can be modified to a certain degree. The second approach is based on processor specification languages.

8.3.2.1 Processor Template Based

Examples of processor template-based approaches are Xtensa [43], Jazz [51], and PEAS [52]. Xtensa [43] is a scalable RISC processor core. Configuration options include the width of the register set, caches, and memories. New functional units and instructions can be added using the Tensilica Instruction Extension (TIE) language. A synthesizable hardware model along with software toolkit can be generated for this class of architectures. Improv's Jazz [51] processor is supported by a flexible design methodology to customize the computational resources and instruction-set of the processor. It allows modifications of data width, number of registers, depth of hardware task queue, and addition of custom functionality in Verilog. PEAS [52] is a GUI-based hardware/software codesign framework. It generates HDL code along with software toolkit. It has support for several architecture types and a library of configurable resources.

8.3.2.2 Specification Language Based

Figure 8.7 shows a typical framework of processor description language-driven HDL generation. Structure-centric ADLs such as MIMOLA are suitable for hardware generation. Some of the behavioral languages (such as ISDL and nML) are also used for hardware generation. For example, the HDL generator HGEN [22] uses ISDL description, and the synthesis tool GO [18] is based on nML. Itoh et al. [53] have proposed a microoperation description based synthesizable HDL generation. It can handle simple processor models with no hardware interlock mechanism or multicycle operations. Mixed languages such as LISA and EXPRESSION capture both structure and behavior of the processor. The synthesizable HDL generation approach based on LISA language [2] produces an HDL model of the architecture. The designer has the choice to generate a VHDL, Verilog, or SystemC representation of the target architecture [2]. The HDL generation methodology presented by Mishra et al. [1] combines the advantages of the processor template-based environments and the language-based specifications using EXPRESSION ADL.

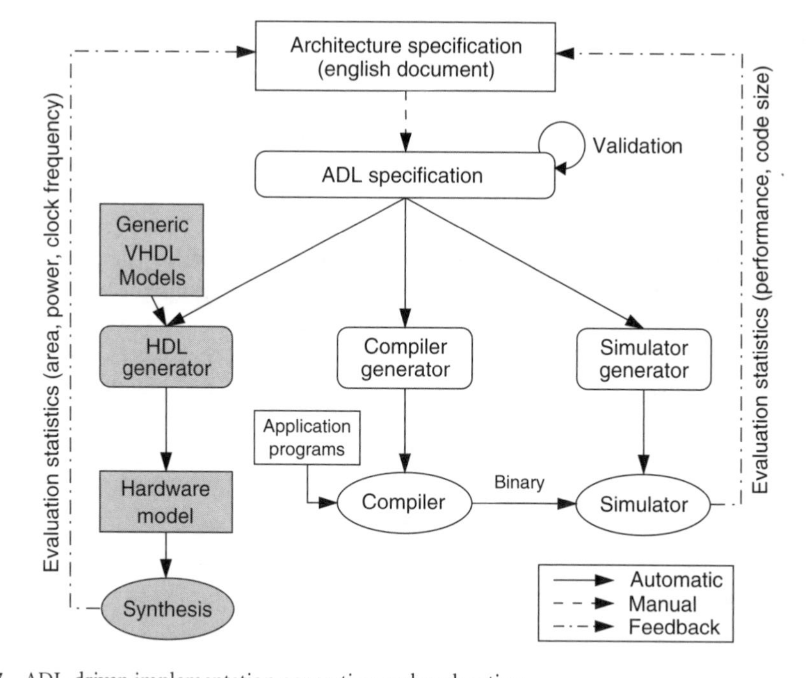

FIGURE 8.7 ADL-driven implementation generation and exploration.

8.3.3 Top-Down Validation

The validation of microprocessors is one of the most complex and important tasks in the current SOC design methodology. Figure 8.8 shows a traditional architecture validation flow. The architect prepares an informal specification of the microprocessor in the form of an English document. The logic designer implements the modules in the RTL. The *RTL design* is validated using a combination of simulation techniques and formal methods. One of the most important problems in today's processor design validation is the lack of a golden reference model that can be used for verifying the design at different levels of abstraction. Thus, many existing validation techniques employ a *bottom–up approach* to pipeline verification, where the functionality of an existing pipelined processor is, in essence, reverse-engineered from its RTL implementation.

Mishra et al. [54] have presented an ADL-driven validation technique that is complementary to these bottom–up approaches. It leverages the system architects knowledge about the behavior of the programmable architectures through ADL constructs, thereby allowing a powerful *top-down approach* to microprocessor validation. Figure 8.9 shows an ADL-driven top–down validation methodology. This methodology has two important steps: validation of ADL specification and specification-driven validation of programmable architectures.

8.3.3.1 Validation of ADL Specification

It is important to verify the ADL specification to ensure the correctness of the architecture specified and the generated software toolkit. Both static and dynamic behavior need to be verified to ensure that the specified architecture is well formed. The static behavior can be validated by analyzing several static properties, such as connectedness, false pipeline and data-transfer paths, and completeness using a graph-based model of the pipelined architecture [55]. The dynamic behavior can be validated by analyzing the instruction flow in the pipeline using a finite state machine (FSM)-based model to verify several important architectural properties such as determinism and in-order execution in the presence of hazards and multiple exceptions [56].

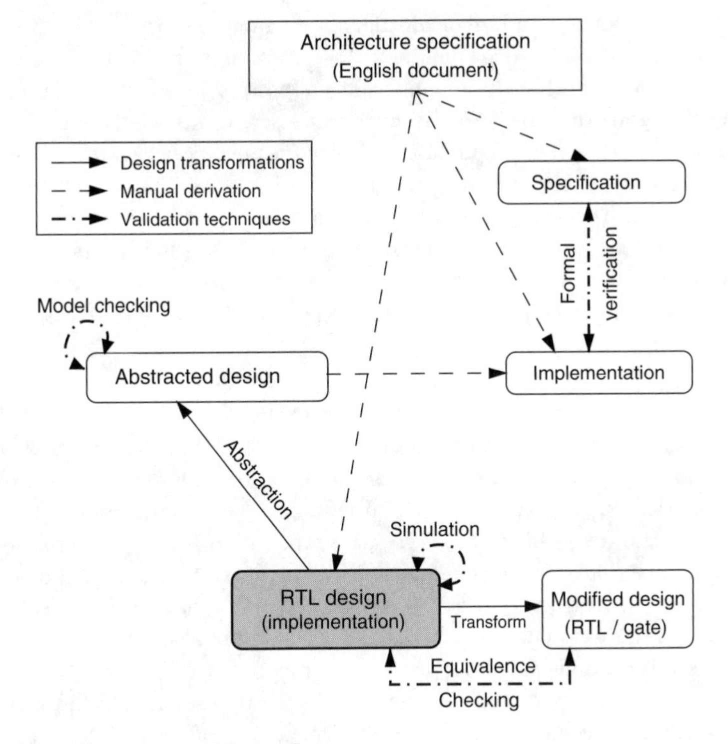

FIGURE 8.8 Traditional bottom–up validation flow.

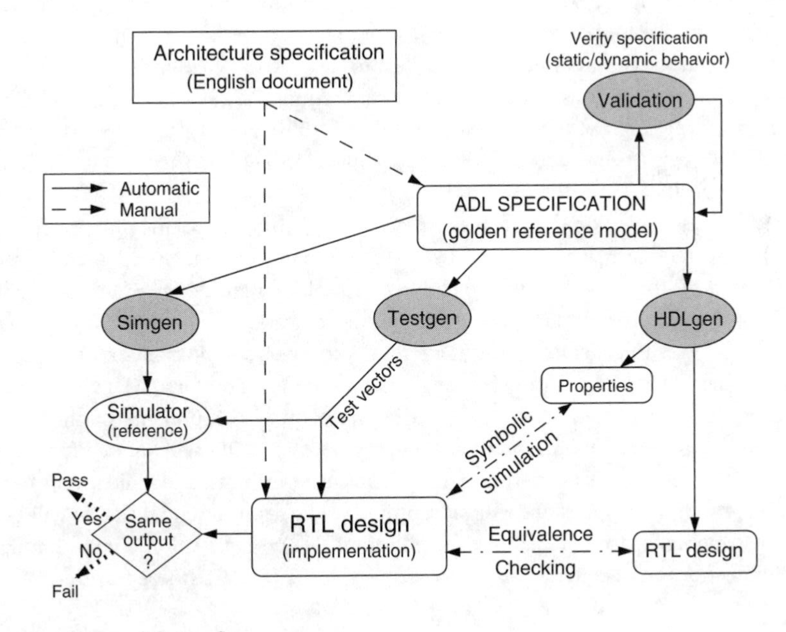

FIGURE 8.9 Top–down validation flow.

8.3.3.2 Specification-Driven Validation

The validated ADL specification can be used as a golden reference model for top–down validation of programmable architectures. The top–down validation approach has been demonstrated in two directions: functional test program generation, and design validation using a combination of equivalence checking and symbolic simulation. Test generation for functional validation of processors has been demonstrated using MIMOLA [12], EXPRESSION [57], and nML [18]. A model checking based approach is used to generate automatically functional test programs from the processor specification using EXPRESSION ADL [57]. It generates graph model of the pipelined processor from the ADL specification. The functional test programs are generated based on the coverage of the pipeline behavior. ADL-driven design validation using equivalence checking has been demonstrated using EXPRESSION ADL [58]. This approach combines ADL-driven hardware generation and validation. The generated hardware model is used as a reference model to verify the hand-written implementation *(RTL design)* of the processor. To verify that the implementation satisfies certain properties, the framework generates the intended properties. These properties are applied using symbolic simulation [58].

8.4 Conclusions

Design and verification of today's complex processors require the use of automated tools and techniques. ADLs have been successfully used in academic research as well as in industry for processor development. The processor is modeled using an ADL. The ADL specification is used to generate software tools including compiler and simulator to enable early design space exploration. The ADL specification is also used to perform other design automation tasks including hardware generation and functional verification.

The early ADLs were either structure-oriented (MIMOLA, UDL/I) or behavior-oriented (nML, ISDL). As a result, each class of ADLs is suitable for specific tasks. For example, structure-oriented ADLs are suitable for hardware synthesis, and unfit for compiler generation. Similarly, behavior-oriented ADLs are appropriate for generating compiler and simulator for instruction-set architectures, and unsuited for generating cycle-accurate simulator or hardware implementation of the architecture. The later ADLs (LISA and EXPRESSION) adopted the mixed approach where the language captures both structure and behavior of the architecture. The existing ADLs are getting modified with the new features and methodologies to perform software toolkit generation, hardware generation, instruction-set synthesis, and test generation for validation of architectures. For example, nML is extended by Target Compiler Technologies to perform hardware synthesis and test generation [18]. Similarly, LISA has been used for hardware generation [2], instruction encoding synthesis [59], and JTAG interface generation [4]. Likewise, EXPRESSION has been used for hardware generation [1], instruction-set synthesis [60], test generation [3], and specification validation [54].

ADLs designed for a specific domain (such as DSP or VLIW) or for a specific purpose (such as simulation or compilation) can be compact and it is possible to automatically generate efficient (in terms of area, power, and performance) tools and hardwares. However, it is difficult to design an ADL for a wide variety of architectures to perform different tasks using the same specification. Generic ADLs require the support of powerful methodologies to generate high-quality results compared to domain or task-specific ADLs. In the future, the existing ADLs will go through changes in two dimensions. First, ADLs will specify not only processor, memory, and co-processor architectures but also other components of the system-on-chip architectures including peripherals and external interfaces. Second, ADLs will be used for software toolkit generation, hardware synthesis, test generation, instruction-set synthesis, and validation of microprocessors. Furthermore, multiprocessor SOCs will be captured and various attendant tasks will be addressed. The tasks include support for formal analysis, generation of RTOS, exploration of communication architectures, and support for interface synthesis. The emerging ADLs will have these features.

References

[1] P. Mishra, A. Kejariwal, and N. Dutt, Synthesis-driven exploration of pipelined embedded processors, *Proceedings of International Conference on VLSI Design*, 2004, pp. 921–926.

[2] O. Schliebusch, A. Chattopadhyay, M. Steinert, G. Braun, A. Nohl, R. Leupers, G. Ascheid, and H. Meyr, RTL processor synthesis for architecture exploration and implementation, *Proceedings of Design Automation and Test in Europe (DATE)*, 2004, pp. 156–160.

[3] P. Mishra and N. Dutt, Functional coverage driven test generation for validation of pipelined processors, *Proceedings of Design Automation and Test in Europe (DATE)*, 2005, pp. 678–683.

[4] O. Schliebusch, D. Kammler, A. Chattopadhyay, R. Leupers, G. Ascheid,and H. Meyr, Automatic generation of JTAG interface and debug mechanism for ASIPs, *GSPx*, 2004.

[5] S. Wang and S. Malik, Synthesizing operating system based device drivers in embedded systems, *Proceedings of International Symposium on Hardware/Software Codesign and System Synthesis (CODES+ISSS)*, 2003, pp. 37–44.

[6] N. Medvidovic and R. Taylor, A framework for classifying and comparing architecture description languages, in *Proceedings of European Software Engineering Conference (ESEC)*, M. J. and H. Schauer, Ed., Springer, Berlin, 1997, pp. 60–76.

[7] Paul C. Clements, A survey of architecture description languages, *Proceedings of International Workshop on Software Specification and Design (IWSSD)*, 1996, pp. 16–25.

[8] M. R. Barbacci, Instruction set processor specifications (ISPS): The notation and its applications. *IEEE Trans. Comp.*, C-30, 24–40, 1981.

[9] W. Qin and S. Malik, *Architecture Description Languages for Retargetable Compilation, in The Compiler Design Handbook: Optimizations & Machine Code Generation*, CRC Press, Boca Raton, FL, 2002.

[10] P. Mishra and N. Dutt, Architecture description languages for programmable embedded systems. *IEE Proceedings on Computers and Digital Techniques*, 2005.

[11] H. Tomiyama, A. Halambi, P. Grun, N. Dutt, and A. Nicolau, Architecture description languages for systems-on-chip design, *Proceedings of Asia Pacific Conference on Chip Design Language*, 1999, pp. 109–116.

[12] R. Leupers and P. Marwedel, Retargetable code generation based on structural processor descriptions, *Design Automat. Embedded Syst.*, 3, 75–108, 1998.

[13] M. Freericks, The nML Machine Description Formalism, Technical Report TR SM-IMP/DIST/08, TU Berlin, Germany, 1993.

[14] G. Hadjiyiannis, S. Hanono, and S. Devadas, ISDL: an instruction set description language for retargetability, *Proceedings of Design Automation Conference (DAC)* , 1997, pp. 299–302.

[15] A. Fauth and A. Knoll, Automatic generation of DSP program development tools, *Proceedings of International Conference Acoustics, Speech and Signal Processing (ICASSP)*, 1993, pp. 457–460.

[16] D. Lanneer, J. Praet, A. Kifli, K. Schoofs, W. Geurts, F. Thoen, and G. Goossens, CHESS: Retargetable code generation for embedded DSP processors, in *Code Generation for Embedded Processors*, Kluwer Academic Publishers, Dordrecht, 1995, pp. 85–102.

[17] F. Lohr, A. Fauth, and M. Freericks, Sigh/sim: An Environment for Retargetable Instruction Set Simulation,. Technical Report 1993/43, Berlin, Germany, 1993.

[18] http://www.retarget.com. *Target Compiler Technologies.*

[19] J. Paakki, Attribute grammar paradigms — a high level methodology in language implementation, *ACM Comp. Surv.*, 27(2):196-256, 1995.

[20] D. Kastner, TDL: A Hardware and Assembly Description Languages, Technical Report TDL 1.4, Saarland University, Germany, 2000.

[21] S. Hanono and S. Devadas, Instruction selection, resource allocation, and scheduling in the AVIV retargetable code generator, *Proceedings of Design Automation Conference (DAC)*, 1998, pp. 510–515.

[22] G. Hadjiyiannis, P. Russo, and S. Devadas, A methodology for accurate performance evaluation in architecture exploration, *Proceedings of Design Automation Conference (DAC)*, 1999, pp. 927–932.

[23] J. Gyllenhaal, B. Rau, and W. Hwu,. HMDES Version 2.0 Specification, Technical Report IMPACT-96-3, IMPACT Research Group, University of Illinois, Urbana, IL, 1996.

[24] The MDES User Manual, http://www.trimaran.org, 1997.

[25] A. Halambi, P. Grun, V. Ganesh, A. Khare, N. Dutt, and A. Nicolau, EXPRESSION: a language for architecture exploration through compiler/simulator retargetability, *Proceedings of Design Automation and Test in Europe (DATE)*, 1999, pp. 485–490.

[26] P. Grun, A. Halambi, N. Dutt, A. Nicolau, RTGEN: an algorithm for automatic generation of reservation tables from architectural descriptions, *Proceedings of International Symposium on System Synthesis (ISSS)*, 1999, pp. 44–50.

[27] A. Halambi, A. Shrivastava, N. Dutt, and A. Nicolau, A customizable compiler framework for embedded systems, *Proceedings of Software and Compilers for Embedded Systems (SCOPES)*, 2001.

[28] A. Khare, N. Savoiu, A. Halambi, P. Grun, N. Dutt, and A. Nicolau, V-SAT: a visual specification and analysis tool for system-on-chip exploration, *Proceedings of EUROMICRO Conference*, 1999, pp. 1196–1203.

[29] http://www.ics.uci.edu/~express. *Exploration framework using EXPRESSION.*

[30] J. Hennessy and D. Patterson, *Computer Architecture: A Quantitative Approach.* Morgan Kaufmann Publishers Inc, San Mateo, CA, 1990.

[31] P. Mishra, M. Mamidipaka, and N. Dutt, Processor-memory co-exploration using an architecture description language, *ACM Trans. Embedded Comput. Syst. (TECS)*, 3, 140–162, 2004.

[32] P. Grun, N. Dutt, and A. Nicolau, Memory aware compilation through accurate timing extraction, *Proceedings of Design Automation Conference (DAC)*, 2000, pp. 316–321.

[33] V. Zivojnovic, S. Pees, and H. Meyr, LISA — machine description language and generic machine model for HW/SW co-design, *IEEE Workshop on VLSI Signal Processing*, 1996, pp. 127–136.

[34] S. Pees, A. Hoffmann, H. Meyr, Retargetable compiled simulation of embedded processors using a machine description language. *ACM Trans. Design Automat. of Electron. Syst.*, 5, 815–834, 2000.

[35] M. Hohenauer, H. Scharwaechter, K. Karuri, O. Wahlen, T. Kogel, R. Leupers, G. Ascheid, H. Meyr, G. Braun, and H. Someren, A methodology and tool suite for c compiler generation from ADL processor models, *Proceedings of Design Automation and Test in Europe (DATE)*, 2004, pp. 1276–1283.

[36] O. Wahlen, M. Hohenauer, R. Leupers, and H. Meyr, Instruction scheduler generation for retargetable compilation. *IEEE Design Test Comput.*, 20, 34–41, 2003.

[37] C. Siska, A processor description language supporting retargetable multi-pipeline DSP program development tools, *Proceedings of International Symposium on System Synthesis (ISSS)*, 1998, pp. 31–36.

[38] T. Morimoto, K. Yamazaki, H. Nakamura, T. Boku, and K. Nakazawa, Superscalar processor design with hardware description language aidl, *Proceedings of Asia Pacific Conference on Hardware Description Languages (APCHDL)*, 1994.

[39] A. Inoue, H. Tomiyama, F. Eko, H. Kanbara, and H. Yasuura, A programming language for processor based embedded systems, *Proceedings of Asia Pacific Conference on Hardware Description Languages (APCHDL)*, 1998, pp. 89–94.

[40] V. Rajesh and R. Moona, Processor modeling for hardware software codesign, *Proceedings of International Conference on VLSI Design*, 1999, pp. 132–137.

[41] ARC Cores. *http://www.arccores.com.*

[42] http://www.axysdesign.com. *Axys Design Automation.*

[43] Tensilica Inc. *http://www.tensilica.com.*

[44] P. Marwedel and G. Goossens, *Code Generation for Embedded Processors*, Kluwer Academic Publishers, Dordrecht, 1995.

[45] J. Zhu and D. Gajski, A retargetable, ultra-fast, instruction set simulator, *Proceedings of Design Automation and Test in Europe (DATE)*, 1999.

[46] R. Cmelik and D. Keppel, Shade: a fast instruction-set simulator for execution profiling. *ACM SIGMETRICS Performance Evaluation Rev.*, 22, 128–137, 1994.

[47] E. Witchel, and M. Rosenblum, Embra: fast and flexible machine simulation, *Measurement and Modeling of Computer Systems*, 1996, pp. 68–79.

[48] A. Nohl, G. Braun, O. Schliebusch, R. Leupers, H. Meyr, and A. Hoffmann, A universal technique for fast and flexible instruction-set architecture simulation, *Proceedings of Design Automation Conference (DAC)*, 2002, pp. 22–27.

[49] M. Reshadi, P. Mishra, and N. Dutt, Instruction set compiled simulation: a technique for fast and flexible instruction set simulation, *Proceedings of Design Automation Conference (DAC)*, 2003, pp. 758–763.

[50] M. Reshadi, N. Bansal, P. Mishra, N. Dutt, An efficient retargetable framework for instruction-set simulation, *Proceedings of International Symposium on Hardware/Software Codesign and System Synthesis (CODES+ISSS)*, 2003, pp. 13–18.

[51] http://www.improvsys.com. *Improv Inc.*

[52] M. Itoh, S. Higaki, Y. Takeuchi, A. Kitajima, M. Imai, J. Sato, and A. Shiomi, PEAS-III: an ASIP design environment, in *Proceedings of International Conference on Computer Design (ICCD)*, 2000.

[53] M. Itoh, Y. Takeuchi, M. Imai, A. Shiomi, Synthesizable HDL generation for pipelined processors from a micro-operation description, *IEICE Trans. Fundamentals*, E00-A(3), March 2000.

[54] P. Mishra, Specification-Driven Validation of Programmable Embedded Systems, Ph.D. thesis, University of California, Irvine, 2004.

[55] P. Mishra and N. Dutt, Automatic modeling and validation of pipeline specifications, *ACM Transactions on Embedded Computing Systems (TECS)*, 3, 114–139, 2004.

[56] P. Mishra, N. Dutt, and H. Tomiyama, Towards automatic validation of dynamic behavior in pipelined processor specifications, *Kluwer Design Automat.Embedded Syst. (DAES)*, 8, 249–265, 2003.

[57] P. Mishra, and N. Dutt, Graph-based functional test program generation for pipelined processors, *Proceedings of Design Automation and Test in Europe (DATE)*, 2004, pp. 182–187.

[58] P. Mishra, and N. Dutt, N. Krishnamurthy, and M. Abadir, A top-down methodology for validation of microprocessors, *IEEE Design Test Comput.*, 21, 122–131, 2004.

[59] A. Nohl, V. Greive, G. Braun, A. Hoffmann, R. Leupers, O. Schliebusch, and H. Meyr, Instruction encoding synthesis for architecture exploration using hierarchical processor models, *Proceedings of Design Automation Conference (DAC)*, 2003, pp. 262–267.

[60] P. Biswas, and N. Dutt, Reducing code size for heterogeneous-connectivity-based VLIW DSPs through synthesis of instruction set extensions, *Proceedings of Compilers, Architectures, Synthesis for Embedded Systems (CASES)*, 2003, pp. 104–112.

[61] M. Hartoog, J. Rowson, P. Reddy, S. Desai, D. Dunlop, E. Harcourt, and N. Khullar, Generation of software tools from processor descriptions for hardware/software codesign, *Proceedings of Design Automation Conference (DAC)*, 1997, pp. 303–306.

9

Embedded Software Modeling and Design

Marco Di Natale
Scuola Superiore S. Anna
Pisa, Italy

Abstract

The development of correct complex software for reactive embedded systems requires automated support for the verification of functional and nonfunctional properties by formal analysis or by simulation and testing. Reduced time-to-market and increased complexity also demand the reuse of components, which also brings the issue of composability of software artifacts. Currently, a language (or a design methodology) that can provide all these desirable features without incurring excessive inefficiency is not available, and a strong semantic characterization of the modeling language or separation of the concerns between the functional and the architecture-level design is the solution advocated by many. This chapter provides an overview of existing models and tools for embedded software, starting from an introduction to the fundamental concepts and the basic theory of existing models of computation, both synchronous and asynchronous. The chapter also features an introduction to commercial languages and tools such as Unified Modeling Language (UML) and Specification and Description Language (SDL), and a quick peek at research work in software models and tools, to give a firm understanding of future trends and currently unsolved issues.

9.1 Introduction

The increasing cost necessary for the design and fabrication of ASICs, together with the need for the reuse of functionality, adaptability, and flexibility, are among the causes for an increasing share of software-implemented functions in embedded projects. Figure 9.1 represents a typical architectural framework for embedded systems, where application software runs on top of a real-time operating system (RTOS) (and possibly a middleware layer) which abstracts from the hardware and provides a common API for the reuse of functionality (such as the OSEK standard in the automotive domain).

Unfortunately, mechanisms for improving the reuse of software at the level of programming code, such as RTOS- or middleware-level APIs, are still not sufficient to achieve the desired levels of productivity, and the error rate of software programs is exceedingly high. Today, model-based design of software bears the promise of a much-needed step-up in productivity and reuse.

The use of abstract software models at the highest possible level in the development process may significantly increase the chances that the design and its implementation are correct. Correctness can be achieved in many ways. Ideally, it should be mathematically proved by formal reasoning upon the model of the system and its desired properties (provided the model is built on solid mathematical foundations and its properties are expressed by some logic function(s)). Unfortunately, in many cases, formal model checking is not possible or simply impractical. In this case, the modeling language should at least provide abstractions for the specification of reusable components, so that software artifacts can be clearly identified with the provided functions or services.

Also, when exhaustive proof of correctness cannot be achieved, a secondary goal of the modeling language should be providing support for simulation and testing. In this view, formal methods can also be used to guide the generation of the test suite and to guarantee coverage. Finally, modeling languages and tools should ensure that the model of the software, after being checked by means of formal proof or by simulation, is correctly implemented in a programming language executed on the target hardware (this requirement is usually satisfied by automatic code generation tools).

Industry and research groups have been working now for decades in the software engineering area looking for models, methodologies, and tools to improve the correctness and increase the reusability of software components. Traditionally, software models and formal specifications have had their focus on behavioral properties and have been increasingly successful in the verification of functional correctness. However, embedded software is characterized by concurrency and resource constraints and by nonfunctional properties, such as deadlines or other timing constraints, which ultimately depend upon the computation platform.

This chapter attempts at providing an overview of (visual and textual) languages and tools for embedded software modeling and design. The subject is so wide, evolving so rapidly, and encompassing so many different issues that only a short survey is possible in the limited space allocated to this chapter. As of today, despite all efforts, existing methodologies and languages fall short in achieving most of the desirable goals and yet they are continuously being extended in order to allow for the verification of at least a subset of the properties of interest.

The objective is to provide the reader with an understanding of what are the principles for functional and nonfunctional modeling and verification, what are the languages and tools available in the market and what can be possibly achieved with respect to practical designs. The description of (some) commercial languages, models, and tools is supplemented with a survey of the main research trends and the

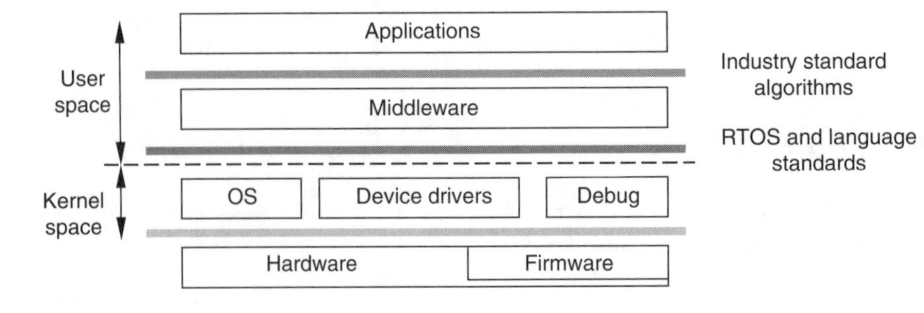

FIGURE 9.1 Common architecture for embedded software.

results that may open new possibilities in the future. The reader is invited to refer to the cited papers in the bibliography section for wider discussion on each issue.

The organization of the chapter is as follows: the introduction section defines a reference framework for the discussion of the software-modeling problem and provides a short review of abstract models for functional and temporal (schedulability) analysis. The second section provides a quick glance at the two main categories of available languages and models purely synchronous as opposed to general asynchronous models. Then, an introduction to the commercial modeling languages Unified Modeling Language (UML) and Specification and Description Language (SDL) is provided. The discussion of what can be achieved with both, with respect to formal analysis of functional properties, schedulability analysis, simulation and testing follows. This chapter also discusses the recent extensions of existing methodologies to achieve the desirable goal of component-based design. Finally, a quick glance at the research work in the area of embedded software design, methods, and tools, closes the chapter.

9.1.1 Challenges in the Development of Embedded Software

According to a typical development process (represented in Figure 9.2), an embedded system is the result of possibly multiple refinement stages encompassing several levels of abstraction, from user requirements, to system testing, and sign-off. At each stage, the system is described using an adequate formalism, starting from abstract models for the user domain entities at the requirements level. Lower-level models, typically developed in the later stages, provide an implementation of the abstract model by means of design entities representing hardware and software components. The implementation process is a sequence of steps that constrain the generic specification by exploiting the possible options (nondeterminism) available from higher levels.

The designer's problem is making sure that the models of the system developed at the different stages do satisfy the properties required from the system, and that low-level descriptions of the system are correct implementations of higher-level specifications. This task can be considerably easier if the models of the system at different abstraction levels are homogeneous, i.e., if the computational models on which they are based share a common semantics and, possibly, a common notation.

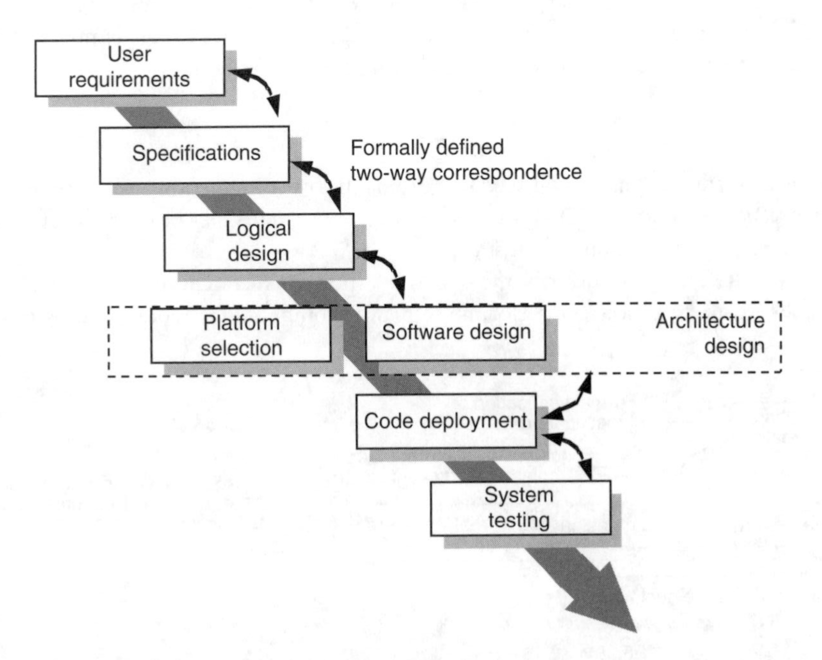

FIGURE 9.2 Typical embedded software development process.

The problem of correct mapping from a high-level specification, employing an abstract model of the system, to a particular software and hardware architecture or *platform* is one of the key aspects of the design of embedded systems.

The separation of the two main concerns of functional and architectural specification and the mapping of functions to architecture elements are among the founding principles of many design methodologies, such as the platform-based design [1] and tools like the Ptolemy and Metropolis frameworks [2,3], as well as of emerging standards and recommendations, such as the UML profile for schedulability, performance, and time (SPT) from the Object Management Group (OMG) [4] and industry-based practices, such as the V-cycle of software development common in the automotive industry [5]. A keyhole view of the corresponding stages is represented in Figure 9.3.

The main design activities taking place at this stage and the corresponding challenges can be summarized as follows:

- *The specification of functionality* is concerned with the development of logically correct system abstractions. If the specification is defined using a formal model, *formal verification* allows checking that the functional behavior satisfies a given set of properties.
- The *system software and hardware* platform components are defined in the *architecture design* level.
- After the definition of logical and physical resources available for the execution of the functional model and the definition of the *mapping of functional model elements onto the platform (architecture elements)* executing them, *formal verification of nonfunctional properties*, such as timing properties, and schedulability analysis usually take place.

Complementing the above two steps, *implementation verification* is the process of checking that the behavior of the logical model, when mapped onto the architecture model, correctly implements the high-level formal specifications.

9.1.2 A Short Introduction to Formal Models and Languages and to Schedulability Analysis

A short review of the most common models of computation (formal languages) proposed by academia and possibly supported by the industry with the objective of formal or simulation-based verification is fundamental for understanding commercial models and languages, and it is also important for understanding present and future challenges (refer to [6,7] for more details).

9.1.2.1 Formal Models

Formal models are formal languages that specify the semantics of computation and communication (also defined as *model of computation* [MOC] [6]). Models of computations may be expressed, for example, by means of a language or automaton formalisms.

A system-level MOC is used to describe the system as a (possibly hierarchical) collection of design entities (blocks, actors, tasks, processes) performing units of computations represented as transitions or

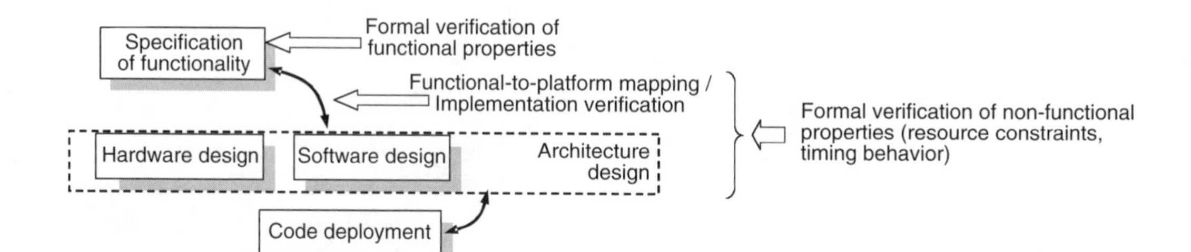

FIGURE 9.3 Mapping formal specifications to a hardware/software platform.

actions, characterized by a state and communicating by means of events (tokens) carried out by signals. Composition and communication rules, concurrency models, and time representation are among the most important characteristics of an MOC.

Once the system specifications are given according to a formal MOC, *formal methods* can be used to achieve design-time verification of *properties* and *implementation* as in Figure 9.3. In general, properties of interest go under the two general categories of *ordered* and *timed executions*:

- Ordered execution relates to verification of event and state ordering. Properties such as *safety*, *liveness*, *absence of deadlock*, *fairness*, or *reachability*, belong to this category.
- Timed execution relates to event enumeration, such as checking that no more that *n* events (including time events) occur between any two events in the system, for example, *timeliness* and some notions of *fairness*.

Verification of desirable system properties may be quite hard or even impossible to achieve by logical reasoning on formal models. Formal models are usually classified according to the *decidability* of properties. Decidability of properties in timed and untimed models depends on many factors, such as the type of logic (propositional or first-order) for conditions on transitions and states, the real-time semantics, including the definition of the time domain (discrete or dense) and the linear or branching time logic that is used for expressing properties (interested readers may refer to [7] for a survey on the subject). In practice [6], decidability should be carefully evaluated. In some cases, even if it is decidable, the problem cannot be practically solved since the required run time may be prohibitive and, in other instances, even if undecidability applies to the general case, it may happen that the problem at hand admits a solution.

Verification of model properties can take many forms. In the *deductive* approach, the system and the property are represented by statements (clauses) written in some logic (for example, expressed in the linear time temporal logic (LTL) [8] or in the branching time Computation Tree Logic (CTL) [9]) and a theorem proving tool (usually under the direction of a designer or some expert) applies deduction rules until (hopefully) the desired property reduces to a set of axioms, or a counterexample is found. In model checking, the system and possibly the desirable properties are expressed using an automaton or some other kind of executable formalism. The verification tool ensures that no executable transition or reachable system state violates the property. To do so, it can generate all the potential (finite) states of the system (exhaustive analysis). When the property is violated, the tool usually produces the (set of) counterexample(s).

The first model checkers worked by constructing the whole structure of states prior to property checking, but modern tools are able to perform verification as the states are produced. This means that the method does not necessarily require the construction of the whole state graph. On the fly model checking and the SPIN [10] toolset provide an instance and an implementation of this approach, respectively.

To give some examples (Figure 9.4), checking a system implementation I against a specification of a property P in case both are expressed in terms of automata (*homogeneous verification*) requires the following steps. The implementation automaton A_I is composed with the complementary automaton A_P expressing the negation of the desired property. The implementation I violates the specification property if the product automaton $A_I \parallel A_P$ has some possible run and it is verified if the composition has no runs. *Checking by observers* can be considered as a particular instance of this method, very popular for synchronous models.

In the very common case in which the property specification consists of a logical formula and the implementation of the system is given by an automaton, the verification problem can be solved algorithmically or deductively by transforming it into an instance of the previous cases, for example, by transforming the negation of a specification formula f_S into the corresponding automaton, and by using the same techniques as in homogeneous verification.

Verification of implementation is usually obtained by exploiting simulation and bisimulation properties.

Following is a very short survey of formal system models, starting with finite-state machines (FSMs), probably the most popular and the basis for many extensions.

In FSMs process behavior is specified by enumerating the (finite) set of possible system states and the transitions among them. Each transition connects two states and it is labeled with the subset of input

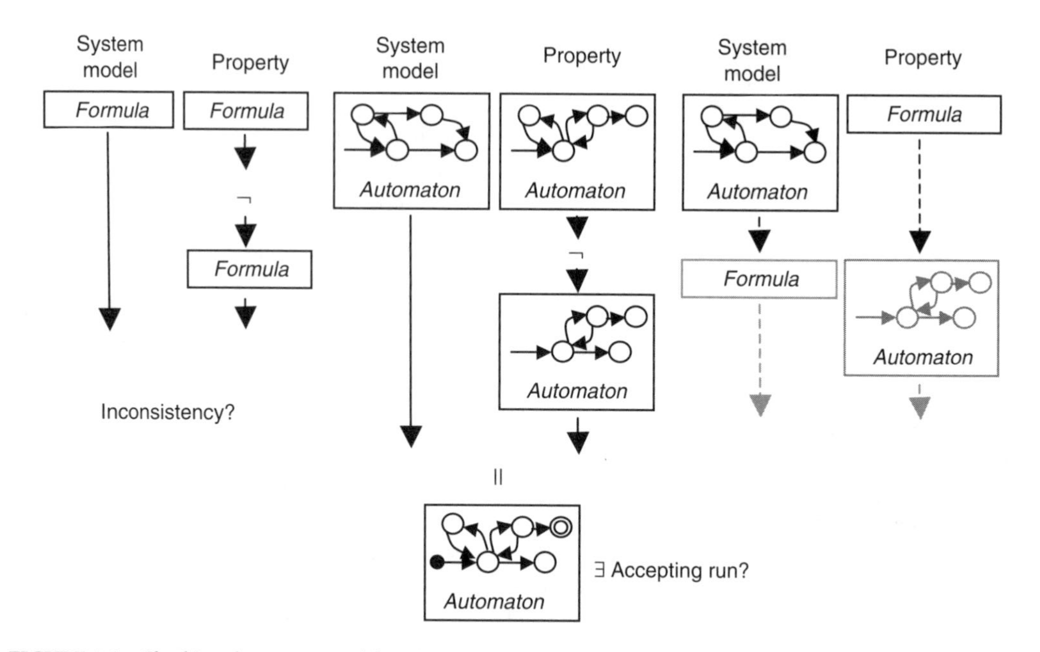

FIGURE 9.4 Checking the system model against some property.

variables (and possibly the guard condition upon their values) that triggers its execution. Furthermore, each transition can produce output variables. In Mealy FSMs, outputs depend on both state and input variables, while in the Moore model, outputs only depend on the process state. Guard conditions can be expressed according to different logics, such as propositional logic, first-order logic, or even Turing-complete programming code.

In the *Synchronous* FSM model, signal propagation is assumed to be instantaneous. Transitions and the evaluation of the next state happen for *all the system components* at the same time. Synchronous languages such as Esterel and Lustre are based on this model. In the *Asynchronous* model, two asynchronous FSMs never execute a transition at the same time except when rendezvous is explicitly specified (a pair of transitions of the communicating FSMs occur simultaneously). The SDL process behavior is an instance of this general model.

Composition of FSMs is obtained by the construction of a product transition system, i.e., a single FSM where the set of states is the cartesian product of the sets of the states of the component machines. The difference between synchronous and asynchronous execution semantics is quite clear when compositional behaviors are compared. Figure 9.5 portrays an example showing the differences in the synchronous and asynchronous composition of two FSMs.

When there is a cyclic dependency among variables in interconnected synchronous FSMs, the Mealy model like any other model where outputs are instantaneously produced based on the input values, may result in a fixed-point problem and possibly inconsistency (Figure 9.6 shows a simple functional dependency). The existence of a unique fixed-point solution (and its evaluation) is not the only problem resulting from the composition of FSMs.

In large, complex systems, composition may easily result in a huge number of states. This phenomenon is known as *state explosion*.

In the *Statecharts* extension to FSMs [11], Harel proposed three mechanisms to reduce the size of an FSM for modeling practical systems: state hierarchy, simultaneous activity, and nondeterminism. In *Statecharts*, a state can possibly represent an enclosed state machine. In this case, the machine is in one of the states enclosed by the super state(s) and concurrency is achieved by enabling two or more state machines to be active simultaneously (and-states, such as lap and off in Figure 9.7).

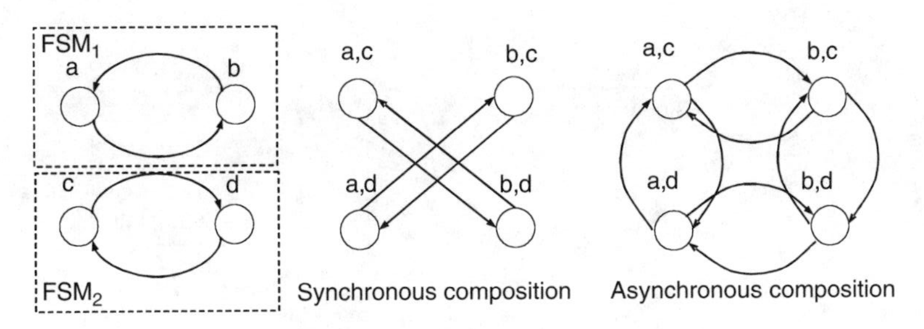

FIGURE 9.5 Composition of synchronous and asynchronous FSMs.

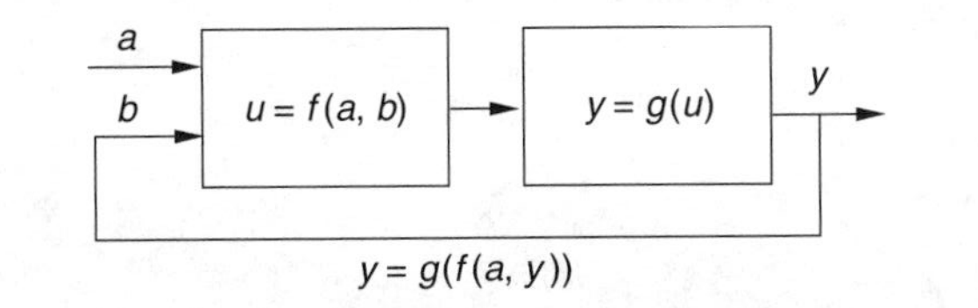

FIGURE 9.6 A fixed-point problem arising from the composition and cyclic dependencies.

In *Petri net* (PN) models, the system is represented by a graph of places connected by transitions. The places represent unbounded channels that carry *tokens*, and the state of the system is represented at any given time by the number of tokens existing in a given subset of places. Transitions represent the elementary reactions of the system. A transition can be executed (fired) when it has a fixed, prespecified number of tokens in its input places. When fired, it consumes the input tokens and produces a fixed number of tokens on its output places. Since more than one transition may originate from the same place, one transition can execute while disabling another one by removing the tokens from the shared input places. Hence, the model allows for non-determinism and provides a natural representation of concurrency by allowing simultaneous execution of multiple transitions (Figure 9.8, left).

The FSM and PN models have originally been developed with no reference to time or time constraints, but the capability of expressing and verifying timing requirements is key in many design domains (including embedded systems). Hence, both have been extended in order to allow time-related specifications. Time extensions differ according to the time model that is assumed. Models that represent time with a discrete-time base are said to belong to the family of discrete-time models, while the others are based on continuous (dense) time. Furthermore, proposals of extensions differ in how time references should be used in the system, whether a global clock or local clocks should be used, and how time should be used in guards on transitions or states, inputs and outputs.

When building the set of reachable states, the time value adds another dimension, further contributing to the state explosion problem. In general, discrete-time models are easier to analyze if compared with dense time models, but synchronization of signals and transitions results in fixed-point evaluation problems whenever the system model contains cycles without delays.

Note that discrete-time systems are naturally prone to an implementation based on the *time-triggered* paradigm, where all actions are bound to happen at multiples of a time reference (usually implemented by means of a response to a timer interrupt), and continuous-time (asynchronous systems) conventionally corresponds to implementations based on the event-based design paradigm, where system actions can happen at any time instant. This does not imply a correspondence between time-triggered systems and synchronous systems. The latter are characterized by the additional constraints that all system components must perform an action synchronously (at the same time) at each tick in a periodic time base.

Many models have been proposed in research literature for time-related extensions. Among those, TPN [12,13] and timed automata (TA) [14] are probably the best known.

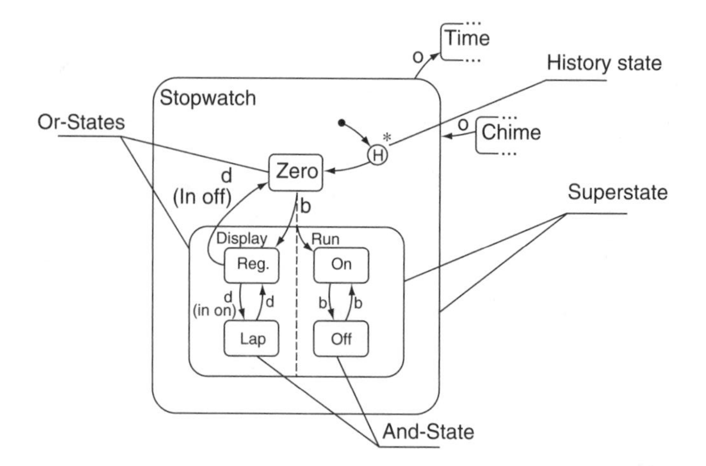

FIGURE 9.7 An example of statechart.

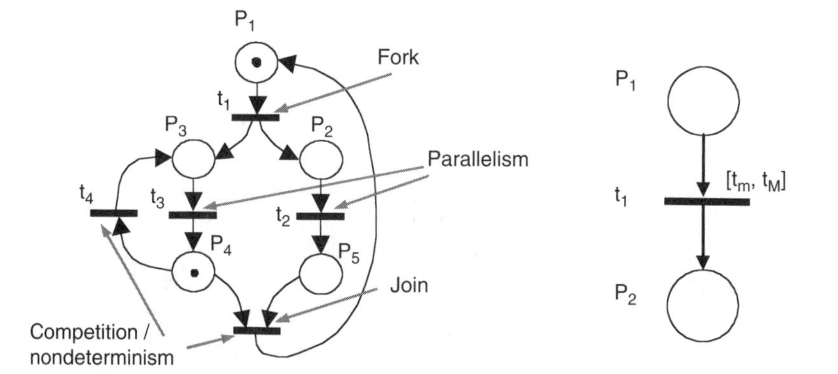

FIGURE 9.8 A sample PN showing examples of concurrent behavior and non-determinism (left) and notations for the time Petri net (TPN) model (right).

Timed automata (an example is given in Figure 9.9) operate with a finite set of locations (states) and a finite set of real-valued clocks. All clocks proceed at the same rate and measure the amount of time that passed since they were started (reset). Each transition may reset some of the clocks and each defines a restriction on the value of the symbols as well as on the clock values required for it to happen. A state may be reached only if the values of the clocks satisfy the constraints and the proposition clause defined on the symbols evaluates to true.

Timed Petri Nets (TDPN) [15] and Time Petri Nets (TPN) are two different, albeit similarly named, models that extend the Petri net formalism allowing for the expression of time-related constraints. The two differ in the way time advances. In timed Petri nets time advances in transitions, thus violating the instantaneous nature of transitions (which makes the model much less prone to analysis). In the TPN model, time advances while token(s) are in places (Figure 9.10). Enabling and deadline times can be associated to transitions, the enabling time being the time a transition must be enabled before firing, and the deadline being the time instant by which the transition must be taken (Figure 9.8, right).

The additional notion of stochastic time allows the definition of the (generalized) stochastic PNs [16,17] used for the purpose of performance evaluation.

Many further extensions have been proposed for both TAs and TPNs. The task of comparing the two models for expressiveness should take into account all the possible variants and is probably not particularly

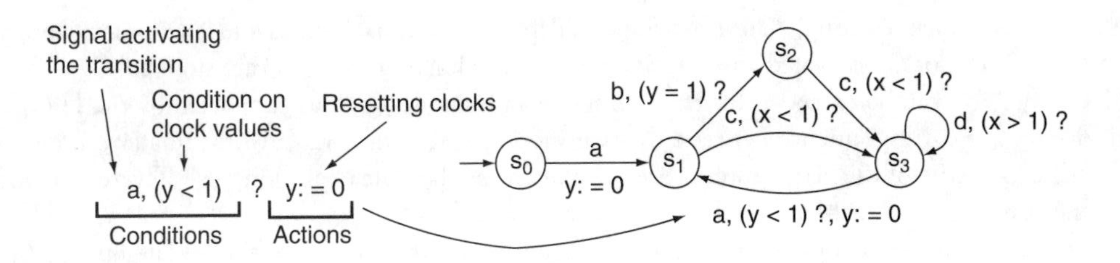

FIGURE 9.9 An example of TA.

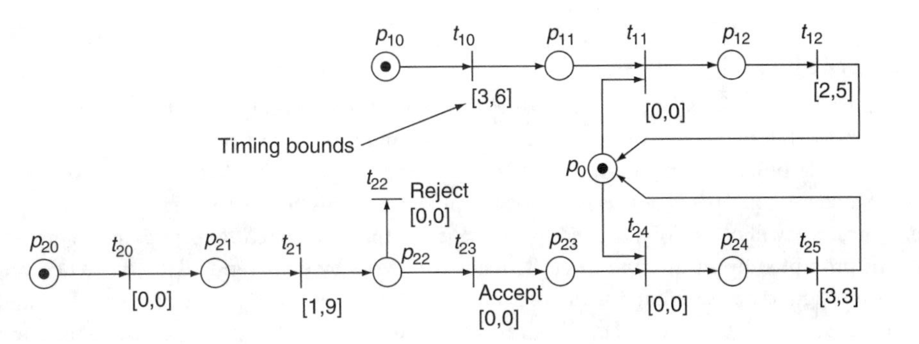

FIGURE 9.10 A sample time Petri net.

interesting in itself. For most problems of practical interest, however, both models are essentially equivalent when it comes to expressive power and analysis capability [18].

A few tools based on the TA paradigm have been developed and are very popular. Among those, we cite Kronos [19] and Uppaal [20]. The Uppaal tool allows modeling, simulation, and verification of real-time systems modeled as a collection of nondeterministic processes with finite control structure and real-valued clocks, communicating through channels or shared variables [20,21]. The tool is free for non-profit and academic institutions.

TAs and TPNs allow the formal expression of the requirements for logical-level resources, timing constraints, and timing assumptions, but timing analysis only deals with abstract specification entities, typically assuming infinite availability of physical resources (such as memory or CPU speed). If the system includes an RTOS with the associated scheduler, the model needs to account for preemption, resource sharing and the non-determinism resulting from them. Dealing with these issues requires further evolution of the models.

For example, in TA, we may want to use clock variables for representing the execution time of each action. In this case, however, only the clock associated with the action scheduled on the CPU should advance, with all the others being stopped.

The hybrid automata model [22] combines discrete transition graphs with continuous dynamical systems. The value of system variables may change according to a discrete transition or it may change continuously in system states according to a trajectory defined by a system of differential equations. Hybrid automata have been developed for the purpose of modeling digital systems interacting with (physical) analog environments, but the capability of stopping the evolution of clock variables in states (first derivative equal to 0) makes the formalism suitable for the modeling of systems with preemption.

Time Petri Nets and TA can also be extended to cope with the problem of modeling finite computing resources and preemption. In the case of TA, the extension consists of the Stopwatch Automata model, which handles suspension of the computation due to the release of the CPU (because of real-time scheduling), implemented in the HyTech [23] for linear hybrid automata. Alternatively, the scheduler is modeled with an extension to the TA model, allowing for clock updates by subtraction inside transitions (besides normal clock resetting). This extension, available in the Uppaal tool, avoids the undecidability of the model where clocks associated with the actions not scheduled on the CPU are stopped.

Likewise, TPNs can be extended to the preemptive TPN model [24], as supported by the ORIS tool [25]. A tentative correspondence between the two models is traced in [26]. Unfortunately, in all these cases, the complexity of the verification procedure caused by the state explosion poses severe limitations upon the size of the analyzable systems.

Before moving on to the discussion of formal techniques for the analysis of time-related properties at the architecture level (schedulability), the interested reader is invited to refer to [27] for a survey on formal methods, including references to industrial examples.

9.1.2.2 Schedulability Analysis

If specification of functionality aims at producing a logically correct representation of system behavior, architecture-level design is where physical concurrency and schedulability requirements are expressed. At this level, the units of computation are the *processes* or *threads* (the distinction between these two operating systems concepts is not relevant for the purpose of this chapter and in the following text, the generic term *task* will be optionally used for both), executing concurrently in response to environment stimuli or prompted by an internal clock. Threads cooperate by exchanging data and synchronization or activation signals and contend for the use of the execution resource(s) (the processor) as well as for the other resources in the system. The physical architecture level is also the place, where the concurrent entities are mapped onto target hardware. This activity entails the selection of an appropriate scheduling policy (for example, offered by an RTOS), and possibly support by timing or schedulability analysis tools.

Formal models, exhaustive analysis techniques, and model checking are now evolving toward the representation and verification of time and resource constraints together with the functional behavior. However, applicability of these models is strongly limited by state explosion. In this case, exhaustive analysis and joint verification of functional and nonfunctional behavior can be sacrificed for the lesser goal of analyzing only the *worst-case timing* behavior of coarse-grain design entities representing concurrently executing threads.

Software models for time and schedulability analysis deal with preemption, physical and logical resource requirements, and resource management policies and are typically limited to a somewhat simplified view of functional (logical) behavior, mainly limited to synchronization and activation signals.

To give an example if, for the sake of simplicity, we limit discussion to single processor systems, the scheduler assigns the execution engine (CPU) to threads (tasks) and the main objective of the real-time scheduling policies is to formally guarantee the timing constraints (deadlines) on the thread response to external events.

In this case, the software architecture can be represented as a set of concurrent tasks (threads). Each task τ_i executes periodically or according to a sporadic pattern and it is typically represented by a simple set of attributes, such as the tuple $(C_i, \theta_i, p_i, D_i)$ representing the worst-case computation time, the period (for periodic threads) or the minimum inter-arrival time (for sporadic threads), the priority, and the relative (to the release time r_i) deadline of each thread instance.

Fixed priority scheduling and rate monotonic analysis (RMA) [28,29] are by far the most common real-time scheduling and analysis methodologies. Rate monotonic analysis provides a very simple procedure for assigning static priorities to a set of *independent periodic tasks* together with a formula for checking schedulability against deadlines.

The highest priority is assigned to the task having the highest rate, and schedulability is guaranteed by checking the worst-case scenario that can possibly happen. If the set of tasks is schedulable in that condition, then it is schedulable under all circumstances. For RMA the critical condition happens when all the tasks are released at the same time instant, initiating the largest busy period (continuous time interval when the processor is busy executing tasks of a given priority level).

By analyzing the busy period (from $t = 0$), it is possible to derive the worst-case completion time W_i for each task τ_i. If the task can be proven to complete before or at the deadline ($W_i \leq D_i$), then it can be guaranteed. The iterative formula for computing W_i (in case $\theta_i \leq D_i$) is

$$W_i = C_i + \sum_{\forall j \in he(i)} \frac{W_i}{\theta_j} C_j$$

where *he(i)* are the indices of those tasks having a priority higher than or equal to p_i.

Rate monotonic scheduling was developed starting from a very simple model, where all tasks are periodic and independent. In reality, tasks require access to shared resources (apart from the processor) that can only be used in an exclusive way, such as, for example, communication buffers shared among asynchronous threads.

In this case, it is possible that one task is blocked because another task holds a lock on the shared resources. When the blocked task has a priority higher than the blocking task, *priority inversion* occurs and finding the optimal priority assignment becomes an NP-hard problem. Real-time scheduling theory settles at finding resource assignment policies that provide at least a worst-case bound upon the blocking time. The priority inheritance (PI) and the (immediate) priority ceiling (PC) protocols [30] belong to this category.

The essence of the PC protocol (which has been included in the RTOS OSEK standard issued by the automotive industry) consists in raising the priority of a thread entering a critical section to the highest among the priorities of all the threads that may possibly request access to the same critical section. The thread returns to its nominal priority as soon as it leaves the critical section. The PC protocol ensures that each thread can be blocked at most once, and bounds the duration of the blocking time to the largest critical section shared between itself or higher priority threads and lower priority threads.

When the blocking time due to priority inversion is bound for each task and its worst-case value is B_i, the evaluation of the worst-case completion time in the schedulability test becomes

$$W_i = C_i + \sum_{\forall j \in he(i)} \frac{W_i}{\theta_j} C_j + B_i$$

9.1.2.3 Mapping the Functional Model into the Architectural Model

The mapping of the actions defined in the functional model onto architectural model entities is the critical design activity where the two views are reconciled. In practice, the actions or transitions defined in the functional part must be executed in the context of one or more system threads. The definition of the architecture model (number and attributes of threads) and the selection of resource management policies, the mapping of the functional model into the corresponding architecture model and the validation of the mapped model against functional and nonfunctional constraints is probably one of the major challenges in software engineering.

Single-thread implementations are quite common and are probably the only choice that allows for (practical) verification of implementation and schedulability analysis, meaning that there exist CASE tools that can provide both according to a given MOC. The entire functional specification is executed in the context of a single thread performing a never-ending cycle where it serves events in a noninterruptable fashion according to the run-to-completion paradigm. The thread waits for an event (either external, like an interrupt from an I/O interface, or internal, like a call or signal from one object or FSM to another), fetches the event and the associated parameters, and, finally, executes the corresponding code.

All the actions defined in the functional part need be scheduled (statically or dynamically) for execution inside the thread. The schedule is usually driven by the partial order in the execution of the actions, as defined by the MOC semantics. Commercial implementations of this model range from the code produced by the Esterel compiler [31] to single-thread implementations by Embedded Coder toolset from Mathworks and TargetLink from DSpace (of Simulink models) [32,33] or the single-thread code generated by Rational Rose Technical Developer [34] for the execution of UML models.

The scheduling problem is much simpler than it is in the multithreaded case, since there is no need to account for thread scheduling and preemption and resource sharing usually result in trivial problems.

On the other extreme, one could define *one thread for every functional block* or every possible action. Each thread can be assigned its own priority, depending on the criticality and on the deadline of the corresponding action. At run time, the operating system scheduler properly synchronizes and sequentializes the tasks so that the order of execution respects the functional specification.

Both approaches may easily prove inefficient. The single-thread implementation suffers from large priority inversion due to the need for completing the processing of each event before fetching the next event in the global queue. The one-to-one mapping of the functions or actions to threads suffers from excessive scheduler overhead caused by the need for a context switch at each action. Considering that the action specified in a functional block can be very short and that the number of functional blocks is usually quite high (in many applications it is of the order of hundreds), the overhead of the operating system could easily prove unbearable.

The designer essentially tries to achieve a compromise between these two extremes, balancing responsiveness with schedulability, flexibility of the implementation, and performance overhead.

9.1.3 Paradigms for Reuse: Component-Based Design

One more dimension can be added to the complexity of the software design problem if the need for maintenance and reuse is considered. To this purpose, component-based and object-oriented (OO) techniques have been developed for constructing and maintaining large and complex systems.

A component is a product of the analysis, design, or implementation phases of the lifecycle and represents a prefabricated solution that can be reused to meet (sub)system requirement(s). A component is commonly used as a vehicle for the reuse of two basic design properties:

- *Functionality.* The functional syntax and semantics of the solution that the component represents.
- *Structure.* The structural abstraction that the component represents. These can range from "small grain" to architectural features, at the subsystem or system level.

The generic requirement for "reusability" maps into a number of issues. Probably the most relevant property that components should exhibit is *abstraction* meaning the capability of hiding implementation details and describing relevant properties only. Components should also be easily adaptable to meet the changing processing requirements and environmental constraints through controlled modification techniques (like *inheritance* and *genericity*). In addition, *composition* rules must be used to build higher-level components from existing ones. Hence, an ideal component-based modeling language should ensure that properties of components (functional properties, such as liveness, reachability, and deadlock avoidance, or nonfunctional properties such as timeliness and schedulability) are preserved or at least decidable after composition. Additional (practical) issues include support for implementation, separate compilations, and imports.

Unfortunately, reconciling the standard issues of software components, such as context independence, understandability, adaptability, and composability, with the possibly conflicting requirements of timeliness, concurrency, and distribution, typical of hard real-time system development is not an easy task and is still an open problem.

Object-oriented design of systems has traditionally embodied the (far from perfect) solution to some of these problems. While most (if not all) OO methodologies, including the UML, offer support for inheritance and genericity, adequate abstraction mechanisms, and especially composability of properties, are still subject of research.

With its latest release, UML has reconciled the abstract interface abstraction mechanism with the common box-port-wire design paradigm. The lack of an explicit declaration of the required interface and the absence of a language feature for structured classes were among the main deficiencies of classes and objects, if seen as components. In UML 2.0, ports allow a formal definition of a required as well as a provided interface. The association of protocol declaration with ports further improves the clarification of the semantics of interaction with the component. In addition, the concept of a structured class allows a much better definition of a component.

Of course, port interfaces and the associated protocol declarations are not sufficient for specifying the semantics of the component. In UML 2.0, OCL can also be used to define behavioral specifications in the form of invariants, preconditions and post-conditions, in the style of the contract-based design methodology (implemented in Eiffel [35]).

9.2 Synchronous vs. Asynchronous Models

The verification of the functional and nonfunctional properties of software demands for formal semantics and a strong mathematical foundation of the models. Many argue that a fully analyzable model cannot be constructed unless shedding generality and restricting the behavioral model to simple and analyzable semantics.

Among the possible choices, the *synchronous reactive* model enforces determinism and provides a sound methodology for checking functional and nonfunctional properties at the price of expensive implementation and performance limitations. Moreover, the synchronous model is built on assumptions (computation times neglectable with respect to the environmental dynamics and synchronous execution) that do not always apply to the controlled environment and to the architecture of the system.

Asynchronous or general models typically allow for (controlled) non-determinism and more expressiveness, at the cost of strong limitations in the extent of the functional and nonfunctional verification that can be performed.

Some modeling languages, such as UML, are deliberately general enough, so that it is possible to use them for specifying a system according to a generic asynchronous or synchronous paradigm, provided that a suitable set of extensions (semantics restrictions) are defined.

By the end of this chapter, it will hopefully be clear how neither of the two design paradigms (synchronous or asynchronous) is currently capable of facing all the implementation challenges of complex systems. The requirements of the synchronous assumption (on the environment and the execution platform) are difficult to meet, and component-based design is very difficult (if not impossible). The asynchronous paradigm on the other hand, results in implementations which are very difficult to analyze for logical and time behavior.

9.3 Synchronous Models

In the *synchronous reactive model*, time advances at discrete instants and the program progresses according to the successive *atomic reactions* (sets of synchronously executed actions), which are performed instantaneously (zero computation time), meaning that the reaction is fast enough with respect to the environment. The resulting discrete-time model is quite natural to many domains, such as control engineering and (hardware) synchronous digital logic design (VHDL).

The composition of system blocks implies product combination of the states and the conjunction of the reactions for each component. In general, this results in a fixed-point problem and the composition of the function blocks is a relation, not a function, as outlined in Section 9.1.

The French synchronous languages *Signal*, *Esterel*, and *Lustre* are probably the best representatives of the synchronous modeling paradigm.

Lustre [36,37] is a declarative language based on the dataflow model where nodes are the main building blocks. In Lustre, each flow or stream of values is represented by a variable, with a distinct value for each tick in the discrete-time base. A node is a function of flows: it takes a number of typed input flows and defines a number of output flows by means of a system of equations.

A Lustre node (an example is given in Figure 9.11) is a pure functional unit except for the `pre` and initialization (−>) expressions, which allow referencing the previous element of a given stream or forcing an initial value for a stream. Lustre allows streams at different rates, but in order to avoid nondeterminism it forbids syntactically cyclic definitions.

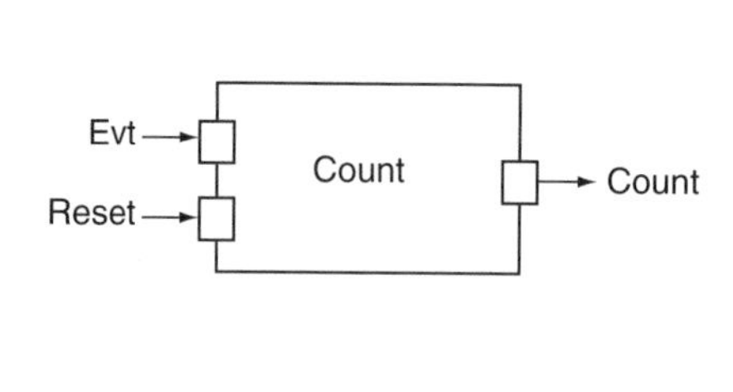


```
Node Count(evt, reset: bool)
    returns(count: int);
Let
    Count = if (true → reset) then 0
             else if evt then pre(count)+1
             else pre(count)
Tel
```

```
Mod60 = Count(second, pre(mod60 = 59));
Minute = (mod60 = 0);
```

FIGURE 9.11 An example of Lustre node and its program.

Esterel [38] is an imperative language, more suited for the description of control. An Esterel program consists of a collection of nested, concurrently running threads. Execution is synchronized to a single, global clock. At the beginning of each reaction, each thread resumes its execution from where it paused (e.g., at a pause statement) in the last reaction, executes imperative code (e.g., assigning the value of expressions to variables and making control decisions), and finally either terminates or pauses, waiting for the next reaction.

Esterel threads communicate exclusively through signals representing globally broadcast events. A signal does not persist across reactions and it is present in a reaction if and only if it is emitted by the program or by the environment.

Esterel allows cyclic dependencies and treats each reaction as a fix-point equation, but the only legal programs are those that behave functionally in every possible reaction. The solution of this problem is provided by *constructive causality* [39], which amounts to checking if, regardless of the existence of cycles, the output of the program (the binary circuit implementing it) can be formally proven to be causally dependent from the inputs for all possible input assignments.

The language allows for conceptually sequential (operator ;) or concurrent (operator ||) execution of reactions, defined by language expressions handling signal identifiers (as in the example of Figure 9.12). All constructs take zero time except `await` and `loop ... each ...`, which explicitly produce a program pause. Esterel includes the concept of preemption, embodied by the `loop ... each R` statement in the example of Figure 9.12 or the `abort` *`action`* `when` *`signal`* statement. The reaction contained in the body of the loop is preempted (and restarted) when the signal R is set. In case of an abort statement, the reaction is preempted and the statement terminates.

Formal verification was among the original objectives of Esterel. In synchronous languages, the verification of properties typically requires the definition of a special program called the *observer*, which observes the variables or signals of interest and at each step decides if the property is fulfilled. A program satisfies the property if and only if the observer never complains during any execution.

The verification tool takes the program implementing the system, an observer of the desired property, and another program modeling the assumptions on the environment. The three programs are combined in a synchronous product, as in Figure 9.13, and the tool explores the set of reachable states. If the observer never reaches a state where the system property is not valid before reaching a state where the assumption observer declares violation of the environment assumptions, then the system is correct. The process is described in detail in [40].

Finally, the commercial package *Simulink* by Mathworks [41] allows modeling and simulation of control systems according to a synchronous reactive model of computation, although its semantics is neither formally nor completely defined. Rules for translating a Simulink model into Lustre have been outlined in [42], while [43] discusses the very important problem of how to map a zero-execution time Simulink semantics into a software implementation of concurrent threads where each computation necessarily requires a finite execution time is discussed.

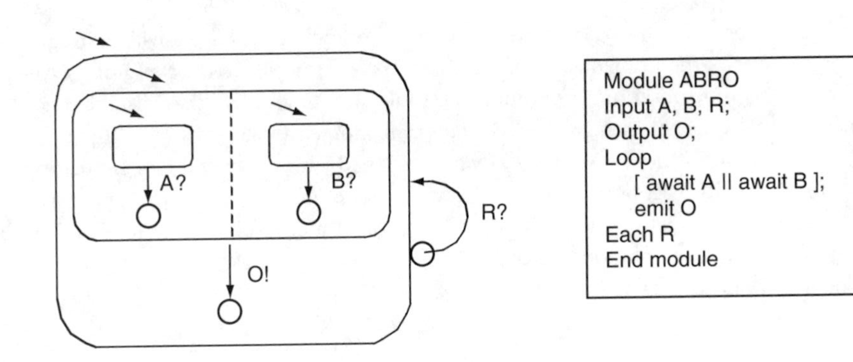

FIGURE 9.12 An example showing features of the Esterel language with an equivalent statechart-like visual formalization.

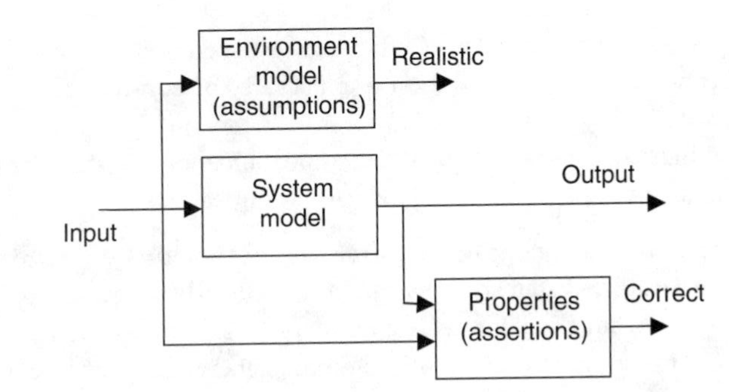

FIGURE 9.13 Verification by observers.

9.3.1 Architecture Deployment and Timing Analysis

Synchronous models are typically implemented as a single task that executes according to an event server model. Reactions decompose into atomic actions that are partially ordered by the causality analysis of the program. The scheduling is generated at compile time, trying to exploit the partial causality order of the functions in order to make the best possible use of hardware and shared resources. The main concern is checking that the synchrony assumption holds, i.e., ensuring that the longest chain of reactions ensuing from any internal or external event is completed within the step duration. Static scheduling means that critical applications are deployed without the need for any operating system (and the corresponding overhead). This reduces system complexity and increases predictability, avoiding preemption, dynamic contention over resources and other non-deterministic operating systems functions.

9.3.2 Tools and Commercial Implementations

Lustre is implemented by the commercial toolset Scade, offering an editor that manipulates both graphical and textual descriptions; two code generators, one of which is accepted by certification authorities for qualified software production; a simulator; and an interface to verification tools such as the plug-in from Prover [44].

The early *Esterel* compilers were developed by Gerard Berry's group at INRIA/CMA and freely distributed in binary form. The commercial version of Esterel was first marketed in 1998 and it is now available from Esterel Technologies, which later acquired the Scade environment. Scade has been used in many industrial projects, including integrated nuclear protection systems (Schneider Electric), flight control software (Airbus A340–600), and track control systems (CS Transport). Dassault Aviation was one of the earliest supporters of the Esterel project, and has long been one of its major users.

Several verification tools use the synchronous observer technique for checking Esterel programs [40]. It is also possible to verify the implementation of Esterel programs with tools exploiting explicit state space reduction and bi-simulation minimization (FC2Tools) and finally, tools can also be used to automatically generate test sequences with guaranteed state/transition coverage.

The very popular *Simulink* tool by Mathworks was developed with the purpose of simulating control algorithms, and has since its inception been extended with a set of additional tools and plug-ins, such as, the Stateflow plug-in for the definition of the FSM behavior of a control block, allowing modeling of hybrid systems, and a number of automatic code generation tools, such as the Realtime Workshop and Embedded Coder by Mathworks and TargetLink by DSpace.

9.3.3 Challenges

The main challenges and limitations that the Esterel language must face when applied to complex systems are the following:

- Despite improvements, the space and time efficiency of the compilers is still not satisfactory.
- Embedded applications can be deployed on architectures or control environments that do not comply with the synchronous reactive model.
- Designers are familiar with other dominant methods and notations. Porting the development process to the synchronous paradigm and languages is not easy.

The efficiency limitations are mainly due to the formal compilation process and the need to check for constructive causality. The first three Esterel compilers used automata-based techniques and produced efficient code for small programs, but they did not scale to large-scale systems because of state explosion. Versions 4 and 5 are based on translation into digital logic and generate smaller executables at the price of slow execution. (The program generated by these compilers wastes time evaluating each gate in every clock cycle.) This inefficiency can produce code 100 times slower than that from the previous compilers [45].

Version 5 of the compiler allows cyclic dependencies by exploiting Esterel's constructive semantics. Unfortunately, this requires evaluating all the reachable states by symbolic state-space traversal [46], which makes it extremely slow.

As for the difficulty in matching the basic paradigm of synchrony with system architectures, the main reasons of concern are:

- the bus and communication lines, if not specified according to a synchronous (time triggered) protocol, and the interfaces with the analog world of sensors and actuators.
- the dynamics of the environment, which can possibly invalidate the instantaneous execution semantics.

The former has been discussed at length in a number of papers (such as [47,48]), giving conditions for providing a synchronous implementation in distributed systems.

Finally, in order to integrate synchronous languages with mainstream commercial methodologies and languages, translation and import tools are required. For example, it is possible from Scade to import discrete-time Simulink diagrams and Sildex allows importing Simulink/Stateflow discrete-time diagrams. Another example is UML with the attempt at an integration between Esterel Studio and Rational Rose, and the proposal for an Esterel/UML coupling drafted by Dassault [49] and adopted by commercial Esterel tools.

9.4 Asynchronous Models

UML and SDL are languages developed in the context of general purpose computing and (large) telecommunication systems respectively. The Unified Modeling Language is the merger of many OO design methodologies aimed at the definition of generic software systems. Its semantics is not completely specified and intentionally retains many variation points in order to adapt to different application domains.

For example, in order to be practically applicable to the design of embedded systems, further characterization (a specialized *profile* in UML terminology) is required. In the 2.0 revision of the language, the system is represented by a (transitional) model where active and passive components, communicating by means of connections through port interfaces, cooperate in the implementation of the system behavior. Each reaction to an internal or external event results in the transition of a *Statechart* automaton describing the object behavior.

The Specification and Description Language (SDL) has a more formal background since it was developed in the context of software for telecommunication systems, for the purpose of easing the implementation of verifiable communication protocols. An SDL design consists of blocks cooperating by means of asynchronous signals. The behavior of each block is represented by one or more (conceptually concurrent) processes. Each process in turn implements an extended FSM.

Until the recent development of the UML profile for Schedulability, Performance and Time (SPT), standard UML did not provide any formal means for specifying time or time-related constraints, or for specifying resources and resource management policies. The deployment diagrams were the only (inadequate) means for describing the mapping of software onto the hardware platform and tool vendors had tried to fill the gap by proposing nonstandard extensions.

The situation with SDL is not very different, although SDL offers at least the notion of global and external time. Global time is made available by means of a special expression and can be stored in variables or sent in messages.

The implementation of asynchronous languages typically (but not necessarily) relies on an operating system. The latter is responsible for scheduling, which is necessarily based on static (design time) priorities if a commercial product is used for this purpose. Unfortunately, as it will be clear in the following, real-time schedulability techniques are only applicable to very simple models and are extremely difficult to generalize to most models of practical interest or even to the implementation model assumed by most (if not all) commercial tools.

9.4.1 Unified Modeling Language

The UML represents a collection of engineering practices that have proven successful in the modeling of large and complex systems, and has emerged as the software industry's dominant OO-modeling language.

Born at Rational in 1994, UML was taken over in 1997 at version 1.1 by the OMG Revision Task Force (RTF), which became responsible for its maintenance. The RTF released UML version 1.4 in September 2001 and a major revision, UML 2.0, which also aims to address the embedded or real-time dimension, has recently been adopted (late 2003) and it is posted on the OMG's website as "UML 2.0 Final Adopted Specification" [4].

The Unified Modeling Language has been designed as a wide-ranging, general-purpose modeling language for specifying, visualizing, constructing, and documenting the artifacts of software systems. It has successfully been applied to a wide range of domains, ranging from health and finance to aerospace and e-commerce, and its domains go even beyond software, given recent initiatives in areas such as systems engineering, testing, and hardware design. A joint initiative between OMG and INCOSE (International Council on Systems Engineering) is working on a profile for Systems Engineering, and the SysML consortium has been established to create a systems modeling language based on UML. At the time of this writing, over 60 UML CASE tools can be listed from the OMG resource page (http://www.omg.org).

After revision 2.0, the UML specification consists of four parts:

- UML 2.0 infrastructure, defining the foundational language constructs and the language semantics in a more formal way than they were in the past.
- UML 2.0 superstructure, which defines the user level constructs.
- OCL 2.0 object constraint language (OCL), which is used to describe expressions (constraints) on UML models.
- UML 2.0 diagram interchange, including the definition of the XML-based XMI format, for model interchange among tools.

The Unified Modeling Language comprises of a metamodel definition and a graphical representation of the formal language, but it intentionally refrains from including any design process. The UML in its general form is deliberately semiformal and even its state diagrams (a variant of statecharts) retain sufficient semantics variation points in order to ease adaptability and customization.

The designers of UML realized that complex systems cannot be represented by a single design artifact. According to UML, a system model is seen under different views, representing different aspects. Each view corresponds to one or more diagrams, which taken together, represent a unique model. Consistency of this multiview representation is ensured by the UML metamodel definition. The diagram types included in the UML 2.0 specification are represented in Figure 9.14, as organized in the two main categories that relate to *structure* and *behavior*.

When domain-specific requirements arise, more specific (more semantically characterized) concepts and notations can be provided as a set of stereotypes and constraints, and packaged in the context of a *profile*.

Structure diagrams show the static structure of the system, i.e., specifications that are valid irrespective of time. Behavior diagrams show the dynamic behavior of the system. The main diagrams are:

- *use case diagram* a high-level (user requirements-level) description of the interaction of the system with external agents;
- *class diagram* representing the static structure of the software system, including the OO description of the entities composing the system, and of their static properties and relationships;
- *behavior diagrams* (including *sequence diagrams* and *state diagrams* as variants of Message Sequence Charts and Statecharts) providing a description of the dynamic properties of the entities composing the system, using various notations;
- *architecture diagrams*, (including *composite* and *component diagrams*, portraying a description of reusable components) a description of the internal structure of classes and objects, and a better characterization of the communication superstructure, including communication paths and interfaces.
- *implementation diagrams*, containing a description of the physical structure of the software and hardware components.

The class diagram is typically the core of a UML specification, as it shows the logical structure of the system. The concept of classifiers (class) is central to the OO design methodology. Classes can be defined as user-defined types consisting of a set of attributes defining the internal state, and a set of operations (signature) that can be possibly invoked on the class objects resulting in an internal transition. As units

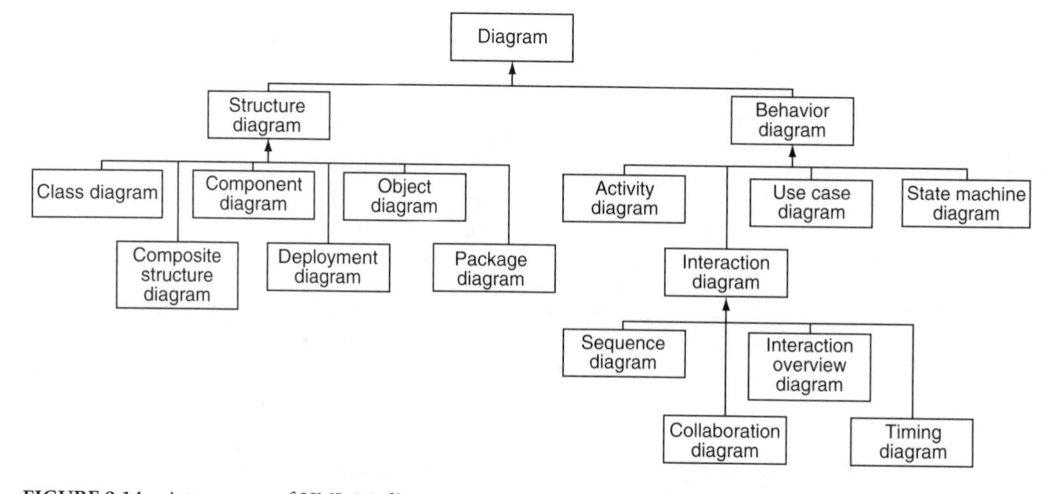

FIGURE 9.14 A taxonomy of UML 2.0 diagrams.

of reuse, classes embody the concepts of encapsulation (or information) hiding and abstraction. The signature of the class abstracts the internal state and behavior, and restricts possible interactions with the environment. Relationships exist among classes and relevant relationships are given special names and notations, such as aggregation and composition, use and dependency. The generalization (or refinement) relationship allows controlled extension of the model by allowing a derived class specification inherit all the characteristics of the parent class (attributes and operations, but also, selectively, relationships) while providing new ones (or redefining the existing).

Objects are instances of the type defined by the corresponding class (or classifier). As such, they embody all of the classifier attributes, operations, and relationships. Several books [50,51] have been dedicated to the explanation of the full set of concepts in OO design. The interested reader is invited to refer to literature on the subject for a more detailed discussion.

All diagram elements can be annotated with constraints, expressed in OCL or in any other formalism that the designer sees as appropriate. A typical class diagram showing dependency, aggregation, and generalization associations is shown in Figure 9.15.

UML 2.0 finally acknowledged the need for a more formal characterization of the language semantics and for better support for component specifications. In particular, it became clear that simple classes provide a poor match for the definition of a reusable component (as outlined in previous sections).

As a result, necessary concepts, such as the means to clearly identify *interfaces* which are provided and (especially) those which are required, have been added by means of the port construct. An interface is an abstract class declaring a set of functions with their associated signature. Furthermore, structured classes and objects allow the designer to specify formally the internal communication structure of a component configuration.

UML 2.0 classes, structured classes, and components are now encapsulated units that model active system components and can be decomposed into contained classes communicating by signals exchanged over a set of *ports*, which models communication terminals (Figure 9.16). A port carries both structural information on the connection between classes or components, and *protocol* information that specifies what messages can be exchanged across the connection. A state machine or a UML Sequence Diagram may be associated to a protocol to express the allowable message exchanges. Two components can interact if there is a connection between any two ports that they own and that support the same protocol in

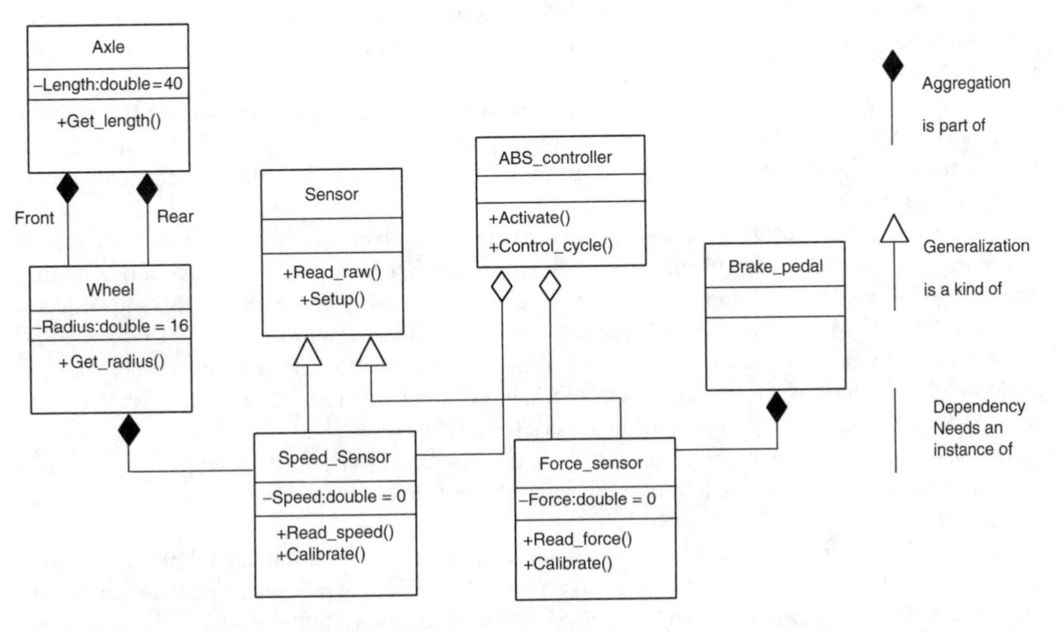

FIGURE 9.15 A sample class diagram.

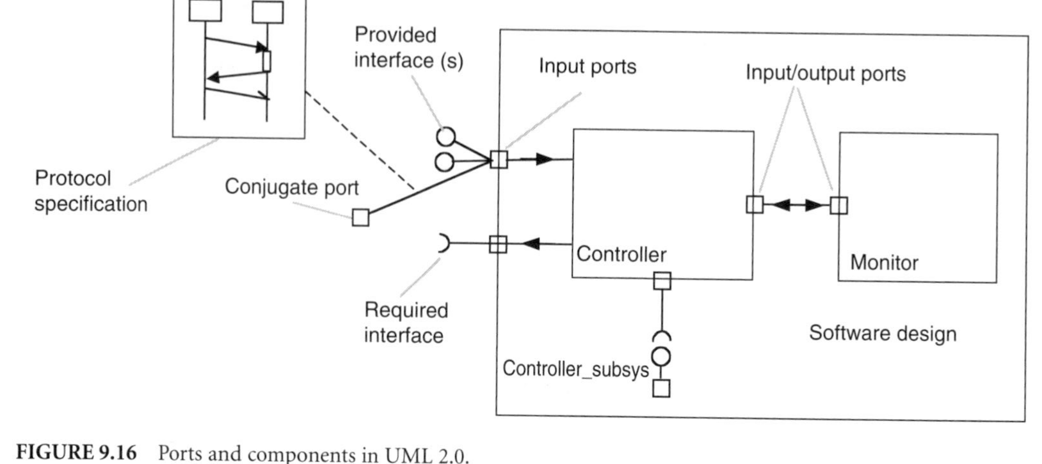

FIGURE 9.16 Ports and components in UML 2.0.

complementary (or *conjugated*) roles. The behavior or reaction of a component to an incoming message or signal is typically specified by means of one or more statechart diagrams.

Behavior diagrams comprise *statechart diagrams, sequence diagrams* and *collaboration diagrams.*

Statecharts [11] describe the evolution in time of an object or an interaction between objects by means of a hierarchical state machine. UML statecharts are an extension of Harel's statecharts, with the possibility of defining actions upon entering or exiting a state as well as actions to be executed when a transition is made. Actions can be simple expressions or calls to methods of the attached object (class) or entire programs. Unfortunately, not only does the Turing completeness of actions prevents decidability of properties in the general model, but UML does not even clarify most of the semantics variations left open by the standard statecharts formalism.

Furthermore, the UML specification explicitly gives actions a run-to-completion execution semantics, which makes them nonpreemptable and makes the specification (and analysis) of typical RTOS mechanisms such as interrupts and preemption impossible.

To give an example of UML statecharts, Figure 9.17 shows a sample diagram where, upon entry of the composite state (the outermost rectangle), the subsystem enters three concurrent (and-type) states, named `Idle`, `WaitForUpdate`, and `Display_all`, respectively. Upon entry into the `WaitForUpdate` state, the variable `count` is also incremented. In the same portion of the diagram, reception of message `msg1` triggers the exit action setting the variable `flag` and the (unconditioned) transition with the associated *call action* `update()`. The `count` variable is finally incremented upon reentry in the state `WaitForUpdate`.

Statechart diagrams provide the description of the state evolution of a single object or class, but are not meant to represent the emergent behavior deriving from the cooperation of more objects; nor are they appropriate for the representation of timing constraints. *Sequence diagrams* partly fill this gap. *Sequence diagrams* show the possible message exchanges among objects, ordered along a time axis. The time points corresponding to message-related events can be labeled and referred to in constraint annotations. Each sequence diagram focuses on one particular scenario of execution and provides an alternative to temporal logic for expressing timing constraints in a visual form (Figure 9.18).

Collaboration diagrams also show message exchanges among objects, but they emphasize structural relationships among objects (i.e., "who talks with whom") rather than time sequences of messages (Figure 9.19).

Collaboration diagrams are also the most appropriate way for representing logical resource sharing among objects. Labeling of messages exchanged across links defines the sequencing of actions in a similar (but less effective) way to what can be specified with sequence diagrams.

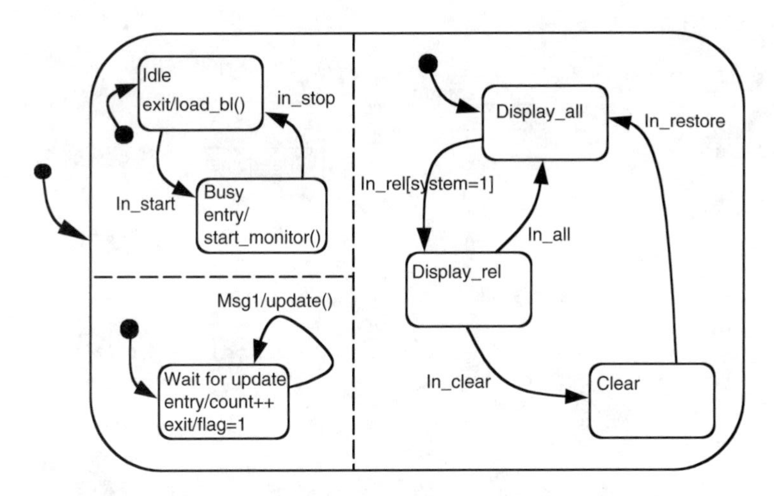

FIGURE 9.17 An example of UML statechart.

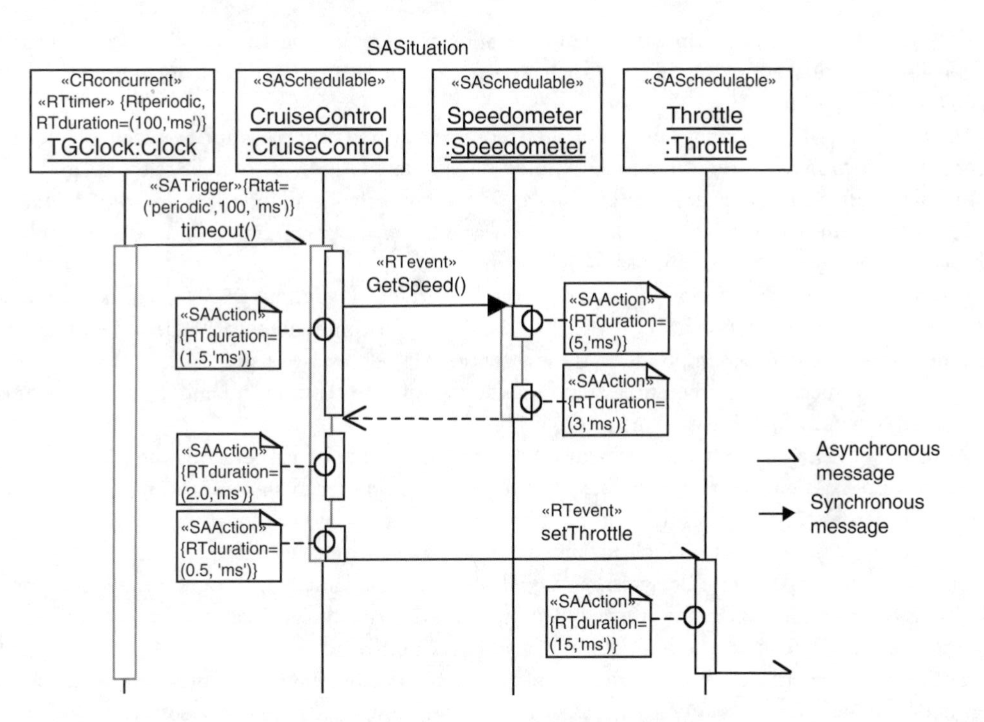

FIGURE 9.18 A sample sequence diagram with annotations showing timing constraints.

Despite the availability of multiple diagram types (or may be because of it), the UML metamodel is quite weak when it comes to the specification of dynamic behavior. The UML metamodel concentrates on providing structural consistency among the different diagrams and provides sufficient definition for the static semantics, but the dynamic semantics is never adequately addressed, up to the point that a major revision of the UML action semantics has become necessary. UML is currently headed in a direction where it will eventually become an executable modeling language, which would for example, allow early verification of system functionality. Within the OMG, a standardization action has been purposely defined with the goal of providing a new and more precise definition of actions. This activity goes under the name of *action semantics for the UML*. Until UML actions are given a more precise semantics, a

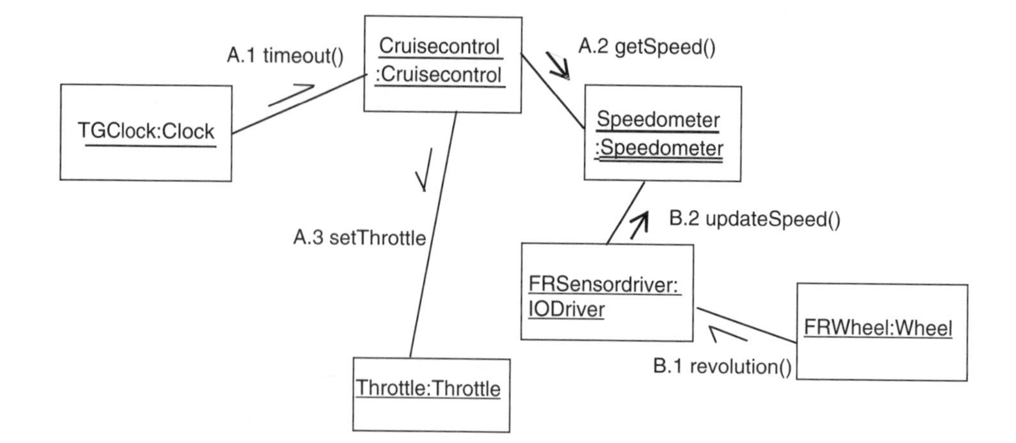

FIGURE 9.19 Defining message sequences in a Collaboration diagram.

faithful model, obtained by combining the information provided by the different diagrams is virtually impossible. Of course, this also nullifies the chances for formal verification of functional properties on a *standard UML model.*

However, simulation or verification of (at least) some behavioral properties and (especially) automatic production of code are features that tool vendors cannot ignore if UML is not to be relegated to the role of simply documenting software artifacts. Hence, CASE tools provide an interpretation of the variation points. This means that validation, code generation, and automatic generation of test cases are tool-specific and depend upon the semantics choices of each vendor.

Concerning formal verification of properties, it is important to point out that UML does not provide any clear means for specifying the properties that the system (or components) are expected to satisfy, nor any means for specifying assumptions on the environment. The proposed use of OCL in an explicit contract section to specify assertions and assumptions acting upon the component and its environment (its users) can hopefully fill this gap in the future.

As of today, research groups are working on the definition of a formal semantic restriction of UML behavior (especially by means of the statecharts formalism) in order to allow for formal verification of system properties [52,53]. After the definition of such restrictions, UML models can be translated into the format of existing validation tools for timed message sequence charts (MSC) or TA.

Finally, the last type of UML diagrams are *Implementation diagrams,* which can be either component diagrams or deployment diagrams. Component diagrams describe the physical structure of the software in terms of software components (modules) related to each other by dependency and containment relationships. Deployment diagrams describe the hardware architecture in terms of processing or data storage nodes connected by communication associations, and show the placement of software components onto the hardware nodes.

The need to express, in UML, timeliness-related properties and constraints, and the pattern of hardware and software resource utilization as well as resource allocation policies and scheduling algorithms found a (partial) response only in 2001 with the OMG issuing a standard SPT profile. The specification of timing attributes and constraints in UML designs will be discussed in Section 9.4.3. Finally, work is currently being conducted in the OMG to develop a test profile for UML 2.0. With the aid of this profile, it will be possible to derive and validate test specifications from a formal UML model.

9.4.1.1 Object Constraint Language

The OCL [54] is a formal language used to describe (constraint) expressions on UML models. An OCL expression is typically used to specify invariants or other types of constraint conditions that must hold for the system. Object constraint language expressions refer to the *contextual instance,* which is the model element to which the expression applies, such as classifiers, like types, classes, interfaces, associations (acting as types), and data types. Also all attributes, association-ends, methods, and operations without side-effects that are defined for these types can be used.

Object constraint language can be used to specify *invariants* associated with a classifier. In this case, it returns a boolean type and its evaluation must be true for each instance of the classifier at any moment in time (except when an instance is executing an operation).

Pre- and *Post-conditions* are other types of OCL constraints that can be possibly linked to an operation of a classifier, and their purpose is to specify the conditions or contract under which the operation executes. If the caller fulfills the precondition before the operation is called, then the called object ensures that the post-condition holds after execution of the operation, but of course only for the instance that executes the operation (Figure 9. 20).

9.4.2 Specification and Description Language

The SDL is an International Telecommunications Union (ITU-T) standard promoted by the SDL Forum Society for the specification and description of systems [55].

Since its inception, a formal semantics has been part of the SDL standard (Z.100), including visual and textual constructs for the specification of both the architecture and the behavior of a system. The behavior of (active) SDL objects is described in terms of concurrently operating and asynchronously communicating abstract state machines (ASMs). SDL provides the formal behavior semantics that UML is currently missing: it is a language based on communicating state machines enabling tool support for simulation, verification, validation, testing, and code generation.

In SDL, systems are decomposed into a hierarchy of *block agents* communicating via (unbounded) channels that carry typed signals. Agents may be used for structuring the design and can in turn be decomposed into sub-agents until leaf blocks are decomposed into *process agents.* Block and process agents differ since blocks allow internal concurrency (subagents) while process agents only have an interleaving behavior.

The behavior of process agents is specified by means of extended finite and communicating state machines (SDL services) represented by a connected graph consisting of states and transitions. Transitions are triggered by external stimuli (signals, remote procedure calls) or conditions on variables. During a transition, a sequence of actions may be performed, including the use and manipulation of data stored in local variables or asynchronous interaction with other agents or the environment via signals that are placed into and consumed from channel queues.

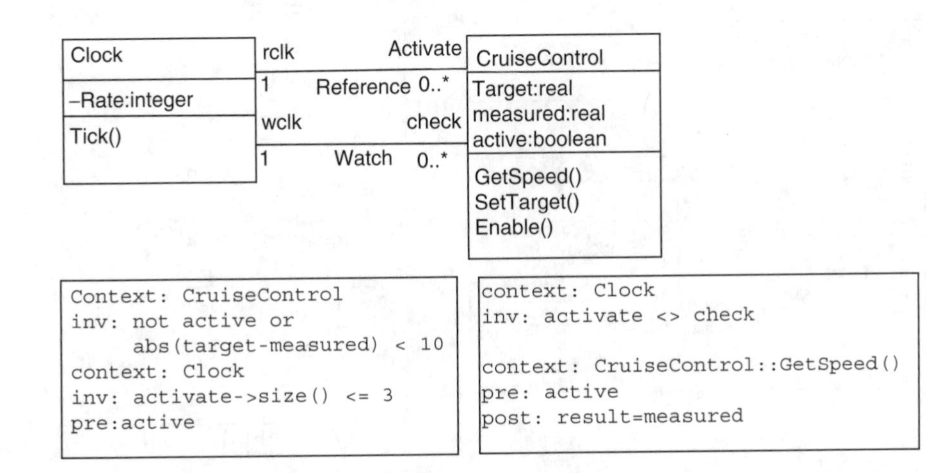

FIGURE 9.20 Examples of OCL adding constraints or defining behavior of operations in a class diagram.

Channels are asynchronous (as opposed to synchronous or rendezvous) FIFO queues (one for each process) and provide a reliable, zero- or finite-delay transmission of communication elements from a sender to a receiver agent.

Signals sent to an agent will be delivered to the input port of the agent. Signals are consumed in the order of their arrival either as a trigger of a transition or by being discarded in case there is no transition defined for the signal in the current state. Actions executed in response to the reception of input signals include expressions involving local variables or calls to procedures.

In summary, an agent definition consists of

- a *behavior specification* given by the agent extended FSM;
- a *structure and interaction diagram* detailing the hierarchy of system agents with their internal communication infrastructure;
- the variables (*attributes*) under the control of each agent;
- the black box or *component view* of the agent defining the interaction points (ports) with the provided and required interfaces.

Figure 9.21 shows an example of SDL notation with five interacting blocks and one leaf block (C2) decomposed into processes. The behavior of the process is specified in terms of its states and reactions as represented in the diagram on the right, where rectangles with concave points represent messages received, and rectangles with convex points denote messages sent. The SDL notation is a representation of the behavior of the equivalent extended state machine (an example is shown on the right side of Figure 9.22).

Figure 9.22 shows a process behavior and a matching representation by means of an extended FSM (right). The figure shows the use of a variable (var1) as a *continuous input* signal for the state S3. A continuous input is an SDL Boolean expression in terms of one or more variables. If the expression evaluates to true and no transition on regular signals is available, the corresponding branch is taken. Note that continuous signals are evaluated after all other signals are checked (i.e., they have the lowest priority).

The state with asterisk reached from the continuous input signal transition is a placeholder, representing the starting state of the transition (S3) itself.

Agents combine behavior and structure in a single entity and fit nicely into the concept of active objects in OO languages (active objects are objects that possess a thread of control and may initiate a reaction).

The latest SDL version, SDL 2000, extends the language by including typical features of OO and component-based modeling techniques, including encapsulation of services by ports (gates) and interfaces, classifiers and associations, and specialization (refinement) of virtual class structure and behavior. In SDL 2000, the user can define classifiers (types in SDL) for blocks, processes, and services, and use instances

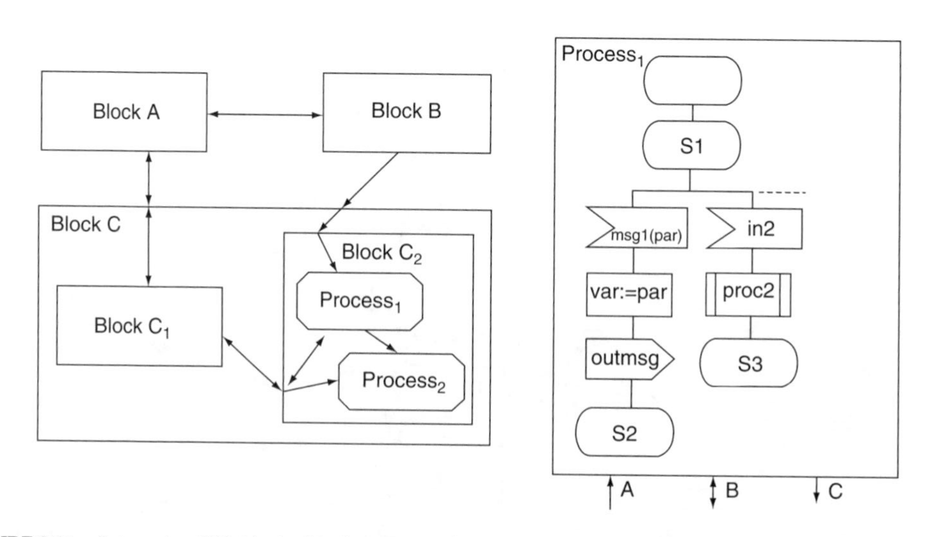

FIGURE 9.21 Interacting SDL blocks, block decomposition, and the description of the behavior of a process.

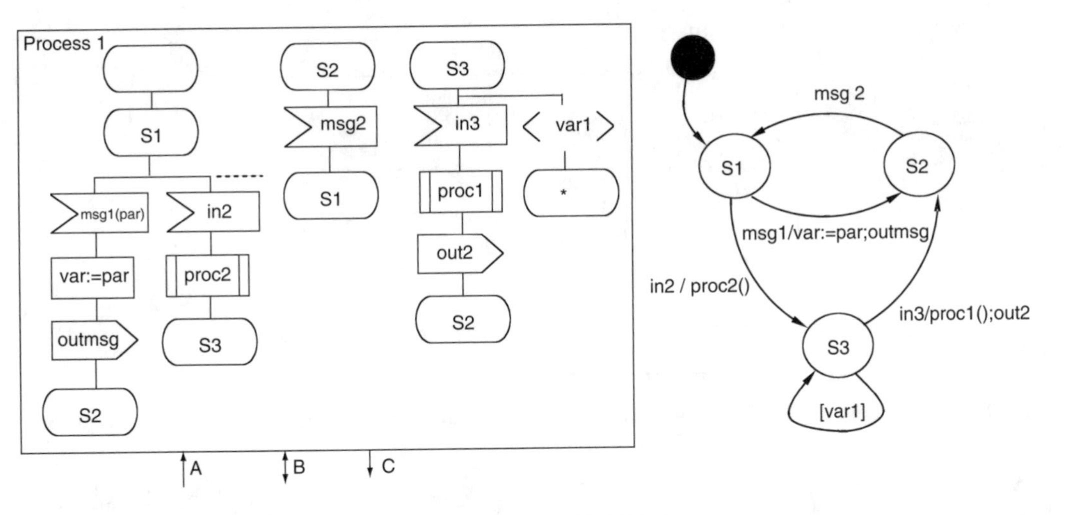

FIGURE 9.22 A process behavior and the corresponding extended FSM.

of types for block, process, service, procedure, and signal description. The specification of behavior for automatic generation of implementations is further improved by the support for exception handling and state machine decomposition.

In SDL 2000, two communicating agents must provide gates with matching interfaces (required or supported) and a connection between the gates must exist.

In the example given in Figure 9.23, an SDL 2000 block exposing two gates (`Bin` of input/output type and the output gate `Bout`) is shown. The block defines two internal signals msg and ack exchanged on the channel connecting the internal processes—`Sender` and `Receiver`.

SDL types can be specialized, leading to subtypes by exploiting the inheritance mechanism. Virtual objects or transitions can be redefined and additional objects or actions can be added to the specification of virtual objects. A redefined transition or object is subject to further specialization (if virtual) unless it is marked as finalized.

SDL offers native mechanisms for representing external (global) time. Time is available by means of the predefined variable *now*, the `now()` primitive and timer constructs. Process actions can set timers, that is the specification of a signal at a predefined point in time, wait and eventually receive a timer expiry signal (as in Figure 9.24). SDL timer timeouts are always received in the form of asynchronous messages and timer values are only meant to be minimal bounds, meaning that any timer signal may remain enqueued for an unbounded amount of time.

In SDL, processes inside a block are meant to be executed concurrently, and no specification for a sequential implementation by means of a scheduler, necessary when truly concurrent hardware is not available, is given. Activities implementing the processes behavior (transitions between states) are executed in a run-to-completion fashion. From an implementation point of view, this raises the same concerns that hold for the implementation of UML models. In SDL, however, the exception mechanism is available to model preemption and interrupt handling.

Other language characteristics make verification of time-related properties impossible: the Z.100 SDL semantics says that external time progresses in both states and actions. However, each action may take an unbounded amount of time to execute, and each process can remain in a state for an indeterminate amount of time before taking the first available transition.

Furthermore, for timing constraints specification, SDL does not include the explicit notion of event, therefore it is impossible to define a time tagging of most events of interest such as transmission and reception of signals, although Message Sequence Charts (MSCs, similar in nature to UML Sequence Diagrams and defined in part Z.120 of the standard [56]) are typically used to fill this gap since they allow the expression of constraints on time elapsing between events.

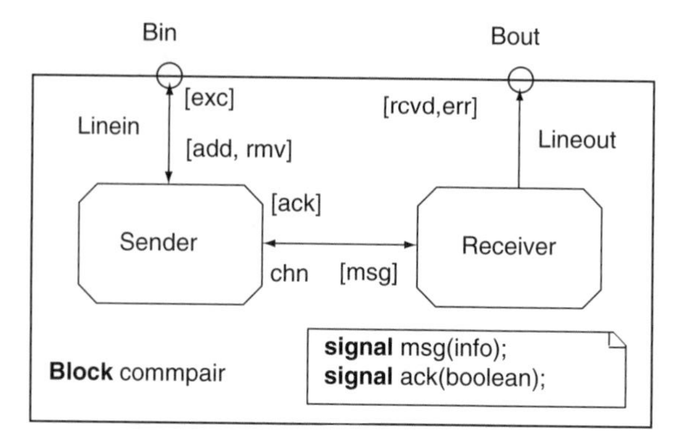

FIGURE 9.23 Specification and description language blocks communicating by ports implementing a signal interface.

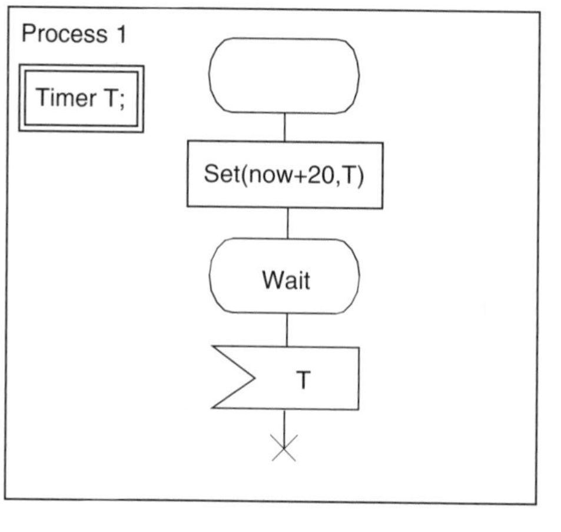

FIGURE 9.24 Example of timer use.

Incomplete specification of time events and constraints prevents timed analysis of SDL diagrams, but the situation is not much better for untimed models. The properties of interest are in general undecidable because of infiniteness of data domains, unbounded message buffers, and the semi-formal modeling style, where SDL is mixed with code fragments inside conditions and actions (the formal semantics of SDL is aimed at code generation rather than at simulation and verification). The untimed verification of properties is only possible in restricted models and seldom performed by commercial or research tools dealing with SDL code. The SDL description, (often restricted to a subset of the language) is provided as an input to transformation tools, which convert it to other formalisms, such as Petri Nets (analyzed by [25, 57]) or TA (for analysis by the Kronos/IF [19,58,59] or Uppaal [20] toolsets).

Research on the verification of SDL designs has been done in academic and industrial laboratories and centers. For example, Regensburger and Barnard [60] describes the verification of a layer of the GSM protocol (6 processes and 1013 reachable states) performed at Siemens by using BDD-based model checking and Bozga et al. [61] report on several case studies related to telecommunication protocols.

9.4.3 Architecture Deployment, Timing Specification, and Analysis

Both UML and SDL have been developed outside the context of embedded systems design and it is clear from previous sections that they are poorly suited to modeling resource allocation and sharing, or to the minimum requirements for timing analysis. In fact, almost nothing exists in standard UML and SDL for modeling (or analyzing) nonfunctional aspects, nor can scheduling or placement of software components on hardware resources be specified and analyzed.

Several proposals exist, most notably the OMG profile for SPT for UML, and others for SDL, which enhance the standard language defining timed models of systems, including time assumptions on the environment and platform-dependent aspects like resource availability and scheduling. Such model extensions allow formal or simulation-based validation of the timing behavior of the software.

The RT-UML profile is currently implemented by commercial tools, and OMG reports that proof of feasibility has been obtained by the interaction with schedulability analysis tools such as TimeWiz by TimeSys and Pert by TriPacific.

9.4.3.1 Unified Modeling Language

The OMG Profile for SPT, voted for adoption in November 2001, aims at substituting a number of proposals for time-related extensions that have appeared in recent years. For example, in order to support better the mapping of active objects into concurrent threads, many research and commercial systems introduced additional nonstandard diagrams. One such example is Artisan's Real-Time Studio, which features the Concurrency diagram, capturing the multitasking and concurrency aspects of the system, and a System Architecture diagram.

The SPT profile defines a comprehensive conceptual framework that extends the UML meta-model with a much broader scope than any other real-time extension and applies to all diagrams. Currently, it consists mostly of a notation framework or vocabulary, with the purpose of providing the necessary concepts for schedulability and performance analysis of (timed) behavioral diagrams or scenarios. However, the SPT inherits from UML the deficiencies related to its incomplete semantics and, at least as of today, it lacks a sufficiently established practice. A major revision of the profile is now being issued by OMG following the request for a profile proposal for real-time embedded systems (MARTES).

The current version of the profile is based on extensions (stereotyped model elements, tagged values and constraints) belonging to three main framework packages, further divided into subpackages (as in Figure 9.25).

Of the three frameworks, the *General Resource Modeling* framework contains the fundamental definitions for modeling *time* in the RTtimeModeling subpackage, generic *resource* specification and usage patterns in the RTresourceModeling profile, and extensions for modeling *concurrency* in the RTconcurrencyModeling profile. The *analysis models* package contains specialized concepts for modeling *schedulability* (RAprofile) and *performance* (PAprofile) analysis. Finally, the *infrastructure models* package deals with real-time middleware specification (RealTimeCORBAModel).

In the SPT, the *time model* provides both continuous and discrete-time models as well as global and local clocks, including drift and offset specifications. The profile allows referencing to time instances (associated to events) of *Time* type, and to the time interval between any two instances of time of *Duration* type in attributes or constraints specifications inside any UML diagram.

The time package not only contains definitions for a formal model of time, but also stereotyped definitions for the two basic mechanisms of timer and clock. Timers can be periodic (as in SDL), they can be set or reset, paused or restarted and, when expired, they send timeout signals. Clocks are specialized periodic timers capable of generating Tick events.

Time values are typically associated to Events, defined in UML as a "specification of a type of observable occurrence" (change of state). A pairing of an event with the associated time instance (time tag) is defined in the SPT profile as a TimedEvent.

The *resource model* package defines schedulability analysis as a quality of service (QoS) matching problem between a client and the resource it uses. The profile provides two interpretations: in the *peer interpretation* the client and the resource coexist at the same level; in the *layered interpretation*, the resources are used to implement the client. This is an attempt at capturing the distinctive aspect of schedulability analysis as a match between the timing (or in general QoS) requirements of *logical entities* in the *logical layer* and the corresponding QoS offers of an implementation or *engineering layer*.

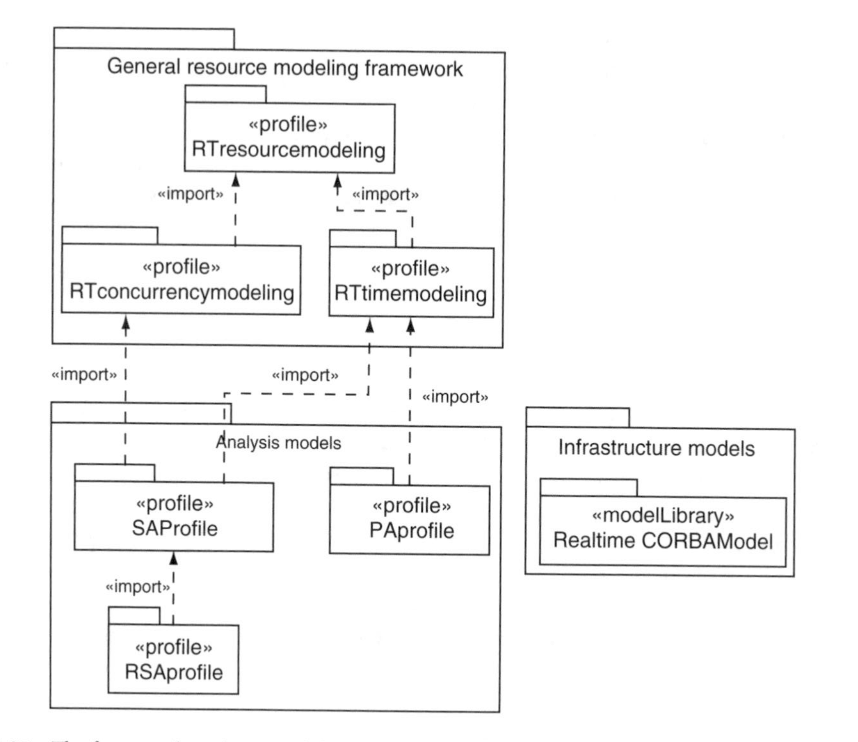

FIGURE 9.25 The framework packages and the subpackages in the OMG SPT.

In the OMG profile, the mapping between the logical entities and the physical architecture supporting their execution is a form of *realization layering* (synonymous of *deployment*). The semantics of the mapping provides a further distinction between the "deploys" mapping, indicating that instances of the supplier are located on the client, an example being the relation between a package and the HW node on which it executes, and the "requires" mapping, which is a specialization indicating that the client provides a minimum deployment environment as required by the supplier, typically in the context of a QoS contract (Figure 9.26).

Although a visual representation of both is possible, it is not clear in which of the existing diagrams it should appear. Hence, the SPT recommendation for expressing the mappings consists of a table of relationships expressing the deploy and requires associations among logical and engineering components.

The QoS match is very simplified view of the schedulability problem with very few practical instances. To give one possible example (Figure 9.26, right) consider the sufficient schedulability condition for rate monotonic, based on processor utilization [28]

$$U = \sum_{i=1,n} \frac{C_i}{\theta_i} \leq U_{\text{lub}} = n(2^{(1/n)} - 1)$$

where

$$U_i = \frac{C_i}{\theta_i}$$

The QoS demand of each thread consists of a demand for utilization U_i to the operating system/scheduling policy and a demand for a target worst-case computation time C_i to the hardware. The QoS offer from the RM scheduling policy is the guaranteed least upper bound on the utilization $U_{\text{lub.}}$ and the QoS matching rule easily follows.

The *schedulability analysis* model is based on stereotyped scenarios (scheduling "situations" in the SPT language). Each Scheduling situation is in practice a sequence diagram, or a collaboration diagram (better

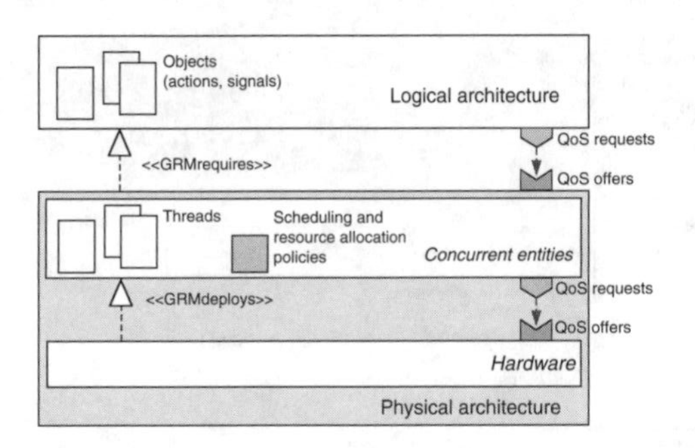

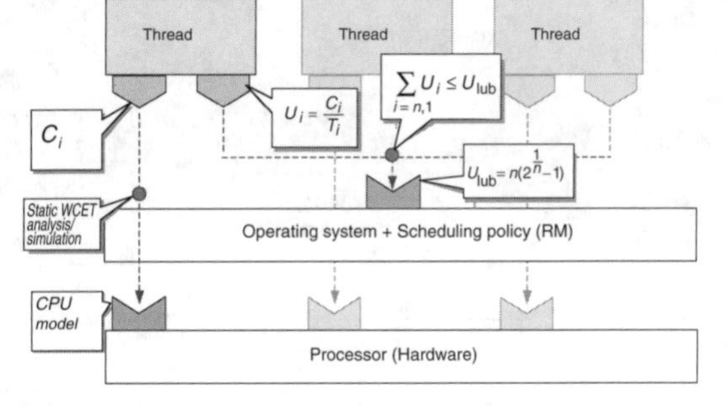

FIGURE 9.26 Schedulability analysis requires creating a correspondence between logical and physical architecture

suited for representing resource sharing), where one or more trigger events result in the actions to be scheduled within the deadline associated with the trigger.

Rate monotonic analysis is the method of choice for analyzing simple models with a restricted semantics, conforming to the so-called *task-centric* design paradigm. This requires updating the current definition of UML actions in order to allow for preemption (which is a necessary prerequisite of RMA). Such a recommendation has been explicitly formulated in the SPT profile and is among the recommended changes in the agenda of the OMG action semantics group.

In this task-centric approach, the behavior of each active object or task consists of a combination of reading input signals, performing computation, and producing output signals (Figure 9.28, left). Each active object can request the execution of actions of other passive objects in a synchronous or asynchronous fashion.

Each task is logically concurrent; it is activated periodically, and handles a single event. Active objects cooperate only by means of pure asynchronous messages, possibly implemented by means of memory mailboxes, a kind of protected (shared resource) object (Figure 9.27).

In the SPT, the TimedEvent that triggers the execution of the task action is identified with the stereotype “SATrigger” and it is the model element to which the deadline constraint (for the execution of the action) and the schedulability response are associated (Figure 9.27).

Unfortunately, the task-centric model, even if simple and effective in some contexts (see also other design methodologies [62–64], where the analysis of UML models is made possible by restricting the semantics of the analyzable models), only allows analysis of simple models where active objects do not interact with each other. In general UML models, each action is part of multiple end-to-end computations (with the associated timing constraints) and the system consists of a complex network of cooperating active objects, implementing state machines, and exchanging asynchronous messages.

In their work [65–67], Saksena, Karvelas, and Wang present an integrated methodology that allows dealing with more general OO models where multiple events can be provided as input to a single thread. According to their model, each thread has an incoming queue of events possibly representing real-time transactions and the associated timing constraints (Figure 12.10).

Consequently, scheduling priority (usually a measure of time urgency or criticality) is attached to events rather than threads. This design and analysis paradigm is called *event-centric design*. Each event has an associated priority (for example, deadline monotonic priority assignment can be used). Event queues are ordered by priority (i.e., threads process events based on their priority) and threads inherit the priority of the events they are currently processing. This model entails a two-level scheduling: the events enqueued as input to a thread need to be scheduled to find their processing order. At the system level, the threads are scheduled by the underlying RTOS (a preemptive priority-based scheduler is assumed).

Schedulability analysis of the general case is derived from the analysis methodology for generic deadlines (possibly greater than the activation rates of the triggering events) by computing response times of actions relative to the arrival of the external event that triggers the end-to-end reaction (transaction) T

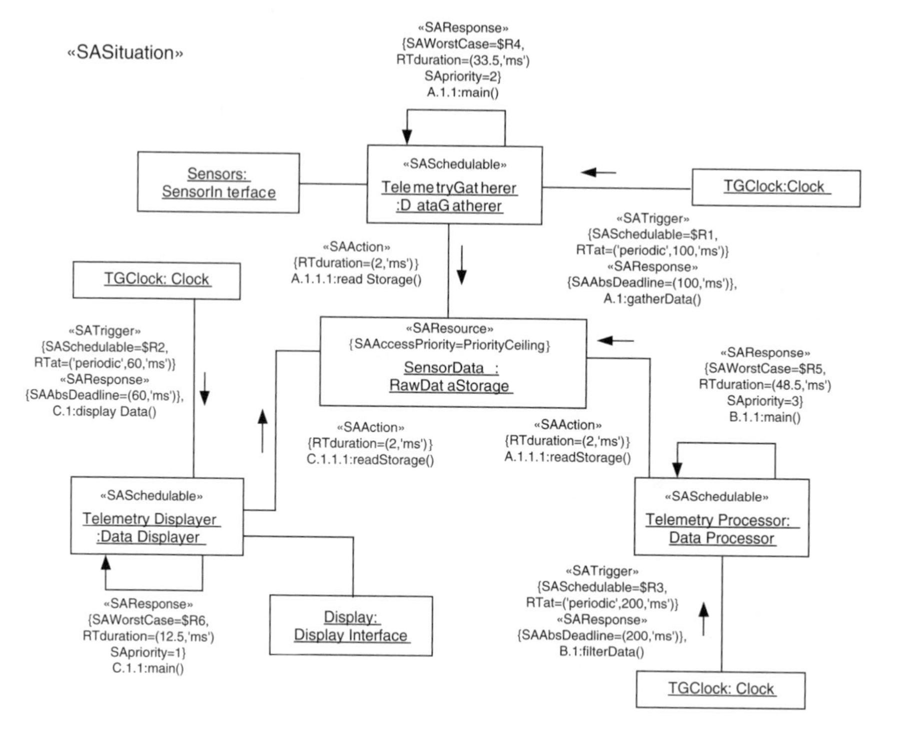

FIGURE 9.27 A sample collaboration instance representing a situation suitable for RMA. (Adapted from the OMG profile document.)

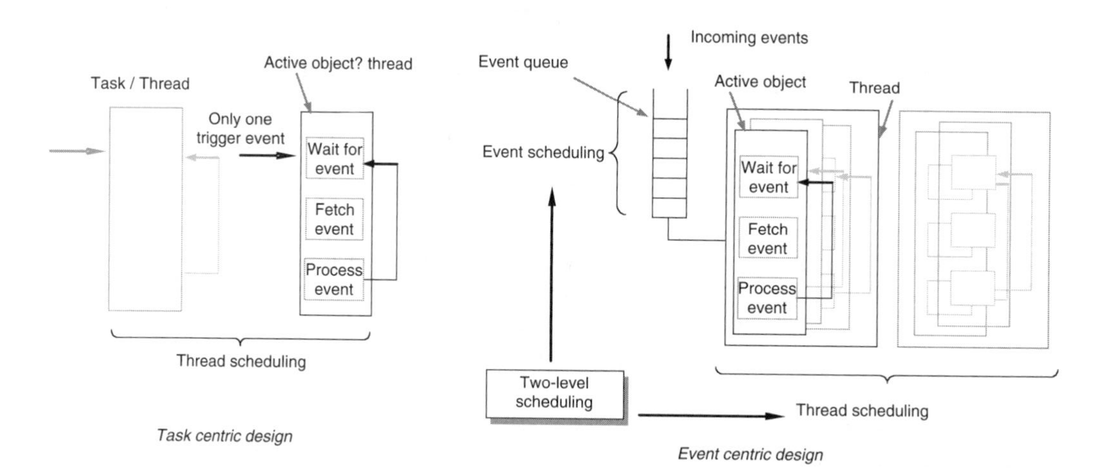

FIGURE 9.28 The dual-scheduling problem in the event-centric analysis.

containing the action. The analysis is based on the standard concepts of critical instant and busy period for task instances with generic deadlines (adapted to the transaction model.)

In [66], many implementation options for the event-centric model are discussed, for example single-thread implementation, multithread implementation with fixed priority or dynamic priority multi-threaded implementation and a formula or procedure for schedulability analysis is provided or at least discussed for each of these models. The interested reader should refer to [66] for details.

From a time analysis point of view, it makes sense to define the set of (nested) actions that are called synchronously in response to an event. This set is called *synchronous set.*

In a *single-thread implementation*, the only application thread processes pending events in priority order. Since there is only one thread, there is only one level of scheduling. Actions inherit their priority from the priority of the triggering event. Also, a synchronously triggered action inherits its priority from its caller. In single-threaded implementations, any synchronous set that starts executing is completed without interruption. Hence, the worst-case blocking time of an action starting at $t = 0$ is bound by the longest synchronous set of any lower priority action that started prior to $t = 0$ and interference can be computed as the sum of the interference terms from other transactions and the interference from actions in the same transaction.

The single-thread model can be analyzed for schedulability (details in [66]) and it is also conformant to the nonpreemptable semantics of UML actions. Most of the existing CASE tools support a single-threaded implementation and some of them (such as Rational Rose RT) support the priority-based enqueuing of activation events (messages).

In *multithreaded implementations*, each event represents a request for an end-to-end sequence of actions, executed in response to its arrival. Each action is executed in the context of a thread. Conceptually, we can reverse the usual meaning of thread scheduling, considering events (and the end-to-end chain of actions) as the main scheduling item and the threads required for the execution of actions as special mutual exclusion (mutex) resources, because of the impossibility of having a thread preempt itself. This insight allows using threads and thread priorities in a way that facilitates response time analysis.

If threads behave as mutexes, then it makes sense to associate with each thread a ceiling priority as the priority at which the highest priority event is served. As prescribed by PI or PCP, *threads inherit the priority of the highest priority event* in their waiting queue and this allows bounding priority inversion. In this case, the worst-case blocking time is restricted to the processing of a lower priority event. Furthermore, before processing an event, a thread locks the active object within which the event is to be processed (this is necessary when multiple threads may handle events that are forwarded to the same object), and a ceiling priority and a priority inheritance rule must be defined for active objects as well as for threads.

For the multithreaded implementation, a schedulability analysis formula is currently available for the aforementioned case in which threads inheriting the priority of the events. The schedulability analysis formula generally results in a reduced priority inversion with respect to the single-thread implementation, but its usefulness is hindered by the fact that (to the author's knowledge) there is no CASE tool supporting generation of code with the possibility of assigning priorities to events and run-time support for executing threads inheriting the priority of the event they are currently handling.

Assigning *static priorities to threads* is tempting (static priority scheduling is supported by most RTOSs) and it is quite a common choice for multithreaded implementations (Rational Rose RT and other tools use this method). Unfortunately, when threads are required to handle multiple events it is not clear what priority should be assigned to them and the computation of the worst-case response time is in general very hard because of the two-level scheduling and because of the difficulty of constructing an appropriate critical instant. The analysis needs to account for multiple priority inversion that arises from all the lower priority events handled by a higher priority thread. However, it may still be possible to use static thread priorities in very simple cases when it is easy to estimate the amount of priority inversion.

The discussion of the real-time analysis problem bypasses some of the truly fundamental problems in designing a schedulable software implementation of a functional (UML) model. The three main degrees of freedom in a real-time UML design subject to schedulability analysis are:

- assigning priorities to events
- defining the optimal number of tasks, and especially
- defining a mapping of methods (or entire objects) to tasks for their execution.

Timing analysis (as discussed here) is only the last stage, after the mapping decisions have been taken.

What is truly required is a set of design rules, or even better, an automatic synthesis procedure that helps in the definition of the set of threads and especially in the mapping of the logical entities to threads and to the physical platform.

The problem of synthesizing a schedulable implementation is actually a complex combinatorial optimization problem. Saksena et al. [67], propose an automatic procedure for synthesizing the three main degrees of freedom in a real-time UML design. The synthesis procedure uses a heuristic strategy based on a decomposition approach, where priorities to events/actions are assigned in a first stage and mapping is performed in a separate stage.

This problem, probably the most important in the definition of a software implementation of real-time UML models, is still an open problem.

9.4.3.2 Specification and Description Language

SDL contains primitives for dealing with global time, but nevertheless, when the specification of time behavior and resource usage is an issue, it is impossible to capture most time specifications and constraints.

For example, SDL does not offer the possibility of defining local time, timers cannot be periodic and it is difficult to express clocks, where periodic ticks are instantaneously consumed or lost. The main reason for these deficiencies is that the SDL notion of time was aimed at code generation, rather than at simulation and verification.

As stated in the previous section, SDL timer values only represent minimal bounds. This means that the SDL semantics provides no means to specify constraints on the time passed in states or actions, whereas the verification or simulation of a property often depends on finite lower and upper bounds. Also, the delays in the transmission of signals are unbounded.

The use of MSC and annotations to express the timing constraints that standard SDL misses is commonplace in commercial tools and research proposals. In particular, some tools allow the expression of duration constraints on actions or other simple constraints in the form of special comments (with the problem that each tool defines its own syntax and semantics.) Other proposals allow the definition of upper and lower time bounds for the delays applied to messages conveyed by a channel, or the definition of the delays probabilities.

Concerning the expressivity problem, a general real-time framework for SDL is presented in [68,69], (the first one is aimed at hardware software codesign). The extension of SDL with more powerful timers and clocks as well as local clocks has been proposed by many (see e.g., [70]). Other proposals for expression of timing requirements can be found in [71,72].

Another problem of SDL is that it provides neither deployment and resource usage information, nor the notion of preemptable or nonpreemptable actions that is necessary for schedulability analysis. One proposal [73] consists in defining a list of resources with an attribute defining their preemptability. Block, Process, and Task definitions need to be complemented with the specification of the resource on which they are executed. Scheduling policies are defined by keywords, such as RMA or Earliest Deadline First (EDF), or by priority rules attached to resources or agents. Priorities may be static or dynamic and in the latter case, they may be specified based on preconditions, or by means of an observer, which explicitly updates priorities depending on the observed states and events.

Some tools, for example time-enhanced versions of ObjectGeode [74] as presented in [75] and Tau [76,77] and the tool and methods based on Queuing SDL [78] propose to attach explicit timing information to SDL constructs such as tasks, and to provide some minimal deployment information. The ObjectGeode simulator, for example, allows associating an execution time (interval) with each action. Schedulability of SDL models is also discussed in [79].

Deficiencies in the specification of time also clearly affect the simulation of SDL models with time constraints. Any rigorous attempt to construct the simulation graph of an SDL system (the starting point for simulation and verification) must account for all possible combinations of execution times, timer expirations, and consumptions. Since no control over time progress is possible, many undesirable executions might be obtained during this exhaustive simulation.

In practice, existing simulation and verification tools make simplifying assumptions on execution and idle times. The usual convention is that statements or entire transactions are atomic; that actions take zero time to execute, and that the system executes immediately whatever it can execute (time passes only where explicitly required by waiting conditions). This option is justified by the fact that it generates the highest degree of determinism, thus reducing the state space by an important factor (making the system practically analyzable).

In [80], an interesting integration of SDL model simulation with resource scheduling is proposed: annotations related to timing constraints and resource usage are translated into scheduler and resource manager models that can be used to simulate the time behavior of the model with respect to schedulability constraints on a standard SDL tool. Performance evaluation of SDL models is the subject of other research works [75, 81, 82]

The verification of time-related properties of SDL models can often be done on a finite control abstraction, which is a system with finite data domains, bounded message buffers, and bounded agent creation. A number of interesting verification problems are decidable on such a finite control abstraction under the condition that in the timed automata obtained by translation, clocks can only be reset to zero, stopped, and restarted, and that the only allowed tests are comparisons of clock values or differences of clock values with constants, which, in practice allows transformation of the model into the standard semantics of TA.

Some SDL tools provide the facility to define observers as a means for defining constraints and properties, but SDL does not provide any standard notation for them.

9.4.4 Tools and Commercial Implementations

There is a large number of specialized CASE tools for UML modeling of embedded systems. Among those, Rhapsody of I-Logix [83], Real-Time Studio of Artisan [84], TAU Generation 2 from Telelogic [76] and Rational Rose RT (now Rational Rose Technical Developer [34], which is in essence a profile implementing ROOM in the UML framework) of IBM (Rational) are probably the most common. In contrast with general-purpose UML tools, all these try to answer the designer's need for automated support for model simulation, verification, and testing. In order to do so, they all provide an interpretation of the dynamic semantics of UML. Common features include interactive simulation of models and automatic generation of code. Nondeterminism is forbidden or eliminated by means of semantics restrictions. Some timing features and design of architecture level models are usually provided, although in nonstandard form if compared to the specifications of the SPT profile. Furthermore, the support for OCL and user-defined profiles is often quite weak.

Formal validation of untimed and timed models is typically not provided, since commercial vendors focus on providing support for simulation, automatic code generation, and (partly) automated testing. Third-party tools inter-operating by means of the standard XMI format provide schedulability analysis and research work on restricted UML semantics models demonstrates that formal verification techniques can be applied to UML behavioral models. This is usually done by transformation of UML models into a formal MOC and subsequent formal verification by existing tools (for both untimed and timed systems). A number of tools, including those previously cited, allow the definition of UML models including real-time constraints, and later possible verification of schedulability properties.

SDL is supported by powerful development environments integrating advanced facilities (like simulation, model checking, test generation, and auto-coding). Tools like ObjectGeode or the Telelogic Tau2 series support many phases of software development, ranging from some restricted form of analysis to implementation (code generation) and on-target deployment. Properties are expressed by MSC or by means of a tool-specific language allowing the definition of observers. No schedulability analysis is currently available from commercial vendors.

The most recent versions of some of these tools attempt to combine various modelling approaches in order to unite the best of both worlds. One example is SDL-UML and the Telelogic TAU series of products [76], where the TTCN language (set to become one of the UML test profile recommended notations) is used to describe formally the test cases and sequences [85].

9.5 Research on Models for Embedded Software

The models and tools described in the previous sections are representatives of a larger number of methods, models, languages, and tools (at different levels of maturity) that are currently being developed to face the challenges posed by embedded software design.

This section provides an insight on other (often quite recent) proposals for solving advanced problems related to the modeling, simulation, and verification of functional and nonfunctional properties, including support for compositionality and possibly for integration of heterogeneous models, where heterogeneity applies to both the execution model and the semantics of the component communications or, in general, interactions.

In particular, the request for heterogeneity arises from the observation that there is no clear winner among the different models and tools and that different parts of complex embedded systems may be better suited to different modeling and analysis paradigms (such as dataflow models for data-processing blocks and FSMs for control blocks). The first objective in this case is to reconcile by unification the synchronous and asynchronous execution paradigms.

The need to handle timing and resource constraints (hence preemption) together with the modeling of system functionality is another major requirement for modern methodologies.

Performance analysis by simulation of timed (control) systems with scheduling and resource constraints is, for example, the main goal of the TrueTime Toolset from the LTH Technical University in Lund [86].

The TrueTime toolset is one example of modeling paradigm where scheduling and resource management policies can be handled as separate modules to be plugged in together with the blocks expressing the functional behavior. TrueTime consists of a set of Simulink blocks that simulate real-time kernel policies as found in commercial RTOSs. The system model is obtained by connecting kernel, network, and ordinary Simulink blocks representing the functional behavior of the control application. The toolbox is capable of simulating the system behavior of the real-time system with interrupts, preemption and scheduling of resources.

The Times tool [87] is another research framework (built on top of Uppaal and available at http://www.timestool.org) attempting an integration of the TA formalism with methods and algorithms for task and resource scheduling. In Times, a timed automaton can be used to define an arbitrary complex activation model for a set of deadline-constrained tasks, to be scheduled by fixed or dynamic (Earliest Deadline First [28]) priorities. Model checking techniques are used to verify the schedulability of the task set, with the additional advantage (if compared with standard worst-case analysis) of providing the possible runs that fail the schedulability test.

Other design methodologies bring further distinction among (at least) three different design dimensions representing:

- The *(functional) behavior* of the system and the timing constraints imposed by the environment and/or resulting from design choices
- The *communication network* and the *semantics of communication* defining interactions on each link
- The *platform* onto which the system is mapped and the timing properties of the system blocks resulting from the binding

The Ptolemy and the Metropolis environments are probably the best-known examples of design methodologies founded on the previous principles.

Ptolemy (http://www.ptolemy.eecs.berkeley.edu/) is a simulation and rapid prototyping framework for heterogeneous systems developed at the Center for Hybrid and Embedded Software Systems (CHESS) in the Department of Electrical Engineering and Computer Sciences of the University of California at Berkeley [2]. Ptolemy (now at version II) targets complex heterogeneous systems encompassing different operations such as signal processing, feedback control, sequential decision-making, and possibly even user interfaces.

Prior to UML 2.0, Ptolemy II had already introduced the concept of actor-oriented design, by complementing traditional object-orientation with concurrency and abstractions of communication between components.

In Ptolemy II, the model of the application is built as a hierarchy of interconnected actors (most of which are domain polymorphic). Actors are units of encapsulation (components) that can be composed hierarchically, producing the design tree of the system. Actors have an interface abstracting their internals and providing bidirectional access to functions. The interface includes ports that represent points of communication, and parameters, which are used to configure the actor operations.

Communication channels pass data from one port to another according to some communication semantics. The abstract syntax of actor-oriented design can be represented concretely in several ways, one example being the graphical representation provided by the Ptolemy II Vergil GUI (Figure 9.29).

Ptolemy is not built on a single, uniform, model of computation, but rather, it provides a finite library of *Directors* implementing different models of computations. A Director, when placed inside a Ptolemy II actor, defines its abstract execution semantics (MOC) and the execution semantics of the interactions among its component actors (its *domain*).

A component that has a well-defined behavior under different models of computation is called a domain polymorphic component, which means that its behavior is polymorphic with respect to the domain or MOC that is specified at each node of the design tree.

The available MOCs in Ptolemy II include:

- Communicating Sequential Processes (CSP, with synchronous rendezvous)
- continuous time (CT, where behavior is specified by a set of algebraic or differential equations)
- discrete events, (DE, consisting of a value and a time stamp)
- Finite State Machines (FSM)
- (Kahn) Process networks (PN)
- Synchronous dataflows (SDF)
- Synchronous/reactive (SR) models and Time-triggered synchronous execution (the Giotto framework [88])

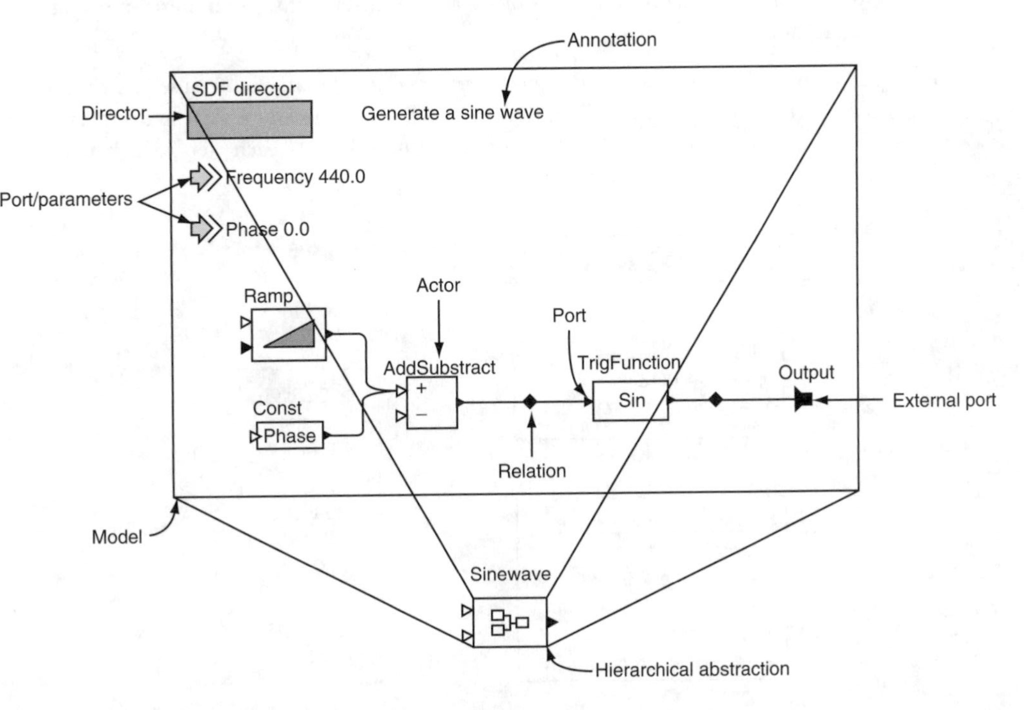

FIGURE 9.29 A Ptolemy actor and its black-box (higher level) representation.

The CT and SR domains have fixed-point semantics, meaning that in each iteration the domain may repeatedly fire the components (execute the available active transitions inside them) until a fixed point is found.

Of course, key to the implementation of polymorphism is proving that an aggregation of components under the control of a domain in turn defines a polymorphic component. This has been proven possible for a large number of combinations of MOCs [89].

In Ptolemy II, the implementation language is Java, and an experimental module for automatic Java code generation from a design is now available at http://ptolemy.eecs.berkeley.edu/. The source code of the Ptolemy framework is available for free from the same web site. Currently, the software has hundreds of active users from industry and academia at various sites worldwide.

The Metropolis environment embodying the platform-based design methodology [1] for design representation, analysis, and synthesis, is under development at the University of California at Berkeley, under the sponsorship of the MARCO Gigascale System Research Center.

In the Metropolis toolset [90], system design is seen as the result of a progressive refinement of high-level specifications into lower-level models, *possibly heterogeneous* in nature. The environment deals with all phases of the design from conception to final implementation.

Metropolis is designed as a very general infrastructure based on a meta-model with precise semantics that is general enough to support existing computation models and accommodate new ones. The metamodel supports not only functionality capture and analysis through simulation and formal verification, but also architecture description and the mapping of functionality to architectural elements (Figure 9.30).

The Metropolis Metamodel (MMM) language provides basic building blocks that are used to describe communication and computation semantics in a uniform way. These components represent *computation*, *communication*, and *coordination* or *synchronization* as follows:

- *Processes* for describing computation
- *Media* for describing communication between processes
- *Quantity Managers* for enforcing a scheduling policy of processes
- *Netlists* for describing interconnection of objects and for instantiating and interconnecting quantity managers and media

A process is an active object (it possesses its own thread of control) that executes a sequence of actions (instructions, subinstructions, function calls, and awaits [91]) and generates a sequence of events. Each process in a system evolves by executing one action after the other. At each step (which is formally

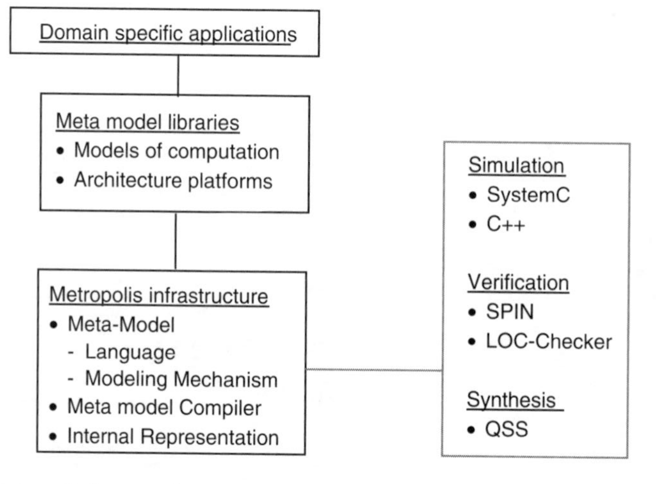

FIGURE 9.30 The Metropolis framework.

described in [91] in terms of a global execution index), each process in the system executes one action and generates the corresponding event. This synchronous execution semantics is relaxed to asynchronicity by means of a special event called NOP that can be freely interleaved in the execution of a process (providing non-determinism).

Each *process* defines its *Ports* as the possible interaction points. *Ports* are typed objects associated with an interface, which declares the services that can be called by the process or called from external processes (bi-directional communication) in a way similar to what is done in UML 2.0.

Processes cannot connect directly to other processes but the interconnection has to go through a *Medium*, which has to define (i.e., implement) the services declared by the interface associated with the ports. The separation of the three orthogonal concepts of Process, Port, and Medium provides maximum flexibility and reusability of behaviors, that is, the meaning of the communication can be changed and refined without changing the computation description that resides in the processes (Figure 9.31).

A Metropolis program can be seen as a variant of a system-encompassing TA where the transition between states is labeled by event vectors (one event per process). At each step, there is a set of event vectors that could be executed to make a transition from the current state to the next state. Unless a suitably defined Quantity Manager restricts the set of possible executions by means of scheduling constraints, the choice among all possible transitions is performed nondeterministically.

Quantity Managers define the scheduling of actions by assigning tags to events. A tag is an abstract quantity from a partially order set (such as, for example, time or execution order). Multiple requests for action execution can be issued to a quantity manager that has to resolve them and schedule the processes in order to satisfy the ordering relation on the set of tags. The tagged-signal model (Lee and Sangiovanni-Vincentelli [92]) is the formalism that defines the unified semantic framework of signals and processes that stands at the foundation of the MMM.

By defining different communication primitives and different ways of resolving concurrency, the user can, in effect, specify different MoCs. For instance, in a synchronous model of computation, all events in an event vector have the same tag.

The mapping of a functional model onto a platform is performed by enforcing a synchronization of function and architecture. Each action in the function side is correlated with an action on the architecture side using synchronization constraints.

The precise semantics of Metropolis allows for system simulation, synthesis and formal analysis. The metamodel includes constraints that represent in abstract form requirements not yet implemented or assumed to be satisfied by the rest of the system and its environment.

Metropolis uses a logic language to capture nonfunctional constraints (time or energy constraints, for example). Constraints are declarative formulas of two kinds: linear temporal logic (LTL) and logic of constraints (LOC).

Although two verification tools are currently provided to check that constraints are satisfied at each level of abstraction, the choice of the analysis and synthesis method or design tool depends on the application domain and the design phase. Metropolis clearly cannot possibly provide algorithms and tools for all possible design configurations and phases, but it provides mechanisms to compactly store and communicate all design information, including a parser that reads metamodel designs and a standard API

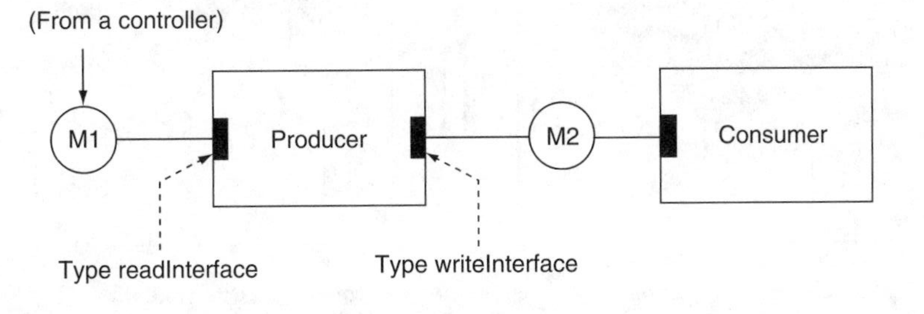

FIGURE 9.31 Processes, Ports, and Media in Metropolis.

that allows developers to browse and modify design information so that they can plug in the required algorithms for a given application domain. This mechanism has been exploited to integrate the Spin software verification tool [10].

Finally, as the last example, separation of behavior, communication or interaction, and execution models are among the founding principles of the component-based design and verification procedure outlined in [93,94]. The methodology is based on the assumption that the system model is a transition system with dynamic priorities restricting non-determinism in the execution of system actions. Priorities define a strict partial order among actions ($a_1 \prec a_2$). A priority rule is a set of pairs $\{(C^j, \prec^j)\}_{j \in J}$ such that $\prec^j$ is a priority order and C^j is a state constraint specifying when the rule applies.

In [95], a component is defined as the superposition of the three models defining its behavior, interactions, and execution model. The transition-based formalism is used to define the behavior, and interaction specification consists of a set of connectors, defining a maximal set of interacting actions and specification of actions that must occur simultaneously in the context of an interaction. The set of actions that are allowed to occur in the system consists of the *complete actions* (Figure 9.32, triangles). Incomplete actions (Figure 9.32, circles) can only occur in the context of an interaction defining a higher-level complete action (as in the definition of $IC[K_1]^+$ in Figure 9.32, which defines the complete action $a_5|a_9$, meaning that incomplete action a_9 can only occur synchronously with a_5). This definition of interaction rules allows for a general model of asynchronous components, which can be possibly synchronized upon a subset of their actions if and when needed. Finally, the execution model consists of a set of dynamic priority rules.

Composing a set of components means applying a set of rules at each of the three levels. The rules defined in [95] define an associative and commutative composition operator. Further, the composition

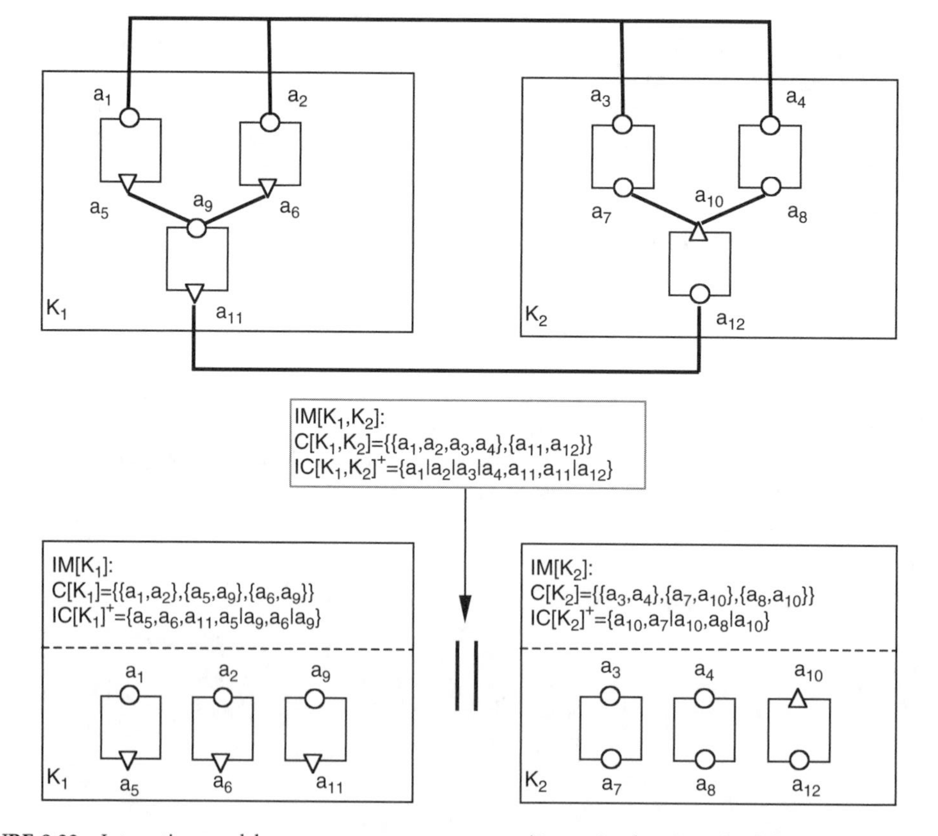

FIGURE 9.32 Interaction model among system components. (From Gossler, G. and Sifakis, J., Composition for component-based modeling. With permission.)

model allows for *composability* (properties of components are preserved after composition) and compositionality (properties of the compound can be inferred from the properties of the components) with respect to deadlock freedom (liveness).

Gossler and Sifakis [95] propose a methodology for analysis of timed systems (and a framework for the composition rules for component-based real-time models) based on the framework represented in Figure 9.33. According to the proposed design methodology, the system architecture consists of a number of layers, each capturing a different set of properties.

At the lowest level, threads (processes) are modeled according to the TA formalism, capturing the functional behavior and the timing properties and constraints.

At the preemption level, system resources and preemptability are accounted for by adding one or more preemption transitions (right-hand side of Figure 9.34), one for each preemptable resource, and mutual exclusion rules are explicitly formulated as a set of constraints acting upon the transitions of the processes.

Finally, the resource management and scheduling policies are represented as additional constraints $K_{pol}=K_{adm} \wedge K_{res}$ where K_{adm} are the admission control constraints and K_{res} are the constraints specifying how resource conflicts are to be resolved.

Once scheduling policies and schedulability requirements are in place, getting a correctly scheduled system amounts to finding a nonempty control invariant K such that $K \Rightarrow K_{sched} \wedge K_{pol}$.

Finally, the last example of research framework for embedded software modeling is the generic modeling environment (GME), developed at Vanderbilt University [96], which is a configurable toolkit offering a meta–meta modeling facility for creating domain-specific models and program synthesis environments. The configuration of a domain-specific metamodel can be achieved by defining the syntactic, semantic, and presentation information regarding the domain. This implies defining the main

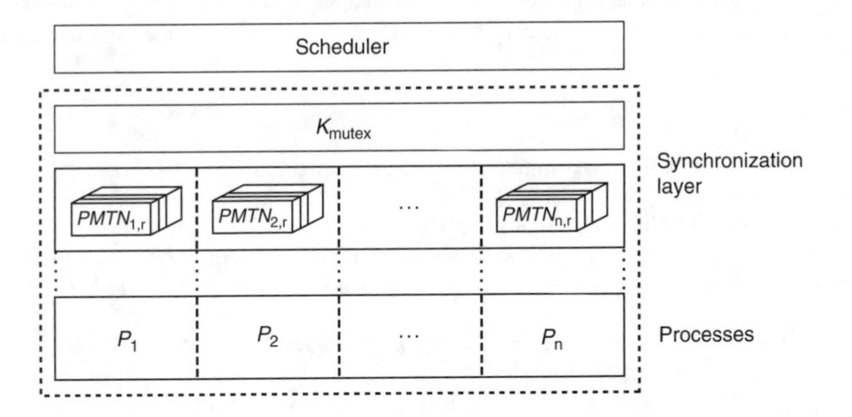

FIGURE 9.33 The architecture of the system model as defined. (From Gossler, G. and Sifakis, J., Composition for component-based modeling. With permission.)

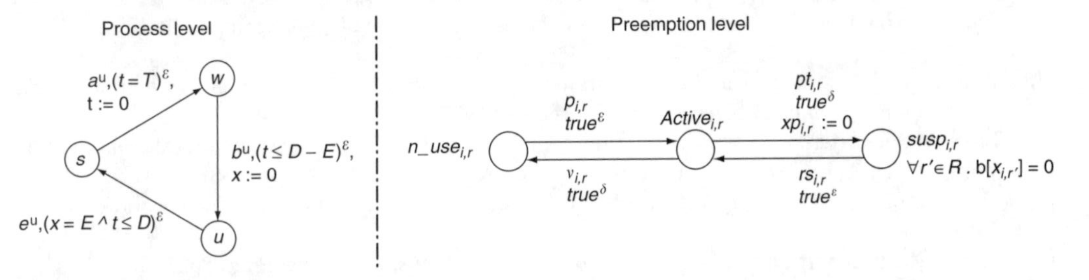

FIGURE 9.34 Process and preemption modeling. (From Gossler, G. and Sifakis, J., Composition for component-based modeling. With permission.)

domain concepts, the possible relationships and the constraints restricting the possible system configurations as well as the visibility rules of object properties.

The vocabulary of the domain-specific languages implemented by different GME configurations is based on a set of generic concepts built into GME itself. These concepts include hierarchy, multiple aspects, sets, references, and constraints. Models, atoms, references, connections, and sets are first-class objects.

Models are compound objects that can have parts and inner structure. Each part in a container is characterized by a role. The modeling instance determines what parts are allowed and in which roles. Models can be organized in a hierarchy, starting with the root module. The aspects provide visibility control. The relationships can be (directed or undirected) connections, further characterized by attributes. The model specification can define several kinds of connections, which objects can participate in a connection and further explicit constraints. The connection only appears between two objects in the same model entity. References help establish connections to external objects as well.

9.6 Conclusions

This chapter discussed the use of software models for the design and verification of embedded software systems. It attempted to classify and a survey existing formal models of computation, following the classical divide between synchronous and asynchronous models and between models for functionality as opposed to models for software architecture specification. Problems like formal verification of system properties, both timed and untimed and schedulability analysis are discussed. This chapter also provided an overview of commercially relevant modeling languages UML and SDL, and discusses recent extensions to both these standards. The discussion of each topic was supplemented with an indication of the available tools that implemented the available methodologies and analysis algorithms. Finally, the chapter contained a short survey of the results of recent researches and a discussion of open issues and future trends.

References

[1] Sangiovanni Vincentelli, A., Defining platform-based design, *EEDesign of EETimes*, February 2002.

[2] Lee, E.A., Overview of the Ptolemy Project, Technical Memorandum UCB/ERL M03/25, July 2, 2003, University of California, Berkeley, CA.

[3] Balarin, F., Hsieh, H., Lavagno, L., Passerone, C., Sangiovanni-Vincentelli, A., and Watanabe, Y., Metropolis: an integrated environment for electronic system design, *IEEE Computer*, April 2003.

[4] *UML Profile for Schedulability, Performance and Time Specification.* OMG Adopted Specification, July 1, 2002, http://www.omg.org.

[5] Beck, T., Current trends in the design of automotive electronic systems, *Proceedings of the Design Automation and Test in Europe Conference*, 2001.

[6] Edwards, S., Lavagno, L., Lee, E.A., and Sangiovanni-Vincentelli, A., Design of embedded systems: formal models, validation and synthesis, *Proceedings of the IEEE*, March 1997.

[7] Alur, R. and Henzinger, T.A., Logics and models of real time: a survey, *Real-Time: Theory in Practice, REX Workshop*, Lecture Notes in Computer Science, Vol. 600, Springer, Berlin, 1991, pp. 74–106.

[8] Pnueli, A., The temporal logic of programs, *Proceedings of the 18th Annual Symposium on the Foundations of Computer Science, IEEE*, November 1977, pp. 46–57.

[9] Emerson, E.A., Temporal and modal logics, in *Handbook of Theoretical Computer Science*, Vol. B, van Leeuwen, J., Ed., Elsevier, Amsterdam, 1990, pp. 995–1072.

[10] Holzmann, G.J., *Design and Validation of Computer Protocols*, Prentice-Hall, Eaglewood cliffs, NJ, 1991.

[11] Harel, D., Statecharts: a visual approach to complex systems, *Sci. Comp. Prog.*, 8, 231–275, 1987.

[12] Merlin, P.M. and Farber, D.J., Recoverability of communication protocols, *IEEE Trans. Comm.*, 24, 36–103, 1976.

[13] Sathaye, A.S. and Krogh, B.H. Synthesis of real-time supervisors for controlled time Petri nets, *Proceedings of 32nd IEEE Conference on Decision and Control*, Vol. 1, San Antonio, 1993, pp. 235–238.

[14] Alur, R. and Dill, D.L., A theory of timed automata, *Theoret. Comput. Sci.*, 126, 183–235, 1994.

[15] Ramchandani, C., Analysis of Asynchronous Concurrent Systems by Timed Petri Nets, Ph. D. thesis, MIT, Cambridge, MA, 1974.

[16] Molloy, M.K., Performance analysis using stochastic Petri nets, *IEEE Trans. Comp.*, 31, 913–917, 1982.

[17] Ajmone Marsan, M., Conte, G., and Balbo, G., A class of generalized stochastic Petri nets for the performance evaluation of multiprocessor systems, *ACM Trans. Comp. Sys.*, 2, 93–122, 1984.

[18] Haar, S., Kaiser, L., Simonot-Lion, F., Toussaint, J., On Equivalence Between Timed State Machines and Time Petri Nets, Rapport de recherche de l'INRIA – Lorraine, 2000.

[19] Yovine, S., Kronos: a verification tool for real-time systems, *Springer Int. J. Software Tools Technol. Transfer*, 1, 123–133, 1997.

[20] Larsen, K.G., Pettersson, P., and Yi, W. Uppaal in a nutshell, *Springer Int. J. Software Tools Technol. Transfer* 1, 134–152, 1997.

[21] Yi, W., Pettersson, P., and Daniels, M., Automatic verification of real-time communicating systems by constraint solving, *Proceedings of the 7th International Conference on Formal Description Techniques*, Berne, Switzerland, 4–7 October 1994.

[22] Henzinger, T.A., The theory of hybrid automata, *Proceedings of the 11th Annual Symposium on Logic in Computer Science (LICS)*, IEEE Computer Society Press, Washington, DC, 1996, pp. 278–292.

[23] Henzinger, T.A., Ho, P.-H., and Wong-Toi, H., HyTech: a model checker for hybrid systems, *Software Tools Technol. Trans.*, 1, 110–122, 1997.

[24] Vicario, E., Static analysis and dynamic steering of time-dependent systems using time Petri nets, *IEEE Trans. Software Eng.*, 27, 728–748, 2001.

[25] The ORIS tool web page: http://www.dsi.unifi.it/~vicario/Research/ORIS/oris.html.

[26] Lime, D. and Roux, O.H., A translation based method for the timed analysis of scheduling extended time Petri nets, *The 25th IEEE International Real-Time Systems Symposium*, December 5–8, Lisbon, Portugal, 2004.

[27] Clarke, E.M. and Wing, J.M., Formal Methods: State of the Art and Future Directions, *Technical Report CMU-CS-96-178*, Carnegie Mellon University (CMU), Sept. 1996.

[28] Liu, C. and Layland, J., Scheduling algorithm for multiprogramming in a hard real-time environment, *J. ACM*, 20, 46–61, 1973.

[29] Klein, M.H. et al., *A practitioner's Handbook for Real-Time Analysis: Guide to Rate Monotonic Analysis for Real-Time Systems*, Kluwer Academic Publishers, Hingham, MA, 1993.

[30] Rajkumar, R., *Synchronization in Multiple Processor Systems, Synchronization in Real-Time Systems: A Priority Inheritance Approach*, Kluwer Academic Publishers, Dordrecht, 1991.

[31] Benveniste, A., Caspi, P., Edwards, S.A., Halbwachs, N., Le Guernic, P., and de Simone, R., The synchronous languages 12 years later, *Proc. IEEE*, 91–1, 64–83, 2003.

[32] Real-Time Workshop Embedded Coder 4.2, web page: http://www.mathworks.com/products/rtwembedded/.

[33] dSPACE Produkte: Production Code Generation Software, web page: www.dspace.de/ww/de/pub/products/sw/targetli.htm.

[34] Rational Rose Technical Developer, web page: http://www-306.ibm.com/software/awdtools/developer/technical/.

[35] Meyer, B., An overview of Eiffel, in *The Handbook of Programming Languages, Vol. 1, Object-Oriented Languages*, Salus, P.H., Ed., Macmillan Technical Publishing, New York, 1998.

[36] Caspi, P., Pilaud, D., Halbwachs, N., and Plaice, J.A., LUSTRE: a declarative language for programming synchronous systems, *ACM Symposium Principles of Programmimg Language (POPL)*, Munich, Germany, 1987, pp. 178–188.

[37] Halbwachs, N., Caspi, P., Raymond, P., and Pilaud, D., The synchronous data flow programming language LUSTRE, *Proc. IEEE*, 79, 1305–1320, 1991.

[38] Boussinot, F. and de Simone, R., The Esterel language, *Proc. IEEE*, 79, 1293–1304, 1991.

[39] Berry, G., The constructive semantics of pure Esterel, 5th *Algebraic Methodology and Software Technology Conference*, Munich, Germany, 1996, pp. 225–232.

[40] Westhead, M. and Nadjm-Tehrani, S., Verification of embedded systems using synchronous observers, in Lecture Notes in Computer Science, Formal Techniques in Real-Time and Fault-Tolerant Systems, Vol. 1135, Springer-Verlag, Berlin, 1996.

[41] The Mathworks Simulink and StateFlow. web page: http://www.mathworks.com.

[42] Scaife, N., Sofronis, C., Caspi, P., Tripakis, S., and Maraninchi, F., Defining and translating a "safe" subset of Simulink/Stateflow into Lustre, *Proceedings of 2004 Conference on Embedded Software, EMSOFT'04*, Pisa, Italy, September 2004, Springer, Berlin.

[43] Scaife, N. and Caspi, P., Integrating model-based design and preemptive scheduling in mixed time- and event-triggered systems, *16th Euromicro Conference on Real-Time Systems (ECRTS'04)*, Catania, Italy, 2004, pp. 119–126.

[44] Prover Technology, http://www.prover.com/.

[45] Edwards, S.A., An Esterel compiler for large control-dominated systems, *IEEE Trans. Comput. -Aided Des. Integr. Circuits Syst.*, 21, 2002, 169–183.

[46] Shiple, T. R., Berry, G., and Touati, H., Constructive analysis of cyclic circuits, *European Design and Test Conference*, 1996.

[47] Benveniste, A., Caspi, P., Le Guernic, P., Marchand, H., Talpin, J.-P., and Tripakis, S., A protocol for loosely time-triggered architectures, *Proceedings of 2002 Conference on Embedded Software, EMSOFT'02*, Sifakis, J. and Sangiovanni-Vincentelli, A., Eds., Lecture Notes in Computer Science, Vol. 2491 , Springer, Berlin, pp. 252–265.

[48] Benveniste, A., Caillaud, B., Carloni, L., Caspi, P., and Sangiovanni-Vincentelli, A., Heterogeneous reactive systems modeling: capturing causality and the correctness of loosely time-triggered architectures (LTTA), *Proceedings of the Conference on Embedded Software, EMSOFT'04*, Buttazzo, G. and Edwards, S., Eds., September 2004, pp. 27–29.

[49] Biannic, Y. L., Nassor, E., Ledinot, E., and Dissoubray, S., UML object specification for real-time software. *RTS Show*, 2000.

[50] Selic, B., Gullekson, G., and Ward, P.T., *Real-Time Object-Oriented Modeling*, Wiley, New York, 1994.

[51] Douglass, B. P., *Doing Hard Time: Developing Real-Time Systems with Objects, Frameworks, and Patterns*, Addison-Wesley, Reading, MA, 1999.

[52] Latella, D., Majzik, I., and Massink, M., Automatic verification of a behavioural subset of UML statechart diagrams using the SPIN modelchecker, *Formal Aspects Comput.* 11, 637–664, 1999.

[53] del Mar Gallardo, M., Merino, P., and Pimentel, E., Debugging UML designs with model checking, *J. Object Technol.*, 1, 2002.

[54] UML 2.0 OCL Final Adopted specification, available at http://www.omg.org/cgi-bin/doc?ptc/2003-10-14.

[55] ITU-T. Recommendation Z.100. Specification and Description Language (SDL). Z-100, Int. Telecom. Union Standard. Sect., 2000.

[56] ITU-T. Recommendation Z.120. Message Sequence Charts. Z-120, Int. Telecom. Union Standard. Sect, 2000.

[57] The PEP tool (Programming Environment based on Petri Nets), Documentation and user guide: http://parsys.informatik.uni-oldenburg.de/~pep/Paper/PEP1.8_doc.ps.gz.

[58] Bozga, M., Ghirvu, L., Graf, S., and Mounier, L., IF: a validation environment for timed asynchronous systems, *Comp. Aided Verification, CAV*, Lecture Notes in Computer Science, Vol. 1855, Springer, Berlin, 2000.

[59] Bozga, M., Graf, S., and Mounier, L., IF-2.0: a validation environment for component-based real-time systems. *Comp. Aided Verification, CAV*, Lecture Notes in Computer Science, Vol. 2404, Springer, Berlin, 2002, pp. 343–348.

[60] Regensburger, F and Barnard, A., Formal verification of SDL systems at the Siemens mobile phone department, *Tools and Algorithms for the Construction and Analysis of Systems. 4th International Conference, TACAS'98*, Lecture Notes in Computer Science, Vol. 1384, Springer, Berlin, 1998, pp. 439–455.

[61] Bozga, M., Graf, S., and Mounier, L., Automated validation of distributed software using the IF environment, *Workshop on Software Model Checking, Electronic Notes in Theoretical Computer Science*, Vol. 55, Elsevier, Amsterdam, 2001.

[62] Gomaa, H. *Software Design Methods for Concurrent and Real-Time Systems*, Addison-Wesley, Reading, MA, 1993.

[63] Burns A. and Wellings, A.J., HRT-HOOD: a design method for hard real-time, *J. Real-Time Sys.*, 6, 73–114, 1994.

[64] Awad, M., Kuusela, J., and Ziegler J., *Object-Oriented Technology for Real-Time Systems: A Practical Approach Using OMT and Fusion*, Prentice-Hall, Eaglewood Cliffs, NJ, 1996.

[65] Saksena, M., Freedman, P., and Rodziewicz, P., Guidelines for automated implementation of executable object oriented models for real-time embedded control systems, *Proceedings IEEE Real-Time Systems Symposium*, December 1997, pp. 240–251.

[66] Saksena, M. and Karvelas, P., Designing for schedulability: integrating schedulability analysis with object-oriented design, *Proceedings of the Euromicro Conference on Real-Time Systems*, Stockholm, June 2000.

[67] Saksena, M., Karvelas, P., and Wang, Y., Automatic synthesis of multi-tasking implementations from real-time object-oriented models, *Proceedings IEEE International Symposium on Object-Oriented Real-Time Distributed Computing*, March 2000.

[68] Slomka, F., Dörfel, M., Münzenberger, R., and Hofmann, R., Hardware/software codesign and rapid-prototyping of embedded systems. *IEEE Design Test Computers, (Special issue: Design Tools for Embedded Systems)*, Vol. 17, 28–38, 2000.

[69] Bozga, M., Graf, S., Mounier, L., Ober, I., Roux, J.-L., and Vincent, D., Timed extensions for SDL, *Proceedings of the SDL Forum 2001*, Lecture Notes in Computer Science, Vol. 2078, Springer, Berlin, June 2001.

[70] Münzenberger, R. Slomka, F. Dörfel, M., and Hofmann, R., A general approach for the specification of real-time systems with SDL, in *Proceedings of the 10th International SDL Forum*, Reed, R. and Reed, J., Eds., Lecture Notes in Computer Science, Vol. 2078, Springer, Berlin, 2001.

[71] Algayres, B., Lejeune, Y., and Hugonnet, F., GOAL: observing SDL behaviors with Geode, *Proceedings of SDL Forum 95.*

[72] Dörfel, M., Dulz, W., Hofmann, R., and Münzenberger, R., SDL and Non-Functional Requirement, Internal Report IMMD7 05/01, University of Erlangen, August 20, 2001.

[73] Mitschele-Thiel, S and Muller-Clostermann, B., Performance engineering of sdl/msc systems, *Computer Networks*, 31,1801–1815, 1999.

[74] Telelogic ObjectGeode, web page: http://www.telelogic.com/products/additional/objectgeode/index.cfm.

[75] Roux, J.L., SDL Performance analysis with ObjectGeode, *Workshop on performance and time in SDL*, 1998.

[76] Telelogic TAU Generation2, web page: http://www.telelogic.com/products/tau/tg2.cfm.

[77] Bjorkander, M., Real-time systems in UML (and SDL), *Embedded System Engineering*, October/November 2000, published by EDA.

[78] Diefenbruch, M., Heck, E., Hintelmann, J., and Müller-Clostermann, B., Performance evaluation of SDL systems adjunct by queuing models, *Proceedings of the SDL-Forum'95*, 1995.

[79] Alvarez, J.M., Diaz, M., Llopis, L.M., Pimentel, E., and Troya, J.M., Deriving hard-real time embedded systems implementations directly from SDL specifications, *International Symposium on Hardware/Software Codesign CODES*, 2001.

[80] Bucci, G., Fedeli, A., and Vicario, E., Specification and simulation of real time concurrent systems using standard SDL Tools, *11th SDL Forum*, Stuttgart, July 2003.

[81] Spitz, S; Slomka, F., and Dörfel, M., SDL* — an annotated specification language for engineering multimedia communication systems, *Workshop on High Speed Networks*, Stuttgart, October 1999.

[82] Malek, M., PerfSDL: interface to protocol performance analysis by means of simulation, *Proceedings of the SDL Forum* 99.

[83] I-Logix Rhapsody, web page: http://www.ilogix.com/rhapsody/rhapsody.cfm.

[84] Artisan Real-Time Studio, web page: http://www.artisansw.com/products/professional_overview.asp.

[85] Telelogic TAU TTCN Suite, web page: http://www.telelogic.com/products/tau/ttcn/index.cfm.

[86] Henriksson, D., Cervin A. and Årzén, K.-E., Truetime: simulation of control loops under shared computer resources, *Proceedings of the 15th IFAC World Congress on Automatic Control*, Barcelona, Spain, July 2002.

[87] Amnell, T., et al., Times — a tool for modelling and implementation of embedded systems, in *Proceedings of 8th International Conference, TACAS* 2002, Grenoble, France, April 8–12, 2002.

[88] Henzinger, T.A., Giotto: a time-triggered language for embedded programming, in *Proceedings of 1st International Workshop on Embedded Software (EMSOFT'2001)*, Tahoe City, CA, USA, Oct. 2001, Vol. 2211, Lecture Notes in Computer Science, Springer, Berlin, 2001, pp. 166–184.

[89] Lee, E.A. and Xiong, Y., System-level types for component-based design, *Proceedings 1st International Workshop on Embedded Software (EMSOFT'2001)*, Tahoe City, CA, October 2001, Lecture Notes in Computer Science, Vol. 2211, Springer, Berlin, 2001, pp. 237–253.

[90] Balarin, F. Lavagno, L. Passerone, C., and Watanabe, Y., Processes, interfaces and platforms. embedded software modeling in Metropolis, *Proceedings of the EMSOFT Conference*, Grenoble, France, 2002, 407–416.

[91] Balarin, F., Lavagno, L., Passerone, C., Sangiovanni Vincentelli, A., Sgroi M., and Watanabe, Y., Modeling and design of heterogeneous systems, Lecture Notes in Computer Science, Vol. 2549, Springer, Berlin, 2002, pp. 228–273.

[92] Lee, E.A. and Sangiovanni-Vincentelli, A., Comparing models of computation, *Proceedings of the IEEE/ACM International Conference on Computer-aided Design*, 1996, pp. 234–241.

[93] Gossler, G. and Sifakis. J., Composition for component-based modeling, *Proceedings of FMCO'02*, November 2002, Leiden, The Netherlands, Lecture Notes in Computer Science, Vol. 2852, Springer, Berlin, pp. 443–466.

[94] Altisen, K., Goessler, G., and Sifakis, J. Scheduler modeling based on the controller synthesis paradigm J.Real-Time Sys. (special issue on Control Approaches to Real-Time Computing), 23, 55–84, 2002.

[95] Gossler, G. and Sifakis, J., Composition for component-based modeling, *Proceedings of the FMCO Conference*, Leiden, The Netherlands, November 5–8, 2002.

[96] Ledeczi, A., Maroti, M., Bakay, A., Karsai, G., Garrett, J., Thomason, IV, C., Nordstrom, G., Sprinkle, J., and Volgyesi, P., The generic modeling environment, *Workshop on Intelligent Signal Processing*, Budapest, Hungary, May 17, 2001.

[97] Jantsch, A., *Modeling Embedded Systems and SoCs-Concurrency and Time in Models of Computation*, Morgan Kaufman, San Fransisco, CA, June 2003, ISBN 1558609253.

10

Using Performance Metrics to Select Microprocessor Cores for IC Designs

Steve Leibson
Tensilica, Inc.
Santa Clara, California

10.1 Introduction

Just as they have become essential building blocks for board-level designs, microprocessors are now absolutely essential components in integrated circuit (IC) design. The reason is simple. Microprocessor cores are the most reusable on-chip components a design team can use because of their easy software programmability and because of the extensive development-tool environment that surrounds any good microprocessor or microprocessor core. Although it is easy to envision using microprocessor cores to implement many tasks on an IC, it is difficult to select an appropriate processor from the many cores now available because contemporary microprocessor cores are complex, multi-function elements. Consequently, there is no single, simple, meaningful way to evaluate the suitability of a microprocessor core for specific embedded tasks on an IC. Many factors must be considered including processor performance, gate count, power dissipation, the availability of suitable interfaces, and the processor's software-development support. This chapter deals with the objective measurement of processor-core performance for the purpose of selection of microprocessor cores for IC design.

Designers measure microprocessor and microprocessor-core performance through benchmarking. Processor chips already realized in silicon are considerably easier to benchmark than processor cores. All major microprocessor vendors offer their processors in evaluation kits, which include complete evaluation boards and software-development tool suites (which include compilers, assemblers, linkers, loaders, instruction-set simulators [ISSs], and debuggers). Consequently, benchmarking chip-level microprocessor performance consists of compiling selected benchmarks for the target processor, downloading and running those benchmarks on the evaluation board, and recording the results.

Microprocessor cores are incorporeal and they are not usually available as stand-alone chips. In fact if these cores were available as chips, they would be far less useful to IC designers because on-chip processors can have vastly greater I/O capabilities than processors realized as individual chips. Numerous wide and fast buses are the norm for microprocessor cores to accommodate instruction and data caches, local memories, and private I/O ports.This is not true for microprocessor chips because these additional buses would greatly increase pin count, which would drive component cost beyond a tolerable level for microprocessor chips. However, extra I/O pins for microprocessor cores are not costly, so they usually have many buses and ports.

Benchmarking microprocessor cores has become increasingly important in the IC-design flow because any significant 21st century IC design incorporates more than one processor core — at least two, often several, and occasionally hundreds (see Figure 10.1). As the number of microprocessor cores used on a chip increases, and as these processor cores perform more tasks that were previously implemented with blocks of logic that were manually designed and coded by hand, measuring processor-core performance becomes an increasingly important task in the overall design of the IC.

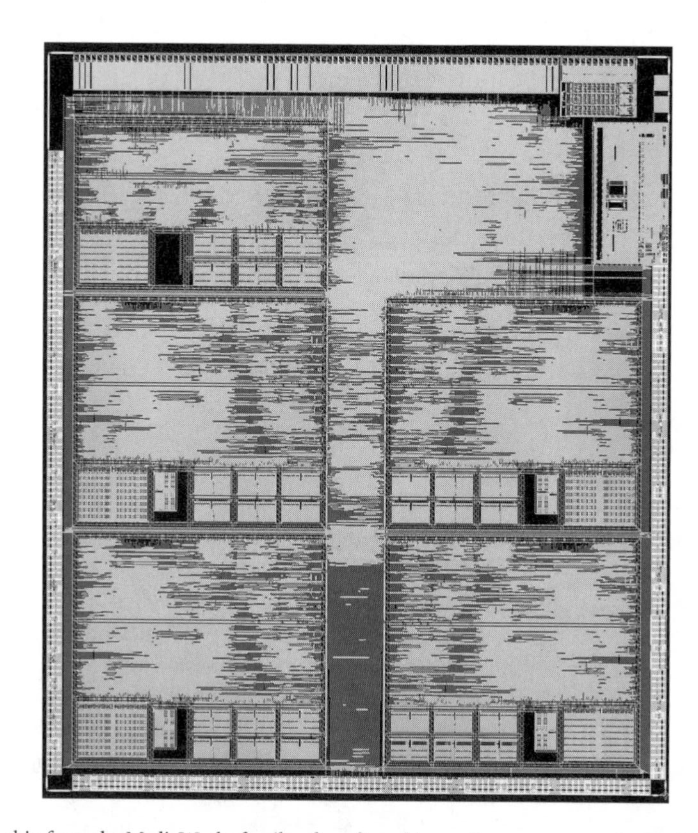

FIGURE 10.1 A chip from the MediaWorks family of configurable media processor products which is a DVD-resolution MPEG video and audio encoder/decoder system on a chip and is intended for use in solid-state camcorders and portable video products. The chip contains five loosely coupled heterogeneous processors. Reproduced with permission.

Measuring the performance of microprocessor cores is a bit more complex than benchmarking microprocessor chips: another piece of software, an ISS, must stand in for a physical processor chip. Each microprocessor and each member of a microprocessor family requires its own specific ISS because the ISS must provide a cycle-accurate model of the processor's architecture to produce accurate benchmark results. Note that it is also possible to benchmark a processor core using gate-level simulation, but this approach is three orders of magnitude slower (because gate-level simulation is much slower than instruction-set simulation) and is therefore used infrequently [1]. The selected benchmarks are compiled and run on the ISS to produce the benchmark results. All that remains is to determine where the ISS and the benchmarks are to be obtained, how they are to be used, and how the results will be interpreted. These are the subjects of this chapter.

10.2 The ISS as Benchmarking Platform

Instruction-set simulators serve as benchmarking platforms because processor cores as realized on an IC rarely exist as a chip. Instruction-set simulators simulate the software-visible state of a microprocessor without employing a gate-level model of the processor so that they run quickly, often 1000 times faster than gate-level processor simulations running on HDL simulators. The earliest ISS was created for the EDSAC (Electronic Delay Storage Automatic Calculator), which was developed by a team led by Maurice V. Wilkes at the University of Cambridge's Computer Laboratory (shown in Figure 10.2). The room-sized EDSAC I was the world's first fully operational, stored-program computer (the first von Neumann machine) and went online in 11949. EDSAC II became operational in 1958.

The EDSAC ISS was first described in a paper on debugging EDSAC programs, which was written and published in 1951 by S. Gill, one of Wilkes' research students [2]. The paper describes the operation of a "tracing simulator", which operates by fetching the simulated instruction, decoding the instruction to save trace information, updating the simulated program counter if the instruction is a branch, or else placing the nonbranch instruction in the middle of the simulator loop and executing it directly, and then

FIGURE 10.2 The EDSAC I, developed at the Computer Laboratory, University of Cambridge, UK, which became the first fully functional stored-program computer in 1948. This computer also had the first instruction-set simulator, which was described in an article published in 1951. Copyright Computer Laboratory, University of Cambridge. Reproduced with permission.

returning to the top of the simulator loop. Thus the first operational processor ISS predates the introduction of the first commercial microprocessor (Intel's 4004) by some 20 years.

Cycle-accurate ISSs, the most useful simulator class for processor benchmarking, compute exact instruction timing by accurately modeling the processor's pipeline. All commercial microprocessor cores have at least one corresponding ISS. They are obtained in different ways. Some core vendors rely on third parties to provide an ISS for their microprocessors. Other vendors offer an ISS as part of their tool suite. Some ISSs must be purchased and some are available in evaluation packages from the processor-core vendor.

To serve as an effective benchmarking tool, an ISS must not only simulate the operation of the processor core, but must also provide the instrumentation needed to provide the critical statistics of the benchmark run. These statistics include cycle counts for the various functions and routines executed, and main- and cache-memory usage. The better the ISS instrumentation, the more comparative information the benchmark will produce.

10.3 Ideal Versus Practical Processor Benchmarks

Integrated-circuit designers use benchmarks to help them pick the "best" microprocessor core for a given task to be performed on the IC. The original definition of a benchmark was literally a mark on a workbench that provided some measurement standard. Eventually, the first benchmarks (carved into the workbench) were replaced with standard measuring tools such as yardsticks. Processor benchmarks provide yardsticks for measuring processor performance. The ideal yardstick would be one that could measure any processor for any task. The ideal processor benchmark would produce results that are relevant, reliable, objective, comparable, and applicable. Unfortunately, no such processor benchmark exists.

In one sense, the ideal processor benchmark would be the actual application code that the processor will run. No other piece of code can possibly be as representative of the actual task to be performed as the actual code that executes that task. No other piece of code can possibly replicate the instruction-use distribution, register and memory use, or data-movement patterns of the actual application code. In many ways however, the actual application code is less than ideal as a benchmark.

First and foremost, the actual application code may not exist when candidate processors are benchmarked, because benchmarking and processor selection must occur early in the project. A benchmark that does not exist is worthless.

Next, the actual application code serves as an extremely specific benchmark. It will indeed give a very accurate prediction of processor performance for a specific task, *and for no other task*. In other words, the downside of a highly specific benchmark is that the benchmark will give a less-than-ideal indication of processor performance for other tasks. Because on-chip processor cores are often used for a variety of tasks, the ideal benchmark may well be a suite of application programs and not just one program.

Yet another problem with application-code benchmarks is their lack of instrumentation. The actual application code has almost always been written to execute the task, not to measure a processor core's performance. Appropriate measurements may require modification of the application code. This modification consumes time and resources, which may not be readily available. Even so, with all of these issues that make the application-code benchmark less than ideal, the application code (if it exists) provides invaluable information on processor-core performance and should be used to help make a processor-core selection whenever possible.

10.4 Standard Benchmark Types

Given that the "ideal" processor benchmark proves less than ideal, the industry has sought "standard" benchmarks that it can use when the target application code is either not available or not appropriate. There are four types of standard benchmarks: full-application or "real-world" benchmarks, synthetic or small-kernel benchmarks, hybrid or derived benchmarks that mix and match aspects of the full-application and synthetic benchmarks, and micro-benchmarks (not to be confused with microprocessor benchmarks).

Full-application benchmarks and benchmark suites employ existing system- or application-level code drawn from real applications, although probably not the specific application of interest to any given IC-design team. These benchmarks may incorporate many thousands of lines of code, have large instruction-memory footprints, and consume large amounts of data memory.

Synthetic benchmarks tend to be smaller than full-application benchmarks. They consist of smaller code sections representing commonly used algorithms. They may be extracted from working code or they may be written specifically as a benchmark. Writers of synthetic benchmarks try to approximate instruction mixes of real-world applications without replicating the entire application.

Hybrid benchmarks mix and match large application programs and smaller blocks of synthetic code to create a sort of torture track (with long straight sections and tight curves) for exercising microprocessors. The hybrid-benchmark code is augmented with test data sets taken from real-world applications. A microprocessor core's performance around this torture track can give a good indication of the processor's abilities over a wide range of situations, although probably not the specific use to which the processor will be put on an IC.

Micro-benchmarks are very small code snippets designed to exercise some particular processor feature or to characterize a particular machine primitive in isolation from all other processor features and primitives. Micro-benchmark results can define a processor's peak capabilities and reveal potential architectural bottlenecks, but peak performance is not a very good indicator of overall application performance. Nevertheless, a suite of micro-benchmarks may approximate the ideal benchmark for certain applications, if the application is a common one with many standardized, well-defined functions to perform.

10.5 Prehistoric Performance Ratings: MIPS, MOPS, and MFLOPS

Lord Kelvin could have been predicting processor performance measurements when he said: "When you can measure what you are speaking about, and express it in numbers, you know something about it; but when you cannot measure it, when you cannot express it in numbers, your knowledge is of a meager and unsatisfactory kind. It may be the beginning of knowledge, but you have scarcely, in your thoughts, advanced to the stage of science". [3]

The need to rate processor performance is so great that at first, microprocessor vendors grabbed any and all numbers at hand to rate performance. These prehistoric ratings are measures of processor performance that lack code to standardize the ratings. Consequently, these performance ratings are not benchmarks. Just as an engine's RPM (revolutions per minute) reading does not give one sufficient information to measure the performance of an internal combustion engine (you need engine torque plus transmission gearing, differential gearing, and tire diameter to compute load), the prehistoric, clock-related processor ratings of MIPS, MOPS, MFLOPS, and VAX MIPS are akin to clock rate: they tell one almost nothing about the true performance potential of a processor.

Before it became the name of a microprocessor vendor, the term "MIPS" was an acronym for "millions of instructions per second." If all processors had the same instruction-set architecture (ISA), then a MIPS rating could possibly be used as a performance measure. However, all processors do not have the same ISA. In fact, microprocessors and microprocessor cores have very different ISAs. Consequently, some processors can do more with one instruction than other processors, as large automobile engines can do more than smaller ones at the same RPM.

This problem was already bad in the days when only CISC (complex-instruction-set computers) processors roamed the Earth. The problem went from bad to worse when RISC (reduced-instruction-set computers) processors arrived on the scene. One CISC instruction would often do the work of several RISC instructions (by design), so that a CISC MIPS rating did not correlate well with a RISC MIPS rating because of the work differential between RISC's simple instructions and CISC's more complex instructions.

The next step in creating a usable processor performance rating was to switch from MIPS to VAX MIPS, which was accomplished by setting the extremely successful VAX 11/780 minicomputer — introduced in 1977 by the now defunct Digital Equipment Corporation (DEC) — as the standard against which all other processors are measured. So, if a processor ran twice as fast as a VAX 11/780 on a set of programs, it was rated at 2 VAX MIPS. The original term "MIPS" then became "native MIPS," so as not to confuse the original ratings with VAX MIPS. DEC referred to VAX MIPS as VUPs (VAX units of performance) to keep things interesting.

Both native and VAX MIPS are inadequate measures of processor performance because they are usually provided without specifying the software (or even the programming language) used to make the measurement. Because different programs have different mixes of instructions, different memory-usage patterns, and different data-movement patterns, the same processor may earn one MIPS rating on one set of programs and quite another rating on a different set. Because MIPS ratings are not linked to a specific benchmark program suite, the "MIPS" acronym has come to stand for "meaningless indication of performance."

A further problem with the VAX MIPS measure of processor performance is that the concept of using a VAX 11/780 minicomputer as the gold performance standard is an idea that is more than a bit long in the tooth in the 21st century. There are no longer many (or any) VAX 11/780s available for running benchmark code and DEC effectively disappeared when Compaq Computer Corp. purchased what was left of the company in January, 1998 following the decline of the minicomputer market. Hewlett-Packard absorbed Compaq in May 2002, submerging DEC's identity even further.

Even more tenuous than the MIPS performance rating is the concept of MOPS, an acronym that stands for "millions of operations per second." Every algorithmic task requires the completion of a certain number of fundamental operations, which may or may not have a one-to-one correspondence with machine instructions. Count these fundamental operations in the millions and they become MOPS. If they are floating-point operations, one gets MFLOPS (millions of floating-point operations per second). Both MOPS and MFLOPS ratings suffer from the same drawback as the MIPS rating: there is no standard software to serve as the benchmark that produces the ratings. In addition, the "conversion factor" for computing how many operations a processor performs per clock (or how many processor instructions constitute one operation) is somewhat fluid as well, which means that the processor vendor is free to develop a conversion factor on its own. Consequently, MOPS and MFLOPS performance ratings exist for various processor cores but they really do not help an IC-design team pick a processor core because they are not true benchmarks.

10.6 Classic Processor Benchmarks (The Stone Age)

Like ISSs, standardized processor performance benchmarks predate the 1971 introduction of Intel's 4004 microprocessor, but barely. The first benchmark suite to attain *de facto* "standard" status was a set of programs known as the Livermore Kernels (also popularly called the Livermore Loops).

10.6.1 The LFK Benchmark

The Livermore Kernels were first developed in 1970 and consist of 14 numerically intensive application kernels written in FORTRAN. Ten more kernels were added in the 1980s and the final suite of benchmarks was discussed in a paper published in 1986 by F. H. McMahon of the Lawrence Livermore National Laboratory (LLNL), Livermore, CA [4]. The Livermore Kernels actually constitute a supercomputer benchmark, measuring a processor's floating-point computational performance in terms of MFLOPS. Because of the somewhat frequent occurrence of floating-point errors in many computers, the Livermore Kernels test both the processor's speed and the system's computational accuracy. Today, the Livermore Kernels are called the Livermore FORTRAN Kernels (LFK).

The LFK are real samples of floating-point computations taken from a diverse workload of scientific applications extracted from operational program code used at LLNL. The kernels were extracted from programs in use at LLNL because those programs were generally far too large to serve as useful benchmark programs; they included hardware-specific subroutines for performing functions such as I/O, memory management, and graphics that were not appropriate for benchmark testing of floating-point performance; and they were largely classified, due to the nature of the work done at LLNL. Some kernels represent widely used, generic computations such as dot and matrix (SAXPY) products, polynomials, first sum and differences, first-order recurrences, matrix solvers, and array searches. Some kernels typify often-used FORTRAN constructs while others contain constructs that are difficult to compile into efficient machine code. These kernels were selected to represent both the best and worst cases of common FORTRAN programming practice to produce results that measure a realistic floating-point performance range by challenging the FORTRAN compiler's ability to produce optimized machine code. Table 10.1 lists the 24 LFK .

A complete LFK run produces 72 timed results, by timing the execution of the 24 LFK three times using different DO-loop lengths.

The LFK are a mixture of vectorizable and nonvectorizable loops, and test the computational capabilities of the microprocessor hardware and the ability of the software tools to create efficient code. The LFK benchmark also tests a processor's vector abilities and the associated software tools' abilities to vectorize code.

At first glance, the LFK benchmark appears to be nearly useless for the benchmarking of embedded microprocessor cores in ICs. It is a floating-point benchmark written in FORTRAN that looks for good vector abilities. FORTRAN compilers for embedded microprocessor cores are also quite rare and unusual. Today, very few real-world applications run tasks like those appearing in the 24 LFK , which are far more suited to research on the effects of very high speed nuclear reactions than the development of commercial, industrial, or consumer products. It is unlikely that the processors in a mobile phone handset or an MP3 music player will ever need to perform 2-D hydrodynamic calculations. Embedded-software developers working in FORTRAN are also rare. Consequently, microprocessor-core vendors are quite unlikely to tout LFK benchmark results for their cores. The LFK benchmarks are far more suited to testing supercomputers such as a Cray XD1 or Intel's ASCI Option Red. However, as on-chip gate counts grow, as microprocessor cores gain SIMD (single-instruction, multiple-data) and floating-point execution units, and as IC designs increasingly tackle tougher number-crunching applications including the implementation of audio and video codecs, the LFK benchmark could become valuable for testing more than just supercomputers.

TABLE 10.1 The 24 Kernels in the Livermore FORTRAN Kernels

LFK Number	Kernel Description
1	An excerpt from a hydrodynamic application
2	An excerpt from an Incomplete Cholesky-Conjugate Gradient program
3	The standard inner-product function from linear algebra
4	An excerpt from a banded linear equations routine
5	An excerpt from a tridiagonal elimination routine
6	An example of a general linear recurrence equation
7	Equation of state code fragment (as used in nuclear weapons research)
8	An excerpt of an alternating direction, implicit integration program
9	An integrate predictor program
10	A difference predictor program
11	A first sum
12	A first difference
13	An excerpt from a 2-D particle-in-cell program
14	An excerpt from a 1-D particle-in-cell program
15	A sample of casually written FORTRAN to produce sub-optimal machine code
16	A search loop from a Monte–Carlo program
17	An example of an implicit conditional computation
18	An excerpt from a 2-D explicit hydrodynamic program
19	A general linear recurrence equation
20	An excerpt from a discrete ordinate transport program
21	A matrix product calculation
22	A Planckian distribution procedure
23	An excerpt from a 2-D implicit hydrodynamic program
24	Finds the location of the first minimum in X

10.6.2 LINPACK

LINPACK is a collection of FORTRAN subroutines that analyze and solve linear equations and linear least-squares problems. Jack Dongarra assembled the LINPACK collection of linear algebra routines at the Argonne National Laboratory, Argonne, IL. The first versions of LINPACK existed in 1976 but the first users' guide was published in 1977 [5,6]. The package solves linear systems whose matrices are general, banded, symmetric-indefinite, symmetric positive definite, triangular, and tridiagonal square. In addition, the package computes the QR and singular value decompositions of rectangular matrices and applies them to least-squares problems. The LINPACK routines are not strictly speaking a benchmark, but they exercise a computer's floating-point capabilities and, as of 2005, Dongarra maintains an online list of computer systems and their LINPACK performance results at www.netlib.org.

Originally, LINPACK benchmarks performed computations on a 100×100 matrix, but with the rapid increase in computing performance that has occurred, LINPACK performance numbers for 1000×1000 arrays now appear on Dongarra's site as well. The original versions of LINPACK were written in FORTRAN but there are now versions written in C and Java as well. Performance is reported in single- and double-precision MFLOPS, which reflects the benchmark's (and the national laboratories') focus on "big iron" machines (supercomputers). The LINPACK benchmarks are not commonly used to measure microprocessor or microprocessor-core performance.

10.6.3 The Whetstone Benchmark

The Whetstone benchmark was written by Harold Curnow of the now defunct CCTA (Central Computer and Telecommunications Agency, the British government agency tasked with computer procurement) to test the performance of a proposed computer. It is the first program to appear in print that was designed as a synthetic benchmark to test processor performance, although it was specifically developed to test the performance of only one computer: the hypothetical Whetstone machine.

The Whetstone benchmark is based on application-program statistics gathered by Brian A. Wichmann at the National Physical Laboratory in England. Wichmann was using an Algol-60 compilation system that compiled Algol statements into instructions for the hypothetical Whetstone computer system, which was named after the small town of Whetstone located just outside the city of Leicester, England, where the compilation system was developed. Wichmann developed statistics on instruction usage for a wide range of numerical computation programs then in use.

An Algol-60 version of the Whetstone benchmark was released in November 1972 and single- and double-precision FORTRAN versions appeared in April 1973. The FORTRAN versions became the first widely used, general-purpose, synthetic computer performance benchmarks. Information about the Whetstone benchmark was first published in 1976 [7].

The Whetstone benchmark suite consists of several small loops that perform integer and floating-point arithmetic, array indexing, procedure calls, conditional jumps, and elementary function evaluations. These loops reside in three subprograms (called p3, p0, and pa) called from a main program. The subprograms call trigonometric (sine, cosine, and arctangent) and other math-library functions (exponentiation, log, and square root). The benchmark authors empirically managed the Whetstone's instruction mix by manipulating loop counters within the modules to match Wichmann's instruction-mix statistics. Empirical Whetstone loop weights range from 12 to 899.

The Whetstone benchmark produces speed ratings in terms of thousands of Whetstone instructions per second (KWIPS), thus using the hypothetical Whetstone computer as the gold standard for this benchmark. In 1978, self-timing versions (written by Roy Longbottom, also of CCTA) produced speed ratings in MOPS and MFLOPS and overall rating in MWIPS.

As with the LFK and LINPACK benchmarks, the Whetstone benchmark focuses on floating-point performance. Consequently, it is not commonly applied to embedded processor cores that are destined to execute integer-oriented and control tasks on an IC because such a benchmark really will not tell the design team much they need to know about the processor(s) in question. However, the Whetstone benchmark's appearance spurred the creation of a plethora of "stone," benchmarks and one of those, the Dhrystone, became the first widely used benchmark to rate microprocessor performance.

10.6.4 The Dhrystone Benchmark

Reinhold P. Weicker, working at Siemens-Nixdorf Information Systems, wrote and published the first version of the Dhrystone benchmark in 1984 [8]. The benchmark's name, "Dhrystone," is a pun derived from the Whetstone benchmark. The Dhrystone benchmark is a synthetic benchmark that consists of 12 procedures in one measurement loop. It produces performance ratings in Dhrystones per second. Originally, the Dhrystone benchmark was written in a "Pascal subset of Ada." Subsequently, versions in Pascal and C have appeared, and the C version is the one most used today.

The Dhrystone benchmark differs significantly from the Whetstone and these differences make the Dhrystone far more suitable as a microprocessor benchmark. First, the Dhrystone is strictly an integer program. It does not test a processor's floating-point abilities because most microprocessors in the early 1980s (and even in the early 21st century) have no native floating-point computational abilities. The Dhrystone does devote a lot of time executing string functions (copy and compare), which microprocessors often execute in real applications.

The original version of the Dhrystone benchmark quickly became successful. One indication of the Dhrystone's success were the attempts by microprocessor vendors to inflate their Dhrystone ratings by "gaming" the benchmark (cheating). Weicker had made no attempt to thwart compiler optimizations when he wrote the Dhrystone benchmark because he reasonably viewed those optimizations as typical of real-world programming. However, version 1 of the Dhrystone benchmark did not print or use the results of its computations allowing "properly" written optimizing compilers using dead-code-removal algorithms to optimize away most of the benchmark by removing the benchmark program's computations that produced never-used results. Clearly, such optimizations go far beyond the spirit of the benchmark. Weicker corrected this problem by publishing version 2.1 of the Dhrystone benchmark in 1988 [9]. At the same time, he acknowledged some of the benchmark's limitations and published his criteria for using the Dhrystone benchmarks to compare processors [10].

Curiously, DEC's VAX also plays a role in the saga of the Dhrystone benchmarks. The VAX 11/780 minicomputer could run 1757 version 2.1 Dhrystones/sec. Because the VAX 11/780 was (erroneously) considered a 1-MIPS computer, it became the Dhrystone standard machine, which resulted in the emergence of the DMIPS or D-MIPS (Dhrystone MIPS) rating. By dividing a processor's Dhrystone 2.1 performance rating by 1757, processor vendors could produce an official-looking DMIPS rating for their products. Thus, DMIPS fuses two questionable rating systems (Dhrystones and VAX MIPS) to create a third, derivative, equally questionable microprocessor-rating system.

The Dhrystone benchmark's early success as a marketing tool encouraged abuse, and this abuse became rampant. For example, vendors quickly (though unevenly) added specialized string routines written in machine code to some of their compiler libraries because accelerating these heavily used library routines boosted Dhrystone ratings, even though the actual performance of the processor did not change at all. Some compilers started in-lining these machine-coded string routines for even better ratings. As the technical marketing teams at the microprocessor vendors continued to study the benchmark, they found increasingly better ways of improving their products' ratings by introducing compiler optimizations that only applied to the benchmark. Some compilers even had pattern recognizers that could recognize the Dhrystone benchmark source code or a "Dhrystone command-line switch" that would cause the compiler to "generate" the entire program using a previously prewritten, precompiled, hand-optimized version of the benchmark program.

It is not even necessary to alter a compiler to produce wildly varying Dhrystone results using the same processors. Using different compiler optimization settings can drastically alter the outcome of a benchmark test, even if the compiler has not been "Dhrystone optimized." For example, Tensilica's Xtensa LX processor core produces different Dhrystone benchmark results that differ by almost 2:1 depending on the setting of the "in-lining" switch of its compiler (shown in Table 10.2).

TABLE 10.2 Performance Difference in Dhrystone Benchmark between In-Lined and Non-In-Lined Code

Xtensa LX Compiler Setting	DMIPS rating (at 150 MHz)	DMIPS/MHz rating
No in-line code	183.187	1.221
In-line code	322.154	2.148

In-lining code is not permissible under Weicker's published rules, but as there is no organization to enforce Weicker's rules, there is no way to ensure that processor vendors benchmark fairly, and no way to force vendors to fully disclose the conditions that produced their benchmark results. Weaknesses associated with conducting Dhrystone benchmark tests and reporting the results highlight a problem: if the benchmarks are not conducted by objective third parties under controlled conditions that are disclosed with the benchmark results, the benchmark results must be suspect even if there are published rules as to a benchmark's use, because there is nobody to enforce the rules and not everyone follows rules when the sales of a new microprocessor are at stake.

Over time, other technical weaknesses have appeared in the Dhrystone benchmark. For example, the actual Dhrystone code footprint is quite small and as microprocessor instruction caches grew, some processors were able to run the Dhrystone benchmark entirely from their caches, which boosted their performance on the Dhrystone benchmark but did not represent real performance gains because most application programs will not fit entirely into a processor's cache. Processors with wider data buses and wide-load and -store instructions that move many string bytes per bus transaction also earn better Dhrystone performance ratings. Although this behavior also benefits application code, it can mislead the unwary who think that the benchmark is comparing the relative computational performance of processors.

One of the worst abuses of the Dhrystone benchmark occurred when processor vendors ran benchmarks on their own evaluation boards and on competitors' evaluation boards, and then published the results. The vendor's own board would have fast memory and the competitors' boards would have slow memory but this difference would not be revealed when the scores were published. Differences in memory speed do affect Dhrystone results, so publishing the Dhrystone results of competing processors without disclosing the use of dissimilar memory systems to compare the processors is clearly dishonest, or at least mean-spirited. Other sorts of shenanigans were used to distort Dhrystone comparisons as well. One of the Dhrystone benchmark's biggest weaknesses is therefore the absence of a governing body or an objective third party that can review, approve, and publish Dhrystone benchmark results.

Despite all of these weaknesses, microprocessor and microprocessor-core vendors continue to publish Dhrystone benchmark results — or at least they have Dhrystone results handily available — because Dhrystone ratings are relatively easy to generate and the benchmark code is very well understood after two decades of use. Vendors also keep Dhrystone ratings handy because the lack of such numbers might raise a red flag for some prospective customers. The Dhrystone's explosive popularity as a tool for making processor comparisons, coupled with its technical weaknesses and the abuses heaped upon the benchmark program, helped to spur the further evolution of processor benchmarks in the 1990s.

10.6.5 The *EDN* Microprocessor Benchmarks

EDN magazine, a trade publication for the electronics industry, was a very early advocate of microprocessor use for general electronic system design, and extensively covered the introduction of new microprocessors starting in the early 1970s. In 1981, just after the first wave of commercial 16-bit microprocessors had been introduced, *EDN* published the first comprehensive article on this subject. The article used a set of microprocessor benchmarks to compare the four leading 16-bit microprocessors of the day: DEC's LSI-11/23, Intel's 8086, Motorola's 68000, and Zilog's Z8000 [11]. The Intel, Motorola, and Zilog processors often competed for system design wins in the early 1980s, and this article provided design engineers with some of the first microprocessor benchmark ratings to be published by an objective third party.

This article was written by Jack E. Hemenway, a consulting editor for *EDN*, and Robert D. Grappel of MIT's Lincoln Laboratory. Hemenway and Grappel summed up the reason that the industry needs objective benchmark results succinctly in their article:

Why the need for a benchmark study at all? One sure way to start an argument among computer users is to compare each one's favorite machine with the others. Each machine has strong points and drawbacks, advantages and liabilities, but programmers can get used to one machine and see all the rest as inferior. Manufacturers sometimes don't help: Advertising and press releases often imply that each new machine is the ultimate in computer technology. Therefore, only a careful, complete and unbiased comparison brings order out of the chaos.

The *EDN* article continues with an excellent description of the difficulties associated with microprocessor benchmarking:

Benchmarking anything as complex as a 16-bit processor is a very difficult task to perform fairly. The choice of benchmark programs can strongly affect the comparisons' outcome so the benchmarker must choose the test cases with care.

Hemenway and Grappel used a set of benchmark programs created in 1976 by a research group at Carnegie-Mellon University (CMU). These benchmarks were first published as a paper, which was presented by the CMU group in 1977 at the National Computer Conference [12]. The tests in these benchmark programs — interrupt handlers, string searches, bit manipulation, and sorting — are very representative of types of tasks microprocessors and processor cores must perform in most embedded applications. The *EDN* authors excluded the CMU benchmarks that tested floating-point computational performance because none of the processors in the *EDN* study had native floating-point resources, something which is true even of most contemporary microprocessor cores.

The seven programs in the 1981 *EDN* benchmark study appear in Table 10.3.

Significantly, the authors of this article allowed programmers from each microprocessor vendor to code each benchmark for their company's processor. Programmers were required to use assembly language to code the benchmarks, which removed the associated compiler from the equation and allowed the processor hardware architectures to compete directly. Considering the state of microprocessor compilers in 1981, this was probably an excellent decision.

The authors refereed the tests by reviewing each program to ensure that no corners were cut and that the programs faithfully executed each benchmark algorithm. The authors did not force the programmers to code in a specific way and allowed the use of special instructions (such as instructions designed to perform string manipulation) because use of these instructions fairly represented actual processor use.

To ensure fairness, the study's authors ran the programs and recorded the benchmark results. They published the results and included both the program size and execution speed. Publishing the program size recognized that memory was costly and limited, a situation that still holds true in IC design. The authors also published the scores of each of the seven benchmark tests separately, acknowledging that a combined score would tend to mask information about a processor's specific abilities.

TABLE 10.3 Programs in the 1981 *EDN* Benchmark Article by Hemenway and Grappel

EDN 1981 Microprocessor Benchmark Component	Benchmark Description
Benchmark A	Simple priority interrupt handler
Benchmark B	FIFO interrupt handler
Benchmark E	Text string search
Benchmark F	Primitive bit manipulation
Benchmark H	Linked-list insertion (tests addressing modes and 32-bit operations)
Benchmark I	Quicksort algorithm (tests stack manipulation and addressing modes)
Benchmark K	Matrix transposition (tests bit manipulation and looping)

In 1988, *EDN* published a similar benchmarking article covering digital signal processors (DSPs) [13]. The article was written by David Shear, one of *EDN*'s regional editors. Shear benchmarked 18 DSPs using 12 benchmark programs selected from 15 candidate programs. The DSP vendors participated in selecting the DSP benchmark programs that were used from the 15 candidate programs. The final set of DSP benchmark programs included six DSP filters, three math benchmarks (a simple dot product and two matrix multiplications), and three fast Fourier transforms (FFTs). Shear used benchmark programs to compare DSPs that are substantially different from the programs used by Grappel and Hemenway to compare general purpose microprocessors because, as a class, DSPs are applied quite differently from general-purpose processors. The 12 DSP benchmark programs of Shear's benchmark set appear in Table 10.4.

Notably, Shear elected to present the performance results of the benchmarks (both execution speed and performance) as bar charts rather than in tabular form, noting that the bar charts emphasized large performance differences while masking differences of a few percent, which he dismissed as "not important."

This article also recognized the potential for cheating by operationalizing the "Three Not-So-Golden Rules of Benchmarking" outlined by *EDN* editor Walt Patstone in 1981, as a follow-up to the article by Hemenway and Grappel [14]:

> Rule 1 — All's fair in love, war, and benchmarking.
> Rule 2 — Good code is the fastest possible code.
> Rule 3 — Conditions, cautions, relevant discussion, and even actual code never make it to the bottom line when results are summarized.

These *EDN* microprocessor benchmark articles established and demonstrated all of the characteristics of an effective, modern microprocessor benchmark:

- Conduct a series of benchmark tests that exercise salient processor features for a class of tasks
- Use benchmark programs that are appropriate to the processor class being studied
- Allow experts to code the benchmark programs
- Have an objective third party check and run the benchmark code to ensure that the experts do not cheat
- Publish benchmark results that maximize the amount of available information about the tested processor
- Publish both execution speed and memory use for each processor on each benchmark program because there is always a tradeoff between a processor's execution speed and its memory consumption

These characteristics shaped the benchmarking organizations and their benchmark programs in the 1990s.

TABLE 10.4 The 12 Programs in EDN's 1988 DSP Benchmark Article

EDN 1988 DSP Benchmark Program	Benchmark Description
Benchmark 1	20-tap FIR filter
Benchmark 2	64-tap FIR filter
Benchmark 3	67-tap FIR filter
Benchmark 4	8-pole canonic IIR filter (4×) (direct form II)
Benchmark 5	8-pole canonic IIR filter (5×) (direct form II)
Benchmark 6	8-pole transpose IIR filter (direct form I)
Benchmark 7	Dot product
Benchmark 8	Matrix multiplication, 2×2 times 2×2
Benchmark 9	Matrix multiplication, 3×3 times 3×1
Benchmark 10	Complex 64-point FFT (radix-2)
Benchmark 11	Complex 256-point FFT (radix-2)
Benchmark 12	Complex 1024-point FFT (radix-2)

10.7 Modern Processor Performance Benchmarks

As the number of available microprocessors mushroomed in the 1980s, the need for good benchmarking standards became increasingly apparent. The Dhrystone experiences showed how useful a standard benchmark could be, and these experiences also demonstrated the lengths (both fair and unfair) to which processor vendors would go to earn top benchmark scores. The *EDN* articles had shown that an objective third party could bring order out of benchmarking chaos. As a result, both private companies and industry consortia stepped forward with the goal of producing "fair and balanced" processor benchmarking standards.

10.7.1 SPEC: The Standard Performance Evaluation Corporation

One of the first organizations to tackle seriously the need for good microprocessor benchmarking standards was SPEC (originally the System Performance Evaluation Cooperative, and now the Standard Performance Evaluation Corporation). A group of workstation vendors including Apollo, Hewlett-Packard, MIPS Computer Systems, and Sun Microsystems working in conjunction with the trade publication *Electronic Engineering Times* founded SPEC in 1988. The consortium produced its first processor benchmark, SPEC89, the following year. SPEC89 provided a standardized measure of compute-intensive microprocessor performance with the express purpose of replacing the existing, vague MIPS and MFLOPS rating systems. Because high-performance microprocessors were primarily used in high-end workstations at the time, and because SPEC was formed as a cooperative by workstation vendors, the SPEC89 benchmark consisted of source code that was to be compiled for UNIX.

The Dhrystone benchmark had already demonstrated that benchmark code quickly rots over time due to rapid advances in processor architecture and compiler technology. (It is no small coincidence that Reinhold Weicker, Dhrystone's creator and one of the people most familiar with processor benchmarking, became a key member of SPEC and has been heavily involved with the ongoing creation of SPEC benchmarks.) The likelihood of "benchmark rot" was no different for SPEC89, so the SPEC organization has regularly improved and expanded its benchmarks, producing SPEC92, SPEC95 (with separate integer and floating-point components called CINT95 and CFP95), and finally SPEC CPU2000 (consisting of CINT2000 and CFP2000).

Table 10.5 and Table 10.6 list the SPEC CINT2000 and CFP2000 benchmark component programs, respectively.

As Table 10.5 and Table 10.6 show, SPEC benchmarks are application-based benchmarks, not synthetic benchmarks. The SPEC benchmark programs are excellent as workstation/server benchmarks because they use the actual applications that are likely to be assigned to these machines. SPEC publishes benchmark performance results for various computer systems on its web site (www.spec.org) and sells its benchmark code.

TABLE 10.5 SPEC CINT2000 Benchmark Component Programs

CINT2000 Benchmark Component	Language	Category
164.gzip	C	Compression
175.vpr	C	FPGA circuit placement and routing
176.gcc	C	C programming language compiler
181.mcf	C	Combinatorial optimization
186.crafty	C	Game playing: chess
197.parser	C	Word processing
252.eon	C++	Computer visualization
253.perlbmk	C	PERL programming language
254.gap	C	Group theory, interpreter
255.vortex	C	Object-oriented database
256.bzip2	C	Compression
300.twolf	C	Place and route simulator

TABLE 10.6 SPEC CFP2000 Benchmark Component Programs

CFP2000 Benchmark Component	Language	Category
168.wupwise	Fortran 77	Physics/quantum chromodynamics
171.swim	Fortran 77	Shallow water modeling
172.mgrid	Fortran 77	Multi-grid solver: 3-D potential field
173.applu	Fortran 77	Parabolic/elliptic partial differential equations
177.mesa	C	3-D graphics library
178.galgel	Fortran 90	Computational fluid dynamics
179.art	C	Image recognition/neural networks
183.equake	C	Seismic wave propagation simulation
187.facerec	Fortran 90	Image processing: face recognition
188.ammp	C	Computational chemistry
189.lucas	Fortran 90	Number theory/primality testing
191.fma3d	Fortran 90	Finite-element crash simulation
200.sixtrack	Fortran 77	High energy nuclear physics accelerator design

As of 2004, The CPU2000 benchmark suite costs $500 ($125 for educational use). A price list for the CPU2000 benchmarks as well as a wide range of other computer benchmarks appears on the SPEC web site.

Because the high-performance microprocessors used in workstations and servers are sometimes used as embedded processors and some of them are available as microprocessor cores for use on SOCs, microprocessor and microprocessor-core vendors sometimes quote SPEC benchmark scores for their products. Use these performance ratings with caution because the SPEC benchmarks are not necessarily measuring performance that is meaningful to embedded applications. For example, no mobile phone is likely to be required to simulate seismic-wave propagation, except for the possible exception of handsets sold in California.

A paper written by Jakob Engblom of Uppsala University compares the static properties of the SPECint95 benchmark programs with code from 13 embedded applications consisting of 334,600 lines of C source code [15]. Engblom's static analysis discovered several significant differences between the static properties of the SPECint95 benchmark and the 13 embedded programs. Significant differences include:

- Variable sizes. Embedded programs carefully control variable size to minimize memory usage. Workstation-oriented software like SPECint95 does not limit variable size nearly as much because workstation memory is relatively plentiful.
- Unsigned data are more common in embedded code.
- Logical (as opposed to arithmetic) operations occur more frequently in embedded code.
- Many embedded functions only perform side effects (such as flipping an I/O bit). They do not return values.
- Embedded code employs global data more frequently for variables and for large constant data.
- Embedded programs rarely use dynamic memory allocation.

Although Engblom's intent in writing this paper was to show that SPECint95 code was not suitable for benchmarking embedded development tools such as compilers, his observations about the characteristic differences between SPECint95 code and typical embedded application code are also significant for embedded processors and processor cores. Engblom's observations underscore the maxim that the best benchmark code is always the actual target application code.

10.7.2 BDTI: Berkeley Design Technology, Inc.

Berkeley Design Technology, Inc. (BDTI) is a technical services company that has focused exclusively on the applications of DSPs since 1991. BDTI helps companies select, develop, and use DSP technology by providing expert advice and consulting, technology analysis, and highly optimized software development services. As part of those services, BDTI has spent more than a decade developing and applying DSP benchmarking tools. As such, BDTI serves as a private third party that develops and administers DSP benchmarks.

BDTI introduced its core suite of DSP benchmarks, formally called the "BDTI Benchmarks," in 1994. The BDTI Benchmarks consist of a suite of 12 algorithm kernels, which represent key DSP operations used in common DSP applications. BDTI revised, expanded, and published information about the 12 DSP algorithm kernels in its BDTI Benchmark in 1999 [16]. The 12 DSP kernels in the BDTI Benchmark appear in Table 10.7.

BDTI's DSP benchmark suite is an example of a specialized processor benchmark. DSPs are not general-purpose processors and therefore require specialized benchmarks. BDTI does not use full applications as does SPEC. The BDTI Benchmark is a hybrid benchmark that uses code (kernels) extracted from actual DSP applications.

BDTI calls its benchmarking methodological approach "algorithm kernel benchmarking and application profiling." The algorithm kernels used in the BDTI benchmarks are functions that constitute the building blocks used by most signal-processing applications. They have been extracted by BDTI from examples of DSP application code. These kernels are the most computationally intensive portions of the donor DSP applications. The resulting DSP kernels are all written in assembly code and must be hand-ported to each new DSP architecture. BDTI believes that the extracted kernel code is more rigorously defined than the large DSP applications such as a V.90 modem or a Dolby AC-3 audio codec, which perform many functions in addition to the core DSP algorithms.

The "application-profiling" portion of the BDTI Benchmarks methodology refers to a set of techniques used by BDTI to either measure or estimate the amount of time, memory, and other resources that an application spends executing various code subsections, including subsections that correspond to the DSP kernels in the BDTI Benchmark. BDTI develops kernel-usage estimates from a variety of information sources including application-code inspections, instrumented code-run simulations, and flow-diagram analysis.

TABLE 10.7 BDTI Benchmark Kernels

BDTI Benchmark Kernel Function	Function Description	Example Applications
Real block FIR	Finite impulse response filter that operates on a block of real (not complex) data	Speech processing (e.g., G.728 speech coding)
Complex block FIR	FIR filter that operates on a block of complex data	Modem channel equalization
Real single-sample FIR	FIR filter that operates on a single sample of real data	Speech processing, general filtering
LMS adaptive FIR	Least-mean-squares adaptive filter, operates on a single sample of real data	Channel equalization, servo control, linear predictive coding
IIR	Infinite impulse response filter that operates on a single sample of real data	Audio processing, general filtering
Vector dot product	Sum of the point-wise multiplication of two vectors	Convolution, correlation, matrix multiplication, multi-dimensional signal processing
Vector add	Pointwise addition of two vectors, producing a third vector	Graphics, combining audio signals or images
Vector maximum	Finding the value and location of the maximum value in a vector	Error control coding, algorithms using block floating-point
Viterbi decoder	Decoding a block of bits that have been convolutionally encoded	Error control coding
Control	A sequence of control operations (test, branch, push, pop, and bit manipulation)	Virtually all DSP applications include some control code
256-point, in-place FFT	Fast Fourier transform converts a time-domain signal into the frequency domain	Radar, sonar, MPEG audio compression, spectral analysis
Bit unpack	Unpacks variable-length data from a bit stream	Audio decompression, protocol handling

BDTI originally targeted its benchmark methodology at realized-in-silicon DSP processors, not processor cores. The key portion of the benchmark methodology that limited its use to processors and not cores was the need to conduct the benchmark tests at actual clock speeds. BDTI did not want processor manufacturers to test this year's processors and then extrapolate the results to next year's clock speeds. However, SOC designers must evaluate and compare processor-core performance long before a chip has been fabricated. Processor-core clock speeds will depend on the IC fabrication technology and cell libraries used to create the SOC. BDTI uses results from a cycle-accurate simulator and worst-case projected clock speeds (based on gate-level processor-core simulation and BDTI's evaluation of the realism of that projection) to obtain benchmark results from processor cores.

BDTI does not release the BDTI Benchmark code. Instead, it works with processor vendors (for a fee) to port the benchmark code. BDTI then acts as the objective third party; conducts the benchmark tests; and measures execution time, memory usage, and energy consumption for each benchmark kernel. BDTI also sells some of the results of its benchmark tests in the form of reports on specific DSP processors and in the form of a book called the *Buyers Guide to DSP Processors* (the 2004 edition of BDTI's DSP buyers guide sells for $2695).

To satisfy people who use one number to make a rough cut of processor candidates, BDTI rolls a processor's execution times on the 12 DSP kernels into one composite number that it has dubbed the BDTImark2000. To prevent the mixing of processor benchmark scores verified with hardware and simulated processor-core scores, BDTI publishes the results of simulated benchmark scores under a different name: the BDTIsimMark2000. Both the BDTImark2000 and the BDTIsimMark2000 scores are available without charge on BDTI's website (www.bdti.com). Figure 10.3 shows BDTImark2000™ and BDTIsimMark2000™ scores for ARM's ARM7 and ARM9E, CEVA CEVA-X, LAS LSI Logic ZSP400 and

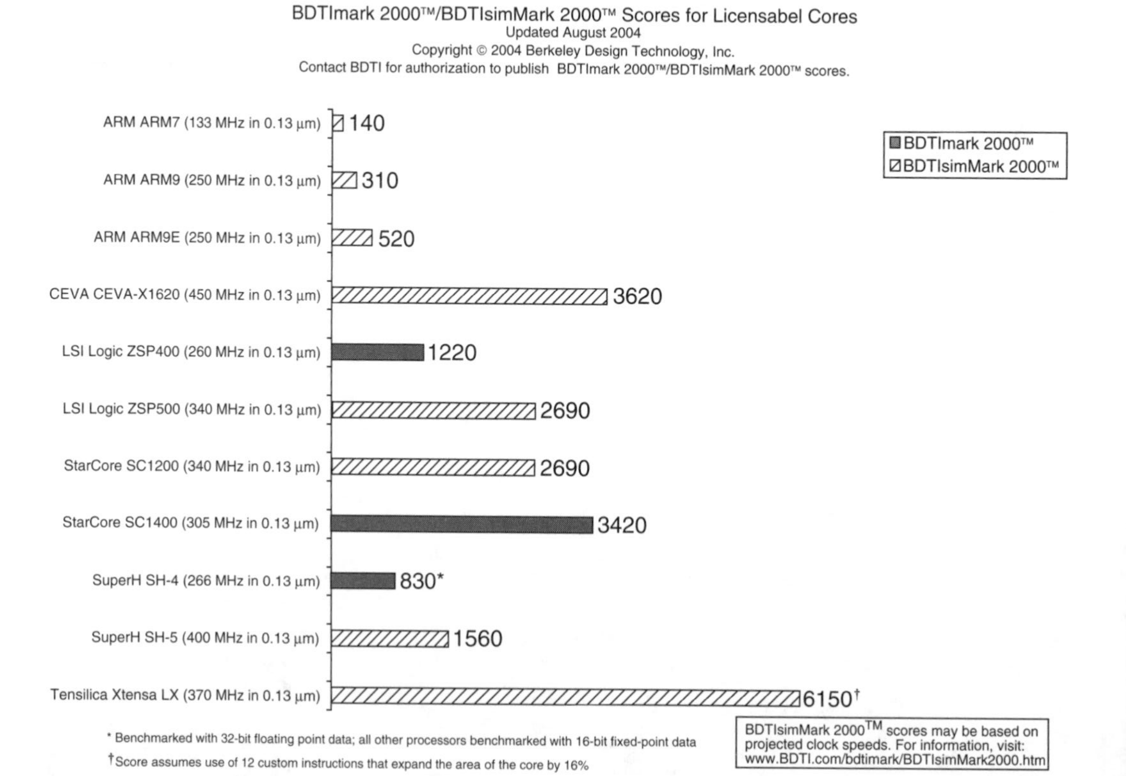

FIGURE 10.3 The BDTImark2000™/BDTIsimMark2000™ provides a summary measure of signal processing speed. For more information and scores see www.BDTI.com. Scores copyright 2004 BDTI.

ZSP500, StarCore SC1200 and SC1400, SuperH SH-4 and SH-5, and Tensilica Xtensa LX microprocessor and DSP cores.

10.7.3 EEMBC (Embedded Microprocessor Benchmark Consortium)

EDN magazine's legacy of microprocessor benchmark articles grew into full-blown realization when *EDN* editor Markus Levy founded the nonprofit, embedded-benchmarking organization called EEMBC (*EDN* Embedded Benchmark Consortium, later dropping the "*EDN*" from its name but not the corresponding "E" from its abbreviation) in 1997. EEMBC's stated goal was to produce accurate and reliable metrics based on real-world embedded applications for evaluating embedded-processor performance. Levy drew remarkably broad industry support from microprocessor and DSP vendors for his concept of a benchmarking consortium which had picked up 21 founding corporate members by the end of 1998: Advanced Micro Devices, Analog Devices, ARC, ARM, Hitachi, IBM, IDT, Lucent Technologies, Matsushita, MIPS, Mitsubishi Electric, Motorola, National Semiconductor, NEC, Philips, QED, Siemens, STMicroelectronics, Sun Microelectronics, Texas Instruments, and Toshiba [17]. EEMBC (pronounced "embassy") spent nearly three years working on a suite of benchmarks for testing embedded microprocessors and introduced its first benchmark suite at the Embedded Processor Forum in 1999. EEMBC released its first certified scores in 2000 and, in the same year, announced that it would start to certify benchmarks run on simulators so that processor cores could be benchmarked. As of 2005, EEMBC has more than 50 corporate members.

The EEMBC benchmarks are contained in six suites loosely grouped according to application. The six suites are: automotive/industrial, consumer, Java GrinderBench, networking, office automation, and telecom. Each of the suites contains several benchmark programs that are based on small, derived kernels extracted from application code. All EEMBC benchmarks are written in C, except for the benchmark programs in the Java benchmark suite, which are written in Java. The six benchmark suites and descriptions of the programs in each suite appear in the Tables 10.8 through 10.13.

EEMBC's benchmark suites with their individual benchmark programs allow designers to select the benchmarks that are relevant to a specific design, rather than lumping all of the benchmark results into one number. EEMBC's benchmark suites are developed by separate subcommittees, each working on one application segment. Each subcommittee selects candidate applications that represent the application segment and dissects each application for the key kernel code that performs the important work. This kernel code coupled with a test harness becomes the benchmark. Each benchmark has published guidelines in an attempt to force the processor vendors to play fair.

However, the industry's Dhrystone experience proved conclusively that some processor vendors will not play fair without a fair and impartial referee, so EEMBC created one: the EEMBC Certification Laboratories, LLC (ECL). Just as major sports leagues hire referee organizations to conduct games and enforce the rules of play, EEMBC employs ECL to conduct benchmark tests and enforce the rules of EEMBC benchmarking. Although a processor vendor ports the EEMBC benchmark code and test harnesses to its own processor and runs the initial benchmark tests, only ECL can certify the results. ECL takes the vendor-supplied code, test harness, and a description of the test procedures used, and then certifies the benchmark results by inspecting the supplied code and rerunning the tests. In addition to code inspection, ECL makes changes to the benchmark code (such as changing variable names) to counteract "over optimized" compilers. Vendors cannot publish EEMBC benchmark scores without ECL certification.

There is another compelling reason for an EEMBC benchmark referee: EEMBC rules allow for two levels of benchmarking "play" (the major league and the minors, to continue the sports analogy). The lower level of EEMBC "play" produces "out-of-the-box" scores. Out-of-the-box EEMBC benchmark tests can use any compiler (in practice, the compiler selected has changed the performance results by as much as 40%) and any selection of compiler switches, but cannot modify the benchmark source code. The "out-of-the-box" results therefore give a fair representation of the abilities of the processor/compiler combination without adding in programmer creativity as a wild card. The higher level of EEMBC play is called "full-fury." Processor vendors seeking to improve their full-fury EEMBC scores (posted as "optimized"

TABLE 10.8 EEMBC Automotive/Industrial Benchmark Programs

EEMBC Automotive/Industrial Benchmark Name	Benchmark Description	Example Application
Angle to time conversion	Compute engine speed and crankshaft angle	Automotive engine control
Basic integer and floating point	Calculate arctangent from telescoping series	General purpose
Bit manipulation	Character and pixel manipulation	Display control
Cache "Buster"	Long sections of control algorithms with pointer manipulation	General purpose
CAN Remote Data Request	Controller Area Network (CAN) communications	Automotive networking
Fast Fourier transform (FFT)	Radix-2 decimation in frequency, power spectrum calculation	DSP
Finite impulse response (FIR) filter	High- and low-pass FIR filters	DSP
Inverse discrete cosine transform (iDCT)	iDCT using 64-bit integer arithmetic	Digital video, graphics, image recognition
Inverse fast Fourier transform (iFFT)	Inverse FFT performed on real and imaginary values	DSP
Infinite impulse response (IIR) filter	Direct-for II, N-cascaded, second-order IIR filter	DSP
Matrix arithmetic	LU decomposition of N×N input matrices, determinant computation, cross product	General purpose
Pointer chasing	Search of a large, doubly linked list	General purpose
Pulse width modulation (PWM)	Generate PWM signal for stepper-motor control	Automotive actuator control
Road speed calculation	Determine road speed from successive timer/counter values	Automotive cruise control
Table lookup and interpolation	Derive function values from sparse 2-D and 3-D tables	Automotive engine control, antilock brakes
Tooth to spark	Fuel-injection and ignition-timing calculations	Automotive engine control

TABLE 10.9 EEMBC Consumer Benchmark Programs

EEMBC Consumer Benchmark Name	Benchmark Description	Example Applications
High pass grey-scale filter	2-D array manipulation and matrix arithmetic	CCD and CMOS sensor signal processing
JPEG	JPEG image compression and decompression	Still-image processing
RGB to CMYK conversion	Color-space conversion at 8 bits/pixel	Color printing
RGB to YIQ conversion	Color-space conversion at 8 bits/pixel	NTSC video encoding

TABLE 10.10 EEMBC GrinderBench Java 2 Micro Edition (J2ME) Benchmark Programs

EEMBC GrinderBench for Micro Java 2 Edition (J2ME) Benchmark Description	Benchmark Name	Example Applications
Chess	Machine -vs.machine chess matches, 3 games, 10 moves/game	Cell phones, PDAs
Cryptography	DES, DESede, IDEA, Blowfish, Twofish encryption and decryption	Data security
kXML	XML parsing, DOM tree manipulation	Document processing
ParallelBench	Multi-threaded operations with mergesort and matrix multiplication	General purpose
PNG decoding	PNG image decoding	Graphics, web browsing
Regular expression	Pattern matching and file I/O	General purpose

TABLE 10.11 EEMBC Networking Benchmark Programs

EEMBC Networking 2.0 Benchmark Name	Benchmark Description	Example Applications
IP Packet Check	IP header validation, checksum calculation, logical comparisons	Network router, switch
IP Network Address Translator (NAT)	Network-to-network address translation	Network router, switch
OSPF version 2	Open shortest path first/Djikstra shortest path first algorithm	Network routing
QoS	Quality of service network bandwidth management	Network traffic flow control

TABLE 10.12 EEMBC Office Automation Benchmark Programs

EEMBC Office Automation Benchmark Name	Benchmark Description	Example Applications
Dithering	Grayscale to binary image conversion	Color and monochrome printing
Image rotation	90-degree image rotation	Color and monochrome printing
Text processing	Parsing of an interpretive printing control language	Color and monochrome printing

TABLE 10.13 EEMBC Telecom Benchmark Programs

EEMBC Telecom Benchmark Name	Benchmark Description	Example Applications
Autocorrelation	Fixed-point autocorrelation of a finite-length input sequence	Speech compression and recognition, channel and sequence estimation
Bit allocation	Bit-allocation algorithm for DSL modems using DMT	DSL modem
Convolutional encoder	Generic convolutional coding algorithm	Forward error correction
Fast Fourier transform (FFT)	Decimation in time, 256-point FFT using Butterfly technique	Mobile phone
Viterbi decoder	IS-136 channel decoding using Viterbi algorithm	Mobile phone

scores on the EEMBC web site) can use hand-tuned code, assembly-language subroutines, special libraries, special CPU instructions, co-processors, and other hardware accelerators. Full-fury scores tend to be much better than out-of-the-box scores, just as application-optimized production code generally runs much faster than code that has merely been run through a compiler. The free-for-all nature of EEMBC full-fury benchmarking rules underscores the need for a benchmark referee.

EEMBC does not publish its benchmark code. Processor vendors gain access to the EEMBC benchmark source code only by joining EEMBC, so EEMBC benchmark results are available for microprocessors and microprocessor cores supplied only by companies that have joined the consortium. With permission from the processor vendor, EEMBC will publish certified EEMBC benchmark-suite scores on the EEMBC website (www.eembc.org). Many processor vendors do allow EEMBC to publish their processors' scores and the EEMBC site already lists more than 300 certified benchmark-suite scores for a wide range of microprocessors and microprocessor cores.

Vendors need not submit their processors for testing with every EEMBC benchmark suite and some vendors do not allow their products' scores to be published at all, although they may share the certified results privately with prospective customers. Vendors tend to publish information about the processors that do well and allow poorer-performing processors to remain shrouded in anonymity.

EEMBC's work is ongoing. To prevent benchmark rot, EEMBC's subcommittees constantly evaluate revisions to the benchmark suites. The Networking suite is already on version 2.0 and the Consumer suite is undergoing revision as of the writing of this chapter. To date, EEMBC has created benchmarks that focus on measuring execution speed but the organization is developing power- and energy-oriented benchmarks because power efficiency and low-power operation have become important processor characteristics.

10.7.4 Modern Processor Benchmarks from Academic Sources

The EEMBC benchmarks have become the *de facto* industry standard for measuring benchmark performance, but two aspects of the EEMBC benchmarks make them less than ideal for all purposes. The EEMBC benchmark code is proprietary and secret. Therefore it is not open to evaluation, analysis, and criticism by industry analysts, journalists, and independent third parties. It is also not available to designers who want to perform their own microprocessor evaluations. Such independent evaluations would be unnecessary if all of the EEMBC corporate members tested their processors and published their results. However, fewer than half of the EEMBC corporate members have published any benchmark results for their processors and few have tested all of their processors [18].

10.7.4.1 UCLA's MediaBench 1.0

There are two sets of university-developed microprocessor benchmarks that can somewhat fill the gaps left by EEMBC. The first is MediaBench, developed by the Computer Science and Electrical Engineering Departments at the University of California at Los Angeles (UCLA) [19]. The MediaBench 1.0 benchmark suite was created to explore compilers' abilities to exploit instruction-level parallelism (ILP) in processors with VLIW (very long instruction word) and SIMD structures. It is targeted at microprocessors and compilers that target new-media applications and consists of several applications culled from image-processing, communications, and DSP applications and a set of input test data files to be used with the benchmark code. (All of the MediaBench files are located at http://cares.icsl.ucla.edu/MediaBench.) Applications used in MediaBench 1.0 appear in Table 10.14.

10.7.4.2 MiBench from the University of Michigan

A more comprehensive benchmark suite called MiBench was presented by its developers from the University of Michigan at the IEEE's 4th Annual Workshop on Workload Characterization in December 2001 [20]. MiBench intentionally mimics EEMBC's benchmark suite. It includes a set of 35 embedded applications in six application-specific categories: automotive and industrial, consumer devices, office automation, networking, security, and telecommunications. Table 10.15 lists the 35 benchmark programs (plus two repeated benchmark tests) in the six categories. All of the benchmark programs in the MiBench suite are available in standard C source code at http://www.eecs.umich.edu/mibench.

TABLE 10.14 MediaBench 1.0 Benchmark Programs

MediaBench 1.0 Application	Application Description	Example Applications
JPEG	Standardized compression method for full-color and gray-scale images	Digital cameras, printers
MPEG	MPEG2 codec for high-quality digital video transmission	Digital television, DVD
GSM	European GSM 06.10 full-rate speech codec	Mobile phone
G.721 Voice Compression	Reference implementations of CCITT G.711, G.721, and G.723 voice compression	Mobile phone
PGP	"Pretty Good Privacy" cryptographic algorithm	Secure data communications
PEGWIT	Publish key encryption and authentication	Secure data communications
Ghostscript	PostScript language interpreter	Graphical display, printing
Mesa	OpenGL 3-D graphics library	Workstation graphics
RASTA	Speech-recognition algorithm	Advanced user interfaces
EPIC	Experimental image-compression algorithm	Digital cameras, printers
ADPCM	Adaptive differential pulse code modulation (audio codec)	Digital telephony

TABLE 10.15 MiBench Benchmark Programs

Auto/Industrial	Consumer Devices	Office Automation	Networking	Security	Telecom
basicmath	JPEG	Ghostscript	Dijkstra	Blowfish encoder	CRC32
bitcount	lame	ispell	patricia	Blowfish decoder	FFT
qsort	mad	rsynth	(CRC32)	PGP sign	IFFT
susan (edges)	tiff2bw	sphinx	(sha)	PGP verify	ADPCM encode
susan (corners)	tiff2rgba	stringsearch	(Blowfish)	Rijndael encoder	ADPCM decoder
susan (smoothing)	tiffdither			Rijndael decoder	GSM encoder
	tiffmedian typeset			sha	GSM decoder

The MiBench automotive and industrial benchmark group tests a processor's control abilities. These routines include:

- *basicmath*: cubic function solving, integer square root, angle conversions
- *bitcount*: counts the number of bits in an integer array
- *qsort*: quicksort algorithm applied to a large string array
- *susan*: image recognition program developed to analyze MRI brain images

The MiBench applications in the consumer devices category include:

- *JPEG encode/decode*: still-image codec
- *tiff2bw*: converts a color tiff image into a black-and-white image
- *tiff2rgba*: converts a color image into one formatted into red, green, and blue
- *tiffdither*: dithers a black-and-white image to reduce its size and resolution
- *tiffmedian*: reduces an image's color palette by taking several color-palette medians
- *lame*: the Lame MP3 music encoder
- *mad*: a high-quality audio decoder for the MPEG1 and MPEG2 video formats
- *typeset*: a front-end processor for typesetting HTML files

Applications in the MiBench office-automation category include:

- *Ghostscript*: a Postscript language interpreter minus a graphical user interface
- *stringsearch*: searches a string for a specific word or phrase
- *ispell*: a fast spelling checker
- *rsynth*: a text-to-speech synthesis program
- *sphinx*: a speech decoder

The set of MiBench networking applications includes:

- *Dijkstra*: computes the shortest path between nodes
- *Patricia*: a routing-table algorithm based on Patricia tries

The networking group also reuses the CRC32, sha, and Blowfish applications from the MiBench suite's security applications group.

The security group of MiBench applications includes:

- *Blowfish encryption/decryption*: a symmetric block cipher with a variable-length key
- *sha*: a secure hashing algorithm
- *Rijndale encryption/decryption*: the cryptographic algorithm used by the AES encryption standard
- *PGP sign/verify*: a public-key cryptographic system called "Pretty Good Privacy"

MiBench's telecommunications application suite includes the following applications:

- *FFT/IFFT*: fast Fourier transform and the inverse transform
- *GSM encode/decode*: European voice codec for mobile telephony

- *ADPCM encode/decode*: a speech-compression algorithm
- *CRC32*: a 32-bit cyclic redundancy check

With its 35 component benchmark applications, MiBench certainly provides a thorough workout for any processor/compiler combination. Both MediaBench and MiBench solve the problem of proprietary code for anyone who wants to conduct a private set of benchmark tests by providing a standardized set of benchmark tests at no cost and with no use restrictions. Industry analysts, technical journalists, and researchers can publish results of independent tests conducted with these benchmarks although none seem to have done so, to date. The tradeoff made with these processor benchmarks from academia is the absence of an official body that enforces benchmarking rules and provides result certification. For someone conducting their own benchmark tests, self-certification may be sufficient. For anyone else, the entity publishing the results must be scrutinized for fairness in the conducting of tests, for bias in test comparisons among competing processors, and for bias in any conclusions.

10.8 Configurable Processors and the Future of Processor-Core Benchmarks

For the past 20 years, microprocessor benchmarks have attempted to show how well specific microprocessor architectures work. As their history demonstrates, the benchmark programs that have been developed have had mixed success in achieving this goal. One thing that has been constant over the years is the use of benchmarks to compare microprocessors and microprocessor cores with fixed ISAs. Nearly all microprocessors realized in silicon have fixed architectures and the microprocessor cores available for use in ASICs and SOCs also had fixed architectures. However, the transmutable silicon of ASICs and SOCs makes it feasible to employ microprocessor cores with configurable architectures instead of fixed ones.

Configurable architectures allow designers to add new instructions and registers to the processor's architecture that boost the processor's performance for specific, targeted applications. Use of configurable processor cores has a profound impact on the execution speed of any program, including benchmark programs. Processor vendors are taking two fundamental approaches to making configurable-ISA processors available to IC designers. The first approach provides tools that allow the designer to modify an existing base processor architecture to boost performance on a target application or set of applications. Companies taking this approach include ARC International, MIPS Technologies, and Tensilica, Inc. The other approach employs tools that compile entirely new processor architectures for each application. Companies taking this approach include ARM, CoWare/LISATek, Silicon Hive, and Target Compilers.

The ability to tailor a processor's ISA to a target application (which includes benchmarks) can drastically increase execution speed of that program. For example, Figure 10.4 shows EEMBC Consumer benchmark suite scores for Tensilica's configurable Xtensa V microprocessor core (labeled Xtensa T1050 in Figure 10.4). The right-hand column of Figure 10.4 shows benchmark scores for a standard processor configuration. The center column shows the improved benchmark scores for a version of the Xtensa V core that has been tailored by a design engineer specifically to run the EEMBC Consumer benchmark programs. Table 10.16 summarizes the performance results of the benchmarks for the stock and tailored processor configurations.

Note that the stock Xtensa V processor core performs on par with, or slightly faster than, other 32-bit RISC processor cores, and that tailoring the Xtensa V processor core for the individual tasks in the EEMBC Consumer benchmark suite by adding new instructions that are matched to the needs of the benchmark computations boosts the processor's performance on the individual benchmark programs from 4.6 × to more than 100 ×. The programmer modifies the benchmark or target application code by adding C intrinsics that use these processor extensions. The resulting performance improvement is similar to results that designers might expect to achieve when tailoring a processor for their target application code, so in a sense, the EEMBC benchmarks are still performing exactly as intended. The optimized Xtensa V EEMBC Consumer benchmark scores reflect the effects of focusing 3 months worth of an engineering graduate student's "full fury" to improve the Xtensa V processor's benchmark performance, optimizing both the processor and the benchmark code.

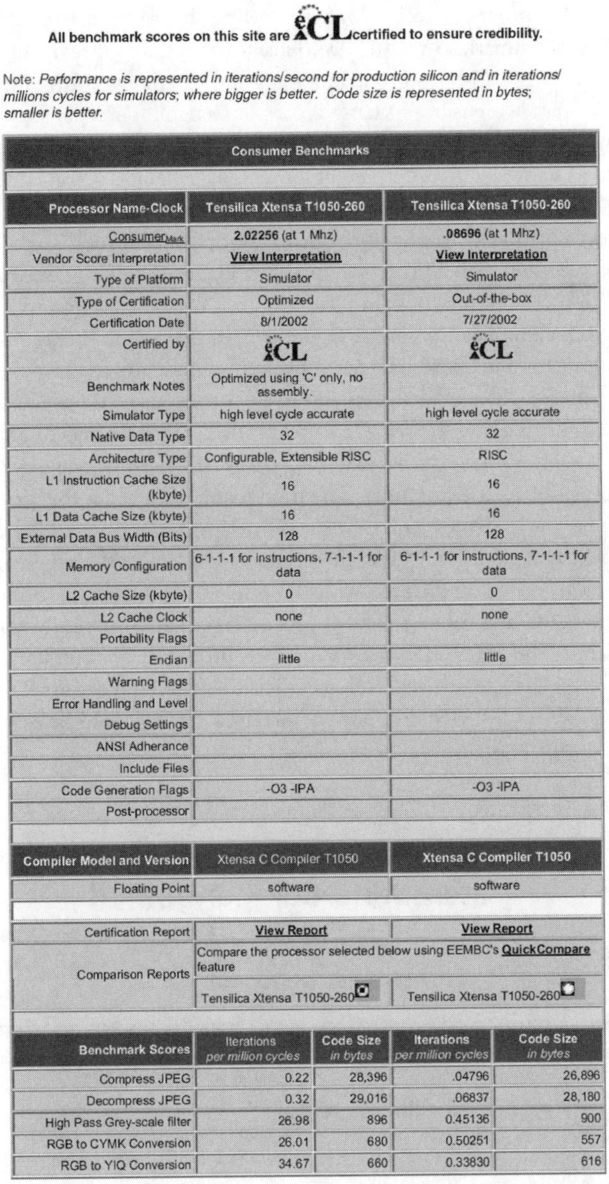

All benchmark scores on this site are ECL certified to ensure credibility.

Note: *Performance is represented in iterations/second for production silicon and in iterations/millions cycles for simulators; where bigger is better. Code size is represented in bytes; smaller is better.*

Consumer Benchmarks		
Processor Name-Clock	**Tensilica Xtensa T1050-260**	**Tensilica Xtensa T1050-260**
ConsumerMark	**2.02256** (at 1 Mhz)	**.08696** (at 1 Mhz)
Vendor Score Interpretation	View Interpretation	View Interpretation
Type of Platform	Simulator	Simulator
Type of Certification	Optimized	Out-of-the-box
Certification Date	8/1/2002	7/27/2002
Certified by	ECL	ECL
Benchmark Notes	Optimized using 'C' only, no assembly.	
Simulator Type	high level cycle accurate	high level cycle accurate
Native Data Type	32	32
Architecture Type	Configurable, Extensible RISC	RISC
L1 Instruction Cache Size (kbyte)	16	16
L1 Data Cache Size (kbyte)	16	16
External Data Bus Width (Bits)	128	128
Memory Configuration	6-1-1-1 for instructions, 7-1-1-1 for data	6-1-1-1 for instructions, 7-1-1-1 for data
L2 Cache Size (kbyte)	0	0
L2 Cache Clock	none	none
Portability Flags		
Endian	little	little
Warning Flags		
Error Handling and Level		
Debug Settings		
ANSI Adherance		
Include Files		
Code Generation Flags	-O3 -IPA	-O3 -IPA
Post-processor		
Compiler Model and Version	Xtensa C Compiler T1050	**Xtensa C Compiler T1050**
Floating Point	software	software
Certification Report	View Report	View Report
Comparison Reports	Compare the processor selected below using EEMBC's QuickCompare feature Tensilica Xtensa T1050-260	Tensilica Xtensa T1050-260

Benchmark Scores	Iterations *per million cycles*	**Code Size** *in bytes*	**Iterations** *per million cycles*	**Code Size** *in bytes*
Compress JPEG	0.22	28,396	.04796	26,896
Decompress JPEG	0.32	29,016	.06837	28,180
High Pass Grey-scale filter	26.98	896	0.45136	900
RGB to CYMK Conversion	26.01	680	0.50251	557
RGB to YIQ Conversion	34.67	660	0.33830	616

FIGURE 10.4 Tailoring of a processor core's architecture for a specific set of programs that can result in significant performance gains. Copyright EEMBC. Reproduced with permission.

TABLE 10.16 Optimized Versus Out-of-the-Box EEMBC Consumer Benchmark Scores

EEMBC Consumer Benchmark Name	Optimized Xtensa T1050 Score (Iterations/M cycles)	Out-of-the-Box Xtensa T1050 Score (Iterations/M cycles)	Performance Improvement
Compress JPEG	0.22	0.04796	4.59 ×
Decompress JPEG	0.32	0.06837	4.68 ×
High pass grey-scale filter	26.98	0.45136	59.77 ×
RGB to CMYK conversion	26.01	0.50251	51.76 ×
RGB to YIQ conversion	34.67	0.33830	102.48 ×

Evolution in configurable-processor technology is allowing the same sort of effects to be achieved on EEMBC's out-of-the-box benchmark scores. In 2004, Tensilica introduced the XPRES Compiler, a processor-development tool that analyzes C/C++ code and automatically generates performance-enhancing processor extensions from its code analysis. Tensilica's Xtensa Processor Generator accepts the output from the XPRES Compiler and automatically produces the specified processor and a compiler that recognizes the processor's architectural extensions as native instructions and automatically generates object code that uses these extensions. Under EEMBC's out-of-the-box benchmarking rules, results from code compiled with such a compiler and run on such a processor qualify as out-of-the-box (not optimized) results. The results of an EEMBC Consumer benchmark run using such a processor appear in Figure 10.5. This figure compares the performance of a stock Xtensa V processor-core configuration with that of an Xtensa LX processor core that has been extended using the XPRES Compiler. (Note that Figure 10.5 compares the Xtensa V and Xtensa LX microprocessor cores running at 260 and 330 MHz, respectively, but the scores are reported on a per-MHz basis, canceling out any performance differences attributable to clock rate. Also, the standard versions of the Xtensa V and Xtensa LX processor cores have the same ISA and produce the same benchmark results.)

The XPRES Compiler boosted the stock Xtensa processor's performance by 1.17 × to 18.7 ×, as reported under EEMBC's out-of-the-box rules. The time required to effect this performance boost was 1 hour. These performance results demonstrate the sort of performance gains designers might expect from automatic processor configuration. However, these results also suggest that benchmark performance comparisons become even more complicated with the introduction of configurable-processor technology and that the designers making comparisons of processors using such benchmarks must be especially careful to understand how the benchmarks tests for each processor are conducted. Other commercial offerings of configurable-processor technology include ARM's OptimoDE, Target Compilers' Chess/Checkers, and CoWare's LISATek compiler. However as far as could be determined, there are no published benchmark results available for these products.

Benchmark score iterations comparison

Consumer benchmarks

Processor name-clock	Tensilica Xtensa T1050 -260	Tensilica Xtensa LX 330 MHz
Consumermark™	0.08696	0.51997
Compress JPEG	**0.04796**	0.056
Decompress JPEG	**0.06837**	0.083
High pass grey-scale filter	**0.45136**	8.661
RGB to CYMK conversion	**0.50251**	7.572
RGB to YIQ conversion	**0.33830**	6.310

Processor name-clock	Tensilica Xtensa T1050 -260	Tensilica Xtensa LX 330 MHz
Consumermark™	1.00	5.98
Compress JPEG	**1.00**	1.17
Decompress JPEG	**1.00**	1.21
High pass grey-scale filter	**1.00**	19.19
RGB to CYMK conversion	**1.00**	15.07
RGB to YIQ conversion	**1.00**	18.65

FIGURE 10.5 Tensilica's XPRES Compiler that can produce a tailored processor core with enhanced performance for a target application program in very little time. Copyright EEMBC. Reproduced with permission.

10.9 Conclusion

As a growing number of processor cores are employed to implement a variety of on-chip tasks, ASIC and SOC designers increasingly rely on benchmarking tools to evaluate processor-core performance. The tool of first recourse is always the actual application code for the target task but this particular tool is not always conveniently at hand. Standard processor benchmark programs run by objective parties must often stand in as replacement tools for processor evaluation. These tools have undergone more than two decades of evolution and their use (and abuse) is now well understood.

Benchmark evolution has not ended. Power-oriented benchmarks are in the wings. The transition to sub-100 nm lithography has placed a new emphasis on the amount of power and energy required to execute on-chip tasks. Benchmarking organizations are developing new processor benchmarks to measure power dissipation on standardized tasks. These new benchmarks will arrive in the next couple of years and will make processor performance benchmarking tools even more valuable to ASIC and SOC design teams.

References

[1] J. Rowson, Hardware/Software Co-Simulation, *Proceedings of the 31st Design Automation Conference (DAC'94)*, San Diego, 1994.
[2] S. Gill, The diagnosis of mistakes in programmes on the EDSAC, *Proceedings of the Royal Society Series A Mathematical and Physical Sciences*, 22, 1951, Cambridge University Press, London, pp. 538–554.
[3] Lord Kelvin (William Thomson), *Popular Lectures and Addresses*, Vol. 1, p. 73 (originally: Lecture to the Institution of Civil Engineers, 3, 1883).
[4] F. McMahon, The Livermore FORTRAN Kernels: A Computer Test of the Numerical Performance Range, Technical Report, Lawrence Livermore National Laboratory, 1986.
[5] J.J. Dongarra, LINPACK Working Note #3, FORTRAN BLAS Timing, Argonne National Laboratory, Argonne, IL, 1976.
[6] J.J. Dongarra, J.R. Bunch, C.M. Moler, and G.W. Stewart, LINPACK Working Note #9, Preliminary LINPACK User's Guide, ANL TM-313, Argonne National Laboratory, Argonne, IL, 1977.
[7] H.J. Curnow and B.A. Wichmann, A synthetic benchmark, *Comput. J.*, 19, 43–49, 1976.
[8] R.P. Weicker, Dhrystone: a synthetic systems programming benchmark, *Commun. ACM*, 27, 1013–1030, 1984.
[9] R.P. Weicker, Dhrystone benchmark: rationale for version 2 and measurement rules, *SIGPLAN Notices*, 1988, pp. 49–62.
[10] R.P. Weicker, Understanding variations in Dhrystone performance, *Microprocessor Report*, 1989, pp. 16–17.
[11] R.D. Grappel and J.E. Hemenway, A tale of four μPs: benchmarks quantify performance, *EDN*, 1, 1981, pp.179–232.
[12] S.H. Fuller, P. Shaman, D. Lamb, and W. Burr, Evaluation of computer architectures via test programs, *AFIPS Conference Proceedings*, Vol. 46, 1977, pp. 147–160.
[13] D. Shear, EDN's DSP Benchmarks, *EDN*, 29, 1988, pp. 126–148.
[14] W. Patstone, 16-bit μP benchmarks — an update with explanations, *EDN*, September 16, 1981, p. 169.
[15] J. Engblom, Why SpecInt95 Should Not Be Used to Benchmark Embedded Systems Tools, *Proceedings of the ACM SIGPLAN 1999 Workshop on Languages, Compilers, and Tools for Embedded Systems (LCTES'99)*, Atlanta, GA, 5, 1999.
[16] *Evaluating DSP Processor Performance*, Berkeley Design Technology, Inc. (BDTI), 1997–2000.
[17] M. Levy, At last: benchmarks you can believe, *EDN*, 5, 1988, pp. 99–108.
[18] T. Halfhill, Benchmarking the Benchmarks, *Microprocessor Report*, 30, 2004.
[19] C. Lee, M. Potkonjak et al., MediaBench: A Tool for Evaluating and Synthesizing Multimedia and Communications Systems, MICRO 30, *Proceedings of the Thirtieth Annual IEEE/ACM International Symposium on Microarchitecture*, 1–3, 1997, Research Triangle Park, NC.
[20] M. Guthaus, J. Ringenberg et al., MiBench: A Free, Commercially Representative Embedded Benchmark Suite, *IEEE 4th Annual Workshop on Workload Characterization*, Austin, TX, 2001.

11

Parallelizing High-Level Synthesis: A Code Transformational Approach to High-Level Synthesis

Gaurav Singh
Virginia Tech
Blacksburg, Virginia

Sumit Gupta
Tensilica Inc.
Santa Clara, California

Sandeep Shukla
Virginia Tech
Blacksburg, Virginia

Rajesh Gupta
University of California, San Diego,
San Diego, California

Abstract

Synthesis is the process of generating circuit implementations from descriptions of what a component does. Synthesis in part consists of refinement, elaboration as well as transformations and optimizations at multiple levels to generate circuits that compare favorably with manual designs. High-level synthesis (HLS) or behavioral synthesis specifically refers to circuit synthesis from algorithmic (or behavioral) descriptions. In this chapter, we focus on the developments in code transformation techniques for HLS over the past two decades, and describe recent progress in coordinated compiler and HLS transformations that seeks to combine effectively advances in parallelizing compiler techniques. We describe how coordinated compiler and HLS techniques can yield efficient circuits through examples of a class of source-level and dynamic transformations. We also describe recent developments in system-level modeling techniques and languages that attempt to raise the level of abstraction in the design process.

11.1 Introduction

The semiconductor industry has seen the widespread acceptance and use of language-level modeling of digital designs over the past decade. Digital designs are modeled using program-like descriptions, often using languages described in previous chapters, instead of schematic or block diagram descriptions of circuit blocks. The typical design process starts with design entry in a hardware description language at the register-transfer level (RTL), followed by logic synthesis.

The increasing willingness of designers to code circuits in a hardware description language and, in limited cases, yield control of exactly where the clock boundaries (i.e., storage elements) are placed in order to meet a given performance target has led system architects to explore at even higher levels of abstractions that allow algorithmic specification of the system. Using higher levels of abstraction reduces the size of the model, thus improving the efficiency of simulation and verification tasks. System architects can explore different algorithmic alternatives and different implementation choices and even hardware–software partitions. Transformations and optimizations applied at higher levels of abstraction have a larger impact on the quality of results of the final synthesized design as compared to applying optimizations at the register-transfer or circuit level.

One key enabling technology for specifying designs and systems in higher levels of abstraction is the ability to generate automatically or synthesize circuits that implement these designs. This process is generally referred to as *high-level synthesis* (HLS) or behavioral synthesis. HLS enables designers to shorten the design times and reduce the possibility of introducing errors or bugs during the process of refining system descriptions to RTL code.

However, current HLS efforts have several limitations: language-level optimizations are few and their impact on final circuit area and performance cannot always be predicted or controlled. The quality of synthesis results is poor, particularly for designs with moderately complex control flow (containing conditionals and loops).

Poor synthesis results can be attributed to several factors. Most of the optimizing transformations that have been proposed over the years are *operation-level* transformations. That is, these transformations typically operate on three-operand computational expressions or instructions [1]. In contrast, *language-* or *source-level* optimizations refer to transformations that require structural and hierarchical information available in the source code to operate. For example, loop transformations such as loop unrolling, loop fusion, and loop-invariant code motion (LICM), use information about the loop structure and the loop index variables and their increments. However, few language-level optimizations have been explored in the context of HLS and their effects on final circuit area and performance are not well understood. Often the effectiveness of transformations has been demonstrated in isolation from other transformations and results are often presented in terms of scheduling results, with little or no analysis of control and hardware costs. Furthermore, designs used as HLS benchmarks are often small, synthetic designs. It is thus difficult to judge if an optimization has a positive impact beyond scheduling results, on large moderately complex designs.

Recently, Gupta et al. [2,3] proposed a new approach to HLS that they termed as *parallelizing high-level synthesis* (PHLS). Parallelizing high-level synthesis seeks to create efficient circuit implementations through a combined strategy of compiler and synthesis transformations. It uses aggressive code parallelizing and code motion techniques to discover circuit optimization opportunities beyond what is possible with traditional HLS. The chief strength of these heuristics is the ability to select the code transformations so as to improve the overall synthesis results.

Parallelizing high-level synthesis addresses the limitations of current HLS approaches by employing techniques and tools that achieve the best compiler optimizations and synthesis results, irrespective of the programming style used in the high-level descriptions. In the past, several compiler techniques have been tried for HLS, albeit with mixed success. This is partly because direct application of compiler techniques does not necessarily optimize hardware synthesis results. The PHLS methodology has shown that for HLS, compiler techniques such as code transformations have to be selected and guided based on their effects on the control, interconnect, and area costs, and not based solely on their impact on instruction schedules. For instance, speculatively executing operations *may* lead to more complex control logic, which in turn can increase the longest combinational path through the design, and thus, increase the clock frequency of the synthesized circuit. Thus, we must examine the overall impact of applying speculation on the synthesized circuit, beyond just looking at scheduling results.

The rest of this chapter is organized as follows: in Section 11.2, we give an overview of the technical background of prior research work in HLS and compilers, and briefly discuss some commercial HLS tool offerings. We also discuss the challenges faced by HLS. We discuss the role of PHLS approach in Section 11.3 and the various transformations used in this approach. The PHLS approach has been validated by implementing a range of PHLS techniques in the *SPARK* software framework, presented in Section 11.4. We conclude this chapter in Section 11.5 with a summary and an outlook for the future of system-level design.

11.2 Background and Survey of the State of the Art

High-level synthesis and compilers are two separate technical streams that have recently merged, leading to the evolution of PHLS methodology. In this section, we provide an overview of prior research work in the HLS and compiler communities.

11.2.1 Early Work in HLS

High-level synthesis techniques have been investigated for two decades now. Over the years, HLS techniques have been chronicled in several books [4–9]. At its core, HLS consists of the following five main tasks:

- *Resource allocation.* This task consists of determining the number of resources that have been allocated or allotted to synthesize the hardware circuit. Typically, designers can specify an allocation in terms of the number of resources of each resource type. Resources consist not only of functional units (like adders and multipliers), but may also include registers and interconnection components (such as multiplexers and buses).
- *Scheduling.* The scheduling problem is to determine the time step or clock cycle in which each operation in the design executes. The ordering between the "scheduled" operations is constrained by the data (and possibly control) dependencies between the operations. Scheduling is often done under constraints on the number of resources (as specified in the resource allocation).
- *Module Selection.* Module selection is the task of determining the resource type from the resource library that an operation executes on. The need for this task arises because of the presence of several resources of different types (and different area and timing) that an operation may execute on. For example, an addition may execute on an adder, an ALU, or a multiply-accumulate (MAC) unit. There are area, performance, and power trade-offs in choosing between these components. Module selection must make a judicious choice between different resources such that a metric like area or timing is minimized.
- *Binding.* Binding determines the mapping between the operations, variables, and data (and control) transfers in the design and the specific resources in the resource allocation. Hence, operations are mapped to specific functional units, variables to registers, and data/control transfers to interconnection components.
- *Control synthesis and optimization.* Control synthesis generates a control unit that implements the schedule. This control unit generates control signals that control the flow of data through the data path (i.e., through the multiplexers). Control optimization tries to minimize the size of the control unit and hence, improve metrics such as area and power.

Each of these tasks are NP-complete and often dependent on each other. To delve deeper into each of these tasks, we refer the reader to [4–6,8,9].

A number of the early work in HLS was focused on scheduling heuristics for data-flow designs. The simplest scheduling approach is to schedule all the operations *as soon as possible* (ASAP) [10–13]. The converse approach is to schedule the operations *as late as possible* (ALAP). Scheduling heuristics such as *urgency scheduling* [14] and *force-directed scheduling* [15] schedule operations based on their urgency and mobility, respectively. Urgency of an operation is the minimum number of control steps from the last step in the schedule that the operation can be scheduled in before a timing constraint is violated. Mobility of an operation is the difference between the ASAP and ALAP start times of the operation.

Another category of heuristics uses a *list scheduling approach* whereby operations are scheduled after they are ordered based on control and data dependencies [16–18]. There are several other types of scheduling approaches that either iteratively reschedule the design [19] or schedule operations along the critical path through the behavioral description [20]. In the MAHA system [20], operations on the critical path are scheduled first and then the critical path is divided into several pipeline stages. Finally, the operations on the off-critical paths are scheduled based on a notion of freedom. This notion of freedom of an operation is similar to the mobility of an operation as defined above.

Resource allocation and binding techniques ranging from clique partitioning [11] to knowledge-based systems [10] have been explored in the past. The optimization goals for resource binding vary from reducing registers and functional units to reducing wire delays and interconnect costs [4,6,10]. Tseng and Siewiorek [11] use clique partitioning heuristics to find a clique cover for a module allocation graph. Paulin and Knight [21] perform exhaustive weight-directed clique partitioning of a register-compatibility graph to find the solution with the lowest combined register and interconnect costs. Stok and Philipsen [22] use a network flow formulation for minimum module allocation while minimizing interconnect. Mujumdar et al. [23] consider operations and registers in each time-step one at a time and use a network flow formulation to bind them.

Since the subtasks of HLS are highly interrelated, there have been several works that have attempted to perform these tasks simultaneously. Approaches using *integer linear programming* (ILP) formulations frequently fall into this category [24–27]. Gebotys and Elmasry [25] present an integer-programming model for simultaneous scheduling and allocation that minimizes interconnect. A 0/1 integer-programming model is proposed for simultaneous scheduling, allocation and binding in the OSCAR system [26]. Wilson et al. [27] presented a generalized ILP approach that gives an integrated solution to the various HLS subtasks. Safir and Zavidovique [28] and Kollig and Al-Hashimi [29] use simulated annealing to search the design space while performing simultaneous scheduling, allocation, and binding.

Pipelining has been the primary technique to improve performance for data-flow design [15,18,30–32]. The Sehwa system [30] automatically pipelines designs into a fixed number of stages to achieve the maximum performance. In the HAL system [15], designs are pipelined, with the help of user-specified pipeline boundaries, so that the resources are distributed uniformly across concurrent control steps. In the Maha system [20], the critical path is repeatedly divided into pipeline stages till the user-specified timing bound is satisfied.

Several optimization techniques such as algebraic transformations, retiming, and code motions across multiplexers have been proposed for improved synthesis results [33–35]. Walker and Thomas [34] proposed a set of transformations that were implemented within the *System architect's workbench* [36]. In this system, the input description is captured in a variant of data-flow graphs known as a *value trace* [37]. The system first applied coarse-level transformations such as function inlining, inverse function inlining, and removal of uncalled functions (or dead code elimination of whole functions). Then the system applies transformations that move operations into and out of select operations (or multiplexers). These code-motion transformations are similar to the speculative code motions used in designs with control flow. Another commonly used flow graph transformation is *tree height reduction* [38,39]. In this technique, the height of an expression tree or number of operations that execute in a chain is reduced by employing the commutative and distributive properties of the operators.

11.2.2 Recent Developments in HLS for Complex Control Flow

The HLS of designs with control flow (conditional branches and loops) has been studied by several groups over the last decade. Path-based scheduling (PBS) [40] is an exact approach that creates a as-fast-as-possible (AFAP) schedule by scheduling each control path independently. However, the order of operations on each control path is fixed and scheduling is formulated as a clique-covering problem on an interval graph; this means that speculative code motions are not supported. An extension of this technique that allowed operation reordering within basic blocks was presented in [41]. On the other hand, tree-based scheduling [42] supports speculative code motion and removes all the join nodes from the design, thus leaving the control-data flow graph (CDFG) looking like a tree. Both these approaches deal with loops by removing the loop's feedback path and scheduling just one iteration of the loop. The PUBSS approach [43] captures scheduling information in the behavioral finite state machine (BFSM) model. The input description (in VHDL) is captured as a network of communicating BFSMs and the schedule is generated using constraint-solving algorithms.

In the CVLS approach from NEC [44–46], a condition vector is generated for each operation based on the conditions under which the operation executes. Operations whose condition vectors are complementary are considered mutually exclusive and are free to be scheduled on the same resource. Speculation and operation duplication across a merge node are supported as well.

Radivojevic and Brewer [47] present an exact symbolic formulation that schedules each control path (or trace) independently and then creates an ensemble schedule of valid, scheduled traces. The traces are validated by using a backtracking procedure that can be computationally expensive. Haynal [48] uses an automata-based approach for symbolic scheduling of cyclic behaviors under sequential timing and protocol constraints. This is an exact approach, but can grow exponentially in terms of internal representation size. The *Waveschedule* approach [49] incorporates speculative execution into HLS to achieve its objective of minimizing the expected number of cycles. Kim et al. [50] transform a CDFG into a data-flow graph (DFG) by exploring various conditional resource sharing possibilities. The DFG is scheduled and then transformed back into a CDFG.

Santos et al. [51–53] and Rim et al. [54] support generalized code motions during scheduling in synthesis systems whereby operations can be moved globally irrespective of their position in the input. Santos creates and evaluates the cost of a set of solutions for the given design and then picks the solution with the lowest cost. This approach supports all types of code motions and employs code motion pruning techniques to reduce the search space and hence the run time of the approach.

Paulin et al. [55] present a hardware–software co-design methodology that uses the Amical tool [56] for synthesizing the control-intensive parts of the design and the Cathedral-2/3 tool [17] for DSP-oriented synthesis. The Amical tool enables the use of large function components in the hardware resource library. The Cathedral-2/3 tool is specifically targeted toward the synthesis of high-throughput application specific units.

The Olympus system [57] is a complete set of HLS, logic-level synthesis and simulation tools. The HLS tool, *Hercules*, accepts an input description in a variant of C called *HardwareC* [58], which introduces the notion of concurrency and timing into C. *Relative scheduling* [59] is used to schedule a design with operations that have unbounded delay.

Prior work on presynthesis transformations has focused on altering the control flow or extracting the maximal set of mutually exclusive operations [60,61]. Li and Gupta [62] restructure the control flow and attempt to extract common sets of operations within conditionals to improve synthesis results. Kountouris and Wolinski [63] perform operation movement and duplication before scheduling and also attempt to detect and eliminate false control paths in the design. Function inlining, where a function call is replaced by the contents of the function, was presented in the context of HLS by Walker and Thomas [34].

11.2.3 Related Work in Compiler and Parallelizing Compiler Approaches

Compiler transformations such as common subexpression elimination (CSE) and copy propagation predate HLS and are standard in most software compilers [1,64]. These transformations are applied as passes on the input program code and for cleanup at the end of scheduling before code generation. Compiler transformations were developed for improving code efficiency for sequential program execution. Their use in digital circuit synthesis has been limited. For instance, CSE has been used for throughput improvement [35], for optimizing multiple constant multiplications [65,66] and as an algebraic transformation for operation cost minimization [67,68].

A converse of CSE, namely, *common subexpression replication* has been proposed to aid scheduling by adding redundant operations [69,70]. *Partial redundancy elimination* [71] inserts copies of operations present in only one conditional branch into the other conditional branch, so as to eliminate common subexpressions in subsequent operations. Janssen et al. [67] and Gupta et al. [72] propose doing CSE at the source level to reduce the effects of the factorization of expressions and control flow on the results of CSE.

In the context of parallelizing compilers, *mutation scheduling* [73] also performs local optimizations such as CSE during scheduling in an opportunistic manner. A range of parallelizing code transformation techniques has also been previously developed for high-level language software compilers (especially parallelizing compilers) [74,75].

Although the basic transformations (e.g., dead code elimination, copy propagation) can be used in synthesis as well, other transformations need to be reinstrumented for synthesis by incorporating ideas of mutual exclusivity of operations, resource sharing, and hardware cost models. Cost models of operations and resources in compilers versus synthesis tools are particularly very different. For example, in compilers, there is generally a uniform push toward executing operations ASAP by speculative code motions since this always lead to improved execution in sequential (and pipelined) processors. Indeed, the optimality claims in *percolation* [76] and *trace scheduling* [74] are based entirely upon maximum movement of operations out of conditional branches. In the context of HLS, such notions of optimality have little relevance because of the interdependency of scheduling and resource utilization in a parallel hardware execution model. In circuit synthesis, code transformations that lead to increased resource utilization may also lead to higher hardware costs in terms of steering logic and associated control circuits. Some of these costs can be mitigated by interconnect minimizing resource binding techniques [77].

The use of code motions as a transformation technique has been explored at length in the context of parallelizing compilers for VLIWs [74,76,78–82]. Trace scheduling [74] uses profiling information to aggregate sequences of basic blocks (or control paths) that execute frequently into *traces*. These traces in the design are scheduled one at a time starting with the most time critical one. Hence, code motions are allowed only within traces and not across traces; in effect, code motions that duplicate or eliminate operations across multiple traces are not supported. Compensation code must be added in a bookkeeping pass applied after scheduling.

Percolation scheduling [76,83,84] applies a set of atomic semantic-preserving transformations to move or *percolate* operations in the design and create a resource-constrained schedule. Given an infinite number of resources, percolation scheduling has been shown to produce optimal schedules. Percolation scheduling suffers from two main limitations: (1) operations are moved linearly by visiting each node on the control path being traversed, and (2) this may lead to duplication of operations in multiple control paths. Trailblazing [80] overcomes the limitations of percolation scheduling by moving operations across large pieces of code with the help of a hierarchically structured intermediate representation [85].

Loop transformations such as loop unrolling and loop pipelining have received some attention in previous HLS works. Early work, such as the Flamel system [13], applied loop unrolling for designs with only data-flow to increase the parallelism of the design. Potasman et al. [86] demonstrated the improvements that can be obtained by applying the parallelizing compiler technique of perfect loop pipelining to data-flow designs. Holtmann and Ernst [87] apply loop pipelining for designs with control flow. In this approach, loops on the program path that have the highest predicted probability of being taken are pipelined, thereby deferring operations belonging to other paths.

11.2.4 High-Level Synthesis Tools in the Industry

Given the focus on HLS for over two decades, several tools and systems for HLS have been built in industry. Below, we list some of them. While technical details about these tools are not always available, our goal is to describe them in sufficient detail to make meaningful comparisons.

(1) *Cynthesizer* from *Forte Design Systems* [88] delivers an implementation path from SystemC to RTL. As shown in Figure 11.1, it includes a combination of behavioral synthesis functions: operation

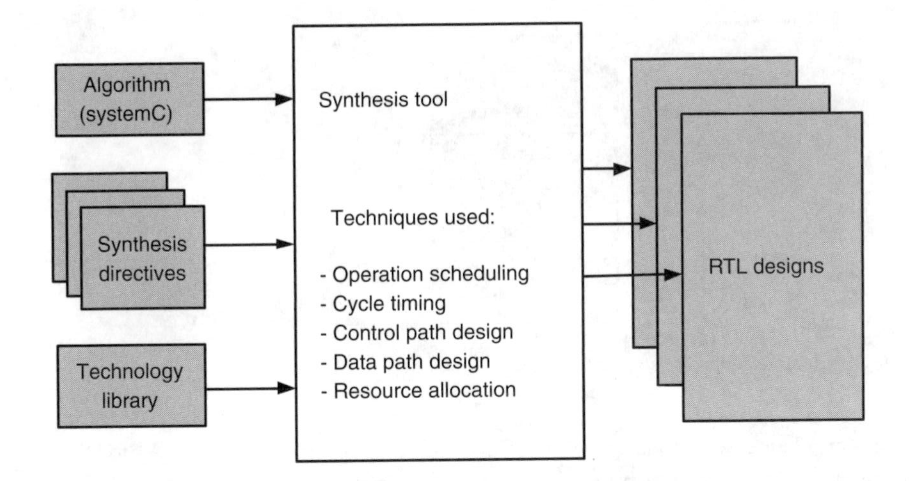

FIGURE 11.1 Forte's HLS tool flow.

scheduling, cycle timing, control and data path design, resource allocation, and RTL generation. Starting from untimed high-level SystemC models, Forte's synthesis tool builds timed RTL hardware implementations based on an external set of directives or constraints specified by the designer, such as clock speed, latency, pipelining, and loop unrolling. As an example, a block of code can be constrained to execute in no more than ten cycles by inserting a latency directive in the SystemC code itself. The block of code can be a straight line code enclosed within braces ({ }), a switch or if/else branch or a loop body.

(2) *PICO Express* from *Synfora* [89] is a tool that synthesizes ANSI-C into RTL. In an ASIC design flow, it takes in the C program code, a data set, and system requirements such as latency, bandwidth, clocking, and area. The other input is Synfora's library (currently characterized for TSMC's 0.13 μm process technology), which enables the tool to calculate area and performance. This tool is based on the PICO-NPA system developed at HP Labs [90]. The PICO-NPA system automatically synthesizes nonprogrammable accelerators (NPAs) to be used as coprocessors for functions expressed as loop nests in C. The NPAs it generates consist of a synchronous array of one or more customized processor datapaths, their controller, local memory, and interfaces. The tool is particularly powerful at finding loop-level parallelism and applying optimizations such as software pipelining, besides instruction-level parallelism. The tool determines the required hardware resources by examining the designer-specified performance requirements. Finally, it schedules operations, after which it allocates resources, and maps the components in Synfora's library.

(3) *Catapult C* from *Mentor Graphics* [91] synthesizes algorithmic descriptions in C++ and provides a graphical user interface to analyze alternative microarchitecture implementations. In addition to providing feedback (in terms of area estimates, latency, and throughput values) for each architectural modification, the micro-architecture solutions can be saved for reexploration at a later time. The tool allows each solution to be reused, thus allowing a single source code representation to be used to derive multiple implementations. The designer can apply constraints on steps such as loop unrolling, pipelining, and merging, array mapping, resource allocation, etc. The design flow through Catapult is as shown in Figure 11.2.

(4) The *Agility compiler* from *Celoxica* [92] provides a synthesis path from SystemC code and outputs electronic design interchange format (EDIF) netlists for FPGA logic and RTL HDL (hardware description language) code for ASIC synthesis. The implemented optimizations include automatic tree balancing, fine-grained logic sharing, and retiming.

(5) *Cyber* from *NEC* [46] is a HLS system used internally within NEC. *Cyber* accepts behavioral descriptions in C or behavioral VHDL and generates RT descriptions in C(BDL), Verilog, VHDL, and NEC's original RT language. C(BDL) is a subset of C with some extensions added to describe

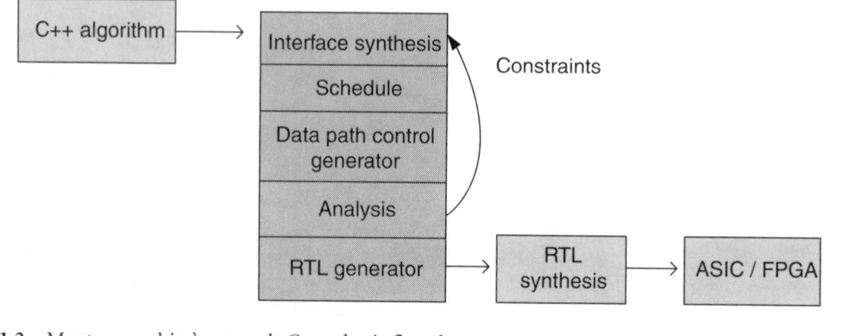

FIGURE 11.2 Mentor graphics's catapult C synthesis flowchart.

hardware. These extensions are provided as specialized comments or pragmas for the Cyber synthesizer. *Cyber* generates bit and cycle-accurate C code at the RT level, which can be compiled with off-the-shelf C compilers. *Cyber* generates a condition vector for each operation based on the conditions under which the operation executes; thus, condition vectors capture control-flow information. Operations whose condition vectors are complementary are considered mutually exclusive and are free to be scheduled on the same resource. Operation can be reordered using speculation and by duplicating operations across merge nodes in the control-flow graph (CFG).

(6) *BlueSpec Compiler* (*BSC*) from *BlueSpec* [93] synthesizes a hardware-centric input description specified in SystemVerilog and generates both the RTL and cycle-accurate C model that can be used for system-level functional verification. The compiler uses the notion of *operation-centric* descriptions [94], wherein the behavior of a system is described as a collection of guarded atomic actions or rules, which operate on the state of the system. Each rule specifies the condition or guard under which it is enabled, and a function that determines the consequent allowable state transition. The atomicity requirement simplifies the task of hardware description by permitting the designer to formulate each rule as if the rest of the system is static. Using this methodology, any legitimate behavior of the system can be expressed as a series of atomic actions on the state.
In the Bluespec methodology, the designer first explicitly instantiates the state elements of the system (registers, FIFOs, memories, etc.). Next, the designer describes the behavior of the system using a collection of rules, which operate on the state of the system. There is a straightforward translation from rules into hardware as shown in Figure 11.3. Assuming all state is accessible (no port contention), each guard and transition function can be implemented easily as combinational logic. A hardware scheduler and control circuit is added so that in every cycle the scheduler dynamically picks one function whose corresponding condition is satisfied and the control circuit updates the state of the system with the result of the selected function. The cycle time in such a synthesis is determined by the slowest guard and the slowest functions. Thus, the challenge in generating efficient hardware from sets of atomic actions is to generate a scheduler, which in every cycle picks a maximal set of rules that can be executed simultaneously. In the BlueSpec compiler, this synthesis is done using term rewriting systems [94].
Early results from BlueSpec show that the method of HLS from the guarded atomic actions is particularly useful for micro-architectural exploration in the design of complex ASICs.

(7) *Behavioral compiler* (BC) from *Synopsys* [8] (now discontinued) is the oldest commercial HLS offering. *BC* accepts the input description in a subset of behavioral VHDL. The VHDL is further restricted by coding guidelines and *pragma* constructs. *BC* uses technology libraries for resource allocation and for mapping functional units to standard cells. *BC* takes the unscheduled input description and generates a schedule for the operations. However, *BC* requires the user to insert some scheduling structure in the code in terms of VHDL *wait* statements. *Wait* statements are used for tasks such as partitioning two sections of the behavioral code that must execute in different clock cycles, or for two consecutive writes to the same variable, or for ensuring that the true and

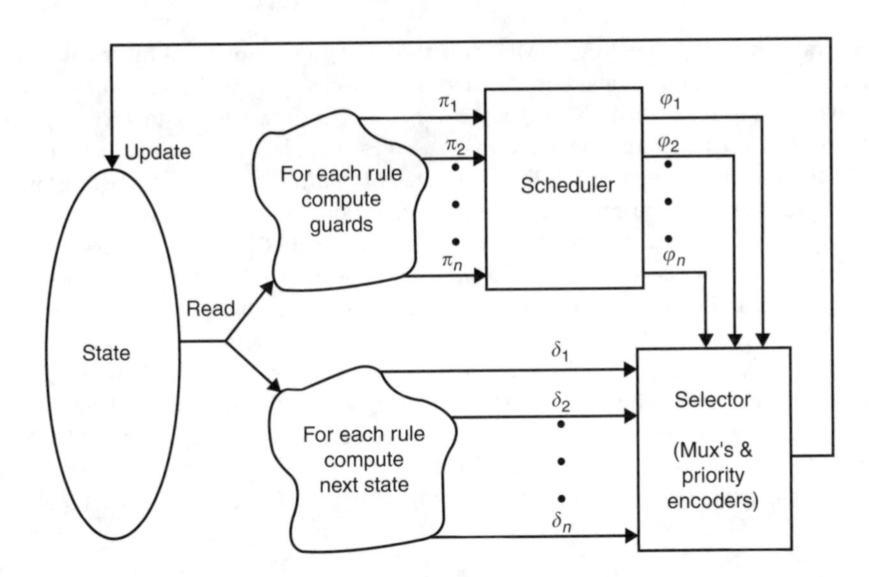

FIGURE 11.3 Synthesis from guarded atomic actions.

false branches of a conditional construct execute in the same number of cycles. The scheduler in *BC* has two different scheduling modes, *cyclefixed* and *superstatefixed*:

- *Cyclefixed mode.* In *cyclefixed* mode, *BC* preserves the cycle-to-cycle behavior of the I/O as defined by the behavioral description. It is used when the designer wants to describe a cycle-accurate description of the design.
- *Superstatefixed mode.* The section of code between two wait statements in the behavioral description is denoted as a *superstate.* In *superstatefixed* mode, *BC* can insert additional clock cycles in a superstate if required. Designers can use the superstatefixed mode to explore the latency versus area trade-offs in their designs.

Apart from these scheduling modes, *BC* can read three other schedule-constraining commands. These commands — *setcycles, setmincycles,* and *setmaxcycles* — can be used to constrain the exact, minimum, and maximum number of cycles between two operations or for a given loop. For resource allocation, *BC* tries to minimize latency by maximizing resource utilization. There is an *extendlatency* scheduling mode in which *BC* minimizes resource allocation at the cost of increased latency. Intermediate designs can be explored using *setcycles* command.

(8) The *XPRES compiler* from *Tensilica* [95,96] creates tailored instructions for the Tensilica Xtensa LX processor from algorithmic specifications in C/C++ code. Thus, it is not a behavioral synthesis tool in its classical sense. However, we mention this tool here since the tool applies parallelizing and vectorizing code transformations on the C/C++ code and generates customized hardware functional units and vector register files and adds them to the base Xtensa processor. The C/C++ code is then reinstrumented using intrinsics to use special instructions that are mapped to these customized functional units. Tensilica has demonstrated the efficacy of this tool through a range of benchmarks [96].

The XPRES compiler uses three techniques for accelerating the application: operator fusion, SIMD vectorization, and using flexible length instruction extensions (FLIX) instructions.

- *Operator fusion* combines a frequently occurring sequence of simple operations into a single instruction. A functional unit is generated corresponding to this new instruction and the code is instrumented to use this instruction. This reduces the number of instructions executed within, say a loop, and also reduces the number of instructions that must be fetched from memory.
- *SIMD vectorization* refers to vectorizing of loops by creating an instruction with multiple identical execution units that operate on multiple data items in parallel. For example, XPRES may

generate a two-operation MAC SIMD instruction that operates on two 16-byte data items simultaneously. Hence, this doubles performance over executing a MAC instruction for each of the two data items sequentially. Note that the rest of the Tensilica software tool will generate vector (or SIMD) functional units and vector register files to implement these instructions.

- The third technique used by XPRES is to create *FLIX instructions* to accelerate code by executing multiple *independent* operations in the same instruction. This is analogous to VLIW instructions in which several operations execute in parallel. These new "FLIX" instructions again execute more operations in fewer cycles, hence, accelerating the application. Note that each of these techniques leads to generation of extra hardware functional units, data path connections, register files, etc.

Thus, XPRES extends a base instruction set architecture with VLIW instructions, vector operations, fusion operations, and vector register files to create an application-specific instruction-set processor optimized for a particular application. Because XPRES is integrated with Tensilicas configurable processor tool chain, the new instructions, operations, and register files are automatically recognized and exploited by the entire software tool chain, including the C/C++ compiler.

Table 11.1 presents the input and output description languages for each of the HLS tools discussed above.

11.2.5 Continuing Challenges in HLS

Despite extensive research in HLS techniques, practical adoption of HLS has suffered from two challenges:

(1) Manual implementations are often superior to HLS solutions due to the ability of designers to apply a large range of RTL and circuit-level optimizations. Commercial HLS tools are often touted for digital signal processing designs with straight-line code. The presence of control flow (conditionals and loops) often leads to poor synthesis results.

(2) The flexibility of high-level and procedural languages such as C, C++, and their variants gives designers the freedom to specify the same algorithm using widely different programming styles and control constructs. This is particularly problematic for behavioral descriptions with complex and nested conditionals and loops, and leads to unpredictable quality of synthesis results. Current HLS tools address this issue by either limiting the input language or by asking designers to conform to preferred coding styles.

(3) During HLS, designers are often given minimal controllability over the transformations that affect synthesis. Designers are, thus, unable to use their design intuition to control and guide the synthesis process.

Speculative code motions have been proposed for improving synthesis results in designs with complex control flow. Whereas these transformations have been shown to be useful for improving schedule lengths, it is not clear if a transformation has a positive impact beyond scheduling, particularly on the control and area costs. Industry experience shows that, often, critical paths in control-intensive designs

TABLE 11.1
High-Level Synthesis Tools

Tools	Input (Behavioral)	Output (RTL)
Forte's cynthesizer	SystemC	SystemC/Verilog
Mentor's catapult	C++/C	Verilog, VHDL
Synfora's PICO express	C	Verilog
Celoxica's agility compiler	SystemC	EDIF netlists/HDL
NEC's cyber	C/VHDL	C(BDL)/Verilog/VHDL/NEC's RTL
BlueSpec's BSC	SystemVerilog	Verilog
Synopsys's behavioral compiler	VHDL	VHDL
Tensilica's XPRES	C++/C	Verilog

pass through the control unit and steering logic. To this end, Rim et al. [54] use an analytical model to estimate the cost of additional interconnect and control caused by code duplication during code motions. Bergamaschi [97] proposes the behavioral network graph to bridge the gap between high-level and logic-level synthesis and aid in estimating the effects of one on the other.

Several approaches use intermediate representations such as CDFGs for design entry. They often use restricted and stylized input descriptions that reduce the appeal of using high-level languages for design entry. Furthermore, language-level optimizations are few and their effects on final circuit area and speed are not well understood.

All these factors continue to limit the acceptance of HLS tools among designers. As mentioned earlier, manual implementations are often superior to HLS solutions because of the range of design optimizations that a system designer has at his/her disposal. While HLS tools have been catching up, just as they have already done so at lower levels of abstraction, a significant gap remains. *Parallelizing high-level synthesis* has been shown to address these factors and has demonstrated through experimental results that it can produce designs that are competitive with manual designs. This can make HLS part of the microelectronic design flow, thus, bringing about the much needed productivity gain required to design increasingly complex circuits.

11.3 Parallelizing HLS

Recent HLS approaches have employed beyond-basic-block code motions such as speculation — derived from the compiler domain — to increase resource utilization. The PHLS methodology systematically incorporates several similar techniques that attempt to increase the scope of code motions and increase available parallelism to improve resource utilization. The transformations in PHLS are structured into two groups: *presynthesis* and *dynamic* transformations. Before discussing these transformations, we briefly describe the intermediate model used to carry out these transformations.

11.3.1 Design Description Modeling

One-key component of the PHLS methodology proposed in the *SPARK* HLS system [98] (described in the next section) is modeling the design description using a novel *three-layered graph representation.* Control and data flow in the input description is captured using a CFG and a DFG, respectively. Additionally, the program structure and hierarchy of the control flow are captured using a hierarchical intermediate representation called *hierarchical task graphs* (HTGs) [85]. The nodes of these — the CFGs, DFGs, and HTGs — form a three-layered graph such that there is a relation between the nodes of each successive layer. Layered graphs are defined as follows:

Definition 11.3.1

A *k-layered graph* is a connected graph in which the vertices are partitioned into k sets $L = l_1, \ldots, l_k$ and edges run between the vertices of successive layers, l_i and l_{i-1}.

The top-level layer consists of the nodes from the HTG, the next layer has nodes from the CFG and the lowest level layer has nodes from the DFG. Data-flow graphs and CFGs have been used extensively in past work: DFGs capture the read-after-write data dependencies between operations and CFG capture the flow of control between basic blocks in the design [4,5]. We present HTGs in the next section.

11.3.1.1 Hierarchical Task Graphs: A Model for Designs with Complex Control Flow

Traditionally, CDFGs have been the most popular intermediate representation for HLS [5]. Control-data flow graphs consist of operation and control nodes with edges for both data flow and control flow. However, to apply the range of coarse- and fine-grain parallelizing compiler transformations proposed in the PHLS methodology, we also require structural information from the source code. Thus, PHLS complements CDFGs with an intermediate representation that also maintains the hierarchical structuring of the design, such as if-then-else blocks and for- and while-loops. This intermediate representation consists of basic blocks encapsulated in HTGs [85,80].

A *HTG* is a directed acyclic graph with unique *Start* and *Stop* nodes such that there exists a path from the *Start* node to every node in the HTG and a path from every node in the HTG to the *Stop* node. Edges in a HTG represent control flow between HTG nodes. Each node in an HTG can be one of the following three types:

1. *Single nodes* are nodes that have no subnodes and are used to encapsulate basic blocks.
2. *Compound nodes* are hierarchical in nature, i.e., they contain other HTG nodes. They are used to represent structures like if-then-else blocks, switch-case blocks or a series of HTGs.
3. *Loop nodes* are used to represent the various types of loops (for, while-do, do-while). Loop nodes consist of a loop head and a loop tail that are single nodes, and a loop body that is a compound node.

The *Start* and *Stop* nodes for all compound and loop HTG nodes are always single nodes. Note that the *Start* and *Stop* nodes of a single node are the node itself.

An example "C" description and its corresponding HTG representation is shown in Figures 11.4(a) and (b). This representation consists of a root HTG (*Htg0*) that has several subnodes. One of them is a If-HTG (compound) node, *Htg2.* This If-HTG has four subnodes of itself. In Figure 11.4(c), we show how the control and DFGs can be overlaid onto the HTG graph. Each basic block is encapsulated in a single node.

Since HTGs maintain a hierarchy of nodes, they are able to retain coarse, high-level information about program structure in addition to operation-level and basic block-level information. This aids in coarse-grain code restructuring (such as that done by loop transformations) and also, in operation movement by reducing the amount of compensation code required. Furthermore, nonincremental moves of operations across large blocks of code are possible without visiting each intermediate node [80].

The layering of the HTG, CFG, and DFG representations of the same example C description is shown in Figure 11.5. Each operation in the DFG has a one-to-one mapping with a basic block in the CFG and each basic block in the CFG is encapsulated in a single HTG node in the HTG representation.

Note that HTGs capture information about the control flow between hierarchical nodes and CFGs capture information about the control flow between basic blocks. Hence, we maintain both graphs since HTGs

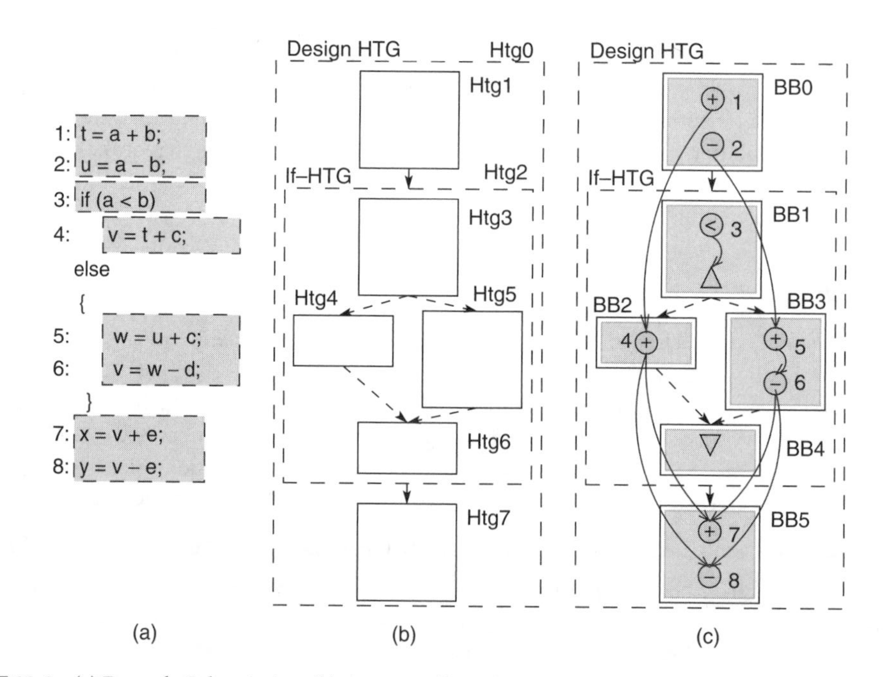

FIGURE 11.4 (a) Example C description; (b) corresponding HTG representation; and (c) HTG representation with the control and DFGs overlaid on top. Each basic block is encapsulated in a single HTG node. Basic block BB4 is added to the CFG as a join node for the If-HTG node.

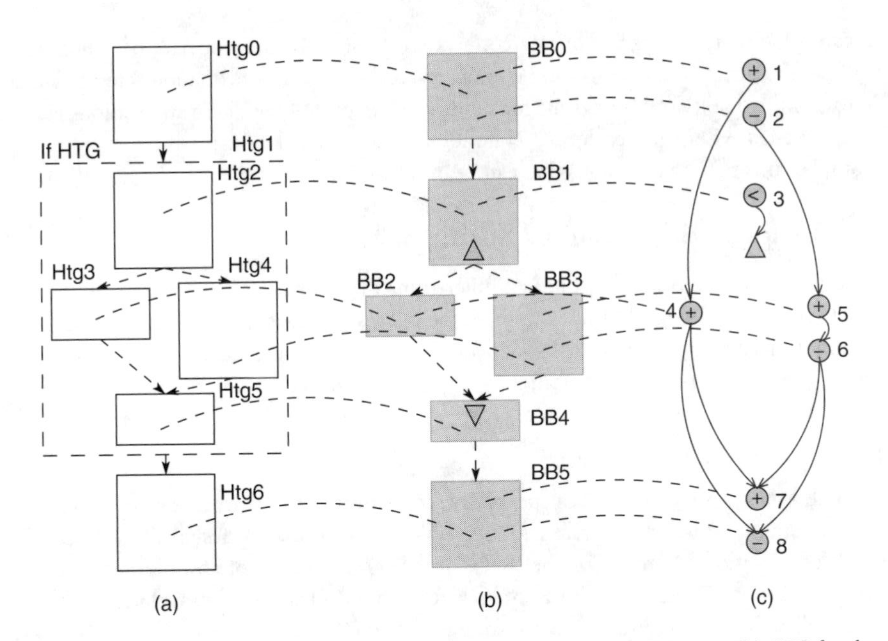

FIGURE 11.5 (a) Hierarchical task graph representation of the example C description; (b) CFG for the example; and (c) DFG for the example. Together these form a three-layered design graph, VQ. The dashed lines denote the edges between the different layers of the design graph.

are an efficient way to traverse the hierarchy of the design (useful for code transformation algorithms), whereas CFGs are efficient for traversing the basic blocks in the design (useful for scheduling algorithms).

11.3.2 Presynthesis Optimizations

Presynthesis optimizations are applied prior to elaboration of the source code into operation-level descriptions. These transformations include: (1) coarse-level code restructuring by function inlining and loop transformations (loop unrolling, loop fusion, etc.), (2) transformations that remove unnecessary and redundant operations such as CSE, copy propagation, and dead code elimination, (3) transformations such as LICM, induction variable analysis and operation strength reduction, that reduce the number of operations within loops and replace expensive operations (multiplications and divisions) with simpler operations (shifts, additions, and subtractions).

Loop transformations such as loop unrolling and loop fusion restructure the code, thereby, exposing many more opportunities for optimizations later on during scheduling. These have the maximum impact on the quality of results and have to be carefully guided, often with designer intervention. The *SPARK* HLS framework described in the next section uses synthesis scripts to guide these transformations [2,3].

Common subexpression elimination is a well-known basic compiler transformation that attempts to detect repeating subexpressions in a piece of code, compute them once and reuse the stored result wherever the subexpression occurs subsequently [1]. Often compilers apply CSE on three-operand instructions. However, Gupta et al. [3,72] have shown that applying CSE on the source code can often expose opportunities for optimizations not otherwise available. In addition to this, CSE can also be applied dynamically during scheduling as explained later in Section 11.3.3.2. *Loop-invariant* operations are computations within a loop body that produce the same results each time the loop is executed [1,64]. These computations can be moved outside the loop body, without changing the results of the code. In this way, these computations will execute only once *before* the loop, instead of for *each* iteration of the loop body.

The PHLS methodology uses a number of compiler transformations such as CSE, LICM, dead code elimination, etc., during the presynthesis stage to remove redundant and expensive operations. Although these transformations have been used extensively in compilers [1,64,99], HLS tools have to take into account the additional control and area (in terms of interconnect) costs of these transformations. Gupta et al. [3,72] have shown that the common subexpression elimination and LICM transformations are particularly effective in improving the quality of HLS results for multimedia applications. These applications are characterized by array accesses within nested loops. The index variable calculation for these array accesses often contains complex arithmetic expressions that can be optimized significantly by CSE and LICM.

11.3.3 Transformations Applied during Scheduling

A range of classical HLS and parallelizing compiler transformations are applied during the scheduling phase of the PHLS methodology. In this chapter, in the interest of space, we only briefly describe the speculative code motions and dynamic transformations. We refer the reader to [3] for details of these techniques and descriptions of other techniques such as chaining of operations across conditional boundaries, multicycle operation scheduling, etc.

11.3.3.1 Speculative Code Motions

The PHLS methodology proposes a set of speculative code motion transformations to alleviate the problem of poor synthesis results in the presence of complex control flow in designs. Effectively, these code motions reorder operations to minimize the effects of the choice of control flow (conditionals and loops) in the input description. These beyond-basic-block code motion transformations attempt to extract the inherent parallelism in designs and increase resource utilization.

Generally, speculation refers to the unconditional execution of operations that were originally supposed to have executed conditionally. However, in the context of PHLS, speculation comes into four forms: (1) speculation, (2) reverse speculation, (3) conditional speculation, and (4) early condition execution. Frequently there are situations when there is a need to move operations *into* conditionals [77,100]. This may be done by *reverse speculation,* where operations before conditionals are moved into *subsequent* conditional blocks and executed conditionally, or this may be done by *conditional speculation,* wherein an operation from after the conditional block is duplicated *up* into *preceding* conditional branches and executed conditionally. Reverse speculation can be cupled with another novel transformation, namely, *early condition execution.* This transformation evaluates conditional checks ASAP. This removes the control dependency on the operations in the conditional branches and thereby makes them available for scheduling.

The movement of operations due to the various speculative code motions is shown in Figure 11.6 using solid arrows. Also shown is the movement of operations across entire hierarchical blocks, such as if-then-else blocks or loops. Since these code motions reorder, speculate, and duplicate operations, they

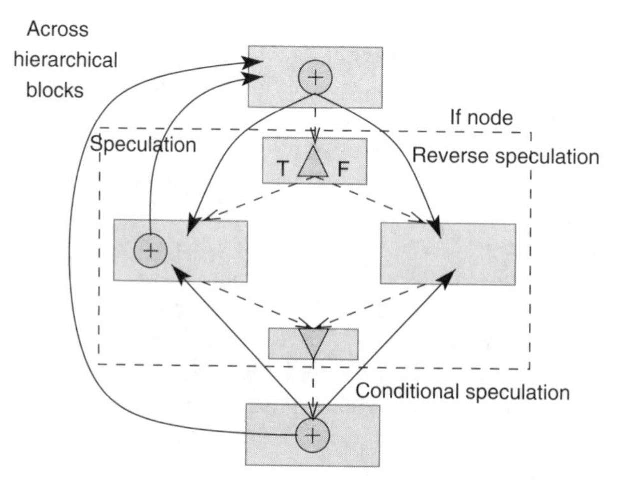

FIGURE 11.6 Various speculative code motions: operations may be speculated, reverse speculated, conditionally speculated or moved across entire conditional blocks.

often create new opportunities for dynamically applying transformations such as CSE and copy propagation, during scheduling.

11.3.3.2 Dynamic Transformations

Dynamic transformations are transformations that are deployed during various HLS tasks, most prominently during operation scheduling and resource allocation. Compiler and synthesis transformations suffer from the classical phase ordering problem, whereby the order in which transformations are applied impacts the effectiveness of the transformations. Dynamic transformations offer a way around this problem by creating opportunities for and applying optimizations during the scheduling and resource allocation phases. Dynamic transformations are particularly effective when combined with speculative code motions.

We have earlier proposed two instances of dynamic transformations in the PHLS methodology: dynamic branch balancing and dynamic CSE. Both these transformations operate during scheduling and are closely interlinked with the speculative code motions. *Dynamic branch balancing* adds scheduling steps in the shorter branches of conditional branch constructs, so that the longest path through the conditional branch is not affected [101]. The additional scheduling steps can then be used to schedule speculatively operations that would otherwise have been scheduled in later time steps. *Dynamic CSE* opportunistically looks for common subexpressions that can be eliminated during scheduling [102]. These are additional opportunities for CSE that are created as a result of code motions and code restructuring, which were unavailable before scheduling.

11.4 The *SPARK* PHLS Framework

The PHLS methodology, along with the various transformations presented in the previous sections has been implemented in the *SPARK* PHLS framework [103]. An overview of this framework is shown in Figure 11.7. *SPARK* provides the ability to control and thus, experiment with the various code transformations using user-defined scripts and command-line options. *SPARK* consists of 100,000+ lines of C++ code and uses a commercial C/C++ front-end from Edison design group (EDG) [104].

One of the main features of *SPARK* is the distinction between code transformation algorithms and the use of these techniques in a given design. This allows the designer to control the transformations applied to the design description by way of synthesis scripts [2]. The scripting ability is used to experiment with new transformations and develop heuristics to guide these transformations, which enables one to find the combination of techniques that leads to the best synthesis results. Various code motions, branch balancing algorithms, methods of calculating priority and other required transformations can be enabled using synthesis scripts. A typical *SPARK* script file has three main sections: the scheduler functions, the list of allowed code motions (code motion rules) and the cost of code motions.

Experiments performed using *SPARK* on moderately complex designs show significant improvements over classical (nonparallelizing) synthesis approaches. Experiments were conducted on a range of multimedia (MPEG-1, MPEG-2, and image processing applications (GIMP, Susan)). The results demonstrated that the various code transformations led to improvements in the performance and controller size of all the designs. The synthesis results for most of the designs follow a similar pattern. By far, the speculative code motions and presynthesis optimizations have the maximum impact on the synthesis results. The overall improvements obtained when all the transformations are applied in the *SPARK* toolkit compared to when only nonspeculative code motions are applied are up to a maximum of 80% and an average of 60% in the total delay through the circuit, with either no increase or a modest increase of 5% in circuit area.

The RTL VHDL generated by *SPARK* for these benchmarks was synthesized using the Synopsys *Design Compiler* logic synthesis tool [105]. The logic synthesis results demonstrated that the improvements in performance seen earlier translate over to the synthesis results of the final netlist. Hence, the total delay through the circuit reduces by up to 50% as the code motions are enabled with negligible impact on critical path length (clock period). However, these aggressive code motions do lead to an increase in area by about 20% when all the code motions are enabled. This area increase is due to the additional steering and control logic caused by increased resource sharing.

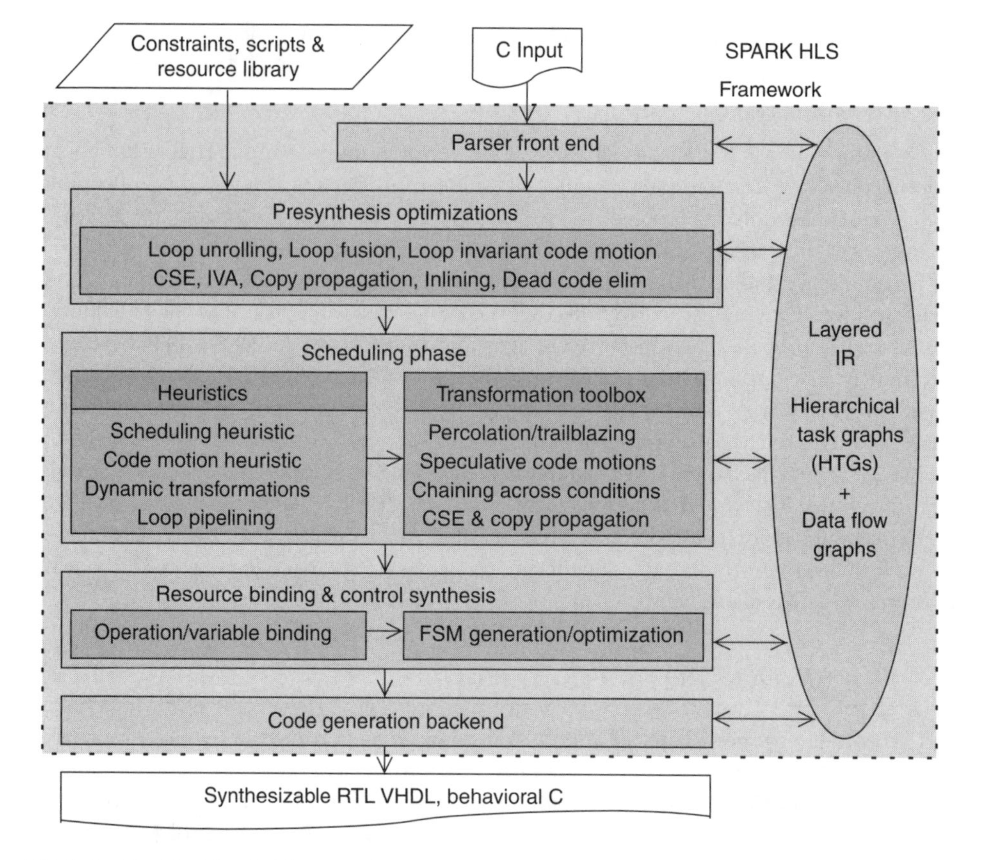

FIGURE 11.7 The *SPARK* PHLS framework.

Dynamic CSE leads to lower area due to fewer operations and the corresponding reductions in interconnect. This ends up counterbalancing the increase in area seen due to the code motions. Thus, enabling all the code motions along with CSE and dynamic CSE, leads to overall reductions in the total delay through the circuit by nearly 70% as compared to only within basic block code motions, while the design area remains almost constant (and sometimes even decreases).

11.5 Summary

The semiconductor industry's needs are shifting from pure ASIC synthesis to hardware–software co-design and direct support for synthesis on reprogrammable (FPGA) or customizable platforms (such as those offered by Tensilica and Arc). System architects are gravitating toward *system-level modeling and design methodologies** that enable them to first specify applications without notions of hardware or software, and then explore different hardware–software partitions and different hardware, communication, and memory architectures. Ideally, software compilers and synthesis tools should then be able to generate the hardware, software, and communication interfaces from these system-level specifications.

The push toward system-level design and synthesis methodologies has led to a renewed interest in HLS, as indicated by a number of startup companies in this domain. In this chapter, we described recent developments in HLS and also reviewed the considerable work done over the past two decades. We focused on a new code transformational, parallelizing approach to HLS that attempts to address the increasing size and

* See chapters on "System-level specification and modeling languages" by J. Buck and "Design and Verification Languages" by S. Edwards in this book for more on system-level design methodologies and languages.

complexity of contemporary designs. The PHLS methodology represents an important step forward in the automated synthesis of hardware components from algorithmic descriptions, so that the quality of results is not compromised by coding style and the use of control constructs. To do so, this methodology applies a coordinated set of coarse- and fine-grain parallelizing compiler and classical HLS transformations. The PHLS methodology has been validated by implementation in the *SPARK* software framework and early results demonstrate the effectiveness of this methodology in producing high-quality synthesized designs.

We believe that code transformational approaches such as the one proposed in the PHLS methodology are key in moving the level of abstraction from RTL hardware design to system-level modeling, design space exploration, refinement, and synthesis, thereby enhancing designer productivity.

References

[1] A. Aho, R. Sethi, and J. Ullman, *Compilers: Principles and Techniques and Tools*, Addison-Wesley, Reading, 1986.

[2] S. Gupta, Coordinated Coarse-Grain and Fine-Grain Optimizations for High-Level Synthesis, PhD thesis, University of California, Irvine, 2003.

[3] S. Gupta, R.K. Gupta, N.D. Dutt, and A. Nicolau, *SPARK: A Parallelizing Approach to the High-Level Synthesis of Digital Circuits*, Kluwer, Dordrecht, 2004.

[4] G. De Micheli, *Synthesis and Optimization of Digital Circuits*, McGraw-Hill, New York, 1994.

[5] D.D. Gajski, N.D. Dutt, A.C.-H. Wu, and S. Y.-L. Lin, *High-Level Synthesis: Introduction to Chip and System Design*, Kluwer, Dordrecht, 1992.

[6] R. Camposano and W. Wolf, *High Level VLSI Synthesis*, Kluwer, Dordrecht, 1991.

[7] D.C. Ku and G. De Micheli, *High Level Synthesis of ASICs Under Timing and Synchronization Constraints*, Kluwer, Dordrecht, 1992.

[8] D.W. Knapp, *Behavioral Synthesis: Digital System Design using the Synopsys Behavioral Compiler*, Prentice-Hall, Englewood Cliff, NJ, 1996.

[9] A.A. Jerraya, H. Ding, P. Kission, and M. Rahmouni, *Behavioral Synthesis and Component Reuse with VHDL*, Kluwer, Dordrecht, 1997.

[10] T.J. Kowalski and D.E. Thomas, The VLSI design automation assistant: what's in a knowledge base, *Design Automation Conference*, 1985.

[11] C.J. Tseng and D.P. Siewiorek, Automated synthesis of data paths in digital systems, *IEEE Trans. Comput-Aided Des., Integrated Circuits and Syst.*, 5, 379–395, 1986.

[12] P. Marwedel, A new synthesis for the MIMOLA software system, *Design Automation Conference*, 1986.

[13] H. Trickey, Flamel: a high-level hardware compiler, *IEEE Trans. Comput-Aided Des., Integrated Circuits and Syst.*, 6, 259–269, 1987.

[14] E. Girczyc, Automatic Generation of Micro-sequenced Data Paths to Realize ADA Circuit Descriptions, PhD thesis, Carleton University, 1984.

[15] P.G. Paulin and J.P. Knight, Force-directed scheduling for the behavioral synthesis of ASIC's, *IEEE Trans. Comput. Aided Des.*, 8, 661–678, 1989.

[16] C.Y. Hitchcock and D.E. Thomas, A method of automatic data path synthesis, *Design Automation Conference*, 1983.

[17] H. De Man, J. Rabaey, P. Six, and L. Claesen, Cathedral-II: A silicon compiler for digital signal processing, *IEEE Des. Test Mag.*, 13, 13–25, 1986.

[18] B.M. Pangrle and D.D. Gajski, Slicer: A state synthesizer for intelligent silicon compilation, ICCAD'86, Santaclara, CA, 1986.

[19] I.-C. Park and C.-M. Kyung, Fast and near optimal scheduling in automatic data path synthesis, *Design Automation Conference*, 1991.

[20] A.C. Parker, J. Pizarro, and M. Mlinar, MAHA: A program for datapath synthesis, *Design Automation Conference*, 1986.

[21] P.G. Paulin and J.P. Knight, Scheduling and binding algorithms for high-level synthesis, *Proceedings of the 26th ACM/IEEE Conference on Design automation*, 1989, pp. 1–6.

[22] L. Stok and W.J.M. Philipsen, Module allocation and comparability graphs, *IEEE International Symposium on Circuits and Systems*, 5, June 1991, pp 2862–2865.

[23] A. Mujumdar, R. Jain, and K. Saluja, Incorporating performance and testability constraints during binding in high-level synthesis, *IEEE Trans. Comp. Aided Des.*, 15, 1212–1225, 1996.

[24] C.T. Hwang, T.H. Lee, and Y.C. Hsu, A formal approach to the scheduling problem in high level synthesis, *IEEE Trans. Comput. Aided Des.*, 10, 464–475, 1991.

[25] C.H. Gebotys and M.I. Elmasry, Optimal synthesis of high-performance architectures, *IEEE J. Solid State Circuits*, 27, 389–397, 1992.

[26] B. Landwehr, P. Marwedel, and R. Doemer, Oscar: Optimum simultaneous scheduling, allocation and resource binding based on integer programming, *European Design Automation Conference*, 1994.

[27] T.C. Wilson, N. Mukherjee, M.K. Garg, and D. K. Banerji, An ILP solution for optimum scheduling, module and register allocation, and operation binding in datapath synthesis, *VLSI Des.*, 3, No.1, 21–36, 1995.

[28] A. Safir and B. Zavidovique, Towards a global solution to high level synthesis problems, *European Design Automation Conference*, March 1990, pp. 283–288.

[29] P. Kollig and B.M. Al-Hashimi, Simultaneous scheduling, allocation and binding in high level synthesis, *Electron. Lett.*, 33, 1516–1518, 1997.

[30] N. Park and A. Parker, Sehwa: A software package for synthesis of pipelines from behavioral specifications, *IEEE Trans. Comput. Aided Des.*, 7, 356–370, 1988.

[31] E. Girczyc, Loop winding — a data flow approach to functional pipelining, *International Symposium of Circuits and Systems*, 1987.

[32] L.-F. Chao, A.S. LaPaugh, and E.H.-M. Sha, Rotation scheduling: A loop pipelining algorithm, *Design Automation Conference*, 1993.

[33] M. Potkonjak and J. Rabaey, Optimizing resource utlization using tranformations, *IEEE Trans. Comput. Aided Des.*, 13, 277–293, 1994.

[34] R. Walker and D. Thomas, Behavioral transformation for algorithmic level IC design, *IEEE Trans. Comput. Aided Des.*, 8, 1115–1128, 1989.

[35] Z. Iqbal, M. Potkonjak, S. Dey, and A. Parker, Critical path optimization using retiming and algebraic speed-up, *Design Automation Conference*, 1993.

[36] D.E. Thomas, E.M. Dirkes, R.A. Walker, J.V. Rajan, J.A. Nestor, and R.L. Blackburn, The system architect's workbench, *Design Automation Conference*, 1988.

[37] M.C. McFarland, The Value Trace: A Database for Automated Digital Design, Technical Report, Carnegie-Mellon University, 1978.

[38] R. Hartley and A.E. Casavant, Tree-height minimization in pipelined architectures, *International Conference on Computer-Aided Design*, November 1989, pp. 112–115.

[39] A. Nicolau and R. Potasman, Incremental tree height reduction for high level synthesis, *Design Automation Conference*, June 1991, pp. 770–774.

[40] R. Camposano, Path-based scheduling for synthesis, *IEEE Trans. Comput. Aided Des.*, 10, 85–93, 1991.

[41] R.A. Bergamaschi, S. Raje, and L. Trevillyan, Control-flow versus data-flow-based scheduling: combining both approaches in an adaptive scheduling system, *IEEE Trans. VLSI Syst.*, 5, 82–100, 1997.

[42] S.H. Huang, Y.L. Jeang, C.T. Hwang, Y.C. Hsu, and J.F. Wang, A tree-based scheduling algorithm for control dominated circuits, *Proceedings of the 30th international Conference on Design Automation*, 1993, pp. 578–582.

[43] W. Wolf, A. Takach, C.-Y. Huang, R. Manno, and E. Wu, The Princeton University behavioral synthesis system, *Design Automation Conference*, 1992, pp. 182–187.

[44] K. Wakabayashi and T. Yoshimura, A resource sharing and control synthesis method for conditional branches, *ICCAD-89*, Nov 1989, pp. 62–65.

[45] K. Wakabayashi and H. Tanaka, Global scheduling independent of control dependencies based on condition vectors, *Design Automation Conference*, 1992, pp. 112–115.

[46] K. Wakabayashi, C-based synthesis experiences with a behavior synthesizer, "Cyber", *Design, Automation and Test in Europe*, 1999, pp. 390–393.

[47] I. Radivojevic and F. Brewer, A new symbolic technique for control-dependent scheduling, *IEEE Trans. Comput. Aided Des.*, 15, 45–57, 1996.

[48] S. Haynal, Automata-Based Symbolic Scheduling, PhD thesis, University of California, Santa Barbara, 2000.

[49] G. Lakshminarayana, A. Raghunathan, and N.K. Jha, Incorporating speculative execution into scheduling of control-flow intensive behavioral descriptions, *Design Automation Conference*, 1998, pp. 108–113.

[50] T. Kim, N. Yonezawa, J.W.S. Liu, and C.L. Liu, A scheduling algorithm for conditional resource sharing — a hierarchical reduction approach, *IEEE Trans. Comput. Aided Des.*, 13, 425–438, 1994.

[51] L.C.V. dos Santos and J.A.G. Jess, A reordering technique for efficient code motion, *Design Automation Conference*, 1999, pp. 296–299.
[52] L.C.V. dos Santos, A method to control compensation code during global scheduling, *Workshop on Circuits, Systems and Signal Processing*, 1997.
[53] L.C.V. dos Santos, Exploiting Instruction-Level Parallelism: A Constructive Approach, PhD thesis, Eindhoven University of Technology, 1998.
[54] M. Rim, Y. Fann, and R. Jain, Global scheduling with code-motions for high-level synthesis applications, *IEEE Trans. VLSI Syst.*, 3, 379–392, 1995.
[55] P. Paulin, J. Frehel, M. Harrand, E. Berrebi, C. Liem, F Nacabal, and J.-C. Herluison, High-level synthesis and codesign methods: an application to a videophone codec, *European Design Automation Conference with EURO-VHDL*, 1995, pp. 444–451.
[56] P. Kission, H. Ding, and A.A. Jerraya, Structured design methodology for high-level design, *Design Automation Conference*, 1994.
[57] G. De Micheli, D.C. Ku, F. Mailhot, and T. Truong, The Olympus synthesis system, *IEEE Des. Test Comput.*, 7, 37–53, 1990.
[58] D. Ku and G. De Micheli, HardwareC — A Language for Hardware Design, Technical Report CSL-TR-90-419, Stanford University, 1988.
[59] D. Ku and G. De Micheli, Relative scheduling under timing constraints, *Design Automation Conference*, 1990, pp. 59–64.
[60] J. Li and R.K. Gupta, HDL optimizations using timed decision tables, *Design Automation Conference*, 1996.
[61] O. Penalba, J.M. Mendias, and R. Hermida, Maximizing conditional reuse by pre-synthesis transformations, *Design, Automation and Test in Europe*, 2002, p. 1097.
[62] J. Li and R.K. Gupta, Decomposition of timed decision tables and its use in presynthesis optimizations, *International Conference on Computer Aided Design*, 1997.
[63] A. Kountouris and C. Wolinski, High level pre-synthesis optimization steps using hierarchical conditional dependency graphs, *Euromicro Conference*, Vol. 1, 1999, pp. 290–294.
[64] S.S. Muchnick, *Advanced Compiler Design and Implementation*, Morgan Kaufmann, San Francisco, 1997.
[65] M. Potkonjak, M.B. Srivastava, and A. Chandrakasan, Multiple constant multiplications: Efficient and versatile framework and algorithms for exploring common subexpression elimination, *IEEE Trans. Comput Aided Des.*, 15, 151–165, 1996.
[66] R. Pasko, P. Schaumont, V. Derudder, S. Vernalde, and D. Durackova, A new algorithm for elimination of common subexpressions, *IEEE Trans. Comput. Aided Des.*, 18, 58–68, 1999.
[67] M. Janssen, F. Catthoor, and H. De Man, A specification invariant technique for operation cost minimisation in flow-graphs, *International Symposium on High-level Synthesis*, 1994, pp. 146–151.
[68] M. Miranda, F. Catthoor, M. Janssen, and H. De Man, High-level address optimisation and synthesis techniques for data-transfer intensive applications, *IEEE Trans. VLSI Syst.*, 6, 677–686, 1998.
[69] D.A. Lobo and B.M. Pangrle, Redundant operator creation: A scheduling optimization technique, *Design Automation Conference*, 1991.
[70] M. Potkonjak and J. Rabaey, Maximally fast and arbitrarily fast implementation of linear computations, *International Conference on CAD*, 1992, pp. 304–308.
[71] R. Kennedy, S. Chan, S.-M. Liu, R. Io, P. Tu, and F. Chow, Partial redundancy elimination in SSA form, *ACM Trans. Progr. Languages Syst.*, 1999.
[72] S. Gupta, M. Miranda, F. Catthoor, and R. Gupta, Analysis of high-level address code transformations for programmable processors, *Design, Automation and Test in Europe*, 2000, pp. 9–13.
[73] S. Novack and A. Nicolau, Mutation scheduling: A unified approach to compiling for fine-grain parallelism, in *Languages and Compilers for Parallel Computing*, Springer-Verlag, London, UK, 1994, 16–30.
[74] J. Fisher, Trace scheduling: A technique for global microcode compaction, *IEEE Trans. Comput.*, 1981.
[75] K. Ebcioglu and A. Nicolau, A global resource-constrained parallelization technique, *3rd International Conference on Supercomputing*, 1989.
[76] A. Nicolau, A Development Environment For Scientific Parallel Programs, Technical Report, Cornell University, 1985.

[77] S. Gupta, N. Savoiu, N.D. Dutt, R.K. Gupta, and A. Nicolau, Conditional speculation and its effects on performance and area for high-level synthesis, *International Symposium on System Synthesis*, 2001, pp. 171–176.
[78] W.W. Hwu et al., The superblock: An effective technique for VLIW and superscalar compilation, *J. Supercomput.*, 7, 1993, pp. 229–248.
[79] R. Gupta and M.L. Soffa, Region scheduling: An approach for detecting and redistributing parallelism, *IEEE Trans. Software Engineering*, 421–431, 1990.
[80] A. Nicolau and S. Novack, Trailblazing: A hierarchical approach to Percolation Scheduling, *International Conference on Parallel Processing*, 1993.
[81] D.W. Wall, Limits of instruction-level parallelism, *International Conference on Architectural Support for Programming Languages and Operating System (ASPLOS)*, 1991.
[82] M.S. Lam and R.P. Wilson, Limits of control flow on parallelism, *International Symposium on Computer Architecture*, 1992, pp. 46–57.
[83] A. Aiken and A. Nicolau, A development environment for horizontal microcode, *IEEE Trans. Software Eng.*, 14, 584–594, 1988.
[84] A. Nicolau, Uniform parallelism exploitation in ordinary programs, *International Conference on Parallel Processing*, 1985.
[85] M. Girkar and C.D. Polychronopoulos, Automatic extraction of functional parallelism from ordinary programs, *IEEE Trans. Parallel Distribut. Syst.*, 3, 166–178, 1992.
[86] R. Potasman, J. Lis, A. Nicolau, and D. Gajski, Percolation based synthesis, *Design Automation Conference*, 1990, pp. 444–449.
[87] U. Holtmann and R. Ernst, Combining MBP-speculative computation and loop pipelining in high-level synthesis, *European Design and Test Conference*, 1995, pp. 550–556.
[88] Forte Design Systems — http://www.forteds.com, Behavioral Design Suite.
[89] Synfora Inc. — http://www.synfora.com/, PICO Express.
[90] R. Schreiber, S. Aditya, S. Mahlke, V. Kathail, B.R. Rau, D. Cronquist, and M. Sivaraman, Pico-npa: High-Level Synthesis of Nonprogrammable Hardware Accelerators, Technical Report HPL-2001-249, HP Labs, October 2001.
[91] Mentor Graphics — http://www.mentor.com/, Catapult C Synthesis.
[92] Celoxica Limited — http://www.celoxica.com, Agility Compiler — Advanced Synthesis Technology For SystemC.
[93] BlueSpec Inc. — http://www.bluespec.com, BlueSpec Compiler (BSC).
[94] J.C. Hoe and A. Arvind, Synthesis of operation-centric hardware descriptions, *IEEE/ACM Intl. Conf. on Computer Aided Design (ICCAD)*, 511–518, 2000.
[95] Tensilica Inc. — http://www.tensilica.com, XPRES Compiler: Xtensa PRocessor Extension Synthesis.
[96] D. Goodwin and D. Petkov, Automatic generation of application specific processors, *International Conference on Compilers, Architectures and Synthesis for Embedded Systems (CASES)*, 2003.
[97] R.A. Bergamaschi, Behavioral network graph unifying the domains of high-level and logic synthesis, *Design Automation Conference*, 1999, pp. 213–218.
[98] SPARK parallelizing high-level synthesis framework website, http://mesl.ucsd.edu/ spark.
[99] D.F. Bacon, S.L. Graham, and O.J. Sharp, Compiler transformations for high-performance computing, *ACM Comput. Surv.*, 26, 345–420, 1994.
[100] S. Gupta, N. Savoiu, S. Kim, N.D. Dutt, R.K. Gupta, and A. Nicolau, Speculation techniques for high level synthesis of control intensive designs, Design Automation Conference, 2001.
[101] S. Gupta, N.D. Dutt, R.K. Gupta, and A. Nicolau, Dynamically increasing the scope of code motions during the high-level synthesis of digital circuits, *Invited Paper in Special Issue of IEE Proc.-Comput. and Digital Tech.: Best of DATE 2003*, 150, 330–337, 2003.
[102] S. Gupta, M. Reshadi, N. Savoiu, N.D. Dutt, R.K. Gupta, and A. Nicolau, Dynamic common subexpression elimination during scheduling in high-level synthesis, *International Symposium on System Synthesis*, 2002, pp. 261–266.
[103] S. Gupta, R.K. Gupta, N.D. Dutt, and A. Nicolau, Coordinated parallelizing compiler optimizations and high-level synthesis, *ACM Trans. Design Autom. Electron. Syst.*, 9, 441–470, 2004.
[104] Edison Design Group (EDG) compiler frontends, http://www.edg.com.
[105] S. Gupta, N.D. Dutt, R.K. Gupta, and A. Nicolau, SPARK: A high-level synthesis framework for applying parallelizing compiler transformations, *International Conference on VLSI Design*, 2003, pp. 461–466.

SECTION III

MICRO-ARCHITECTURE DESIGN

12

Cycle-Accurate System-Level Modeling and Performance Evaluation

Marcello Coppola
STMicroelectronics
Grenoble, France

Miltos D. Grammatikakis
ISD S.A.
Athens, Greece

12.1 Introduction

The number of transistors on a chip grows exponentially, pushing technology toward highly integrated systems-on-a-chip (SoC) [1]. Existing design tools fail to exploit all the possibilities offered by this technological leap, and shorter than ever time-to-market trends drive the need for innovative design methodology and tools. It is expected that a productivity leap can only be achieved by focusing on higher levels of abstraction, enabling optimization of the top part of the design where important algorithmic and architectural decisions are made, and massive reuse of predesigned system and block components.

Detailed VHDL or Verilog models are inadequate for system-level description due to poor simulation performance. Advanced system-level modeling may lead from a high-level system model derived from initial specifications, through successive functional decomposition and refinement, to implementing an optimized, functionally correct, unambiguous protocol, i.e., without deadlock conditions or race hazards. A popular open-source C++ system-level modeling and simulation library is OSCI's SystemC [2]. SystemC consists of a collection of C++ classes describing mainly hardware concepts and a simulation kernel implementing the runtime semantics. It provides all basic concepts used by HDLs, such as fixed-point data types, modules, ports, signals, time, and abstract concepts such as interfaces, communication channels, and events. SystemC 2.0 allows the development and exchange of fast system models, providing seamless integration of tools from a variety of vendors [3]. A detailed system-level modeling framework can rely on a SystemC-based C++ Intellectual Property (IP) modeling library, a powerful simulation engine, a runtime and test environment, and refinement methodology. Two commercial tools for domain-independent SystemC-based design have been designed previously.

- In September 1998, the Felix initiative of the Alta Group of Cadence Design Systems delivered the Cierto Virtual Component Co-design tool suite (VCC 2.0) [4] VCC supports both top-down design based on communication refinement, and bottom-up reuse-oriented design. When a design is fully refined and its performance meets system requirements, the designer may export an implementation. Hardware elements are ready for HDL simulation, floor planning, and logic synthesis, while software models are mapped to processor-specific code for linking to a selected real-time operating system (RTOS) configured with the appropriate interrupt handlers, counter/timer initialization, and static schedulers. Cadence claims that only VCC permits a consistent level of design abstraction, even after detailed refinement. In addition, VCC generates complete test benches for verifying implementations using popular coverification tools. These tools combine a simulation of the hardware description, with software running on a processor simulator. Today, Cadence has withdrawn VCC from the market.
- In June 2001, Synopsys formally announced the Cocentric System Studio [5] and Cocentric SystemC Compiler [6]. Today, the first one is called just System Studio and the second has been withdrawn by Synopsys from the market. The tool is a design entry and modeling environment that utilizes block and state diagramming techniques using graphical abstractions to represent concurrency, reactive control, and dynamic switching concepts. System Studio allows the user to execute models either on a time- or an event-driven simulation engine. The Cocentric SystemC Compiler integrates behavioral and RTL synthesis engines to transform SystemC descriptions into gate-level netlists, including support for resource sharing, multicycle/pipelined operations, FSM generation, and a “MemWrap” graphical interface for mapping to specific types of memories. In addition, “Bcview” provides a graphical cross-linking capability to source code.

Despite current system design efforts, there is not yet a complete and efficient SystemC-based development environment capable of:

- providing parameterized architecture libraries, synthesis, and compilation tools for fast user-defined creation and integration of concise, precise, and consistent SystemC models;
- enabling IP reuse at the system and block level, which results in significant savings in project costs, time scale, and design risk [7,8]; and
- supporting efficient design space exploration.

In this chapter, we focus on general methodologies for cycle-accurate system-level modeling and performance evaluation. More specifically, in Section 12.2, we consider system modeling, design flow, and design space exploration. We first introduce abstraction levels, and outline general system-level design methodology, including *top-down and bottom-up protocol refinement, hierarchical modeling* (i.e., decomposing system functionality into subordinate modules [blocks and subblocks]), *orthogonalization of concerns* (i.e., separating module specification from architecture, and behavior from communication), and *communication protocol layering.* Then, we examine design space exploration, a key process to enhancing design quality.

In Section 12.3, we outline SystemC-based system-level modeling objects, such as module, clock, intramodule memory, intramodule synchronization, and inter-module communication channels. Then we discuss, with examples, back-annotation issues, especially back-annotating time for transactional clock-accurate modeling. We also examine a more complex example of back-annotating time on a user-defined SystemC-based hierarchical finite-state machine (`HFsm`). This object provides two descriptions of model behavior; using *states, conditions, actions, and delays* as in traditional (flat) finite-state machines, or *states and events.*

In Section 12.4, we focus on system-level modeling methodologies for collecting performance statistics from modeling objects, including automatic extraction of statistical properties. Our description focuses on the efficient, open-source on-chip communication network (OCCN) framework that includes an object-oriented C++ library built on top of SystemC [2]. Advanced high-level performance modeling environments may be based on advanced system-level monitoring activities. Thus, we examine the design of integrated system-level tools, *generating, processing, presenting, and disseminating* system monitoring information.

In Section 12.5, we discuss open system-level modeling issues related to asynchronous modeling, parallel and distributed system-level simulation, and interoperability with other design tools. By resolving these issues, SystemC-based system-level modeling could achieve a higher degree of productivity in contrast to other similar libraries currently available in the market.

12.2 System Modeling and Design Methodology

We examine fundamental concepts in system-level design, including abstraction level, system-level modeling methodology, and design exploration.

12.2.1 Abstraction Levels

A fundamental issue in system design is model creation. A model is a concrete representation of IP functionality. In contrast to component IP models, a *virtual platform prototype* refers to system modeling. Virtual platform enables integration, simulation, and validation of system functionality, reuse at various levels of abstraction, and design space exploration for various implementations and appropriate hardware/software partitioning. A virtual prototype consists of:

- *Hardware modules,* including peripheral IP block model (e.g., I/O, timers, Audio, Video code, or DMA), processor emulator via instruction set simulator (ISS) (e.g., ARM V4, PowerPC, ST20, or Stanford DLX), and communication network with internal or external memory (e.g., bus, crossbar, or network-on-chip).
- *System software,* including hardware dependent software, models of RTOS, device drivers, and middleware.
- *Environment simulation,* including application software, benchmarks, and stochastic models.

Notice that a virtual prototype may hide, modify, or omit system properties. As shown in Figure 12.1, abstraction levels in system modeling span multiple levels of accuracy ranging from functional to transactional cycle- and bit-accurate to gate level. Each level introduces new model details [9]. We now provide an intuitive description for the most used abstraction levels from the most abstract to the most specific.

- *Functional models* have no notion of resource sharing or time, i.e., functionality is executed instantaneously and the model may or may not be bit-accurate. This layer is suitable for system concept validation and functional partitioning between control and data. This includes definition of abstract data types, specification of hardware or software (possibly RTOS) computation, communication, and synchronization mechanisms, and algorithm integration to high-level simulation using Ada, Matlab, ODAE Solver, OPNET, SDL, SIMSCRIPT, SLAM, SystemC, or UML technology.
- *Transactional behavioral models* (denoted simply as transactional) are functional models mapped to a discrete-time domain. Transactions are atomic operations with their duration stochastically determined, i.e., a number of clock cycles in a synchronous model. Although general transactions on buses cannot be modeled, transactional models are fit for modeling pipelining, RTOS introduction, basic communication protocols, test-bench realization, and preliminary performance estimation.
- *Transactional clock-accurate models* (denoted transactional CA) map transactions to a clock. Thus, synchronous protocols, wire delays, and device access times can be accurately modeled. Using

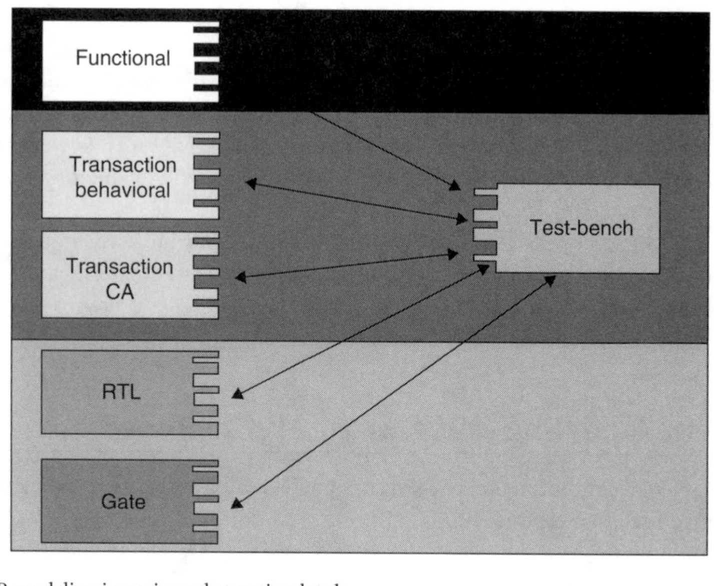

FIGURE 12.1 IP modeling in various abstraction levels.

discrete-event systems, this layer allows for simple, generic, and efficient cycle-accurate performance modeling of abstract processor core wrappers (called bus functional models), bus protocols, signal interfaces, peripheral IP blocks, instruction set simulators, and test benches. Time delays are usually back-annotated from RTL models, since transactional CA models are not always synthesizable.

- *RTL models* correspond to the abstraction level from which synthesis tools can generate gate-level descriptions (or netlists). Register-transfer-level systems are usually visualized as having two components: data and control. The data part consists of registers, operators, and data paths, while the control part provides the time sequence of signals that evoke activities in the data part. Data types are bit-accurate, interfaces are pin-accurate, and register transfer is time-accurate. Propagation delay is usually back-annotated from gate models.
- *Gate models* are described in terms of primitives, such as boolean logic with timing data and layout configuration. For simulation reasons, gate models may be internally mapped to a continuous-time domain, including currents, voltages, noise, clock rise, and fall times. Storage and operators are broken down into logic implementing the digital functions corresponding to these operators, while timing for individual signal paths can be obtained. Thus, according to the above, an embedded physical SRAM memory model may be defined as:
  - a functional model in a high-level programming language such as Ada, C, C++, or Java;
  - a transactional CA model, allowing validation of its integration with other components;
  - implementation-independent RTL logic described in VHDL or Verilog;
  - a vendor gate library described using NAND, flip-flop schematics; or,
  - a detailed and fully characterized mask layout at the physical level, depicting rectangles on chip layers and geometrical arrangement of I/O and power locations.

12.2.2 General System Modeling Methodology

Optimal system modeling methodology is a combination of stepwise protocol refinement, hierarchical modeling, orthogonalization of concerns, and communication layering techniques. Stepwise protocol refinement is achieved through a top-down and bottom-up approach.

- In *bottom-up refinement*, IP-reuse-oriented integration with optimal evaluation, composition, and deployment of prefabricated or predesigned IP block and system components drive the process.

- In *top-down stepwise refinement*, emphasis is placed on specifying unambiguous semantics, capturing desired system requirements, optimal partitioning of system behavior into simpler behaviors, and gradually refining the abstraction level down to a concrete low-level architecture by adding details and constraints in a narrower context, while preserving desired properties. Top-down refinement allows the designer to explore modeling at different level of abstractions, thus trading model accuracy with simulation speed. This continuous re-arrangement of existing IP in ever-new composites is a key process to new marketing ideas. It also allows for extensive and systematic reuse of design knowledge and application of formal correctness techniques. The concrete architecture obtained through formal refinement must satisfy:
- *relative correctness*, i.e., it must logically imply the abstract architecture, and
- *faithfulness*, i.e., no new properties could be derived.

While formal refinement is hard to achieve since there are no general proof techniques, relative refinement is based on *refinement patterns* consisting of a pair of architectural schemas that are relatively correct with respect to a given mapping. By applying refinement patterns, e.g., a general optimized state transformation [10], we can systematically and incrementally transform an abstract architecture to an equivalent lower-level form. Notice that formal architectural transformation is related to providing an augmented calculus with annotations for properties such as correctness, reliability, and performance [11]. If architectural components eventually become explicit formal semantic entities, then architectural compatibility can be checked in a similar way to type-checking in a high-level programming language [12].

Hierarchy is a fundamental relation for modeling conceptual and physical processes in complex systems of organized complexity based either on a top-down analytical approach, i.e., the divide and conquer paradigm, or a bottom-up synthesis approach, i.e., the design-reuse paradigm. Hierarchical modeling based on simpler, subordinate models controlled by high-level system models enables a better encapsulation of the design's unique properties, resulting in efficient system design with simple, clean, pure, self-contained, and efficient modular interfaces. For example, hierarchical modeling of control, memory, network, and test bench enables the system model to become transparent to changes in sub-block behavior and communication components. A systematic framework supporting hierarchical modeling may be based on the SystemC module object, and the inheritance and composition features of the C++ language [13,14]. Thus, SystemC modules consist of SystemC submodules in nested structures. Moreover, SystemC facilitates the description of hierarchical systems by supporting module class hierarchy very efficiently since simulation of large, complex hierarchical models imposes *no* impact on performance over the corresponding nonhierarchical models.

Stepwise transformation of a high-level behavioral model of an embedded system into actual implementation can be based on the concept of *orthogonalization of concerns* [15]:

- Separation of function specification from final architecture, i.e.,, *what* are the basic system functions for system-level modeling vs. how the system organizes hardware and software resources in order to implement these functions. This concept enables system partitioning and hardware/software codesign by focusing at progressively lower levels of abstraction.
- Separation of communication from computation (called behavior) enables plug-and-play system design using communication blocks that encapsulate and protect IP cores. This concept reduces ambiguity among designers, and enables efficient design space exploration.

Layering simplifies individual component design by defining functional entities at various abstraction levels and implementing protocols to perform each entity's task. Advanced inter-module communication refinement may be based on establishing distinct communication layers,thus, greatly simplifying the design and maintenance of communication systems [16,17]. We usually adopt two layers, namely:

- The *communication layer* which provides a generic message API, abstracting away the fact that there may be a point-to-point channel, a bus, or an arbitrarily complex network-on-chip.
- The *driver layer* which builds the necessary communication channel adaptation for describing high-level protocol functionality, including compatible protocol syntax (packet structure) and semantics (temporal ordering).

12.2.3 Design Space Exploration

In order to evaluate the vast number of complex architectural and technological alternatives, the architect must be equipped with a highly parameterized, user-friendly, and flexible design space exploration methodology. This methodology is used to construct an initial implementation from system requirements, mapping modules to appropriate system resources. This solution is subsequently refined through an iterative-improvement strategy based on reliability, power, performance, and resource contention metrics obtained through domain- or application-specific performance evaluation using *stochastic analysis models and tools*, and *benchmarks* for common applications, e.g., commercial networking or multimedia. The proposed solution provides values for all system parameters, including configuration options for sophisticated multiprocessing, multithreading, pre-fetching, and cache hierarchy components.

After generating an optimized mapping of behavior onto architecture, the designer may either manually decompose hardware components to the RTL of abstraction, or load system-level configuration parameters onto available behavioral synthesis tools, such as the Synopsys coreBuilder and coreConsultant. These tools integrate a pre-loaded library of configurable high-level (soft) IPs, e.g., STMicroelectronics' `STbus` [18,19], VSIA's `VCI` [20], OCP [9], and generic interface prototypes. The target is to parameterize IP in order to generate the most appropriate synthesis strategy automatically. Now, design flow can proceed normally with routing, placement, and optimization by interacting with tools, such as Physical Compiler, Chip Architect, and PrimeTime.

12.3 Back-Annotation of System-Level Modeling Objects

In this section, we consider back-annotation of system-level models, focusing especially on hardware modeling objects.

12.3.1 System-Level Modeling Objects

In this section, we examine a high-level modeling environment consisting of a SystemC-based C++ modeling library and associated simulation kernel. A proprietary runtime and test environment may be provided externally [21]. Thus, our high-level modeling environment provides user-defined C++ building blocks (macro functions) that simplify system-level modeling, enable concurrent design flow (including synthesis), and provide the means for efficient design space exploration and hardware/software partitioning.

High-level SystemC-based modeling involves hardware objects as well as system and application software components. For models of complex safety critical systems, software components may include calls to an RTOS model [22]. An RTOS model abstracts a real-time operating system by providing generic system calls for operating system management (RTOS kernel and multitask scheduling initialization), task management (fork, join, create, sleep, activate), event handling (wait, signal, notify), and real-time modeling. SystemC 3.0 is expected to include user-defined scheduling constructs on top of the core SystemC scheduler, providing RTOS features such as thread creation, interrupt, and abort. During software synthesis, an RTOS model may be replaced with a commercial RTOS.

Focusing on hardware components and on-chip communication components, we will describe system-level modeling objects (components) common in SystemC models. Most objects fall into two main categories: *active* and *passive modeling objects*. While active objects include at least one thread of execution, initiating actions that control system activity during the course of simulation, passive objects do not have this ability and only perform standard actions required by active objects. Notice that communication objects are neither active nor passive. Each hardware block is instantiated within a SystemC class called `sc_module`. This class is a container of a user-selected collection of SystemC clock domains, and active, passive, and communication modeling objects including:

- *SystemC and user-defined data types* providing low-level access, such as `Bits` and `Bytes`, allowing for bit-accurate, platform-independent modeling.

- *User-defined passive memory objects* such as `Register`, `FIFO`, `LIFO`, `circular FIFO`, `Memory`, `Cache` as well as user-defined collections of externally addressable hierarchical memory objects.
- *SystemC and user-defined intramodule communication and synchronization objects* based on message passing, such as `Monitor`, or concurrent shared memory such as `Mutex`, `Semaphore`, `Conditional Variables`, `Event Flag`, and `Mailbox`. This includes `Timer` and `Watchdog Timer` for generating periodic and nonperiodic time-out events.
- *Active control flow objects* such as SystemC processes (`SC_Method`, asynchronous `Thread`, and clocked `Thread`), and `HFsm`. Intramodule interaction including computation, communication, and synchronization is performed either by local processes or by remote control flow objects defined in other modules.
- *inter-module communication object* usually based on message passing, e.g., Message [23].
- *user-defined inter-module point-to-point and multi-point communication channels and corresponding interfaces*, e.g., peripheral, basic, and advanced VCI [20], Amba bus (`AHB`, `APB`) [24], ST Microelectronics' proprietary Request Acknowledge (`RA`), Request Valid (`RV`), and `Stbus` (type1, type2 and type3) [18,19]. These objects are built on top of SystemC in a straightforward manner. For example, OCCN provides a library of objects with the necessary semantics for modeling on-chip communication infrastructure [23].

12.3.2 Back-Annotating Time Delay in System Models

When high-level views and increased simulation speed are desirable, systems may be modeled at the transactional level using an appropriate abstract data type, e.g., processor core, video line, or network communication protocol-specific data structure. With transactional modeling, the designer is able to focus on IP functionality rather than on details of the physical interface, including data flows, FIFO sizes, or time constraints. For efficient design space exploration, transactional models (including application benchmarks) must be annotated with a number of architectural parameters.

- Realizability parameters refer to system design issues that control system concurrency such as peripheral, processor, memory, network interface, router, and their interactions, e.g., packetization, arbitration, packet transfer, or instruction execution delays; RTOS models, e.g., context switch or operation delays; and VLSI layout, e.g., time-area tradeoffs, clock speed, bisection bandwidth, power consumption models, pin count, or signal delay models.
- Serviceability parameters refer to reliability, availability, performability, and fault-recovery models for transient, intermittent, and permanent hardware/software errors. When repairs are feasible, fault recovery is usually based on detection (through checkpoints and diagnostics), isolation, rollback, and reconfiguration. While reliability, availability and fault-recovery processes are based on two-state component characterization (faulty or good), performability metrics evaluate degraded system operation in the presence of faults, e.g., increased latency due to packet congestion when there is limited loss of network connectivity.

Time delay is only one of many possible annotations. In order to provide accurate system performance measurements, e.g., power consumption, throughput rates, packet loss, latency statistics, or QoS requirements, all transactional models involving computation, communication, and synchronization components must be back-annotated with an abstract notion of time. Thus, timing characteristics, obtained either through hardware synthesis of cycle-accurate hardware models, or by profiling software models on an RTOS-based embedded processor with cycle-accurate delays associated to each operation in its instruction set, can be back-annotated to the initial TLM model for transactional cycle-accurate (CA) modeling. The cycle-accurate system architecture model can be used for system-level performance evaluation, design space exploration, and also as an executable specification for the hardware and software implementation.

However, architectural delays are not always cycle-accurate but cycle-approximate. For example, for computational components mapped to a particular RTOS, or large communication transactions mapped to a particular shared bus, it is difficult to estimate accurately thread delays, which depend on precise system configuration and load. Similarly, for deep submicron technology, wire delays that dominate protocol timings cannot be determined until layout time. Thus, in order to avoid system revalidation for deadlock and data race hazards, and ensure correct behavior independent of computation, communication, and synchronization delays, one must include all necessary synchronization points and interface logic in the transactional models. Analysis using parameter sweeps helps estimate sensitivity of system-level design due to perturbations in the architecture, and thus examine the possibility of adding new features in derivative products [1,25].

For most SystemC-based hardware objects, time annotation is performed statically, during instantiation time. Notice that time delays also occur due to the underlying protocol semantics. As an example, we now consider memory and communication channel objects; `HFsms` are discussed in the next section:

- Time is annotated during instantiation of memory objects by providing the number of clock cycles (or absolute time) for respective read/write memory operations. The timing semantics of the memory access protocol implemented within an active control flow object may contribute additional dynamic delays. For example, a *congestion delay* may be related to allowing a single-memory operation to proceed within a particular time interval, or a *large packet delay* may be due to transferring only a small amount of information per unit time.
- Time is annotated during instantiation of the processes connected to the communication channels through timing constraints implemented within an active control flow object, blocking calls to intramodule communication and synchronization objects, and SystemC `wait` or `wait_until` statements. Furthermore, blocking calls within inter-module communication objects instantiated within the SystemC channel object or adherence to specific protocol semantics may imply additional dynamic delays, e.g., compatibility in terms of the size of communicated tokens; this parameter may be annotated during communication channel instantiation.

12.3.3 An Example in Back-Annotation: Hierarchical Finite-State Machine (HFsm)

Finite-state machines are very important for modeling dynamic aspects of hardware systems. Most hardware systems are highly state-dependent, since their actions depend not only on their inputs, but also on what has previously happened. Thus, states represent a well-defined system status over an interval of time, e.g., a clock cycle. In cycle-accurate simulation, we assume that system events occur at the beginning (or the end) of a clock cycle. In this way, we can capture the right behavior semantics, and we can back-annotate hardware objects in a very detailed way.

Hierarchical finite-state models define complex, control-dominated systems as a set of smaller subsystems, by allowing the model developer to consider each subsystem in a modular way. Scoping is used for types, variable declaration, and procedure calls. Hierarchy is usually disguised into two forms, namely:

- Structural hierarchy which enables the realization of new components based on existing ones. A system is a set of interconnected components, existing at various abstraction levels.
- *Behavioral hierarchy* which enables system modeling using either *sequential*, e.g., procedure, or recursion, or *concurrent computation*, e.g., parallel or pipelined decomposition.

As asynchronous and clocked threads defined in SystemC, an `HFsm` modeled in C++/SystemC is a primary and powerful unit for capturing description of dynamic behavior within a SystemC model. `HFsm` may be used to capture formally the dynamic behavior of a hardware module, e.g., a circuit representing an embedded system controller or a network protocol. For simple control tasks, asynchronous or clocked threads in SystemC and `HFsm` implementations can be made equivalent. Although clocked threads are usually more efficient from the simulation time point of view, an `HFsm` implementation is more general, and improves clarity and efficiency when specifying complex control tasks. It also allows one to back-annotate system-level models in a more detailed way, thus making cycle accuracy easier to obtain. Since outer HFsm states are typically behavioral generalizations of inner ones, HFsm facilitates the implementation of policy mechanisms such as signaling exceptions to a higher scope and pre-emptive events. It also avoids pitfalls in implementations of traditional (flat) finite-state machine implementations, such as:

- Exponential explosion in the number of states or transitions when composing sub-states, hence the system-level model is characterized using less redundant timing information.
- Difficulty in modeling complex hierarchical control flow, with transitions spanning different levels of the behavioral hierarchy.
- Difficulty in handling hierarchical group transitions, or global control signals specified for a set of states, e.g., reset and halt.
- No support for concurrent, multithreaded computation.

HFsm are based on Harel statecharts [26] and at least 20 variants [27].

12.3.3.1 Flat HFsm Description Using States, Conditions, Actions, and Delays

With a simple, commonly used user–interface description, a flat HFsm is represented using states, conditions, delays, and actions. States are used to identify HFsm status, conditions are used to drive actions, and delay times are used to postpone actions. Composite transitions are defined as a quadruple formed by the next state, a condition, an action, and a delay. A flat HFsm represented graphically is shown in Figure 12.2.

An equivalent representation can be given in tabular form.

- State S1={{C1, A1, t1, S2}, {NULL}};
- State S2={{C3, A3, t3, S3}, {NULL}};
- State S3={{C4, A4, t4, S1}, {C3, A3, t5, S3}, {NULL}};

Let us consider the state S3. In this state we have two transitions. In the first one, S1 is the next state, C4 is the condition, A4 is the action, and t4 is a constant delay. The semantics of this transition are as follows. If the HFsm is in state S3 and condition C4 is true, then HFsm performs action *A4* and transfers control to state S1 after a delay of *t4* clock cycles. Thus, HFsm designer could re-define this order, e.g., condition-action-delay.

Furthermore, if HFsm conditions were *not* allowed to make blocking function calls, then the delay associated in performing the corresponding action would always be *deterministic* and could possibly be estimated by the model designer. *Dynamic delays*, e.g., delays due to concurrent access to shared resources, could be handled at the beginning of the action routine by placing special blocking function calls to intramodule communication and synchronization objects.

12.3.3.2 HFsm Description using States and Events

For more complex examples, hierarchical states may have "parents" or "children", placed at different levels. This approach is common with machines providing various operating modes, i.e., machine features

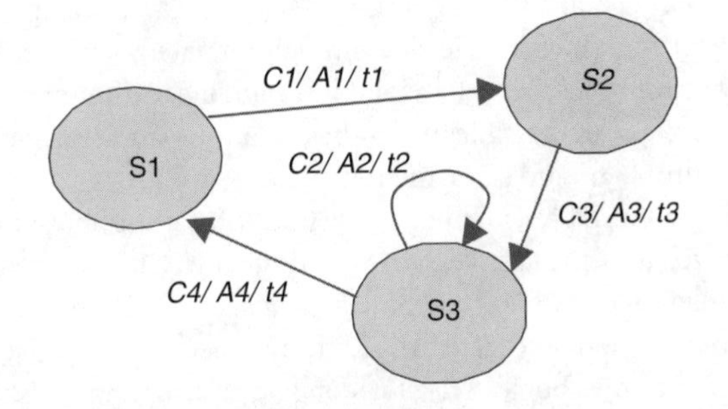

FIGURE 12.2 Graphical illustration of a flat, clocked finite-state machine.

that define multiple, alternative configurations. Then, `Hfsm` is described optimally using discrete-event systems, i.e., a finite set of states, a finite set of event categories, start-up and regular state transitions mapping state-event pairs to other states, and actions associated with transitions. This description is natural, abstract, powerful, and efficient.

- *Initial start-up transitions* originate during initial transitions from a `top` superstate. These transitions invoke entry and start-up actions (e.g., setting state parameters) for the state associated with the transition as well as for intermediate superstates, if a transition crosses two or more levels of hierarchy.
- *Regular transitions* occur at the current state during normal `Hfsm` processing. They are triggered by events selected either randomly or through external stimula from a user-defined enumerated event list. These transitions cause entry (and exit) actions upon entering (respectively, exiting) an `HFsm` state. Depending on the position in the state hierarchy of the new vs. the current state, entry and exit actions are executed in order, either from the least deeply nested to the most deeply nested state, or *vice versa.* Entry and exit actions always precede processing actions possibly associated with the newly entered hierarchical state.

With this description, a SystemC `wait` or `wait_until` statement can be associated with each state event. Moreover, blocking calls to intramodule communication and synchronization objects may account for additional dynamic delays.

12.4 Automatic Extraction of Statistical Features

System-level modeling is an essential ingredient of system design flow. Data and control flow abstraction of the system hardware and software components express not only functionality, but also performance characteristics that are necessary to identify system bottlenecks. For example, virtual channels are used to avoid both network and protocol deadlock, and also to improve performance, and provide quality of service. While for software components it is usually the responsibility of the user to provide appropriate performance measurements, for hardware components and interfaces it is necessary to provide a statistical package that hides internal access to the modeling objects. This statistical package can be used for evaluating system performance characteristics. The statistical data may be analyzed using visualization software, e.g., the open source Grace tool or dumped to a file for subsequent data processing, e.g., via an electronic spreadsheet or a specialized text editor.

12.4.1 Statistical Classes for System-Level Modeling

Considering the hardware system modeling objects previously proposed, we observe that dynamic performance characteristics, e.g., latency, throughput, packet loss, resource utilization, and possibly power consumption (switching activity) are definitely required from:

- *inter-module communication and synchronization objects* (`Message`) reflecting communication channel performance metrics for throughput, latency, or packet loss; and
- *intramodule passive memory objects* (`Register`, `FIFO`, `LIFO`, circular `FIFO`, `Memory`, `Cache`) reflecting memory performance metrics for throughput, latency, packet loss, buffer size, and hit ratio. Although similar metrics for certain joint intramodule communication and synchronization objects, e.g., `Mailboxes` are possible, these objects may be implemented in hardware in various ways, e.g., using control logic and static memory.

Assuming a common 2-D graphic representation, a statistical API for the above metrics can be based on a function `enable_stat(args)` enabling the data-capturing activity. Its arguments specify a distinct name of the modeling object, the absolute start and end time for statistics collection, the title and legends for the x and y axes, the time window for windowed statistics, i.e., the number of consecutive points averaged for generating a single statistical point, and a boolean flag for stopping or restarting statistics during simulation.

Since complex systems involve both time- (instant) and event-driven (duration) statistics, we may provide two general monitoring classes, collecting instant and duration measurements from system components with the following functionality.

In *time-driven simulation*, signals usually have instantaneous values. During simulation, these values can be captured at a specific time by calling the following function:

- `stat_write (double time, double value).`

In *event-driven simulation*, recorded statistics for events must include arrival and departure time (or duration). Since the departure time is known later, the interface must be based on two functions `stat_event_start` and `stat_event_end`. Thus, first the user invokes an operation

- `a = int stat_event_start(double arrival_time)`

to record the arrival time and save the unique location of the event within the internal table of values in a local variable `a`. Then, when the event's departure time is known, this time is recorded within the internal table of values at the correct location, by calling the `stat_event_end` function with the appropriate departure time.

- `void stat_event_end(double departure_time, a).`

The `stat_event_start/end` operations may take into account memory addresses in order to compute duration statistics for consecutive operations, e.g., consecutive read/write accesses (or enqueue/dequeue) operations corresponding to the same memory block (or packet) for a `Register`, `FIFO`, `Memory`, `Cache`, or inter-module communication object, i.e., `Message`. Thus, a modified `StatDurationLib` class performs the basic `stat_event_start/end` operations using (transparently to the user) internal library object pointers. Then, upon an event arrival, we invoke the command

- `void stat_event_start(long int MAddr, double arrival_time)`

in order to record the following in the next available location within the internal table:

- the current memory address (`MAddr`),
- the arrival time (`arrival_time`), and
- an undefined (-1) departure time.

Then, upon an event departure, we invoke the command

- `void stat_event_end(long int MAddr, double departure_time)`

to search for the current `MAddr` in the internal table of values, and update the entry corresponding to the formerly undefined, but now defined departure time.

For duration statistics in 2-D graphic form, the *y*-axis point may indicate time of occurrence of a read operation performed on the same address as a previous cache write operation whose time of occurrence is shown at the corresponding point on the *x*-axis. Event duration, e.g., latency for data access from the same memory block can be obtained by subtracting these two values. For example, using `Grace`™ this computation can be performed very efficiently using the `Edit data sets: create_new (using Formula)` option. Furthermore, notice that the write-read mapping is one to one, i.e., it is first written and then it is read, while a *reverse access*, i.e., read before write, causes an error.

As explained before, statistics collection may either be performed by the user, or directly by the hardware modeling objects, e.g., bus channels, such as `APB`, `AHB`, or `StBus`, using library-internal object pointers. In the latter case, software probes are inserted into the *source code of library routines*, either manually by setting sensors and actuators, or more efficiently through the use of a monitoring segment which automatically compiles the necessary probes. Software probes share resources with the system model, thus offering low cost, simplicity, flexibility, portability, and precise application performance measurement in a timely, frictionless manner.

Furthermore, based on the previous statistical functions, we can derive application-specific statistics for the following system modeling objects (similar to instant and duration classes, `enable_stat(args)` functions are provided for initializing parameters):

- *average throughput over a specified time window* of `Register`, `FIFO` objects, `Memory`, `Cache`, and inter-module communication objects (`Message`),
- *cumulative average throughput over consecutive time windows* of `Register`, `FIFO` objects, `Memory`, `Cache`, and inter-module communication objects (`Message`),
- *instant value of counter-based objects* such as `FIFO`, `LIFO`, and circular `FIFO` objects; this class can also be used to compute the instant `Cache` hit ratio (for read/write accesses) and cell loss probability with binary instantaneous counter values,
- *average instant value over a specified time window of counter-based objects;* this class can also be used to compute the average `Cache` hit ratio and cell loss probability,
- *cumulative average value over consecutive time windows* of counter-based objects; this class can also be used to compute the cumulative average `Cache` hit ratio and cell loss probability,
- *latency* statistics for consecutive read/write (or enqueue/dequeue) accesses on `Register`, `FIFO`, `Memory`, `Cache`, and inter-module communication objects (`Message`).

In addition to the previously described classes which cover all basic cases, it is sometimes necessary to combine statistical data from different modeling objects, e.g., from hierarchical memory units, in order to compare average read vs. write access times, or for computing cell loss in ATM layer communication. For this reason, we need new user-defined *joint or merged statistics* built on top of time- and event-driven statistics. Special parameters, e.g., boolean flags for these joint detailed statistic classes can lead to more detailed statistics.

We have discussed automatic extraction of statistical properties from hardware modeling objects. This statistical approach may also be extended to software modeling. As an example, the `enable_stat_` construct for obtaining write throughput statistics (read throughput is similar) from a FIFO object `buffer` is listed below. This function call will generate Figure 12.3.

```
//Enable statistics in [0,500], with time window = 1 sample
enable_stat_throughput_read("buffer", 0, 500, 1,
      "Simulation Time", "Average Throughput for Write Access");
```

For more details on the implementation of statistical classes, and the use of online or off-line `Grace`-based statistical graphs, the reader is referred to the OCCN user manual and the statistical test benches that accompany the OCCN library [23].These test benches are compatible with both Solaris and Linux operating systems.

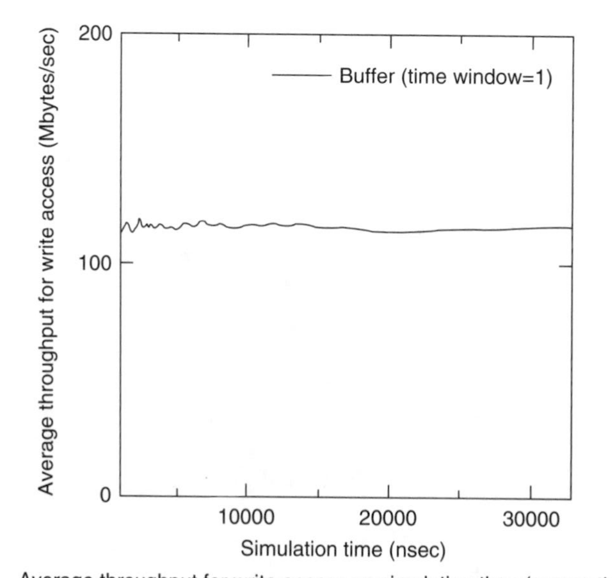

FIGURE 12.3 Performance results using Grace for a transport layer protocol.

12.4.2 Advanced System Monitoring

Conventional text output and source-code debugging are inadequate for monitoring and debugging complex and inherently parallel system models. Current tools, such as the Synopsys System Studio generate vcd files for signal tracing, or build relational databases in the form of tables for data recording, visualization, and analysis. Advanced high-level performance modeling environments may be based on advanced system-level monitoring activities. Although these activities may correspond to distinct monitoring phases occurring in time sequence, potentially there is a partial overlap between them.

- *Generation* refers to detecting events, and providing *status and event reports* containing monitoring traces (or histories) of system activity.
- *Processing* refers to functions that deal with monitoring information such as filtering, merging of traces, correlation, analysis, validation, and model updating. These functions convert low-level monitoring data to the required format and level of detail.
- *Presentation* refers to displaying monitoring information in an appropriate form.
- *Dissemination* concerns the distribution of selected monitoring reports to system-level developers and external processing entities.

In addition to the above main activities, implementation issues relating to intrusiveness and synchronization are crucial to the design and evaluation of monitoring activities.

12.4.2.1 Generation of Monitoring Reports: Status and Event Reports

In order to describe the dynamic behavior of an object (or group of objects) over a period of time, status and event reports are recorded in time order as *monitoring traces.* A *complete monitoring trace* contains all monitoring reports generated by the system since the start of the monitoring session, while a *segmented trace* is a sequence of reports collected during a limited period of time, e.g., due to overflow of a trace buffer or deliberate halting of trace generation. A monitoring trace may also be used to generate nonindependent traces based on various logical views of objects or system activity. For each trace we need to identify the reporting entity, the monitored object, the type of the report, as well as user-defined parameters, e.g., start and end time, time-window, priority, and size. We also provide browsing or querying facilities (by name or content) or runtime adjustments, e.g., examining the order of event occurrences or readjusting the interval between event occurrences.

A *status report* contains a subset of system state information, including object properties, such as time stamp, status, and identity. Appropriate criteria define the sampling rate, the reporting scheme, and the contents of each report. For example, the report may either be generated periodically, i.e., based on a predetermined finite-state machine or Thread schedule, or on demand, i.e., upon receiving a request for solicited reporting. The request may itself be periodic, i.e., via polling or on a random basis.

System events may be detected immediately upon occurrence, or after a delay. For example, signals on an internal system bus may be monitored in real time, while alternatively, status reports may be generated, stored, and processed in order to detect events at a later time. Event detection may be internal to the monitored object, i.e., typically performed as a function of the object itself, or external, i.e., performed by an external agent who receives status reports and detects changes in the state of the object. Once the occurrence of an event is detected, an *event report* is generated. In general, an event report contains a variable number of attributes such as reporting entity, monitored object, event identifier, type, priority, time of occurrence, state of the object immediately before and after event occurrence, application-specific state variables, time stamps, text messages, and possibly pointers to detailed information.

12.4.2.2 Processing of Monitoring Information

A system-level model may generate large amounts of monitoring information. Design exploration is successful, if the data can be used to identify design problems and provide corrective measures. Thus, the data collection phase is split into four different phases.

Validation of monitoring information provides data consistency checks, possibly specified in a formal language, ensuring that the system is monitored correctly. This includes:

- *sanity tests* based on the validity of individual monitoring traces, e.g., by checking for correct token values in event fields, such as an identity or time stamp, and
- *validation of monitoring reports against each other*, e.g., by checking against known system properties, including temporal ordering.

Filtering reduces the amount of monitoring data to a suitable rate and level of detail. For example, filter mechanisms reduce the complexity of displayed process communications by:

- incorporating a variable report structure
- displaying processes down to a specified level in the module hierarchy
- masking communication signals and data using filter dialogs and
- providing advanced filter functionality for displaying only tokens with pre-determined values

Analysis processes monitoring traces based on special user-selected criteria. Since analysis is application-dependent, it relies on sophisticated stochastic models involving combinatorics, probability theory, and Markov chain theory. *Analysis* techniques enable:

- simple data processing, e.g., average, maxima, variance statistics of particular state variables,
- statistical trend analysis for forecasting using data analysis packages such as Maple, Grace, SAS, or S from AT & T,[28,29] and *correlation* (merging, combination, or composition) of monitoring traces which raises the abstraction level.

Together with filtering, it prevents the users from being overwhelmed by an immense amount of detailed information, and helps identify system bottlenecks. Thus, for example, events generated by sensors or probes may be combined using `and`, `or`, and `not` operators to provide appropriate high-level reliability metrics. Since correlation of system-level monitoring information is a very challenging task, a relational database, such as mini-SQL, that includes selection, projection, and join operators is sometimes useful.

12.4.2.3 Presentation of Monitoring Information

Various visualization techniques such as simple textual representation, time-process diagrams, and animation may be provided.

Textual representation (ascii) increases its expressive power by providing appropriate indentation, color, and highlighting to distinguish information at different abstraction levels. Events may be displayed in a causal rather than temporal order by including parameters such as event type, name of process initiating the event, name of process(es) handling the event, and contents of transmitted messages.

A *time-process diagram* is a 2-D diagram illustrating the current system state and the sequence of events leading to that state. The horizontal axis represents events corresponding to various processes, while the vertical one represents time. In synchronous systems, the unit of time corresponds to actual time, while in asynchronous systems, it corresponds to the occurrence of an event. In the latter case, the diagram is called a *concurrency map*, with time dependencies between events shown as arrows. An important advantage of time-process diagrams is that monitoring information may be presented on a simple text screen or a graphical one.

An *animation* captures a snapshot of the current system state. Both textual and graphical event representations e.g., input, output, and processing events, can be arranged in a 2-D display window. Graphical representations use various formats such as icons, boxes, Kiviat diagrams, histograms, bar charts, dials, *X–Y* plots, matrix views, curves, pie charts, and performance meters. Subsequent changes in the display occur in single step or continuous fashion, and provide an animated view of system evolution.For online animation, the effective rates at which monitoring information is produced and presented to the display must be matched. For each abstraction level, animation parameters include enable/disable event monitoring or visualization, clock precision, monitoring interval, or level of detail, and viewing/printing statistics.

12.4.2.4 Dissemination of Monitoring Information

Monitoring reports would have to reach designers, managers, or processing entities. Thus, dissemination schemes range from very simple and fixed, to very complex and specialized. Selection criteria contained within the subscription request are used by the dissemination system to determine which reports (and with what contents, format, and frequency) should be delivered.

12.4.2.5 Other Important Monitoring Design Issues

Specifications for the actual graphical user interface of the monitoring tool depend on the designer's imagination, experience, and time. We present here desirable features of a system-level monitoring interface:

- *Visualization at different abstraction levels* which enables the user to observe behavior at various abstraction levels on a per object, per block, or per system basis. Thus, the user may start observation at a coarse level and progressively (or simultaneously) focus on lower levels. At each level, application performance and platform-specific metrics may be presented in appropriate easy-to-read charts and graphs. The communication volume (reliability or power consumption) may be visualized by adjusting the width or color of the lines interconnecting (respectively) the boxes representing the corresponding modules,
- A *history function* which visualizes inherent system parallelism by permitting the user to
  - scroll the display of events forwards or backwards in time by effectively changing a simulated system clock, and
  - control the speed at which system behavior is observed using special functions, e.g., to start, stop, pause, or restart an event display, perform single-step or continuous animation, and allow for real-time or slow-motion animation.
- *Visibility of interactions* which enables the user to visualize dynamically contents of a particular communication message, object data structure, module or driver configuration, or general system data such as log files or filtering results.
- *Placement of monitoring information* which greatly enhances visibility and aids human comprehension. Placement may either be automatic, i.e., using computational geometry algorithms, or manual by providing interface functions, e.g., for moving or resizing boxes representing objects, system events, or coupled links representing process communication.
- *Multiple views* use multiple windows representing system activities from various points of view, thus providing a more comprehensive picture of system behavior.
- *Scalable monitoring* focuses on monitoring large-scale models, with tens or hundreds of thousands of objects. Large-scale monitoring could benefit from efficient message queues, or nonblocking and blocking concurrent queues [30,31] that achieve a high degree of concurrency at low implementation cost compared to other general methods [32,33].

  In addition, the monitoring system designer should focus on the issues of intrusiveness and synchronization occurring in distributed system design.
- Intrusiveness refers to the effect that monitoring may have on the monitored system due to sharing common resources, e.g., processing power, communication channels, and storage space. Intrusive monitors may lead not only to system performance degradation, e.g., due to increased memory access, but also to possible deadlock conditions and data races, e.g., when evaluating symmetric conditions which result in globally inconsistent actions.
- Distributed simulation can support time- and space-efficient modeling of large modular systems at various abstraction levels, with performance orders of magnitude better than existing commercial simulation tools [34]. Distributed systems are more difficult to monitor due to increased parallelism among processes or processors, random and nonnegligible communication delays, possible process failures, and unavailable global synchronized time. These features cause interleavings of monitoring events that might result in different output data from repeated runs of deterministic distributed algorithms, or different views of execution events from various objects

[35–38]. However, these problems do not occur with SystemC scheduling (and early SystemC-based parallel efforts), since current versions of the kernel are sequential, offering simulated parallelism and limited event interleaving.

12.5 Open System-Level Modeling Issues

Cycle-accurate transactional system-level modeling is performed at an abstraction level higher than traditional hardware description languages, thus improving design flow and enhancing design quality by offering an increased design space, efficient simulation, and block- and system-level reuse. We have described efforts toward a general, rich, and efficient system-level performance evaluation methodology that allows multiple levels of abstraction, enables hardware/software codesign, and provides interoperability between several state-of-the-art tools. Activities that could enhance this methodology include

- System reliability and fault-tolerance models [39].
- Power-consumption models [40–42].
- Asynchronous modeling [43], e.g., by modifying the SystemC kernel, and providing waves, concurrency map data structures, and system snapshots.
- Graphical visualization [44,45,46], including appropriate GUIs for:
  - *interactive model design* based on importing ready-to-use library modules, e.g., `HFsms`, or reusable IP block or system components,
  - *simulation control*, such as saving into files, starting, pausing, and restarting simulation as well as displaying waveforms by linking to graphical libraries, and dumping or changing simulation and model parameters during initialization or runtime [47],
  - *advanced monitoring features*, based on generation, processing, presentation, and dissemination activities, and
  - *platform-specific performance metrics*, such as simulation efficiency, and computation, and communication load, which help improve simulation performance in parallel platforms, e.g., through automatic data allocation, latency hiding, or dynamic load balancing.
- Executable model specification by customizing existing tools, such as Doxygen or Doc++.
- Interoperability between libraries and tools, e.g., ISS, system and hardware description languages, and simulation interfaces which can provide long-term reuse opportunities. This could be achieved with middleware, e.g., by inventing an appropriate interface definition language.
- System validation and verification — existing block- and system-level verification tools are usually external to the system-level modeling library and tightly linked to actual implementation and this causes a complete rebuilt of the verification environment, even if only very few architectural parameters change.

References

[1] Carloni, L.P. and Sangiovanni-Vincentelli, A.L., Coping with latency in SoC design,. *IEEE Micro* (Special Issue on Systems on Chip), 22–25, 24–35, 2002.

[2] SystemC, http://www.systemc.org

[3] SystemC 2.0, *User's Guide*, 2001, see http://www.systemc.org and http://www-ti.informatik.uni-tuebingen.de/~systemc

[4] Krolikoski, S., Schirrmeister, F., Salefski, B., Rowson, J., and Martin, G., Methodology and Technology for Virtual Component Driven Hardware/Software Co-Design on the System Level, Paper 94.1, ISCAS 99, Orlando, FL, May 30–June 2, 1999.

[5] SystemC Modeling with the Synopsys Cocentric SystemC Studio, Synopsys, 2002. See http://www.synopsys.com/products/cocentric_studio/cocentric_studio.html

[6] SystemC Synthesis with the Synopsys Cocentric SystemC Compiler, Synopsys, 2002, see http://www.synopsys.com/products/cocentric_systemC/cocentric_systemC.html

[7] Ferrari, A. and Sangiovanni-Vincentelli, A., System design: traditional concepts and new paradigms, *Proceedings of the International Conference on Computer Design*, 1999, pp. 2–13.

[8] Rowson, J.A. and Sangiovanni-Vincentelli, A.L., Interface-based design, *Proceedings of the Design Automation Conference*, 1997, pp. 178–183.

[9] Haverinen, A., Leclercq, M., Weyrich, N., and Wingard, D., SystemC-based SoC communication modeling for the OCP protocol, white paper submitted to SystemC, 2002. Also see http://www.ocpip.org/home

[10] Moriconi, M., Qian, X., and Riemenschneider, R.A., Correct architecture refinement, *IEEE Trans. Software Eng.*, SE-21, 356–372, 1995.

[11] Medvidovic, N. and Taylor, R.N., A framework for classifying and comparing architecture description languages, *Proceedings of the Software Engineering Conference*, Lecture Notes in Computer Science, Vol. 1301, Spinger, Berlin, 1997, pp. 60–76.

[12] Allen, R. and Garlan, D., A formal basis for architectural connection, *ACM Trans. Software Eng. Methodol.*, 6, 213–249, 1997.

[13] Charest, L., Aboulhamid, E.M., and Tsikhanovich, A., Designing with SystemC: multi-paradigm modeling and simulation performance evaluation, *Proceedings of the HDL Conference*, 2001, pp. 33–45.

[14] Virtanen, S., Truscan, D., and Lilius, J., SystemC based object oriented system design, *Proceedings* of the *Forum on Design Languages*, 2001, pp. 1–4.

[15] Gajski, D.D., Zhu, J., Doemer, A., Gerstlauer, S., and Zhao, S., *SpecC: Specification Language and Methodology*, Kluwer, Dordrecht, 2000. Also see, http://www.specc.org

[16] Brunel, J.-Y., Kruijtzer, W.M., Kenter, H.J.H.N., Petrot, F., Pasquier, L., de Kock, E.A, and Smits, W.J.M., Cosy communication IP's, *Proceedings of the Design Automation Conference*, 2000, pp. 406–409.

[17] Coppola, M., Curaba, S., Grammatikakis, M.D., Locatelli, R., Marrucia, G., and Papariello, F., OCCN: a NoC modeling framework for design exploration, *J. Syst. Arch. — Special Issue on Network-on-Chip*, Vol. 50, Elsevier, North Holland, 2004, pp. 129–163.

[18] Carey, J., *STbus Superhighway: Type 3*, internal document, ST Microelectronics, March 2001.

[19] Scandurra, A., *STbus C++ Class*, internal document, ST Microelectronics, December 2000.

[20] VSI Alliance, http://www.vsi.org/

[21] Lyoyd, A., *Runtests Lite User Guide*, internal document, ST Microelectronics, June 2000.

[22] Melkonian, M., Get by without an RTOS. Embedded Systems, 13 (10), September 2000, available from http://www.embedded.com/2000/0009/0009feat4.htm

[23] Coppola, M., Curaba, S., Grammatikakis, M., Maruccia, G., and Papariello, F., *The OCCN user manual*, unpublished report, 2004. Available from http://occn.sourceforge.net

[24] Amba Bus, Arm, http://www.arm.com

[25] Carloni, L.P., McMillan, K.L., and Sangiovanni-Vincentelli, A.L., Theory of latency-insensitive design, *IEEE Trans. Computer-Aided Design of Integrated Circuits & Syst.*, 20–29, 1059–1076, 2001.

[26] Harel, D., Statecharts: a visual formalism for complex systems, *Sci. Comput. Program.*, 8, 231–274, 1987.

[27] von der Beeck, M., A comparison of statecharts variants, *Proceedings of the Formal Techniques in Real Time and Fault Tolerant Systems*, Lecture Notes in Computer Science, Vol. 863, Spinger, Berlin, 1994, pp. 128–148.

[28] Zimmerman, C., *The Quantitative Macroeconomics and Real Business Cycle*, for statistical software see: http://dge.repec.org/software.html

[29] Grace see http://plasma-gate.weizmann.ac.il/Grace/

[30] Hunt, G.C., Michael, M., Parthasarathy, S., and Scott, M.L., An efficient algorithm for concurrent priority queue heaps, *Inf. Proc. Lett.*, 60, 151–157, 1996.

[31] Michael, M. and Scott, M.L., Simple, fast and practical non-blocking and blocking concurrent queue algorithms, *Proceedings of the ACM Symposium on Princ. Distrib. Comput.*, 1996 , pp. 267–275.

[32] Prakash, S., Yann-Hang, L., and Johnson, T., A non-blocking algorithm for shared queues using compare-and-swap, *IEEE Trans. Comput.*, C-43, 548–559, 1994.

[33] Turek, J., Shasha, D., and Prakash, S., Locking without blocking: making lock-based concurrent data structure algorithms nonblocking, *Proceedings of the ACM Symposium on Princ. Database Systems*, 1992, pp. 212–222.

[34] Grammatikakis, M.D. and Liesche, S., Priority queues and sorting for parallel simulation, *IEEE Trans. Soft. Eng.*, SE-26, 401–422, 2000.
[35] Chandy, K. M. and Lamport, L., Distributed snapshots: determining global states of distributed systems, *ACM Trans. Comp. Syst.*, 3, 63–75, 1985.
[36] Christian, F., Probabilistic clock synchronization, *Distr. Comput.*, 3, 146–158, 1989.
[37] Fidge, C.J. Partial orders for parallel debugging, *Proceedings of the ACM Workshop Parallel Distr. Debug.*, 1988, pp. 183–194.
[38] Lamport, L., Time, clocks and the ordering of events in distributed systems, *Comm. ACM*, 21, 558–564, 1978.
[39] Grammatikakis, M.D., Hsu, D.F., and Kraetzl, M., *Parallel System Interconnections and Communications*, CRC Press, Boca Raton, FL, 2000.
[40] Benini, L. and De Micheli, G., System-level power optimization: techniques and tools, *ACM Trans. Autom. Electr. Syst.*, 5, 115–192, 2000.
[41] Benini, L., Ye, T.T., and De Micheli, G., Packetized on-chip interconnect communication analysis for mpsoc, *Proceedings of the Design Automation & Test in Europ Conf.*, 2003.
[42] Ye, T.T., Benini, L., and De Micheli, G., Analysis of power consumption on switch fabrics in network routers, *Proceedings of the Design Automation Conf.*, 2002.
[43] Mueller, W., Ruf, J., Hoffmann, D., Gerlach, J., Kropf, T., and Rosenstiehl, W., The simulation semantics of SystemC, *Proceedings of the Design, Automation & Test in Europe Conf.*, 2001, pp. 64–71.
[44] Koutsofios, E. and North, S.C., *Dot User Manual*, AT&T Bell Labs, Murray Hill, NJ, 1993. See http://www.cs.brown.edu/cgc/papers/dglpttvv-ddges-97.ps.gz
[45] Reid, M., Charest, L., Tsikhanovich, A., Aboulhamid, E.M., and Bois, F., Implementing a graphical user interface for SystemC, *Proceedings of the HDL Conf.*, 2001, pp. 224—231.
[46] Sinha, V., Doucet, F., Siska, C., Gupta, R., Liao, S., and Ghosh, A., YAML: a tool for hardware design, visualization, and capture, *Proceedings* of the *International Symp. Syst. Synthesis*, 2000, pp. 9–17.
[47] Charest, L., Reid, M., Aboulhamid, E.M., and Bois, G., A methodology for interfacing open source SystemC with a third party software, *Proceedings of the Design, Automation & Test in Europe Conference*, 2001, pp. 16–20.

13

Micro-Architectural Power Estimation and Optimization

Enrico Macii
Politecnico di Torino
Torino, Italy

Renu Mehra
Synopsys, Inc.
Mountain View, California

Massimo Poncino
Politecnico di Torino
Torino, Italy

Abstract

As power consumption has become one of the major limiting factors in current electronic systems, this chapter introduces innovative methodologies for successfully dealing with power estimation and optmization during the early stages of the design process. In particular, the presentation offers an insight into state-of-the-art techniques for power estimation at the micro-architectural level, describing how power consumption of components like data-path macros, glue and steering logic, memory macros, buses, interconnect, and clock wires can be efficiently modeled for fast and accurate power estimation. Then, the focus shifts to power optimization, covering the most popular classes of techniques, such as those based on clock-gating, on exploitation of common-case computation, and on threshold and supply voltage management. Ad-hoc optimization solutions for specific components, such as on-chip memories and global buses are also briefly discussed for the sake of completeness.

Most of the aforementioned approaches to micro-architectural power estimation and optimization have now reached a significant level of maturity, and are thus finding their way into commercial CAD tools that are currently hitting the EDA market. Strengths and limitations of the design technology that is at the basis of such tools will be discussed in detail throughout this chapter.

13.1 Introduction

Low-power consumption has become one of the most important features of modern electronic devices, due to several factors, ranging from the increased market share of mobile and portable systems for telecom and computing to the increased cost of implementing, packaging and manufacturing circuits working at high speed and temperature. As a consequence, techniques and tools that enable tight power consumption control during design are required.

Pioneering work on low-power design techniques has focused on gate and transistor levels where, due to the available information on the structure and the macroscopic parameters of the devices, accurate power estimates are expected and satisfactory methods for both estimation and optimization are available.

The increased complexity of modern designs, facilitated by the advent of aggressively scaled technologies, and the augmented pressure of time-to-market constraints, called for modifications to the way ICs are designed. This resulted in a substantial shift upward in the design paradigm and in the supporting tool flow. Today, electronic systems are specified and then developed starting from the micro-architectural (also called register-transfer, RT) level, in which the basic design entities are no longer elementary objects such as transistors or logic gates, but rather complex blocks capable of performing nontrivial functions. Typical components at the micro-architectural level include data-path macros, such as adders and multipliers, storage elements, such as registers and memory banks, communication resources (e.g., buses) and steering elements (e.g., multiplexers, codecs). Along this line, which defines the RT level as the common entry point for digital design, technologies, methodologies, and tools for low-power design applicable at the RTL have come up in the last decade, now providing the industrial design community with integrated EDA solutions for power analysis, estimation, and optimization starting from this level of abstraction.

As electronic technology keeps evolving, so does EDA technology; higher levels of abstraction are currently being proposed as possible starting points for new design development. For instance, in the system-on-chip (SoC) domain, hardware–software combined specification, design and synthesis are becoming common practice, and new methodologies and tools are being investigated. It is fair to say, though, that fully automated solutions are still not commercially available, especially that concerning power-oriented design. As of today, the RTL is thus the highest level of abstraction for which extensive EDA support is guaranteed.

The objective of this chapter is to provide a comprehensive overview of the most advanced, yet well-established EDA solutions for power estimation and optimization. For the reasons highlighted above, the focus of the discussion will be restricted to the RT level. The chapter will start by providing some background information concerning power consumption in CMOS circuits; it will continue with a brief overview of the architectural template assumed by most of the power modeling, estimation, and optimization approaches, which constitute the core of the chapter and which are the subjects of Sections 13.4 and 13.5.

13.2 Background

CMOS is, by far, the most common technology used for manufacturing digital circuits. There are three major sources of power dissipation in a CMOS circuit [1]:

$$P = P_{SW} + P_{SC} + P_{L} \tag{13.1}$$

P_{SW}, called *switching* or *dynamic power*, is due to charging and discharging capacitors driven by the gates in the circuit. P_{SC}, called *short-circuit power*, is caused by the short-circuit currents that arise when pairs of PMOS/NMOS transistors are conducting simultaneously. Finally, P_{L}, called *leakage* or *static* or *stand-by power*, originates from subthreshold currents caused by transistors with low threshold voltages and from gate currents caused by reduced thickness of the gate oxide.

For older technologies (e.g., 0.25-μm), P_{SW} was dominant. For deep-submicron processes, P_{L} becomes more important. For instance, in ASIC designs, leakage power accounts for around 5–10% of the total power budget at 180 nm, and this fraction grows to 20–25% at 130 nm and to 35–50% at 90 nm. Therefore, leakage power minimization must be addressed from the design standpoint, and not just at the technology or process level, as was done in the past.

Design methods for leakage power control are currently the subject of intensive investigation; approaches based on variable-threshold (VTCMOS), dual-threshold (DTCMOS), and multi-threshold CMOS devices (MTCMOS); the insertion of (possibly distributed) sleep transistors; the adoption of multi-voltage gates (MVCMOS); and the application of reverse and forward body biasing are some examples of promising solutions for reducing the impact of leakage in nanometer circuits. Yet, most of the methods and tools for low-power design in use today are still primarily targeting the minimization of the dynamic component of the power; this is because research in this domain has been carried out for a much longer time, and the available solutions have now reached a significant degree of maturity. As a consequence, a significant part of this chapter will be focused on techniques addressing switching power modeling, estimation, and optimization.

Switching power for a CMOS gate working in a synchronous environment is modeled by the following equation:

$$P_{Sw} = \frac{1}{2} C_L V_{dd}^2 f_{Ck} E_{Sw} \tag{13.2}$$

where C_L is the output load of the gate, V_{dd} the supply voltage, f_{Ck} the clock frequency, and E_{Sw} the *switching activity* of the gate, defined as the probability of the gate's output of making a logic transition during one clock cycle.

Reductions of P_{Sw} are achievable by combining the minimization of the four parameters in the equation above.

Historically, supply voltage scaling has been the most used approach to power optimization, since it normally yields considerable savings owing to the quadratic dependence of P_{Sw} on V_{dd}. The major shortcoming of this solution, however, is that lowering the supply voltage affects circuit speed. As a consequence, both design and technological solutions must be applied in order to compensate the decrease in circuit performance introduced by reduced voltage. In other words, speed optimization is applied first, followed by supply voltage scaling, which brings the design back to its original timing but with a lower power requirement.

A similar problem, i.e., performance decrease, is encountered when power optimization is obtained through frequency scaling. Techniques that rely on reductions of the clock frequency to lower power consumption are thus usable under the constraint that some performance slack does exist. Although this may seldom occur for designs considered in their entirety, it happens quite often that some specific units in a larger architecture do not require peak performance for some clock/machine cycles. Selective frequency scaling (as well as voltage scaling) on such units may thus be applied, at no penalty in the overall system speed.

Optimization approaches that have a lower impact on performance, yet allowing significant power savings, are those targeting the minimization of the *switched capacitance* (i.e., the product of the capacitive load and the switching activity). Static solutions (i.e., applicable at design time) handle switched capacitance minimization through area optimization (that corresponds to a decrease in the capacitive load) and switching activity reduction via exploitation of different kinds of signal correlations (temporal, spatial, and spatio-temporal). Dynamic techniques, on the other hand, aim at eliminating power wastes that may be originated by the application of certain system workloads (i.e., the data being processed).

Static and dynamic optimizations can be achieved at different levels of design abstraction. Actually, addressing the power problem from the very early stages of design development offers enhanced opportunities to obtain significant reductions of the power budget and to avoid costly redesign steps. Power-conscious design flows must then be adopted; these require, at each level of the design hierarchy, the exploration of different alternatives, as well as the availability of power estimation tools that could provide accurate feedback on the quality of each design choice.

As discussed in the introduction, in this chapter, we will devote our attention to the RT level, as it represents the actual front-end for automated design today. More specifically, although RTL descriptions may contain components of different kinds (see the next section for more details on the architectural template), we will primarily concentrate on estimation and optimization of the portion of the architecture that is normally designed with the help of automatic tools (e.g., data-path, controller, interconnect and clock tree). Modeling, estimation and optimization techniques for components with peculiar characterisitics, such as memories and buses will also be surveyed in order to make the picture as complete as possible.

13.3 Architectural Template

The definition of a "micro-architecture" goes through the assumption of an *architectural template,* which serves as a reference for all micro-architectural descriptions. There are two main advantages in assuming such a template; first, it allows one to infer a fine-grain partition of the entire design into *objects* of manageable size, for which customized models and specific optimizations can be devised. Second, by assuming that all micro-architectural descriptions will map onto this template, the chance of reusing models and optimization techniques increases.

A traditional micro-architectural template views a design as the interaction of a *data-path* and a *controller,* which fits well to the so-called Finite State Machine with Data-path FSMD model [2]. Variants to the base FSMD model concern the structure of the controller (e.g., sparse logic implementation vs. wired logic), the structure of the interconnect (e.g., number, type, and size of the buses), or the supported arithmetic operations (e.g., number and type of the available data-path units).

Figure 13.1 shows a possible architecture of a typical FSMD that exposes its basic building blocks: the controller (shown on the left) and the data-path (right), consisting of a register file (or equivalently, a set of sparse registers), a memory, various interconnection buses, and some data-path units (integer and/or floating point). Besides the blocks themselves, the interconnection between them (not explicitly depicted in the figure) is also a source of power consumption; these wires can be global or local, depending on whether they span the entire architecture or not. Examples of the former classes are the clock signal and the global buses; examples of the latter are the signals connecting the units to the global wires. The most important of all wires is the clock that constitutes a significant source of consumption, because of its large load and high activity. While most of the local wires are usually hidden inside the primitive blocks, global wires need a special treatment because they cannot be captured otherwise.

The architectural template introduced above is sometimes defined as *structural RTL,* with the purpose of emphasizing the explicit notion of structure it contains, as well as the RT level nature of the operations that take place during execution.

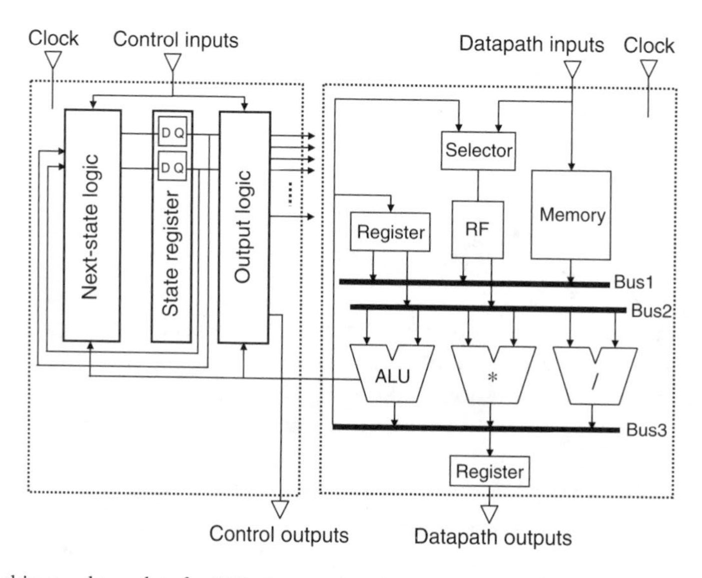

FIGURE 13.1 Architectural template for RTL power estimation.

The key issue is that fitting a micro-architectural description to this template allows us to restrict the granularity of the power models and of the optimizations to that of the following main building blocks: *controller, data-path units, memory, buses,* and *wires.* Registers may fall either in the class of memory devices (for a register-file implementation style) or in that of data-path units (for "sparse" style). This approach is followed by most micro-architectural power estimation and optimization frameworks that have been proposed in the recent literature [3–10].

The FSMD template of Figure 13.1 also nicely fits micro-architectural descriptions specified using hardware description languages (e.g., VHDL or Verilog). They are in fact described as a state machine (the controller), whose states consist of a series of micro-architectural operations (e.g., assignments, arithmetic, or logic operations) corresponding to data-path operations. It is worth emphasizing, however, that such descriptions, after parsing by standard HDL compilers, tend to lose some of their structural semantics and are partly flattened into a netlist of finer-grain primitives such as abstract logic gates [11], in which only memory macros are preserved.

13.4 Micro-Architectural Power Modeling and Estimation

The problem of power estimation at the RTL amounts to building a *power model* that relates the power consumption of the target design to suitable quantities. In formula,

$$P = P(X_1, \ldots, X_n)$$

where X_i, $i = 1, \ldots, n$, represent the *n model parameters.* The construction of the model shown above implies addressing several issues that call for the development of various modeling alternatives, and are discussed in detail in the sequel.

13.4.1 Modeling Issues

13.4.1.1 Model Granularity

This issue has already been addressed in Section 13.3, where the definition of an architectural template has been shown to determine the granularity of the power models; power estimation of a design requires thus the construction of power models for the following classes of building blocks: *controller, data-path units, memory, buses, and wires.*

13.4.1.2 Model Parameters

The choice of the parameters to be included in the model is constrained by the fact that they must be quantities that are observable at the abstraction level at which they are used. Under the architectural model of Section 13.3, the abstract model of switching power as shown in Equation (13.2) translates into the following high-level expression:

$$P_{\text{total}} = k \sum_{\forall \text{component } i} A_i C_i \tag{13.3}$$

where A_i and C_i denote the *switching activity* and the *physical capacitance* of component *i*, respectively; *k* lumps the fV_{dd}^2 terms, which can be considered as scaling factors at the RTL.

Equation (13.3) decouples the problem of building a power model into that of building a model for activity and a model for capacitance, for each type of component. In general, activity and capacitance models will depend on different parameters because they are affected by different physical quantities.

Activity is easier to model, because at the RTL it is a well-defined quantity. Activity models rely on *activity parameters,* such as bit-wise (referred to specific input or output signals of a component) or word-wise (regarding input or output values of a component) transition and static probabilities. Most activity models proposed in the literature use these probability measures as parameters. Other choices for activity parameters may include variants of transition probabilities such as *transition density* [12], defined as an average (over time) switching rate or various *correlation measures.*

Modeling physical capacitance is a more problematic task than modeling activity. The term "physical" suggests the difficulty in linking capacitance to quantities observable at the RTL. In spite of that, RTL capacitance models can be derived with a reasonable degree of accuracy. They all rely on the intuitive observation that capacitance will be roughly related to the number of "objects" (gates or similar lower-level primitives) of the target component. In other words, physical capacitance at the RTL is approximated by *complexity,* and we thus speak of *complexity models,* based on *complexity parameters.*

Complexity parameters that are available at the RTL are restricted to the *width* of a component (i.e., its number of inputs and outputs) or the *number of states* (which is relevant only to the controller, since notion of state is explicit.) Any complexity parameter different from the above would require some additional information derived from back-annotation of physical information of previous implementations.

13.4.1.3 Model Semantics

Models can be distinguished in terms of how the values that they return can be interpreted. The most intuitive option is to assume that models express *average power,* which is commonly used as a metric to track battery lifetime or average heat dissipation. In this case, the semantics of the model is that of a single figure to represent the consumption of the target description. Average power models are called *cumulative power models* [13].

However, the notion of *cycle* intrinsic to RTL descriptions allows us to obtain a power model with a richer semantic by simply changing the way we collect statistics. The first step in this direction consists of expanding the model of Equation (13.3) into

$$P_{\text{total}} = k \sum_{\forall \text{cycle } j} P_j = k \sum_{\forall \text{cycle } j} \sum_{\forall \text{component } i} A_{ij} C_{ij} \tag{13.4}$$

where P_j denotes the power consumption at cycle *j,* which can be obtained by summing the power consumption for each component (as in Equation [13.3]), this time using activities and capacitances of component *i* at each cycle *j.* The semantics of the model of Equation (13.4) is *cycle-accurate,* because it allows to track cycle-by-cycle (total) power.

The use of a cycle-accurate model affects the choice of the parameters. For example, transition or static probabilities are not suitable quantities anymore, since they are intrinsically "average." Conversely, cycle accurate models should use cycle-based activity measures, such as the *number of bit toggles* between consecutive patterns (i.e., the *Hamming distance)* [14,15] or *the values* of consecutive input patterns [13,16].

A cycle-accurate model provides several advantages over a cumulative one. First, it allows one to go beyond the bare evaluation of average power and can be used to perform sophisticated analysis of power consumption over time, such as reliability, noise, or IR drop analysis. In addition, a cycle-accurate model is more accurate than a cumulative one, but not just because it provides a set of power values rather than a single one. In fact, the relation between input statistics and power is nonlinear: average consumption is usually different from the consumption associated to average input statistics, especially when power consumption varies significantly over time. Therefore, even when average power is the objective, averaging the series of cycle-by-cycle values will yield a more accurate estimate than a model of average power. On the negative side, cycle-accurate models require significantly larger storage than cumulative ones.

13.4.1.4 Model Construction

Concerning model construction, there are two options to build an RTL power model: a *top-down* (or *analytical)* approach, and a *bottom-down* (or *empirical)* one [3].

Top-down approaches relate activity and capacitance of an RTL component to the model parameters through a closed formula. The term "top-down" refers to the fact that the model is derived directly from the micro-architectural description, and is not based on lower-level information. For this reason, such formula usually has a *physical interpretation.* Analytical models are particularly useful either when dealing with a newly designed circuit for which no information of previous implementations is available, or when the implementation of the circuit follows some predictable template, which can be exploited to force some specific relation between the model parameters. A memory is a typical example of an entity for which an analytical model is suitable: its internal organization is quite fixed (cell array, bit and wordlines, decoders, MUXes, and sense amps), thus allowing accurate modeling based on "internal" parameters [17,18].

If we exclude these special cases, however, top-down models are not very accurate, since their links to the implementation (e.g., technology and synthesis constraints) are quite weak.

Bottom-up approaches, conversely, are based on "measuring" the power consumption of existing implementations, from which the actual power model is derived. Typically, the template of the power model (i.e., the parameters and a set of coefficients used to weigh the parameters) is defined up front; statistical techniques are then used to fit the model template to the measure of power values. This approach is known as *macromodeling,* and has proved to be a very accurate and robust methodology for RTL power estimation, and can be considered the state-of-the-art solution. Section 13.4.2 will be devoted to the detailed description of the macromodeling flow.

13.4.1.5 Model Storage

The issue of model storage is concerned with the *shape* of the model. Since models express a mathematical relation between power and a set of parameters, the problem amounts to that of representing such a relation. The two options are to store it as an equation or as a look-up table, corresponding to the choice of representing a relation as a continuous function (equation-based models), or a discrete function approximated by points (look-up table-based models). These two types of models differ in their storage requirements and *robustness,* that is, the sensitivity of the model to the conditions (i.e., the experiments) used for its construction. In that sense, robustness is an issue only for empirical models. Concerning storage requirements, equation-based models are much more compact than table-based ones. In general, an equation will only require the storage of the coefficients of the model, as opposed to a full table. In addition, the accuracy of a table-based model is proportional to its size (the denser the table, the higher the accuracy), whereas the accuracy of an equation-based model is independent of the model size.

13.4.2 RTL Power Models

In this section we will review the most relevant results on power modeling of data-path units, controller, and wires. On the other hand, as already mentioned in Section 13.2, we will not consider components such as memories and communication buses.

13.4.2.1 Power Modeling of Data-Path Units

There are three main reasons for which achieving accurate power estimates for data-path units must resort to empirical models. First, because of the variety of available data-path units, a distinct analytical model for each type of component would be required. Second, analytical models, because of their functional meaning, would not be able to capture differences due to various implementation styles of a given component (e.g., a ripple-carry vs. a carry-lookahead adder). Third, accuracy of analytical models would be too low for units that have poorly regular structures (e.g., dividers or floating-point operators).

Besides being able to cope with these issues, empirical models allow one to devise a single-model template that can be used for all types of data-path units. Empirical power models are commonly called *macromodels.* The term has been borrowed from the statistical domain to denote the fact that such models have a "coarse" level of detail, and are used to relate quantities pertaining to different abstraction levels, such as RTL parameters, to the actual measured power. We thus talk of "macromodeling" when defining the process of building an empirical power model for a generic component.

Before reviewing the most important results in data-path macromodeling, it is worth summarizing the flow that is normally used for building such macromodels.

13.4.2.1.1 Macromodeling Flow

Macromodeling involves four major steps, as depicted in Figure 13.2:

1. *Choice of model parameters.* Although this step applies to non-empirical models also, it is especially important for macromodels, since it defines the parameters space and it affects the complexity of the following steps.
2. *Design of the training set.* The training set is a representative subset of the set of all possible input vector pairs that will be used to determine the model. Key points in the choice of the training set are its size (i.e., the total number of vector pairs) and its statistical distribution. The former is responsible for the simulation time, while the latter impacts accuracy; a bad distribution of the training set may offset the advantage of a large number of vector pairs. What defines a "good" distribution depends on the chosen parameters. A requirement for the training set is that it should span the domain of all the model parameters as much as possible. When one or more domains are not sufficiently covered by the training set, we say that the model is *insufficiently trained.* For instance, if the parameter of the model is the switching activity, the choice of random patterns as a training set would not be a good one, since only a very small portion of the activity domain (the one around a switching activity of 0.5) would be exercised.
3. *Characterization.* This step consists of using the training set to generate a set of points in the *(power, parameters)* space. For each element in the training set (a vector pair), a corresponding value of power is obtained by means of a low-level power simulator (gate or transistor-level).
4. *Model extraction.* This step consists of deriving the model from the set of measurements obtained in the previous step. The actual calculation depends on how the model is stored. For equation-based models, a least-mean-square (LMS) regression engine is typically applied to the sample measurements; if a look-up table is used, extraction consists of collecting the power values for each of the discrete points of the parameters space.

13.4.2.1.2 Macromodels for Data-Path Units

The literature on power macromodeling for data-path units is quite rich, and it includes solutions that cover all points of the power model space discussed in Section 13.4.1, with different accuracy/effort tradeoffs. In spite of the vast amount of available material, it is quite easy to identify the key results in the power macromodeling domain.

The technique described by Powell and Chau [19] can be considered as the first proposal of a power macromodel for data-path units; it is actually just a capacitance macromodel, because it assumes fixed activity, that is, no activity parameters are included. Landman and Rabaey [20] recognized that modeling activity is essential, by observing that even "random" data do not exhibit true randomness on all bits,

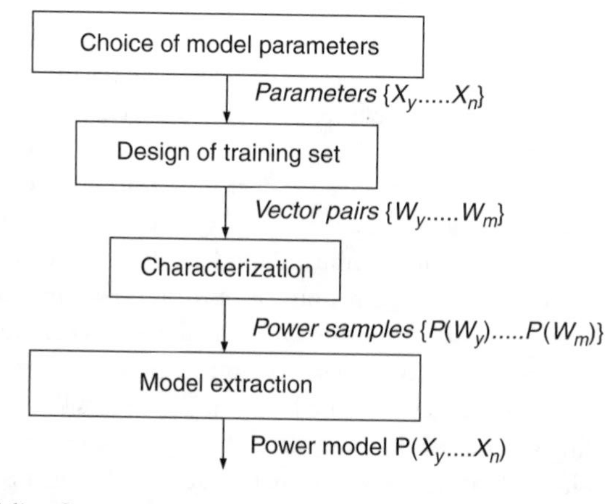

FIGURE 13.2 Macromodeling flow.

because of sign extension bits due to reduced dynamic ranges of typical values. Their model, called *dual-byte type* (DBT), separates data bits in two regions (sign and random), with distinct activity parameters. Since each unit may have different combinations of input and output regions depending on their functionalities, each unit must have a distinct model.

Gupta and Najm [21] proposed a power macromodel that can be considered as the state of the art; it consists of a three-dimensional look-up table of power coefficients, whose dimensions are P_{in}, the average input probability, D_{in}, the average input *transition density*, and D_{out}, the average output transition density. Characterization is based on quantizing each dimension into equally spaced intervals.

Figure 13.3 shows the three dimensions of the look-up table in the case of a quantization interval of 0.2. Notice how the relation between D_{in} and P_{in} constrains the number of feasible points in the (D_{in}, P_{in}) parameters' space.

Under this model, characterization involves the generation of a number of input streams for each feasible value of parameters D_{in} and P_{in}. Values of parameter D_{out}, instead, are extracted from the resulting output stream (obtained through simulation), and are discretized using a binning technique.

The three-dimensional look-up table model provides two main advantages with respect to the DBT model (or its variants). First, the use of a look-up table with discretized dimensions significantly improves the robustness of the estimates. Second, since parameters are normalized values, the same model template can apply to any data-path unit, without the need of (1) customizing it to its functionality, and (2) including an explicit "capacitance" (i.e., complexity) parameter (thus, without the need of being parameterized).

Many macromodels published in the literature build around this three-dimensional table-based model, by either improving some specific aspects that help to increase accuracy (e.g., interpolation schemes and local accuracy improvements) [22–25], or by generalizing it into a parameterized model (i.e., where the bit-width of the operators becomes a parameter, as in the DBT model) [26].

13.4.2.2 Power Modeling of Controllers

Control logic typically has a smaller impact than the data-path; however, its dissipation is not negligible for several reasons. First, some designs are control-intensive, and may contain many interacting controllers whose impact on power may be sizable. Second, data-path units are often hand-crafted while controllers are synthesized; then, they use silicon area less efficiently than data-path. Third, controllers are usually active for a large fraction of the operation time, while the data-path can be partially or completely idle.

What makes controller power modeling more challenging than data-path units is the fact that controllers are usually specified in a very abstract fashion even at the RT level and they are specific to each

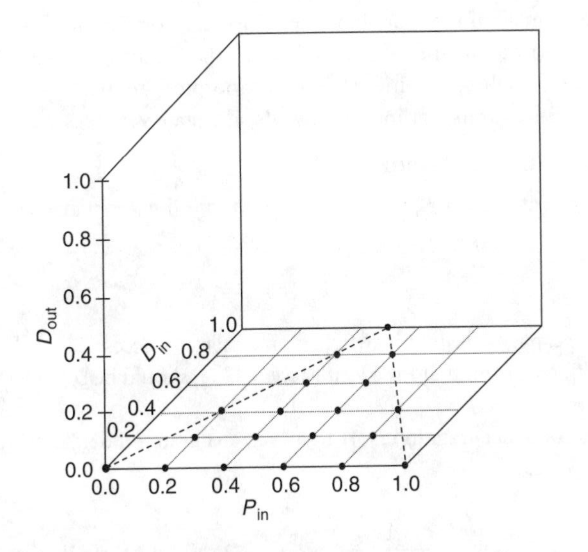

FIGURE 13.3 Three-dimensional look-up able macromodel.

design. In contrast, data-path units are usually instantiated from a library that can be precharacterized once and for all. Some controller implementation styles employ regular structures, like PLAs or ROMs, that can be precharacterized with good accuracy. On the other hand, the most common choice for controllers is to synthesize them as sparse logic. Because of its irregular structure, this makes power modeling considerably harder.

One additional difficulty lies in the fact that the description of controllers is more abstract than that of data-path components because states are usually specified symbolically. Thus, controller synthesis goes through two steps, namely state assignment and logic synthesis.

Power modeling of controllers at the micro-architectural level can be classified into *pre-state assignment* [27,28] and *post-state assignment* [4,6,29] depending on the target description of the controller. The former is generally less accurate, and can be used to provide upper and lower bounds to power consumption. The latter tries to give actual estimates of average power. Strictly speaking however, RTL descriptions of controllers typically fall in the former category.

Due to the poor link between an abstract, state-based description and the actual implementation, power models for controllers that have some practical value are empirical. The analytical model presented in [27] provides in fact theoretical bounds for the switching in FSMs, but the correlation to actual power figures is intrinsically very weak. For controllers, macromodeling is thus more than just an option. The difference with respect to data-path components, however, is that the model template is not so intuitive, because it is difficult to estimate the correlation between parameters and the actual power consumption. Therefore, the choice of the parameters is even more critical than in the case of data-path units. Possible parameters for controllers include:

- *Static behavioral* parameters (i.e., parameters that can be obtained from the behavioral specification): the number of inputs, the number of outputs and the number of states.
- *Dynamic behavioral* parameters (i.e., parameters that can be obtained from functional simulation): the average input and output signal and transition probabilities.
- *Static structural* parameters (i.e., parameters that are available after state encoding): the number of state variables.
- *Dynamic structural* parameters (i.e., parameters that can be obtained from RTL simulation): the average transition probability of the state variables.

A framework for the exploration of all possible models of order M with up to N terms is described in [30]; it defines a third-order model with four terms that yields an average relative error of about 30% and relative standard deviation of the error of about the same magnitude. The model uses only a subset of the above parameters, but from all of the four above-described categories. This model has been shown to be far more accurate and robust than other models, including the "intuitive" model, that is, the one that uses a linear combination of switching activities (for inputs, outputs, and state variables) weighted by their corresponding cardinalities (number of inputs, outputs, and state variables).

13.4.2.3 Power Modeling of Memories

As for any other component, memory power can be fitted into the template of Equation (13.2):

$$P_{\text{mem}} = \frac{1}{2} C_{\text{mem}} V_{\text{dd}}^2 f_{Ck} E_{\text{Sw}}$$

where C_{mem} denotes the capacitance that is switched on a memory access. Since the capacitance for a read access is, in general, different from that for a write access, C_{mem} should be considered as an average capacitance.

For this reason, it is more accurate to resort to a cycle-accurate model, and express average power as follows:

$$P_{\text{mem}} = \frac{1}{2} \frac{1}{N_{\text{cycles}}} (N_{\text{read}} C_{\text{read}} + N_{\text{write}} C_{\text{write}}) f_{Ck} V_{\text{dd}}^2 \qquad (13.5)$$

where N_{cycles} is the total number of cycles, N_{read} (N_{write}) the number of read (write) accesses, and C_{read} (C_{write}) the corresponding capacitances.

Memories, and in particular SRAM arrays, unlike generic RTL blocks, lend themselves to a relatively easy modeling of capacitance, owing to their well-defined internal structure which is characterized by high regularity. In other words, it is feasible to adopt an analytical model that expresses the total memory power as the sum of the various components: read/write circuitry, decoders, cells, bitlines, wordlines, MUXes, and sense amplifiers [17,31,32]. These models are very accurate, but suffer from being strongly dependent on technology; in fact, they require the knowledge of parameters such as the capacitance of the bitline or the wordline, or the capacitance of a minimum-sized transistor, which cannot always be easily accessible.

A simpler, yet effective model for micro-architectural estimation is an empirical one, where *all* model parameters are chosen in order to be easily available at a higher level of abstraction (e.g., the number of words) and the relation to the technology is established in a characterization run. For example, a typical capacitance model can be expressed as

$$C_{mem} = a + bW + cN + dWN$$

where W is the number of rows and N the number of columns of the memory array [4]. A similar model that also included the number of words as parameters was used in [33]. The characterization proceeds very similarly to the conventional macromodeling flow, in which the values of a, b, c, and d are determined by means of LMS regression. If we stick to the template of Equation (13.5), separate characterizations for read and write capacitances are required.

13.4.2.4 Power Modeling of Wires

If we assume that the power consumption of data-path units and controller is based on empirical macromodels, their power estimates include the contribution of (internal) wires. There are then two categories of wires that need a custom power model: *Global* wires, such as reset and clock, and *inter-component* wires, such as those connecting RTL blocks.

For both types of wires similar considerations apply: in fact, power is consumed by wires when charging and discharging the corresponding capacitances. Therefore, the model of Equation (13.3) where A_i and C_i are the switching probability and the parasitic capacitance of the ith net, respectively well fits wires also. Since factor A_i is available for both global wires and inter-component wires, the problem of modeling wiring power reduces to that of modeling wiring capacitance, or, equivalently, wiring length. Wiring capacitances are unknown at the RTL, but realistic estimates can be obtained based on structural information, area estimates, and low-level wiring models. The problem of estimating the total wire length at high levels of abstraction is quite well understood; it is based on variants of Rent's rule [34–36] which relate the length of the interconnect to macroscopic parameters that can be easily inferred from a high-level specification, such as the number of I/O pins.

The main difference between the power models for global wires and inter-component wires lies in their "context." Area-based estimates work reasonably well for global wires, since they span the entire design. As a matter of fact, typical power models for global wires directly relate wire length to a power k of design area (often $k = 0.5$), and the latter to a power q of the number of I/O pins [20].

For inter-component wires, conversely, their length is more weakly correlated to the entire design area. The solution adopted is very similar, yet on a smaller scale—that is, rather than referring to the entire design, each wire refers to the components it connects. In this case, wire length is made proportional to a power of the component's area, which is related to a power of the number of I/O pins of the component [37].

Figure 13.4 shows the conceptual topology of an inter-component wire. The output capacitance C_{out} of a component is implicitly accounted for by its power model. Conversely, the input capacitance C_{in} is not included in the power model, since power estimates provided by low-level simulation represent the power drawn from the supply net, while input capacitors are directly charged by the primary input lines. The total wire capacitance should thus take into account the fanout of a wire (known at the RTL), both externally

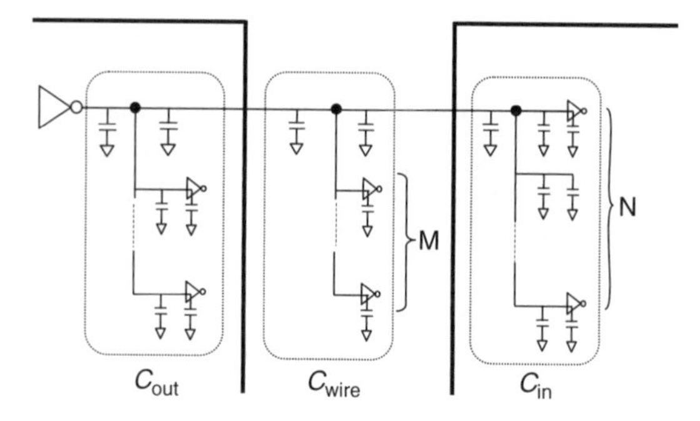

FIGURE 13.4 Hierarchical topology of an inter-component wire.

(C_{wire}) and internally (C_{in}) by splitting each contribution into a fixed term (the "stem" capacitance) and a quantity proportional to the number of fanouts (the "branch" capacitance).

It is worth emphasizing that in ultra-deep-submicron technologies, where wires dominate chip area, the estimation of wire length requires some early floorplanning information in order to be reliable. Such information can then be back-annotated into the RTL description to guide wire length estimation based on the empirical models discussed above.

13.4.2.5 Power Modeling of Buses

At the micro-architectural level, buses are relatively straightforward entities consisting of a set of wires. However, we believe they deserve a separate treatment from the generic wires we have considered in Section 13.4.2.4, since they are usually seen as a single-interconnection resource that will eventually be routed together as *a set* of wires; this physical "grouping" of bus wires is also reflected by typical HDL descriptions, where buses are represented as arrays of signals.

Buses may be shared (that is, they may not just be point-to-point connections) and, mostly, are single-master buses relying on a simple protocol that defines the physical (i.e., the signaling scheme) and the "data-link" (i.e., the binary representation of the transmitted "values") layers. More complex on-chip bus architectures that include features such as support for multi-master and complex protocols (i.e., those required by modern multi-processor SoC architectures) tend to be categorized as hardware blocks rather than as a wiring infrastructure, because of the high amount of hardware control they require [38]. This trend is witnessed by the wide availability of synthesizable IP blocks that can be modularly used by RTL designers and automatically synthesized (e.g., Synopsys' DesignWare IPs for AMBA buses [39]). In this section, we will focus on single-master buses that are typical in non-core-based designs.

Specifically, with reference to the template of Figure 13.1, buses are the resources that are used to connect blocks of two types (possibly in a shared manner): computational units and memories.

A conceptual model of the average power consumed by a bus consisting of *n* lines can be obtained by simply summing up the contribution of each wire, according to the model of Equation (13.2):

$$P_{bus} = \frac{1}{2}\sum_{i=1}^{n}(C_L V_{dd}^2 f_{Ck} E_{Sw}^i)$$

where C_L is, in this case, the capacitance of a single bus line, and E_{Sw}^i the switching activity of the *i*th bus line. The equation is based on the assumption that all lines have roughly the same capacitance. This model would be reasonably accurate if the bus lines were routed independently of each other, possibly on different metal layers. Conversely, the "grouped" nature of buses makes the model unrealistic, because it considers each wire as if it were isolated from the others, thus completely ignoring the contribution due to *coupling capacitances,* that is, the mutual capacitive effect of two neighboring wires.

Figure 13.5 shows a simplified view of the capacitances switched between two adjacent wires. C_L is the ground capacitance considered in the model (also called *self*-capacitance), while C_C denotes the coupling

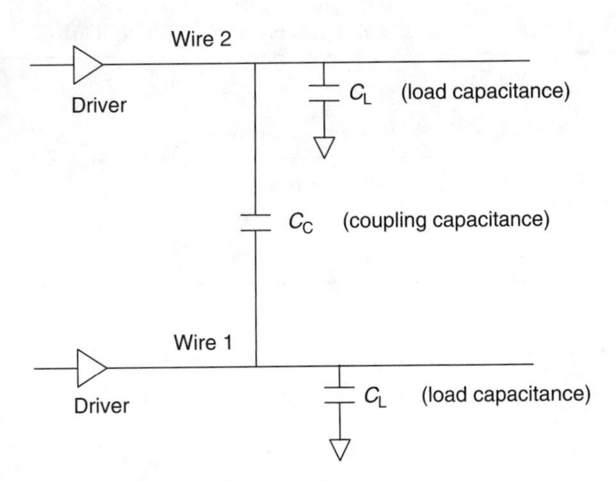

FIGURE 13.5 Capacitances involved in bus line switching.

capacitance. In deep-submicron technologies, the magnitude of C_C far exceeds the self-capacitance: Electrical-level simulations for 0.13 μm technologies have shown that C_C is more than three times the value of C_L [40], and that this factor will further increase in future technologies.

In view of the discussion above, accounting for the contribution of C_C in on-chip buses is mandatory; this can be done by augmenting the basic power model as follows [41]:

$$P_{bus} = \frac{1}{2}\left(\sum_{i=1}^{n}(E^i_{Sw}C_L + E^i_{Sw}C_C)f_{Ck}V^2_{dd}\right) \tag{13.6}$$

E^i_{Sw} denotes the *coupling switching activity*, i.e., a quantity related to the simultaneous switching of two adjacent lines. In fact, when the transitions on two adjacent lines, *a* and *b*, are aligned in time, there are only two transition pairs that cause C_C to switch: (1) when both *a* and *b* switch to different final values; (2) when one of the two lines switches, while the other one does not, and their final values are different.

Table 13.1 shows the normalized switched capacitance for a two-line bus, when all capacitive effects are considered. λ represents the ratio C_C/C_L. In the table, only $0 \rightarrow 1$ transitions are counted as power-dissipating transitions on C_L.

The table clearly shows that increasingly larger values of λ will tend to emphasize the importance of the energy due to switching of the coupling capacitance.

13.4.3 Power Estimation

The modeling technology discussed in the previous sections can be successfully used to enhance state-of-the-art RTL-to-physical design flows with power estimation capabilities.

Assuming that the design to be estimated is described by an FSMD, as defined in Section 13.3, the estimation procedure consists of three basic steps [8], as shown in Figure 13.6.

The first operation implies identifying and separating the data-path components from each other and from the FSM that represents the control. This is needed in order to enable the power estimator to generate the power models for each component in the FSMD.

The FSMD then needs to be simulated. To this purpose, an RTL simulator is used to trace all the internal signals that define the boundaries between the various components in the FSMD. This is of fundamental importance for the evaluation of the power models that is needed to complete the estimation procedure.

Finally, the actual power estimation must take place. The design hierarchy is traversed at the top level, from the inputs toward the outputs, and for each component, model construction and model evaluation

TABLE 13.1 Normalized Switched Capacitance on Two Adjacent Bus Lines

		(b^t, b^{t+1})			
		$0 \to 0$	$0 \to 1$	$1 \to 0$	$1 \to 1$
(a^t, a^{t+1})	$0 \to 0$	0	$1+\lambda$	0	0
	$0 \to 1$	$1+\lambda$	2	$1+2\lambda$	1
	$1 \to 0$	0	$1+2\lambda$	0	λ
	$1 \to 1$	0	1	λ	0

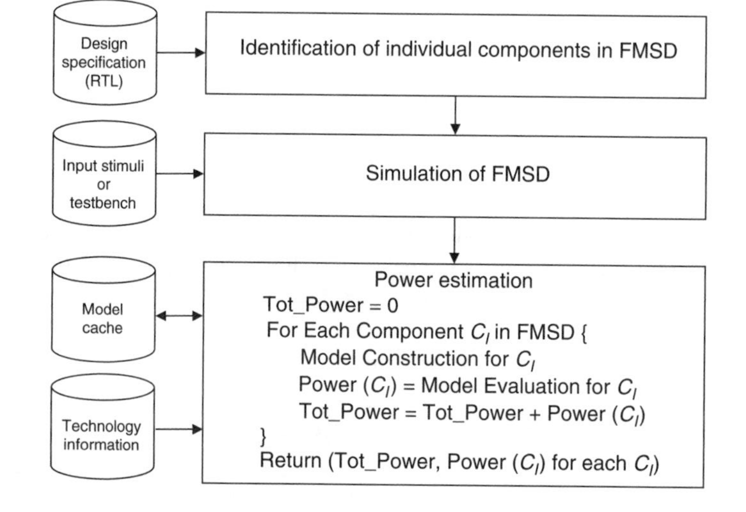

FIGURE 13.6 Micro-architectural power estimation flow.

are carried out. Model construction entails building the proper model according to the type of component being considered. A caching strategy may be used to limit the number of times models are built; more specifically, after the model for a given component is built, it is stored in a cache, so that it can possibly be reused at later times. Model evaluation, on the other hand, requires that the parameter values obtained during the RTL simulation phase are plugged into each model to get the actual power values.

The total power information for the design is then obtained by summing up the contribution of the model of each component; both a total power budget and a power breakdown can thus be reported to the user.

13.5 Micro-Architectural Power Optimization

This section presents some of the most popular micro-architectural power optimization techniques used in designs today. We first deal with data-path components and controllers; in this case, we classify the existing solutions into three categories:

- Those based on clock-gating, whose objective is to stop the clock for some cycles of operation in order to achieve a reduction of the switching component of the power.
- Those based on exploitation of common-case computation, whose objective is to optimize switching power consumption for the most common operation conditions.
- Those based on dynamic management of threshold and supply voltages, whose goal is that of reducing either leakage or switching power, or both by operating the logic in a multi-V_{th}/multi-V_{dd} regime.

For each class of approaches we discuss the basic idea as well as the algorithms used and challenges faced in applying these techniques in automated flows. We then move to techniques applicable to memories and buses. Owing to the peculiarty of these components, their optimization is treated separately, toward the end of this section.

13.5.1 Clock-Gating

Clock-gating dynamically shuts off the clock to portions of a design that are "idle" for some cycles. The theory of this technique has been investigated extensively in the past [42–44], and clock-gating is now considered the most successful and widely adopted solution for switching power reduction in real designs [45–48].

In some cases, it may be possible to shut off the clock to an entire block in the design, thereby saving large amounts of power when the block is not functioning. Perfect cases for this are when a block is used only for a specific mode of operation, e.g., the receiver and transmitter parts in a transceiver may not be active at the same time, and the receiver can be shut off during transmit stages, or vice versa.

It is also possible to gate the clock of a single register or set of registers. For instance, synchronous, load-enabled registers are usually implemented using a clocked D-type flip-flop and a recirculating multiplexor, with the flip-flop being clocked at every cycle as shown in Figure 13.7a.

In the gated-clock version of Figure 13.7b, the register does not get the clock signal in the cycles when no new data are loaded, therefore reducing switching power. Savings are further enhanced by the removal of the multiplexor. Gating a single-bit register, however, has the associated penalty of power consumption in the clock-gating logic. The key is then to amortize such a penalty over a large number of registers, by saving the flip-flop clocking power and the multiplexor power of all of them using a single clock-gating circuit.

The transparent latch is used to guarantee that spurious glitches in the enable signal occurring when the clock is high are not propagated to the clock input of the register. The latch freezes the output at the rising edge of the clock, and ensures that the new enable signal, EN1, at the AND gate is stable when the clock is high. Meanwhile, the enable signal can time-borrow from the latch, so that it has the entire clock period avaliable to propagate.

We cover advanced methods for detecting clock-gating opportunities in Section 13.5.1.1; next, we present issues associated with implementing clock-gating in a typical automatic flow, along with possible solutions in Sections 13.5.1.2–13.5.1.7.

13.5.1.1 Advanced Clock-Gating

The clock-gating conditions that depend on the enable signal of the register bank, which can be identified by means of topological inspection and analysis of the RTL description, can be extended by considering the functional behavior of the circuit. In particular, it is possible to augment the opportunities for stopping the clock that feeds a given register anytime the output of such a register is not observable, that is,

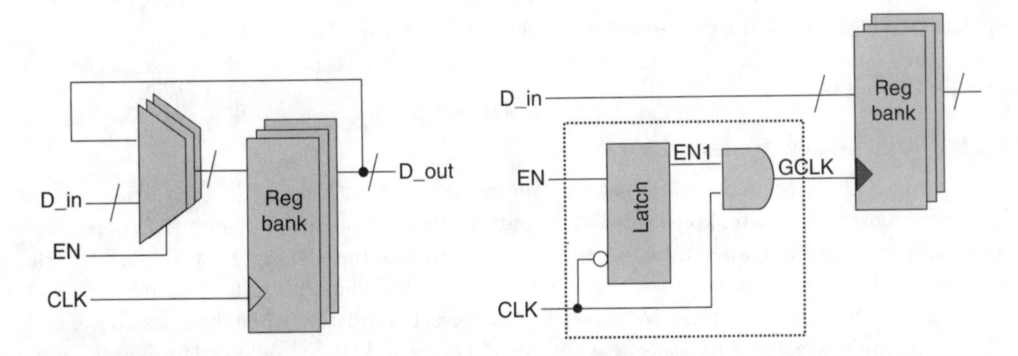

FIGURE 13.7 Clock-gating: (a) load-enabled register implementation; (b) gated-clock implementation.

the value of the output is not used by the gates in the register's fanout cone. Calculating these additional options for clock-gating entails finding the *observability don't care* (ODC) conditions for the output of the register; Babighian et al. [49] propose to extract idle conditions from a RTL netlist, by focusing on control signals that drive steering modules (e.g., multiplexers, tristate buffers, and load-enabled registers). Referring, for instance, to a multiplexer, all but one data inputs are unobservable at any given time depending on the value(s) that are carried out by the control signal(s), so that the logic values of the unobservable branches become irrelevant to the correct operation of the circuit. Hence, the clock signals of the registers in the fanin cone of an unobservable multiplexer branch can be gated without compromising the functionality.

The goal of the approach of Babighian et al. [49] is to improve the effectiveness of clock-gating by creating an activation function that can stop the clock of a set of registers for a significant fraction of cycles when the register outputs are unobservable. This is done in two steps. First, by performing ODC computation, which is based on a backward traversal of the data-path in decreasing topological order, considering only ODCs created by steering modules. Second, by generating the activation function, which implies the synthesis of the ODC function and the addition of clock-gating logic to the RTL netlist. Even though ODC computation is simple for a single-steering module, ODC expressions can become quite large if the netlist has many levels of steering modules and many fanout points. Therefore, traversal of only a limited number of levels in the netlist is allowed.

Once the ODC expression is computed, the corresponding logic must be instantiated in order to drive properly the clock-gating logic. The main difficulty in this step is due to the fact that ODC conditions masking register in clock cycle T may be used to gate their clock in cycle T-1. In other words, the clock-gating logic may need to be active in the clock cycle immediately before the register becomes unobservable. Unfortunately, the control signals at the inputs of the ODC functions are generated one clock cycle too late.

If the control signals are available directly as outputs of the registers, the instantiation of the clock-gating logic is relatively straightforward. Logic gates implementing the ODC expressions are inserted and their inputs are connected to the inputs of the registers. In real-life designs, however, the control inputs of the steering modules seldom come directly from the registers; instead, they are often generated by additional logic. In this case, the entire cone of logic between registers and control signals should be duplicated and connected at the inputs of the registers, and ODC computation gates should then be connected at the outputs of the duplicated cones. Clearly, the addition of this extra logic may represent a nonnegligible overhead.

Most of today's commercial synthesis tools deal with this issue by restricting the type of activation function used to gate the clock. In practice, they just detect ODCs generated locally to the registers by assuming the register's outputs to be always observable by the environment. This corresponds to considering, as ODC function, the complement of the enable signal, which feeds the clock-gating logic.

By detecting clock-gating conditions only when enable signals are present, no precomputation of clock-gating conditions is required and thus no duplication is needed. Notice that ODC-based clock-gating subsumes traditional automatic clock-gating as a very special limit case (i.e., backward traversal is completely avoided). Thus, the clock-gating conditions computed by the ODC-based approach are guaranteed to be a superset of those targeted by tools that introduce clock-gating only for load-enabled registers (see Figure 13.7).

13.5.1.2 Clock-Skew Issues

A problem in latch-based architectures comes from the fact that clock skew between the latch and the AND gate can result in glitches at the gated-clock output. This is explained in Figure 13.8. In particular, Figure 13.8a shows the case when the clock arrives much earlier at the AND gate than at the latch. Here, the clock skew between the latch and the AND gate should be less than the clock-to-output delay of the latch for the circuit to function properly. Figure 13.8b depicts the situation when the clock arrives earlier at the latch. In this case, the clock skew between the AND gate and latch should be less than the sum of the setup time of the latch and the input-to-output delay of the latch to function properly. Therefore, the

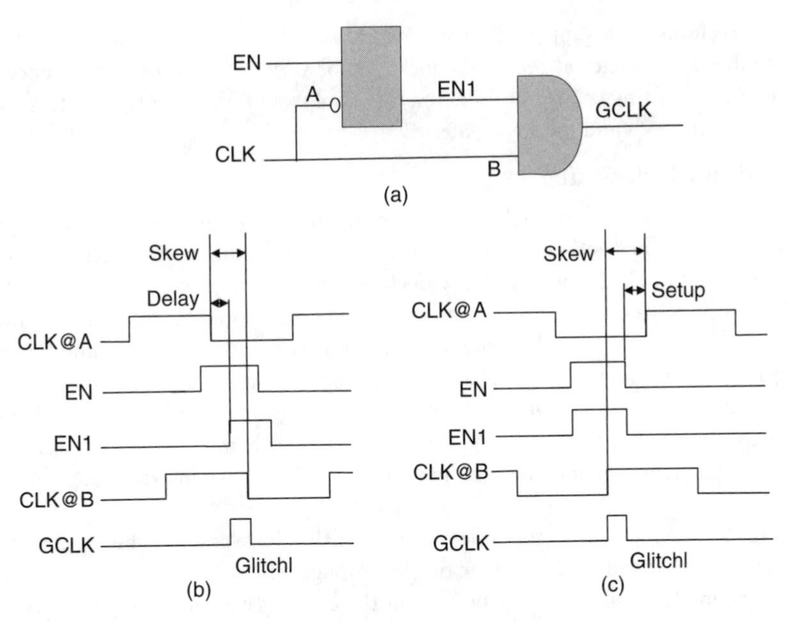

FIGURE 13.8 Clock skew within clock-gating logic: (a) clock-gating logic; (b) Glitches due to positive clock skew between latch and AND gate; (c) glitches due to negative clock skew between latch and AND gate.

clock-skew between the latch and AND gate, C_s, should be carefully controlled according to the following equation:

$$-(s + d_{in}) < C_s < d_{clk}$$

where s is the setup time of the latch, d_{in} the input-to-output delay of the latch, C_s the difference in clock arrival time between the latch and the AND gate (the clock arrival time at the AND gate minus the clock arrival time at the latch), and d_{clk} the clock-to-output delay of the latch.

Depending on the relative placement of the latch and the AND gate, these requirements may pose very stringent constraints on the clock-tree synthesis (CTS) tool.

The best way to control the relative timing of the two clock signals is to keep the entire structure in a single cell, called the *integrated clock-gating* (ICG) cell. The cell should be designed specifically for clock-gating, with the explicit requirements discussed above. Most technology libraries today do include the ICG cell as part of their primitives. Another way to address this issue is to ensure that the latch and the AND gate are close to each other during the placement phase of the design, by placing hard constraints on the distance between them. This makes it simpler for CTS tools to reduce the clock skew between them during the clock routing phase.

13.5.1.3 Clock Latency Issues

In order to maximize the power savings, a single clock-gating cell may be used to gate several registers, if the activation function is common to all of them. But the gated-clock may not have enough strength to drive all these registers, calling for a clock tree at its output. If a clock tree is introduced between the ICG cell and the registers it controls, the clock signal at the gating logic arrives much before the clock signal at the registers, and the activation signal must be ready before the clock arrives at the gating logic. This applies strict timing constraints on the activation signal, which must be addressed during synthesis.

13.5.1.4 Clock-Tree Synthesis

In the presence of massively gated-clocks, CTS tools must automatically address the presence of gated-clocks, both combinational and sequential, and also the integrated clock-gating cell on the clock line.

Clock tree synthesis tools must support different relative latency requirements at different points in the clock tree, since the clock latency at the gated-clocks can be very different from the latency at the registers. If integrated clock-gating cell is not used, the CTS tool would need to provide stringent control of clock skew between the latch and the AND gate.

13.5.1.5 Physical Clock-Gating

Physical clock-gating simultaneously takes into account the factors mentioned in Sections 13.5.1.2–13.5.1.4, namely skew, latency and CTS issues. There exists a spectrum of clock-gating approaches with regard to the placement of clock-gating cells into a clock tree. Designers often opt to place the clock-gating cells as close as possible to the final placement of their corresponding registers, as shown in Figure 13.9a. This placement can be enforced during physical synthesis by specifying a bound for the proximity of the clock-gating cells to the registers. Some advantages of this approach are that it makes it easier to estimate the latency from the clock-gating cell and it also increases the amount of available slack for the arrival of the activation signal. The impact on the clock tree is fairly minimal since the clock-gating cells are placed close to the registers and can eliminate the need for buffer insertion after clock-gating cell insertion.

A disadvantage to this approach is that it leaves most of the clock-tree switching even when branches are leading to registers that will have the clock blocked by a gated-clock. In order to save as much power as possible, it is desirable to gate as many buffers on the clock tree as possible. This is difficult for the designer to do without knowledge of the actual physically induced timing constraints.

In a physically aware clock-gating system, CTS works in conjunction with placement and clock-gating to determine an optimized placement and insertion of the clock-gating cells into the clock tree. This information is used to balance the delay on the activation signal with the amount of potential power saved by placing the clock-gating cell closer to the root of the clock tree, as shown in Figure 13.9b.

13.5.1.6 Clock Tree Planning

Cell placement, clock-gating cell insertion and CTS can also be decoupled, provided that the process of clock tree planning is started after cell placement and it does not interfere with clock routing. The objective of the approach in [50,51] is to build a power-optimal gated-clock tree structure fully compatible with state-of-the-art physical design tools to perform detailed clock routing and buffering. Thus, the output of the proposed clock tree planning methodology is not a completely routed clock tree; instead, it is a clock netlist (including clock-gating cells and related control logic) and constraints that, provided as input to CTS tools, lead to a low-power gated-clock tree, while still accounting for all nonpower-related requirements (e.g., controlled skew and low crosstalk-induced noise).

The methodology consists of three steps:

- Calculation of the clock-gating activation functions
- Generation of the clock tree logical topology
- Instantiation and propagation of the clock-gating cells and related logic

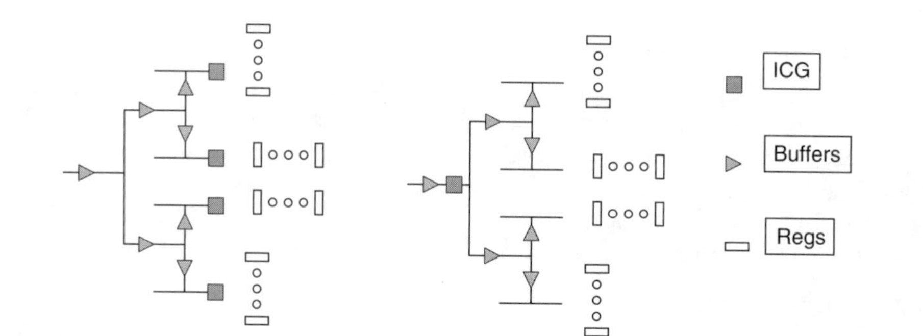

FIGURE 13.9 Integrated clock-gating cell placement: (a) close to registers; (b) buffering after ICG cell insertion.

The gated-clock activation functions for all the RTL modules are computed first. Next, according to the activation functions and the physical position of the registers, the logical topology of the clock tree is planned. This entails balancing the reduction in clock-switching activity against clock and activation function capacitive loads. Clock-gating cells are then inserted into the clock-tree topology and propagated upward in the tree whenever this is convenient, thus balancing the clock power consumption against the power of the gated-clock subtree. The information about the gated-clock tree is finally passed to the back-end portion of the flow, which will take care of clock-tree routing and buffering.

13.5.1.7 Testability Issues

Clock-gating reduces test coverage of the circuit since gated-clock registers are not clocked unless the activation signal is high. During test or scan modes, test vectors need to be loaded into the registers, and hence they must be clocked irrespective of the value of the activation signal. One way to address this, investigated from the theoretical standpoint in [52], is to include a control point or control gate at the activation signal, as shown in Figure 13.10a. This allows clock-gating signal, EN, to be overridden during the scanning in or out of vectors by the test-mode signal. In this way, during the test clock cycles, the clock signal is not gated by the activation signal, EN, and the register can be tested to see if it holds the correct state. Further, the test-mode signal is held at logic "1" during test mode, making any stuck-at fault on the activation signal unobservable. If full observability is required, this signal must be explicitly made observable by tapping it into an observability exclusive or (XOR) tree, as shown in Figure 13.10b.

A growing concern around scan-based testing is the power consumed during the scanning in and out of test vectors [53,54]. The changing register values during scanning can create activity levels that are much higher than those experienced during "normal" operation and lead to "good" chips failing during testing.

13.5.2 Exploitation of Common-Case Computation

It is well known that in complex digital architectures, some functionalities are far more exercised than others; this is due, mainly, to the fact that the data to be processed (i.e., the workload) may not have equiprobable distribution. Optimizing the common-case computation has thus become an established practice in high-throughput design [55], where variable-latency units replace fixed-latency ones to improve the overall system performance by adapting the latency of the data path to the length of the computation to be executed. Architectural retiming [56], speculative completion [57], and telescopic units [58,59] are examples of application, even in an automatic fashion, of the optimization paradigm based on exploitation of common-case computation.

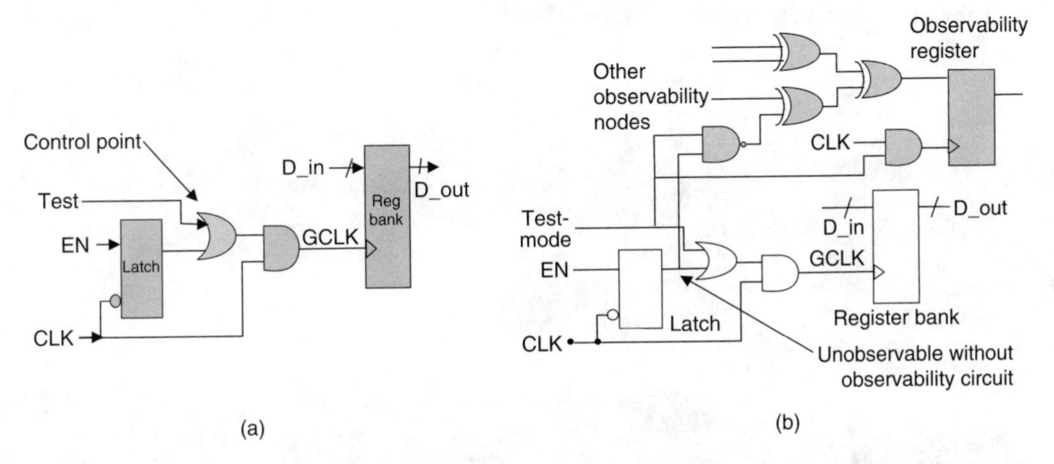

FIGURE 13.10 Testability issues with clock-gating: (a) adding controllability; (b) improving observability.

This idea has been extended recently to the case of power minimization; more specifically, as switching power depends on what the units of a design are which are activated by the input data as well as on the type of data to be processed, it may be possible to come up with variable-power architectural solutions which guarantee minimum power demand for the most probable execution conditions [60].

Figure 13.11 shows a possible architecture implementing the common-case optimization approach.

Block *A* supports the full functionality of the design while block *B* only covers the most common subset of it. *A* and *B* work in mutual exclusion, owing to the latch-enabled registers placed on the primary inputs of the two blocks. Based on the next datum to be processed, block *SEL* is in charge of selecting whether *A* or *B* should be activated in the next clock cycle. As *B* is much smaller than *A*, any time *B* is active, power is reduced. On the other hand, when *A* has to compute, there may be a penalty in power, as the whole design is larger (thus there are more gates that have to switch) if compared to block *A* alone. Let q be the probability of block *B* to be active, let P_A and P_B be the average power consumed by blocks *A* and *B*, respectively, when they are computing, and let P_{Ov} be the power overhead due to block *SEL* and the additional logic needed to make the architecture working. Actual power savings are achieved if

$$(1 - q)P_A + qP_B + P_{Ov} < P_A \qquad (13.7)$$

Equation (13.7) tells us that the architecture exploiting the common-case computation obtains power reductions over block *A* alone for high values of probability q. Clearly, high q usually implies a larger *SEL* and a larger *B* (intuitively, what happens is that as q increases, some of the functionality of *A* is incorporated also into *B*). The challenge is thus that of designing the smallest possible *SEL* and *B* blocks that maximize the value of q and limit the impact on area and delay penalty.

In the sequel we review some solutions proposed in the recent literature that can be considered as practical actuation, with a slightly different flavor, of the design framework discussed above.

13.5.2.1 Operand Isolation

The concept at the basis of *operand isolation* [61] is illustrated in Figure 13.12.

In Figure 13.12a, we observe that the output of the multiplier is only used when the control signals to the multiplexors, SEL_0 and SEL_1, are both high. In cycles when either of the control signals is low, if the multiplier inputs change, the multiplier performs computation but its result is not used. The wasted power may be substantial if these idle cycles occur for long periods of time. Figure 13.12b shows operand isolation applied to the multiplier. First, the activation signal AS is created to detect the idle cycles of the multiplier. AS is high in the "active" clock cycles when the multiplier output is being used, and otherwise low. This signal is used to isolate the multiplier by freezing its inputs during idle cycles using a set of gates, called *isolation logic*. In Figure 13.12b, the isolation logic consists of AND gates, but OR gates or latches may also be used. Using AND/OR gates avoids the introduction of new sequential elements and reduces the impact on the rest of the flow. Also, AND/OR gates are cheaper and tend to give better power savings overall.

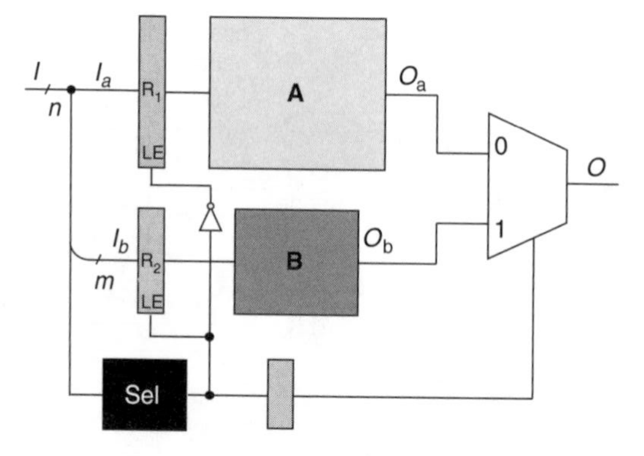

FIGURE 13.11 Exploitation of common-case computation.

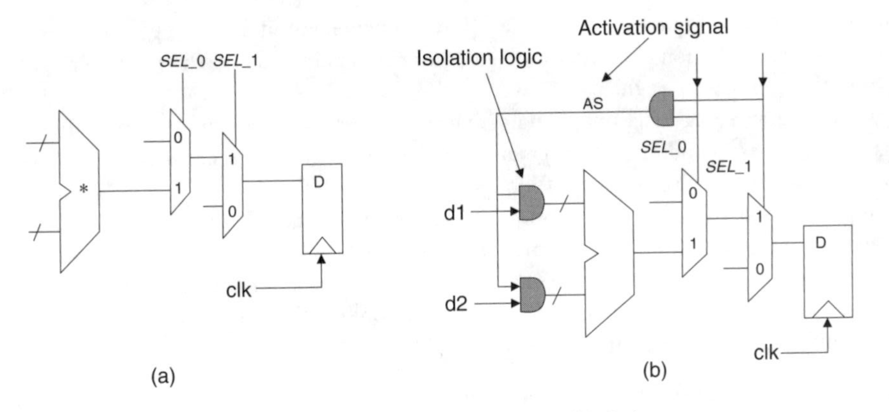

FIGURE 13.12 Operand isolation: (a) original circuit; (b) after operand isolation.

Operand isolation saves power by reducing switching in the operator being isolated, but it also introduces timing, area, and power overhead from the additional circuitry for the activation signal and the isolation logic. This overhead must be carefully evaluated against the power savings obtained to ensure a net power saving without too much delay or area penalty.

13.5.2.2 Precomputation

Precomputation [62,63] relies on duplication of part of the logic with the purpose of precomputing the circuit output values one clock cycle before they are required, and then using these values to reduce the total amount of switching in the circuit during the next clock cycle. Knowing the output values one clock cycle in advance allows the original logic to be turned off during the next time frame, thus eliminating any charging/discharging of internal capacitances.

The size of the logic that precalculates the output values must obviously be kept under control, since its contribution to the total power balance may offset the savings achieved by blocking the switching inside the original circuit. Several variants to the basic architecture can be adopted to take care of this problem. In particular, it may sometimes be convenient to resort to partial, rather than global, shutdown, i.e., to select for precomputation only a (possibly small) subset of the circuit inputs.

As an example, consider Figure 13.13a; the combinational block, *A*, implements an *N*-input, single-output Boolean function, *f*, and it has the I/O pins connected to registers R_1 and R_2. A possible precomputation architecture is depicted in Figure 13.13b.

Key elements of the precomputation architecture are the two N-input, single-output predictor functions, called g_1 and g_0, whose behavior is required to satisfy the following constraints:

$$g_1 = 1 \Rightarrow f = 1$$

$$g_0 = 1 \Rightarrow f = 0$$

The consequence is that if at the present clock cycle either g_1 or g_0 evaluates to 1, the load-enable (LE) signal goes to 0, and the inputs to block *A* at the next clock cycle are forced to retain the current values. Hence, no gate output transitions inside block *A* occur, while the correct output value for the next time frame is provided by the two registers located on the outputs of g_1 and g_0.

As mentioned earlier, the choice of the predictor functions is a difficult task. Perfect prediction requires $g_1 \equiv f$ and $g_0 \equiv f'$. However, this solution would not give any advantage in terms of power consumption over the original circuit, since it would entail the triplication of block *A*, and thus it would cause the same number of switchings as before, but with an area three times as large as the original network. Consequently, the objective to be reached is the realization of two functions for which the probability of their logical sum (i.e., $g_1 + g_0$) being 1 is as high as possible, but for which the area penalty due to their implementations is very limited. Also, the delay of the implementation of g_1 and g_0 should be given some attention, since the prediction circuitry may be on the critical path and, therefore, it may impact the performance of the optimized design.

One way of guaranteeing functions g_1 and g_2 to be much less complex than function f, thus implying a marginal area overhead in addition to a remarkable power savings, consists of making the two predictor functions depend on a limited number of inputs as compared to f.

Precomputation-based power optimization has been shown to be effective in the case of designs with pipelined structure. On the contrary, it seems to be hardly applicable to the case of sequential circuits with feedback. The reason for this is that the precomputation functions never attempt to stop the present-state inputs, which represent the majority of the inputs to the combinational logic for sequential circuits with a realistic number of memory elements (i.e., flip-flops).

13.5.2.3 Computational Kernel Extraction

It is known that, when in their steady state, complex sequential circuits tend to run through a limited set of states. Once such a set, called *computational kernel* [64], is extracted from a given circuit specification, it can be successfully used for various types of optimization, including power minimization.

Given a sequential circuit with the traditional topology shown in Figure 13.14a, the paradigm proposed in [64] for improving its power dissipation is based on the architecture depicted in Figure 13.14b.

The essential elements of the architecture are the following:

- The combinational portion of the original circuit (block *C*)
- The computational kernel (block *K*)
- The selector function (block *SEL*)

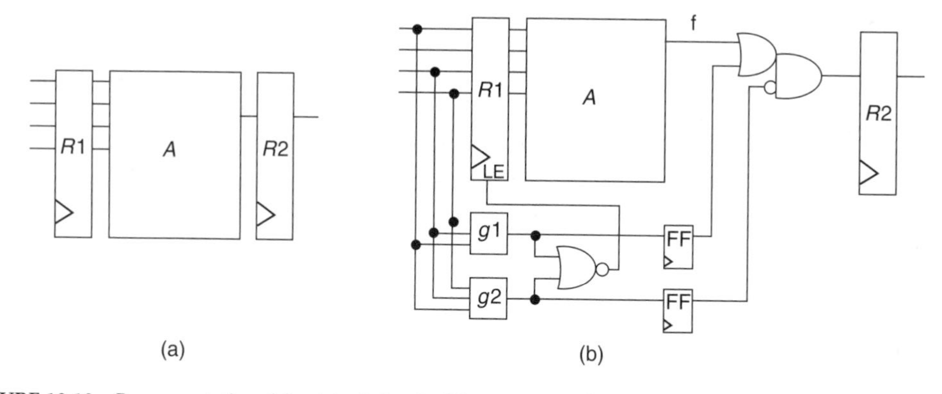

FIGURE 13.13 Precomputation: (a) original circuit; (b) precomputed architecture.

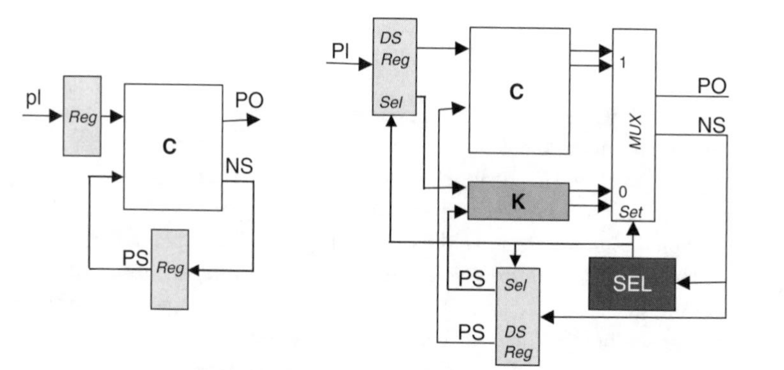

FIGURE 13.14 Computational kernel extraction: (a) original circuit; (b) kernel-based architecture.

- The dual-state registers (block *DS Reg*)
- The output multiplexor (block *MUX*)

The computational kernel can be seen as a "dense" implementation of the circuit it has been extracted from. In other terms, *K* implements the core functions of the original circuit, and because of its reduced complexity, it usually implements such functions in a faster and less power consuming way.

The purpose of the selector function *SEL* is that of deciding what logic block, between *C* and *K*, will provide the output value and the next state in the following clock cycle. To take a decision, *SEL* examines the values of the next-state outputs at clock cycle *n*. If the output and next-state values in clock cycle $n + 1$ can be computed by the kernel *K*, then *SEL* generates the value 1. Otherwise, it generates the value 0. The output of block *SEL* is connected to the multiplexor that selects which block produces the output and the next-state, as well as to the control input of the dual-state registers, which steer the primary and state inputs to the appropriate combinational block (either *C* or *K*). The optimized implementation is functionally equivalent to the original one.

The scheme in Figure 13.14b is just one among several possible architectures. The peculiar feature of this solution concerns the topology of the selection logic. In particular, the choice of having a selection function that only depends on the next-state outputs is dictated by the need of obtaining a small implementation. Reducing the size of the support of *SEL*, i.e., not including the primary inputs, is one way of pursuing this objective.

Fundamental for a successful application of the kernel-based optimization paradigm is the procedure adopted for kernel extraction. If the circuit is described by means of its state transition graph, the kernel can be determined exactly through symbolic (i.e., binary decision diagram (BDD-based)) procedures similar to those employed for FSM reachability analysis. On the other hand, when the state transition graph is too large to be managed, there are two options. If a gate-level description of the circuit is available, then block *K* can be iteratively synthesized by means of implication analysis followed by redundancy removal. Otherwise (i.e., if only a functional description of the circuit is available), kernel identification and synthesis can be performed by resorting to simulation of typical input traces and by then running probabilistic analysis and resynthesis of the state transition graph based on the results of the simulation. Obviously, in case approximate kernel extraction is adopted, the savings that can be obtained are usually more limited.

13.5.3 Managing Voltages

As discussed in Section 13.2, power consumption in CMOS designs heavily depends both on the operating supply voltage (switching power) and on the transistors' threshold voltage (leakage power).

The two variables are not independent, and switching speed constitutes the link. Starting from the 0.5 μm technology node, supply voltage levels have been scaling at approximately 1 V/0.1 μm. As we are now below the 100 nm feature size, the reduced operating voltages are forcing threshold voltages down to 0.25 V and below, in order to preserve speed. This has had a major impact on the leakage current of the transistors built into these technologies. As Equation (13.8) shows, the subthreshold leakage current grows exponentially as the threshold voltage decreases:

$$I_{sub} = I_0(e^{(-V_{th}/S)}(1 - e^{-qV_{ds}/kT})) \quad (\text{at } V_{gs} = 0) \tag{13.8}$$

For approximately every 65–85 mV decrease in threshold voltage (depending on temperature), there is an order of magnitude increase in subthreshold leakage current.

13.5.3.1 Managing Threshold Voltages

Silicon foundries have started to offer multiple threshold devices at the same process node to address the need to control leakage current and enabling designers to trade leakage and performance [65]. Along with the standard V_{th}, a low and high V_{th} transistor may be offered where the low-V_{th} device may have order of magnitude higher leakage than the standard V_{th} device, and the high-V_{th} device may have an order of magnitude lower leakage than the standard. This reduction in leakage is not free though, and it comes at the expense of the speed of the device — there could be a 20% to 2*x* delay penalty between the standard and the high-V_{th} devices — and an increase in the cost of the fabrication process.

Synthesis algorithms can be used to optimize leakage by using high-threshold voltages while still meeting the timing requirements using low V_{th} devices on the more critical paths. The threshold voltage of a transistor can be dynamically changed by varying its back bias. The change in V_{th} is roughly proportional to the square root of the back-bias voltage. As threshold voltages drop below 0.25 V, variable back biasing may gain more appeal. A distinct advantage of this approach is that during periods when heavy processing is needed, V_{th} can be reduced, thus speeding up the cells. When the cells are in a slower drowsy or idle mode, the threshold voltage can be raised, thus lowering the leakage. One significant impact of using variable back-biasing is that there are two new terminals for each cell that now need to be routed. A common ASIC design practice is to create cells that tie the *n-well* regions to V_{dd}, and the *p-well* regions to ground. In the physical implementation, these are simply pre-defined contacts designed into the cell that are connected as part of the power and ground routes. To enable back-biasing, new voltage lines are routed to control the bias. These can be to individual cells or, more likely, to regions that contain multiple cells sharing the same *well* and a common tie-cell to control the *well bias* [66,67].

13.5.3.2 Managing Supply Voltages

13.5.3.2.1 Power Gating

For applications like cell phones that have long shut-down periods, the power consumption is dominated by the leakage power consumption during off periods. *Power gating* addresses this issue by shutting off the power supply to blocks that are not in use. This not only eliminates the switching power dissipation in the block, but also hugely saves the leakage power dissipation when the block is shut down. In the simplest case of power gating, the voltage level across the chip is the same, but different power supply grids are used in different parts of the chip. Besides both leakage and switching power savings, this helps to control the IR drop on the power grids. These design techniques create interesting switching on the chip. When one block is shut down, the voltage on its output ports may drift to undefined values causing large leakage and unexpected functionality in the gates that it drives. Therefore, it is necessary to isolate all the output ports of a block that is shut down. Further, if a block is shut down, it might be important for it to *remember* some of its previous state values. The registers used to store these values need to be specially designed to be powered by a secondary supply that will allow them to retain their values during shut down and restore it when the primary supply is up [68,69].

Figure 13.15 shows a typical floorplan of a power-gated chip with isolation cells and retention registers.

EDA tools need to insert automatically isolation cells and retention registers for powered-down blocks. To complete the entire flow, it is important to be able to represent all this information in the source description (the RTL), and infer both isolation and retention logic during RTL synthesis. Since the power-down behavior directly impacts the functionality of the chip, it is important to be able to verify this functionality both with simulation tools and formal verification tools.

To enable power gating in a complete automated design flow, place, route and back-end optimization tools need to be aware of the different power supply regions. Cells powered by each supply need to be placed in separate areas, and signals should be routed so that they do not traverse areas serviced by a different power supply, since subsequent repeater/buffer insertion on these signals will not be legal.

With complex power-gating schemes, the role of power and power-network analysis tools becomes more important. Turning the voltage ON or OFF to a block can cause large transients on the power grid, affecting other blocks on the chip. These effects should be accounted for by the power analysis tools.

13.5.3.2.2 Multi-V_{dd} Design

Over the last few years, more and more chips are using multivoltage design in one form or the other, and EDA tools need to support the various special needs of these design styles [70,71]. For instance, the power-gating technique discussed earlier can be thought of as the simplest version of multivoltage design, where the ON voltage level is the same for all the blocks, but can be changed to 0 V during shut down.

A more complex methodology uses different voltages for different blocks on the chip, which are powered with separate power supplies that can be independently shut down [72].

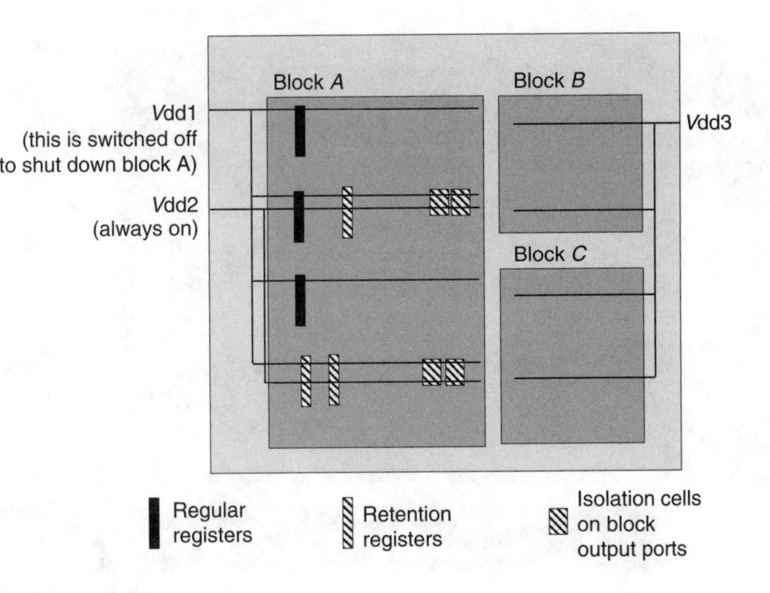

FIGURE 13.15 Applying power-gating.

An example of a multivoltage chip is shown in Figure 13.16.

Variations of multivoltage design styles include dynamic V_{dd} scaling, where on-chip controls are used to switch the voltage on the different parts of the chip to different predefined levels. Adaptive scaling takes this a step further, where high-level control feedback loops can be used to change the V_{dd} values of different blocks based on the speed/power requirements for the chip at any given time.

Along with the issues associated with power gating, multivoltage design has other requirements from EDA tools. First, tools need to understand the impact of voltage levels on timing and power of library cells (hence, on the design) and need to provide both analysis and optimization capabilities that accurately account for the different V_{dd} levels. Secondly, the implementation tools should be be able to insert automatically level shifters to be able to adapt signals from one level to another. Instantiation and optimization of level shifter cells is key to multivoltage design [73]. The tools have to manage cells that have more than one supply rail, and circuitry that can vary or completely shut down the supply voltage to a block.

Clock tree generators need to account for buffers that operate at different voltages to provide clock signals to each block, and the router needs to account for buffer placement in the context of different voltage regions on the chip. Routing a feed-through signal via a region may now require the insertion of level shifters in order to drive the signal adequately. Analysis tools need to understand these different situations — tracking all of these new voltage-based modes — and provide useful feedback to the designer.

An additional major impact on the design flow is the need to treat the supply line as another variable. For most previous mainstream designs, logical netlists only specified the input and output connections between gates. V_{dd} pins for all the cells (as also the V_{ss} pins) were attached to a single power (ground) network after the place and route step was done. In a multivoltage design, different cells are connected to different power (ground) networks and this information needs to be managed in the entire design flow starting from the RT level, to be able to simulate correctly and verify the design right from the start.

A final design implication that optimization and analysis tools must account for is the signal integrity impact of driving some lines at higher voltages than others. The higher-voltage lines can cause larger spikes in neighboring low-voltage lines than other lower-voltage aggressors, which impacts timing analysis, power, and the routing of lines on the chip.

13.5.4 Memory Power Optimization

The power model of Equation (13.5) exposes the two quantities that can be targeted to reduce memory power, namely, the number of (read or write) accesses and the (read or write) capacitance. The former are

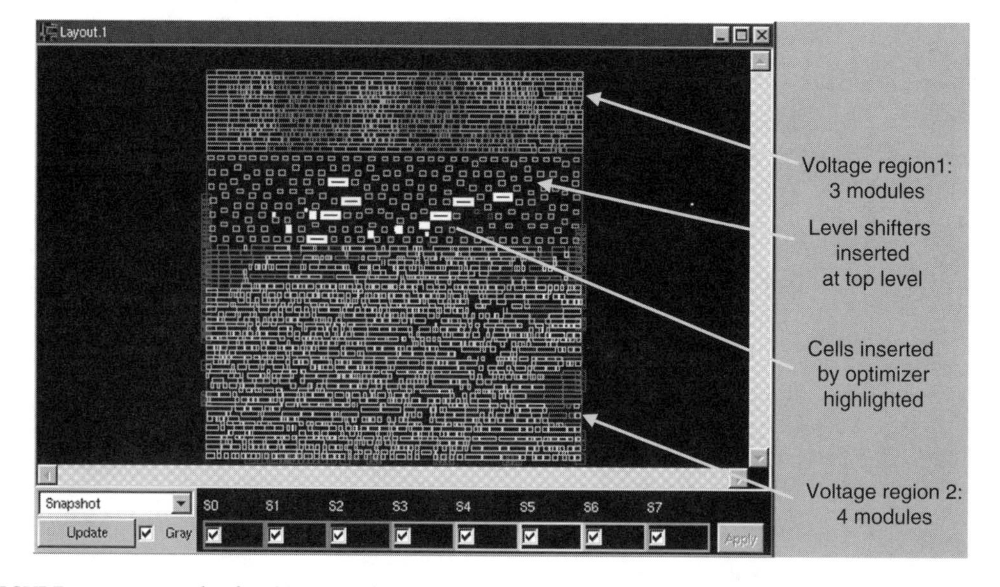

FIGURE 13.16 Example of multivoltage chip.

actually dictated by the scheduling of the operations resulting from register-transfer level design or high-level synthesis and should be considered as fixed. Attention should then focus on capacitance minimization.

Apparently, the read or write capacitances are related to physical parameters and are also uniquely defined for a given technology. In practice, C_{read} and C_{write} should be regarded as *average* read or write capacitances, that is, they should represent the cost of an *average* read or write access. This subtle difference opens the way to a class of optimization techniques that modify the *organization* of the memory, which can be seen as the RTL counterpart of techniques that, in a more general macro-architectural context, aim at optimizing the *memory hierarchy* of a system. The most well-known example is the use of a cache between processor and memory, whose net effect is that of reducing the *average cost* of accessing memory.

Such a reduction of average cost (i.e., capacitance) can be achieved by either augmenting the hierarchy vertically (i.e., adding levels to the hierarchy), or by growing it horizontally (i.e., adding support structures in parallel to existing hierarchy levels), or both [74,75]. In an RTL context, the cycle-accurate timing information contained in the description rules out the vertical transformations, since they affect the average cycle time of a memory. We can reasonably assume that, in a micro-architecture, only one level of memory hierarchy does (typically) exist. The latter option is therefore the only viable one that allows a relatively seamless integration in a micro-architecture. Horizontal modification of a memory can be restated in terms of the usual common-case paradigm discussed in Section 13.5.2.

In this specific situation, the common case is represented by the memory cells that are accessed most. The idea is that of "isolating" this common case from the average case; since both are identified by memory cells, this amounts to instantiating a memory block "in parallel" to the main memory. This is shown in the example in Figure 13.17.

The left side of the figure depicts the original memory, consisting of $W = 1000$ words, while on the right-hand side, the W memory locations are mapped onto two different memory blocks of sizes $W_1 = 200$ and $W_2 = 800$. Assume that for a given workload the memory cells in the monolithic block Mem are accessed a total of $N = 10000$ times. For the partitioned architecture, the common case, i.e., the most frequently accessed memory cells are placed into the smaller block Mem_1 (with $N_1 = 8000$ accesses) and the least frequently accessed locations are mapped to the larger block Mem_2 (with $N_2 = 2000$ accesses).

We realistically assume that the power cost for accessing a cell in a memory block is directly related to the size of the block itself (the larger the block, the higher the power cost). Specifically, for our example, we assume the power access cost for cells in Mem to be $C_{Mem} = 0.5$, the cost for cells in Mem_1 to be $C_{Mem1} = 0.1$, and the cost for cells in Mem_2 to be $C_{Mem2} = 0.4$.

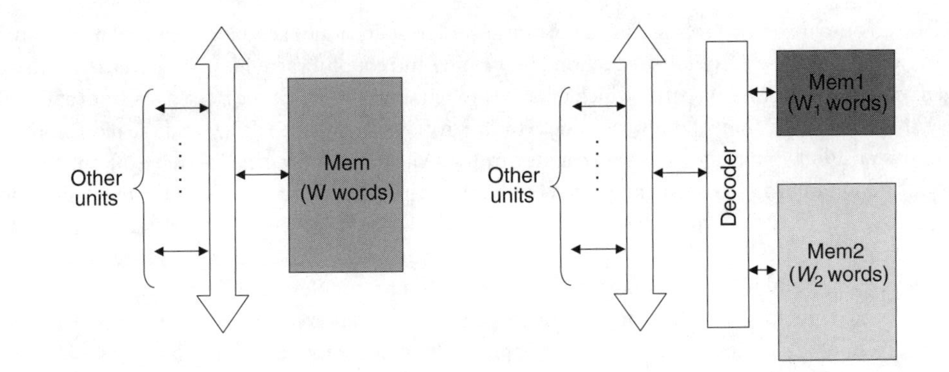

FIGURE 13.17 Common-case optimization applied to memory.

The average power for the monolithic memory is $P_{\text{mono}} = P_{\text{Mem}} = \dfrac{NC_{\text{Mem}}}{N} = 0.5$. For the partitioned memory, neglecting the decoder overhead, the power is $P_{\text{part}} = P_{\text{Mem1}} + P_{\text{Mem2}} = \dfrac{N_1 C_{\text{Mem1}} + N_2 C_{\text{Mem2}}}{N} = \dfrac{8000 \cdot 0.1 + 2000 \cdot 0.4}{10000} = 0.16$. The partitioned scheme is more convenient as N_1 gets larger w.r.t. N_2, meaning that the most frequently accessed locations should be mapped as much as possible onto smaller blocks.

The generic template illustrated in Figure 13.17 lends itself to numerous variants, depending for instance on whether the subblocks are overlapping or not, on the choice of the number and of the size of the subblocks, on the possibility of having noncontiguous partitions (i.e., addresses could be *relocated* within blocks), on the architecture and implementation style of the decoder, on the organization and routing of the address and data buses, on the placement of the subblocks, etc.

The common-case principle applied to memories requires proper extra logic to drive the accesses to the correct memory bank (the block generically denoted as "encoder" in the figure) whose power and performance overhead constrains the type of partitioning scheme allowed. Some memory partitioning variants, such as those proposed in [76], provide significant power reductions with a very limited hardware overhead that can be easily tolerated into an RTL design.

It is important to mention that, although the delay of the decoder is on the critical path, it usually does not affect the overall cycle time, since the access times of the subblocks are smaller than that of the monolithic memory; it suffices that $(t_{\text{decoder}} + \max_i(t_{\text{block}i})) \leq t_{\text{mem,mono}}$ in order to guarantee a seamless integration of the partitioned scheme ($t_{\text{mem,mono}}$ is the access time of the monolithic memory, and $t_{\text{block}i}$ is the access time of the generic subblock after partitioning).

13.5.5 Bus Power Optimization

Among the various parameters that appear in the model of Equation (13.6), only switching activity can be exploited at the micro-architectural level in order to reduce power. While supply voltage and frequency are somehow assigned up-front, capacitance values (both switching and coupling) can only be reduced during physical design through proper wire sizing (for self-capacitance), spacing, or shielding (for coupling capacitance).

Reducing the activity factors amounts to modifying the binary values that are transmitted to the bus, in other terms, *encoding* the values. For correct operations, clearly, both encoding and decoding are required at each end of the bus.

Since we are dealing with hardware components, encoding/decoding must be implemented through specialized hardware blocks. This obvious requirement poses serious limitations on the encoding that can be applied on an on-chip bus: codecs will in fact consume power, which if not kept under control may easily offset the power gain achieved from the reduction of switching and coupling activities. Moreover, the codec also adds delay to the corresponding paths, which must be limited since the cycle time is fixed up-front at the RTL. There is therefore a clear tradeoff between codec complexity (power, but also delay) and the reduction in the number of transitions. Since the latter are weighted by the switched capacitance, this trade-off can be cleanly expressed as a minimum value (switching and coupling) of capacitance that represents the break-even point of the power cost function; beyond this minimum value, the encoding scheme becomes advantageous.

This trade-off drastically limits the type of encoding to be chosen. The existing literature on low-power bus encoding is vast and it consists, in most of the cases, of quite complex encoding schemes that are more suitable for off-chip buses or for long global on-chip buses rather than for short data-path buses. Moreover, technology scaling plays against the application of encoding to on-chip buses: as wire capacitance progressively increases with respect to that of the cells, the average length of the wires must be decreased, thus making it more and more difficult to amortize the cost of the codec.

13.5.5.1 Bus-Encoding Schemes

Bus encoding can be applied at the physical level *(signal encoding)* or at the data-link level *(data encoding)*. The former consists of modifying (in time or space) the way the binary 0's and 1's are represented. For instance, a 1 could be encoded as a $0 \rightarrow 1$ transition. Data encoding, instead, consists of modifying the way the binary patterns are represented. For example, a word could be represented by adding one parity bit. The application of the two types of encoding is not mutually exclusive. However, signal encoding may have an impact on the technology used to implement the design. For instance, a typical signal encoding technique consists of reducing the voltage swing on (possibly some) of the bus wires. To make this possible, the technology on which the system will be implemented must be able to support that voltage level. For this reason, signal-encoding techniques are more related to the physical level of abstraction, and will not be analyzed further. Conversely, data encoding is more general, since the only underlying assumption is that of tolerating the insertion of additional hardware to perform the encoding.

The problem of defining a code that minimizes the number of (self and/or coupling) transitions enables a theoretical formulation that has led to solutions that mix results from information and probability theory. Although surveying these techniques is out of the scope of this chapter, a rough classification may help in understanding the problem better. Bus-encoding techniques can be categorized based on two dimensions:

- *The amount of redundancy allowed.* Some encoding schemes rely on spatial or temporal redundancy. Spatial redundancy implies the addition of extra bus lines, whereas temporal redundancy implies the addition of extra cycles to the bus transfers.
- *The amount of knowledge on the statistics of the transmitted data.* Some schemes assume *a priori* knowledge of the statistical properties of the information transmitted on the bus, which can be exploited to customize the encoding functions to the most typical behavior. An example is that of address buses, in which, although the specific patterns sent on the bus are not known, there exists a high degree of correlation between them, because of the sequential execution of a program.

Redundant codes are very popular in literature. One of the most referenced ones is the *bus-invert* (BI) scheme [77], in which the transmitter computes the Hamming distance between the word to be sent and the previously transmitted one. If the distance is larger than half of the bus width, the word to be transmitted is inverted, i.e., complemented. The information about inversion is carried by an additional wire that is used at the receiver end to restore the original data.

The BI scheme has some interesting properties. First, the worst-case number of transitions of an n-bit bus is $n/2$ at each cycle. Second, if we assume that data are uniformly randomly distributed, it is possible to show that the average number of transitions with this code is lower than that of any other encoding scheme with just one redundant line. Moreover, the basic 1-bit redundant BI code has the property that the average number of transitions per line increases as the bus gets wider, and asymptotically converges to 0.5, which is also the average switching per line of an unencoded bus, and is already close to this value for 32-bit buses. This shortcoming has spun a number of variants of the basic BI scheme, based on the partitioning of the bus into smaller blocks and on the use of bus inversion on each block independently. Since the trivial application of this partitioned variant on an m-block bus would require m control lines, these methods have tried to reduce this additional complexity (see, for example, the work in [78,79]).

Addition of redundancy is not very desirable at the RTL. Temporal redundancy obviously alters the timing of the operations, thus giving rise to performance issues, while spatial redundancy may require the modification of the bus interface to support the extra connections, which may not be feasible when connecting synthesizable IP blocks with predefined I/O.

Discarding redundancy drastically limits the spectrum of applicability of bus encoding. In particular, if no assumption on the statistical properties of the transmitted data can be made, results from information theory show that it is not possible to reduce the number of transitions [80]. Some irredundant codes proposed in the literature bypass this theoretical limitation by building statistical information online over a given timing window [81,82]. These adaptive schemes require however a significant hardware overhead that is not generally affordable at the RTL.

When some knowledge of the statistical properties is available *a priori*, more effective encodings can be devised. The most realistic option at the RTL is the case of address buses, for which there exists a high degree of correlation between consecutive addresses; in particular, addresses generated by processors typically exhibit a high degree of *sequentiality;* this is particularly true for data-dominated applications, where the few control structures only occasionally break the sequentiality of the address stream.

Some authors have suggested the adoption of Gray coding [83] for address buses. This code achieves its asymptotic best performance of a single transition per emitted address when infinite streams of consecutive addresses are considered. This average can be lowered to asymptotic zero transitions, at the price of adding some spatial redundancy [84]. Alternatively, the asymptotic zero-transition behavior can be achieved without any redundancy by exploiting the decorrelating characteristics of the XOR function, when applied to consecutive bus patterns. In this way, the values on the bus are encoded using a transition signaling scheme [85]. Notice that the decorrelation implies an operation of a signal with its previous copy and may not be feasible at the RTL for timing reasons. Conversely, Gray encoding is a fully combinational transformation and it is potentially feasible to insert Gray codecs into an RTL description, should the cycle time constraints allow it.

So far, we have discussed techniques that aim at reducing switching activity. When addressing coupling activity, things are even more critical, because the encoding involves pairs of wires, thus making the hardware overhead required by the codec more complex by construction. In addition, solutions based on encoding the data to reduce coupling activity tend to ignore the most important implication of coupling, that is, its impact on timing due to *crosstalk*. Minimizing the number of simultaneous transitions on adjacent wires may reduce power consumption of coupling capacitance, yet it does not reduce crosstalk by itself. Crosstalk mainly affects signal integrity, and even a significant reduction of crosstalk-induced power is of little interest for designers, if it does not guarantee the proper functionality of the circuit. Therefore, a solution to the problem must be consistent with performance-oriented crosstalk reduction techniques. In other terms, since crosstalk is mainly a capacitive effect, the only way to reduce it is by reducing the capacitance that causes it, and let power reduction come as a by-product. This makes the power optimization an issue to be dealt with during the physical design step, with the application of techniques ranging from (static or dynamic) wire permutation [86–88] to nonuniform wire spacing [89], from the insertion of shielding lines [90–92] to different combinations of the approaches above [93,94]. Solutions based on encoding of the data, however, may still be applied on top of capacitance reduction techniques, should the timing constraint allow it.

13.6 Conclusions

This chapter has reviewed the basic principles of power modeling, estimation, and optimization for digital CMOS circuits described at the micro-architectural level. The most common and successful modeling solutions for various types of micro-architectural components, such as data-path macros, controllers, memories, wires and buses have been discussed in detail, offering a comprehensive overview of the state of the art in this domain. The flow for power estimation that makes use of the various models has then been illustrated.

The focus has then been shifted to power optimization technologies; here, solutions based on clock-gating, exploitation of common-case computation and dynamic V_{th}/V_{dd} management, memory partitioning, and bus encoding have been explored, pointing out the main context in which they can be more suitably applied. Emphasis has been put on design automation aspects of most of the techniques considered, with the objective of making this chapter of practical use not only to IC designers and architects, but also to EDA engineers.

References

[1] A.P. Chandrakasan, S. Sheng, and R.W. Brodersen, Low-power CMOS digital design, *IEEE J. Solid State Circuits,* 27, 473–484, 1992.

[2] D.D. Gajski, N.D. Dutt, A.C.-H. Wu, and S.Y.-L. Lin, *High-Level Synthesis: Introduction to Chip and System Design,* Kluwer Academic Publishers, Dordrecht 1992.

[3] P. Landman, High-level power estimation, *ISLPED-96: ACM/IEEE International Symposium on Low Power Electronics and Design,* Monterey, CA, August 1996, pp. 29–35.

[4] P. Landman and J. Rabaey, Activity-sensitive architectural power analysis, *IEEE Trans. Comput. Aided Des. Integr. Circuits Syst,* 15, 571–587, 1996.

[5] P. Landman, R. Mehra, and J. Rabaey, An integrated CAD environment for low-power design, *IEEE Des. Test Comput.,* 13, 72–82, 1996.

[6] A. Raghunathan, S. Dey, and N. Jha, Register-Transfer Level Estimation Techniques for Switching Activity and Power Consumption, *ICCAD-96: IEEE/ACM International Conference in Computer-Aided Design,* San Jose, CA, November 1996, pp. 158–165.

[7] S. Katkoori and R. Vemuri, Architectural power estimation based on behavioral profiling, *J. VLSI Des.,* 7, 255–270, 1998.

[8] A. Bogliolo, I. Colonescu, R. Corgnati, E. Macii, and M. Poncino, An RTL power estimation tool with on-line model building capabilities, *PATMOS-01: IEEE International Workshop on Power and Timing Modeling, Optimization and Simulation,* Yverdon-les-Bains, Switzerland, September 2001, pp. 2.3.1–2.3.10.

[9] S. Ravi, A. Raghunathan, and S. Chakradhar, Efficient RTL power estimation for large designs, *IEEE International Conference on VLSI Design,* New Delhi, India, January 2003, pp. 431–439.

[10] D. Helms, E. Schmidt, A. Schulz, A. Stammermann, and W. Nebel, An improved power macro-model for arithmetic datapath components, *PATMOS-02: IEEE International Workshop on Power and Timing Modeling, Optimization and Simulation,* Sevilla, Spain, September 2002, pp. 16–24.

[11] M. Bruno, A. Macii, and M. Poncino, A statistical power model for non-synthetic RTL operators, *PATMOS-03: IEEE International Workshop on Power and Timing Modeling, Optimization and Simulation,* Torino, Italy, September 2003, pp. 208–218.

[12] F. Najm, Transition density: a new measure of activity in digital circuits, *IEEE Trans. Comput. Aided Des. Integr. Circuits Syst.,* 12, 310–323, 1993.

[13] Q. Qiu, Q. Wu, C.-S. Ding, and M. Pedram, Cycle-accurate macro-models for RT-level power analysis, *IEEE Trans. VLSI Syst.,* 6, 520–528, 1998.

[14] H. Mehta, R.M. Owens, and M.J. Irwin, Energy characterization based on clustering, *DAC-33: ACM/IEEE Design Automation Conference,* Las Vegas, NV, June 1996, pp. 702–707.

[15] S. Gupta and F. Najm, Energy-per-cycle estimation at RTL, *ISLPED-99: ACM/IEEE International Symposium on Low-Power Electronics and Design,* San Diego, CA, August 1999, pp. 16–17.

[16] L. Benini, A. Bogliolo, M. Favalli, and G. De Micheli, Regression models for behavioral power estimation, *PATMOS-96: IEEE International Workshop on Power and Timing Modeling, Optimization and Simulation,* Bologna, Italy, October 1996, pp. 125–130.

[17] D. Liu and C. Svensson, Power consumption estimation in CMOS VLSI chips, *IEEE J. Solid State Circ.,* 29, 663–671, 1994.

[18] E. Schmidt, G. Jochens, L. Kruse, F. Theeuwen, and W. Nebel, Memory power models for multilevel power estimation and optimization, *IEEE Trans. VLSI Syst.*, 10, 106–109, 2002.

[19] S. Powell and P. Chau, Estimating power dissipation of VLSI signal processing chips: the PFA technique, *VLSI Signal Processing*, 4, 250–259, 1990.

[20] P. Landman and J. Rabaey, Architectural power analysis: the dual-bit type model, *IEEE Trans.VLSI Syst.*, 3, 173–187, 1995.

[21] S. Gupta and F. Najm, Power macromodeling for high level power estimation, *DAC-34: ACM/IEEE Design Automation Conference*, Anaheim, CA, June 1997, pp. 365–370.

[22] M. Barocci, L. Benini, A. Bogliolo, B. Ricco, and G. De Micheli, Look-up table power macro-models for behavioral library components, *IEEE Alessandro Volta Memorial Workshop on Low-Power Design*, Como, Italy, March 1999, pp. 173–181.

[23] R. Corgnati, E. Macii, and M. Poncino, Clustered table-based macromodels for RTL power estimation, *GLS-VLSI-99: IEEE/ACM Great Lakes Symposium on VLSI*, Ann Arbor, MI, March 1999, pp. 354–357.

[24] A. Bogliolo, E. Macii, V. Mihailovici, and M. Poncino, Combinational characterization-based power macro-models for sequential macros, *PATMOS-99: IEEE International Workshop on Power and Timing Modeling, Optimization and Simulation*, Kos, Greece, October 1999, pp. 293–302.

[25] M. Anton, I. Colonescu, E. Macii, and M. Poncino, Fast characterization of RTL power macro-models, *ICECS-01: IEEE International Conference on Electronics, Circuits and Systems*, La Valletta, Malta, September 2001, pp. 1591–1594.

[26] A. Bogliolo, R. Corgnati, E. Macii, and M. Poncino, Parameterized RTL power models for combinational soft macros, *IEEE Trans. VLSI Syst.*, 9, 880–887, 2001.

[27] A. Tyagi, Entropic bounds on FSM switching, *ISLPED-96: ACM/IEEE International Symposium on Low Power Electronics and Design*, Monterey, CA, August 1996, pp. 323–327.

[28] D. Marculescu, R. Marculescu, and M. Pedram, Theoretical bounds for switching activity analysis in finite-state machines, *ISLPED-98: ACM/IEEE International Symposium on Low Power Electronics and Design*, Monterey, CA, August 1998, pp. 36–41.

[29] S. Katkoori and R. Vemuri, Simulation-based architectural power estimation for PLA-based controllers, *ISLPED-96: ACM/IEEE International Symposium on Low Power Electronics and Design*, Monterey, CA, August 1996, pp. 121–124.

[30] L. Benini, A. Bogliolo, E. Macii, M. Poncino, and M. Surmei, Regression-based RTL power models for controllers, *GLS-VLSI-00: ACM/IEEE Great Lakes Symposium on VLSI*, Evanston, IL, March 2000, pp. 147–152.

[31] U. Ko and P. T. Balsara, Characterization and design of a low-power, high-performance cache architecture, *IEEE International Symposium on VLSI Technology, Systems and Applications*, May 1995, pp. 235–238.

[32] M.B. Kamble and K. Ghose, Analytical energy dissipation models for low power caches, *ISLPED-97: ACM/IEEE International Symposium on Low Power Design*, Monterey, CA, August 1997, pp. 143–148.

[33] S.L. Coumeri and D.E. Thomas, Memory modeling for system synthesis, *IEEE Trans. VLSI Syst.*, 8, 327–334, 2000.

[34] B.S. Landman and R.L. Russo, On a pin vs. block relationship for partitions of logic graphs, *IEEE Trans. Comput.*, 20, 1469–1479, 1971.

[35] S. Sastry and A.C. Parker, Stochastic models for wireability of analysis of gate arrays, *IEEE Trans. Comput. Aided Des.Integr. Circuits Syst.*, 5, 52–65, 1986.

[36] F. J. Kurdahi and A.C. Parker, Techniques for area estimation of VLSI layouts, *IEEE Trans. Comput.-Aided Des. Integr. Circuits Syst.*, 8, 81–92, 1989.

[37] C. Anton, A. Bogliolo, P. Civera, I. Colonescu, E. Macii, and M. Poncino, RTL estimation of steering logic power, *PATMOS-00: IEEE International Workshop on Power and Timing Modeling, Optimization and Simulation*, Gottingen, Germany, September 2000, pp. 36–45.

[38] L. Benini and G. De Micheli, Networks on chips: a new SoC paradigm, *IEEE Comput.*, 35, 70–78, 2002.

[39] Synopsys Design Ware Library, *http://www.synopsys.com.*

[40] R. Ho, K.W. Mai, and M.A. Horowitz, The future of wires, *Proceedings of the IEEE,* Vol. 89, No. 4, April 2001, pp. 490–504.

[41] P. Sotiriadis and A. Chandrakasan, A bus energy model for deep submicron technology, *IEEE Trans. VLSI Syst.,* 10, 341–350, 2002.

[42] L. Benini, P. Siegel, and G. De Micheli, Automatic Synthesis of Gated Clocks for Power Reduction in Sequential Circuits, *IEEE Des. Test Comput.,* 11, 32–40, 1994.

[43] L. Benini and G. De Micheli, Transformation and synthesis of FSMs for low power gated clock implementation, *IEEE Trans. Comput. Aided Des. Integr. Circuits Syst.,* 15, 630–643, 1996.

[44] L. Benini, G. De Micheli, E. Macii, M. Poncino, and R. Scarsi, Symbolic synthesis of clock-gating logic for power optimization of synchronous controllers, *ACM Trans. Des. Automat. Electron. Syst.,* 4, 351–375, 1999.

[45] M. Gowan, L.L. Biro, and D.B. Jackson, Power considerations in the design of the Alpha 21264 microprocessor, *DAC-35: ACM/IEEE Design Automation Conference,* San Francisco, CA, June 1998, pp. 726–731.

[46] A. Correale, Overview of the power minimization techniques employed in IBM powerPC 4xx embedded controllers, *ISLPD-95: ACM International Symposium on Low Power Design,* Dana Point, CA, August 1995, pp. 75–80.

[47] V. Tiwari, D. Singh, S. Rajgopal, G. Mehta, R. Patel, and F. Baez, Reducing power in high-performance microprocessors, *DAC-35: ACM/IEEE Design Automation Conference,* San Francisco, CA, June 1998, pp. 732–737.

[48] Z. Khan, and G. Mehta, Automatic clock gating for power reduction, *SNUG-99: Synopsys Users Group,* San Jose, CA, March 1999.

[49] P. Babighian, L. Benini, and E. Macii, A scalable algorithm for RTL insertion of gated clocks based on observability don't cares computation, *IEEE Trans. Comput. Aided Des. Integr. Circuits Syst.,* 24, 29–42, 2005.

[50] L. Benini, M. Donno, A. Ivaldi, and E. Macii, Clock-tree power optimization based on RTL clock-gating, *DAC-40: ACM/IEEE Design Automation Conference,* Anaheim, CA, June 2003, pp. 622–627.

[51] M. Donno, E. Macii, and L. Mazzoni, Power-aware clock tree planning, *ISPD-04: ACM/IEEE International Symposium on Physical Design,* Phoenix, Arizona, April 2004, pp. 138–147.

[52] L. Benini, M. Favalli, and G. De Micheli, Design for testability of gated-clock FSMs, *EDTC-96: IEEE European Design and Test Conference,* Paris, France, March 1996, pp. 589–596.

[53] B. Pouya and A. Crouch, Optimization for vector volume and test power, *ITC-00: IEEE International Test Conference,* Atlantic City, NJ, October 2000, pp. 873–881.

[54] J. Saxena, K. Butler, and L. Whetsel, An analysis of power reduction techniques in scan testing, *ITC-01: IEEE International Test Conference,* Baltimore, MD, October 2001, pp. 670–677.

[55] J.L. Hennessy and D.A. Patterson, *Computer Architecture: A Quantitative Approach,* Morgan Kaufmann Publishers, San Francisco, CA, 1996.

[56] S. Hassoun and C. Ebeling, Architectural retiming: pipelining latency-constrained circuits, *DAC-33: ACM/IEEE Design Automation Conference,* Las Vegas, NV, June 1996, pp. 708–713.

[57] S. Nowick, Design of a low-latency asynchronous adder using speculative completion, *IEE Proc.–Comput. Digit. Tech.,* 143, 301–307, 1996.

[58] L. Benini, G. De Micheli, E. Macii, and M. Poncino, Telescopic units: a new paradigm for performance optimization of VLSI designs, *IEEE Trans. Comput. Aided Des. Integr. Circuits Syst.,* 17, 220–232, 1998.

[59] L. Benini, G. De Micheli, A. Lioy, E. Macii, G. Odasso, and M. Poncino, Automatic synthesis of large telescopic units based on near-minimum timed supersetting, *IEEE Trans. Comput.,* 48, 769–779, 1999.

[60] G. Lakshminarayana, A. Raghunathan, K. S. Khouri, N. K. Jha, and S. Dey, Common case computation: a new paradigm for energy and performance optimization, *IEEE Trans. Comput. Aided Des. Integr. Circuits Syst.,* 23, 33–49, 2004.

[61] M. Munch, B. Wurth, R. Mehra, J. Sproch, and N. Wehn, Automating RT-level operand isolation to minimize power consumption in datapaths, *DATE-00: IEEE Design Automation and Test in Europe,* Paris, France, March 2000, pp. 624–631.
[62] M. Alidina, J. Monteiro, S. Devadas, A. Ghosh, and M. Papaefthymiou, Precomputation-based sequential logic optimization for low power, *IEEE Trans. VLSI Syst.,* 2, 426–436, 1994.
[63] J. Monteiro, S. Devadas, and A. Ghosh, Sequential logic optimization for low power using input-disabling precomputation architectures, *IEEE Trans. Comput. Aided Des. Integr. Circuits Syst.,* 17, 279–284, 1998.
[64] L. Benini, G. De Micheli, A. Lioy, E. Macii, G. Odasso, and M. Poncino, Synthesis of power-managed sequential components based on computational kernel extraction, *IEEE Trans. Comput. Aided Des. Integr. Circuits Syst.,* 20, 1118–1131, 2001.
[65] S. Svilan, J.B. Burr, and G.L. Tyler, Effects of elevated temperature on tunable near-zero threshold CMOS, *ISLPED-01: ACM/IEEE International Symposium on Low Power Electronics and Design,* Huntington Beach, CA, August 2001, pp. 255–258.
[66] K. Flautner, D. Flynn, and M. Rives, A combined hardware–software approach for low-power SoCs: applying adaptive voltage scaling and the vertigo performance-setting algorithms, *DesignCon,* Paper SA2-3, Santa Clara, CA, January 2003.
[67] S. Martin, K. Flautner, T. Mudge, and D. Blaauw, Combined dynamic voltage scaling and adaptive body biasing for lower power microprocessors under dynamic workloads, *ICCAD-02: ACM/IEEE International Conference on Computer Aided Design,* San Jose, CA, November 2002, pp. 721–725.
[68] S. Shigematsu, S. Mutoh, Y. Matsuya, Y. Tanabe, and J. Yamada, A 1-V high-speed MTCMOS circuit scheme for power-down application circuits, *IEEE J. Solid State Circuits,* 32, 861–869, 1997.
[69] V. Zyuban and S. Kosonocky, low power integrated scan-retention mechanism, *ISLPED-02: ACM/IEEE International Symposium on Low Power Electronics and Design,* Monterey, CA, August 2002, pp. 98–102.
[70] K. Usami, M. Igarashi, F. Minami, T. Ishikawa, M. Kawakawa, M. Ichida, and K. Nogami, Automated low-power technique exploiting multiple supply voltages applied to media processor, *IEEE J. Solid State Circuits,* 33, 463–472, 1998.
[71] L. Wei, K. Roy, and V. De, Low power low voltage CMOS design techniques for deep submicron ICs, *VLSI-00: International Conference on VLSI Design,* January 2000, pp. 24–29.
[72] D.E. Lackey, S. Gould, T.R. Bednar, J. Cohn, and P.S. Zuchowski, Managing power and performance for system-on-chip designs using voltage islands, *ICCAD-02: ACM/IEEE International Conference on Computer Aided Design,* San Jose, CA, November 2002, pp. 195–202.
[73] F. Ishihara, F. Sheikh, and B. Nikolic, Level conversion for dual supply systems, *ISLPED-03: ACM/IEEE International Symposium on Low Power Electronics and Design,* Seoul, Korea, August 2003, pp. 164–167.
[74] A. Macii, L. Benini, and M. Poncino, *Memory Design Techniques for Low-Energy Embedded Systems,* Kluwer Academic Publishers, Dordrecht, 2002.
[75] P. Panda and N. Dutt, *Memory Issues in Embedded Systems-on-Chip Optimization and Exploration,* Kluwer Academic Publishers, Dordrecht, 1999.
[76] L. Benini, L. Macchiarulo, A. Macii, and M. Poncino, Layout-driven memory synthesis for embedded systems-on-chip, *IEEE Trans. VLSI Syst.,* 10, 96–105, 2002.
[77] M. Stan, and W.P. Burleson, Bus-invert coding for low-power I/O, *IEEE Trans. VLSI Syst.* 3, 49–58, 1995.
[78] Y. Shin, S.-I. Chae, and K. Choi, Partial bus-invert coding for power optimization of application-specific systems, *IEEE Trans. VLSI Syst.* 9, 377–383, 2001.
[79] U. Narayanan, K.-S. Chung, and K. Taewhan, Enhanced bus invert encodings for low-power, *ISCAS-02: IEEE International Symposium on Circuits and Systems,* Vol. 5, Scottsdale, AZ, May 2002, pp. 25–28.
[80] S. Ramprasad, N. Shanbhag, and I. Hajj, Signal coding for low power: fundamental limits and practical realizations, *IEEE Trans. Circuits Syst. II: Analog Digit. Signal Process.,* 46, 923–929, 1999.

[81] L. Benini, A. Macii, E. Macii, M. Poncino, and R. Scarsi, Architectures and synthesis algorithms for power-efficient bus interfaces, *IEEE Trans. Comput. Aided Des. Integr. Circuits Syst.*, 19, 969–980, 2000.

[82] M. Mamidipaka, D. Hirschberg, and N. Dutt, Low power address encoding using self-organizing lists, *ISLPED-01: ACM/IEEE International Symposium on Low Power Electronics and Design*, Huntington Beach, CA, August 2001, pp. 188–193.

[83] H. Mehta, R. M. Owens, and M. J. Irwin, Some issues in gray code addressing, *GLS-VLSI-96: IEEE/ACM Great Lakes Symposium on VLSI*, Ames, IA, March 1996, pp. 178–180.

[84] L. Benini, G. De Micheli, E. Macii, D. Sciuto, and C. Silvano, Asymptotic zero-transition activity encoding for address busses in low-power microprocessor-based systems, *GLS-VLSI-97: IEEE/ACM Great Lakes Symposium on VLSI*, Urbana-Champaign, IL, March 1997, pp. 77–82.

[85] Y. Aghaghiri, F. Fallah, and M. Pedram, Irredundant address bus encoding for low power, *ISLPED-01: ACM/IEEE International Symposium on Low Power Electronics and Design*, Huntington Beach, CA, August 2001, pp. 182–187.

[86] Y. Shin and T. Sakurai, Coupling-driven bus design for low-power application-specific systems, *DAC-38: ACM/IEEE Design Automation Conference*, Las Vegas, NV, June 2001, pp. 750–753.

[87] L. Macchiarulo, E. Macii, and M. Poncino, Low-energy encoding for deep-submicron address buses, *ISLPED-01: ACM/IEEE International Symposium on Low Power Electronics and Design*, Huntington Beach, CA, August 2001, pp. 176–181.

[88] J. Henkel and H. Lekatsas, A2BC: adaptive address bus coding for low-power deep sub-micron designs, *DAC-38: ACM/IEEE Design Automation Conference*, Las Vegas, NV, June 2001, pp. 744–749.

[89] L. Macchiarulo, E. Macii, and M. Poncino, Wire placement for crosstalk energy minimization in address buses, *DATE-02: IEEE Design Automation and Test in Europe*, Paris, France, March 2002, pp. 158–162.

[90] A. Vittal and M. Sadowska, Crosstalk reduction for VLSI, *IEEE Trans. Comput. Aided Des. Integr. Circuits Syst.*, 16, 290–298, 1997.

[91] H. Kaul, D. Sylvester, and D. Blauuw, Active shields: a new approach to shielding global wires, *GLS-VLSI-02: ACM/IEEE Great Lakes Symposium on VLSI*, New York, NY, March 2002, pp. 112–117.

[92] S. Salerno, E. Macii, and M. Poncino, Crosstalk energy reduction by temporal shielding, *ISCAS-04: IEEE International Symposium on Circuits and Systems*, Vol. 2, Vancouver, BC, Canada, May 2004, pp. 749–752.

[93] R. Arunachalam, E. Acar, and S. Nassif, Optimal shielding/spacing metrics for low power design, *ISVLSI-03: IEEE Annual Symposium on VLSI*, Tampa, FL, February 2003, pp. 167–172.

[94] S. Salerno, E. Macii, and M. Poncino, Combining wire swapping and spacing for low-power deep-submicron buses, *GLS-VLSI-03: IEEE/ACM Great Lakes Symposium on VLSI*, Washington, DC, April 2003, pp. 198–202.

14 Design Planning

Ralph H.J.M. Otten
Eindhoven University of Technology
Eindhoven, Netherlands

14.1 Introduction

In the past decade, *design automation* faced a sequence of what were called *closure problems.* "How to achieve wireability in placement of components or modules on a chip," and "how to allocate resources in order to optimize schedules" are among the early ones. In the 1990s timing closure was a dominant challenge; that is "how to ensure timing convergence with minimal size." Toward the end of that decade so many additional characteristics and constraints, such as (static and active) power, signal integrity, and electromagnetic compliance, had to be considered that industry started to speak of the ever elusive *design closure* (For an analysis of these characteristics and constraints, see the chapter "Design Closure", by John Cohn, and Peter Osler in volume II of this book). Yet, that is today's challenge of design automation: "how to specify a function to be implemented on a chip, feed it to an *EDA* tool, and get, without further interaction, a design that meets all requirements of functionality, speed, size, power, yield and other costs."

14.1.1 Wiring Closure

Around 1980, designers realized that the complexity of chips in the decade to come forced them to use more than just a router and occasionally a placer to find a starting point. At the same time, it was obvious that the two tasks were heavily dependent on each other. Routing without a placement was inconceivable, while at the same time a placement might render any routing infeasible. This "phase problem" between placement and routing was tackled with error-prone spacing repairs and wasteful reservations, which turned out to be inadequate. Thus the quest for *wiring closure* was born.

The answer came with the introduction of *floorplans,* data structures that capture relative positions of objects rather than co-ordinates. Once such a "topology" was generated and stored, optimizations could convert it into geometry, often a dissection of the rectangular chip into rectangular regions for the modules. The restriction to rectangular shapes was convenient because the optimization could be organized as trading off dimensions in order to reserve an adequate shape for each module while achieving an optimal overall contour. Under a mild structural constraint this required only polynomial time.

In a sense, floorplanning is simply a generalization of placement. Whereas placement was the manipulation of geometrically fixed objects to arrange them in a configuration without overlaps, floorplanning handled objects for which the shape did not have to be predetermined. It lent itself much better to top-down approaches where the shape and sometimes even the precise size of the modules were unknown. Certain restraints enabled *stepwise refinement* in that the tree representation of the floorplan was forced to be a refinement of a given hierarchy tree. These features caused a strong association of the floorplan concept with hierarchical design.

The salient feature of floorplans is that they allow designers to perform early analysis on their design decisions so that performance can be improved without resorting to lengthy iterations (in its original context floorplanning was presented as an iteration-free approach to chip design!). It thus enables *silicon virtual prototyping,* where sophisticated assessment procedures are integrated with hierarchical floorplan design to adapt modules in a timely manner to their (preliminary) environments, while respecting intellectual property reuse.

Today it can safely be stated that placement has been replaced by floorplanning, followed by a legalization step. Many modern back-end tools start with generating point configurations (with or without size-dependent spacing) or overlapping geometrical objects, mostly by analytic or stochastic optimizers that can handle complex but flat incidence structures with library elements of fixed shape. Once the optimum for this configuration is established, legalizers ensure that all overlap is removed, and that routability is enhanced. Ironically, this was exactly what early fully automatic floorplanners did [1]. Other back-end tools start by partitioning a netlist while giving relative positions to the blocks. Details about the algorithms can be found in the chapter "Digital Layout — Placement" by Andrew Kahng and Sherief Reda in volume II of this book.

In Section 14.2, we work from the definition of a floorplan to the many models and configurations that have served to capture relative positions in one way or another. We emphasize the importance of meaningful and easy to handle metrics for evaluating floorplans, although the in-depth treatment and comparison is left to other chapters in this book.

14.1.2 Timing Closure

With wiring as the main closure problem of the 1980s, timing closure became the target of the next decade. Two timing closure approaches emerged in the middle of the 1990s, both founded on the observation that gate delay can be kept constant under load variation by sizing [2]. Applying this rule to the whole design flat is one of them, and it has been commercialized successfully. There is however a flaw in this approach: as interconnects get longer, wire resistance cannot be neglected anymore. Unfortunately, keeping the delay constant by sizing is a model valid only when there is no resistance between the driving gate and the capacitive load(s). Line buffering is a patch that often helps, but cannot rescue the method. The other approach is by *wire planning.* It allocates delay to *global* interconnect first, then assigns budgets to modules, and applies the constant delay method to each of these modules. The assumptions of this method are that interconnect delay is linear with its length, and the size of the individual models is small compared to the whole design. The latter is used in two ways: interconnect delay within the modules cannot be reduced by buffer insertion, and modules can be treated as points during the global delay allocation.

If interconnect delay is proportional to interconnect length, path delay is insensitive to module position as long as detours are avoided. Interconnect linear with length is achieved by optimal segmentation of wires with buffers. To achieve minimal delay, buffers should be optimized in size. This fixes line impedance, thus giving up degrees of freedom to keep module delay within the budget. Adapting line impedance changes the delay per unit length. Fortunately, delay change by size variation around the optimum is small, and by minimizing area during the budget assignment, slack is created that can be used to compensate for additional interconnect delay.

What remains is to implement an algorithm for time budget assignment. This implies trading module size for speed: the smaller the module, the slower the signal propagation through it. If we know the relation between the two performance characteristics for every module, a minimization of the total area under timing constraints is easily formulated. Every path imposes a constraint on the budget assignment, and every budget assignment to a module implies a certain amount of area. The mathematical program obtained in this way can be efficiently solved if the trade-off function is convex, i.e., there is an algorithm with runtime polynomial in the size of the tableau.

There may however be exponentially many paths in a graph, and the size of the tableau could be exponential in the size of the graph. What is needed is a formulation that is polynomial in the size of the graph. Such a formulation will be presented in Section 14.3.3. Before that, we will review constant delay synthesis, and summarize the assumptions of wire planning. We conclude with a Section that shows how mild violations of these assumptions can be accommodated.

14.1.3 Design Closure

Up until 2000, the EDA industry has concentrated on tools and techniques for achieving closure with respect to one target design parameter. Moreover, these parameters were mainly related to that part of the trajectory that synthesized mask level data from an *rtl* description or even lower levels. After wiring and timing, power is likely to become the closure problem of this decade.

Today's chip synthesis requires tools and methodologies for manipulating designs at higher levels of abstraction to meet a large variety of performance constraints. What is needed is a formal system to handle trade-offs between performance characteristics over many levels of abstraction. Such a system will be indispensable to achieve *design closure*, in the sense alluded to at the beginning of this chapter. It should free design trajectories from the need for iterations and interactions to which users resort when their tools fail in one or more aspects to achieve closure, not knowing whether the process will converge or not, and in the latter case not even being sure whether a solution within specifications exists.

The approach hinges on a generalization of the concept of a performance characteristic, and a set of operations over such characteristics that extend and reduce performance spaces without ever losing candidates for optimal solutions with respect to any monotonic cost function. Of course, the key is the effectiveness in keeping the search space small, and the efficiency of the algorithms which do that.

After analyzing successes of the past in achieving closure, and pointing to the efforts of today, we formulate a number of observations that are the basis for the construction of such a formal system, an algebra for trade-offs, the topic of Section 14.4.

14.2 Floorplans

Formally, a floorplan is a data structure that captures the relative positions of objects which get their final shape by optimizing some objective without violating constraints.

This definition does not imply any underlying hierarchical structure. It simply generalizes the notion of placement: the manipulation of objects with fixed geometrical features in order to allocate them without any overlap in an enclosing plane region. The generalization refers to both the uncertainty or flexibility of the shape of some or all of the objects, and the possibility of overlap when these objects have a shape or area associated with them.

It is evident that hierarchical approaches that want to find regions for allocating subsets of modules almost always must resort to floorplanning in some way, because it is neither always wise nor often feasible to predetermine the shape of all modules in a hierarchy. This obvious need for floorplanning in approaches that want to preserve, completely or partly, properties of earlier partitioning, caused an almost complete identification of floorplanning with hierarchical layout design. The elegant formulation of stepwise refinement for layout synthesis, where preserving the hierarchical structure of a functional design occurs automatically by adoptation of a structural restraint that expanded, ordered and labeled the original hierarchy tree, planted this idea firmly in the minds of back-end tool developers of the late 1980s.

This is easily understood if we see that a supermodule's floorplan (that is the relative positions of its modules) was derived from:

- a preliminary shape of the supermodule
- a preliminary environment (e.g., pin positions)

- shape information for the modules (e.g., shape constraints)
- an incidence structure (i.e., a netlist)
  and possibly
- timing information of external signals
- module delay information
- path delay information.

The result was subsequently used in

- creating a preliminary environment for the supermodule's modules
- assessing wire space requirements
- estimating the delays.

All these are controlled by the (given or derived) hierarchy.

Figure 14.1 illustrates this, and essentially invites the recursive application of the central floorplan function. The terminology, the creation, and the use of module environments, and the explicit mention of hierarchy in the last item, seems to bind floorplanning exclusively to hierarchical design. However, a flat design with a single supermodule and a single level of leaf cells (undivided modules) is perfectly consistent, and was always the most important practical application of the floorplanning concept. It has rendered classical constructive placement obsolete, because of its flexibility and relative efficiency.

14.2.1 Models

In this section, we look at three classes of models for floorplans, not necessarily in chronological order of origin, but fairly representative of what has been conceived over three decades for capturing relative positions of modules.

14.2.1.1 Graphs

The oldest floorplans are graphs. Drawing a rectangle dissection already produces a graph, where line segments are interpreted as edges and the points where these come together (mostly *T-junctions,* sometimes *crossings)* as vertices. It should be noted that different dissections may lead to the same graph. Special properties that such a graph should have is that its edges can be divided into two classes, *h*- and *v-edges,* and it should possess a plane representation where these edges are represented by orthogonal line segments, one orientation per class. This graph is called the *floorplan graph* (Figure 14.2a).

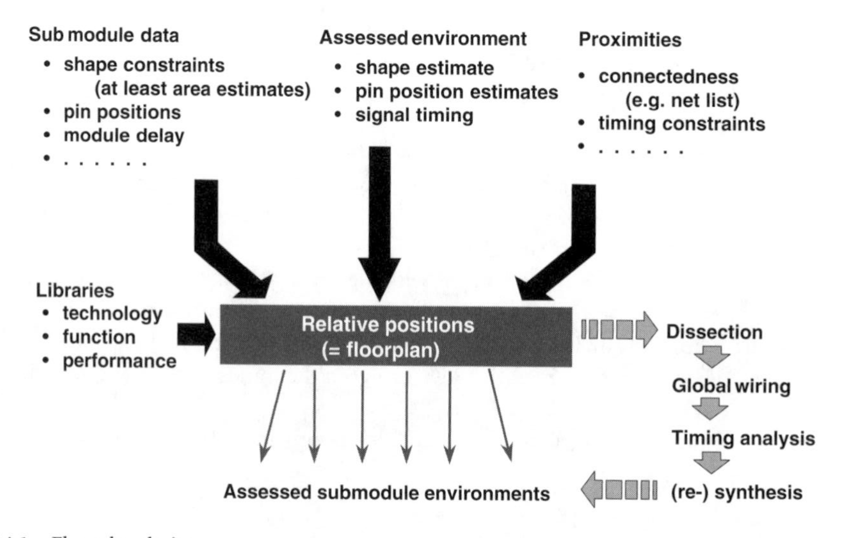

FIGURE 14.1 Floorplan design.

The first formal floorplanning procedure started with a graph in which the edges represented desired adjacencies. The vertices represent the objects, that is modules, but in the first applications, *rooms.* Obviously, not every such graph can be converted into a rectangle dissection. They should be planar, but that is not enough. In addition, they should have a dual that has a plane representation with the two orthogonal sets of line segments, as described above. Graphs with such a dual are called *Grason graphs* in honor of the first scientist to publish a floorplanning procedure [3]. This pioneer also gave a characterization of Grason graphs, but efficient algorithms for recognizing Grason graphs and constructing an associated rectangle dissection came some 15 years later [4–7].

On the basis of the division into h- and v-edges, we can also divide the edges of the Grason graph, yielding two graphs: the *Grason-h-graph* and the *Grason-v-graph.* By directing these graphs consistently with the relative positions of the rectangles of their vertices, we obtain the *Grason-h-digraph* (Figure 14.2b) and the *Grason-v-digraph.*

Equivalent to this digraph pair (or the colored Grason graph), and also to the colored floorplan graph (i.e., with explicitly given h- and v-edgesets) is the *channel digraph* (Figure 14.2d) where each channel (i.e., maximal line segment) has a vertex and each T-junction has an arc pointing to the "bar of the T" [8]. This graph has rarely been used though it has some characteristic properties and retains generality, where the much more popular *polar digraphs* are not capable of capturing all adjacencies.

Although known already in the first half of the last century, these polar graphs were introduced [9] into layout synthesis in 1970 by Tatsuo Ohtsuki. The *polar h-digraph* (Figure 14.2c) takes the maximal horizontal

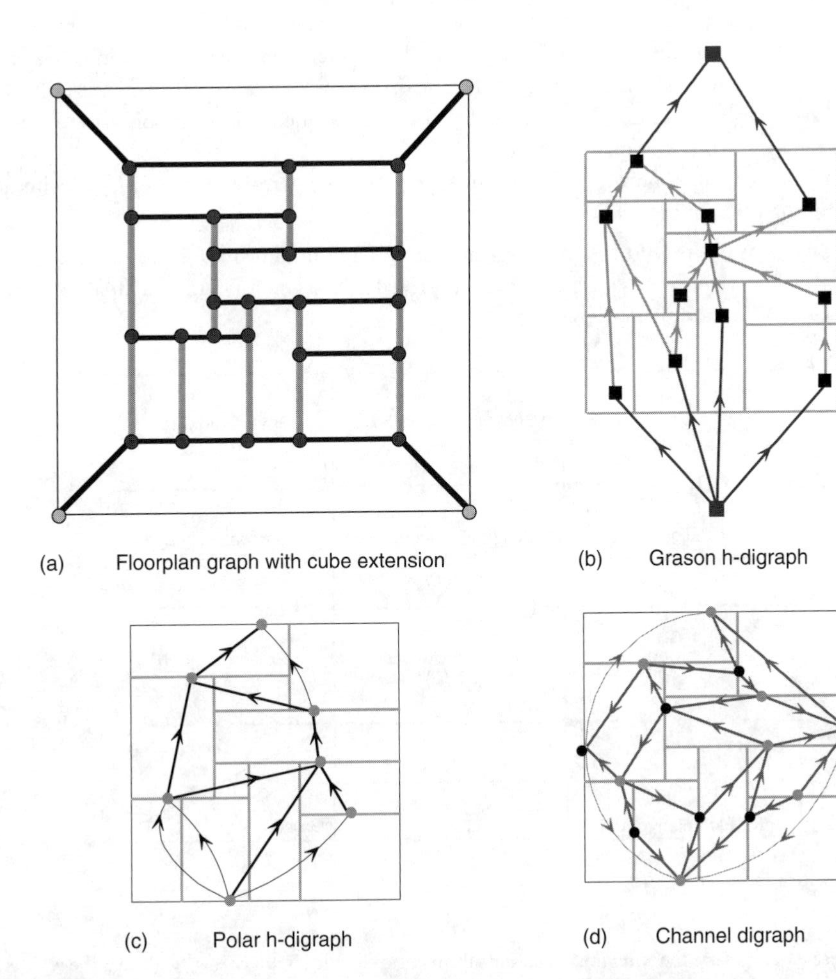

FIGURE 14.2 Graphs.

line segments as its vertices, and has an arc for every rectangle. It does capture incidence of rectangles to line segments, but adjacency between rectangles is not exactly covered. Consequently, the *Grason h-digraph* is a subgraph of the line graph of the *polar h-digraph.*

The relations between the graphs acting as floorplans for the same rectangle dissection are brought together in the table of Figure 14.3, where "D" stands for dualization and "L" for forming the line digraph. The upward arrow toward the polar digraphs and superset signs indicate the loss of information when going to polar graphs. Polar graphs are nevertheless the most popular graph models because they can be transformed into sets of linear equalities that must be satisfied by the rectangle dimensions of any compatible dissection*. This forms a good starting point for sizing using mathematical programming.

14.2.1.2 Point Configurations

Plane point sets are by now the most commonly used floorplans. If the only information that is used is the sequence of the points (representing the modules) after projecting them on an axis, then it is better to speak of a sequence pair, because all distance information is left unused.

Methods that try to embed a distance space into a euclidean space, and finally into a two-dimensional euclidean space often use eigenvalues to find the plane with maximum spread and to lose as little distance information as possible when projecting on that plane. Depending on the distance metric certain statements about optimality can be made. However, the euclidean distance is not really a useful metric and projection may force modules apart that should and can be kept together. Eigenvalue methods are good mainly for recovering structure in low-dimensional configurations.

With the rise of annealing as a competitive method for improving complex configurations, it was soon considered for floorplan design [10]. The question was how to represent relative positions in such a way that total wiring length was calculated very quickly. The first answer was a pair of sequences, which meant that the moves were transpositions of two elements in one such sequence, but the points were then spaced according to their area (not the square root of area!).

Sequence pairs obtained a more specific meaning when the two permutations of modules implied the following meaning for relative positions:

- A module m_i is to the left of module m_j if it precedes that module in both sequences.
- A module m_i is above a module m_j if it precedes that module in the first sequence, and if it follows that module in the second sequence.

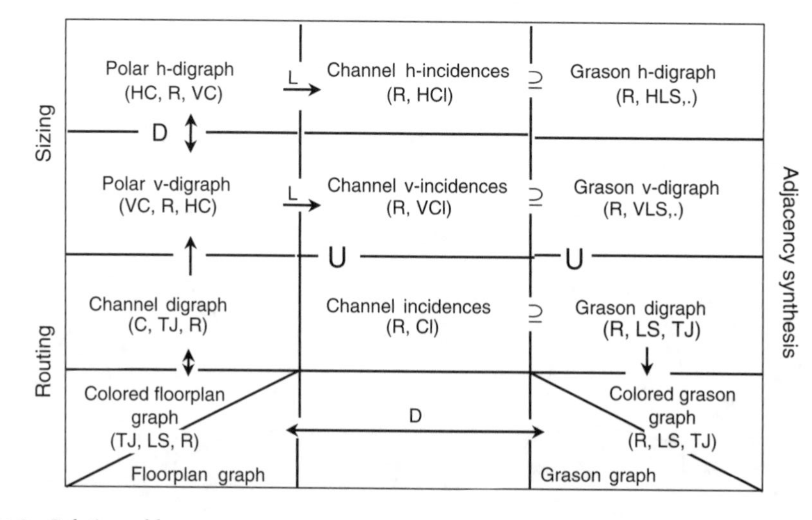

FIGURE 14.3 Relation table.

*In fact, these equalities are the Kirchhoff equations from basic electric circuit theory, with e.g., the widths as current intensities, and the heights as voltage differences between the nodes.

Till now sequence pairs are only generated by stochastic algorithms such as annealing. Evaluations are time consuming, although polynomial or even linear, which causes long runtimes, because these evaluations are placed in the inner loop. This makes such algorithms impractical. In favor of sequence pairs, it should be noted that they can easily handle empty rooms without explicitly encoding them in the sequences. This is not possible, for example, with polar graphs!

14.2.1.3 Trees

Trees for representing rectangle configurations have become popular recently, as shown by the appearance of *O-*, *bi-*, *star* and *stair trees.* These representations compete in their efficiency to solve the packing problem for their compatible rectangle configurations. These packing problems are easy to solve for rectangles with fixed shape and orientation. An example is the *labeled ordered tree* (Figure 14.4) from which a packing can be computed in linear time [11]. Among all possible trees with the adequate number of vertices there is one that produces the smallest possible packing. There are however very many possible trees though their number is bounded by the *Catalan number* times the number of permutations, and this number is asymptotically smaller than $(n!)^2$, the number of sequence pairs.

14.2.2 Design

In many ways, one can argue that floorplan design has not been solved adequately. The various algorithms for generating floorplans modeled by graphs are either confined to subspaces that do not have to contain optimum or even near-optimum plans, or produce in essence a number of configurations that cannot be bounded by a polynomial in the number of modules.

14.2.2.1 Adjacency-Based Floorplan Design

The first thoughts on floorplan design were directed toward realizing the required or desired adjacencies. These adjacencies form a graph, and if this graph is a Grason graph, a rectangle dissection with these adjacencies exists. Linear-time algorithms[†] for testing graphs for this property exist. In general, these adjacencies do not form a Grason graph, and seldom a subgraph of a Grason graph. The adjacency requirements therefore have to be pruned to such a subgraph and subsequently extended to a suitable Grason graph, either explicitly or implicitly. A generic procedure is given in Figure 14.5.

Graphs with rectangular duals (even graphs with duals, when combinatorially defined) have to be planar. Procedures that try to realize adjacencies therefore often start with heuristics for planarizing the

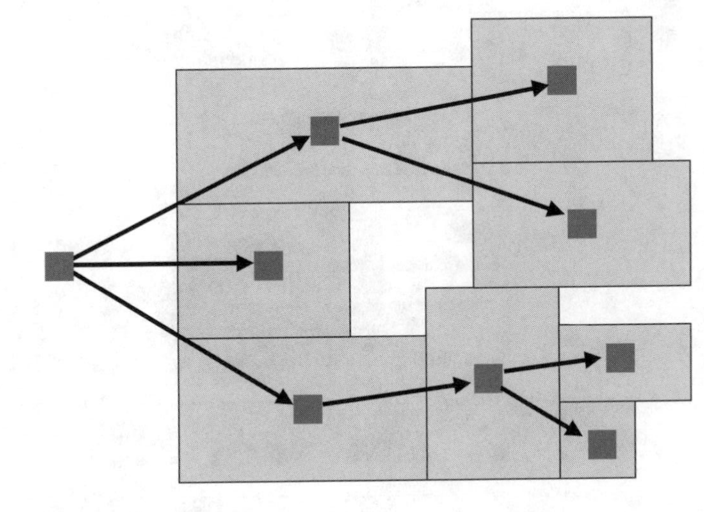

FIGURE 14.4 Labeled ordered tree.

[†]Do not trust the textbooks on this! Consult the original articles.

corresponding graph. A straightforward way for doing that is by removing "weak" adjacencies. Removing enough edges will planarize a graph sooner or later. But adding nodes also can have this effect. These "phony" rectangles (because nodes are supposed to represent rectangles) are called wiring nodes in [12]. Other heuristics can be found, but after the first stage in Figure 14.5 the graph is supposed to be planar, so it has a dual and of course a plane representation. But does it have a rectangular dual? Not for a representation with *complex triangles* - that is, representations with circuits of three edges with other edges inside. If that is the case there are a number of options:

- modify the representation
- remove the complex triangles
- triangulate avoiding complex triangles.

In [13], all options are exploited although in the reverse sequence. The remaining freedom can be used for assigning the corners.

The resulting graph has a rectangular dual, and therefore a corresponding rectangle dissection can be found. A constructive procedure for generating such a dissection in an efficient way was published in [5]. An elegant method, based on the observation that construction is in essence assigning T-junctions, was *complete matching* in [14].

A flow such as in Figure 14.5 does not consider shape requirements (not even size requirements) in its floorplan generation. Shape constraints are taken into account only afterward in a floorplan optimization step, mostly a tableau of inequalities holding the floorplan (e.g., by the equations derived from polar graphs) and the shape constraints. This often leads to unacceptable results, the main reason why pure adjacency-based floor-planning was never successful. The approach in the next section is much better equipped to take shapes into account.

14.2.2.2 Proximity-Based Floorplan Design

Point configurations can be generated efficiently depending on the objective. Eigenvalue methods are very fast but are bound to use a euclidean metric. Besides, they usually use projection to lower the

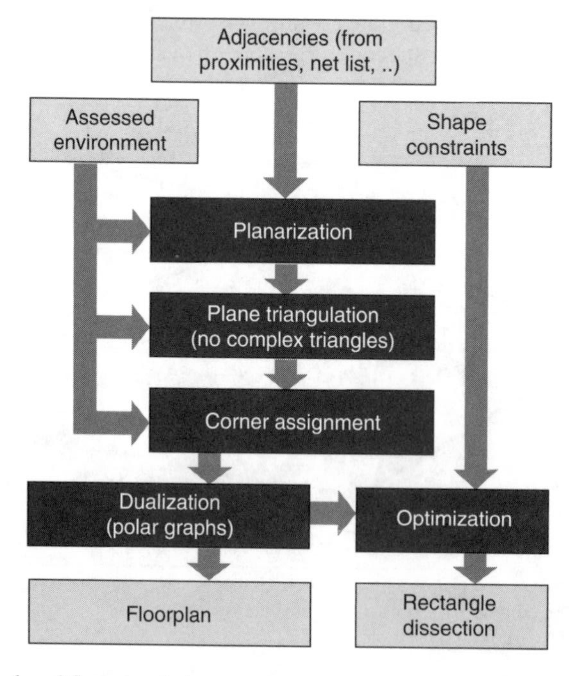

FIGURE 14.5 Adjacency-based floorplan design.

dimensionality of the solution, which may reduce distances, leaving other modules unnecessarily far apart. Repeated probing based on eigenvalues makes the generation slow.

Probabilistic techniques, such as annealing, genetic algorithms, and evolution, need very fast evaluations, which makes them resort to crude measures. For fixed rectangles, techniques such as in Section 14.2.1.3 can be of advantage, but they cannot handle more flexible shape constraints.

Currently the best results seem to be obtained with analytic optimizers of the quadratically convergent Newton-type, followed by legalizers. They are usually applied to flat designs, that is designs without any hierarchical structure, regardless of the number of objects to be treated. Since these analytic optimizers do not produce nonoverlapping configurations of the objects to be allocated, they are classical examples of floorplanners, although the literature often calls them (global) placement procedures, which is only justified when combined with a valid legalizer.

In any way, critical for floorplanners is the metric by which the results are evaluated. In the chapter "Digital Layout - Placement", by Andrew Kahng and Sherief Reda, this issue will be addressed.

14.2.3 Constrained Floorplans

Much more can be said about floorplanning, especially about the misconceptions and misinterpretations surrounding it. A particular topic that attracts many scientists (but few tool developers) is the avoidance of special constraints. Early in its history the restraint of preserving sliceability was recommended [1]. It was exactly that constraint that made floorplanning fit as a glove with stepwise refinement from a given hierarchy. Floorplans with that property could be represented as a tree, but there were other and more important properties. The fact that slicing structures have the minimum number of "channels" may have had its day as well as the conflict-free routing sequence of those channels. But the fact that many algorithms that are at least *NP*-hard for general floorplan structures become tractable for slicing structures, remains of paramount importance. Among these algorithms is of course the floorplan optimization problem [15]. Later it was proven by Stockmeyer [16] that even the simple special case of fixed module geometries and free orientations was *NP*-hard! Given a slicing tree and the more general shape constraints defined in [17] makes floorplan optimization efficiently solvable. Also, labeling a given tree to find the optimum slicing tree with that topology can be done in polynomial time [18]. And last but not the least, given a point configuration (but actually a pair of sequences), the optimal compatible slicing structure can be found in polynomial time. "Compatible" means that every slice in the floorplan corresponds exactly with a rectangular set in the point configuration [19].

Considering the place and the nature of floorplanning capable of handling uncertainties, flexibility, and preliminary positions, there is not much sense in paying any price for also including nonslicing floorplan topologies [20, 21].

14.3 Wireplans

A wireplan is an incidence structure of modules and global wires. A global wire is one that can be sped up by buffer insertion. The latter implies that its (minimum) length and layer are known. For the modules, a trade-off between speed and size should be available. If timing constraints are given for the inputs and outputs, time budgets can be assigned to the modules so that the total size is minimal. The modules can then be designed using the concept of constant delay so that they do not exceed their budget.

14.3.1 Constant Delay

14.3.1.1 Delay Models

In [2], the observation was made that sizing a gate proportional to its load keeps the gate delay constant. This confirms the delay model introduced in [22] that reads as

$$\tau = g\,\frac{C_{\mathrm{L}}}{C_{\mathrm{in}}} + p$$

where p is the inherent parasitic delay and g the computational effort. The latter depends on the function, the gate topology, and the relative dimensioning of the transistors. Both p and g are size-independent, i.e., increasing the gate size with an arbitrary factor does not affect these parameters. The quotient C_L/C_{in} is called the *restoring effort* of the gate.

14.3.1.2 Sizing formulation

Constant delay synthesis starts from a network of gates where each gate has a fixed delay assigned to it. In order to realize this delay, its *restoring effort* has to have the appropriate value. Satisfying these requirements for all gates in the network simultaneously leads to a set of linear equations, as can be seen from Figure 14.6.

Using the notations in that figure the set can be written as

$$\mathbf{c} = \mathbf{q} + \mathrm{N}\mathbf{f}^{\mathrm{D}}{}_{\mathbf{C}} \quad \text{or} \quad (\mathrm{I} - \mathrm{N}\mathbf{f}^{\mathrm{D}})\,\mathbf{c} = \mathbf{q} \tag{14.1}$$

The algebraic and numerical aspects of this set have been analyzed in [23] where a detailed derivation can be found as well.

The sizing procedure is summarized in Figure 14.7. It shows that if synthesis produces, beside a netlist, also a vector **f** of reciprocal values of restoring efforts (thereby fixing the delay of each gate), solving set Equation (14.1) yields the vector **c** of input capacitances (and implicitly the gate sizes). Placement of these sized gates may reveal changes in the wire capacitances, and consequently in the vector **q**. Iterating the calculation converges to a set of sizes and imposed capacitances consistent with the placement.

14.3.1.3 Limits

The gate model of Section 14.3.1.1, which underlies the procedure in Section 14.3.1.2, assumes negligible resistance between the gate and its load. For relatively short wires in the network this assumption is certainly acceptable. Modules of up to 100,000 gates can be safely dimensioned in this way. For larger modules wires may get so long that wire resistance is no longer negligible. In today's chips, it is even said that interconnect dominates performance. It is not possible to derive an equally simple and numerically robust iteration method when resistance has to be accounted for. Moreover, improving speed by sizing does not work when there is line resistance. Line segmentation and buffering is effective up to a certain level. This limits the validity of constant delay synthesis (Figure 14.8).

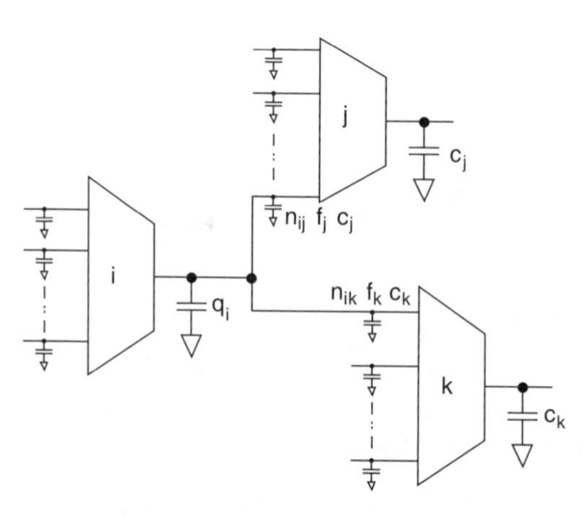

FIGURE 14.6 Composing the load capacitance of gate i : it consists of an "imposed" capacitance q_i, and the input capacitances in its fanout, which are proportional to the respective load capacitances.

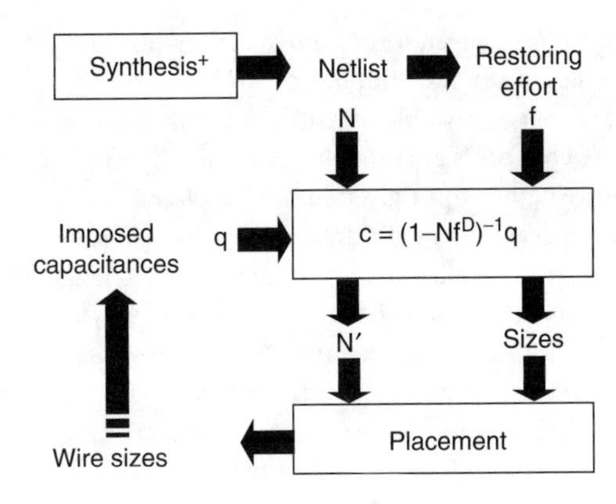

FIGURE 14.7 Computational procedure for sizing the gates.

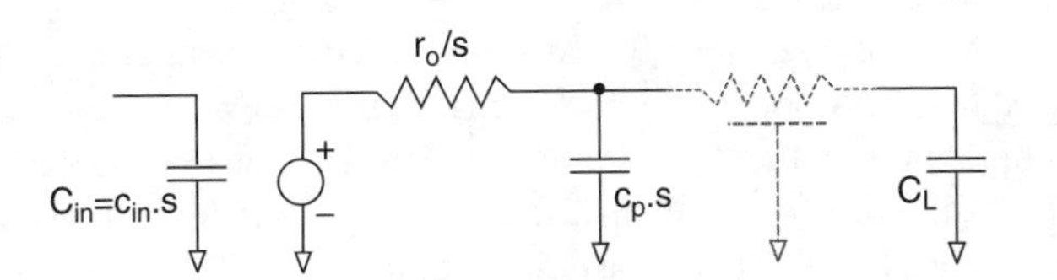

FIGURE 14.8 Wire resistance, here de-emphasized, is neglected in the model underlying the constant delay synthesis.

14.3.2 Wire Planning

14.3.2.1 Problem Statement

Larger networks will have wires with nonnegligible wire resistance; if some wires have to connect widely separated points this can only be avoided in very regular designs by only-neighbor connections. To extend the methodology of the previous section to these designs, such wires have to be treated appropriately. Not uncommon in complex designs is dividing the whole into parts that can be handled by the sizing procedure. Such a partitioning often comes naturally in functional design. The time plan can then identify connections that allow for (or have to have) delays that exceed a critical value. If this value is taken to be the delay of a segment in an optimally buffered interconnection, the thus identified interconnect can be characterized as wires with a length that enables delay reduction by segmentation and buffering. It has been shown that the critical value only depends on the chosen technology and not, for example, on the layer to which the wire is going to be assigned [23]‡.

The approach is to allocate the delay of such identified interconnect first, and then divide the remaining time budget over the modules. These modules do not contain such interconnect and can be sized with the procedure described in Section 14.3.1.2, and thus be held within the budget assigned to that module. The problem to be solved is therefore: given a network of modules connected by *global* interconnect and timing requirements on paths in that network, assign size and delay to each module so that a network can be produced that satisfies the timing requirements. Of course, the relation between size and delay of each module has to be known. This relation will be discussed in Section 14.3.2.3. But first, possible properties of global interconnect will be exploited.

14.3.2.2 Wire Delay

Optimally segmented wires have a delay linear in the length [23]. And if wire delay is linear with its length, its contribution to the path delay between two points on a chip does not depend on the position

‡The delay of a segment in an optimally buffered line is indeed a process constant while the length of a segment-depends on the layer. That length does not depend on the buffer sizes!

of the modules on this path. Even stronger, if the path is detour-free, the wire delay on a path is fixed: it only depends on the coordinates of the two endpoints.

Of course, it may in general not be possible to realize the paths of a network in a detour-free manner when the coordinates of these endpoints are given.[§] An easy net-by-net criterion has been formulated in [23], answering the question whether for a given endpoint placement, a detour-free realization exists. Even a point placement can be derived straightforwardly if that criterion is tested (and satisfied) for all nets. The given endpoint placement is called *monotonic* in that case. Also it has been demonstrated there that every acyclic functional network has a realization that makes every endpoint placement monotonic. For logic networks a synthesis procedure has been implemented that preserves monotonicity [25].

Here we assume that an acyclic network is given with the coordinates of its endpoints and that it can be realized without path detours. In Section 14.3.4, we discuss what can be done if detours cannot be avoided and extra wire delay is incurred. The endpoints can be the location of connections to the outside, entries and exits of a supermodule containing all modules in the network, the positions of preplaced registers, etc.

14.3.2.3 Module Size and Delay

The relation between the size and delay of a module is essentially a trade-off: speeding up a module is usually paid for with additional area. Plotted in an area versus delay quadrant, the realization values will be in a region bounded below for both performance characteristics by a monotonically decreasing function with the points of interest on that function. In practice, that function can be well approximated by a convex curve. In [26] this curve is approximated by finding a best fit for $(delay - c)\ (area + b) = a$ through the identified *Pareto* points among the design points.

14.3.3 Time Budgetting

14.3.3.1 Problem Formulation

Assume that a wireplan is given with a monotonic placement of the endpoints (in the sense of Section 14.3.2.2). This means that there exists a point placement for its modules enabling simultaneous detour-free routing of all paths between inputs and outputs. The Manhattan or L_l length of any input–output path ($L_{i,o}$) can therefore be made equal to the L_l distance between the input and output terminal pins of that path. The wire delay on such a "monotonic" path is proportional to this length, regardless of the position of the modules on this path, because the delay of optimally buffered wires is linear with length, and the summed delay of the wire segments between the modules is the summed length of the segments multiplied by some constant. The main conclusion is that we can calculate the wire delay of a monotonic path directly from the input and output pin positions. Because module positions are not needed to do that delay calculation, it is possible to do time budgeting even before relative module positions are known.

To obtain the total path delay, the delay of its modules has to be added to the wire delay. That delay depends on the implementation of the modules. Let P^{jk} be the set of all paths from input j to output k; in a given wire plan and p a single path in this set. Then the constraints imposed by the timing requirements read as follows:

$$\forall_{j,k}\forall_{p\in P_{jk}}\left[D_{W_{jk}}+\sum_{m\in p}D_m(A_m)\leq \mathrm{T}_{\mathrm{req}_{jk}}\right]$$

These constraints simply state that for each single path between two pins, the wire delay plus the summed module delay should not exceed a given timing constraint. $T_{\mathrm{req}_{jk}}$ is an upper bound on the delay between pins j and k. $D_{W_{jk}}$ is the wire delay on the path, and thus equal to L_l distance between both pins multiplied by a (technology/layer dependent) constant. The minimum (Pareto) delay of a module m is approximated by $D_m(A_m)$, a function of its size (area) A_m.

[§] In practice, the overwhelming majority of two-point routes are without detours as experiments in [24] show: only 1.37% were detoured!

The first question that can be posed is whether the wire plan can be realized in a footprint of a given size, while satisfying the timing requirements. It can be answered by minimizing total area, ΣA_i, subject to the given constraints. If the obtained minimal area is larger than the area of the footprint, the wire plan does not fit. Otherwise, not only is the answer to the question obtained, but also the available "slack" area that can be used to adapt modules to the wire characteristics.

The method to use for solving the above optimization problem depends on the delay model, that is on the module delay as a function of module area. Linearizing the D_m curves, and using linear program solvers yields a method which can be polynomial in the size of the program, but the linearizations need to be sufficiently coarse. Other time budgeting techniques have appeared in the literature. Among the more popular techniques, we have to count the *zero-slack algorithm* of Nair et al. [27] and its variants, the *knapsack of approach* of Karkowski [28] and the convex optimization of Sapatnekar, et al. [29].

This optimization problem can be solved by *geometric program* solvers if both object function and constraints have the form $f(t) = \Sigma c_j \Pi t_i^{aij}$. Choosing for the module delay

$$D(s) = d/s + r.$$

achieves that. The function captures the essence of common area-delay trade-offs. D is the delay as a function of a size factor s, making the area of a module $A(s) = a_0 s$. For each module d, r and a_0 are given. Now, the resulting program for n modules is

$$\text{minimize} \sum_{i=1}^{n} a_{0i} s_i$$

with constraints of the following form:

$$\sum_{j \in p} \left[d_j s_j^{-1} + r_j \right] + D_{Wjk} \leq T_{\text{req}_{jk}}$$

A certain class of geometric programs (including the example above) has *posynomial* functions for both object function and constraints $C_j > 0$. By applying the transformation $t_i = e^{xi}$, both object function and constraints of the posynomial formulation become convex, and can be solved in polynomial time. Further, general convex solvers can be used. The number of constraints depends on the number of paths. Unfortunately, the latter is exponential in the number of edges, resulting in a program of unmanageable size.

14.3.3.2 Complexity Reduction

The number of *constraints* is determined by the total number of input–output paths. The number of terms in a constraint is the number of modules in the associated path. Note that the number of terms is far greater than the number of variables (module sizes), since modules may be on several paths. The number of terms is of interest here, and renders the straightforward formulation impractical already for relatively small designs.

It is easy to see that the number of paths in a graph can be exponentional with regard to the number of edges, but there may be some hope that in practical designs the number of paths is small. This is not the case, and an approach based on considering all paths is out of the question.

However, all the necessary information can be captured by maintaining a number of variables at each node. The delay from a primary input i to an internal node n depends on the *maximum* delay of the paths from i to n. If s of those subpaths exists, s constraints appear for each subpath from n to an output. Since only maximum delays count, only the delay of the slowest (sub)path is of interest. To use that observation, we introduce at each node, for each primary input in its fanin cone, a variable representing the maximum module delay on the subpath from the input to this node. Using the fact that we have a directed, acyclic graph, this value can easily be calculated as the sum of the delay of this node and the *maximum* of the values of its predecessors.

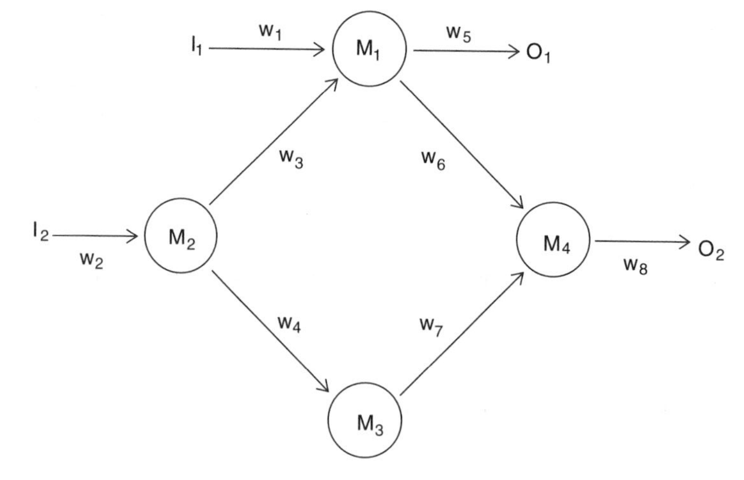

FIGURE 14.9 A circuit of modules.

For example, in Figure 14.9, there are two paths between I_2 and O_2 . With w_x denoting the wire delay of segment x, and M_y the module delay of module y, we find (in the straight forward formulation) as constraints for these path

$$\omega_2 + M_2 + \omega_3 + M_1 + \omega_6 + M_4 + \omega_8 \leq T_{\mathrm{req},I2O2}$$
$$\omega_2 + M_2 + \omega_4 + M_3 + \omega_7 + M_4 + \omega_8 \leq T_{\mathrm{req},I2O2}$$

Using monotonicity and linear wire delay

$$\omega_2 + \omega_3 + \omega_6 + \omega_8 = \omega_2 + \omega_4 + \omega_7 + \omega_8 = Dw_{I2,O2}$$

we obtain the form of section 14.3.3.1

$$M_2 + M_1 + M_4 + Dw_{I2,O2} \leq T_{req,I2O2}$$
$$M_2 + M_3 + M_4 + Dw_{I2,O2} \leq T_{req,I2O2}$$

In the formulation of this section, introducing the variable AT_n^I to denote the maximum module delay from input I to node n, we find

$$AT_{M_4}^{I_2} + Dw_{I2,O2} \leq T_{\mathrm{req}} I_2 O_2$$

$$AT_{M_4}^{I_2} = M_4 + max\,(AT_{M_4}^{I_2}, AT_{M_3}^{I_2})$$

$$AT_{M_3}^{I_2} = M_3 + max\,(AT_{M_2}^{I_2})$$

$$\vdots$$

$$AT_{M_2}^{I_2} = M_2 + 0$$

Although the savings in efficiency are not noticeable in such a small example, it is obvious that they are tremendous in large examples. In fact, the number of constraints becomes equal to the number of nodes plus the product of the number of edges and the number of primary inputs. Each node has for each primary input in the fanin cone a variable representing the module delay up to here. The number of these variables is bound by the number of primary inputs. For each edge, a constraint is generated for each of those variables at the source node, linking the delay of a node with the delay of its successor. For each node, an additional constraint representing its own delay is generated. Both kinds of constraints are simple and have only two terms.

This shift from paths to nodes makes the *timing graph* from [30] a more convenient model. It is in essence the line digraph of the original: the arrows become nodes, and two arrows that have a head–tail connection at a module node yield an arrow in the timing graph. The conversion is illustrated in Figure 14.10.

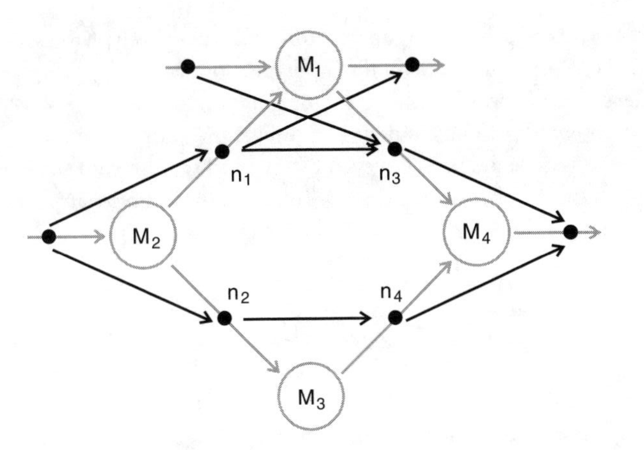

FIGURE 14.10 The line graph of the digraph in Figure 14.9.

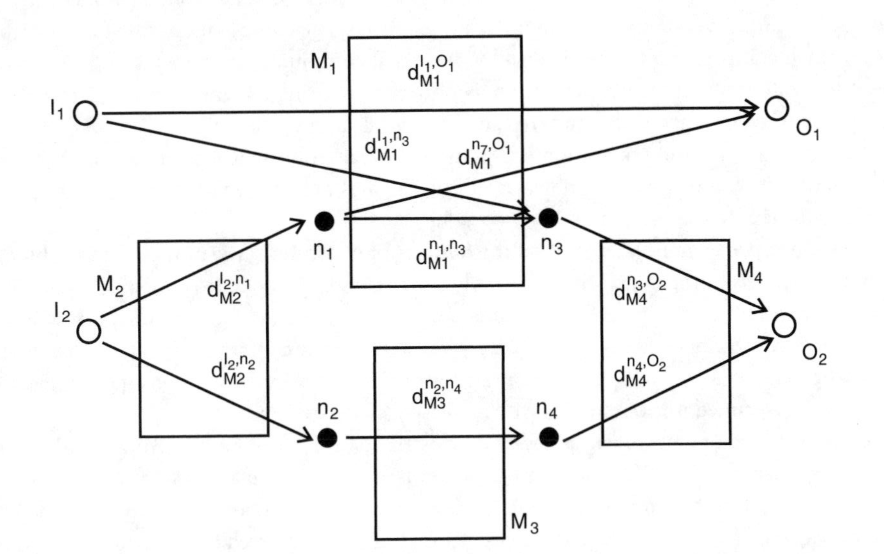

FIGURE 14.11 The timing graph of the digraph in Figure 14.9.

In this model, we can distinguish different delays between different module pin pairs and associate them with the corresponding arrow. Figure 14.11 shows the timing graph that corresponds with the wire plan of Figure 14.9.

As an alternative to [30], we have for each node as many variables AT as there are primary inputs in its input cone. In Figure 14.11, node n_3 has two variables associated, $AT^{I_2}_{n_3}$ and $AT^{I_2}_{n_3}$. In this way, all input–output delays can be taken into account. The complexity of the problem is now, as is the case with static timing analysis, polynomial in the size of the graph.

The problem formulation after this complexity reduction becomes

$$\text{minimize} \sum_M A_M$$

subject to

$$\forall I \in C(m)\ \forall (m,n) \in E \left[AT^I_n \geq AT^I_m + d^{mn}_M(A_M) \right] \quad \text{(E)} \tag{14.2}$$

$$\forall I \in PI\ \forall o \in Po \left[AT^I_n \leq T_{req,IO} - Dw_{IO} \right]$$

where *PI* and *PO* are the sets of primary inputs and outputs, respectively, *E* the set of edges in the timing graph, and *C(m)* the set of primary inputs in the input cone of *m*.

Moving from the straightforward path-based formulation of Section 14.3.3.1 to the nodebased approach of this section enables an efficient area optimization under timing constraints. But there are some more techniques to reduce the problem even further and save on runtime.

A simple reduction in the number of constraints is obtained by not restricting the formulation to variables derived from looking at maximum delays from the inputs. The first set of constraints in (E) contains

$$\sum_m \gamma(m)|C(m)|$$

constraints, g(m) being the outdegree of node m. If the generation of constraints is done with reference to the primary outputs, the total number of constraints is determined by output cones and indegrees. This may yield a quite different and sometimes considerably lower number of constraints. Although a graph may be analyzed for this comparison in a straightforward manner, the tableau of constraints may be generated for both cases after which the smaller one is used. The computation time for generating such a tableau is small compared to the time taken by the actual computation.

A more intricate improvement in efficiency is obtained by using a technique called *pruning* [31]. It reduces problem size, degeneracy, and redundancy without sacrificing accuracy. As illustrated by Figure.14.12, the basic pruning operation is a graph transformation that replaces a node with touching arcs, and replaces it with other arcs. The variables on the arcs are such that the associated optimization problem is equivalent to the original one.

Generally, solver performance depends on the number of constraints, the number of variables, and the total number of terms in the constraints. Pruning affects these numbers. It is possible to assign to each node a *gain*, a measure of the benefit in case this node was pruned, taking these effects into account. If, for example, two constraints with four variables are replaced by one constraint with six variables, this may or may not be beneficial, depending on the used solver. This is reflected by a positive or negative gain. Nodes are greedily pruned until no nodes with positive gains exist anymore.

In [31], only the numbers of variables and constraints are taken into account, and given equal weights, but one can also consider the number of terms, and tailor the associated weights to the solver used. In the original formulation, only one *AT* variable resides at each node, while one for each primary input may also exist. Therefore, the pruning procedure has to be adjusted to calculate gains for *AT variables* rather than nodes. In this way, an *AT* at node *n* may be pruned for primary input I_1, but not for primary input I_2. Another way of looking at this is that *for each primary input*, a timing graph is being constructed, consisting of the input and its fanout cone, and to which the original pruning procedure can be applied. Then, the resulting optimization program is simply the sum of object functions and the concatenation of the associated constraints. The pruning procedure requires only one graph traversal, and results not only in a dramatically more compact formulation, but also one that is numerically better conditioned.

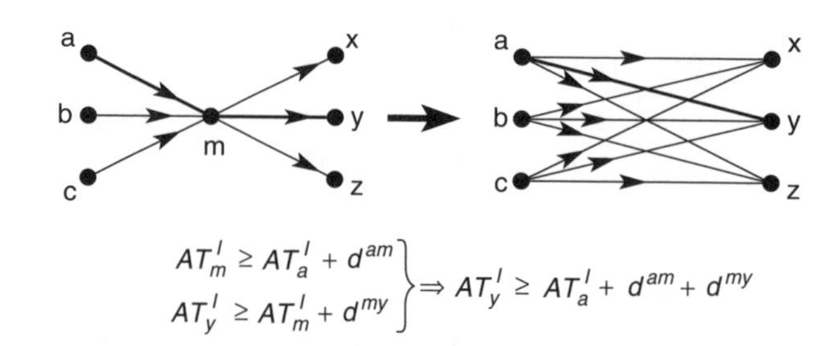

FIGURE 14.12 The basic pruning operation with constraints for the bold path.

14.3.4 Robustness

The time budgets are based on assumptions concerning path lengths, monotonicity, and trade-offs. In real life, however, some of these assumptions may be hard to meet later on: one may, for instance, sometimes need to deviate from monotonicity. Another uncertainty is the accuracy of the trade-off models. More generally, one can say that in a noniterative design flow, one needs a certain amount of "slack separation"[¶] in order to obtain "robustness" with respect to uncertainties later on. Fortunately, slack on the majority of paths can be introduced at very low cost.

First, the formulation itself ensures a certain amount of slack: wire delay is calculated as the path length times some constant. When the modules are realized, however, they will occupy space, effectively reducing wire length, and thus delay. Second, the formulation ensures that all input–output pairs will have a critical path: a path whose delay equals the constraint. This implies that every module is on a critical path. It does not however imply that every wire segment is. This offers the possibility of detouring these wires. Third, if simple bounds on area are used, this may also result in slack. Finally, wire delay is calculated for ideally buffered wires. If logic is pulled out of the module, and spread over the wires, buffers may be replaced by "useful" gates. Therefore, wire delay estimation is kept conservative.

If the slack inherent in the formulation is still not sufficient, Westra et al. [26] introduces a way to provide additional slack at little area cost. It also has the advantage that *truely critical paths* that limit circuit performance most are revealed, making it clear where the main thrust of design effort should focus.

14.4 A Formal System For Trade-Offs

It is interesting to look back on how wiring closure was achieved, though keeping in mind that in those days wiring was realized in two or three layers and allocated to channels between modules. Critical was the generalization of placement, concerned with the arrangement of geometrically fixed objects without overlap in the plane (possible bounded or preferably small) into floorplan design, where only relative positions of objects are being fixed. Accepting slicing as a structural restraint then guaranteed sufficient wiring space that could be treated uniformly with the modules during sizing, and yielded a conflict-free sequence of channel routing problems. In addition, it rendered several optimizations tractable, including sizing which was in essence a trade-off between the dimensions of height and width. These dimension pairs are partially ordered by a *dominance relation:* one pair dominates another if neither of its dimensions is larger than the corresponding dimension of the other pair.

Similar observations can be made with respect to timing closure. The essential shift there is from sizing with speed as an arbitrary outcome to timing constraints with size beyond control. For smaller modules the concept of constant delay enabled closure. However, the underlying assumption that the delay of a gate can be kept constant by varying its size linearly with its load is only valid when resistance between the gate and its load can be neglected. This required planning of all wires for which resistance could not be neglected. They take a considerable but constant part of the timing budget leaving the remainder to be divided over the modules. Again this comes down to trade-offs, now between size and speed, and a corresponding *dominance relation.*

After timing, the primary concern for many of today's *SoC* devices is power consumption, which if appropriate paradigm shifts can be found, will lead to similar trade-offs between speed and power. The problem however is that performance characteristics can no longer be dealt with by considering them in isolated pairs or triples. What is needed is a formal system that allows us to manipulate a design without losing potentially optimal final solutions. Such a system is the topic of this section, but before introducing it we observe the following.

[¶] *Negative slack* means constraints are violated, *(positive) slack* means there is room to tighten a constraint, and *slack separation* is the difference between the most critical path and other paths.

Closure has up to now been achieved by Pareto-style trade-offs, where every realization implies a combination of values of performance characteristics. These values are linearly ordered, though the realization points are partially ordered by dominance. Only points that are not dominated by others can be optimal under monotone objectives. At the end, the best one with respect to such an objective is chosen. The two examples above enable hierarchical application of the methods.

Today, it is of paramount importance to avoid premature commitment while being unaware of the consequences, because of the many relevant performance characteristics that interact with different levels of design. Choosing the best one at a certain level may preclude optimal ones in later stages. Yet, multidimensional Pareto analysis may be, and mostly is, very complex.

14.4.1 Realization Spaces

A performance characteristic is a set of realization values that indicate a quality aspect of a product or design step. Performance characteristics that quickly come to mind in chip design are *size*, *delay*, and *power*. Considered by themselves they are of the type "the smaller the better." Without loss of generality we will assume all characteristics to be of that type. Another property of these classic performance characteristics is that they are totally ordered.[||] That is, for every pair of realizations we can say whether one is better than the other with respect to such a parameter. The performance characteristics measurable in the final product often share this property. Although performance characteristics without that property do not easily come to mind, we will not adopt such a constraint for performance characteristics. We assume that they are ordered, i.e., there is a reflexive, transitive, and anti-symmetric relation associated with any performance characteristic, but not all of its values are comparable. Any trade-off between totally ordered performance characteristics may produce such a relation. Power and delay, for example, form pairs that are so ordered by dominance: such a pair dominates another pair if it is at least as good in both aspects. Obviously, two pairs each of which improves upon the other in some aspect are not comparable. If we consider a number of realizations of a given product specification and compare them on the basis of power consumption and speed, we are in fact looking at a number of pairs that are *partially ordered*. We call such a partially ordered set of realizations with respect to some performance characteristic a *realization space*, and its ordering relation *dominance*, i.e., a realization *dominates* another realization if it is at least as good with respect to every concerned performance characteristic. Abstractly, a *realization space* is simply a set Q with a partial order $\preccurlyeq_Q$, and we say $q \in Q$ *dominates* $q' \in Q$ whenever $q \preccurlyeq_Q q'$.

Let us consider all performance characteristics that might be relevant for a design. Some of these characteristics will be measurable in the final result. Others might be absorbed during the design process after they have played a role in certain stages of that process. It may even be so that certain characteristics have not been encountered at all upon completion because early decisions precluded them. That is, at each stage of the design process, realization spaces concerned with subsets of the performance characteristics are being manipulated and transformed into other realization spaces. It is important to note here that when realization spaces are combined to form a new space they induce a new partial ordering. In other words, the result is also a realization space. We no longer distinguish spaces concerned with basic performance characteristics and results that are associated with partial orders!

Each element (or point) in a realization space S represents a (partial) realization of a system. For convenience, we write them as vectors, $\bar{c} = (c_1, c_2, \ldots, c_n)$, where each c_i is a realization value of a characteristic Q_i. Then $\bar{a}$ in S *dominates* $\bar{b}$ in S ($\bar{a} \preccurlyeq_S \bar{b}$) if $a_i \preccurlyeq_{Q_i} b_i$ for all i. Dominance still expresses the fact that the underlying realization is in no aspect worse than the other. It can be raised to subsets of a realization space to say that any point in the dominated subset can be matched or improved upon by some point in the dominating subset: A subset A of a realization space dominates another subset B if for every element $\bar{b} \in B$ there is an element $\bar{a} \in A$ such that $\bar{a} \preccurlyeq_C \bar{b}$; this is written as $A \preccurlyeq B$. If A and B dominate each other, they are called *Pareto-equivalent* which is denoted as $A \equiv B$. Pareto-equivalence between two subsets means that neither subset contains a point that cannot be matched by a point in the other subset.

[||] A performance characteristic whose ordering is total (or linear) is called a basic performance characteristic here.

Realization points that are strictly dominated by other points in the same set can be removed without losing interesting realization options. In general, we do not want to maintain larger sets than necessary. This brings about the notion of minimality: a subset C of a realization space is *Pareto-minimal* if the removal of any of its elements yields a subset that is not Pareto-equivalent to C. That is, no element dominates another element in a Pareto-minimal subset. Or in other words, the Pareto-minimal subsets are *anti-chains* of the partial order.

Note that two Pareto-equivalent subsets need not be the same. However, when they are Pareto-minimal they have to be identical! More disturbing is the fact that not all realization spaces have a unique Pareto-equivalent subset that is Pareto-minimal. Whenever it exists for a realization space S we denote it by $\min(S)$. A sufficient requirement for such a *Pareto-frontier* is that every chain in the space has a smallest element (i.e., the space is *well-ordered*). This implies that *every finite realization space has a Pareto-equivalent subset that is Pareto-minimal.* In the present context, we mostly have only discrete points obtained by profiling in the realization spaces. From that perspective, the implication is certainly satisfactory. We assume in the rest of this section that all realization spaces are well-ordered.

14.4.2 Operators

Of course, the unbridled combination of the performance characteristics yields immense, unmanageable spaces. Essential, therefore, is an efficient algorithm that takes a realization space and returns a so-called *Pareto frontier.* This potentially reduces realization spaces, and might, if applied intelligently, keep a design flow manageable. Other operations may be defined provided that no potentially optimal candidate realizations are lost. Operations should preferably *preserve minimality,* i.e., applying the operation on minimized sets produces a minimal resulting set. However, a number of obviously necessary operations do not have that property. In such cases, we have to settle for *optimum preserving,* meaning that minimizing the result of such an operation on minimized operands, is the same as minimizing the result of that same operation on the unminimized operands.

14.4.2.1 Minimization

We denote by min(C) the unique Pareto-minimal set of a realization spaces C. It contains all realizations from that space that we would be interested in practice because all other realizations are dominated by some realization in min(C).

We call the operation to obtain that subset *minimization.* It is the key operation in the system. It would be nice if other operations when applied to minimized realization spaces would not require minimization to obtain the Pareto-minimal equivalent space. More precisely, if we have an operator Ψ that takes n spaces as its operands, we would like

$$\min(\Psi\,(C_1, \ldots C_n)) = \Psi\,(\min(C_1), \ldots, \min(C_n)) \tag{14.3}$$

We say then that Ψ *preserves minimality.* In that case minimization after applying operator Ψ is unnecessary. The result is always minimal when the operands have been minimized, which saves time in computing the results of the operation.

Unfortunately, some indispensable operations will not preserve minimality. That does not mean that it is not advantageous to apply the operation on minimized spaces. The computational effort spent on these generally smaller spaces is almost always considerably less than when the unreduced spaces are the arguments. If only a Pareto-equivalent subset is obtained, that is

$$\Psi\,(C_1, \ldots C_n) \equiv \Psi\,(\min\,(C_1), \ldots, \min C_n)) \tag{14.4}$$

Subsequent minimization then yields the same realization space as when the operation was applied to $C_1, \ldots, C_n$. So by minimizing operands, which are often intermediate results in the design process, before applying Ψ, no optimal realizations are lost. Every practical operator should *support minimization* in that sense, for otherwise all realizations have to be analyzed to identify the optimal ones.

If Pareto-equivalence is a congruence with respect to the operator Ψ, that is

$$\forall_{1 \leq i \leq n}[C_i' \equiv C_i] \rightarrow [\Psi\ (C_i', \ldots, C_n') \equiv \Psi\ (C_1, \ldots, C_n)]$$

then the operation necessarily supports minimization. This means that to prove that minimization support it suffices to show that the operator preserves dominance:

$$\forall_{1 \leq i \leq n}[C_i' \preccurlyeq C_i] \rightarrow [\Psi\ (C_1', \ldots, C_n') \preccurlyeq \Psi\ (C_1, \ldots, C_n)]$$

14.4.2.2 Free Product

When designing subsystems independently with their own realization spaces and tradeoffs, they have to be put together later, which means that their realization spaces have to be combined. The corresponding operator is called *free product.*

Let C_1 and C_2 be realization spaces each with their own performance characteristics. The free product $C_1 \bullet C_2$ consists of the points $\bar{c}_1 \cdot \bar{c}_2$, the concatenation of the "vectors" of each point $\bar{c}_1 \in C_1$ and each point $\bar{c}_2 \in C_2$

Obviously, not only is Pareto-equivalence a congruence with respect to free product, the operator even preserves minimality, which was to be expected when its purpose is considered.

14.4.2.3 Union

To enable a choice between realizations resulting from different design paths, the operator *union* is introduced. It can be thought of as the realization points of both spaces being brought into one and the same space spanned by the same set of performance characteristics as the two individual spaces. The notation is $C_1 \sqcup C_2$.

The operator does not preserve minimality. Obviously, a design trajectory may produce a realization that is dominated by one resulting from the other design path. So, minimization may result in a subset strictly included in its operand. But minimization is supported, because it can be seen as easily that $C_1 \sqcup C_2 \equiv \min(C_1) \sqcup \min(C_2)$. It is even true that Pareto-equivalence is a congruence with respect to the union operator.

14.4.2.4 Constraints

When we have a set of realizations, some of which are invalid because of additional constraints, then we can apply a constraint to filter out only those realizations that satisfy the constraint. Such constraints can be formally expressed as a set D of acceptable realizations or equivalently as a proposition on the realization space, identifying the acceptable realizations. Application of the constraint D to a realization space C leaves those realizations of C satisfying the constraint. We denote the new space as $C \sqcap D$.

When defined in that general way, constraints present a problem. It may easily happen that a constraint filters out a realization that dominated one that was not filtered out. Minimization cannot be supported in this way, because that operator may have removed such dominated realization points, while they may be part of the minimized result of the constraint applied to the unminimized space. The constraints in our algebra — we will call them *safe constraints* — will never have a dominating realization passing while the dominating realization fails the test. More precisely, a constraint D is called a safe constraint when for every $\bar{a}$ and $\bar{b}$ with $\bar{a} \preccurlyeq \bar{b}$ $\bar{b} \in D$ implies that $\bar{a} \in D$. The notation for applying a safe constraint D on a realization space C is $C \breve{\sqcap} D$.

14.4.2.5 Projection

In certain design stages, certain performance characteristics may no longer be relevant for future decisions. Projection can then be used to remove such a dimension of the realization space. In general, minimality is not preserved by projection, but dominance is. Therefore congruence and hence minimization support follow.

14.4.3 Cost Functions

The selection of a final realization point often happens with a cost function. Such a cost function should be one which never selects a point that is dominated under the space's partial order by another point. If that were possible, the whole formal system constructed here would be pointless, because the purpose was to manipulate and form realization spaces without losing any candidates for optimal realizations. In the process, strictly dominated realizations are being removed when reducing a space to the Pareto frontier. To ensure that cost functions are in concord with the motivation for building this formal system, we require *monotonicity*, i.e., whenever $\bar{c}_1 \preccurlyeq \bar{c}_2$ the cost function f has to satisfy $f(\bar{c}_1) \leq f(\bar{c}_2)$.

Thus defined, it requires cost functions to assign real numbers to every point in the final realization space. This is perfectly in order when the real objective is like that, as for example the chip size (or area). Often the selection is not so unambiguous, and designers resort to weighted combinations. Setting these weights is more often than not a trial-and-error affair. Yet the outcome may depend heavily on the exact setting. In essence, it is a consequence of restricting the range of cost functions to a basic performance characteristic. This is not necessary. If we allow the range to be any performance characteristic, that is any partially ordered set, then a simple adaptation of the definition of monotonicity preserves the essential property cost functions should have.

Monotone functions from a realization space C into a performance characteristic are then defined to be functions satisfying

$$\forall_{C_1, C_2 \in C}[\bar{c}_1 \preccurlyeq \bar{c}_2 \rightarrow \bar{c}_2 f(\bar{c}_1 \preccurlyeq f(\bar{c}_2)]$$

Thus defined they still can only select realizations in the Pareto-minimal set of the space as optimal. Moreover, any member of that set can be selected by such a function. That is, for every point in the Pareto-minimal set of a realization space, there is a monotone cost function that will select that point as minimal. This means that further reduction of spaces is not possible without giving up the guarantee that no potentially optimal candidate realizations are lost, if the cost function is not known. With the generalized cost function however a partial selection of candidates is possible, leaving fewer realizations to consider in subsequent steps.

14.4.4 Concluding Remarks

This section describes the algebra (C, min, $\bullet$, $\sqcup, \downarrow, \breve{\sqcap}$), where C is the set of all possible realization spaces that can be encountered during the design of a chip. The algebra was introduced in [32] where it mainly served as a formal system for runtime reconfiguration. The manipulations and calculations then have to take place on mostly resource-constrained devices. Here it is used for design-time exploration with a wide choice in computer power and lesser constraints on compute time. Not surprisingly, different algorithms may be chosen in these two application areas.

In the latter application, the algebra is meant for supporting the design while it forms realization spaces by combining subsystems and their realization spaces. The number of realizations, also when restricted to Pareto-minimal sets, can grow very fast, and it is therefore crucial to keep the space sizes small. Reducing them to their Pareto-minimal spaces is the most important means here, but pruning the space may still be necessary. The challenge is then to find the best approximation of realization space by one with a limited number of points [33, 34].

To date the most efficient minimization algorithm has time complexity $O(N (\log N)^d)$ [35], where d is the number of dimensions of the realization space and N the number of realizations. The number of points for which the algorithm is faster than simpler algorithms such as the *simple cull* in [36], having a complexity of $O(N^2)$, is quite high. This makes the former algorithm the choice only for large spaces, while *simple cull* and hybrid algorithms [36] are recommendable for smaller numbers of realizations. A similar effect is observed for data structures for storing sets of realizations. The quad-tree data structure [37, 38] is shown to be more efficient than linear lists in [38], but here also, the gain is achieved for large numbers of data points. Yukish[36] shows that when keeping points lexicographically sorted, normalized linear tables can be advantageous for computing the union and intersection of sets.

References

[1] R.H.J.M. Otten, Automatic floorplan design, *Proceedings of the 19th Design Automation Conference*, Las Vegas, USA, 1982, pp. 261–267.

[2] J. Grodstein, E. Lehman, H. Harkness, B. Grundmann, and Y. Watanabe, A delay model for logic synthesis of continuously-sized networks, *Proceedings of ICCAD*, November 1995.

[3] J. Grason, *A Dual Linear Graph Representation for Space-filling Location Problems of the Floor-Planning Type*, MIT Press, Cambridge, Massachusetts, USA, 1970.

[4] K. Kozminsky and E. Kinnen, Rectangular duals of planar graphs, *Networks*, 15, 145–157, 1985.

[5] J. Bhasker and S. Sahni, Representation and generation of rectangular dissections, *Proceedings of the 23rd Design Automation Conference*, Las Vegas, USA , 1986, pp. 108–114.

[6] J. Bhasker and S. Sahni, A linear time algorithm to check for the existence of a rectangular dual of a planar triangulated graph , *Networks*, 17, 307-317, 1987.

[7] R. Schmid, Synthese von Floorplans auf der Basis von Dualgraphen mit rechtwinkliger Einbettung, Ph.D. Thesis, Universität Karlsruhe, West Germany, 1987.

[8] U. Flemming, Representation and generation of rectangular dissections, *Proceedings of the 15th Design Automation Conference*, Las Vegas, USA, 1978, 138–144.

[9] T. Ohtsuki, N. Sugiyama, and H. Kawanishi, An optimization technique for integrated circuit lay-out design *Proceedings of ICCST*, Kyoto, Japan, 1970, pp. 67–68.

[10] R.H.J.M. Otten, L.P.P.P. van Ginneken, Floorplan design using annealing, *Proceedings of ICCAD*, SantaClara, 1984.

[11] T. Takahashi, A new encoding scheme for rectangle packing problem, *Proceedings ASP-DAC 2000*, Yokohama, Japan, 2000, pp. 175-178.

[12] W.R.Heller, G. Sorkin, and K. Maling, The planar package planner for system designers, *Proceedings of the 19th Design Automation Conference*, Las Vegas, USA, 1982, pp. 663–670.

[13] S. Tsukiyama, K. Koike, and I. Shirakawa, An algorithm to eliminate alle complex triangles in a maximal planar graph for use in VLSI floorplan, *Proceedings ISCAS*, Philadelphia, PA, USA, 1986, pp. 321-324.

[14] S.M. Leinwand and Yen-Tai Lai, An algorithm for building rectangular floor-plans, *Proceedings of the 21st Design Automation Conference*, Albuquerque, USA, 1984, pp. 663-664.

[15] R.H.J.M. Otten and P.S.Wolfe, Algorithm for wiring space assignment in slicing floorplans, Technical Report Y0882-0682, IBM Thomas J. Watson Research Center, Yorktown Heights, USA, 1982.

[16] L. Stockmeyer, Optimal orientations of cells in slicing floorplan designs, *Inform. Control*, 57, 91-101, 1983.

[17] R.H.J.M. Otten, Efficient floorplan optimization, *Proceedings of ICCD*, Port Chester, USA, 1983, pp. 499–502.

[18] W. Keller, F. Schreiner, B. Schurman, E. Siepmann, and G. Zimmerman, Hierarchisches Top Down Chip Planning , Report of the University of Kaiserslautern, West Germany, 1987.

[19] L.P.P.P. van Ginneken and R.H.J.M. Otten, Optimal slicing of plane point placements, *Proceeding of the European Design Automation Conference*, Glasgow, 1990, pp. 322–326.

[20] Young, F.Y. and D. F. Wong, How good are slicing floorplans?, *INTEGRATION: VLSI J.* 23, 61-73, 1997.

[21] Lai, M. and D.F. Wong, Slicing Tree is a complete floorplan representation, *Proceedings of the Design, Automation and Test in Europe Conference (DATE)*, March 2001.

[22] I. Sutherland and R. Sproull, The theory of logical effort: designing for speed on the back of an envelope, in *Advanced Research in VLSI*, C. Séquin, Ed., MIT Press, Cambridge, 1991.

[23] R.H.J.M. Otten, A design flow for performance planning, in *Architecture Design and Validation Methods*, E. Borger, Ed., Springer, Berlin, 2000.

[24] J. Westra, C. Bartels, and P.R. Groeneveld, Probabilistic congestion prediction, *Proceedings of the International Symposium on Physical Design*, 2004, pp. 204-209.

[25] W. Gosti, A. Narayan, R. K. Brayton, and A. L. Sangiovanni-Vincentelli, Wireplanning in Logic Synthesis, *Proceedings of ICCAD*, 1998.

[26] J. Westra, D.-J. Jongeneel, R.H.J.M. Otten, and C.Visweswariah, Time budgeting in a wire planning context, *Proceedings of DATE,* 2003.

[27] R. Nair, C.L. Berman, P.S. Hauge, and E.J. Yoffa, Generation of performance constraints for layout, *IEEE Transactions on CAD,* Vol. 8, August 1989.

[28] I. Karkowski, Circuit delay optimization as a multiple choice linear knapsack problem, *Proceedings of the European Design Automation Conference (EDAC),* March 1993, pp. 419-423.

[29] S.S. Sapatnekar, V.B. Rao, P.M. Vaidya, and S.M. Kang, An exact Solution to the transistor sizing problem for CMOS circuits using convex optimization, *IEEE Trans. CAD,* 12, 1621–1634, 1993.

[30] A.R. Conn, I.M. Elfadel, W.W. Molzen, Jr., P.R. O'Brien, P.N. Strenski, C. Visweswariah, and C.B. Whan, Gradient-based optimization of custom circuits using a static-timing Formulation, *Proceedings of the 36th Design Automation Conference,* June 1999, p. 452–459, 1999.

[31] C. Visweswariah and A.R. Conn, Formulation of static circuit optimization with reduced size, degeneracy and redundancy by timing graph manipulation, *Proceedings of the International Conference on Computer-Aided Design (ICCAD),* 1999, p.244.

[32] M. Geilen, A.A. Basten, B. Theelen, and R.H.J.M. Otten, An algebra of Pareto points, pp. 88–97, *Proceedings of ASCD 2005.*

[33] C. Mattson, A. Mullur, and A. Messac, Minimal representations of multiobjective design space using a smart Pareto filter, *Proceedings of the 9th AIAA/ISSMO Symposium on Multidisciplinary Analysis and Optimization,* 2002.

[34] E. Zitzler, L.Thiele, M. Laumanns, C. Fonseca, and V.G.D.Fonseca, Performance assessment of multiobjective optimizers: an analysis and a review, *IEEE Trans. Evol. Comput.,* 7,117-132, 2003.

[35] J. Bentley, Multidimensional command-and-conquer, *Commun. ACM,* 23, 214–229, 1980.

[36] M. Yukish, Algorithms to Identify Pareto Points in Multidimensional Data Sets, Ph. D. Thesis, Pennsylvania State University, August 2004.

[37] R. Finkel and J. Bentley, Quad trees: a data structure for retrieval on composite keys, *Acta Inform.,* 4, 1–9, 1974.

[38] M. Sun and R.E.Steuer, Quad trees and linear lists for identifying nondominated criterion vectors, *ORSA J. Comput.,* 8, 367–375, 1996.

[39] L. Cheng, L. Deng, and M.D.F. Wong, Floorplan Design for 3-D VLSI design, *Proceedings of the IEEE Asia South Pacific Design Automation Conference (ASP-DAC),* January 2005.

SECTION IV

LOGICAL VERIFICATION

15

Design and Verification Languages

Stephen A. Edwards
Columbia University
New York, New York

Abstract

After a few decades of research and experimentation, register–transfer dialects of two standard languages — Verilog and VHDL — have emerged as the industry standard starting point for automatic large-scale digital integrated circuit synthesis. Writing register-transfer-level descriptions of hardware remains a largely human process and hence the clarity, precision, and ease with which such descriptions can be coded correctly has a profound impact on the quality of the final product and the speed with which the design can be created.

While the efficiency of a design (e.g., the speed at which it can run or the power it consumes) is obviously important, its correctness is usually the paramount issue, consuming the majority of the time (and hence money) spent during the design process. In response to this challenge, a number of the so-called verification languages have arisen. These have been designed to assist in a simulation-based or formal verification process by providing mechanisms for checking temporal properties, generating pseudorandom test cases, and for checking how much of a design's behavior has been exercised by the test cases.

Through examples and discussion, this chapter describes the two main design languages — VHDL and Verilog — as well as SystemC, a language currently used to build large simulation models; System Verilog, a substantial extension of Verilog; and OpenVera, e, and PSL, the three leading contenders for becoming the main verification language.

15.1 Introduction

Hardware description languages (HDLs) are now the preferred way to enter the design of an integrated circuit, having supplanted graphical schematic capture programs in the early 1990s. A typical design methodology in 2005 starts with some back-of-the-envelope calculations that lead to a rough architectural design.

This architecture is refined and tested for functional correctness by implementing it as a large simulator written in C or C++ for speed. Once this high-level model is satisfactory, it is passed on to designers who recode it in a register–transfer-level (RTL) dialect of VHSIC (Very High-Speed Integrated Circuit) Hardware Description Language (VHDL) or Verilog — the two industry-dominant HDLs. This new model is often simulated to compare it to the high-level reference model, then fed to a logic synthesis system such as Synopsys' Design Compiler, which translates the RTL into an efficient gate-level netlist. Finally, this netlist is given to a place-and-route system that ultimately generates the list of polygons that will become wires and transistors on the chip.

None of these steps, of course, is as simple as it might sound. Translating a C model of a system into RTL requires adding many details, ranging from protocols to cycle-level scheduling. Despite many years of research, this step remains stubbornly manual in most designs. Synthesizing a netlist from an RTL dialect of an HDL has been automated, but is the result of many years of university and industrial research, as are all the automated steps after it.

Verifying that the C or RTL models are functionally correct presents an even more serious challenge. At the moment, simulation remains the dominant way of raising confidence in the correctness of these models, but has many drawbacks. One of the more serious is the need for simulation to be driven by appropriate test cases. These need to exercise the design, preferably the difficult cases that expose bugs, and be both comprehensive and relatively short since simulation takes time.

Knowing when simulation has exposed a bug and estimating how complete a set of test cases actually is, are two other major issues in a simulation-based functional verification methodology. The verification languages discussed here attempt to address these problems by providing facilities for generating biased, constrained random test cases, checking temporal properties, and checking functional coverage.

15.2 History

Many credit Reed [1] with the first HDL. His formalism, simply a list of Boolean functions that define the inputs to a block of flip-flops driven by a single clock (i.e., a synchronous digital system), captures the essence of an HDL: a semi-formal way of modeling systems at a higher level of abstraction. Reed's formalism does not mention the wires and vacuum tubes that would actually implement his systems, yet it makes clear how these components should be assembled.

In the five decades since Reed, both the number and the need for HDLs has increased. In 1973, Omohundro and Tracey [2] could list nine languages and dozens more have been proposed since.

The main focus of HDLs has shifted as the cost of digital hardware has dropped. In the 1950s and 1960s, the cost of digital hardware remained high and was used primarily for general-purpose computers. Chu's CDL [3] is representative of languages of this era: it uses a programming-language-like syntax; has a heavy bias toward processor design; and includes the notions of arithmetic, registers and register transfer, conditionals, concurrency, and even microprograms. Bell and Newell's influential ISP (described in their 1971 book [4]) was also biased toward processor design.

The 1970s saw the rise of many more design languages [5, 6]. One of the more successful was ISP'. Developed by Charles Rose and his student Paul Drongowski at Case Western Reserve in 1975 to 1976, ISP' was based on Bell and Newell's ISP and was used in a design environment for multiprocessor systems called N.mPc [7]. Commercialized in 1980 (and since owned by a variety of companies), it enjoyed some success, but starting in 1985, the Verilog simulator (and accompanying language) began to dominate the market.

The 1980s brought Verilog and VHDL, which remain the dominant HDLs to this day (2004). Initially successful because of its superior gate-level simulation speed, Verilog started life in 1984 as a proprietary language in a commercial product, while VHDL was designed at the behest of the US Department of Defense as a unifying representation for electronic design [8].

While the 1980s was the decade of the widespread commercial use of HDLs for simulation, the 1990s brought them an additional role as input languages for logic synthesis. While the idea of automatically synthesizing logic from an HDL dates back to the 1960s, it was only the development of multilevel logic synthesis in the 1980s [9] that made them practical for specifying hardware, much as compilers for software require optimization to produce competitive results. Synopsys was one of the first to release a commercially successful logic synthesis system that could generate efficient hardware from RTL Verilog specifications and by the end of the 1990s, virtually every large integrated circuit was designed this way.

Hardware description languages continue to be important for providing inputs for synthesis and modeling for simulation, but their importance as aids to validation has recently grown substantially. Long an important part of the design process, the use of simulation to check the correctness of a design has become absolutely critical, and languages have evolved to help perform simulation quickly, correctly, and judiciously.

Clearly articulated in features recently added to System Verilog, it is now common to generate automatically simulation test cases using biased random variables (e.g., to generate random input sequences in which reset occurs very infrequently), check that these cases thoroughly exercise the design (e.g., by checking whether certain values or transitions have been overlooked), and check whether invariants have been violated during the simulation process (e.g., making sure that each request is followed by an acknowledgment). Hardware description languages are expanding to accommodate such methodologies.

15.3 Design Languages

15.3.1 Verilog

The Verilog hardware description language [10–12] was designed and implemented by Phil Moorby at Gateway Design Automation in 1983 to 1984 (see Moorby's history of the language for more details [6]). The Verilog product was very successful, buoyed largely by the speed of its "XL" gate-level simulation algorithm. Cadence bought Gateway in 1989 and largely because of the pressure from the competing, open VHDL language, made the language public in 1990. Open Verilog International (OVI) was formed shortly thereafter to maintain and promote the standard. IEEE adopted it in 1995, and ANSI in 1996.

The first Verilog simulator was event-driven and very efficient for gate-level circuits, the fashion of the time, but the opening of the Verilog language in the early 1990s paved the way for other companies to develop more efficient compiled simulators, which traded up-front compilation time for simulation speed.

Like tree rings, the syntax and semantics of the Verilog language embodies a history of simulation technologies and design methodologies. At its conception, gate- and switch-level simulations were in fashion, and Verilog contains extensive support for these modeling styles that are now little used. (Moorby had worked with others on this problem before designing Verilog [13].)

Like many HDLs, Verilog supports hierarchy for structural modeling, but was originally designed assuming modules would have at most tens of connections. Hundreds or thousands of connections are now common, and Verilog-2001 [11] added a more succinct connection syntax to address this problem.

Procedural or behavioral modeling, once intended mainly for specifying testbenches, was pressed into service first for RTL specifications, and later for the so-called behavioral specifications. Again, Verilog-2001 added some facilities to enable this (e.g., always @* to model combinational logic procedurally) and SystemVerilog has added additional support (e.g., always_comb, always_ff).

The syntax and semantics of Verilog have always been a compromise between modeling clarity and simulation efficiency. A "reg" in Verilog, the variable storage class for behavioral modeling, is exactly a shared variable. This means it simulates very efficiently (e.g., writing to a reg is just an assignment to memory), but also means that it can be misused (e.g., when written to by two concurrently running processes) and misinterpreted (e.g., its name suggests a memory element such as a flip-flop, but it often represents purely combinational logic).

The book by Thomas and Moorby [14] has long been the standard text on the language (Moorby was the main designer), and the language reference manual [10], since it was adopted from the original Verilog simulator user manual, is surprisingly readable. Other references include Palnitkar [15] for an overall description of the language, and Mittra [16] and Sutherland [17] for the programming language interface (PLI). Smith [18] compares Verilog and VHDL. French et al. [19] present a clever way of compiling Verilog simulations and also discuss more traditional ways.

15.3.1.1 Coding in Verilog

A Verilog description is a list of modules. Each module has a name; an interface consisting of a list of named ports, each with a type, such as a 32-bit vector, and a direction; a list of local nets and regs; and a body that can contain instances of primitive gates such as ANDs and ORs, instances of other modules (allowing hierarchical structural modeling), continuous assignment statements, which can be used to model combinational datapaths, and concurrent processes written in an imperative style.

Figure 15.1 shows the various modeling styles supported in Verilog. The two-input multiplexer circuit in Figure 15.1(a) can be represented in Verilog using primitive gates (Figure 15.1(b)), a continuous assignment (Figure 15.1(c)), a user-defined primitive (a truth table, Figure 15.1(d)), and a concurrent process (Figure 15.1(e)). All of these models exhibit roughly the same behavior (minor differences occur when some inputs are undefined) and can be mixed freely within a design.

One of Verilog's strengths is its ability to also represent testbenches within the model being tested. Figure 15.1(f) illustrates a testbench for this simple mux, which applies a sequence of inputs over time and prints a report of the observed behavior.

Communication within and among Verilog processes takes place through two distinct types of variables: *nets* and *regs*. Nets model wires and must be driven either by gates or by continuous assignments. Regs are exactly shared memory locations and can be used to model memory elements. Regs can be assigned only by imperative assignment statements that appear in *initial* and *always* blocks. Both nets and regs can be single bits or bit vectors, and regs can also be arrays of bit vectors to model memories. Verilog also has limited support for integers and floating-point numbers. Figure 15.2 shows a variety of declarations.

The distinction between regs and nets in Verilog is pragmatic: nets have slightly more complicated semantics (e.g., they can be assigned a capacitance to model charge storage and they can be connected to multiple tri-state drivers to model buses), but regs behave exactly like memory locations and are therefore easier to simulate quickly. Unfortunately, the semantics of regs makes it easy to inadvertently introduce nondeterminism in the language (e.g., when two processes simultaneously attempt to write to the same reg, the result is undefined). This will be discussed in more detail in the next section.

Figure 15.3 illustrates the syntax for defining and instantiating models. Each module has a name and a list of named ports, each of which has a direction and a width. Instantiating such a module consists of giving the instance a name and listing the signals or expressions to which it connected. Connections can be made positionally or by port name, the latter being preferred for modules with many (perhaps ten or more) connections.

Continuous assignments are a simple way to model both Boolean and arithmetic datapaths. A continuous assignment uses Verilog's comprehensive expression syntax to define the function to be computed, and its semantics are such that the value of the expression on the right of a continuous expression is always copied to the net on the left (regs are not allowed on the left of a continuous assignment). Practically, Verilog simulators implement this by recomputing the expression on the right whenever any variable it references changes. Figure 15.4 illustrates some continuous assignments.

Behavioral modeling in Verilog uses imperative code enclosed in *initial* and *always* blocks that write to reg variables to maintain state. Each block effectively introduces a concurrent process that is awakened by an event and runs until it hits a delay or a wait statement. The example in Figure 15.5 illustrates basic behavioral usage.

Figure 15.6 shows a more complicated behavioral model, in this case a simple state machine. This example is written in a common style where the combinational and sequential parts of a state machine are written as two separate processes. The first process is purely combinational. The @(a or b or state) directive triggers its execution when signals a, b, or state change. The code consists of a multiway choice — a case statement — and performs procedural assignments to the o and nextState registers. This illustrates one of the odder aspects

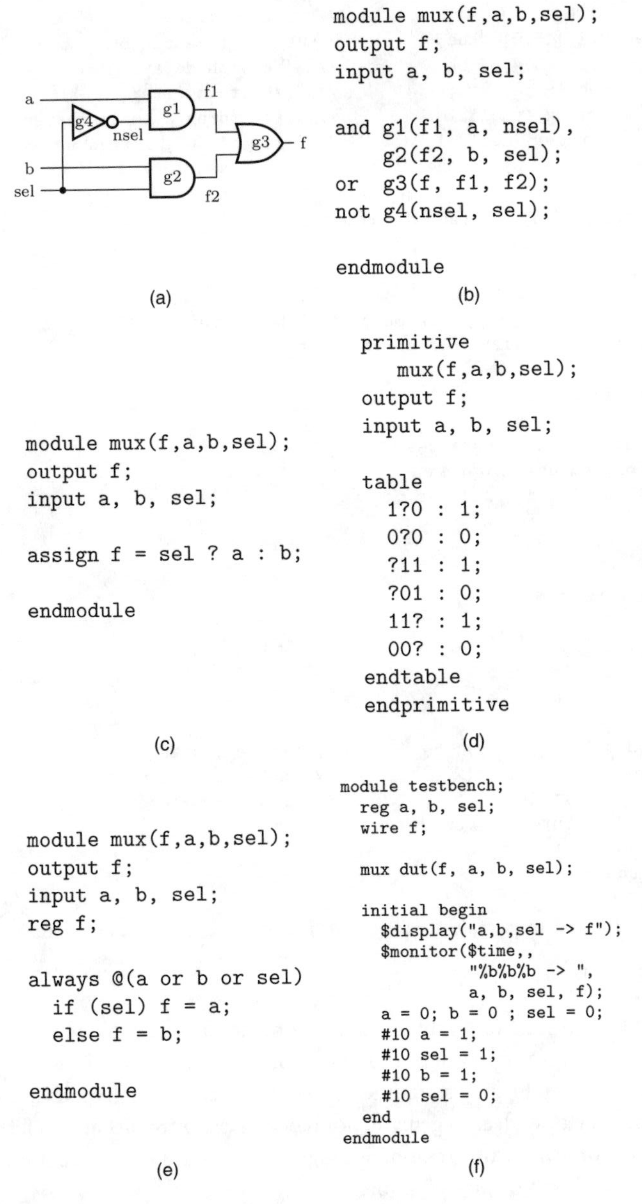

FIGURE 15.1 Verilog examples: (a) a multiplexer circuit; (b) the multiplexer described as a Verilog structural model; (c) the multiplexer described using a continuous assignment; (d) a user-defined primitive for the multiplexer; (e) the multiplexer described with imperative code; (f) a testbench for the multiplexer.

of Verilog modeling: both are declared reg, yet neither corresponds to the output of a latch or flip-flop. This reflects the Verilog requirement that procedural assignment can only be performed on regs.

The second process models a pair of flip-flops that holds the state between cycles. The @(posedge clk or reset) directive makes the process sensitive to the rising edge of the clock or a change in the reset signal. At the positive edge of the clock, the process captures the value of the nextState variable and copies it to state.

The example in Figure 15.6 illustrates the two types of behavioral assignments. The assignments used in the first process are the so-called blocking assignments, written =, and take effect immediately. Nonblocking assignments are written <= and have somewhat subtle semantics. Instead of taking effect

```
wire a;                   // Simple wire
tri [15:0] dbus;          // 16-bit tristate bus
tri #(5,4,8) b;           // Wire with delay
reg [-1:4] vec;           // Six-bit register
trireg (small) q;         // Wire stores a small charge
integer imem[0:1023];     // Array of 1024 integers
reg [31:0] dcache[0:63];  // A 32-bit memory
```

FIGURE 15.2 A sampling of Verilog *net* and *reg* definitions.

```
module mymod(out1, out2, in1, in2);
output out1;        // Outputs first by convention
output [3:0] out2;  // four-bit vector
input in1;
input [2:0] in2;

// Module body: instances,
// continuous assignments,
// initial and always blocks

endmodule

module usemymod;
reg a;
reg [2:0] b;
wire c, e, g;
wire [3:0] d, f, h;

mymod m1(c, d, a, b);              // simple instance
mymod m2(e, f, c, d[2:0]),         // instance with part-select input
      m3(.in1(e), .in2(f[2:0]),   // connect-by-name
         .out1(g), .out2(h));

endmodule
```

FIGURE 15.3 Verilog structure: an example of a module definition and another module containing three instances of it.

immediately, the right-hand sides of nonblocking assignments are evaluated when they are executed, but the assignment itself does not take place until the end of the current time instant. Such behavior effectively isolates the effect of nonblocking assignments to the next clock cycle, much like the output of a flip-flop is only visible after a clock edge. In general, nonblocking assignments are preferred when writing to state-holding elements for exactly this reason. See Figure 15.7 and the next section for a more extensive discussion of blocking vs. nonblocking assignments.

15.3.1.2 Verilog Shortcomings

Compared to VHDL, Verilog does a poor job at protecting users from themselves. Verilog's regs are exactly shared variables and the language permits all the standard pitfalls associated with them, such as races and nondeterministic behavior. Most users avoid such behavior by following certain rules (e.g., by restricting assignments to a shared variable to a single concurrent process), but Verilog allows more dangerous usage. Tellingly, a number of EDA companies exist solely to provide lint-like tools for Verilog that report such poor coding practices. Gordon [20] provides a more detailed discussion of the semantic challenges of Verilog.

Nonblocking assignments are one way to ameliorate most problems with nondeterminism caused by shared variables, but they too can lead to bizarre behavior. To illustrate the use of shared variables, consider a three-stage shift register. The implementation in Figure 15.7(a) appears to be correct, but in fact may not behave as expected because the language says the simulator is free to execute the three *always* blocks in any

```
module add8(sum, a, b, carryin);
output [8:0] sum;
input [7:0] a, b;
input carryin;

// unsigned arithmetic
assign sum = a + b + carryin;
endmodule

module datapath(addr_2_0, icu_hit, psr_bm8, hit);
output [2:0]  addr_2_0
output        icu_hit
input         psr_bm8;
input         hit;
wire   [31:0] addr_qw_align;
wire   [3:0]  addr_qw_align_int;
wire   [31:0] addr_d1;
wire          powerdown;
wire          pwdn_d1;

// part select, vector concatenation is {}
assign addr_qw_align = { addr_d1[31:4], addr_qw_align_int[3:0] };

// if-then-else operator
assign addr_offset = psr_bm8 ? addr_2_0[1:0] : 2'b00;

// Boolean operators
assign icu_hit = hit & !powerdown & !pwdn_d1;

// ...

endmodule
```

FIGURE 15.4 Verilog modules illustrating continuous assignment. The first is a simple eight-bit full adder producing a nine-bit result. The second is an excerpt from a processor datapath.

```
module behavioral;
reg [1:0] a, b;

initial begin
  a = 'b1;
  b = 'b0;
end

always begin
  #50 a = ~a;   // Toggle a every 50 time units
end

always begin
  #100 b = ~b;   // Toggle b every 100 time units
end

endmodule
```

FIGURE 15.5 A simple Verilog behavioral model. The code in the *initial* block runs once at the beginning of simulation to initialize the two registers. The code in the two *always* blocks runs periodically: once every 50 and 100 time units.

order when they are triggered. If the processes execute top-to-bottom, the module becomes a one-stage shift register, but if they execute bottom-to-top, the behavior is as intended.

Figure 15.7(b) shows a correct implementation of the shift register that uses nonblocking assignments to avoid this problem. The semantics of these assignments is such that the value on the right-hand side

```
module FSM(o, a, b, reset);
output o;
reg o;  // declared reg only because o is assigned procedurally
input a, b, reset;
reg [1:0] state, nextState; // only "state" holds state

// Combinational logic block: sensitive to changes on all inputs and
// outputs o and nextState always assigned

always @(a or b or state)
 case (state)
    2'b00: begin
      o = a & b;
      nextState = a ? 2'b00 : 2'b01;
    end
    2'b01: begin
      o = 0; nextState = 2'b10;
    end
    default: begin
      o = 0; nextState = 2'b00;
    end
 endcase

// Sequential block: sensitive to clock edge

always @(posedge clk or reset)
  if (reset)
    state <= 2'b00;
  else
    state <= nextState;

endmodule
```

FIGURE 15.6 A Verilog behavioral model for a state machine illustrating the common practice of separating combinational and sequential blocks.

of the assignment is captured when the statement runs, but the actual assignment of values is only done at the "end" of each instant in time, i.e., after all three blocks have finished executing. As a result, the order in which the three assignments are executed does not matter and therefore the code always behaves like a three-stage shift register.

However, the use of nonblocking assignments can also be deceptive. In most languages, the effect of an assignment can be felt by the instruction immediately following it, but the delayed-assignment semantics of a nonblocking assignment violates this rule. Consider the erroneous decimal counter in Figure 15.8(a). Without knowing the subtle semantics of Verilog nonblocking assignments, the counter would appear to count from 0 to 9, but in fact it counts to 10 before being reset because test of o by the *if* statement gets the value of o from the previous clock cycle, not the results of the o <= o + 1 statement. A corrected version is shown in Figure 15.8(b), which uses a local variable count to maintain the count, blocking assignments to touch it, and finally a nonblocking assignment to o to send the count outside the module.

Coupled with the rules for register inference, the circuit implied by the counter in Figure 15.8(b) is actually just fine. While both o and count are marked as reg, only the count variable will actually become a state-holding element.

15.3.2 VHDL

Although VHDL and Verilog have grown to be nearly interchangeable, they could not have had more different histories. Unlike Verilog, VHDL was deliberately designed to be a standard HDL after a lengthy, deliberate process. As Dewey explains [8], VHDL was created at the behest of the U. S. Department of Defense in response to the desire to incorporate integrated circuits (specifically very high-speed integrated

```
module bad_sr(o, i, clk);
output o;
input i, clk;
reg a, b, o;

always @(posedge clk) a = i;
always @(posedge clk) b = a;
always @(posedge clk) o = b;

endmodule
```

(a)

```
module good_sr(o, i, clk);
output o;
input i, clk;
reg a, b, o;

always @(posedge clk) a <= i;
always @(posedge clk) b <= a;
always @(posedge clk) o <= b;

endmodule
```

(b)

FIGURE 15.7 Verilog examples illustrating the difference between blocking and nonblocking assignments. (a) An erroneous implementation of a three-stage shift register that may or may not work depending on the order in which the simulator chooses to execute the three *always* blocks. (b) A correct implementation using nonblocking assignments, which make the variables take on their new values after all three blocks are done for the instant.

```
module bad_counter(o, clk);
output o;
input clk;
reg [3:0] o;

always @(posedge clk) begin
  o <= o + 1;
  if (o == 10)
    o <= 0;
end

endmodule
```

(a)

```
module good_counter(o, clk);
output o;
input clk;
reg [3:0] o;
reg [3:0] count;

always @(posedge clk) begin
  count = count + 1;
  if (count == 10)
    count = 0;
  o <= count;
end

endmodule
```

(b)

FIGURE 15.8 Verilog examples illustrating a pitfall with nonblocking assignments: (a) an erroneous implementation of a counter, which counts to 10, not 9; (b) A correct implementation using a combination of blocking and nonblocking assignments.

circuits (VHSICs)) in military hardware, hence the name of the program from which VHDL evolved. Starting with a summer study at Woods Hole, Massachusetts in 1981, requirements for and scope of the language were first established; then, after a bidding process, a contract to develop the language was awarded in 1983 to three companies: Intermetrics, which was the prime contractor for Ada, the software programming language developed for the U. S. military in the early 1980s; Texas Instruments; and IBM. Dewey and de Geus [21] describe this history in more detail.

The VHDL language was created in 1983 and 1984, essentially concurrent with Verilog, and first released publicly in 1985. Interest in an IEEE standard HDL was high at the time, and VHDL was eventually adapted and adopted as IEEE standard 1076 in 1987 [22] and revised in 1993 [23]. Verilog, meanwhile, remained proprietary until 1990. The standardization and growing popularity of VHDL was certainly instrumental in Cadence's decision to make Verilog public.

The original objectives of VHDL [8] were to provide a means of documenting hardware (i.e., as an alternative to imprecise English descriptions) and of verifying it through simulation. As such, a VHDL simulator was developed along with the language.

VHDL is vast, complicated, and has a verbose syntax obviously derived from Ada. While its popularity as a means of formal documentation is questionable, it has succeeded as a modeling language for hardware simulation, and, like Verilog, more recently as a specification language for RTL logic synthesis.

Although there are many features absent in Verilog that are unique to VHDL (and vice versa), in practice most designers use a nearly identical subset because this is what the synthesis tools accept. Thus although the syntax of the two languages differs greatly, and the semantics appear different, their usage has converged to a common core.

Unlike Verilog, VHDL has spawned a plethora of books discussing its proper usage. Basic texts include Lipsett et al. [24] (one of the earliest), Dewey [25], Bhasker [26], Perry [27], and Ashenden [28, 29]. More advanced is Cohen [30], who suggests preferred idioms in VHDL, and Harr and Stancluescu [31], who discuss using VHDL for a variety of modeling tasks, not just RTL.

15.3.2.1 Coding in VHDL

Like Verilog, VHDL describes designs as a collection of hierarchical modules. But unlike Verilog, VHDL splits them into interfaces — called entities — and their implementations — called architectures. In addition to named input and output ports, entities also define compile-time parameters (generics), types, constants, attributes, use directives, and others.

Figure 15.9 shows code for the same two-input multiplexer roughly equivalent to Verilog examples in Figure 15.1. Figure 15.9(a) is the entity declaration for the multiplexer, which defines its input and output ports. Figure 15.9(b) is a purely structural description of the multiplexer; it defines internal signals, the interface to the Inverter, AndGate, and OrGate components, and instantiates four of these gates. The name of the architecture, "structural" is arbitrary; it is used to distinguish among various architectures. The Verilog equivalent in Figure 15.1(b) did not define the internal signals, relying instead on Verilog's rule of automatically considering undeclared signals to be single-bit wires. Furthermore, the Verilog example used the built-in gate-level primitives; VHDL itself does not know about logic gates, but can be taught about them.

Figure 15.9(c) illustrates a dataflow model for the multiplexer with each logic gate made explicit. VHDL does have built-in logical operators. Figure 15.9(d) shows an even more succinct implementation, which uses the multi-way *when* conditional operator.

Finally, Figure 15.9(e) shows a behavioral implementation of the mux. It defines a concurrently running process sensitive to the three mux inputs (a, b, and c) and uses an if-then-else statement (VHDL provides most of the usual control-flow statements) to select between copying the a and the b signal to the output d.

One of the design philosophies behind the VHDL was to maximize its flexibility by making most things user-definable. As a result, unlike Verilog, it has only the most rudimentary built-in types of (e.g., Boolean variables, but nothing to model four-valued logic), but has a much more powerful type system that allows such types to be defined. The Bit used in the examples in Figure 15.9 is actually a predefined part of the standard environment, i.e.,

```
type BIT is ('0', '1');
```

which is a character enumeration type whose two values are the characters 0 and 1 (VHDL is case insensitive; Bit and BIT are equivalent.)

There are advantages and disadvantages to this approach. While it allows more things to be added to the language later, it also makes for a few infelicities. For example, the predicate in the *if* statement in Figure 15.9(e) must be written "c = '1'" instead of just "c" because the argument must be of type Boolean, not Bit. While not a serious issue, it is yet another thing that contributes to VHDL's verbosity.

Figure 15.10 is a more elaborate example showing an implementation of the classic traffic light controller from Mead and Conway [32]. This is written in a synthesizeable dialect, using the common practice of separating the output and next-state logic from the state-holding element. Specifically, the first process is sensitive only to the clock signal. The *if* statement in the first process checks for an event on the clock (VHDL signals have a variety of attributes; *event* is true whenever the value has changed) and the clock being high, i.e., the rising edge of the clock. The second process is sensitive only to the inputs and present state of the machine, not the clock, and is meant to model

```
entity mux2 is
  port (a, b, c : in Bit; d : out Bit);
end;
```

(a)

```
architecture structural of mux2 is

  signal cbar, ai, bi : Bit;       -- internal signals

  component Inverter               -- component interfaces
    port (a:in Bit; y: out Bit);
  end component;
  component AndGate
    port (a1, a2:in Bit; y: out Bit);
  end component;
  component OrGate
    port (a1, a2:in Bit; y: out Bit);
  end component;

begin
  I1: Inverter port map(a => c, y => cbar); -- connect-by-name
  A1: AndGate  port map(a, c, ai);          -- connect-by-position
  A2: AndGate  port map(a1 => b, a2 => cbar, y => bi);
  O1: OrGate   port map(a1 => ai, a2 => bi, y => d);
end;
```

(b)

```
architecture dataflow1 of mux2 is
  signal cbar, ai, bi : Bit;
begin
  cbar <= not c;
  ai   <= a and c;
  bi   <= b and cbar;
  d    <= ai or bi;
end;
```

(c)

```
architecture dataflow2 of mux2 is
begin
  d <= a when c = '1' else
       b;
end;
```

(d)

```
architecture behavioral of mux2 is
begin
  process(a, b, c)  -- define sensitivity list
  begin
    if c = '1' then
      d <= a;
    else
      d <= b;
    end if;
  end process;
end;
```

(e)

FIGURE 15.9 VHSIC Hardware Development Language (VHDL) code for a two-input multiplexer: (a) the entity definition for the multiplexer; (b) a structural implementation instantiating primitive gates; (c) a dataflow implementation with an expression for each gate; (d) a direct dataflow implementation. (e) a behavioral implementation.

combinational logic. It illustrates the multi-way conditional *case* statement, constants, and bit vectors. It employs types (i.e., std_ulogic and std_ulogic_vector) and operators from the ieee.std_logic_1164 library, an IEEE standard library [33] for modeling logic that can represent unknown values ("X") as well as 0s and 1s.

```
library ieee;
use ieee.std_logic_1164.all;

entity tlc is
  port (
    clk            : in  std_ulogic;
    reset          : in  std_ulogic;
    cars           : in  std_ulogic;
    short          : in  std_ulogic;
    long           : in  std_ulogic;
    highway_yellow : out std_ulogic;
    highway_red    : out std_ulogic;
    farm_yellow    : out std_ulogic;
    farm_red       : out std_ulogic;
    start_timer    : out std_ulogic);
end tlc;

architecture imp of tlc is
signal current_state, next_state : std_ulogic_vector(1 downto 0);
constant HG : std_ulogic_vector := "00";
constant HY : std_ulogic_vector := "01";
constant FY : std_ulogic_vector := "10";
constant FG : std_ulogic_vector := "11";
begin

P1: process (clk)    -- Sequential process
begin
  if (clk'event and clk = '1') then
    current_state <= next_state;
  end if;
end process P1;

-- Combinational process: sensitive to input changes
P2: process (current_state, reset, cars, short, long)
begin
  if (reset = '1') then
    next_state <= HG;
    start_timer <= '1';
  else
    case current_state is
      when HG =>
        highway_yellow <= '0'; highway_red <= '0';
        farm_yellow    <= '0'; farm_red    <= '1';
        if (cars = '1' and long = '1') then
          next_state <= HY; start_timer <= '1';
        else
          next_state <= HG; start_timer <= '0';
        end if;
      when HY =>
        highway_yellow <= '1'; highway_red <= '0';
        farm_yellow    <= '0'; farm_red    <= '1';
        if (short = '1') then
          next_state <= FG; start_timer <= '1';
        else
          next_state <= HY; start_timer <= '0';
        end if;
      when FG =>
        highway_yellow <= '0'; highway_red <= '1';
        farm_yellow    <= '0'; farm_red    <= '0';
        if (cars = '0' or long = '1') then
          next_state <= FY; start_timer <= '1';
        else
          next_state <= FG; start_timer <= '0';
        end if;
      when FY =>
        highway_yellow <= '0'; highway_red <= '1';
        farm_yellow    <= '1'; farm_red    <= '0';
        if (short = '1') then
          next_state <= HG; start_timer <= '1';
        else
          next_state <= FY; start_timer <= '0';
        end if;
      when others =>
        next_state     <= "XX"; start_timer <= 'X';
        highway_yellow <= 'X'; highway_red    <= 'X';
        farm_yellow    <= 'X'; farm_red       <= 'X';
    end case;
  end if;
end process P2;
end imp;
```

FIGURE 15.10 The traffic-light controller from Mead and Conway [32] implemented in VHDL, illustrating the common practice of separating combinational and state-holding processes.

15.3.2.2 VHSIC Hardware Description Language (VHDL) Shortcomings

One shortcoming of VHDL is its obvious verbosity: the use of begin/end pairs instead of braces, the need to separate entities and their architectures, the need to spell out things like ports, its lengthy names for standard logic types (e.g., std_ulogic_vector), and its requirement of enclosing Boolean values and vectors in quotes. Many of these are artifacts of its roots in the Ada language, another fairly verbose language commissioned by the U. S. Department of Defense, but others are due to questionable design decisions. Consider the separation of entity/architecture pairs. While separating these concepts is a boon to abstraction and simplifies the construction of simulations of the same system in different configurations (e.g., to run a simulation using a gate-level architecture in place of a behavioral one for more precise timing estimation), in practice most designers only ever write a single architecture for a given entity and such pairs are usually written together.

The flexibility of VHDL has advantages and disadvantages. Its type system is much more flexible than Verilog's, providing things such as aggregate types and overloaded functions and operators, but this flexibility comes with a need for standardization and also tends to increase the verbosity of the language. Some of the need for standardization was recognized early, resulting in libraries such as the widely supported IEEE 1164 library for multivalued logic modeling. However, a standard for signed and unsigned arithmetic on logic vectors was slower in coming (it was eventually standardized in 1997 [34]), prompting both Synopsys and Mentor to each introduce similar but incompatible and incomplete versions of such a library.

Fundamentally, many of the problems stem from a desire to make the language too general. Aspects of the type system suffer from this as well. While the ability to define new enumerated types for multi-valued logic modeling is powerful, it seems a little odd to require virtually every VHDL program (since its main use has long been specification for RTL synthesis) to include one or more standard library. This also leads to the need to be constantly comparing signals to the literal '1' instead of just using a signal's value directly, and requires a user to carefully watch the types of subexpressions.

15.3.3 SystemC

SystemC is a relative latecomer to the HDL wars. Developed at Synopsys in the late 1990s, primarily by Stan Liao, SystemC was originally called Scenic [35] and was intended to replace Verilog and VHDL as the main system description language for synthesis (see Arnout [36] for some of the arguments for SystemC). SystemC is not so much a language as a C++ library along with a set of coding rules, but this is exactly its strength. It evolved from the common practice of first writing a high-level simulation model in C or C++, refining it, and finally recoding it in RTL Verilog or VHDL. SystemC was intended to smooth the refinement process by removing the need for a separate HDL.

SystemC, especially in its original incarnation, can be thought of as a dialect of C++ for modeling digital hardware. Like Verilog and VHDL, it supports hierarchical models whose blocks consist of input/output ports, internal signals, concurrently running imperative processes, and instances of other blocks. The SystemC libraries make two main contributions: an inexpensive mechanism for running many processes concurrently (based on a lightweight thread package; see Liao et al. [35]), and an extensive set of types for modeling hardware systems, including bit vectors and fixed-point numbers. A SystemC model consists of a series of class definitions, each of which define a block.

Methods defined for such a class become concurrently running processes, and the constructor for each class starts these processes running by passing them to the simulation kernel. Simulating a SystemC model starts by calling the constructors for all blocks in the design, then invoking the scheduler which is responsible for executing each of the concurrent processes as needed.

The computational model behind earlier versions of SystemC was cycle-based instead of the event-driven model of Verilog and VHDL. This meant that the simulation was driven by a collection of potentially asynchronous, but periodic clocks. Later versions (SystemC 2.0 and higher) adopted an event-driven model much like VHDLs. Also added were many constructs useful for higher-level modeling, such as abstract channels and interfaces, which has pushed the use of SystemC toward system-level modeling and away from its origins as hardware description.

SystemC books have only appeared recently. Grötker et al. [37] provide a good introduction to SystemC 2.0. Bhasker [38] is also an introduction. The volume edited by Müller et al. [39] surveys more advanced SystemC modeling techniques.

15.3.3.1 Coding in SystemC

Figure 15.11 shows a small SystemC model for a 0 to 99 counter driving a pair of seven segment displays. It defines two modules (the decoder and counter *structs)* and an sc_main function that defines some internal signals, instantiates two decoders and a counter, and runs the simulation while printing out what it does.

The two modules in Figure 15.11 illustrate two of the three types of processes possible in SystemC. The decoder module is the simpler one: it models a purely combinational process by defining a method (called, arbitrarily, "compute") that will be invoked by the scheduler every time the number input changes, as indicated by the "sensitive << number;" statement beneath the definition of compute as an SC_METHOD.

In the second module, counter is an SC_CTHREAD process: a method (here called "tick") that is invoked in response to a clock edge (here, the positive edge of the clk input, as defined by the SC_CTHREAD(tick, clk.pos()); statement) and can suspend itself with the wait() statement. Specifically, the scheduler resumes the method when the clock edge occurs, and the method runs until it encounters a wait() statement, at which point its state is saved and control passes back to the scheduler.

This example illustrates only a very small fraction of the SystemC type libraries. It uses unsigned integers (sc_uint), bit vectors (sc_bv), and a clock (sc_clock). The nonclock types are wrapped in sc_signals, which behave like VHDL signals. In particular, when an SC_CTHREAD method assigns a value to a signal, the effect of this assignment is felt only after all the processes triggered by the same clock edge have been run. Thus, such assignments behave like blocking assignments in Verilog to ensure that the nondeterministic order in which such processes are invoked (the scheduler is allowed to invoke them in any order) does not affect the ultimate outcome of simulating the system.

15.3.3.2 SystemC Shortcomings

Like many languages, the most common use of SystemC has diverged from its designers' original intentions i.e., as an input for hardware synthesis in the case of SystemC 1.0. A number of synthesis tools for the language do exist, but since version 2.0, SystemC has been primarily (and quite successfully) used for system modeling. This does mean, however, that it does not solve the "separate language for synthesis problem."

A big disadvantage of SystemC is that C++ was never intended for modeling digital hardware and as a result, is even more lax about enforcing rules than Verilog. The syntax, similarly, is somewhat awkward and relies on some very tricky macro preprocessor definitions. On detailed models, the simulation speed of a good compiled-code Verilog or VHDL simulator may be better, although SystemC is much faster for higher-level models. For such systems which consist of complex processes, SystemC should be superior since the simulation becomes nearly a normal C++ program. However, the context-switching cost in SystemC is higher than that of a good Verilog or VHDL simulator when running a more detailed model, so a system with many small processes would not simulate as quickly.

Another issue is the ease with which a SystemC model can inadvertently be made nondeterministic. Although carefully following a discipline of only communicating among processes through signals will ensure the simulation is nondeterministic, any slight deviation from this will cause problems. For example, library functions that use a hidden global variable may cause nondeterminism if called from different processes. Accidentally holding state in an SC_METHOD process (easily done if class variables are assigned) can also cause problems since such methods are invoked in an undefined order.

Many argue that nondeterministic behavior in a language can be desirable for modeling nondeterministic systems, which certainly exist and need to be modeled. However, the sort of nondeterminism in a language such as SystemC or Verilog creeps in unexpectedly and is difficult to use as a modeling tool.

```
#include "systemc.h"
#include <stdio.h>

struct decoder : sc_module {
  sc_in<sc_uint<4> > number;
  sc_out<sc_bv<7> > segments;

  void compute() {
    static sc_bv<7> codes[10] = {
      0x7e, 0x30, 0x6d, 0x79, 0x33,
      0x5b, 0x5f, 0x70, 0x7f, 0x7b };
    if (number.read() < 10)
      segments = codes[number.read()];
  }

  SC_CTOR(decoder) {
    SC_METHOD(compute);
    sensitive << number;
  }
};

struct counter : sc_module {
  sc_out<sc_uint<4> > tens;
  sc_out<sc_uint<4> > ones;
  sc_in_clk clk;

  void tick() {
    int one = 0, ten = 0;
    for (;;) {
      if (++one == 10) {
        one = 0;
        if (++ten == 10) ten = 0;
      }
      ones = one;
      tens = ten;
      wait();
    }
  }

  SC_CTOR(counter) {
    SC_CTHREAD(tick, clk.pos());
  }
};

int sc_main(int argc, char *argv[])
{
  sc_signal<sc_uint<4> > ones, tens;
  sc_signal<sc_bv<7> > ones_segments, tens_segments;
  sc_clock clk;

  decoder decoder1("decoder1");
  decoder1(ones, ones_segments);
  decoder decoder2("decoder2");
  decoder2(tens, tens_segments);

  counter counter1("counter1");
  counter1(tens, ones, clk);

  for (int i = 0 ; i < 12 ; i++) {
    sc_start(clk, 1);
    printf("%d %d %x %x\n",
           (int)tens.read(), (int)ones.read(),
           (int)(sc_uint<7>)tens_segments.read(),
           (int)(sc_uint<7>)ones_segments.read());
  }

}
```

FIGURE 15.11 A SystemC model for a two-digit decimal counter driving two seven-segment displays.

For the simulation of a nondeterministic model to be interesting, there needs to be some way of seeing the different possible behaviors, yet a nondeterministic artifact such as an SC_METHOD process that holds state provides no mechanism for ensuring that it is not, in fact, predictable. As a result, a designer has a hard time answering whether a model of a nondeterministic system can exhibit undesired behavior, even through a careful selection of test cases.

15.4 Verification Languages

Thanks to dramatic improvements in integrated circuit fabrication technology, it is now possible to build very large, complex integrated circuits. Hardware description languages, logic synthesis, and automated place-and-route technology have similarly made it possible to design such complicated systems. Unfortunately, technology to validate these designs, i.e., to identify design errors, has had a hard time keeping pace with these trends.

The time required to validate a design now greatly outstrips the time required to design or fabricate it. Although many novel techniques have been proposed to address the validation problem, simulation remains the preferred method. So-called formal verification techniques, which amount to efficient exhaustive simulation, have been gaining ground, but suffer from capacity problems.

Simulation applies a stimulus to a model of a design to predict the behavior of the fabricated system. Naturally, there is a trade-off between highly detailed models that can predict many attributes, say, logical values, timing, and power consumption, and simplified models that can only predict logical behavior but execute much faster.

Because the size of the typical design has grown exponentially over time, functional simulation, which only predicts the logical behavior of a synchronous circuit at clock-cycle boundaries, has become the preferred form of simulation because of its superior speed. Furthermore, designers have shied away from more timing-sensitive circuitry such as gated clocks and transparent latches because they require more detailed simulation models and are therefore more costly to simulate.

Simulation-based validation raises three important questions: what the stimulus should be, whether it exposes any design errors, and whether the stimulus is long and varied enough to have exposed "all" design errors. Historically, these three questions have been answered manually, i.e., by having a test engineer write test cases, check the results of simulation, and make some informed guesses about how comprehensive the test suite actually is.

A manual approach has many shortcomings. Writing test cases is tedious and the number necessary for "complete" verification grows faster than the size of the system description. Manually checking the output of simulation is similarly tedious and subtle errors can be easily overlooked. Finally, it is difficult to judge quantitatively how much of a design has really been tested.

More automated methodologies, and ultimately languages, have evolved to address some of these challenges, although the verification problem remains one of the most difficult. Biased random test case generation has become standard practice, although it has only supplemented manual test case generation, not completely supplanted it. Designer-inserted assertions, long standard practice for software development, have also become standard for hardware, although the sort of assertions needed in hardware are more complicated than the typical "the argument must be nonzero" sort of checking that works well in software. Finally, automated "coverage" checking, which attempts to quantify how much of a design's behavior has been exercised in simulation, has also become standard.

All of these techniques, while an improvement, are not a panacea. While biased random test case generation can quickly generate many interesting tests, it provides no guarantees of completeness, meaning bugs may go unnoticed. Because they must often check a temporal property (e.g., "acknowledge arrives within three cycles of every request"), good assertions in hardware systems are much more difficult to write than those for software (which most often check data structure consistency) and again, there is no way to know when enough assertions have been added, and it is possible that the assertions themselves have flaws (e.g., they let bugs by). Finally, test cases that achieve 100% coverage can also let bugs by because the criteria for coverage are necessarily weak. Coverage typically checks what states particular variables have been in, but it cannot consider all combinations because they grow exponentially quickly with design size. As a result, certain important combinations may not be checked even though coverage checks report "complete coverage."

While the utility of biased random test generation and coverage metrics is mostly limited to simulation, assertion specification techniques are useful for, and have been heavily influenced by, formal verification. Pure formal techniques consider all possible behaviors by definition and therefore do not require explicit test cases (implicitly, they consider all possible test cases) and also do not need to consider coverage. But knowing what behavior is unwanted is crucial for formal techniques, whose purpose is to either expose unwanted behavior or formally prove it cannot occur.

Recently, a sort of renaissance has been occurring in verification languages. Temporal logics, specifically linear temporal logic (LTL) and computation tree logic (CTL), form the mathematical basis for most assertion checking, but their traditional mathematical syntax is awkward for hardware designers. Instead, a number of more traditional computer languages, which combine a more human-readable syntax for the bare logic with a lot of "syntactic sugar" for more naturally expressing common properties, have been proposed for expressing properties in these logics. Two industrial efforts from Intel (ForSpec) and IBM (Sugar) have emerged as the most complete ones.

Meanwhile, some EDA companies have produced languages designed for writing testbenches and checking simulation coverage. Vera, originally designed by Systems Science and since acquired by Synopsys, and e, designed and sold by Verisity, acquired in 2005 by Cadence, have been the two most commercially successful. Bergeron [40] discusses how to use these two languages.

All four of these languages have recently undergone extensive crossbreeding. Vera has been made public, rechristened Open Vera, had Intel's ForSpec assertions grafted onto it, and added almost in its entirety to SystemVerilog. Sugar, meanwhile, has been adopted by the Accelera standards committee, rechristened the Property Specification Language (PSL), and has also been added in part to SystemVerilog. Verisity's e has changed the least, having only recently been made public, but is being standardized by the IEEE in 2005 (P1647).

There are obvious advantages in having a single industry-standard language for assertions, so it seems likely that most of these languages will eventually disappear, but as of this writing (2004), there is no obvious winner.

The sections that follow describe languages that are currently public. As mentioned above, most started as proprietary in-house or commercial.

15.4.1 Open Vera

OpenVera began life around 1995 as Vera, a proprietary language implemented by Systems Science, Inc. mainly for creating testbenches for Verilog simulations. As such, its syntax was heavily influenced by Verilog. VHDL support was added later. Synopsys bought the company in 1998, rechristened it OpenVera, and released the language to the public in 2001.

OpenVera is a concurrent, imperative language designed for writing testbenches. It executes in concert with a Verilog or VHDL simulator and can both provide stimulus to the simulator and observe the results. In addition to the usual high-level imperative language constructs, such as conditionals, loops, functions, strings, and associative arrays, it provides extensive facilities for generating biased random patterns (designed to be applied to the hardware design under test) and monitoring what values particular variables take on during the simulation (for checking coverage).

In the process of making OpenVera public, Synopsys added the assertion specification capabilities of Intel's ForSpec language, making it easy to check whether certain behaviors ever appear during the simulation.

In a further nod to crossbreeding, around 2003 much of Vera was incorporated into System Verilog 3.1. In particular, its style of generating biased random variables and checking for behavior coverage have been adopted more or less verbatim.

The following is a quick overview of OpenVera 1.3, the latest version as of this writing (September 2004). References for the language include the OpenVera language reference manual, available from the OpenVera website, and Haque et al. [41].

OpenVera has three main facilities that separate it from more traditional programming languages: biased random variable generation subject to constraints, monitoring facilities for reporting coverage of state variables, and the ability to specify and check temporal assertions.

Figure 15.12 shows a simple OpenVera program that demonstrates the language's ability to generate biased random variables. First, a Java/C++-like class is defined containing the 16-bit field addr and the 32-bit field data. These are marked rand, meaning their values will be set by a call to the randomize() method implicitly defined for every class.

Following the definition of the fields, a constraint (named arbitrarily word_align) is defined for objects of this class. Such constraints restrict the possible values the randomize() method may assign to various fields. This particular constraint simply restricts the two lowest bits of addr to be zero, i.e., aligned on a four-byte boundary.

The "demo" program creates a new Bus object then, for 50 times, invokes randomize() to generate a new set of "random" values (in fact, they are taken from a pseudorandom sequence guaranteed to be the same each time the program runs) for the two fields in the bus object that conform to the given constraint.

Overly constrained or inconsistent constraints may lead randomize() to fail. It signals this with a non-OK return value (OK and FAIL are reserved words in Vera), here assigned to the local integer variable result. For this simple example, randomization will always succeed. The constraint solver guarantees that it will only fail if there is no consistent solution to the supplied constraints.

This example of course only illustrates a fraction of OpenVera's facilities for biased random variable generation. It can also add constraints on-the-fly, impose set membership constraints, follow user-defined distributions, impose conditional constraints, impose constraints between variables, selectively randomize and disable randomization of user-specified variables, and generate random sequences from a language specified by a grammar.

OpenVera has extensive support for so-called functional coverage checking, which can monitor state variables and state transitions (unlike software coverage, which usually monitors which statements and branches have been executed). It uses a bin model: each bin represents a particular state or transition and when a matching activity occurs, the counter for that bin is incremented. The number of bins that remain empty after simulation therefore gives a rough idea of what behavior has yet to be exercised.

The coverage_group construct defines a type of monitor. Each specifies a set of variables to monitor and an event that triggers a check of the variables, typically the positive edge of a clock. Like a class, these constructs must be explicitly instantiated and there may be multiple copies of each, especially useful when a coverage group has parameters. Figure 15.13 illustrates a simple coverage group.

Like the earlier example, Figure 15.13 shows only the most basic coverage functionality. Open Vera can also monitor cross coverage, i.e., the combinations of values taken by two or more variables; selectively

```
class Bus {
  rand bit[15:0] addr;
  rand bit[31:0] data;

  constraint world_align { addr[1:0] == '2b0; }
}

program demo {
  Bus bus = new();
  repeat (50) {
    integer result = bus.randomize();
    if (result == OK)
      printf("addr = %16h data = %32h\n", bus.addr, bus.data);
    else
      printf("randomization failed: overconstrained.\n");
  }
}
```

FIGURE 15.12 A simple Vera program illustrating its ability to generate biased random variables and its mix of imperative and object-oriented styles. From the OpenVera 1.3 LRM.

```
bit clk;
bit [15:0] addr;
bit [7:0] data;

coverage_group MyChecker {
  sample_event = @(posedge clk);
  sample addr, data;
}

program demo_coverage {
  MyChecker checker = new();
  ...
}
```

FIGURE 15.13 A simple Vera program containing a single coverage group that monitors which values appear on the addr and data state variables. From the OpenVera 1.3 LRM.

disable coverage checking, e.g., during system reset; allow the user to specify bins explicitly and the values mapped to them; and monitor transition coverage, i.e., combinations of values taken by the same variable in successive cycles.

OpenVera's assertions, which were adapted from Intel's ForSpec language, provide a way to check temporal properties such as "an acknowledge signal must occur within three clock cycles after any request." Because of its source, the syntax for assertions is a little unusual.

Figure 15.14 illustrates the assertion language. It defines a checker called "mychecker" that takes a single integer parameter *p*, an enable signal, a clock, and an eight-bit bus, and looks for the sequence *p* (4), 6, 9, 3 on the bus. The clock construct defines a collection of events (patterns) synchronized to the positive edge of the clock signal. Events e0 through e3 are very simple properties: they look for particular patterns on the result signal. The myseq pattern is the interesting one: it looks for the appearance of the four events separated by a single clock cycle (a one-cycle delay is written 2 #1). The assert directive means to check for the myseq event and report an error otherwise.

Finally, the bind directive indicates where to instantiate the checker, names the particular instance of it myinst, passes the parameter 4, and connects the checker to the enable, clock, and result signals.

Once more, the example in Figure 15.14 barely scratches the surface. Sequences can also contain consecutive and nonconsecutive repetition, explicit delays, simultaneous and disjoint sequences, and sequence containment.

15.4.2 The e Language

The e language was developed by Verisity as part of its Specman product as a tool for efficiently writing testbenches. Like Vera, it is an imperative object-oriented language with concurrency, the ability to generate constrained random values, mechanisms for checking functional (variable value) coverage, and a way to check temporal properties (assertions). Books on e include Palnitkar [42] and Iman and Joshi [43].

The syntax of e is a little unusual. First, all code must be enclosed in <' and '>' symbols, otherwise it is considered a comment. Unlike C, e declarations are written "name : type." The syntax for fields in compound types (e.g., structs) includes particles such as % and !, which indicate when a field is to be driven on the device-under-test and not randomly computed, respectively.

Figure 15.15 shows a fragment of an e program that defines an abstract test strategy for a very simple microprocessor, specifically how to generate instructions for it. It illustrates the type system of the language as well as the utility of constraints. It defines two enumerated scalar types, opcode and reg, and defines the width of each. The struct instr defines a new compound type (instr) that represents a single instruction. First is the op field, which is one of the opcodes defined earlier. It, the op1, and the op2 fields are marked with %, indicating that they should be considered by the pack built-in procedure, which marshals data to send to the simulation.

The kind field is also an enumerated scalar, but is used here as a type tag. It is not marked with %, which means that its value will not be included when the structure is packed and sent to the simulation.

```
/* Checks for the sequence p, 6, 9, 3 */
unit mychecker
  #(parameter integer p = 0)
  (logic en, logic clk, logic [7:0] result);

  clock posedge (clk) {
    event e0: (result == p);
    event e1: (result == 6);
    event e2: (result == 9);
    event e3: (result == 3);
    event myseq: if (en) then (e0 #1 e1 #1 e2 #1 e3);
  }

  assert myassert: check(myseq, "Missed a step.");
endunit

/* Watch for the sequence 4, 6, 9, 3 on the outp bus */
bind instances cnt_top.dut:
  mychecker myinst #(4) (enable, clk, outp);
```

FIGURE 15.14 A sample of OpenVera's assertion language. It defines a checker that watches for the sequence p, 6, 9, 3 and creates an instance of it that checks for the sequence 4, 6, 9, 3 on the outp bus. From the OpenVera 1.3 LRM.

The two when directives define two subtypes, i.e., "reg instr" and "imm instr." Such subtypes are similar to derived classes in object-oriented programming languages. Here, the value of the kind field, which can be either imm or reg, determines the subtype.

The two keep directives impose constraints between the kind field and the opcode, ensuring, e.g., that ADD and SUB instructions are of the reg type. Although these constraints are simple, e is able to impose much more complicated constraints on the values of fields in a struct.

The final when directive further constrains the JMP and CALL instructions, i.e., by restricting what values the op2 field may take for these instructions.

The extend sys directive adds a field named instrs to the sys built-in structure, which is the basic environment. The leading ! makes the system create an empty list of instructions, which will be filled in later.

Figure 15.16 illustrates how the definition of Figure 15.15 can be used to generate tests that exercise the ADD and ADDI instructions. It first adds constraints to the instr class (the template for instructions defined in Figure 15.15) that restrict the opcodes to either ADD or ADDI, then imposes a constraint on the top level (sys) that makes it generate exactly ten instructions. Running the source code of Figure 15.15 and Figure 15.16 together makes the system generate a sequence of ten pseudorandom instructions.

15.4.3 Property Specification Language

The property specification language, PSL, evolved from the proprietary Sugar language developed at IBM. Its focus is narrower than either OpenVera or e, since its goal is purely to specify temporal properties to be checked in hardware designs, but is more disciplined and has more formal semantics.

Beer et al. [44] provide a good introduction to an earlier version of the language, which they explain evolved over many years. It has been used within IBM in the RuleBase formal verification system since 1994 and was also pressed into service as a checker generator for simulators in 1997. Accelera, an EDA standards group, adopted it as their formal property language in 2002. Cohen [45] provides instruction on the language.

PSL is based on CTL [46], a powerful but rather cryptic temporal logic for specifying properties of finite-state systems. It is able to specify both safety properties ("this bad thing will never happen") as well as liveness properties ("this good thing will eventually happen"). Liveness properties can only be checked formally because they make a statement about all the possible behaviors of a system while safety properties can also be tested in simulation. Linear Temporal Logic (LTL), a subset of CTL, expresses only safety properties and can therefore be turned into checking automata meant to be run in concert with a simulation

```
Instruction encoding for a very simple processor
<'
type opcode : [ ADD, SUB, ADDI, JMP, CALL ] (bits: 4);
type reg : [ REG0, REG1, REG2, REG3 ] (bits: 2);

struct instr {
  %op  : opcode;      // Four-bit opcode
  %op1 : reg;         // Two-bit operand
  kind : [imm, reg]; // Whether instruction is immediate or register

  when reg instr { %op2 : reg; }  // Second operand register
  when imm instr { %op2 : byte; } // Second operand and immediate byte

  // Constrain certain instructions to be register, immediate
  keep op in [ ADD, SUB ] => kind == reg;
  keep op in [ ADDI, JMP, CALL ] => kind == imm;

  // Constrain the second operand for JMP and CALL instructions
  when imm instr {
    keep opcode in [JMP, CALL] => op2 < 16;
  }
};

extend sys {
  !instrs : list of instr; // Add a non-generated field called instrs
};
'>
```

FIGURE 15.15 e Code defining instruction encoding for a simple eight-bit microprocessor. After an example in the Specman tutorial.

to look for unwanted behavior. PSL carefully defines which subset of its properties are purely LTL and are therefore candidates for use in simulation-based checking.

PSL is divided into four layers. The lowest, Boolean, consists of instantaneous Boolean expressions on the signals in the design under test. The syntax of this layer follows that of the HDL to which PSL is being applied, and can be Verilog, SystemVerilog, VHDL, or GDL. For example, a[0:3] & b[0:3] and a(0 to 3) and b(0 to 3) represent the bit-wise *and* of the four most significant bits of vectors a and b in the Verilog and VHDL flavors, respectively.

The second layer, temporal, is where PSL gets interesting. It allows a designer to state properties that hold across multiple clock cycles. The always operator, which states that a Boolean expression holds in every clock cycle, is one of the most basic. For example, always ! (ena & enb) states that the signals ena and enb will never be true simultaneously in any clock cycle.

More interesting are operators that specify delays. The next operator is the simplest one. The property always (req -> next ack) states that in every cycle that the req signal is true, the ack signal is true in the next cycle. The -> symbol denotes implication, i.e., if the expression to the left is true, that on the right must also be true. The next operator can also take an argument, e.g., always req -> next [2] ack means that ack must be true two cycles after each cycle in which req is true.

PSL provides an extended form of regular expressions convenient for specifying more complex behaviors. Although it would be possible to write always (req -> next (ack -> next ! cancel)) to indicate that ack must be true after any cycle in which req is true, and cancel must be false in the cycle after that, it is much easier instead to write always {req ; ack ; ! cancel}. This illustrates a basic principle of PSL: most operators are actually just "syntactic sugar;" the set of fundamental operators is quite small.

PSL draws a clear distinction between "weak" operators that can be checked in simulation (i.e., safety properties) and "strong" operators, which express liveness properties and can only be checked formally. Strong operators are marked with a trailing exclamation point (!), and some operators come in both strong and weak varieties.

```
<'
extend instr {
  keep opcode in [ADD, ADDI];
  keep op1 == REG0;
  when reg instr { keep op2 == REG1; }
  when imm instr { keep op2 == 0x3; }
};

extend sys {
   keep instrs.size() == 10;
};
'>
```

FIGURE 15.16 e Code that uses the instruction encoding of Figure 15.15 to randomly generate ten instructions. After an example in the Specman tutorial.

The eventually! operator illustrates the meaning of strong operators. The property always (req -> eventually! ack) states that after req is asserted, ack will always be asserted eventually. This is not something that can be checked in simulation: if a particular simulation saw req but did not see ack, it would be incorrect to report that this property failed because running that particular simulation longer might have produced ack. This is the fundamental difference between safety and liveness properties. Safety states something bad never happens; liveness states something good eventually happens.

Another subtlety is that it is possible to express properties in which time moves backward through a property. A simple example is always ((a && next [3] (b)) -> c, which states that when a is true and b is true three clock cycles later, c is true in the first cycle, i.e., when a is true. While it is possible to check this in simulation (for each cycle in which a is true, remember whether c is true and look three clock cycles later for b), it is more difficult to build automata that check such properties in simulation.

The third layer of PSL, the verification layer, instructs a verification tool what tests to perform on a particular design. It amounts to a binding between properties defined with expressions from the Boolean and temporal layer, and modules in the design under test. The following simple example:

```
vunit ex1a(top_block.i1.i2) {
  A1: assert never (ena && enb);
}
```

declares a "verification unit" called ex1a, binds it to the instance named top_block. i1. i2 in the design under test, and declares (the assertion named A1) that the signals ena and enb in that instance are never true simultaneously.

In addition to assert, verification units may also include assume directives, which state the tool may assume a given property; assume_guarantee, which both assumes and tests a particular property; restrict, which constrains the tool to only consider those behaviors that have the given property; cover, which asks the tool to check whether a certain property was ever observed; and fairness, which instructs the tool to only consider paths in which the given property occurs infinitely often, e.g., only when the system does not wait indefinitely.

The fourth, modeling layer of PSL essentially allows Verilog, System Verilog, VHDL, or GDL code to be included inline in a PSL specification. The intention here is to allow additional details about the system under test to be included in the PSL source file.

15.4.4 System Verilog

Recently, many aspects of the Vera, Sugar, and ForSpec verification languages have been merged into Verilog, along with the higher-level programming constructs of Superlog [47] (which were taken nearly verbatim from C and C++) to produce SystemVerilog [12]. As a result, Verilog has become the English of the HDL world: voraciously assimilating parts of other languages and making them its own.

```
class Bus;
  rand bit[15:0] addr;
  rand bit[31:0] data;

  constraint world_align { addr[1:0] = 2'b0; }
endclass

initial begin
  Bus bus = new;

  repeat (50) begin
    if (bus.randomize() == 1)
      $display("addr = %16h data = %h\n",
               bus.addr, bus.data);
    else
      $display("overconstrained: no satisfying values exist\n");
  end
end

typdef enum { low, mid, high } AddrType;

class MyBus extends Bus;
  rand AddrType atype; // Additional random variable

  // Additional constraint on address: still word-aligned}
  constraint addr_range {
    (atype == low ) -> addr inside { [0:15] };
    (atype == mid ) -> addr inside { [16:127] };
    (atype == high) -> addr inside { [128:255] };
  }
endclass

task exercise_bus;
  int res;

  // Restrict to low addresses
  res = bus.randomize() with { atype == low; };

  // Restrict to particular address range
  res = bus.randomize()
        with { 10 <= addr && addr <= 20 };

  // Restrict data to powers of two
  res = bus.randomize() with { data & (data - 1) == 0 };

  // Disable word alignment
  bus.word_align.constraint_mode(0);

  res = bus.randomize with { addr[0] || addr[1] };

  // Re-enable word alignment
  bus.word_align.constraint_mode(1);
endtask
```

FIGURE 15.17 Constrained random variable constructs in System Verilog. The example starts with a simple definition of a Bus class that constrains the two least-significant bits of the address to be zero, then invokes the randomize method to randomly generate address/data pairs and print the result. Next is a refined version of the Bus class that adds a field taken from an enumerated type that further constrains the address depending on its value. The example ends with a task that illustrates various ways to control the constraints. After examples in the SystemVerilog LRM [12].

The C- and C++-like features added to SystemVerilog read like the list of features in those languages. SystemVerilog adds enumerated types, record types *(structs)*, *typedefs*, type casting, a variety of operators such as +=, operator overloading, control-flow statements such as break and continue, as well as object-oriented programming constructs such as classes, inheritance, and dynamic object creation and deletion.

```
enum { red, green, blue } color;

bit [3:0] adr, offset;

covergroup g2 @(posedge clk);
  Hue:    coverpoint color;
  Offset: coverpoint offset;

  // Consider (color, adr) pairs, e.g.,
  // (red, 3'b000), (red, 3'b001), ..., (blue, 3'b111)
  AxC:    cross color, adr;

  // Consider (color, adr, offset) triplets
  // Creates 3 * 16 * 16 = 768 bins
  all:    cross color, adr, Offset;
endgroup

g2 g2_inst = new; // Create a watcher

bit [9:0] a; // Takes values 0--1023

covergroup cg @(posedge clk);

  coverpoint a {
    // place values 0--63 and 65 in bin a
    bins a = { [0:63], 65 };

    // create 65 bins, one for 127, 128, ..., 191
    bins b[] = { [127:150], [148:191] };

    // create three bins: 200, 201, and 202
    bins c[] = { 200, 201, 202 };

    // place values 1000--1023 in bin d
    bins d = { [1000:$] };

    // place all other values (e.g., 64, 66, .., 126, 192, ...) in their own bin
    bins others[] = default;
  }

endgroup

bit [3:0] a;

covergroup cg @(posedge clk);
  coverpoint a {
    // Place any of the sequences 4 -> 5 -> 6, 7 -> 11, 8 -> 11, 9 -> 11, 10 ->11,
    // 7 -> 12, 8 -> 12, 9 -> 12, and 10 -> 12 into bin sa.
    bins sa = (4 => 5 => 6), ([7:9],10 => 11,12);

    // Create separate bins for 4 -> 5 -> 6, 7 -> 10, 8 -> 10, and 9 -> 10
    bins sb[] = (4 => 5 => 6), ([7:9] => 10);

    // Look for the sequence 3 -> 3 -> 3 -> 3
    bins sc = 3 [* 4];

    // Look for any of the sequences 5 -> 5, 5 -> 5 -> 5, or 5 -> 5 -> 5 -> 5
    bins sd = 5 [* 2:4];

    // Look for any sequence of the form 6 -> ... -> 6 -> ... -> 6
    // where "..." represents any sequence that excludes 6
    bins se = 6 [-> 3];
  }
endgroup
```

FIGURE 15.18 SystemVerilog covers constructs. The example begins with a definition of a "covergroup" that considers the values taken by the color and offset variables as well as combinations. Next is a covergroup illustrating the variety of ways "bins" may be defined to classify values for coverage. The final covergroup illustrates SystemVerilog's ability to look for and classify sequences of values, not just simple values. After examples in the System Verilog LRM [12].

```
// Make sure req1 or req2 is true if we are in the REQ state}
always @(posedge clk)
  if (state == REQ)
    assert (req1 || req2);

// Same, but report the error ourselves
always @(posedge clk)
  if (state == REQ)
    assert (req1 || req2)
    else
      $error("In REQ; req1 || req2 failed (\%0t)", $time);

property req_ack;
  @(posedge clk) // Sample req, ack at rising clock edge
    // After req is true, between one and three cycles later,
    // ack must have risen.
    req ##[1:3] $rose(ack);
endproperty

// Assert that this property holds, i.e., create a checker
as_req_ack: assert property (req_ack);

// The own_bus signal goes high in 1 to 5 cycles,
// then the breq signal goes low one cycle later.
sequence own_then_release_breq;
  ##[1:5] own_bus ##1 !breq
endsequence

property legal_breq_handshake;
  @(posedge clk)      // On every clock,
  disable iff (reset) // unless reset is true,
  // once breq has risen, own_bus should rise and breq should fall.
  $rose(breq) |-> own_then_release_breq;
endproperty

assert property (legal_breq_handshake);
```

FIGURE 15.19 SystemVerilog assertions. The first two *always* blocks check simple safety properties, i.e., that req1 and req2 are never true at the positive edge of the clock. The next property checks a temporal property: that ack must rise between one and three cycles after each time req is true. The final example shows a more complex property: when reset is not true, a rising breq signal must be followed by own_bus rising between one and five cycles later and breq falling.

At the highest level, it also adds strings, associative arrays, concurrent process control (e.g., fork/join), semaphores, and mailboxes, giving it features only found in concurrent programming languages such as Java.

Perhaps the most interesting features added to SystemVerilog are those directly related to verification. Specifically, SystemVerilog includes constrained, biased random variable generation, user-defined functional coverage checking, and temporal assertions, much like those in Vera, e, PSL, and ForSpec.

Figure 15.17 illustrates some of the random test-generation constructs in SystemVerilog, which were largely taken from Vera. Compare this with Figure 15.12.

Figure 15.18 illustrates some of the coverage constructs in SystemVerilog. In general, one defines "covergroups," which are collections of bins that sample values on a given event, typically a clock edge. Each covergroup defines the sorts of values it will be observing (e.g., values of a single variable, combinations of multiple variables, and sequences of values on a single variable) and rules that define the "bins" each of these values will be placed in. In the end, the simulator reports which bins were empty, indicating that none of the matching behavior was observed. Again, much of this machinery was taken from Vera (cf. Figure 15.13).

Figure 15.19 shows some of SystemVerilog's assertion constructs. In addition to signaling an error when an "instantaneous" condition does not hold (e.g., a set of variables are taking on mutually incompatible values), SystemVerilog has the ability to describe temporal sequences such as "ack must rise between one and five cycles after req rises" and check whether they appear during simulation. Much of the syntax comes from PSL/Sugar.

15.5 Conclusions

VHDL and Verilog remain the dominant HDLs and are likely to be with us for a long time, although perhaps they will become what assembly language has become to programming: a part of the compilation chain, but not generally written manually. Both have deep flaws, but these can be largely avoided by adhering to coding conventions, and in practice are fairly practical design entry vehicles.

The future of the verification languages discussed in this chapter is less certain. Clearly, there is a need to automate the validation process as much as possible, and these languages do provide useful assistance in the form of biased constrained pseudorandom test case generation, temporal property assertions, and coverage estimates. However, none has clearly proven itself essential to modern IC design, and the plethora of variants and derivatives suggests that their evolution is not complete. An even more serious question is whether the complexity of these languages, especially in their specification of temporal properties, creates more problems than it solves.

The fundamental burdens of specifying digital hardware and verifying its correctness will continue to fall on design and verification languages. Even if those in the future bear little resemblance to those described here, the current crop forms a strong foundation on which to build.

References

[1] I.S. Reed, Symbolic synthesis of digital computers, in *Proceedings of the ACM National Meeting*, Toronto, Canada, September 1952, pp. 90–94.

[2] W.E. Omohundro and J. H. Tracey, Flowware — a flow charting procedure to describe digital networks, in *Proceedings of the First International Conference on Computer Architecture (ISCA)*, Gainesville, FL, December 1973, pp. 91–97.

[3] Y. Chu, An ALGOL-like computer design language, *Commn. ACM*, 8, 607–615, 1965.

[4] C. Gordon Bell and A. Newell, *Computer Structures: Readings and Examples*, McGraw-Hill, New York, 1971.

[5] Y. Chu, D.L. Dietmeyer, J.R. Duley, F.J. Hill, M. R. Barbacci, C.W. Rose, G. Ordy, B. Johnson, and M. Roberts, Three decades of HDLs: Part I, CDL through TI-HDL. *IEEE Design Test Comput.*, 9, 69–81, 1992.

[6] D. Borrione, R. Piloty, D. Hill, K.J. Lieberherr, and P. Moorby, 2nd Three decades of HDLs: Part II, Colan through Verilog, *IEEE Design Test Comput.*, 9, 54–63, 1992.

[7] F.I. Parke, An introduction to the N.mPc design environment, in *Proceedings of the 16th Design Automation Conference*, San Diego, CA, June 1979, pp. 513–519.

[8] Al Dewey, VHSIC hardware description (VHDL) development program, *Proceedings of the 20th Design Automation Conference*, Miami Beach, FL, 1983, pp. 625–628.

[9] R.K. Brayton, G.D. Hachtel, and A.L. Sangiovanni-Vincentelli, Multilevel logic synthesis, *Proc. IEEE*, 78, 264–300, 1990.

[10] *IEEE Standard Hardware Description Language Based on the Verilog Hardware Description Language (1364–1995)*, IEEE Computer Society, New York, NY, 1996.

[11] *IEEE Standard Verilog Hardware Description Language (1364–2001)*, IEEE Computer Society, New York, NY, September 2001.

[12] Accelera, 1370 Trancas Street #163, Napa, CA 94558. *SystemVerilog 3.1a Language Reference Manual: Accellera's Extensions to Verilog*, May 2004.

[13] P.L. Flake, P.R. Moorby, and G. Musgrave, An algebra for logic strength simulation, *Proceedings of the 20th Design Automation Conference*, Miami Beach, FL, June 1983, pp. 615–618.

[14] D.E. Thomas and P.R. Moorby, *The Verilog Hardware Description Language*, (5th ed.), Kluwer, Boston, MA, 2002.

[15] S. Palnitkar, *Verilog HDL: A Guide to Digital Design and Synthesis,* Prentice-Hall, Upper Saddle River, NJ, 1996.

[16] S. Mittra, *Principles of Verilog PLI,* Kluwer, Boston, MA, 1999.

[17] S. Sutherland, *The Verilog PLI Handbook,* Kluwer, Boston, MA, 1999.

[18] D.J. Smith, VHDL & Verilog compared & constrasted — plus modeled examples written in VHDL, Verilog, and C, in *Proceedings of the 33rd Design Automation Conference,* Las Vegas, Nevada, June 1996, pp. 771–776.

[19] R.S. French, M.S. Lam, J.R. Levitt, and K. Olukotun, A general method for compiling event-driven simulations, in *Proceedings of the 32nd Design Automation Conference,* San Francisco, CA, June 1995, pp. 151–156.

[20] M.J.C. Gordon, The semantic challenge of Verilog HDL, in *Proceedings of the Tenth Annual IEEE Symposium on Logic in Computer Science (LICS),* San Diego, CA, June 1995.

[21] A. Dewey and A.J. de Geus, VHDL: toward a unified view of design, *IEEE Design Test Comput.*, 9, 8–17, 1992.

[22] *IEEE Standard VHDL Reference Manual (1076–1987),* The Institute of Electrical and Electronics Engineers (IEEE), New York, NY, 1988.

[23] *IEEE Standard VHDL Language Reference Manual (1076–1993),* IEEE Computer Society, New York, NY, 1994.

[24] R. Lipsett, C.F. Schaefer, and C. Ussery, *VHDL: Hardware Description and Design,* Kluwer, Boston, MA, 1989.

[25] A.M. Dewey, *Analysis and Design of Digital Systems with VHDL,* Brooks/Cole Publishing (Formerly PWS), Pacific Grove, CA, 1997.

[26] J. Bhasker, *A VHDL Synthesis Primer.* 2nd ed., Star Galaxy Publishing, Allentown, PA, 1998.

[27] D.L. Perry, *VHDL,* 3rd ed., McGraw-Hill, New York, 1998.

[28] P.J. Ashenden, *The Designer's Guide to VHDL,* Morgan Kaufmann, San Francisco, CA, 1996.

[29] P.J. Ashenden, *The Student's Guide to VHDL,* Morgan Kaufmann, San Francisco, CA, 1998.

[30] B. Cohen, *VHDL Coding Styles and Methodologies,* , 2nd ed., Kluwer, Boston, MA, 1999.

[31] R.E. Harr and A.G. Stanculescu, Eds., *Applications of VHDL to Circuit Design,* Kluwer, Boston, MA, 1991.

[32] C. Mead and L. Conway, *Introduction to VLSI Systems,* Addison-Wesley, Reading, MA, 1980.

[33] *IEEE Standard Multivalue Logic System for VHDL Model Interoperability (Std_logic_1164),* The Institute of Electrical and Electronics Engineers (IEEE), New York, NY, 1993.

[34] *IEEE Standard VHDL Synthesis Packages (1076.3–1997),* The Institute of Electrical and Electronics Engineers (IEEE), New York, NY, 1997.

[35] S. Liao, S. Tjiang, and R. Gupta, An efficient implementation of reactivity for modeling hardware in the Scenic design environment, in *Proceedings of the 34th Design Automation Conference,* Anaheim, CA, June 1997, pp. 70–75.

[36] G. Arnout, SystemC standard, *Proceedings of the Asia South Pacific Design Automation Conference (ASP-DAC),* Yokohama, Japan, January 2000, pp. 573–578.

[37] T. Grötker, S. Liao, G.Martin, and S. Swan, *System Design with SystemC,* Kluwer, Boston, MA, 2002.

[38] J. Bhasker, *A SystemC Primer,* 2nd ed., Star Galaxy Publishing, Allentown, PA, 2004.

[39] W. Müller, W. Rosenstiel, and J. Ruf, Eds., *SystemC: Methodologies and Applications,* Kluwer, Boston, MA, 2003.

[40] J. Bergeron, *Writing Testbenches: Function Verification of HDL Models,* 2nd ed., Kluwer, Boston, MA, 2003.

[41] F. Haque, J. Michelson, and K. Khan, *The Art of Verification with Vera,* Verification Central, www.verificationcentral.com, 2001.

[42] S. Palnitkar, *Design Verification with e,* Prentice-Hall, Upper Saddle River, NJ, 2003.

[43] S. Iman and S. Joshi, *The e Hardware Verification Langauge,* Kluwer, Boston, MA, 2004.

[44] I. Beer, S. Ben-David, C. Eisner, D. Fisman, A. Gringauze, and Y. Rodeh, The temporal logic Sugar, in *Proceedings of the 13th International Conference on Computer-Aided Verification (CAV), Lecture Notes in Computer Science,* Paris, France, Springer, Berlin, 2001, Vol. 2102, pp. 363–367.

[45] B. Cohen, S. Venkataramanan, and A. Kumari, *Using PSL/Sugar for Formal Verification,* VhdlCohen Publishing, Palos Verdes Peninsula, CA, 2004.

[46] E.M. Clarke and E.A. Emerson, Design and synthesis of synchronization skeletons using branching time temporal logic, in *Proceedings of the Workshop on Logic of Programs, Lecture Notes in Computer Science,* Yorktown Heights, New York, May 1981. Springer, Berlin, Vol. 131, pp. 52–71.

[47] P.L. Flake and S.J. Davidmann, Superlog, a unified design language for system-on-chip, *Proceedings of the Asia South Pacific Design Automation Conference (ASP-DAC),* Yokohama, Japan, January 2000, pp. 583–586.

16

Digital Simulation

John Sanguinetti
Forte Design Systems, Inc.
San Jose, California

16.1 Introduction

Logic simulation is the primary tool used for verifying the logical correctness of a hardware design. In many cases logic simulation is the first activity performed in the process of taking a hardware design from concept to realization. Modern hardware description languages are both simulatable and synthesizable. Designing hardware today is really writing a program in the hardware description language. Performing a simulation is just running that program. When the program (or model) runs correctly, then one can be reasonably assured that the logic of the design is correct, *for the cases that have been tested in the simulation.*

Simulation is the key activity in the design verification process. That is not to say that it is an ideal process. It has some very positive attributes:

1. It is a natural way for the designer to get feedback about his design. Because it is just running a program — the design itself — the designer interacts with it using the vocabulary and abstractions of the design. There is no layer of translation to obscure the behavior of the design.
2. The level of effort required to debug and then verify the design is proportional to the maturity of the design. That is, early in the design's life, bugs and incorrect behavior are usually found more quickly. As the design matures, it takes longer to find the errors.
3. Simulation is completely general. Any hardware design can be simulated. The only limits are time and computer resources.

On the negative side, simulation has two drawbacks, one of which is glaring:

1. There is (usually) no way to know when you are done. It is not feasible to completely test, via simulation, all possible states and inputs of any nontrivial system.
2. Simulation can take an inordinately large amount of computing resources, since typically it uses a single processor to reproduce the behavior of many (perhaps millions of) parallel hardware processes.

Every design project must answer the question "have we simulated enough to find all the bugs?" and every project manager has taped out his design knowing that the truthful answer to that question is either "no" or "I don't know." It is this fundamental problem with simulation that has caused so much effort to be spent looking for both tools to help answer the question, and formal alternatives to simulation.

Code coverage, functional coverage, and logic coverage tools have all been developed to help gauge the completeness of simulation testing. None are complete solutions, though they all help. Formal alternatives have been less successful. Just as in the general software world where proving programs correct has proven intractable, formal methods for verifying hardware designs have still not proven general enough to replace simulation. That is not surprising, since it is the same problem.

The second drawback motivates much of the material in this chapter. That is, simulation is always orders of magnitude slower than the system being simulated. If a hardware system runs at 1 GHz, a simulation of that system might run at 10–1000 Hz, depending on the level of the simulation and the size of the system. That is a slowdown from 10^6 to 10^8! Consequently, many people have spent considerable time and effort finding ways to speed up logic simulation.

Considering both the advantages and disadvantages of logic simulation, it is really quite a good tool for verifying the correctness of a hardware design. Despite its drawbacks, simulation remains the first choice for demonstrating correctness of a design before fabrication, and its value has been well established.

16.1.1 Levels of Abstraction

Because simulation is a general technique, a hardware design can be simulated at different levels of abstraction. Often it is useful to simulate a model at several levels of abstraction in the same simulation run. The commonly used levels of abstraction are gate level, register transfer level (RTL), and behavioral (or algorithmic) level. However, it is possible to incorporate lower levels like transistor level or even lower physical levels as well as higher levels such as transaction level or domain-specific levels. For this discussion, we will restrict our attention to behavioral, RTL, and gate levels, with the understanding that other levels are completely compatible with the techniques described here.

16.1.2 Discrete Event Simulation

"we decided to focus on simulation ..., because that's the only really interesting thing to do with a computer" (Alan Kay, Second West Coast Computer Faire, 1978)

Logic simulation is a special case of the more general discrete event simulation methods, which were initially developed in the 1960s [5]. Discrete event simulation is a method of representing the behavior of a system over time, using a computer. The system may be either real or hypothetical, but its state is assumed to change over time due to some combination of external stimulus and internal state. The fundamental idea is that the behavior of any system can be decomposed into a set of discrete instants of time at which things happen. Those instants are called events, and the "things that happen" are state changes. This is very analogous to the way we digitize continuous physical phenomena, like audio sampling. In essence, we digitize a time-based process by dividing it up into discrete events. It is easy to see that with a fine enough granularity, one can get an adequately accurate representation for just about any purpose.

The basic operation of a discrete event simulation is given in Figure 16.1.

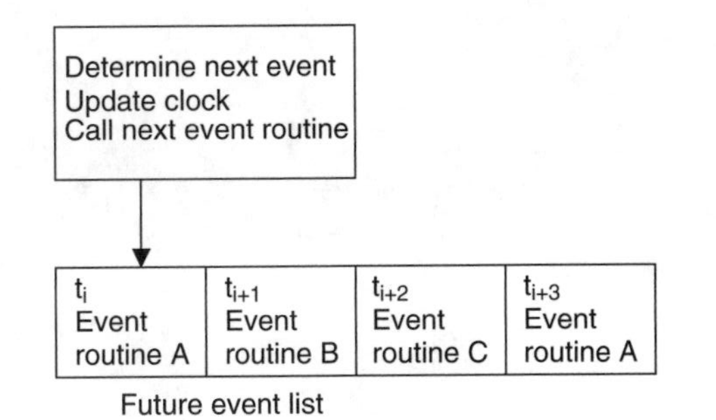

FIGURE 16.1 Discrete event simulation basic operation.

Each event in the system is represented by an event routine. An event routine is some code to be executed to represent the action at that event, which usually amounts to a state change and a determination of the occurrence of one or more future events. The set of possible events is usually (though not always) fixed, but the number of occurrences of each type of event can be variable.

The list of events shown in Figure 16.1 is a list of those events that are scheduled to happen at known times in the future ($t_{i+1}, t_{i+2}, \ldots$). The entries on this list contain the time at which the event will occur, and the type of event it is; that is, what event routine should be called to realize that event's behavior. It is common to have more than one instance of any particular type of event scheduled to occur at various times on this list. This list is called the future event list.

16.2 Event- vs. Process-Oriented Simulation

The basic discrete event simulation paradigm was in use in the early 1960s (and probably earlier), and several subroutine libraries were developed for FORTRAN and Algol to support it. It was soon realized that there were more natural ways to write simulation models of many systems, and that led to the development of specialized simulation languages, particularly GPSS, SIMSCRIPT, and Simula [3]. Of these, GPSS is still used today to model systems as extensive as the U.S. air traffic control system. Simula was the first object-oriented programming language and introduced the concepts of classes and objects.

Both GPSS and Simula took a process-oriented view of modeling. That is, instead of writing a separate routine for each event, one would write a routine that represented a process which might include several events within it taking place over a nonzero period of simulated time. The routine would have wait statements in it to indicate that some time would pass between one event and the next. This did not affect the underlying simulation mechanism, but it did change the way the simulation model was written. A process orientation is particularly useful for hardware simulation at a behavioral level, while an event orientation works for RTL and gate levels. The most popular hardware description languages, Verilog [6,7], VHDL [2], and SystemC [4], are process-oriented simulation languages which are indirectly descended from Simula.

16.3 Logic Simulation Methods and Algorithms

Logic simulation is essentially a process of computing a state trajectory of the system over time. The system's state is defined by the state variables, which are the storage elements in the circuit. We usually think of these as the registers and latches in the design. How the state, taken as a whole, changes over time is called the state trajectory.

A simple state trajectory as a function of time can be written as follows:

$$f(m, n, t+1) = m_t + n_t$$

where m is a state variable and n is an input.

It is obvious how to compute this function:

$$
\begin{aligned}
&\text{Do } (i = 1, \text{endtime}) \\
&\quad t = t + 1 \\
&\quad m = m + n \\
&\quad n = \text{new input} \\
&\text{enddo}
\end{aligned}
$$

This example illustrates how a function is simulated over time. There is a loop, which consists of updating the system time or clock, evaluating the logic components, and updating the state variables. Compare this loop to the process of Figure 16.1. There is only one event routine, and it gets executed repeatedly at a regular time interval.

Now we can expand this example to two functions,

$$f(m, n, t + 1) = m_t + n_t$$
$$g(x, n, t + 1) = x_t \wedge n_t$$

Evaluating these two functions together would look like this:

$$
\begin{aligned}
&\text{Do } (i = 1, \text{endtime}) \\
&\quad t = t + 1 \\
&\quad m = m + n \\
&\quad x = x \wedge n \\
&\quad n = \text{new input} \\
&\text{enddo}
\end{aligned}
$$

This set of state functions when taken together, make up a system with a state variable, which is a duple (m, x). Running this simulation computes the trajectory of the state vector (m, x) over the simulation interval.

Notice that the way we have defined the functions f and g, the value of n used in the computation is the value of n at time t, even though the time of the evaluation is $t + 1$. This leads to a further refinement of the simulation process. In real hardware, state variables (registers or latches) do not get updated instantaneously. That is, the computation even if it is a simple assignment, takes some nonzero amount of time. So the simulation loop could be rewritten as in Figure 16.2.

```
do (i=1, endtime)
  t = t + 1
  // evaluate the functions
  t_m = m + n
  t_x = x ^ n
  t_n = new input
  // update the variable values
  m = t_m
  x = t_x
  n = t_n
enddo
```

FIGURE 16.2 Basic simulation loop.

While we have not represented the actual delays involved in the computation, we have abstracted them into a behavior that takes into account their effect. We now have the typical event-processing loop in a digital logic simulation: advance the clock, evaluate all the logic functions, and update the variable values.

Note that this has incorporated an abstraction of physical behavior over time, since we are moving time from one discrete moment to another, and we are assuming that nothing interesting happens in between those two instants. We also assumed that m, x, and n do not change their values instantaneously, but they do change before the next time instant $(t + 1)$. Note that they all change together after all the evaluation has been done. This is usually called a "delta cycle," meaning it is a set of events that happen at time t, but after all the evaluation events. That is, the new-value assignments happen at $t + \delta t$.

A further complication arises when time delays must be introduced into a computation. That is, the function being computed may look like this:

$$f(m, n, t) = g(m_{t-k}, n_{t-k})$$

That is, the new value does not get updated into the state variable until k time units after the evaluation. To handle this, the simulation loop would look like Figure 16.3.

It is common practice to replace the array *t_m*() in Figure 16.3 with a linked list, which is ordered by the index t. In order to make this work, the delay is usually incorporated into the new-value computation, and the new value is put on a future update list at time $t + k$. The future update list is analogous to the future event list, and in fact does not need to be separate from it. The simulation loop would then look like Figure 16.4.

Here we have assumed that the future update list has values only for the variable m. Generally, there may be many different state variables whose new values may be saved on the future update list, so each entry

```
do (i=1, endtime)
  t = t + 1
  // evaluate the functions
  t_m(t+k) = m + n
  t_n = new input
  // update the variable values
  m = t_m(t)
  n = t_n
enddo
```

FIGURE 16.3 Simulation loop with update step.

```
do (i=1, endtime)
  t = t + 1
  // evaluate the functions
  t_m= m + n
  put_on_future_update_list(t_m, t+k)
  t_n = new input
  // update the variable values
  m = take_off_future_update_list(t)
  n = t_n
enddo
```

FIGURE 16.4 Simulation loop with update list.

on the list must identify the state variable that the value is associated with. Note also that the new value is always at the front of the list, if the list is ordered by increasing time t. That is, at any given time t, there are no values on the list with time $<t$ (those would be values in the past), so all the variable updates can be found at the front of the list. Of course, there may be no values on the list associated with time t, that is, the first entry on the list may have time $> t$, in which case no variable would be updated at time t.

16.3.1 Synchronous and Asynchronous Logic

The example above shows synchronous, or sequential, logic. That is, each variable's new value is computed at a regular interval $(t, t + 1, \ldots)$. There is also asynchronous, or combinational, logic that might be present in a system to be simulated. The salient characteristic of asynchronous logic is that the function is computed continuously. That is, when the inputs change, the output changes immediately, at least as compared to the granularity of the simulation time. For example, in an and gate followed by an inverter, the two inputs are *and*ed into the inverter whose output is created with no (apparent) delay (Figure 16.5).

This would be represented by $d = \sim(c \,\&\, b)$. In terms of our simulation loop, asynchronous logic functions have their outputs updated during the evaluation phase, not delayed until the update phase. That is, there is no temporary variable created to hold the new value.

It now is apparent that we need some way of distinguishing between the kind of assignment that is immediate and the previous, synchronous, kind, which is delayed. A simulation language typically has the ability to represent instantaneous change as well as delayed change.

16.3.2 Propagation

Combinational logic leads to a further complication of our simulation loop, namely, where do we do the evaluation of a combinational expression? Combinational logic must be evaluated whenever one of its constituent inputs changes. Sequential logic can be evaluated regularly, at time intervals corresponding to a clock signal. That is why it was natural to write the simulation loop with the evaluations happening immediately after updating the time variable. But where do we put combinational logic update?

The answer can be found by looking at what would cause a combinational expression to be evaluated. That is, when would its inputs change? If we look at a typical circuit that has both combinational and sequential logic, it might look like Figure 16.6.

Looking at this, we see that the input to a combinational expression is either the value of a state variable (a register) or the output of another combinational expression. Going back to the simulation loop, the logical place to put the evaluation of combinational expressions is after the update of state variables. If our combinational logic consisted of

$$c = m \mid a$$
$$d = \sim(c \,\&\, b)$$

Then the loop would look like Figure 16.7.

The simulation loop as it is described in Figure 16.7 is now sufficient to simulate circuits that have the form of Figure 16.6, but with an important caveat. That caveat is that the combinational expressions must be ordered such that each one is evaluated only after all of its inputs have been updated. If a circuit does not contain any combinational feedback loops, then it is possible to satisfy this ordering requirement. Ordering the expressions is called *levelizing*.

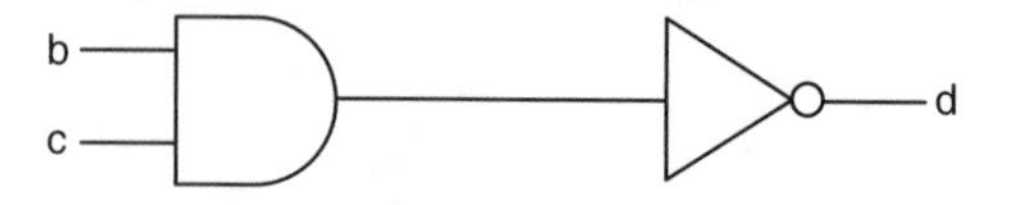

FIGURE 16.5 Example asynchronous logic.

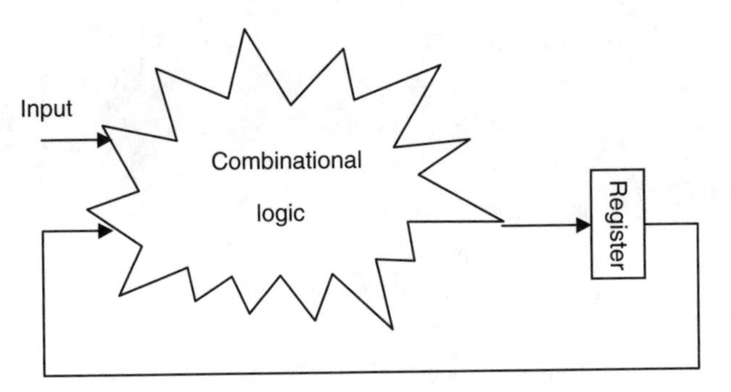

FIGURE 16.6 Example with synchronous and asynchronous logic.

```
do (i=1, endtime)
  t = t + 1
  // evaluate the state variables
  t_m = m + n
  put_on_future_update_list(t_m, t)
  t_n = new input
  // update the variable values
  m = take_off_future_update_list(t)
  n = t_n
  // update all combinational expressions
  c = m | a
  d = ~(c & b)
enddo
```

FIGURE 16.7 Simulation loop with asynchronous logic.

If the delays for the sequential variables are all zero, which is a common case, then this simulation loop is quite efficient, and in fact many logic simulators have been created with just this simulation loop. The drawback of this approach however, is that all of the expressions in the circuit must be evaluated on every loop iteration, i.e., at each time instant. When very much of the circuit remains unchanging for long periods, this can be quite wasteful, as expressions that do not change get continually reevaluated.

To improve the simulation efficiency in this common case, we can return to the basic event scheduling idea of discrete event simulation, and only evaluate an expression when one or more of its inputs has changed. This is called *propagation* of a value from an output to an input, or in language terms, from a left-hand side of an assignment to a right-hand side. The simulation loop now would look like Figure 16.8.

Just as we introduced a future update list, we can also create a propagation list, where the entry on the list is the expression whose inputs have changed. Rewriting the simulation loop using that, we get Figure 16.9.

Notice here that the propagate list contains a pointer or other identifier of the expression that needs to be evaluated. Notice also that the expressions that depend on any variable are given by the function $c(v)$. The function $c(v)$ is static, since in logic simulation, the expressions that depend on any variable are known at compile time. Finally, notice that it is a simple efficiency improvement to only propagate those values that have changed. That is, if the new value of a variable is the same as the old value, there is no need to reevaluate any dependent expressions. With that enhancement, we have a complete simulation loop for a logic simulator.

```
do (i=1, endtime)
  t = t + 1
  // evaluate the state variables
  t_m = m + n
  put_on_future_update_list(t_m, t)
  t_n = new input
  // update the variable values
  m = take_off_update_event_list(t)
  n = t_n
  // propagate m and n to combinational expressions
  c = m | a
  // propagate c to combinational expressions
  b = ~(c & d)
  // propagate b to combinational expressions
  ...
enddo
```

FIGURE 16.8 Simulation loop with propagation.

```
do (i=1, endtime)
  t = t + 1
  for (all state variables m)
    t_m= f(.)   // evaluate the state variables
    put_on_future_update_list(m, t_m, t)
  endfor
  // update the variable values
  while (u = take_off_future_update_list(t) != empty)
    u.v = u.t_v
    put_on_propagate_list(c(u.v), t+u.d)
  endwhile
  // propagate changed values to combinational expressions
  while (c = take_off_propagate_list(t) != empty)
    c.v = eval(c.expr)
    put_on_propagate_list(c(c.v), t+c.d)
  endwhile
enddo
```

FIGURE 16.9 Simulation loop with propagation list.

The only remaining necessary generalization is time. We have dealt with time as simply an ordered set of instants, t_1, t_2, … This would be sufficient if each time instant was able to be mapped onto a real time line, for example, if each t_i corresponded to clock cycle *i*. However, in the general case, there may be nonuniform time intervals at which events happen. Just as there could be a delay in updating the state variables with new values, there could be delays updating combinational variables with new values (this would represent gate delays). So both the update event list and the propagate list would be ordered by time, and each event on either list would include a time at which it was to occur. Now, when all the events at the current time *t* are finished, the main loop would be iterated upon and time would be advanced to the nearest time in the future from either of the two lists (Figure 16.10).

The entry which is put on the future event list is the sequential expression, which will be evaluated at a given time in the future (often a clock-cycle boundary). This now has all the elements of a logic simulation loop, sometimes also called the scheduling loop or the simulation kernel. People often talk about a "simulation engine" in an attempt to make simulators sound more sophisticated than they really are. The scheduler is about the only thing that could be called a "simulation engine," and as we have seen, conceptually it is pretty simple.

16.3.3 Processes

We have described a logic simulation model as a collection of synchronous state variables (registers) and asynchronous combinational expressions. This corresponds to a typical RTL-style description of a logic circuit. We might write a description of such a circuit as follows:

at rising edge of clock:

$reg_i = newval_i$

$reg_j = newval_j$

...

asynchronous:

$newval_i = f(reg_i, reg_j, \ldots)$

$newval_j = g(reg_i, reg_j, \ldots)$

...

This maps directly onto the simulation loop described in Figure 16.10. However, this is a pretty low level of description, and it would be laborious to write circuit models this way routinely. By introducing the

```
do (while t < endtime)
  t = min(future_event_list, future_update_list, propagate_list)
  while (m = take_off_future_event_list(t) != empty)
    t_m= eval(m.expr)   // evaluate the state variable in this event
    put_on_future_update_list(m.v, t_m, t+m.d)
    put_on_future_event_list(m.v, m.expr, t+m.next)
  endwhile
  // update the variable values
  while (u = take_off_future_update_list(t) != empty)
    u.v = u.t_v
    put_on_propagate_list(c(u.v), t+u.d)
  endwhile
  // propagate changed values to combinational expressions
  while (c = take_off_propagate_list(t) != empty)
    c.v = eval(c.expr)
    put_on_propagate_list(c(c.v), t+c.d)
  endwhile
enddo
```

FIGURE 16.10 Simulation loop with future event, update, and propagate lists.

idea of a process, we can gain expressive power and still use the same simulation mechanics. A process would look like this:

```
process
    wait (rising edge of clock)
    reg_i <= f (reg_i, reg_j, ...)
    reg_j <= g (reg_i, reg_j, ...)
    ...
endprocess
```

Here we have combined the combinational logic with the sequential logic. Note that we have used a special symbol <=, to indicate that the assignment is a delayed assignment. This is not strictly necessary, as we could infer that any assignment to a state variable should be a delayed assignment. Verilog uses a special assignment symbol, VHDL and SystemC do not.

We could extend the process by allowing time to pass during the execution of the process, as in Figure 16.11.

This process represents a communication protocol with another process whereby data are exchanged using a ready/valid handshake, and the output happens ten clock-cycles after the input. It is straightforward to map this process onto our simulation loop by decomposing the process into events that get put onto the future event list. Each of the events corresponds to a wait in the process. This is left as an exercise for the reader.

The notion of a process becomes more powerful when a system has multiple processes, which can operate on common variables. Processes are inherently independent. That is, the events in two processes may or may not have any ordering relationship between them. Typically, all processes are assumed to begin at the beginning of simulation (time 0), but the events they wait on may be different and unrelated. Events within processes may be synchronized by means of common events or variables, but they do not have to be.

It is easy to see how arbitrarily complicated behavior can be described with multiple processes, and it is straightforward to map the events in each process into our simulation structure. Indeed, it is fairly easy to construct models that cannot be physically realized using these structures. While it may not be apparent what the value of describing unrealizable systems is, this power can be very useful when modeling the environment that the system is subjected to. This is usually called the *testbench*, while the target system is called the *design under test*. It is very useful to be able to describe both in the same simulation environment.

```
process
    while (forever)
        while (!valid)
            wait (rising edge of clock)
        d <= data.in
        ready <= 0
        wait (rising edge of clock)
        for (i=0; i<10; i++)
            wait (rising edge of clock)
        dataout <= f(d)
        ready <= 1
    endwhile
endprocess
```

FIGURE 16.11 An example process.

16.3.4 Race Conditions

A race condition can occur in a concurrent system when the behavior of the system depends on the order of execution of two events that are logically unordered. The most common cause of this is when one process modifies a variable and another reads the same variable at the same simulated time. This will not happen with state variables when delayed assignment is used, but it can happen with combinational variables, or with state variables if delayed assignment is not used. VHDL took the approach of making all assignments to state variables delayed, while Verilog did not. Thus, it is easier to write a model with race conditions in Verilog than in VHDL. There is an efficiency cost to delayed assignment of course, which is one of the reasons that VHDL simulators are typically slower than Verilog simulators.

16.4 Impact of Languages on Logic Simulation

The three major hardware description languages, Verilog, VHDL, and SystemC, are the primary languages used for hardware simulation today. All of them have the richness required to represent the vast majority of hardware designs. Verilog has more low-level capabilities than VHDL or SystemC, while SystemC has more high-level capabilities than VHDL or Verilog. They have common features that enable efficient simulation of hardware constructs, primarily hardware data types, hardware-oriented hierarchy, and hardware-oriented timing and synchronization.

16.4.1 Data Types

One of the ways that hardware description languages differ from other programming languages is the data types they offer. Logic signals are either 1 or 0, or in some cases neither (i.e., undriven or floating). But real hardware can be more complicated, so it is often convenient to be able to model signals as a range of strengths. That is, a strong signal can override a weak signal. It is also convenient for simulation purposes to include an unknown value (x), which indicates that a value is either uninitialized, or driven by conflicting values. This range of possible values is most useful when modeling at low levels, which is why Verilog has the richest set of signal values and SystemC has the least.

All hardware description languages have data types that allow the explicit specification of bit widths. That is, it can be specified that a variable is n bits wide, where the maximum value of n is usually some large number. This is useful when describing buses and collections of signals that are to be treated as a single variable. There are also operations to go along with these data types, like concatenation and subset selection. Operations on these data types cause the simulator to do more work than would be done in a normal C program, since the underlying computer must use several instructions, to accomplish them, rather than a single native instruction.

16.4.2 Variables

Verilog, VHDL, and SystemC make a distinction between variables which are state variables and those which are combinational variables. This makes the simulator's job easier because the classification is made by the programmer. In fact, a simulator could determine which variables are state variables and which are not by context. Verilog goes even further and requires the programmer to indicate which assignments are delayed assignments and which are not.

16.4.3 Hierarchy

The organization of a model in a hardware description language consists of a collection of modules in a tree structure. This corresponds with the way hardware is built. Signals are passed between modules in the tree by means of input and output ports. As far as the simulation goes, the port connections between parent and child are the same as combinational logic assignments. That is, for an input port, the left-hand side of the assignment is the port variable in the child module, and the right-hand side is the port variable in the parent module. By including these constructs in the language, the user does not have to write so much code, but the simulator still has to do the same amount of work.

16.4.4 Time Control

Hardware description languages offer a clock-based model of synchronization and event control. This makes synchronous logic circuits easy to write. Again, it does not make the simulator do any less work. The main characteristic is the definition of events in relation to one or more common signals, which are usually interpreted as a system clock signal, or a set of clock signals. Thus, a process can wait on the rising or falling edge of a signal, or either edge. The common way of indicating that two processes execute at the same time is to have both wait on the same edge of a common signal. At the RTL, where registers are explicitly instantiated, it is common to have hundreds or thousands of registers all triggered by the rising edge of a clock. This can be taken advantage of by the simulator to reduce the overhead of these events. Note however, that there is nothing inherently different about a clock signal from any other signal. The only difference is how it is used.

A Verilog description of a clock signal is shown in Figure 16.12a, a register using it is shown in Figure 16.12b, and a process using it is shown in Figure 16.12c.

16.4.5 Combinational Logic

A distinctive feature of hardware description languages from a simulation point of view, is the inclusion of a separate construct to describe combinational or continuous logic. As previously described, combinational logic is composed of assignments to variables that are done reactively. That is, the assignment is evaluated and performed whenever any of its constituent variables changes. We saw earlier how propagation is handled in the simulator. Combinational expressions are evaluated only as a result of propagation. Their use is a natural way to represent hardware at the RTL, but their use imposes complications for the simulator, since a poor choice of when to evaluate them can have dramatic consequences on the running time of the simulation. Figure 16.13a shows a Verilog continuous assignment, and Figure 16.13b shows the same variable assignment written as a process. Note that in both cases, the expression evaluation and assignment will be performed whenever one of the right-hand side variables changes.

```
reg clk;
clk = 0;
always #10
  clk = ~clk;
```

(a)

```
reg [7:0]  state, newstate;
always @(posedge clk)
  state <= newstate;
```

(b)

```
always begin
  while ( go == 0)
    @(posedge clk) ;   // wait for go
  count = 0;
  while (count < 10)   // wait for 10 cycles
    @(posedge clk) count = count + 1;
  done <= 1;           // raise flag
  @(posedge clk)
    done <= 0;   // drop flag after one cycle
end
```

(c)

FIGURE 16.12 Description and use of a clock signal in verilog.

```
assign var = a + (b ^ c);

        (a)

always @(a or b or c)
    var = a + (b ^ c);

        (b)
```

FIGURE 16.13 Logic written as a) Continuous assignment; b) process.

16.5 Logic Simulation Techniques

Simulation speed is defined as the ratio of simulation time to simulated time. Simulation time is the real time required to execute the simulation model, while simulated time is the time represented in the model. Because an essentially unbounded amount of simulation is required to verify the correct behavior of a complex digital design, simulation speed is very important. It is interesting to note that in logic simulation, the simulation speed ratio is often called the slow-down, since the simulation takes longer than the real system, usually by orders of magnitude. However, in general system simulation, the simulation speed ratio is often called a speedup, since the simulated time scale may be very large.

The underlying mechanics of logic simulation are as described above. Except in the case of cycle-based simulation as described below, all simulators use a scheduler that functions pretty much the same. The scheduler is responsible for selecting the next event to execute and transferring control to it. There are well-known algorithms for implementing schedulers, and all mainstream logic simulators have reasonably well-optimized schedulers.

Nevertheless, different simulators can have vastly different performances on the same model written in the same language (or the same model written in a different language). The difference in execution efficiency between simulators is due to how the event routines are executed, how operations are executed, and especially how events are scheduled. It is not uncommon for two different simulators to execute a different number of events — differing by a factor of 2 or 3 — for the same model, and still yield the same results.

16.5.1 Interpreted Simulation

Most early logic simulators were interpreters. That is, the simulator read in the model source, built some internal data structures, and encoded the event routine operations in a unique instruction set. Then, as the simulation was run, the event routines were interpreted by a special piece of code, the interpreter, which "executed" those custom instructions. This is a technique that has been used for many years to translate and execute programs in a variety of languages. In general, interpreted execution offers a good opportunity for debugging the program, since the interpreter can relate any errors directly to the source of the program. However, there is an efficiency cost. Interpreted execution is slow, because the interpreter has to do a lot of work for each instruction. As a result, there are few modern logic simulators that are interpreters.

16.5.2 Compiled Code Simulation

Logic simulators have been categorized as "compiled code" simulators to distinguish them from interpreters. A better description would be simply "compiler." [1] A compiled code simulator is nothing more than a compiler for the simulation language it implements. Fundamentally, it differs from an interpreter in how the event routines are executed. While an interpreter executes the event routines by executing a sequence of operations represented as higher-level instructions, a compiler prepares the event routines so that they can be executed directly by the host machine. That is, an event routine is compiled into machine code so that it can be called as a subroutine. Essentially, the compiler has to do more work initially, but it produces event routines that execute much faster. It is common for compiled code to execute one to two orders of magnitude faster than interpreted code. That difference in efficiency has been observed across many different programming languages and many different underlying machine architectures. It is no surprise that compiled logic simulators typically run between 10 and 20 times faster than interpreted logic simulators on the same models.

It is common for compiled logic simulators to emit event routine code in C and then use the host machine's C compiler to produce machine code. The C code emitted is nothing more than an intermediate form of the compiled program, with the C compiler serving as the last phase of the compiling process. In general, there is little difference in efficiency between compiled simulators that use the host C compiler to produce machine instructions and those that emit machine instructions directly.

It is worth noting that the relative efficiency of interpreters and compilers is dependent on the level of abstraction of the model being simulated. The lower the level, the smaller the difference. The fastest gate-level simulator, even today, is still an interpreted Verilog simulator.

16.5.3 Cycle-Based Simulation

In the general case, updating a variable value may have a time delay which results in an event being scheduled and then executed by the scheduling loop. In a typical synchronous logic model, every clock cycle could have many intermediate events within the cycle when different variables are updated. In most cases however, the same state behavior will be observed at the clock boundaries if those intermediate events are collapsed to two events, one at the clock edge and one immediately after. This is generally called cycle-based simulation. Since the state behavior is all that is important when verifying logical correctness, this simplification is appealing. The question then becomes, how much faster can you simulate a model using this abstraction?

A number of different cycle-based techniques have been tried. The obvious way to do it was mentioned previously, where every expression in the model is ordered and executed once in every cycle. This is sometimes called ubiquitous cycle-based simulation. The big advantage this has is that there is no overhead of putting events on the future event list or taking them off. In an RTL logic simulation, this overhead can amount to as much as 40% of the execution time. For some designs this static scheduling has proven effective, but the big drawback it has is that if the model has few state variables changing in each cycle, there is a lot of wasted work done. This technique is seldom used for commercial logic simulators now.

However, the ubiquitous cycle-based technique is amenable to hardware acceleration. The scheduling loop is simple enough in that it is relatively easy to program an FPGA-based hardware device to perform the simulation. Products that do this are called emulators, and they can perform logic simulations several orders of magnitude faster than software simulators.

16.5.4 Level of Abstraction and Speed

As noted above, while not limited to these levels, the primary levels of abstraction that logic simulation is concerned with are gate level, RTL, and behavioral level. From a simulation perspective, the main difference between these levels is the number of events that are executed. Between RTL and gate level, the primary abstraction is the width of operands and results. At the RTL, multiple-bit variables, or *vectors* in Verilog terminology, are operated on as a unit. At gate level, vectors are typically split into their individual bits. Thus, at gate level, every logical operation is performed by a hardware element, which requires an event to produce its output. If the timing of the design is important for the simulation, then simulating each physical component is required, since each component may have a different delay.

At the RTL, operations are typically performed as aggregates on their input variables, producing vectors as results. It is easy to see that *and*ing two 16-bit vectors to produce a 16-bit result can be done with one event at RTL, while it will take 16 events at gate level. Since the event routine itself takes about the same amount of time to *and* two 1-bit inputs as two 16-bit inputs, the gate-level simulation will run 16 times slower than the RTL simulation. It is common for the same design to simulate an order of magnitude faster at RTL than at gate level.

Moving up to behavioral level, the primary abstraction is the reduction of clocked events. That is, in behavioral code, variable values are computed without regard to the mechanics of the computation. A computation that may take several cycles in the ultimate hardware will be done in just one event. Synchronization in behavioral code is typically done via communication signals rather than fixed-cycle counts. Consequently, the simulation can be done with significantly fewer events at behavioral level than at RTL. Again, the difference between simulation time for a behavioral model and the same design at the RTL can be an order of magnitude.

16.5.5 Co-Simulation Methods

There are a variety of hardware simulation languages that have been invented and used over the years. In addition to the primary HDLs, Verilog, VHDL, and SystemC, other languages have been created to write testbench code in, most prominently e and Vera. In addition, code written in other more general-purpose languages (C, PERL, etc.) is often incorporated into a simulation model. While these other languages are mainly used for modeling the parts of the system that make up the environment for the hardware design, there is no difference between them and the hardware design as far as the simulator is concerned. An event is an event, whether in the hardware model or the environment model.

Integrating this multiplicity of languages into a single simulation model can be a problem. Integrating a general-purpose programming language with a simulation language is not especially hard, since there is no concept of time in general-purpose languages. Thus, all one really needs to do is allow an escape mechanism so that code written in the general-purpose language can be called in an event routine. Providing an API so that the general-purpose language code can schedule events and get called as an event routine is pretty straightforward. All mainstream logic simulators provide this capability.

Integrating a simulator with another logic simulator is a more difficult problem. Ideally, a logic simulator would be able to understand all the required simulation languages and compile a model written in all of them into a single model. Practically, logic simulators are created to handle one language, and integration with other languages is done at a coarser level of granularity.

The simplest way to integrate models written in two or more languages is to compile the components with separate compilers, and then put them together by coordinating their schedulers. This is called *co-simulation*. That is, each simulator runs as it normally would, keeping its own scheduler and future event list as well as other local data, and synchronizing through the schedulers. This can be done as shown in Figure 16.14. The points marked as <==n==> are points at which the schedulers must synchronize with each other. All schedulers must have finished the preceding section before any can proceed to the next section.

The only remaining difficulty is that propagation may be required between parts of the model that reside in different simulators. That is, a Verilog module could have a variable whose value depends on a variable in

```
do (forever)                                          do (forever)
  t = next t                          <==1==>           t = next t
  while (m = future_event_list(t))                      while (m = future_event_list(t))
    execute event routine m                               execute event routine m
  endwhile                            <==2==>           endwhile
  // update the variable values                         // update the variable values
  while (u = update_list(t))                            while (u = update_list(t))
    u.v = u.t_v                                           u.v = u.t_v
    put_on_propagate_list(c(u.v), t+u.d)                  put_on_propagate_list(c(u.v), t+u.d)
  endwhile                            <==3==>           endwhile
  // propagate changed values                           // propagate changed values
  while (c = propagate_list(t))                         while (c = propagate_list(t))
    c.v = eval(c.expr)                                    c.v = eval(c.expr)
    put_on_propagate_list(c(c.v), t+c.d)                  put_on_propagate_list(c(c.v), t+c.d)
  endwhile                            <==4==>           endwhile
enddo                                                 enddo
```

FIGURE 16.14 Synchronization points in co-simulation.

a VHDL module. It is easy to see how to do this — just provide a way for the VHDL model to propagate a value onto the Verilog model's propagate list. Actually doing it in a concise and automatic way is more problematic, and this problem has been solved in a variety of ways through a combination of language features (to identify a variable as external) and an API. Sometimes, this ability is simply restricted to module ports.

16.5.6 Single-Kernel Simulators

Co-simulation with multiple schedulers works, but the coarse granularity of the synchronization can impose a substantial overhead if there are many propagation events across the simulator boundary. In general, the API calls between simulators involve a fair amount of overhead as each one must establish its environment on every call. It would be more efficient if the model components written in all the languages could use the same scheduler. This is the approach taken by the so-called single-kernel simulators.

In essence, simulators that handle multiple languages with a single scheduler have separate compilers for each language but they all use the same scheduler. That avoids the need for synchronization between different schedulers, and can save a substantial amount of overhead. There are commercial single-kernel simulators for Verilog and VHDL, and Verilog, VHDL, and SystemC.

16.6 Impact of HVLs on Simulation

Hardware verification languages (HVLs) came into vogue as special-purpose languages for writing testbenches. They provide convenient means of generating stimulus for a hardware model, and also provide useful abstractions for modeling the hardware model's environment. For instance, it is generally easier to write a communication protocol that provides correct, variable input in an HVL than it is to write the same thing in an HDL. From the simulator's point of view, the mechanics of dealing with an HVL are the same as co-simulation. Indeed, an HVL is just another flavor of simulation language. Simulators that integrate an HVL and an HDL into a single scheduler are becoming more common.

16.7 Summary

In this chapter, we have covered many of the details of digital logic simulation. Logic simulation is simply a special case of discrete event simulation, which has a long history in general system modeling. The speed of simulation is proportional to the level of detail in the simulation model, which in turn is determined by the level of abstraction at which the model is expressed. In logic simulation, the three common levels of abstraction are gate level, RTL, and behavioral level. There are techniques like cycle simulation used to speed up simulation, as well as techniques like co-simulation used to improve the hardware modeling capability. Ultimately, logic simulation is the most general technique available to verify that a hardware design does what it is intended to do.

References

[1] A. Aho, R. Sethi, and J. Ullman, *Compilers Principles, Techniques, and Tools*, Addison-Wesley, Reading, 1986.

[2] P. Ashenden, *The Designer's Guide to VHDL*, Morgan Kaufmann Publishers, San Francisco, CA, 1995.

[3] G.M. Birtwistle, O.-J. Dahl, B. Myhrhaug, and K. Nygaard, *Simula Begin*, Auerbach Publishers, Philadelphia, PA, 1973.

[4] T. Grötker, S. Liao, G. Martin, and S. Swan, *System Design with SystemC*, Kluwer Academic Publishers, Dordrecht, 2002.

[5] M.H. McDougall, Computer system simulation: an introduction. *Comput. Surv.*, 2, 1970, pp. 191–209.

[6] S. Palnitkar, *Verilog HDL, A Guide to Digital Design and Synthesis*, 2nd ed., Sunsoft Press, Mountain View, CA, 2003.

[7] D.E. Thomas and P. Moorby, *The Verilog Hardware Description Language*, Kluwer Academic Publishers, Dordrecht, 1991.

17

Using Transactional-Level Models in an SoC Design Flow

Alain Clouard
STMicroelectronics
Crolles, France

Frank Ghenassia
STMicroelectronics.
Crolles, France

Laurent Maillet-Contoz
STMicroelectronics.
Crolles, France

Jean-Philippe Strassen
STMicroelectronics
Crolles, France

17.1 Introduction

Multimillion gate circuits currently under design with the latest CMOS technologies include not only hardwired functionalities but also embedded software, running most often on more than a single processor. This complexity is driving up the need for extensions to the traditional register-transfer level (RTL) to layout design and verification flow. Indeed, these circuits are complete systems: system-on-chip (SoC).

Systems-on-chip, as the name implies, are complete systems composed of processors, busses, hardware accelerators, I/O peripherals, analog/RF devices, memories, and embedded software. Less than a decade ago, these components were assembled on boards; nowadays, they can be embedded into a single circuit.

This skyrocketing complexity results in two major consequences: (1) mandatory reuse of many existing Intellectual Property (IPs) to avoid redesigning the entire chip from scratch for every new generation; and (2) employment of embedded software to provide major parts of the expected functionality of the chip. Allowing software development to start very early in the development cycle is therefore of paramount importance in reducing time-to-market. Meanwhile, real-time requirements are key parameters of the specifications, especially in the application domains targeted by STMicroelectronics (e.g., automotive, multimedia, telecom). It is therefore equally important to be able to analyze the expected real-time behavior of a defined SoC architecture. Another crucial issue is the functional verification of IPs that compose the system, as well as verification of their integration. The design flow must support an efficient verification process to reduce development time and to avoid silicon re-spins that could jeopardize the return on investment of the product under design. We will also describe how the proposed approach strengthens the verification process.

At STMicroelectronics, one approach that addresses these two issues is to extending the CAD solution made available to product divisions, Unicad, beyond the RTL entry point; this extension is referred to as the system-to-RTL flow. As the current ASIC flow mainly relies on three implementation views of a design: layout, gate, and RTL levels; the extended flow adds two new views: transaction-level model (TLM) and algorithmic.

In the remainder of this chapter, we first introduce the system-to-RTL design and verification flow with its objectives and benefits. We then focus on the TLM view with a description of the modeling style and also discuss the application of these models to support critical activities of platform-based design: early embedded software development, early architecture analysis, and providing a reference model for the functional verification. Finally, we present how we used this approach in the design of a multimedia multiprocessor platform.

17.2 Related Work

Transaction-level model has become a topic of much greater interest in the industry for the last couple of years, since the delivery of SystemC 2.0 [1]. It has rapidly raised hopes of delivering the long-awaited benefits of a system-design approach for industrial designs.

There are various possible "system-level" abstraction levels of a hardware and software SoC, ranging from paper specification of the functionality before hardware/software partitioning, down to synthesizable RTL of the complete hardware and source code of embedded software (in C typically) [2,3]. Transaction-level model is an intermediate level between these two extremes, and is most often based on C++ — especially using SystemC as base set of C++ classes [4].

Even the most complex SoC platforms can be modeled using a TLM approach [5,6]. Actually there have been various abstraction levels referred to as TLM. We will describe them, following an ordering from less abstract to most abstract. Some tool frameworks encompass several of these levels for mixed-level simulation [7]. The following is a bottom-up description for easy understanding by hardware-oriented readers. However, a real project flow would be essentially top-down, including some reuse.

17.2.1 BCA or CA-TLM

Transaction-based modeling is sometimes understood as only using a C++ variable to contain the value represented by a bus signal for a word with a given bus width, at a given bus cycle. Such modeling of transfers between IPs in an SoC may be called cycle-accurate TLM (CA-TLM) but is more suitably called bus cycle-accurate (BCA) modeling, and is very close to RTL modeling [8]. It requires the IP behaviors to interface with the bus through a bus interface that includes protocol finite-state machines (FSMs) that are difficult to code, debug, and maintain consent with an evolving design of synthesizable VHDL/Verilog RTL.

17.2.2 Protocol-Phase-Aware TLM

A slightly higher level of abstraction, yet sometimes referred to as TLM, is modeling only the phase, but not the cycles of a bus protocol [9]. A transaction is composed of several subtransactions, e.g., bus request, bus grant, transfer single word, or burst transfer. However, the data transferred are structured according to bus width and bus protocol. In such modeling, the modeling engineer must, as in the BCA approach, select a given protocol, understand it, and manage interfaces between his IP behaviors and the bus protocol.

17.2.3 Our TLM and PV

Our definition of TLM, as described later in the chapter, raises the level of abstraction compared to BCA or protocol-phase-aware TLM modeling. Yet it is a model of the SoC hardware architecture after hardware/software partitioning. In our TLM, the data of a transaction can be any structure of words and bits (e.g., as little granularity as a bus half-word transferred between two peripheral registers, or much larger granularity — a full VGA image transferred between two memory buffers) that is meaningful as a transfer between two IPs, at a time determined by an event from the circuit (system) specification. IP models can be pure C/C++ along with register-accurate interfaces accessed using a simple TLM transport application programming interface (API). This level of TLM modeling has proven to be extremely useful to create reference models for functional verification of hardware, and also for SoC simulation platforms for embedded software developers.

Optionally, the transaction definition of our TLM definition can be refined into a structure that is bus-protocol aware (bus width, burst capability, lock, etc.), but this is optional and useful only for SoC architects for fine analysis of arbitrations and other characteristics of the SoC interconnect.

Our TLM definition is slightly different from the one described in [10] because we define an abstraction that models an SoC architecture with *no requirement for* timing [11]. To our knowledge, this abstraction level is unique, and is the foundation for the numerous benefits of our methodology. Timing remains, however, a vital part of the system specification; and a timing-complementary view of TLM is widely used for timing-behavior analysis as described in [12–15]. We also describe how timing can be managed in our methodology without compromising the untimed TLM models.

17.2.4 Higher-Level Models

Our definition of TLM, being register-accurate for the IP interfaces but not necessarily protocol-accurate for the transfers between IPs, is an intermediate between the protocol-phase aware TLM on the one hand, and more abstract models on the other.

More abstract models, e.g., untimed functional models [16], have some high-level architectural features such as queues but no registers and no address map, preventing execution of the embedded software. There exist variants such as timed functional models [16,17], abstract virtual component (VC) models [18], or timing extensions of multithread graphs [19], which may be useful for architects but are not ready for use by embedded software developers. Even higher abstraction models, including for example, communicating sequential processes that only have links between behaviors without any hardware architectural concept [20], may be useful.

17.2.5 TLM/PV — At the Heart of System-to-RTL Flow

From our experience, the TLM/PV level is the most useful one in which to invest modeling effort. This is because it can be created either by hardware-dependent software developers, SoC architects, SOC RTL integration engineers, or hardware functional verification engineers with C++ knowledge. Then TLM/PV models can be used by any or all of these categories of hardware and software engineers, with or without the timed extension to the base TLM model.

For instance, let us describe how TLM/PV addresses functional verification. Most efforts in this domain, as described in [21–23], have been focused on the measurement of test coverage. Our TLM practice provides a complementary benefit enabling the automatic generation of expected results that can then be compared with the output generated by the design under test. A similar approach is in embedding an HDL cycle-accurate (CA) reference model into the test-bench so that outputs can be compared cycle by cycle. Our approach is different because we use an untimed reference model that leads to great savings in the model development and maintenance. For functional verification, a similar technique using an untimed model is sometimes applied for processor verification [24]. We generalize this approach for any digital hardware IP.

17.2.6 TLM Standardization and Tools

Our TLM methodology and associated SystemC-based classes have been submitted in a joint proposal in May 2003 to the TLM working group of the Open SystemC Initiative (OSCI, www.systemc.org), and the OSCI standardization work has been using it intensively. The OSCI standardization process with various partners in the OSCI TLM WG also benefited from inputs by the OCP-IP TLM WG [25]. Various papers reflect the work done in the OSCI TLM WG [2,9].

Currently, various CAD vendors have started offering TLM support in their tools and libraries, either with an untimed, system event-based approach , which we promote as our main modeling style , or complementing it with protocol-aware APIs [9].

17.2.7 Conclusion on Related Work

In this chapter, we uniquely address the rationalization of an industrial ASIC/SoC design flow around the real deployment of a new higher abstraction level, i.e., TLM. This abstraction enables the systematic practice of three key activities of an SoC design: embedded software development, architecture analysis, and functional verification.

17.3 Overview of the System-to-RTL Design Flow

Our system-to-RTL flow depends on three abstraction levels (see Figure 17.1):

- The *SoC functional view* models the expected behavior of the circuit without taking into account how it is implemented.
- The *SoC architecture view* (SoC-A or SoC TLM platform) captures all information required to program embedded software on the processors in the circuit.
- The *SoC microarchitecture view* (SoC-MA or SoC RTL platform) captures all information that enables CA simulation of the SoC. Most often, it is modeled at the RTL using VHDL or Verilog. Such models are almost always available because they are used as input for logic synthesis.

The three views have complementary objectives that avoid the conflict between the need for accurate descriptions, which leads to undesirably slow simulation speed and the need for simulating real-time embedded application software.

The *SoC functional view*, as implied by its name, specifies the expected behavior of the circuit perceived by the user. It is an executable specification of the SoC function that is normally composed of algorithmic software. It contains neither architecture nor address mapping information. Performance figures are

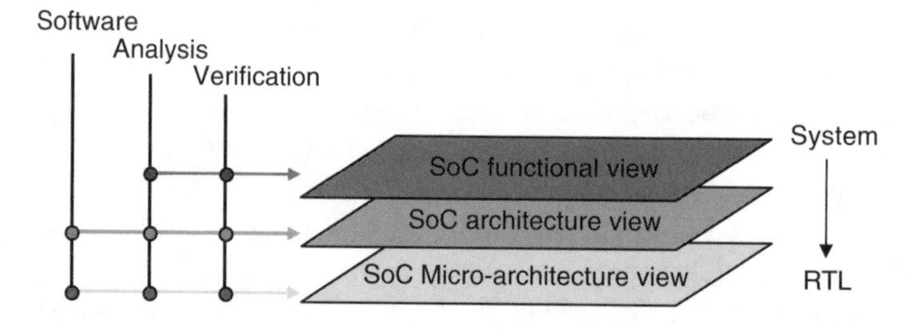

FIGURE 17.1 Overview of the design flow.

specified separately, very often as a paper specification.

Farther down in the flow is the *SoC architecture view* that focuses on:

- *Early firmware development.* It allows the hardware-dependent software to be validated before running on a prototype.
- *SoC architecture exploration.* In the early phase of the design cycle, architects have to define the architecture that will best fit the customer requirements. The early availability of the TLM platform is a means to obtain the quantitative figures needed for determining the appropriate architectural decisions.
- *Functional verification.* It provides a reference model for the verification engineer. This "golden" model is used to generate tests that can be applied to the implementation model in order to verify whether the functionality matches the expected behavior.

The *SoC microarchitecture view* at the end of the system-to-RTL design flow is intended for:

- *Low-level embedded software debug and validation in the real hardware simulation environment.* Our goal is to enable the debugging of device drivers, and their integration into the target operating system before the first breadboard or even the hardware emulator is available.
- *SoC functional verification.* We aim at verifying that the IPs, once integrated, still provide the expected functionality, and that parallel block execution and communication do not corrupt the IP behavior. Certain blocks are processors that run software. The verification activity must take this fact into account and validate all possible interactions between the hardware and software components.
- *SoC microarchitecture validation.* In order to sustain the real-time requirements of the application, the embedded software is usually optimized and the hardware is configured accordingly. In case of insufficient performance, the SoC architecture may be upgraded to match the requirements. Any part requiring cycle accuracy will count on an RTL view.

A discussion of these activities is available [26].

The three abstract views can be integrated gracefully into the SoC development process as explained below. The functionality is independent of the architecture so that development of its corresponding view can be initiated as soon as there is a need for a product specification.

While software and architecture teams are using the same reference—SoC TLM—for firmware development and coarse-grain architecture analysis, the RTL development takes place in parallel, leading to an SoC RTL platform. Hence, concurrent hardware/software engineering is achieved as depicted in Figure 17.2. By the time the SoC RTL platform is available, some tasks closely related to the hardware implementation are ready to start: fine-grain architecture analysis, hardware verification, and low-level software integration with hardware. These tasks are conducted concurrently with the emulation setup, the synthesis, and back-end implementation of the standard ASIC design flow.

Finally, as the first breadboard is ready, the software (drivers, firmware, and a simplified application) will have been completed with good confidence and the SoC will have gone through a thorough verification process. This will certainly increase the probability of achieving first-time silicon and system success.

Note that we might have to mix SoC architecture and microarchitecture models in a single simulation model. Such a mixture of abstraction levels requires a standard mechanism that provides a methodology with an automated assembly tool, i.e., design flow automation (see Section 17.7). This standard automation will not only ensure the compatibility of the different abstraction models within the existing tools, but it also gives us a benefit with a gain in time.

Currently, one difficulty with the proposed approach is that in the absence of synthesis tools that can generate these views from a reference model, coherence with the other two views must be maintained throughout the circuit development. To address this issue, we reuse the functional subset of "system" test vectors across all views, to ensure their conformity with the expected functionality. Alternatively, we could run the same embedded software (without modification) on different simulation environments. Identical behavior proves the compatibility of the underlying models. Obviously, the coherence is ensured up to

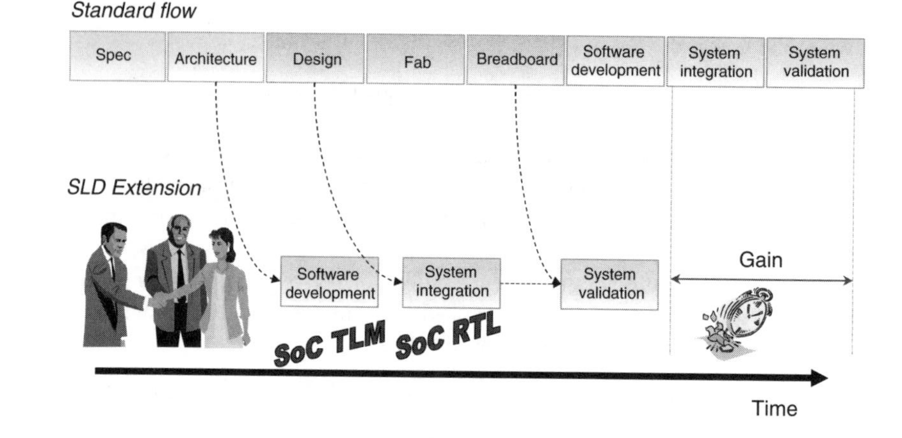

FIGURE 17.2 Concurrent hardware/software engineering.

the level of functional coverage that the tests provide. Also, as mentioned earlier, TLM models may be used as executable specifications of an SoC for RTL verification.

17.4 TLM — A Complementary View for the Design Flow

17.4.1 TLM Modeling Concepts

When addressing the need for early embedded software development and architecture analysis, engineers list the following requirements as barriers for this activity:

- *Speed.* Models used for the activities above involve millions of simulated cycles in certain cases; it is thus unacceptable to wait for even a day to complete a simulation run, especially in the case of interactive sessions.
- *Accuracy.* Certain analysis requires full-CA models to obtain reliable results. The minimum requirement is that the model should be detailed enough to run the related embedded software.
- *Lightweight modeling.* Any modeling activity besides the RTL coding (which is anyway mandatory for silicon) must be simple enough so that the modeling cost is not higher than the resulting benefits to the SoC project management.

Complex designs usually start with the development of an algorithmic model (e.g., based on a dataflow simulation engine for DSP-oriented design). The advantage of an algorithmic model is that it simulates extremely quickly, since it captures the algorithm free of the implementation details. However, such a model has no notion of the hardware and software components; it does not model registers or system synchronizations related to the SoC architecture. Thus, this model cannot fulfill our need of enabling embedded software execution.

With the emerging C-based dialects that support hardware concepts, it is often believed that writing CA models in a C-based environment could meet all the stated requirements. This approach has several major drawbacks [8,25,27–29]:

- Much information captured in such a model is not available in the IP documentation, but is maintained only in the designer's mind ... and in the RTL source code. As a result, the RTL designer has to spend considerable time to update the CA model development, or the modeling engineer is forced to reverse-engineer the RTL code. Either way ends up being tedious and time-consuming, without solving the issue.

- The CA model is merely an order of magnitude faster than the equivalent RTL simulation … very close to the speed of a cycle-based VHDL/Verilog model.

Not only is the speed too low to run a significant amount of embedded software in a reasonable time frame, the development cost is also very high compared to the benefits.

Meanwhile, we have noticed that architects and software engineers do not require cycle accuracy for all of their activities. For instance, software development may not require cycle accuracy until engineers are ready to work on software optimization; earlier in the process, they do not need high simulation speed coupled with cycle accuracy, e.g., verification of cycle-dependent behavior.

Based on this experience and new understanding, we have adopted the "divide and conquer" approach by providing two complementary environments:

- An SoC TLM platform for early use with a relatively lightweight development cost. This abstraction level sits between the CA bit-true model and the untimed algorithmic model. This is, in our opinion, an adequate trade-off between speed and accuracy with a complementary SoC RTL platform (at least 1000 times faster than RTL simulation in our experience, as depicted in Figure 17.3).
- An SoC RTL platform for fine-grained CA simulations at the cost of slower simulation speed and later availability.

Let us now define some terms in order to understand the TLM modeling approach. A system is a mixture of hardware and software that provides functionality with adequate performance to its user. An SoC is a particular system in which both hardware and software are embedded in a single circuit. A TLM system model is composed of a set of *modules* that:

- Execute part of the expected system behavior through one or more concurrent *threads.*
- Communicate data to each other (and, potentially, to a test-bench) to perform the system functionality. Each set of data exchanges is a *transaction.* Two consecutive transactions may transfer data sets of various sizes. The size corresponds to the amount of data being exchanged between two system synchronizations.
- Synchronize with each other (and potentially with the test-bench). System *synchronization* is an explicit action between at least two modules that need to coordinate some behavior distributed over the modules that depends on this synchronization. Such coordination is essential to assure predictable system behavior. A synchronization example is an interrupt raised by a DMA to notify a transfer completion.

The simulation speedup achieved by a TLM model vs. the equivalent RTL model correlates with the ratio of the average number of cycles between two system synchronizations. An explicit system synchronization is compulsory for the TLM model to behave properly because it is the only mechanism available for synchronizing the different processes in the system. Our experience shows that all systems comply with this rule. Figure 17.3 provides approximate ratios in terms of simulation speedup and modeling efforts.

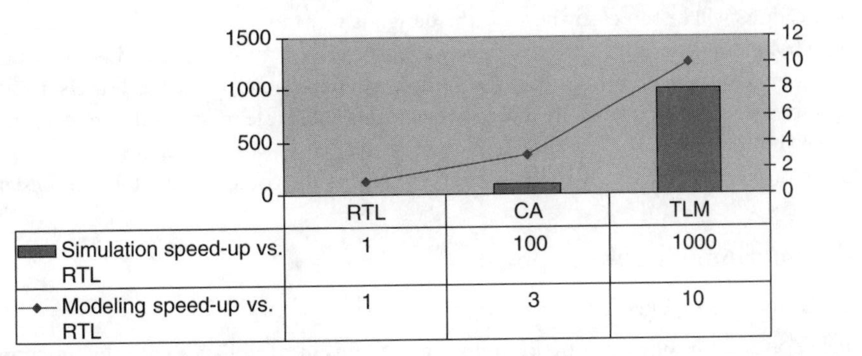

	RTL	CA	TLM
Simulation speed-up vs. RTL	1	100	1000
Modeling speed-up vs. RTL	1	3	10

FIGURE 17.3 Efficiency of modeling strategies.

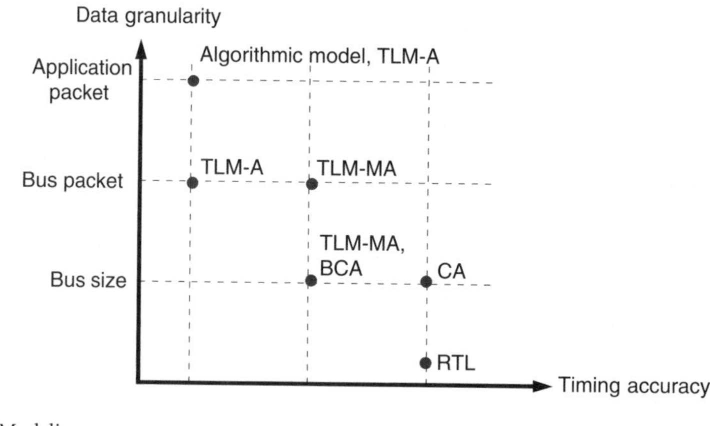

FIGURE 17.4 Modeling accuracy.

Figure 17.4 gives a better picture for the different models that we are going to quote throughout this chapter according to their timing accuracy and data granularity.

All of the models included in Figure 17.4 have bit-true behavior except the algorithmic specification model. TLM-A is the TLM architecture model that gives an untimed SoC architectural view as presented earlier. For some particular purposes, it may be necessary to refine a TLM model to include some information related to microarchitecture. We label this refined, timed model as TLM-MA, i.e., TLM microarchitecture model. To ensure satisfactory efficiency, we provide a method to compile and build the TLM IP source code, without any modification, as a TLM-A or TLM-MA model. In this chapter, unless explicitly stated, the term TLM refers to TLM-A. Further down in abstraction is the BCA model, a timing-accurate model for communication structure and interface. Following that is the CA model. Finally, we have the RTL model, the most precise timed model for a system module in the range shown.

17.4.2 Embedded Software Development

Software development (at least debugging and validation) starts only if it can be executed on the target platform. The traditional approach considers the physical prototype, e.g., an emulator or FPGA board prototype, as the starting point for software development. The advantage is earlier availability before the chip has been fabricated, whereas the drawback is the fact that hardware development is almost complete. As a result, if the software execution uncovers any need to modify the hardware at this stage, it will be be very costly.

Hardware/software coverification could start executing software on the target-simulated hardware earlier. Any small hardware modification is still possible if this technique is employed with the first top-level VHDL/Verilog netlist. The downside is the dependence on VHDL/Verilog models for simulation; this results in a slow execution speed running only a reduced part of the embedded software. Any major hardware modifications will be too costly because the design is way too advanced.

An SoC TLM platform can be delivered shortly after the architecture is specified. As a result, it is available for software development very early in the hardware design phase. A true hardware/software codevelopment is therefore attained. Hardware and software teams cooperate in line with the "contract" between them — the TLM platform. The software developers regard the TLM model as the reference standard on which to run their code while the hardware designers use it as the golden model for their design.

17.4.3 Functional Verification

17.4.3.1 Block-Level Verification

Functional verification assures that the IP implementation is in compliance with the functional specification. Using the paper specification, a verification engineer defines test scenarios that will be applied to the RTL IP model. Frequently, the engineer needs to determine "manually" the expected result of each scenario.

The TLM model is an actual functional specification of an IP. It is also referred to as the executable specification of the IP since it only captures the intended behavior perceived by the user, but excludes any internal implementation details. In other words, it models the architecture of the IP but not its microarchitecture.

Therefore, a TLM model can replace the manual process undertaken by the verification engineer to generate expected results. As demonstrated in Figure 17.5, the test scenario AVP (architecture verification pattern) is applied to the TLM model to generate automatically the expected output (i.e., reference output). A common question is: how can it be ensured that the executable specification, the TLM, is compliant with the written specification? And thus, what is the level of confidence that its resulted output can be used as a reference? There is no answer to this question. Also, the same question can be asked of the output results manually generated by the verification engineer. No warranty exists that can grant that the engineer has not misunderstood the specification or that he has not made a mistake while generating the reference output.

Note that an implicit assumption for TLM IP verification is the existence of a TLM IP model prior to the availability of an RTL model. Legacy IP may not comply with this criterion since it is normally delivered as an RTL model only. In this case, a TLM model needs to be developed based on the RTL reference model. The test scenarios and vectors are extracted and recorded from the RTL test-bench. A TLM test-bench will then load the transaction traces and signal values to drive the TLM IP model developed later. Once the TLM view is validated against its RTL counterpart, it is ready to be integrated into an SoC platform for software development and architecture analysis.

17.4.3.2 SoC System-Level Verification

In a typical design flow, any independently verified IPs must be verified in an SoC environment as well. Functional integration tests are developed to verify the following features:

- *Memory map.* To prevent registers from being mapped at the same addresses and; guarantee that hardware and software teams use the same mapping.
- *Data consistency.* To ensure that data generated by an IP matches the input data (in terms of format and content) of another IP that will reuse the generated data.
- *Concurrency.* To ensure that concurrent execution of all IPs is controlled for deterministic expected behavior.

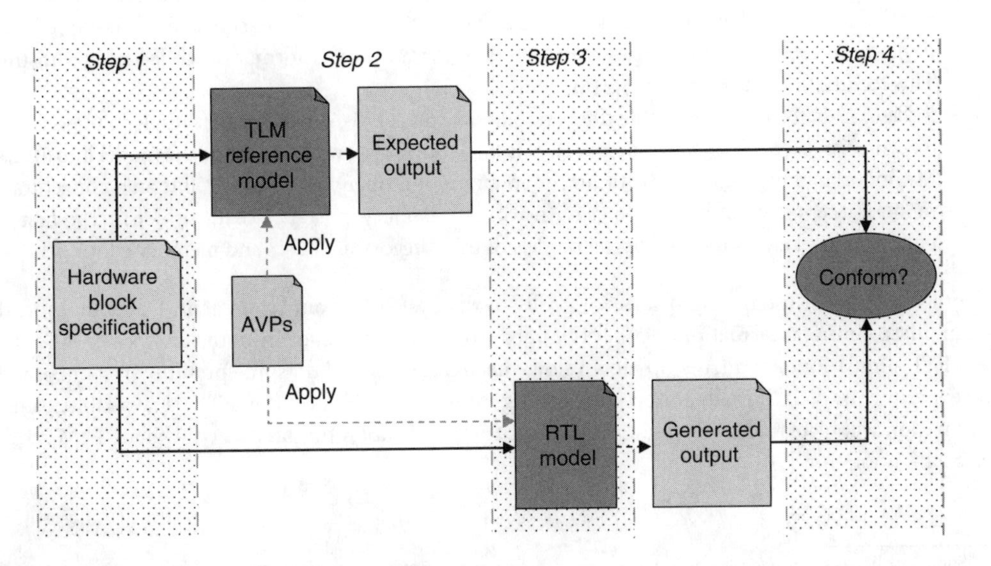

FIGURE 17.5 Functional verification using a TLM model.

The same approach as for block-level verification holds here. The SoC TLM platform is used to generate expected system reference output for a given scenario.

17.4.4 Early System Integration

System integration is the assembly of different system components to create a properly functioning system. A stepwise process with successive partial integrations throughout the development of the hardware and software components is adopted. TLM models participate in such an integration process in the following manner:

- Software is executed on the reference model of the hardware IP for a common understanding of the hardware specification.
- Hardware RTL implementation is checked against the reference model used by software developers.
- Earlier integration of hardware components can take place by mixing TLM and RTL views of IPs instead of waiting for the release of all RTL blocks.
- Software can be executed on a partial RTL platform by using TLM models for IPs whose RTL code is not available yet. It enables earlier hardware/software integration.

Abstraction-level adaptors are required for interfacing TLM and RTL models in this sort of mixed simulation. Thus, reusable bus functional models are developed for different communication protocols such as STBus or AMBA.

The process of system integration could consume considerable effort since it involves diverse parts from different points of a given design. Front-end design flow automation could be very helpful in reducing this effort (see Section 17.7).

17.4.5 Performance Analysis and Verification

A system must be carefully controlled such that its circuit respects the real-time constraints stated in the initial system definition and specification. Every design flow step is verified against this condition. A system performance analysis and verification will achieve this purpose. The system performance can be studied once the timing information is included. The following two cases may arise:

- *Timed TLM-A model.* It may be sufficient to add delays to execution paths of a TLM-A model to mimic its timing behavior. The delay computation may vary from a fixed value to a complex function whose returned value varies over time and depends on information collected at run-time. This technique is illustrated in Section 17.6.
- *Timed TLM-MA model.* A simple time annotation on a TLM-A model might result in unacceptable inaccuracies, particularly for complex IPs. It is therefore indispensable to refine the model into a TLM-MA model and then sum up the timing annotations. It is often necessary to refine an architectural IP model if it represents a shared resource. An arbitration policy must replace the simple first-in first-out policy* of the TLM-A model. Typical models of this category are buses and memory controllers.

Both cases model the timing delays with calls to the SystemC "wait" statement. Depending on the design phase, timing information is either obtained from the architect or from the analysis of the equivalent RTL simulation. For a true top–down flow, the architect, based on some preliminary benchmarking experiments, will initially insert the timing information. Once the RTL is available, timing information will be extracted from the RTL simulation for annotation into the equivalent TLM model, i.e., back-annotation [30].

* Such a policy is optimized for simulation speed because it removes the need for a delta cycle.

17.5 TLM Modeling Application Programming Interface

17.5.1 Modeling Environment

As described in depth in the literature [31], our modeling environment should support modularity, concurrency, hierarchy, separation of behavior from communication, and hardware and algorithmic data types. After evaluation of C/POSIX and synchronous language environments, which had some drawbacks with regard to our specific TLM goal, we evaluated and selected SystemC [1,4] — a language that complies with most of these requirements. In addition, as it is becoming a de facto industry standard, SystemC is supported by most CAD tools, hence facilitating a smooth transition to successive steps of the design flow (see Section 17.8).

Each module of an SoC TLM platform is characterized by:

- Bit-true behavior of the block
- Register-accurate block interface
- System synchronizations managed by the block

For the integration of modules into a platform, SystemC does not offer the inherent ability to perform transactions among several modules connected to a single channel based on a destination identifier (e.g., address map in case of bus-based architectures). To address this, we created a user-defined interface along with the associated bus channels (see the next section).

The computation model derived from the previous TLM definition implies that all system synchronizations are potentially rescheduling points in a simulation environment for accurate concurrent modeling. System synchronizations could be modeled by fixed means (e.g., event, signal) or data exchanges (e.g., polling). The simulated system will behave correctly only if all these potential system synchronizations cause a call to the simulation kernel, thus enabling its scheduler to activate other modules. SystemC events and signals comply with this requirement. The simulator is however unable to distinguish data exchanges of transactional models from implicit synchronizations. A conservative approach must therefore be adopted to allow the kernel to reschedule on each transaction completion.

Once the TLM platform is integrated, embedded software can be executed either as a native compilation on the simulation host for higher simulation speed, or as a cross-compilation on the embedded processor architecture for an instruction set simulator (ISS) execution.

17.5.2 Modeling Application Programming Interface

This section discusses the definition of an API for IP models to support data exchange among the different IPs. The general characteristics follow:

- *IP-A model reuse.* Reuse an IP-A model in an SoC-A model for simulation without any code modification, regardless of its microarchitectural implementation.
- *Port.* Enable emission and reception of data to/from communication structures.
- *Communication/behavior separation.* Separate IP behavior from the communication mechanism such that various communication schemes could be tried with the same IP-A model.
- *Information locality.* Information should be captured locally by the model that owns it.
- *Simulation efficiency.* The API should not result in significant simulation degradation.
- *Intuitive API.* Minimize learning curve and potential misuse through an intuitive API.
- *Abstraction and protocol adaptors.* Minimize the number of these models that must be developed.
- *Automation.* Should automate activities such as monitoring, adaptor insertion, code generation, and consistency check at minimum tool development cost.

There are several important differences between IP-A and IP-MA models. First, the granularity of the data sent and received by an IP-A model corresponds to the model function, while IP-MA models exchange data as managed by the communication layer. Second, an IP-A model only expects to read and write data in the communication control layer whereas an IP-MA model expects to control the handshaking mechanism required by a data transfer. Third, no timing information is necessary for an IP-A model since deterministic execution relies on a defined order of system event occurrences. In contrast, an IP-MA model captures details of data exchange and the internal IP structure to provide an accurate timing annotation.

17.5.2.1 TLM API: TLM Transport

The TLM API enables the transport of transactions among the IPs for information exchange. The information conveyed by a transaction includes the initiator dataset, target dataset, initiator access port, and target access port. A corresponding transaction structure follows:

```
class tlm_transaction        {
     tlm_port *m_initiator_port;          //Initiator port pointer
     tlm_port *m_target_port;             //Target port pointer
     tlm_transaction_data *m_data;        //Transaction data pointer
     tlm_metadata *m_metadata;            //Metadata pointer
     ...                     };
```

Transactions are exchanged among the different IPs via the ports. Such a transaction is atomic, and the IP will resume its execution when the transaction is completed. The TLM interface is defined as a subclass of `sc_interface`. Then, this TLM interface defines a function named `transport` that will be called to serve a transaction. A C++ function prototype looks like:

```
tlm_status transport(tlm_transaction &transaction);
```

where the parameter `transaction` is a reference to the transaction structure, and the return value provides transaction status information.

This API is considered as a generic transport layer for TLM. The actual content of the transaction is defined as another layer on top of it. In other words, a protocol is defined based on this generic API wherein the actual data type being transferred is specified. Such specific data are derived from the empty class, `tlm_transaction_data`. Subsequently, IP models can be developed based on this protocol. This idea is demonstrated in Figure 17.6.

17.5.2.2 TLM API: TLM Protocol

Different protocols could be defined based on the transport API to reflect the rules required for data exchange. Structures corresponding to the following TLM protocols are available as examples:

- *TAC* for supporting the interconnection of IP-A models. A transaction is either a read or a write operation.
- *AMBA* for support of IP-MA models based on the AMBA protocol. Single transfers, bursts, and locked transfers are supported.
- *STBus* for support of IP-MA models based on the STBus protocol. Request/acknowledge and separation between requests and responses are supported.
- *OCP* for support of IP-MA models based on the OCP protocol. Features are very similar to the STBus.

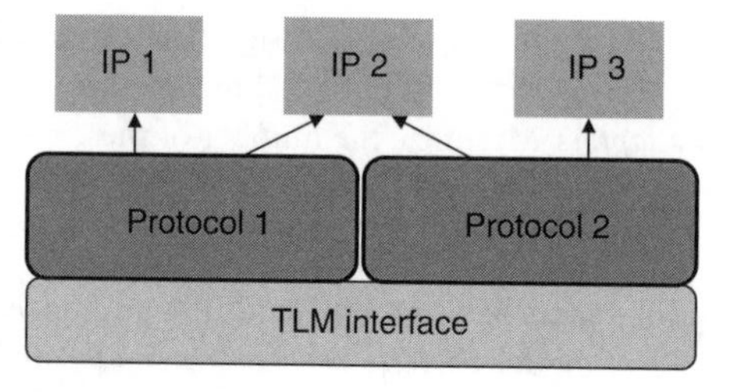

FIGURE 17.6 TLM layers.

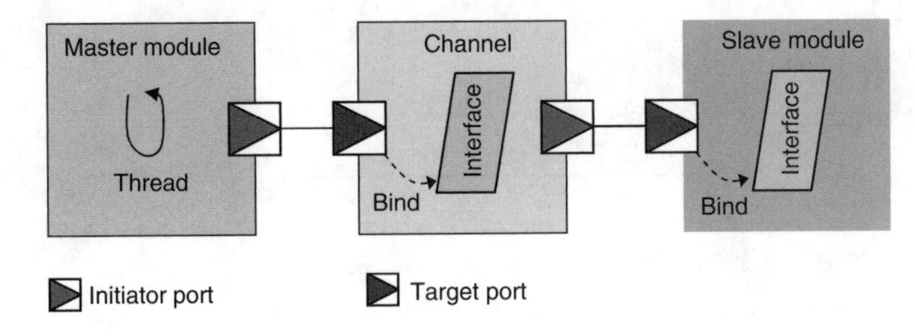

FIGURE 17.7 TLM modeling API.

Figure 17.7 illustrates the foundation of our TLM modeling API. Communication protocols that require more than one exchange to transfer data can be decomposed into multiway transfers. The request is initiated when the port is accessed, and acknowledged when the `transport` function returns on the initiator side. An identical scheme holds for the response. Such a decoupling between requests and responses enables separate pipelining of requests and responses.

17.6 Example of a Multimedia Platform

We present here a model of an MPEG4 coder–decoder (CODEC) based on our TLM TAC protocol. An SoC architecture model is indispensable for a CODEC because an efficient simulation model is required for embedded software development and interactive debugging.

17.6.1 System Model

The MPEG4 platform employs a distributed multiprocessing architecture with two internal buses and four application-specific processors (ASIP), on which are distributed the embedded software, a general-purpose host processor managing the application-level control, and seven hardware blocks. Consequently, there is much software parallelism with explicit synchronization elements in the software. Each processing block is dedicated to a particular part of the CODEC algorithm.

The hardware and software are partitioned according to the complexity and flexibility needed by the CODEC algorithm. All operators work at the macroblock level (the video unit for the CODEC). Descriptions of some major system blocks follow (see Figure 17.8 for the platform block diagram).

Multisequencer (MSQ). Multisequencer is an RISC processor for video pipeline management. It is mainly firmware with a hardware scheduler for fast context switches and process management.

Multichannel controller (MCC). Multichannel controller is a hardware with microprogrammed DMA. It arbitrates all the present requests from the operators and performs I/O with external memory. Scheduling is done for each request where a request is a memory burst with variable size. An I/O request is considered as a single transaction even if it may take several cycles to execute.

VLIW Image predictor (VIP). VLIW image predictor is a mix of hardware and firmware for performing motion compensation. Control is done by firmware while processing is done by hardware with a special VLIW instruction set. The instruction-set is modeled.

Encoder (COD). Encoder is a pure hardware. It computes the difference between the acquired and predicted image, DCT, zigzag, quantization and run/level coding.

In terms of system behavior, the external processor posts commands to the mailbox, HIF. The MSQ (programmed in C) takes the command into consideration and is in charge of the internal control of the CODEC. Consequently, the MSQ activates the different internal hardware or programmable blocks to achieve the coding or decoding of the video flow by reading status registers and writing command registers for each block. All the operators are pipelined and communicate with the system memory through a

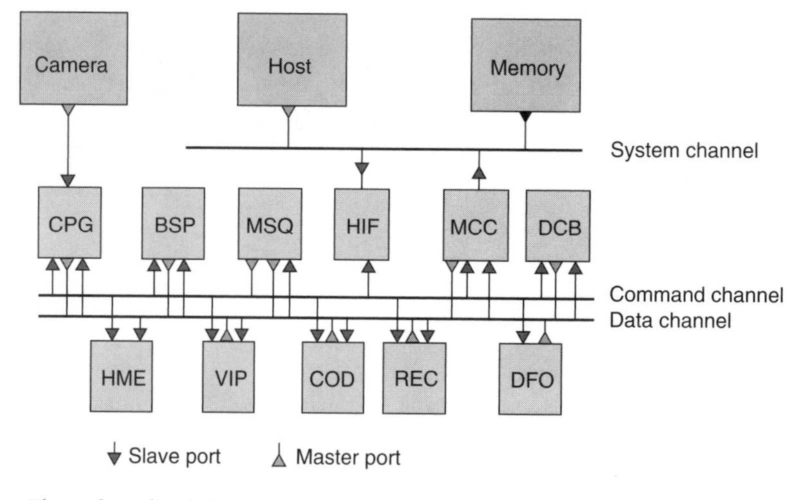

FIGURE 17.8 The multimedia platform block diagram.

memory controller. This memory controller receives requests from the internal operators and generates transactions on the top-level bus to access the system memory. It communicates with the internal operators through well-identified input/output registers that contain the values to be stored into and loaded from the memory.

The two internal channels on the platform, command and data, serve different purposes. For those modules that need to give access to their control register, their ports are bound to the command channel, whereas for those modules that handle data communication, their ports are bound to the data channel. The MSQ is a particular case since it has a double role: it generates read/write operations to control the system behavior, and hence its command master port is connected to the command channel; it also initiates transactions to the memory, and hence its communication port is bound to the data channel.

17.6.2 Design Choices

The requirement of being able to simulate the embedded software prevents us from developing a pure functional and sequential circuit model. Each computing block is modeled as a SystemC module with its own processes and associated synchronization elements.

17.6.2.1 Programmable Modules

This multiprocessor design is composed of C-programmed modules— pure hardware blocks as well as mixed blocks that encompass hardware and programmable operators. There are three categories with specific modeling strategies as stated below.

Software block. We rely on the sequential aspect of the codes for MSQ, BSP, HME, and VIP. All firmware is written in C. This enables the code to be natively compiled on the workstation and to communicate with the TLM model through an I/O library (via simple stubs to call C++ from C). The specific built-in instructions of the processors are modeled as C functions. One part of the C model is used directly for ROM code generation using retargetable C cross-compilers. An ISS is integrated in the environment to run the cross-compiled application software on the SoC host. Once the software is written, it remains unchanged despite its environment: TLM or RTL simulations, emulation, or application board.

Hardware block. A high-level model is written in SystemC. It is a functional, bit-true model with the memory transactional representation. We do not characterize the internal block structure. Instead, we model the input/output of the block, the synchronization, and the internal computation at the functional level. The complete SystemC model is used as a reference for RTL validation.

Let us consider an example. The coder block (COD) is a hardware pipeline with five operations: Delta, DCT, Zigzag, Quantization, and Run/Level. A FIFO is inserted among all these operations; the computation, controlled by an FSM, is pipelined; and a DMA manages the inputs and outputs. The RTL model is fully representative of this architecture. Developing such a model requires at least 3-person-months effort for a senior designer. In contrast, the corresponding transactional model consists of the input data acquisition (acquired and predicted blocks in our example), computes results with a C function and the resulting output. The gain is twofold: (1) easier and faster modeling; and (2) greater simulation speed. The corresponding effort is only about 1-person-week.

Mixed hardware/Software block. Mixed hardware/software blocks are modeled as a mix of the two previous categories.

17.6.3 System Integration Strategy

Some features of the modeling environment applied to the MPEG4 platform are illustrated in this section. We will focus our attention only on the fundamental aspects of our approach.

System synchronization in the platform is twofold. First, the MSQ is in charge of the global control of the platform. It is responsible for activating the coding/decoding tasks according to the current system status. A task will then be executed only if the relevant data are available; otherwise it will be suspended. Several tasks may be activated at the same time due to the internal operation pipeline. Synchronization is achieved by writing or reading the command and status registers of the different operators. Second, data exchanges between the operators and the memory are blocking operations. The platform synchronization scheme ensures that an operator will resume its computation only when the previous transaction is completed from the system point of view.

We model the data exchange with arbitrary sizes in the platform with respect to the system semantics of the exchange. For instance, transactions between the camera and grabber are line-per-line image transfer while transactions between MSQ and other operators are 32-bits wide.

17.6.4 Experimental Results — Simulation Figures

We compare the performance figures obtained for the MPEG4 platform in terms of code size and execution speed on different complementary environments as shown in Figure 17.9.

The modeling choices of the IP bring a significant gain in terms of model size: TLM models are 10 times smaller than RTL. As a result, they are easier to write and faster to simulate. The TLM simulation speed is 1400 times faster than an RTL model on a SUN Ultra 10 workstation of 330 MHz, 256 MB memory.

The RTL model simulates a coded image in an hour while the TLM model in SystemC only requires 2.5 sec per coded image. Before the Accellera SCE-MI interface [32] is available on the Mentor Celaro hardware emulator, an ad-hoc co-emulation transactional interface has been implemented through the C API; it provides both controllability and observability of the emulated design (clock control, memory and register load/dump, and programmable built-in logic analyzer). On the hardware side, synthesizable models have been developed for the camera, the memory, and several RTL transactors [32], that translate a data packet from a transaction into a BCA data exchange. On the software side, the host TLM model has been extended for HW debug and performance evaluation purposes. The coemulation requires 35 sec per coded image, i.e., a system clock at about 40–60 kHz. This speed (more than 30 times faster than a cycle-based coemulation) enables us to run the software developed onto the SoC TLM platform without any external or synthesizable CPU core being required for the host modeling.

The performances obtained for the TLM environment are well suited for embedded software development, since it runs a significant test-bench of 50 images in a couple of minutes on an Ultra 10 with full source level debugging facilities.

17.7 Design Flow Automation

Systems-on-chips are composed of a large number of IP blocks. They are interconnected with complex communication structures such as layered buses and crossbars. In addition, as we have discussed earlier,

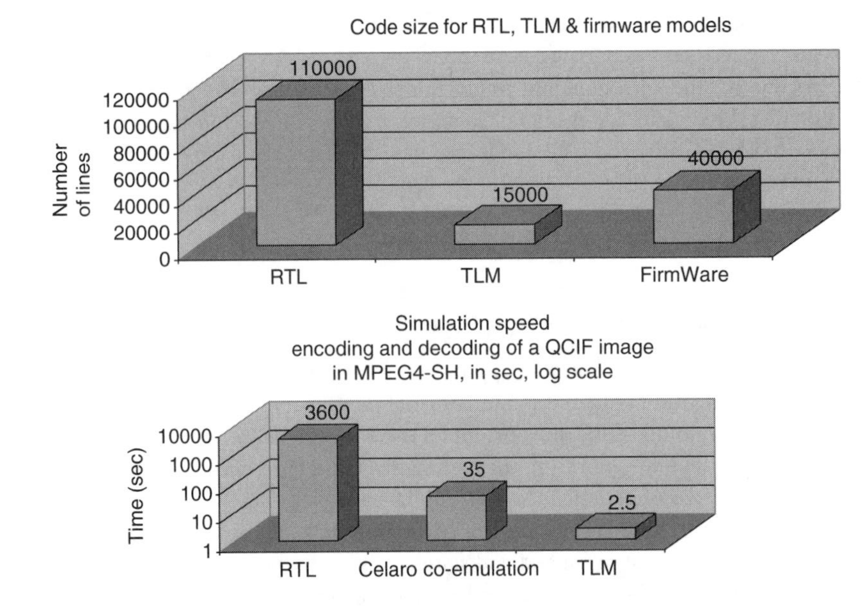

FIGURE 17.9 Performance figures of the MPEG4 platform.

different abstraction models are frequently mixed in a single simulation platform. The creation of such hardware platforms is a tedious and error-prone procedure with many repetitive tasks. In addition, low-level embedded software is extremely dependent on the hardware configuration.

Because of this complexity, design flow automation is a vital approach to reduce the time-to-market of an SoC product. We propose a platform automation tool that automates all the repetitive tasks in creating an SoC platform while leaving room for users to intervene manually in the procedure wherever necessary. This section discusses the idea of such a tool that aims to support system-to-RTL activities, and relies on the emerging SPIRIT† standard [33].

17.7.1 Overview of Design Flow Cockpits

The design and verification flow for an SoC encompasses the traditional ASIC flow as well as other activities related to the embedded software development and (micro-) architecture definition.

The design flow is split into two main categories:

- *Front-end cockpit.* Technology-independent tools and methods such as embedded software, architecture definition, and functional verification.
- *Back-end cockpit.* Technology-dependent tools and methods requiring information on the target technology. This includes logic synthesis as well as place-and-route activities.

The distinction made above does not, however, suggest that technology-independent tools would not rely on any implementation-dependent information (thus being technology-dependent). It simply means that there are far fewer dependencies. These dependencies must be identified and the related information must be exchanged between the front-end and back-end cockpits.

An obvious example is the clock frequency. Front-end activities depend on a target clock frequency to run CA (RTL or BCA) simulations. This frequency will remain an assumption until the design has gone through the physical implementation. Front-end activities that rely on this number should then be validated with the final number provided by tools of the back-end cockpit. Figure 17.10 illustrates the front-end and back-end cockpits for an SoC design flow.

† Structure for Packaging, Integrating and Re-using IP within Tool-flows.

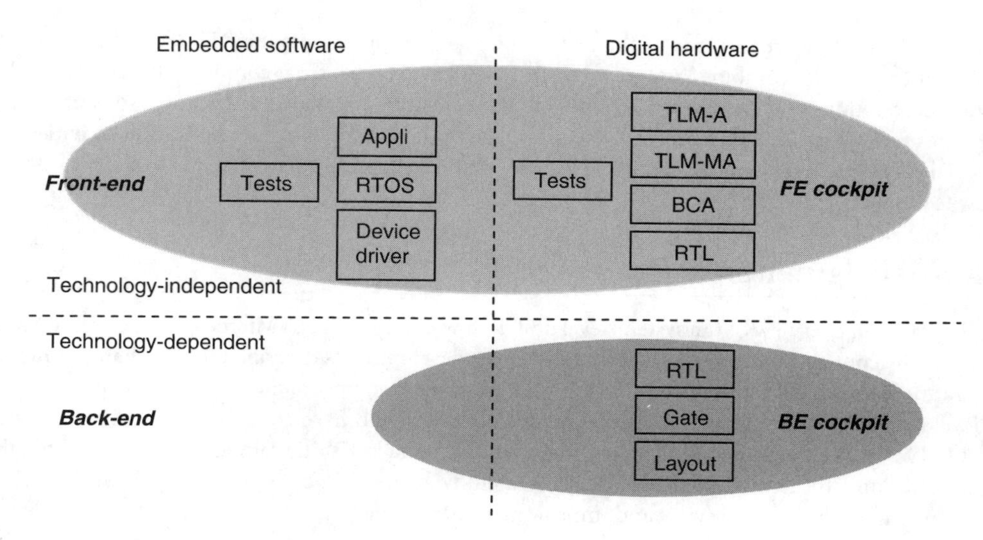

FIGURE 17.10 Design flow cockpits.

17.7.2 Front-End Cockpit

The front-end cockpit automates activities related to:

- embedded software development
- architecture definition and validation
- IP and SoC functional verification

The automation process relies on the availability of reusable IPs, for both their hardware, and associated software such as drivers. The meta-description of these IPs based on the SPIRIT standard will enable the automation tool to automate their usage in the design flow.

The main expected features of the automation tool are:

- Generation of SoC models at various levels of abstraction, namely TLM, BCA, and RTL, out of a single description of an SoC architecture. An editor, for instance a schematic editor, can be provided to edit the architecture.
- Assembly of all low-level embedded software that is necessary to boot the platform and initialize the peripherals.
- Assembly of specified software. For example, diagnostic routines can be assembled and executed to verify whether the generated platform behaves as expected.

Getting an IP meta-description ready for such an automation tool entails a repetititive effort. An industry standard is thus being developed through the SPIRIT Consortium to define a standard mechanism for describing and handling IP. This standard enables automated IP configuration and integration.

Within the SPIRIT-defined Design Environment, an IP meta-description or meta-data is written as an XML file. It could then be imported or exported by any SPIRIT-compliant CAD tool — that is, the design flow automation tool. The automation tool will configure an IP based on its XML file and subsequently generate a design by assembling the different IPs. This tool also has the ability to invoke any potential plug-in required by the configuration and generation of the IPs.

17.8 Conclusion

As highlighted in this chapter, we have introduced a new abstraction level in our SoC design flow to address the needs of early embedded software development, architecture analysis, and functional verification.

[16] W. Rosenstiel, and J. Ruf, Y. Vanderperren, M. Pauwels, W. Dehaene, A. Berna, and F. Ozdemir, A SystemC based system on chip modelling, and Design methodology. Chapter 1, *SystemC: Methodologies and Applications*, W. Müller, W. Rosenstiel, and J. Ruf, Eds., Kluwer Academic Publishers, Dordrecht, 2003.
[17] M. Baleani, A. Ferrari, A. Sangiovanni-Vincentelli, and C. Turchetti. HW/SW codesign of an engine management system, *Proceedings of Design, Automation and Test in Europe conference*, Paris, France, March 2000, pp. 263–267.
[18] G. Martin, H. Chang, L. Cooke, M. Hunt, A. McNelly, and L. Todd, *Surviving the SOC Revolution — A Guide to Platform-Based Design*, Kluwer Academic Publishers, Dordrecht, 1999, chap. 4.
[19] P. Thoen and F. Catthoor, *Modeling, Verification and Exploration of Task-Level Concurrency in Real-Time Embedded Systems*, Kluwer Academic Publishers, Dordrecht, 2000.
[20] C. Hoare, *Communicating Sequential Processes (CSP)*, International Series in Computer Science, Prentice-Hall, New York, 1985.
[21] M. Abrahams and J. Barkley, RTL verification strategies, *IEEE WESCON'98*, Anaheim, California, 1998, pp. 130–134.
[22] S. Asaf, E. Marcus, and A. Ziv, Defining coverage views to improve functional coverage analysis, *Proceedings of the 41st Annual Conference on Design Automation*, San Diego, California, June 2004, pp. 41–44.
[23] TransEDA web site: http://www.transeda.com
[24] F. Casaubieilh, A. McIsaac, M. Benjamin, M. Bartley, F. Pogodalla, F. Rocheteau, M. Belhadj, J. Eggleton, G. Mas, G. Barrett, and C. Berthet, Functional verification methodology of chameleon processor, *Proceedings of the 33rd Annual Conference on Design Automation*, Las Vegas, Nevada, 1996, pp. 421–426.
[25] A. Haverinen, M. Leclercq, N. Weyrich, and D. Wingard, SystemC-based SoC Communication Modeling for the OCP Protocol, 2002, http://www.ocpip.org
[26] C. Chevallaz, N. Mareau, and A. Gonier, Advanced methods for SoC concurrent engineering, *Proceedings of Design, Automation and Test in Europe Conference (DATE'02)*, 2002, Paris, France, pp. 59–63.
[27] J. Gerlach and W. Rosenstiel, System level design using the SystemC modeling platform, *Proceedings of the Forum on Specification & Design Languages (FDL'00)*, Tüübingen, Gernamy, 2000.
[28] L. Semeria and A. Ghosh, Methodology for hardware/software co-verification in C/C++, *Proceedings of the High-Level Design Validation and Test Workshop (HLDVT '99)*, San Diego, California, 1999.
[29] A. Fin, F. Fummi, M. Martignano, and M. Signoretto, SystemC: a homogenous environment to test embedded systems, *Proceedings of the IEEE International Symposium on Hardware/Software Codesign (CODES'01)*, Copenhagen, Denmark, 2001, pp. 17–22.
[30] A. Clouard, G. Mastrorocco, F. Carbognani, A. Perrin, and F. Ghenassia, Towards bridging the precision gap between SoC transactional and cycle-accurate, *Proceedings of Design, Automation and Test in Europe Conference (DATE'02)*, Paris, France, 2002.
[31] D. Gajski, J. Zhu, R. Domer, A. Gerstlauer, and S. Zhao, *SpecC: Specification Language and Methodology*, Kluwer Academic Publishers, Boston/Dordrecht/London, 2000.
[32] Accellera Interface Technical Committee, 2002, http://www.eda.org/itc
[33] SPIRIT Consortium web site: http://www.spiritconsortium.org

18

Assertion-Based Verification

Erich Marschner
Cadence Design Systems
Berkeley, California

Harry Foster
Jasper Design Automation
Mountain View, California

18.1 Introduction

Functional verification is a process of confirming that the *intent* of the design was preserved during its implementation. Hence, this process requires two key components: a *specification of design intent* and a *design implementation.* Yet historically, describing design intent in a fashion useful to the verification process has been problematic. For example, typical forms of specification are based on natural languages, which certainly do not lend themselves to any form of automation during the verification process. Furthermore, ambiguities in the specification often lead to misinterpretation in the design and verification environments. The problem is compounded when a verification environment cannot be shared across multiple verification processes (that is, the lack of interoperability between the specification and the various verification environments for simulation, acceleration, emulation, or formal verification).

In this chapter, we introduce a new approach to addressing the functional verification challenge, known as *assertion-based verification,* which provides a unified methodology for unambiguously specifying design intent across multiple verification processes using assertions. Informally, an assertion is a statement of design intent that can be used to specify design behavior [1]. Assertions may specify internal design behaviors (such as a specific FIFO structure) or external design behavior (such as protocol rules for a design's interfaces or even higher-level requirements that span across design blocks). One key characteristic of assertions is that they allow the engineer to specify *what* the design is supposed to do at a high level of abstraction, without having to describe the details of *how* the design intent is to be implemented. Thus, this abstract view of the design intent is ideal for the verification process — whether we are specifying high-level requirements or lower-level internal design behavior.

18.1.1 Controllability and Observability

Fundamental to the discussion of assertion-based verification is an understanding of the concepts of *controllability* and *observability.* Controllability refers to the ability to influence an embedded finite state-machine,

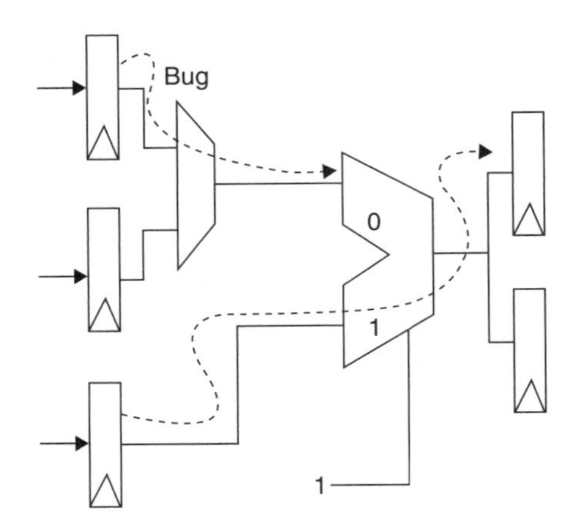

FIGURE 18.1 Poor observability misses bugs.

structure, or specific line of code within the design by stimulating various input ports. Note that while in theory, a simulation testbench has high controllability of the design model's input ports, during verification it can have very low controllability of an internal structure within the model. Observability, in contrast, refers to the ability to observe the effects of a specific internal finite state-machine, structure, or stimulated line of code. Thus, a testbench generally has limited observability if it only observes the external ports of the design model (because the internal signals and structures are often hidden from the testbench).

To identify a design error using the testbench approach, the following conditions must hold (i.e., evaluate to True):

1. The testbench must generate proper input stimulus to activate (i.e., sensitize) a bug.
2. The testbench must generate proper input stimulus to propagate all effects resulting from the bug to an output port.

It is possible, however, to set up a condition where the input stimulus activates a design error that does not propagate to an observable output port (Figure 18.1). In these cases, the first condition cited above applies; however, the second condition is absent.

Embedding assertions in the design model increases observability. In this way, the verification environment no longer depends on generating proper input stimulus to propagate a bug to an observable port. Thus, any improper or unexpected behavior can be caught closer to the source of the bug, in terms of both time and location in the design intent.

While assertions help solve the observability challenge in simulation, they do not help with the controllability challenge. However, by adopting an assertion-based, constraint-driven simulation environment, or applying formal property checking techniques to the design assertions, we are able to address the controllability challenge.

18.2 History

In this section, we discuss different approaches to solving the functional specification challenge. We begin by briefly introducing propositional temporal logic, which forms the basis for modern property specification languages. Building on this foundation, we then present a historical perspective for various forms of assertion specification.

18.2.1 Reasoning about Behavior

Logic, whose origins can be traced back to ancient Greek philosophers, is a branch of philosophy (and mathematics) concerned with reasoning about a model. In a classical logic system, we state a proposition, and then deduce (or infer) if a given model satisfies our proposition, as illustrated in Figure 18.2.

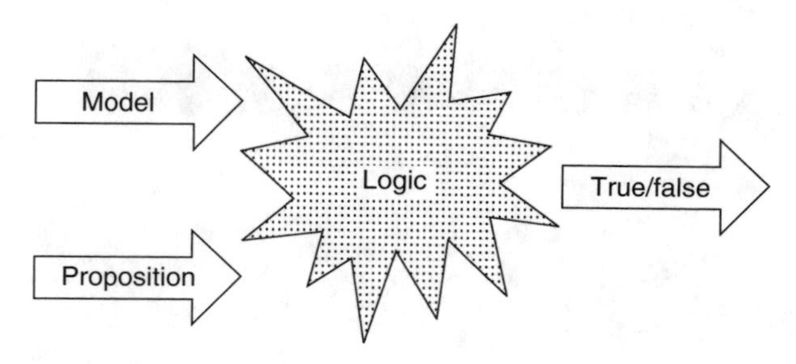

FIGURE 18.2 Classical logic system.

For example, consider the following set of propositions:

- *The moon is a satellite of the earth.*
- *The moon is rising (now).*

If we let the universe be our model, then using classical logic we can check whether our set of propositions holds for the given model.

Classical logic is good for describing *static* situations. However, it is unsuitable for describing dynamic behavior (i.e., situations involving time). Returning to the previous example, it would not be possible to express the following proposition in classical logic, since it involves time:

- *The moon will rise again and again.*

Note that our interest in functional verification of hardware systems requires that we use a logic that is expressive enough to describe properties of reactive systems. For a reactive system, components of the system concurrently maintain ongoing interaction with their environment, as well as other concurrent components of the system. Hence, in the next section, we discuss a more expressive logic that involves time.

18.2.1.1 Propositional Temporal Logic

In this section, we build a foundation for understanding propositional temporal logic by introducing a few basic concepts. The advantage of using temporal logic to specify properties of reactive systems is that it enables us to reason about these systems in a very simple way. That is, temporal logic eliminates the need to explicitly specify the time relationships between system events which are represented as Boolean formulas that describe states of the design. For example, instead of writing the property expression involving time explicitly, such as

```
∀t.!(grant1(t) & grant2(t))
```

which specifies for all values of time `t`, `grant1` and `grant2` are mutually exclusive, we simply write in a temporal property language, such as the Accellera Property Specification Language (PSL):

```
always !(grant1 & grant2)
```

which states that `grant1` and `grant2` should not hold at the same time.

In temporal logic, we define a computational path π as an infinite sequence of states:

$$\pi = s_0, s_1, s_2, s_3, \ \ldots$$

that represents a forward progression of time and a succession of states, s_i. When proving a temporal assertion, we may assume that a point in time or given state along the path has a unique future or successor state (for example, $s_0 \rightarrow s_1$), in which case the assertion is checked on a given path or trace of execution (for example, a single simulation trace). Thus, each possible computational path of a system is

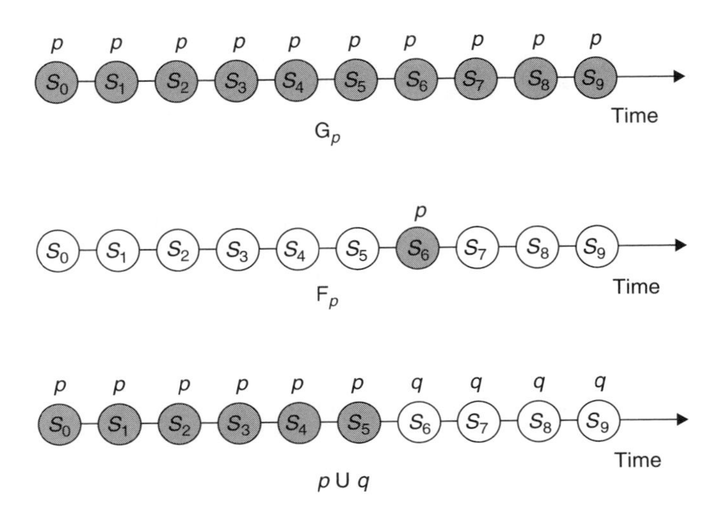

FIGURE 18. 3 Temporal formula path operator semantics.

considered separately. We refer to this form of logic as *linear-time temporal logic*, and Linear Temporal Logic (LTL) is one example [2]. Alternatively, we may assume that each point in time or given state along a path may split into multiple futures or successor states (for example, $s_0 \rightarrow s_1$ and $s_0 \rightarrow s_5$). In that case, all computational paths are considered concurrently, and are usually represented as a tree of infinite computational paths. We refer to this form of logic as *branching-time temporal logic*, and Computation Tree Logic (CTL) is one example [3].

CTL and LTL play essential roles in formal hardware verification for branching-time temporal logic and linear-time temporal logic, respectively. To show how these two types of logic differ, we first introduce CTL* [4]. CTL* contains operators for reasoning about paths of computation, such as the *path formula* operators **G** (always), **F** (eventually), **U** (until) and **X** (next), and operators for reasoning on branching paths of execution, that is, the *state formula* operators **A** (for all paths of execution) and **E** (for some path of execution). In addition to these quantifiers, any Boolean compositions of CTL* formulas are CTL* formulas as well.

For any temporal formulas p and q, as illustrated in Figure 18.3, the temporal formula $\mathbf{G}\ p$ specifies that the temporal formula p holds for all states of a path π (or simply, p always holds). The temporal formula $\mathbf{F}\ p$ specifies that p holds for some future state of a path π. The temporal formula $p\ \mathbf{U}\ q$ specifies that the temporal formula p holds in all states of a path π until q holds in some future state of π.

The temporal formula $\mathbf{A}f$ specifies that for all paths π starting from the current state, f holds. The temporal formula $\mathbf{E}f$ means that there is a path π starting from the current state for which f holds.

As seen in the previous paragraphs, CTL* can be separated into *state formulas* (involving **A** and **E**) and *path formulas* (involving **G**, **F**, **X**, and **U**). Any atomic proposition p (or Boolean expression) over *state formulas* is by definition a *state formula*. In addition, existential quantification over path formulas (for example, $\mathbf{E}f$, where **E** is the existential quantifier and f a *path formula*) is also by definition a *state formula*.

Any *state formula* is a *path formula*, as are Boolean compositions of *path formulas*. In addition, path formulas can be composed using the temporal operators $\mathbf{X}f$ and $f_1\ \mathbf{U}\ f_2$.

Note that the **F** operator can be thought of as an alias for the unary form of the *until* operator (for example, $\mathbf{F}p = true\ \mathbf{U}\ p$), and that **G** and **F** are dual (that is, $\mathbf{G}p$ is equivalent to $\neg\mathbf{F}\neg p$). The rationale behind the first alias is that *eventually* (**F**) is equivalent to waiting vacuously until p is valid; and the rationale behind the second formula is that saying "*p always valid*" is equivalent to saying that "*it is not the case that ¬p will be valid in the future.*" Similarly, it is not difficult to show that **E** and **A** are dual.

Note that in CTL* we do not make any restriction on the order in which temporal and branching operators appear in a valid formula. As a result, **FG** and **AG** are valid CTL* formulas. The first formula (**FG**) states that *eventually our proposition will be valid forever*. The second formula (**AG**) states that *for all paths starting from the current state, our proposition will always be valid.*

Now that we have presented CTL*, we can restrict this logic to CTL and LTL.

- *CTL:* A CTL formula is a CTL* formula beginning with a branch quantifier (**A** and **E**), restricting that temporal operators (**F**, **G**, **U**, and **X**) be proceeded by branch quantifiers. For example, the formula **AG** p is a valid CTL formula, but **FG** p is not.
- *LTL:* An LTL formula is the subset of CTL* formulas obtained by simply restricting the valid formulas to path formulas. For example, the formula **FG** p is a valid LTL formula, but **EG** p is not.

If we consider an implicit universal quantifier for all paths (that is, **A**) in front of an LTL formula, we can see that certain behaviors, such as **A**(**GF** p), cannot be represented by CTL, although it is a valid LTL formula. Similarly **AG**(**EF** p) is a valid CTL formula, but not a valid LTL formula. While the first formula says that for all states of all paths, eventually p holds (a fairness constraint), the second formula says that for all branches of all states, at least in one of the paths, eventually p holds.

Note that neither LTL nor CTL can express a property that involves counting, such as

- *p is asserted in every even cycle.*

In the next section, we introduce extended regular expressions, which allow us to express behavior that involves counting.

We have not thoroughly discussed the semantics of CTL*, CTL, and LTL in this chapter, but we have given a short introduction on this subject to subsidize the remainder of this chapter. We refer the reader to [5] for a more complete definition of these logics' semantics and complexities.

18.2.1.2 Extended Regular Expressions

Regular expressions provide a convenient way to define sets of computations (that is, a temporal pattern used to match various *sequences* of states). Extended regular expressions are regular expressions extended with conjunctions and negation. Hence, the computational path defined by an extended regular expression can be combined to form the building blocks used to specify hardware design assertions. Regular expressions can express counting *modulo n* type behavior through the use of the * operator. Hence, regular expressions allow us to specify properties that cannot be described by either CTL or LTL, such as `{`true; p}`[*], which states that *p is asserted in every even cycle.*

Note that not all properties can be expressed with extended regular expressions. For example, the property *eventually p holds forever* cannot be expressed using extended regular expressions, yet this property can be expressed in LTL (that is, **FG** p).

18.2.2 Assertion Languages

Assertions may be expressed either declaratively or procedurally. A declarative assertion is always active, and is evaluated concurrently with other components in the design. A procedural assertion, on the other hand, is a statement within procedural code, and is executed sequentially in its turn within the procedural description. Hence, declarative properties are natural for specifying block-level interfaces, as well as systems. Similarly, a procedural assertion is convenient for expressing implementation-level properties that must hold in the context of procedural code. A key difference between the declarative assertion and the procedural assertion is that the declarative assertion concurrently monitors the assertion expression, while the procedural assertion only validates the assertion expression during sequential visits through the procedural code.

In the following sections, we discuss various languages and techniques in use today for expressing assertions, including: the VHSIC Hardware Description Language (*VHDL*), the *Open Verification Library* (OVL), and *SystemVerilog Assertions* (SVA). A comprehensive discussion of Accellera PSL follows in Section 18.3.

18.2.2.1 VHSIC Hardware Description Language (VHDL) Assertions

The concept of assertions was introduced quite early in the development of VHDL. VHDL was originally developed in the early 1980s, by a team consisting of Intermetrics, Inc., IBM, and Texas Instruments, under a DoD-sponsored effort to create a design and documentation language for VHSIC-class electronic designs. The initial requirements for this language were defined in June 1981 during a workshop in Woods Hole, Massachusetts. The requirements published in the proceedings of that workshop [6] included a requirement to support exception handling, but they did not explicitly mention assertions. However, by January 1983, when these requirements were included in the Request for Proposal (RFP) for the VHDL effort [7], they had been extended to include many of the so-called "Steelman" requirements developed earlier for the Ada programming language [8], including a requirement for assertions that was almost identical to the assertion requirement for Ada (section 3.8.6 Assertions):

> It shall be possible to include assertions in programs. If an assertion is false when encountered during simulation, it shall raise an exception. It shall also be possible to include assertions, such as the expected frequency for selection of a conditional path, that cannot be verified. [Note that assertions can be used to aid optimization and maintenance.]

Ironically, the original Ada requirement did not lead to an explicit assertion construct in Ada, yet the derivative requirement for VHDL resulted in the creation of not just one, but two kinds of assertions in VHDL: a sequential assertion statement, which can appear within sequential (i.e., procedural) code in a process or subprogram; and a concurrent assertion statement, which can appear in a block, and acts like an independent process.

VHDL assertions involve only combinational expressions; no temporal operators are available in VHDL, other than the very limited capability provided by certain signal-valued attributes such as S'delayed. Even so, VHDL assertions quickly proved to be a very useful mechanism for expressing invariants that are expected to hold within a design, such as a relationship among variables that is expected to hold at a given point in the execution of algorithmic code (expressed with a sequential assertion statement) or a relationship among signals that is expected to hold at all times (expressed with a concurrent assertion statement).

The simulation semantics for VHDL assertions, which are defined in the VHDL Language Reference Manual [9], ensure that an assertion will issue an error message any time it is executed and the asserted expression evaluates to False. However, error detection during simulation is not the only value of such assertions. They also can play a significant role in documentation of code and therefore maintenance of the code over time. Furthermore, as formal equivalence checking developed as an alternative to simulation, VHDL assertions were adapted to that verification method as well. For example, in the early to mid-1990s, VHDL assertions were used in one formal verification tool [10] to specify both "axioms" or assumptions about the environment, and assertions about the design that needed to be verified.

At the time of this writing, work is in progress to extend VHDL assertions further by incorporating PSL v1.1 (see Section 18.3) into the next version of the VHDL standard. This extension will enable PSL temporal operators, properties, sequences, and directives to be used alongside the existing VHDL concurrent assertion statements to express much more complex behavioral requirements.

18.2.2.2 Open Verification Library

One of the challenges in creating an assertion-based methodology is ensuring consistency within a design project. Any inconsistencies between multiple stakeholders involved in the design and verification process can become so problematic that the benefits the assertions provide during the verification process are overshadowed by an unmanageable methodology. For example, an assertion-based methodology needs to provide a consistent manner of:

- controlling assertions (for example, enabling and disabling assertions)
- reporting assertion violations
- specifying reset

Aside from methodology consistency, another challenge related to assertion adoption has been the effort required to teach engineers new languages that specify assertions.

OVL was developed as a means to overcome these challenges within an assertion-based methodology [11]. Actually, OVL is not an assertion language, and lacks the expressiveness of languages such as PSL or SystemVerilog Assertions (SVA). On the contrary, OVL is a library of simulation monitors written in both Verilog and VHDL. Hence, OVL could be classified as a declarative form of assertion specification. During the RTL development process, engineers select and then instantiate OVL monitors into the RTL model. Each assertion monitor is then used to check for a specific Boolean or temporal violation during the verification process. The following is an example of a Verilog instantiated OVL monitor that checks for the case when `grant1` and `grant2` are not mutually exclusive:

```
assert_always mutex (clk, reset_n, !(grant1 & grant2));
```

OVL monitors address assertion-based methodology considerations by encapsulating a unified and systematic method of reporting, and a common mechanism for enabling and disabling assertions during the verification process. The reporting and enable/disable features use a consistent process, which provides uniformity and predictability within an assertion-based methodology.

OVL offers a wide set of monitors, enabling the engineer to capture a large class of assertions such as the simple `assert_always` Boolean invariant shown in the previous example, one-hot checking, as well as multicycle temporal checks such as `assert_next` and `assert_change`.

18.2.2.3 SystemVerilog Assertions

Unlike VHDL, the IEEE-1364 Verilog language never contained a Boolean assertion construct. It was not uncommon for designers to capture assertions in their Verilog RTL in an ad hoc fashion using the Verilog `$display()` system task. The problem with this approach is that not all `$display()` calls could be treated or identified as assertions, which means that any assertions specified using a `$display()` could not be verified by formal verification tools. In addition, this approach of specifying assertions required a significant amount of extra modeling to express sequences or other complex temporal behaviors.

SystemVerilog Assertions was developed to provide engineers the means to describe complex behaviors about their designs in a clear and concise manner. SVA is based on LTL built over sublanguages of regular expressions. Although SVA lacks the full expressiveness of LTL (that is, it only supports *globally* (G) in the form of the SystemVerilog `always` or an implicit *globally* of a concurrent assertion), most engineers will find SVA sufficient to express most common assertions required for hardware design.

SVA supports two types of assertions: *procedural* (or immediate) and *declarative* (or concurrent). Both the types of assertion are intended to convey the intent of the design engineer and to identify the source of a problem as quickly and directly as possible. In addition, engineers can use SVA to define temporal correctness properties and coverage events.

The following is an example of a SystemVerilog concurrent assertion used to check for mutual exclusion of grant signals:

```
property mutex;
  @(posedge clk) disable iff (!reset_n) (!(grant1 & grant2));
endproperty

assert property (mutex);
```

At the time of this writing, the Accellera 3.1a SystemVerilog LRM [12] has moved to the IEEE for standardization under the auspices of the IEEE's Corporate Standards Group (CAG) as IEEE P1800. In 2004, great effort was taken within Accellera to ensure semantics alignment between SVA with PSL. However, there are a few syntactical differences between SVA and PSL due to the divergent objectives of these two languages. Most notably, SVA was designed to provide an embedded assertion capability directly within the SystemVerilog language, while PSL was designed as a standalone assertion language that harmoniously works across multiple hardware description languages (HDLs), such as SystemVerilog, Verilog, and VHDL.

18.3 State of the Art

The following section introduces principles and concepts behind Accellera PSL [13].

18.3.1 Property Specification Language Principles

PSL is a comprehensive language that includes both LTL and CTL constructs. This enables PSL to support various kinds of formal verification flows, including event- and cycle-driven simulation and various algorithms for formal verification (model checking).

Most tools are based on a single algorithm and therefore cannot support all of PSL. In particular, most simulation tools are characterized by a single trace and monotonically advancing time, which preclude support for many CTL-based properties and make it difficult to support some LTL-based properties efficiently. The subset of PSL that can be supported efficiently in both simulation and formal verification is known as the "Simple Subset" and is defined by a short list of simple syntactic restrictions.

The temporal semantics of PSL are formally defined. This formal definition ensures precision; it is possible to understand precisely what a given PSL construct means, and does not mean. The formal definition also enables careful reasoning about the interaction of PSL constructs, or about their equivalence or difference. The formal definition has enabled validation of the PSL semantic definition, either manually or through the use of automated reasoning [14].

For practical application, it is imperative that PSL statements be grounded in the domain to which they apply. This is accomplished in PSL by building on expressions in the HDL used to describe the design that is the domain of interest. At the bottom level, PSL deals with Boolean conditions that represent states of the design. By using HDL expressions to represent those Boolean conditions, PSL temporal semantics are connected to, and smoothly extend, the semantics of the underlying HDL.

18.3.2 Basic PSL Concepts

PSL is defined in *layers* (Figure 18.4). The Boolean layer is the foundation; it supports the temporal layer that uses Boolean expressions to specify behavior over time. The verification layer consists of directives that specify how tools should use temporal layer specifications to verify functionality of a design. Finally, the modeling layer defines the environment within which verification is to occur.

PSL is also defined in various *flavors* (Figure 18.4). Each flavor corresponds to a particular underlying HDL. Currently defined flavors include VHDL, Verilog, and SystemVerilog. A SystemC flavor has been added in the recently completed IEEE 1850–2005 standard [14a].

Since PSL works with various underlying HDLs, it must adapt to the various forms of expression allowed in each language, and the various notions of datatype that exist in each language. To that end,

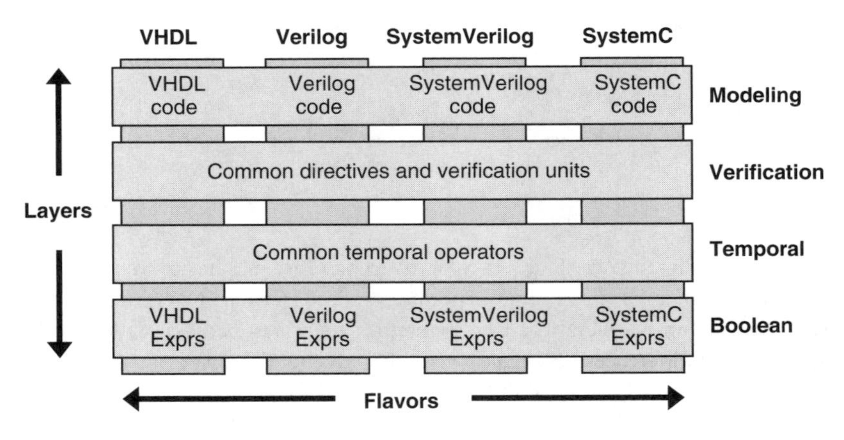

FIGURE 18.4 PSL layers and flavors.

PSL defines a number of expression type classes — Bit, Boolean, BitVector, Numeric, and String — and specifies how native and user-defined datatypes in VHDL, Verilog, SystemVerilog (and now, SystemC) all map to those type classes.

18.3.2.1 Boolean Layer

The Boolean layer consists of expressions that belong to the Boolean type class, i.e., those whose values are, or map to, True or False. These expressions represent conditions within the design, e.g., the state of the design, or the values of inputs, or a relationship among control signals. Typically, this includes any expression that is allowed as the condition in an 'if' statement in a given HDL. PSL extends this set of expressions with a collection of built-in functions that check for rising or falling signals, signal stability, and other useful utilities.

18.3.2.2 Temporal Layer

The temporal layer involves time, and Boolean expressions that *hold* (i.e., are True) at various points in time. This includes expressions that hold pseudo-continuously (i.e., at every point in time considered by the verification tool) as well as expressions that hold at selected points in time, such as those points at which a clock edge occurs, or an enabling condition is True. Temporal operators enable the specification of complex behaviors that involve multiple conditions over time.

One class of temporal operators is used to construct sequential extended regular expressions (SEREs) or *sequences*. A sequence describes behavior in the form of a series of conditions that hold in succession. A sequence may specify that a given condition repeats for a minimum, maximum, or even unbounded number of times, and it may specify that two or more subordinate sequences overlap or hold in parallel. If the behavior described by a sequence matches the behavior of the design starting at a given time, then the sequence holds at that time. PSL also supports *endpoints*, which are identical to sequences except that they hold at the time in which the behavior completes.

For example, consider the following sequence declarations, which describe a simple handshake-based bus protocol. In this case, a default clock declaration specifies the times at which successive expressions are evaluated:

```
default clock = (posedge clk);
sequence GetBus = { req[+]; (req && ack) };
sequence HoldBus = { (req && ack)[*] };
sequence RlsBus = { (!req && ack); !ack };
sequence ReadOp = { rwb && ardy; !drdy[*0:2]; drdy };
sequence ReadT = { GetBus; {HoldBus && ReadOp}; RlsBus };
```

The first three define the series of control conditions involved in arbitrating for control of the bus. Sequence GetBus describes behavior in which a request signal is high for one or more cycles, followed by a cycle in which the request signal is still high and an acknowledge signal goes high. Sequence HoldBus describes behavior in which both request and acknowledge signals stay high for an indefinite length of time. Sequence RlsBus describes behavior in which the request signal drops, and in the next cycle the acknowledge signal drops. Similarly, sequence ReadOp describes the control signals involved in a read operation on the bus: in the first cycle, a read operation is indicated and "address ready" signal is asserted; following that, "data ready" is asserted after a delay of 0, 1, or 2 cycles. Sequence ReadT combines all of these to represent a read transaction in which the GetBus sequence occurs first, then the ReadOp sequence occurs in parallel with the HoldBus sequence, and finally the RlsBus sequence occurs.

A second class of temporal operators provides a means of expressing conditional behavior using implication. This includes Boolean implication (a −> b) as well as suffix implication, in which the antecedent may be a sequence rather than a simple Boolean. In the latter case, the consequent holds at the end of the antecedent sequence. A third class of temporal operators consists of a set of English keywords that describe temporal relationships. These include **always, never, next, eventually!, until, before,** and slight variations thereof. Temporal operators may be combined with Booleans and sequences to describe behaviors or *properties* of a design.

For example, consider the following property declarations, which describe certain characteristics of the same simple bus protocol:

```
property BusArbitrationCompletes =
    always GetBus |=> {HoldBus; RlsBus};

property ReqSteady =
    always rose(req) -> next req until ack;

property AckSteady =
    always rose(ack) -> next !req before !ack;

property AckWithin (constant N) =
    always rose(req) -> next {{ack} within [*N]};
```

The first declaration defines a property BusArbitrationCompletes: if sequence GetBus occurs, then it will always be followed by sequence HoldBus and then sequence RlsBus. The second declaration defines property ReqSteady: if signal req rises, then it will stay high until signal ack is high. Similarly, the third declaration defines property AckSteady: if signal ack rises, then signal req will go low before ack goes low. The final declaration defines parameterized property AckWithin: after req rises, ack will occur within *N* cycles, where *N* is to be provided later.

18.3.2.3 Verification Layer

The verification layer specifies how sequences and properties are expected to apply to the design, and therefore how verification tools should attempt to verify the design using those sequences and properties. PSL *directives* indicate what to do with a given property or sequence. The **assert** directive says that a given property is expected to hold for the design, and that this should be checked during verification. The **assume** directive says that a given property should be taken for granted — it can be used in verifying other properties, but it need not be checked. The **restrict** directive is similar; it says that that a given sequence should be taken for granted. Typically, **assert** directives apply to the design being verified, while **assume** and **restrict** directives are used to constrain the environment within which the design is being verified. Additional directives (in particular, the **cover** directive, described below) provide additional capabilities.

The Verification layer also includes *verification units*, which enable packaging a collection of related PSL declarations and directives so they can be applied as a group to a particular part of the design. A verification unit may be bound to a particular instance within the design, or to a module used within the design. This causes the PSL directives to apply to that instance, or to all instances of that module, respectively. Unbound verification units may be defined to create reusable packages of PSL definitions. One verification unit may *inherit* another verification unit, which makes the definitions of the latter available for use in the former.

18.3.2.4 The Modeling Layer

The Modeling layer consists of HDL code used to model the environment of the design and/or to build auxiliary state machines that simplify the construction of PSL sequences or properties. Modeling the environment of the design under verification is primarily of interest for formal verification (see Chapter 20 *Formal Property Verification*). Building auxiliary state machines applies to both formal verification and simulation, and is addressed here.

Counters are one class of auxiliary state machines that are often necessary in specifying behavior. The modeling layer of PSL allows the underlying HDL to be used for constructing such a counter within a verification unit, so that it is kept separate from the design itself, yet is still available for use in a PSL assertion. For example, consider a byte-serial bus protocol that specifies the number of bytes being transferred as the first byte following the packet header. Assuming the Verilog flavor of PSL, the following verification unit might be written to specify the behavior of this protocol:

```
vunit SerialProtocol (serial_interface_module) {
   integer count, crc;

   always @(posedge clk)
     if (header)
       begin
       count ⇐ currentbyte;
       crc ⇐ 0;
       end
   else if (count > 0)
       begin
       crc ⇐ compute_crc(crc, currentbyte);
       count ⇐ count - 1;
       end

endpoint header = {currentbyte == `PKT_HDR; 1};
sequence data = {(count>0)[*]; crc == currentbyte};

assert always header -> next data;

}
```

In this example, verification unit SerialProtocol contains Verilog code (in Courier font) that acts as an auxiliary state machine supporting the assertion at the end of the verification unit. The Verilog code maintains a counter, loaded from the first byte after the header, and decrements the counter as successive data bytes appear. In parallel, it computes a cyclic redundancy check (CRC) for the packet, to compare against the CRC embedded at the end of the packet.

This example illustrates reference to a PSL endpoint within the HDL modeling layer code. Endpoint "header" is a sequence consisting of the header byte of a packet followed by the packet size. When this sequence has been recognized, it triggers the loading of the packet size into the counter in the Verilog code. The assertion also uses this endpoint as the indication that a packet has started, which implies that the data must follow (and that the CRC computed by the Verilog code must match the byte following the data in the packet).

Although PSL is defined as a separate language, it is also possible to embed PSL directly into HDL code. One approach involves a convention adopted by many PSL tools in which PSL code is embedded in HDL comments, following the comment delimiter and the keyword "psl." Another approach involves incorporation of PSL directly into the underlying HDL, as is planned for the next version of IEEE 1076 VHDL. Embedded assertions are of most use for designers who want to add assertions as they go, to document their assumptions. External assertions (in verification units) are often of more use to verification engineers, who want to verify blocks without modifying the source code of those blocks.

18.3.3 Future Direction

The current interest in assertion-based verification is focused on the use of assertions in simulation. Assertions are usually added after the fact, to increase observability of events within the design so that bugs can be detected more closely to their source. But this approach is only beginning to take advantage of the power of assertion languages. Over time, we can expect to see increased use of the coverage capabilities of assertion languages to enable coverage-driven verification. At the same time, taking advantage of the formal semantics of assertion languages such as PSL, formal verification of designs will become more common. Eventually, we can expect to see assertion languages used to thoroughly specify the behavior of design IP blocks and their interfaces [15], with such specifications developed before, rather than after, the design is done. This may even lead to use of assertion languages as a vehicle for design, if appropriate methods for assertion-based design synthesis or component selection can be developed.

In parallel with increasing application of assertion languages, development of those languages will continue to increase their capabilities and applicability within a variety of design languages and environments. Much of this development is occurring within IEEE standards activities. The IEEE 1800 SystemVerilog working group is pursuing integration of IEEE 1364 Verilog and Accellera 3.1a SystemVerilog, and in the process has refined SystemVerilog Assertions further. The IEEE 1850 PSL working group has completed the definition of IEEE Std 1850-2005 Property Specification Language (PSL), which refines and enhances Accellera PSL v1.1.

References

[1] H. Foster, A. Krolnik, and D. Lacey, *Assertion-Based Design*, 2nd ed., Kluwer Academic Press, Dordrecht, 2004.

[2] A. Pnueli, The temporal logic of programs, *Proceedings of 18th IEEE Symposium on Foundations of Computer Science*, 1977, pp. 46–57.

[3] E.M. Clarke and E.A. Emerson, Design and synthesis of synchronization skeletons using branching time temporal logic, *Proceedings of Workshop on Logic of Programs*, Lecture Notes in Computer Sciences, Springer, Berlin, Vol. 131, 1981, pp. 52–71.

[4] E.A. Emerson and J.Y. Halpern, "Sometimes" and "not never" revisited: on branching vs. linear time, *Proceedings of 10th ACM Symposium on Principles of Programming Languages*, 1983, pp. 127–140.

[5] T. Kropf, *Introduction to Formal Hardware Verification*, Springer, Berlin,1998.

[6] G.W. Preston, Report of IDA Summer Study on Hardware Description Language (IDA Paper P-1595), Institute of Defense Analyses, October 1981.

[7] Department of Defense Requirements for Hardware Description Languages, January 1983.

[8] Department of Defense Requirements for High Order Computer Programming Languages ("Steelman"), June 1978, at http://www.adahome.com/History/Steelman.

[9] IEEE Standard 1076-2002, *IEEE Standard VHDL Language Reference Manual*, IEEE, 17 May 2002.

[10] VFormal User Guide, COMPASS Design Automation, August 1995.

[11] H. Foster and C. Coelho, Assertions targeting a diverse set of verification tools, *Proceedings of 11th Annual International HDL Conference*, March 2001.

[12] Accellera, 1370 Trancas Street, #163, Napa, CA 94558. *SystemVerilog Language Reference Manual*, 2004. http://www.eda.org/sv/SystemVerilog_3.1a.pdf.

[13] Accellera, 1370 Trancas Street, #163, Napa, CA 94558. *Property Specification Language, Reference Manual*, 2004. www.eda.org/vfv/docs/PSL-v1.1.pdf.

[14] M.J.C. Gordon, Validating the PSL/Sugar semantics using automated reasoning, in *Formal Aspects of Computing*, Vol. 15, Springer, London, 2003, pp. 406–421.

[14a] IEEE Std 1850-2005, *IEEE Standard for Property Specification Lanuage (PSL)*, IEEE, New York, September 2005.

[15] E. Marschner, B. Deadman and G. Martin, IP reuse hardening via embedded sugar assertions, in *Proceedings of the IP-Based System-on-Chip Design Workshop*, Grenoble, France, October 30–31, 2002.

19 Hardware Acceleration and Emulation

Ray Turner
Cadence Design Systems, San Jose, California

Mike Bershteyn
Cadence Design Systems, Inc. Cupertino, California

19.1 Introduction

According to a Collett International North American survey, 82% of silicon respins are due, at least in part, to functional errors. Thus, comprehensive functional verification is key to reducing developmental costs and delivering a product on time. Functional verification of a design is most often performed using logic simulation and prototyping. There are advantages and disadvantages to each, and often both methods are used. Logic simulation is easy, accurate, flexible, and low cost. But it is not fast enough for large designs and it is much too slow to run application software against the hardware design. Field-programmable gate array (FPGA)-based prototypes are fast and inexpensive. But the time required to implement a large design into several FPGAs can be very long and it is error-prone. Changes to fix design flaws also take a long time to implement and may require board wiring changes. Since FPGA prototypes have little debugging capability, probing signals inside the FPGAs in real time is very difficult, if not impossible, and recompiling FPGAs to move probes takes too long. Clearly, simulation should be used earlier in the verification process when bugs and fixes are frequent, and prototyping should be used at the end of the development cycle when the design is basically complete, and speed is needed to get sufficient testing to uncover any remaining system-level bugs. Prototyping is also popular for testing software.

Simulation acceleration can address the performance shortcomings of simulation to an extent. Here, the design is mapped into a hardware accelerator to run faster, and the testbench (and any behavioral design code) continues to run on the simulator on the workstation. A high bandwidth, low latency channel connects the workstation to the accelerator to exchange signal data between testbench and design. By Amdahl's law, the slowest device in the chain will determine the speed achievable. Normally, this is the testbench in the simulator. With a very efficient testbench (written in C or transaction-based), the channel may become the bottleneck.

In-circuit emulation (ICE) greatly reduces the long time required to implement and change designs typically seen with FPGA prototyping, and provides a comprehensive, efficient debugging capability. While it takes weeks or months to implement an FPGA prototype, it takes only days to implement emulation. And design changes take a few hours or less. Emulation does this at the expense of running speed and cost compared to FPGA prototypes. Looking at emulation from the other direction, it improves on acceleration's performance by substituting "live" stimulus for the simulated testbench. This stimulus can come from a target system (the product being developed) or from test equipment. At 10,000 to 1,00,000 times the speed of simulation, emulation alone delivers the speed necessary to test application software while still providing a comprehensive hardware debug environment. Later we will discuss creation of emulation environments.

It is worth noting that simulation and prototyping involve two different styles of execution. Simulation executes the register transfer level (RTL) code serially while a prototype executes fully in parallel. This leads to differences in debugging. In simulation, you set a breakpoint and stop simulation to inspect the design state, interact with the design, and resume simulation. You can stop execution "midcycle" as it were, with only part of the code executed. You can see any signal in the design and the contents of any memory location at any time. You can even back up (if you have saved a checkpoint) and rerun. With a prototype, you use a logic analyzer for visibility, so you can see only a limited number of signals, which you determine ahead of time (by clipping on probes). The target does not stop when the logic analyzer triggers, so each time you change the probes or trigger condition, you have to reset the environment and start again from the beginning. Acceleration and emulation are more like prototyping and silicon in terms of RTL execution and debugging, since the entire design executes simultaneously as it will in the silicon. Since the same hardware is often used to provide both simulation acceleration and ICE, these systems provide a blend of these two very different debugging styles.

Another difference between simulation, and acceleration and emulation is a consequence of accelerators using hardware for implementation — they are fundamentally two-state machines — acting the way the silicon will when fabricated. Thus, they are not useful for analyzing X-state initialization or strength resolution. Accelerators also do not model precise circuit timing, hence they will probably not find any race conditions or set-up and hold time violations. These tasks are properly carried out during simulation or with a static timing analysis tool. A key distinction between an emulator and an FPGA prototyping system is that the emulator provides a rich debug environment, while a prototyping system has little to no debug capability and is primarily used after the design is debugged, to create multiple copies for system analysis and software development.

Logic Simulation	Acceleration	In-Circuit Emulation	FPGA Prototype
Accurate	Faster	Fast enough for embedded software	Inexpensive
Flexible	Handles large designs	Real world stimulus	Fast enough for embedded software
Not fast enough for large designs	Debugs via simulator UI	Short implementation time from RT level	Long implementation time
Not fast enough to test embedded software	Not fast enough for embedded software	Full debugging capabilities	Little, if any debug capability
			Error-prone
	Setup time for porting your testbench	Slower than FPGA prototype	Hard to change design
		Higher cost than FPGA prototype	

19.1.1 Hardware-Accelerated Verification Systems

In the 1980s, IBM developed the Engineering Verification Engine (EVE) family of simulation acceleration hardware for internal use. In 1988, Quickturn delivered the first commercial emulator when FPGA technology had reached sufficient capacity to allow their implementation. In 1997, Quickturn (now Cadence) delivered a processor-based emulator based on custom silicon.

19.1.1.1 FPGA-Based Systems

FPGAs would seem to be an obvious choice for implementing an emulator. And, if the design would fit into 2 to 5 FPGAs, they would be. But, the number of gates an FPGA can hold drops dramatically as a design is partitioned into multiple FPGAs. Literally, hundreds of FPGAs are required to implement a general-purpose accelerator/emulator that has comprehensive debugging capability. When a design is partitioned into so many FPGAs, the FPGA I/O pin count rather than the vendor-rated gate count determines the capacity of the FPGA. Rent's Rule may be used to calculate this [10–12].

FPGA-Based Emulation Systems

Mentor Celaro
Mentor (formerly Ikos and VMW) VStation
Cadence (formerly Verisity and Axis) Xtreme
Cadence (formerly Quickturn) RPM, Mars, Enterprise, and System Realizer
Cadence Mercury and Mercury^{Plus}

19.1.1.2 Processor-Based Systems

Processor-based emulators consist of a massive array of Boolean processors able to share data with one another, running at very high speed. The software technology consists of partitioning a design among the processors and scheduling individual Boolean operations in the correct time sequence and in an optimal way. Initially, performance did not match FPGA-based emulators, but compile times of less than an hour, and the elimination of timing problems that plagued FPGA-based systems made the new technology appealing for many use models, especially simulation acceleration. Future generations of this technology eventually surpassed FPGA systems in emulation speed while retaining the huge advantage in compilation times — and without a farm of a hundred PCs for compilation. Advances in software technology extended the application of processor-based emulators to handle asynchronous designs with any number of clocks. Other extensions supported 100% visibility of *all* signals in the design, visibility of all signals at *any* time from the beginning of the emulation run, and dynamic setting of logic analyzer trigger events without recompilation. At the same time that the emulation speed of FPGA-based systems was decreasing (due to heavily multiplexing pins), new generations of processor-based systems not only increased emulation speed, but also proved scalable in capacity to hundreds of millions of gates.

Processor-Based Emulation Systems

Processor-Based Emulation Systems
Synopsys Arkos
Tharas Hammer
Cadence (formerly Quickturn) CoBALT, and CoBALT^{Plus}
Cadence Palladium and Palladium II

Comparison of General Characteristics of FPGA-Based and Processor-Based Emulators

	FPGA-Based Emulators	Processor-Based Emulators
Technology	Uses commercial off-the-shelf components	Requires custom silicon
Compile speed from RTL	Takes many hours on many CPUs	10–30 million gates/h on 1 CPU
Emulation speed	150–800 kHz	400–1500 kHz
Predictable, reliable compiles	Fair	Excellent
Maximum capacity	30–120 million gates	256 million gates
Ability to handle asynchronous designs	Good	Good
Power requirements	50–150 W/million gates	100–250 W/million gates
Partitioning problems	Not found until individual FPGA compiles fail	Found early in process
Timing issues in emulation model	Potential for setup and hold issues and subtle, hard to find timing problems	Avoided completely
Size	0.1–0.3 ft^3/million gates	0.25–1 ft^3/million gates
Ability to make small changes quickly	Must recompile entire FPGA(s)	Can compile and download incremental changes to processor

19.2 Emulator Architecture Overview

19.2.1 Small-Scale Emulation and Logic Prototyping with FPGA

Field-programmable gate arrays provide a convenient and inexpensive mechanism for prototyping low complexity logic designs. In fact, recently the logic gate and memory capacity of the FPGA has grown so much that the high-end FPGA parts are roughly equivalent in capacity to a median size application specific integrated circuit (ASIC), that is, about 2 to 5 million gates (usable capacities of FPGAs is typically ¼ to $^1/_{10}$ the rated capacity). Similarly, modern FPGA can operate at speeds up to hundreds of megahertz which is also close to an average ASIC clock rate. These factors have resulted in the growth in popularity of FPGA prototyping for small- to medium-size logic designs.

In the majority of cases, the designers of such prototypes create and customize them independently, based on the requirements and characteristics of the systems in which their logic designs are intended to operate. The external inputs and outputs are connected either to a production board or to a prototype board specifically designed for that purpose. A minimal set of tools and infrastructure required for prototyping is supplied by FPGA vendors for whom such prototyping activities represent a significant market. Such tools typically consist of the compilation software that converts the logic design into an FPGA programming bitstream, and the means for debugging the design once compiled into an FPGA.

For example, Xilinx (www.xilinx.com) offers a comprehensive ISE design environment that covers logic synthesis, placement, routing, and timing closure for all the FPGA parts available from the supplier. After the design is mapped and downloaded into a part, the *ChipScope*™ debugger can be used to observe the internal signals under specified triggering conditions. This debugger acts as an embedded logic analyzer and consists of the logic core embedded into the user's design and the software to operate the debugger and display the results.

Similarly, Altera (www.altera.com) offers the *QuartusII*™ software kit for design compilation and the *SignalTap*™ embedded logic analyzer that is used with this company's FPGA devices.

As much as FPGA vendors attempt to simplify prototyping, there are still some difficulties that the developers of such a prototype must overcome. For example, FPGA devices are volatile. Their configurations must be stored in nonvolatile memory and downloaded at powerup. The facilities for that purpose need to be created. In many cases, the verification environment for the design is implemented in software. For that, the FPGA needs to be interfaced with the workstation that runs the software. These issues create a business opportunity for a number of vendors of small FPGA-based rapid prototyping systems that are typically implemented on boards, which contain the target FPGA and the facilities for design download and workstation interfaces. The use of such systems shortens the time required for prototype bringup in comparison with the completely custom-designed setups. FPGA prototyping system companies include Eve, Simpod, Hardi, and ExpressIP.

Things rapidly become more complicated when the design size grows beyond the capacity of one FPGA. Although multiple FPGA prototyping systems are also available, they all suffer from difficult problems of partitioning, creating signal connectivity, and multi-FPGA timing closure.

19.2.2 Large-Scale Emulation with FPGA Arrays

19.2.2.1 FPGA Arrays with Direct Connection Pattern

Over the last 15 years, FPGA density has increased much faster than the density of an average very Large Scale Integrated Circuit (VLSI). Because of that, early emulation systems faced the problem of modeling the target design with multiple FPGA parts. Many more FPGA devices were necessary to implement a logic design intended for an average ASIC than would be needed today.

The systems of the early 1990s had already identified the major difficulties of large-scale FPGA array emulation: partitioning, timing closure, and interconnect schemes. One distinct approach used for FPGA interconnect was a regular pattern of direct FPGA to FPGA connections. Although moderately successful, this scheme significantly magnified the difficulties of timing closure and partitioning because many signals had to visit several FPGA parts in order to reach their destinations.

19.2.2.2 FPGA Arrays with Partial Crossbar

This architecture, highly successful in the 1990s, implements the interconnection of FPGA parts using a number of special crossbar devices [1,15]. Each such device is a fully programmable crossbar with N terminals. Assuming that the FPGA array consists of K parts each having M external inputs and outputs, every crossbar device will connect to at most N/K inputs or outputs of each FPGA. This ratio, commonly called the "richness" of a partial crossbar, determines the routability of the interconnection scheme. The number of pins of a single crossbar device defines the limit to which a one-level partial crossbar can be scaled. Further capacity increase is accomplished using the multilevel schemes, where the crossbar devices are again connected in a partial crossbar pattern. It has been shown that with clever use of the internal symmetry of FPGA pins, even low richness values allow most practical logic designs to be routed successfully. Hybrid schemes that combine direct connections with partial crossbar have also been implemented [14].

The problem of multi-FPGA timing closure has also been addressed. This problem in FPGA-based emulators primarily manifested itself with hold time violations. Unlike the insufficient setup time which can be corrected simply by reducing the operating speed of an emulator, hold time violations affect the functionality of the design and must be removed. A number of approaches have been proposed, which utilize partitioning constraints and clock signal duplications [2] or special circuit modification [3,4,15].

The problem that presented the FPGA array emulators with their biggest challenge was that of multi-FPGA partitioning. As commercial FPGA devices were increasingly architected to contain functionally complete designs or modules, they did not maintain a pin/gate ratio that would result from automatic partitioning of the large logic design into many (often hundreds of) parts. As a consequence, the usable gate utilization was typically way below the level that a single FPGA could offer. As FPGA capacitycontinued to grow much faster than their pin count, the problem was aggravated with every new FPGA generation.

19.2.2.3 FPGA Arrays with Time-Multiplexed Connections

As an answer to the challenge of insufficient FPGA pin count, several time division multiplexing (TDM) schemes have been proposed, which divide the bandwidth available at an FPGA input/output pin among several signals that need to be transmitted between the parts.

A virtual wire scheme [5,13,16], though defined in general terms, has been primarily applied to direct FPGA connectivity architecture. It schedules the transmission of signals between FPGA parts as well as the storage of signal values at design flip/flops, in terms of the cycles T_V of a single clock signal which is distributed throughout the emulator. After each logical signal that crosses an FPGA partition boundary is assigned to both the physical pin and a T_V cycle, the total number of the T_V cycles necessary to advance the circuit to the next state is determined. This number essentially constitutes a TDM factor M for this design, which will be equal to the ratio of the design clock cycle (in a simplifying assumption that it was a single-clock synchronous design to start with) to T_V. Circuits are later built in the FPGA programmable logic that implement and control TDM and signal evaluation schedules. It is important to note that virtual wire technology assumes that the design can be statically scheduled (although exceptions are made for asynchronous signals which are not to be multiplexed at all) and that the execution of exactly M T_V cycles assures its correct transition to the next state. This methodology effectively transforms the original circuit to a synchronous equivalent, which operates with a clock period T_V. It has to assure that the signal setup time requirements for such synchronous circuit are satisfied and therefore no combinatorial path exists with the propagation time larger than T_V.

A different scheme [6], though again defined in general terms, has been primarily used with the partial crossbar architecture. This scheme is different in that it makes no attempt to transform the circuit to its synchronous equivalent, and it makes no assumptions about the signal propagation times inside the FPGA. All it does is the high-speed multiplexing of the FPGA inputs and outputs with the maximal speed permitted by the partial crossbar, printed circuit board, etc. The multiplexing period T_V does not depend at all on the design being emulated and is in fact constant, as is the multiplexing factor M. On the other hand, there are no guarantees that exactly M T_V cycles would be sufficient for advancing the design state. Signal transitions are propagated freely until they reach the FPGA boundary where they are synchronized to the TDM clock. As in the virtual wires technique, some signals can be excluded from multiplexing (and the clocks always are) so that the asynchronous behavior can be emulated more closely. The crossbar parts have the ability to demultiplex signals before switching them. Therefore, the fact that some signals share a physical wire does not impose any constraints on the way these signals should be routed.

For each solution, it has been argued that it has advantages over the other in the areas of emulation speed, support for asynchronous design styles, as well as the overall cost and the ease of adoption. However, it is important to point out the similar difficulties encountered by both. The virtual wires technology allows the increase of effective bandwidth of the FPGA pins only to the extent that T_V is less than the design state evaluation time. In order to keep the emulation speed from going down, it is necessary therefore to keep T_V low. However, to satisfy setup time requirements, T_V must be larger than any combinatorial path delay. Controlling these delays in an FPGA, while keeping compilation time low, is notoriously difficult. Besides, the delays are known only after the FPGA compilation is finished while the signal schedule has to be determined before this compilation starts. As a result, pin bandwidths cannot be increased so that they keep up with the ever-growing gate/pin ratio. Although the technique by Sample et al. [6] does not explicitly require controlling setup time, in the end it suffers from similar limitations.

There is an intriguing observation one can make while considering the multi-FPGA systems with TDM. A configurable logic structure (typically, a lookup table) that produces a signal which is subject to time multiplexing is severely underused. While its output needs to be valid only during one T_V period, this output is in fact being produced continuously. Thus on average, only one out of

M configurable logic structures could be in actual use at any given time. Conversely, we could use the configurable logic much more efficiently if it could compute different signals during different T_V cycles. Had this been the case however, it would effectively turn a lookup table into a processor that executes instructions over time.

19.2.3 Processor Arrays for Emulation

A processor-based hardware emulator [7] consists of bit-wide processors which execute a different Boolean function of N variables every T_V cycle. The input values on which the Boolean functions operate are supplied by sets of multiplexers that switch every T_V cycle as well. The Boolean functions and the control codes for the input multiplexers (addresses) together constitute instructions that are stored in the instruction memory. The depth of such memory determines the TDM factor M that applies both to the processors and to the interconnect structures. In this approach, every processor effectively performs the work that M FPGA lookup tables formally had.

In the semiconductor implementations of a processor-based emulator (Figure 19.1), routing structures are significantly simplified compared with FPGA. In place of a large variety of wire segments that could be arbitrarily combined with programmable switches, processor-based emulation chips contain a two-level routing structure [8] that recognizes processor cluster-level routing in addition to the uniform chip-level routing. Because of that, signal propagation delays are uniform. Processor-based emulators offer tighter control over the signal timing than an FPGA-based emulator's virtual logic can allow, and corresponding low duration of T_V cycle and high values of M.

To better understand how a processor-based emulator works, it is useful to briefly review how a logic simulator works. Recall that a computer's arithmetic-logic unit (ALU) can perform basic

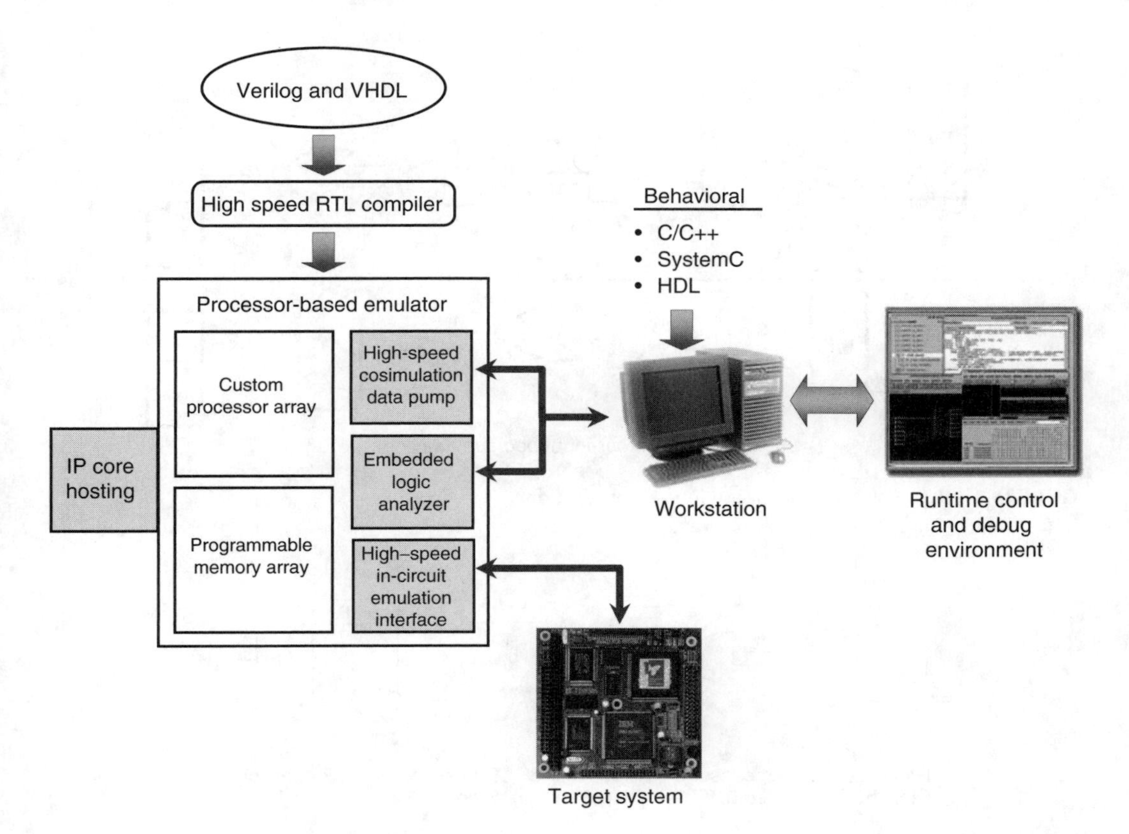

FIGURE 19.1 Processor-based emulator architecture.

Boolean operations on variables, e.g. AND, OR, NOT and that a language construct such as "always @ (posedge Clock) Q = D" forms the basis of a flip-flop. In the case of gates (and transparent latches), simulation order is important. Signals race through a gate chain schematically "left-to-right" so to speak, or "top-to-bottom" in RTL source code. Flip-flops (registers) break up the gate chain for ordering purposes.

One type of simulator, a levelized compiled logic simulator, performs the Boolean equations one-at-a-time in the correct order. (Time delays are not relevant for *functional* logic simulation.) If two ALUs were available, you can imagine breaking the design up into two independent logic chains and assigning each chain to an ALU, thus parallelizing the process and reducing the time required, perhaps to one half. A processor-based emulator has from tens of thousands to hundreds of thousands of ALUs, which are efficiently scheduled to perform all the Boolean equations in the design in the correct sequence. The following series of drawings (starting with Figure 19.2 and Figure 19.3) illustrate this process.

Step 1: Reduce Boolean Logic to Four-Input Functions

The following sequencing constraint set applies:

- The flip-flops must be evaluated first
- S must be calculated before M
- M must be calculated before P
- Primary inputs B, E, and F, must be sampled before S is calculated
- Primary inputs G, H, and J, must be sampled before M is calculated
- Primary inputs K, L, and N, must be sampled before P is calculated

Note that primary input A can be sampled at any time after the flip-flops.

Step 2: Scheduling Logic Operations among Processors and Time Steps

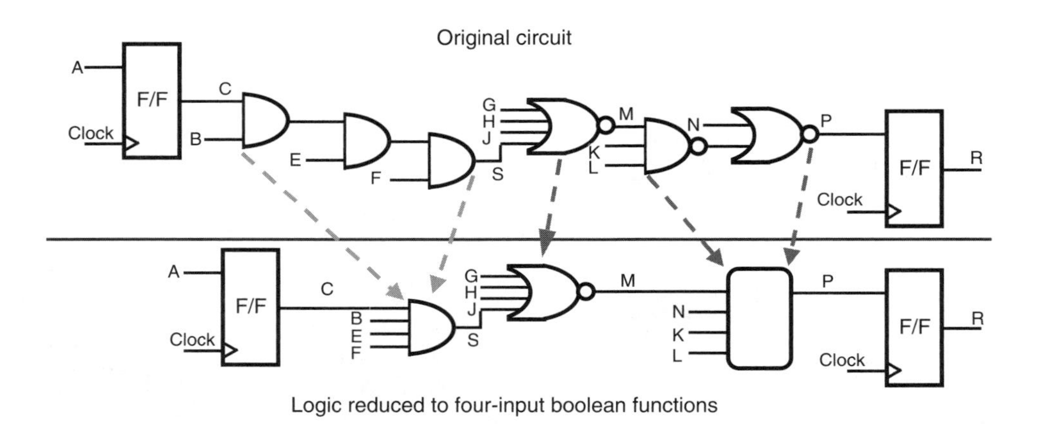

FIGURE 19.2 Result of reducing logic to four-input functions.

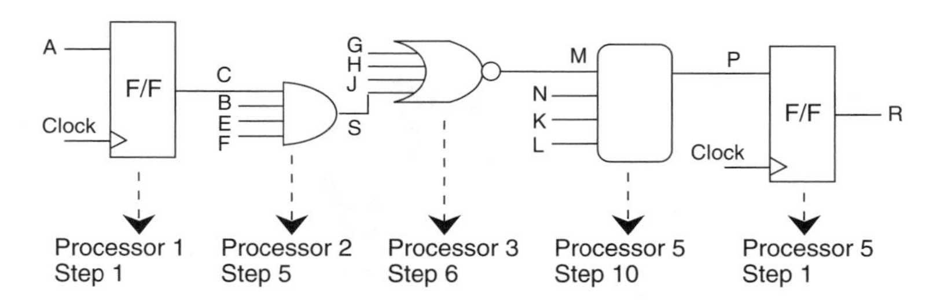

FIGURE 19.3 Result of scheduling logic.

Time Step	Processor 1	Processor 2	Processor 3	Processor 4	Processor 5
1	Calculate C	Sample B	Sample G		Calculate R
2		Sample E	Sample H		Sample N
3		Sample F	Sample J		
4					Sample K
5		Calculate S			Sample L
6			Calculate M		
7					Receive M
8					
9					
10	Sample A				Calculate P

One possible scheduling is shown above. An *emulation cycle* consists of running all the steps for a complete modeling of the design . Large designs take typically from 60 to 300 steps to schedule, depending on whether the compiler optimizes for speed or capacity.

In addition to the Boolean equations representing the design logic, the emulator must also efficiently implement memories, encrypted "soft" intellectual property (IP), and support physical IP "bonded-out" cores. For simulation acceleration it must also have a fast, low-latency channel to connect the emulator to a simulator on the workstation, since behavioral code cannot be synthesized into the emulator. For in-circuit operation, the emulator needs to support connections to a target system and test equipment. Finally, to provide visibility into the design for debugging, the emulator contains a logic analyzer. The logic analyzer will be described later.

The balance of this chapter will focus primarily on processor-based emulators as they represent the state-of-the-art currently, with comparative information where FPGA-based emulator capabilities are significantly different.

Key Characteristics of Processor-Based Accelerator/Emulator

- Maps any design style; automatically maps multiple clock domains, gated clocks, asynchronous loops
- Fast compilation without needing dozens of workstations
- High running speed
- Complete debug capabilities without recompiling design
- Architecture scales to 200 million gates and beyond
- Several hundred megabytes of design memory per million gates
- Ten thousand I/O pins or more for in-circuit applications
- Many simultaneous users
- No performance or capacity penalty for 100% debug visibility
- Built-in logic analyzer with complex triggering capability
- Flexible licensing
- Hard and soft IP support

19.3 Design Modeling

As was demonstrated above, the further we move along the line from a simple FPGA-based prototype to a complicated, highly scalable emulation system, the less the actual model execution mechanism resembles the target design implementation in silicon. In other words, for emulation, the user's design needs to be modeled. Below, we shall cover the major reasons and methods of design modeling.

19.3.1 Tri-State Bus Modeling

Tri-state busses are modeled with combinatorial logic. When none of the enables are on, emulators give the user a choice of "pull-up," "pull-down," or "retain-state." In the latter case, a latch is inserted into the design to hold the state of the bus when no drivers are enabled. In case multiple enables are on, for pull-up and retain-state logic, 0 will "win," and for pull-down logic, 1 will win. (Note that this is a good place to use assertions.)

19.3.2 Asynchronous Loop Breaking

Emulators do not model the precise gate-level timing of silicon. FPGA-based emulators have random "gate" delays and processor-based emulators have delays of 0. Because of the 0 delay characteristic of processor-based emulators, asynchronous loops are broken automatically by a delay flip-flop during compilation. However, by allowing the user to specify where loop breaks should occur, emulator performance may be enhanced, since the performance is related to the length of long combinatorial paths.

19.3.3 Clock Handling in Processor-Based Emulators

As described earlier, clocking was one of the prime sources of unreliability in FPGA-based emulators. Processor-based emulators completely avoid this problem. They can generate all the clocks necessary for a design or they can accept externally generated clocks. It is much more convenient to have the emulator generate all the design clocks needed — and it runs faster. To provide the maximum possible emulation speed while retaining the asynchronous accuracy required, processor-based emulators provide two methods of handling asynchronous design clocks: *aligned edge* and *independent edge* (Figure 19.4).

19.3.4 Independent Edge Clocking

Since emulators do not model design timing, but rather are *functional* equivalents, the exact timing between asynchronous clock edges is not relevant. It is only necessary that clock edges that are not simultaneous in "real life" are emulated independently. With independent edge clocking (Figure 19.5 and Figure 19.6), an emulation cycle is scheduled for every edge of every clock unless an edge is naturally coincident with an already scheduled edge. This is very similar to an event-driven simulator.

19.3.5 Aligned Edge Clocking

Aligned edge clocking is based on the fact that although many clocks in a design *happen* to have noncoincident edges because of their frequencies, proper circuit operation does not *depend* on the edges being independent (Figure 19.7). In this case, while proper frequency relationships are maintained, clock edges are aligned to the highest frequency clock, thereby reducing the number of emulation cycles required (compared to independent edge clocking), and increasing emulation speed.

In aligned edge clocking, an emulation cycle is first scheduled for each edge of the fastest clock in the design. Then, all other clocks are scheduled relative to this clock, with the slower clock edges "aligned" to the next scheduled emulation cycle. Note that no additional emulation cycles have been added for the second and third (slower) clocks (Figure 19.8 and Figure 19.9). Thus emulation speed is maintained. Also note that while edges are moved to following fastest clock edge, frequency relationships are maintained which could be essential for proper circuit operation. Giving the user the flexibility to switch between

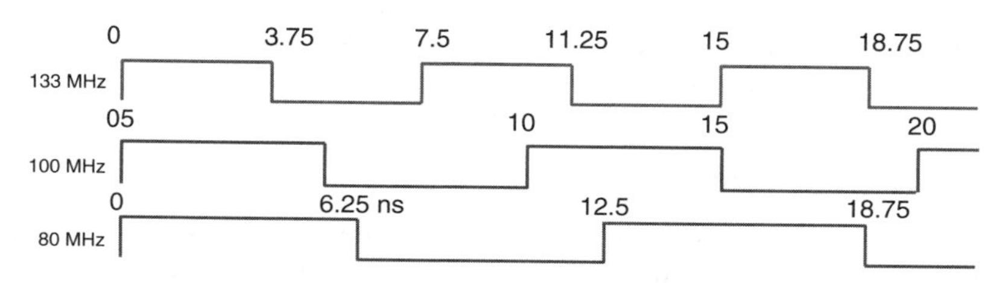

FIGURE 19.4 Clocking example — three asynchronous clocks.

aligned edge clocking and independent edge clocking in processor-based emulators provides both high asynchronous accuracy and the fastest possible emulation speed for a wide variety of design styles.

Comparison of Clocking Techniques in Processor-Based Emulator

	Independent Edge Clocking	Aligned Edge Clocking
Preserves frequency relationships	Yes	Yes
Matches simulation	Yes	May not
Maintains speed with more clocks	No	Yes
Maintains edge independence	Yes	No

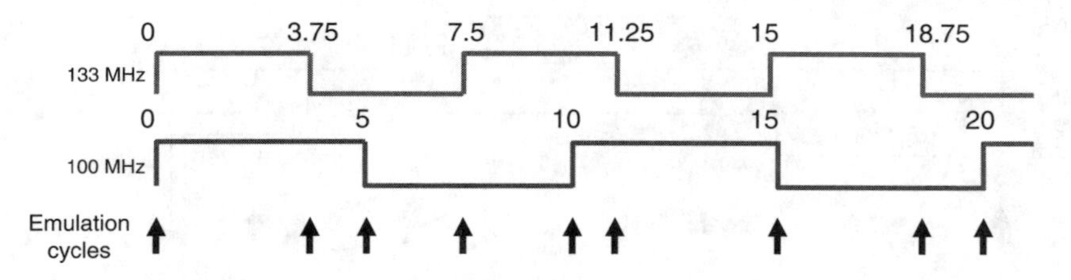

FIGURE 19.5 Independent edge clocking of two asynchronous clocks.

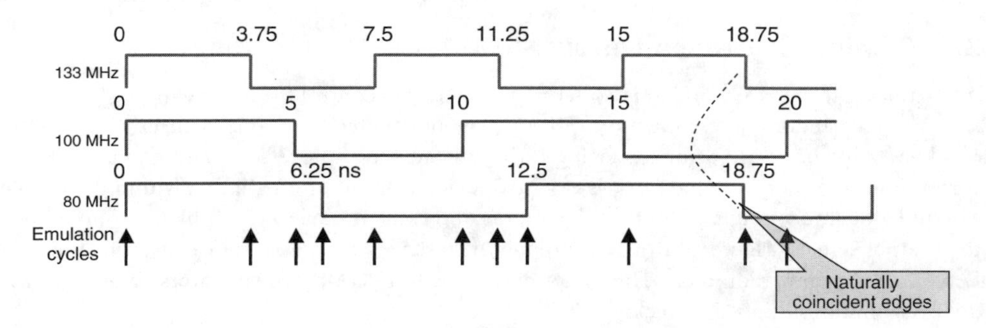

FIGURE 19.6 Adding a third asynchronous clock with independent edge clocking.

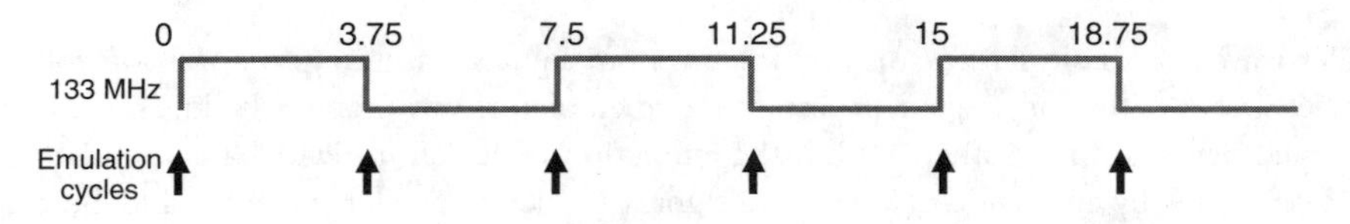

FIGURE 19.7 Aligned edge clocking starts by assigning an emulation cycle for each edge of the fastest clock.

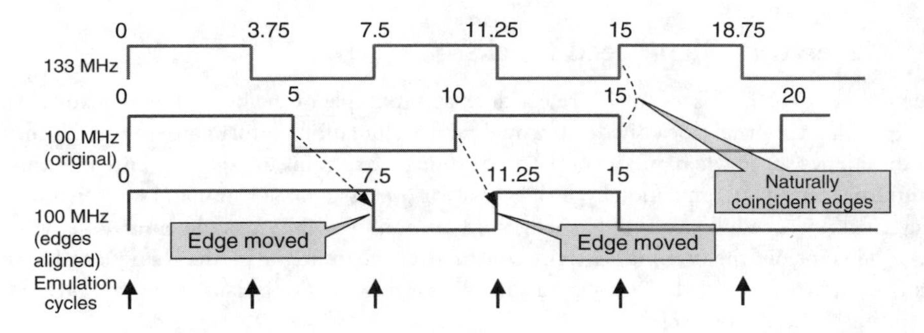

FIGURE 19.8 Add second clock aligning edges to fastest clock.

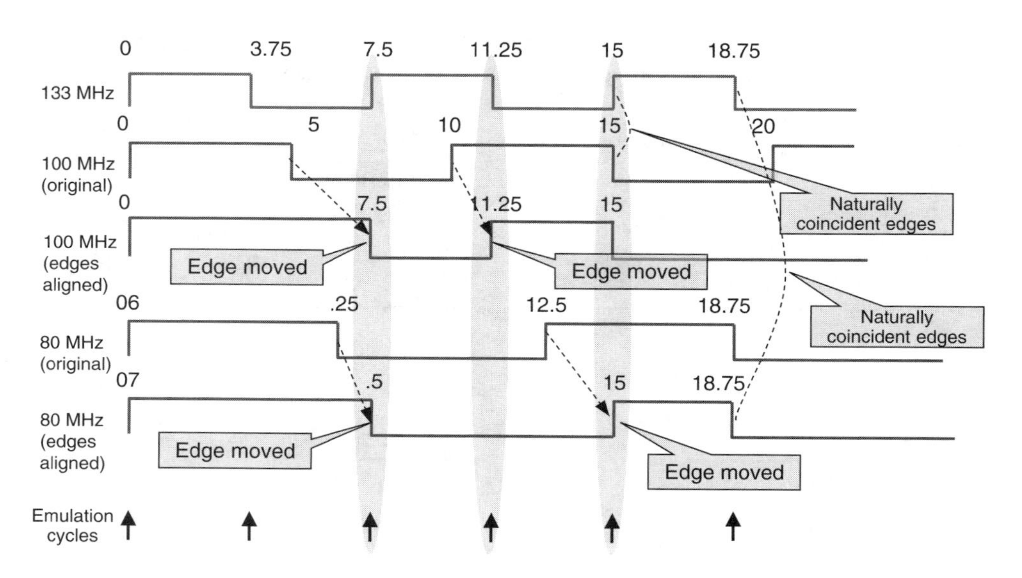

FIGURE 19.9 Add additional clocks by aligning edges to fastest clock.

19.3.6 Timing Control on Output

When interfacing to the real world, it is sometimes necessary to control the relative timing of output signals. A dynamic random access memory (DRAM) memory interface is one such example — all the address lines must be stable before asserting Write-Enable. Since processor-based emulators schedule logic operations to occur in sequence, it is easy to add a constraint on the timing within the emulation cycle on individual (or groups of) output signals, to control the timing to a very high resolution relative to other output signals (Figure 19.10). The compiler then schedules this output calculation at the appropriate point in the emulation cycle. This is not possible with FPGA-based emulators since they have no control over timing within a design clock.

19.3.7 Timing Control on Input

In a similar way, and for similar reasons, timing on input signals (sampling) may be controllable by the user to meet specific situations. Again, processor-based emulators can simply schedule specific input pins to be sampled before or after others within the emulation cycle. With FPGA-based emulators the user must "tweak" timing by adding delays to certain signals — including a large guard-band, because FPGA-based emulators cannot control absolute timing along different logic paths. For FPGA-based emulators, this is a hit-or-miss proposition and may vary from compile to compile.

19.3.8 Generating High-Speed Clocks

Occasionally, it may be necessary to generate a clock at a multiple of the highest design clock. An example is a chip with an internal clock divider. It would reduce emulation performance to run the entire emulation at this higher speed when only a couple of flip-flops are running at this rate. Since there are usually over a hundred steps in an emulation cycle, it is easy for processor-based emulators to generate a higher frequency clock by scheduling multiple changes on an output within a single emulation cycle (Figure 19.11). As shown below, the user may exercise control over the timing of all the changes and create fairly complex clock signals, if needed. FPGA-based emulators cannot do this because they have no control over timing within the emulation cycle.

19.3.9 Handling IP

Intellectual property can be an important part of a design. For acceleration and emulation design, IP may be supplied in two basic forms: (1) as physical silicon — a "bonded-out core," or (2) as a synthesizable RTL source file. Note that if IP is supplied in behavioral form, accelerated cosimulation must be used, with the behavioral code running on the workstation, which will reduce verification performance.

19.3.9.1 Physical IP

Most emulators and accelerators provide a mechanism for incorporating physical IP into the design model. The IP component will be mounted on an IP block of some sort and inserted into the emulator. Some emulation vendors also supply turn-key IP blocks with various components mounted, e.g., a large, programmable FPGA with a flash card reader on the front panel, or an ARM IP block which mates with the ARM-supplied IP tile and accepts ARM programming files (Figure 19.12).

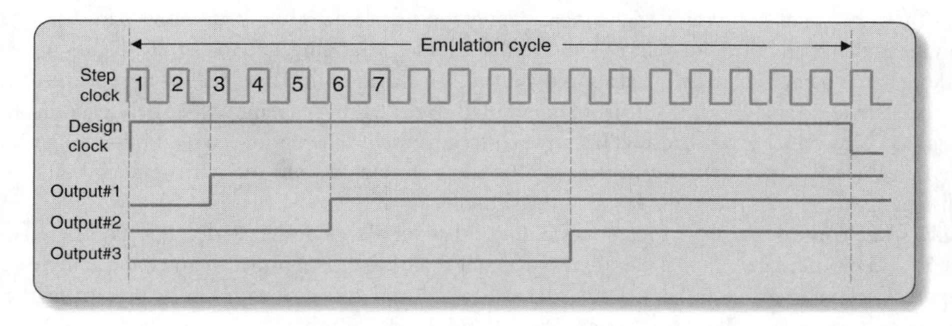

FIGURE 19.10 Processor-based emulators can adjust output timing with high precision.

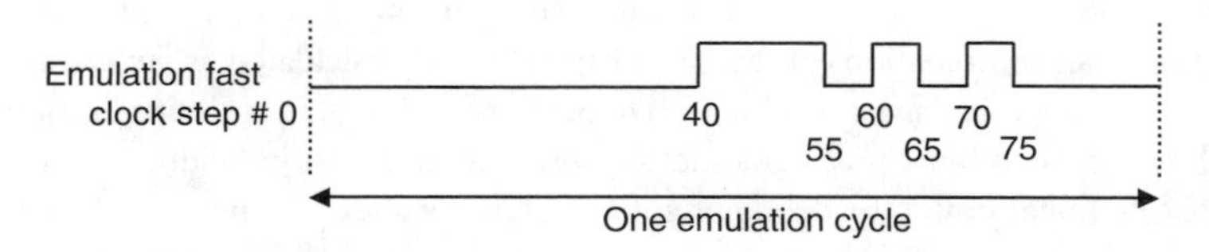

FIGURE 19.11 Special, high frequency clocks may be generated by processor-based emulators.

FIGURE 19.12 Physical IP prototype block with TriMedia core mounted.

The user may also mount a physical IP on their target system, but it must be noted that these connections may be treated differently when running vector regressions.

19.3.9.2 Source Code IP

RTL IP source files may be supplied and are included along with the user's design files in the compilation flow. RTL files for IP may be encrypted by the IP vendor for confidentiality reasons. In such a case, the user cannot inspect the encrypted block or any signals inside the block However, the IP vendor may provide access to defined signals the user might reasonably need in order to debug designs using the IP. An example is a microprocessor register set, which may be needed for debugging purposes. In other respects, the IP block is handled as part of the design.

19.4 Debugging

19.4.1 Logic Analyzer and Signal Trace Display

Most of the time spent verifying designs is on debugging problems. Thus, productivity in debugging is paramount in delivering time-to-market. Core to this capability is being able to rapidly display any signal in the design, and being able to trace back rapidly through a circuit to find the cause of the observed anomaly. The built-in logic analyzer is the heart of an emulator's debugging capability. The debugger displays in an integrated way, signal waveforms, RTL source code, generated schematic views, and the design hierarchy. Signal values at a specific point in time can be back-annotated onto the schematic and RTL source code displays to accelerate the user's grasp of the situation.

To give some context, consider that an emulator with a very large design with 20 million signals, running at 500 kHz would generate 60 Terabytes of signal trace data *per minute*! In any logic analyzer, there is a finite amount of memory for signal trace data storage. Emulators deal with this by presenting the user with trade-offs between width (number of signals) and depth (number of clock cycles). Emulators also use several techniques to "extend" the logic analyzer trace capacity greatly. Some of these techniques are covered under various patents granted or applications pending.

Emulators only need to capture design inputs and the outputs of flip-flops and memories. The values of other signals are Boolean combinations of these and can be quickly calculated, *when needed.* This capability would be necessary in any event since combinatorial logic in the design is optimized into n-input Boolean functions and thus some combinatorial signals in the design do not exist individually in the emulator.

Oftentimes, the engineer doing the debugging may not know ahead of time which signals will need to be probed to debug a problem. This is especially true when verification is performed by a separate team of engineers. In the course of debugging in such situations, the engineer may need to see *any* signal in the design to trace the problem back to the root cause. Emulators often allow the user to trade-off display speed, logic analyzer depth, and the support of dynamic target systems to get the best debugging capabilities for their particular needs.

In contrast to 100% signal visibility, *Dynamic Probing* provides a very deep logic analyzer with a smaller number of probes. Users must specify prior to running — and also prior to compilation with FPGA-based emulators — the signals they wish to see. Much as with simulation, adding a signal to the list allows display of that signal going forward in time, but not backwards in time. ("Forward" refers to clock cycles that have not yet been emulated, "backward" refers to clock cycles that have already been emulated.) Processor-based emulators accomplish this rapid change of signals to be probed (<1 sec to change 1000 signal probes) by utilizing otherwise "idle" time steps in the processor array to route signals to available logic analyzer memory channels. The compiler reserves a small percentage of the processor array bandwidth for this purpose when using Dynamic Probing, allowing new signals to be probed without changing the "program" for emulating the design. Note that in FPGA-based emulators, when not using 100% visibility, recompiling the entire design with the new probes (or at least a few FPGAs) is necessary — which changes the timing of the design model, sometimes resulting in different functional behavior or reduced operating speed.

For any of the above vision choices, emulators often also provide the ability to extend the depth of signal tracing (number of samples) to infinity — from the beginning of emulation, showing "*all the signals, all the time*" — but with some trade-offs. To support "infinite" trace data, the emulator is stopped periodically (as the physical logic analyzer becomes full) to upload the necessary data to the workstation. Then emulation continues. This approach also requires a use model, which tolerates periodic stopping of the clocks to the target system (typically a static target system). The advantage is that the logic analyzer now has infinite depth, and is able to display signal trace data from the beginning of the emulation run for any or *all* signals in the design. The trade-off is that stopping the clocks to upload the logic analyzer data results in reduced net emulation speed — usually 2 to 3 times slower. However, the user can turn on and off the infinite trace recording at run time so that the reduced speed is only experienced during times of interest. For example, one could boot a PC at full speed and then turn on infinite tracing to study a particular problem. Thus, in spite of having a finite, practical amount of logic analyzer trace memory, emulators can provide a rich set of vision choices from which the user can select a suitable set of trade-offs for their use model.

19.4.2 Definition of Trigger Condition

Emulators allow the user to specify complex trigger conditions for the logic analyzer. In the course of debugging a problem, the need to change the trigger condition is often encountered. Processor-based emulators support changing the trigger condition instantly, without needing to recompile the entire design. FPGA-based emulators may require recompiling one or more FPGAs in order to change trigger conditions when the change involves new signals.

19.4.3 Design Experiments with Set, Force, and Release Commands

Design Experiments allow the user to make certain design changes instantly, without recompilation, and see the result. This can be useful

- To disable sections of the circuit that are interfering with the section you want to inspect
- To force signals on or off to enable/disable signal effects
- To inject changes into the circuit, including complex time sequences, that duplicate the proposed fix and observe their effect

This allows the user to experiment with ideas without having to recompile the design. Design experiments are accomplished using the Set, Force, and Release commands, possibly included in a script for time sequencing. *Set* instantly sets a flip-flop to a state of 1 or 0, but then allows normal circuit operation to change it. *Force* and *Release* are complimentary commands; *Force* will freeze a signal to 1 or 0 until a *Release* command is given for that signal.

19.5 Use Models

Emulators and accelerators can be used in a number of ways to accelerate functional verification.

19.5.1 Accelerated, Signal-Based CoSimulation

Using a logic simulator such as Cadence Incisive, in this use model achieves faster simulation. The synthesizable RTL is moved into the emulator, where it runs several orders of magnitude faster than on the workstation. Behavioral code (e.g., testbench and memory models) continues to be run on the workstation by the simulator. Overall performance is limited by the communication channel between emulator and workstation and also by the testbench execution on the simulator. Most users in this mode prefer to debug in the familiar simulation environment, providing visibility into the signals in the RTL code in the emulator.

19.5.2 C/C++ Testbench

Similar to cosimulation, this use model has a C testbench (and possibly models) instead of a hardware description language (HDL) testbench. The C code is interfaced to the emulator through an application programming interface (API). This mode can provide higher performance than HDL cosimulation because C code executes faster than HDL in a simulator. C testbenches are becoming more popular because C is often used for system-level modeling before designing the ASIC and some of the test routines may be reusable for the ASIC testbench. Debugging in this mode involves using the emulator debug environment (which the user prefers to look like the simulator) as well as a C debugger.

19.5.3 Transaction-Based Acceleration

Transaction-Based Acceleration (TBA) is a form of cosimulation in which *transactions* are sent to the emulator rather than bit-by-bit signal exchanges. This reduces the traffic between workstation and emulator, allowing higher performances to be achieved (where the channel is the limiting factor). Transactors are designed in synthesizable RTL and put into the emulator with the design. These transactors map the transactions from the workstation into bit-by-bit operations for the design in the emulator. TBA has the additional advantage of raising the level of debugging of the design from individual logic levels to transactions which are more meaningful to the designer. There is a significant issue of acquiring a rich library of transaction IP that can be (or has been) synthesized. The Accellera committee (www.accellera.org) has standardized an interface between simulators and emulators as the Standard Co-Emulation Modeling Interface (SCE-MI) which, if supported, allows mix-and-match use of emulators/accelerators with simulators for transaction-based testbenches.

19.5.4 Vector Regression

This use model applies large sets of vectors for regression testing small changes in the design. The vectors may come from simulation, testers, or other sources. Typically, they are used close to tape-out when small changes are made to the logic and the user must ensure that nothing was "broken" by the change. Comparison of resulting vectors against known good results may be done either in the emulator or the workstation (depending on emulator capabilities). Emphasis is on fast execution and pass/fail results.

19.5.5 Embedded Testbench

In this use model, the testbench (and the entire design) is written in synthesizable RTL code and the design and testbench are run in the emulator at full emulation speeds. Few users are willing to write a testbench in synthesizable RTL because of the extra effort and the existing investment in behavioral testbench code. This use model is most often applied to processor chips, because performance is valued so highly as to make the extra testbench effort worthwhile. When the ASIC contains a processor core, embedded testbench mode can be applied using a simple synthesizable testbench and processor object code executing out of emulator memory.

19.5.6 In-Circuit Emulation

This model of operation connects the emulator with a user's *target system* — a prototype of the system the user is designing. The emulator typically replaces the ASIC(s) being designed for the target system, allowing system-level and software testing prior to silicon availability. Because the emulator runs much slower than the ASIC being designed, slow-down solutions must be implemented to match the real-world interfaces to the emulator. Clocks may be derived in the target system or in the emulator. In some simple environments, the target system may be a piece of test equipment that generates and verifies test data — network testers are a typical example of this.

19.6 The Value of In-Circuit Emulation

In-circuit emulation provides the highest performance and delivers the following value for verification:

1. Tests the design with "live" data in real world environment
   a. Finding bugs you did not imagine test cases for
   b. Reducing the number of test cases you have to write
   c. Reducing your risk of silicon respin
2. Allows software testing to begin before first silicon, even before tape-out
   a. Hardware/software interaction uncovers interface bugs
   b. Supports software verification with live data
   c. Enables optimal hardware/software trade-offs
   d. Saves months off software delivery cycle
3. Delivers products to market sooner
4. Reduces risk of shipping products with bugs requiring a product recall
5. Avoids performance reduction as design size grows compared to simulation

19.7 Considerations for Successful Emulation

Creating a successful ICE environment involves a few additional considerations relative to simulation or acceleration.

19.7.1 Creating an In-Circuit Emulation Environment

Emulation generally runs at 200 kHz to 1 MHz while the actual system you are designing will run much faster. When the emulator must interface with real world, full speed devices, an emulation environment which bridges the speed difference between the real-world speeds and emulation speeds must be created. There are a variety of ways this might be accomplished. Sometimes it is as simple as using a tester that can be slowed down. For example, for Ethernet interfaces, usually the PHY is bypassed. A Spirent *SmartBits* tester can slow down the Ethernet bit stream to emulation speed (both transmit and receive).

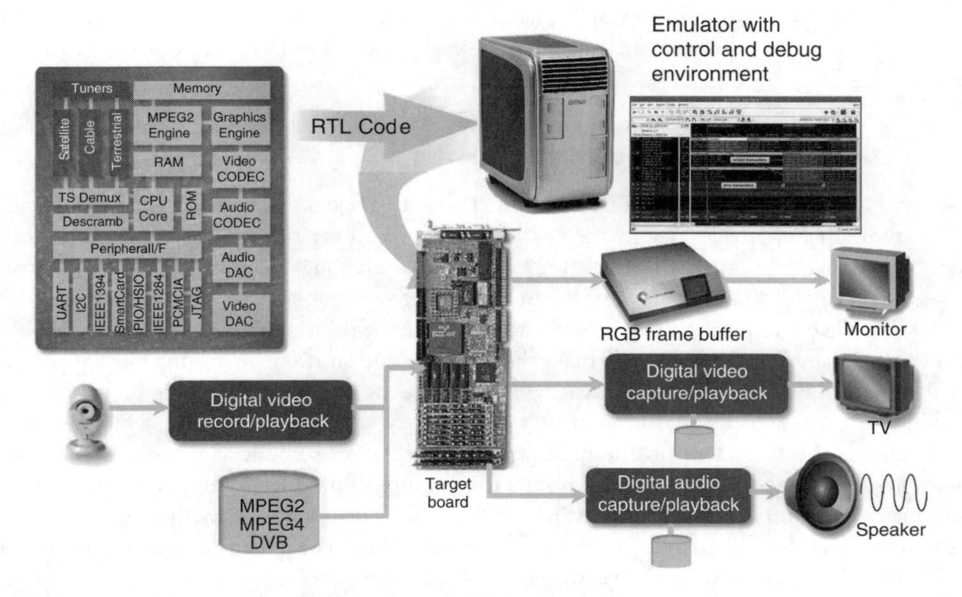

FIGURE 19.13 Emulation environment for set-top box design.

Alternatively, you can use an Ethernet bidirectional packet buffer, such as a Cadence SpeedBridge™, between multiple real-world connections and slowed-down emulation speed Ethernet ports.

Another emulation environment example is the verification of an ASIC for a video graphics card. The red-green-blue (RGB) data will be produced at emulation speeds, but the monitor cannot sync the image at these speeds. You can use a video frame buffer, such as the Cadence RGB SpeedBridge, which accepts the RGB data at emulation speeds and puts the pixel data into a frame buffer. The frame buffer is then read out at normal scan rates and drives a monitor. In this way, the user views the image produced by the emulated design but the frame update rate is reduced. Such an environment is common as it allows the user to "see" the results of their design.

A set-top box emulation environment is shown in Figure 19.13. The RF circuitry is dropped — emulation is for doing functional verification of the digital portion of the design. The rest of the ASIC is compiled into the emulator and the emulator plugged into the ASIC socket on the PC board. A digital video recorder/player is used to deliver the digital data stream representing the demodulated RF. This data stream could come from some standard data sets for motion picture expert group (MPEG) compression testing or it could be data that the user recorded from a video camera and microphone. The video frame buffer is used to display (nearly) still images. A second digital video recorder/player captures video data at emulation speed, records it to hard disk, and then plays it back at full speed. Being able to view motion is a critically important factor in MPEG decoder work, since the MPEG process does not reproduce the exact bit stream put through it (termed "lossy"). Hence, decoder algorithms are evaluated subjectively. In a similar way, audio is produced via a digital audio recorder/player which buffers and replays audio at full speed so that you can hear the quality of reproduction.

19.7.2 Debug Issues with Dynamic Targets

Some emulators temporarily suspend the clocks during operation, either to capture or to calculate and display signal data. While the emulator is doing this, it is not available to emulate with the target system. As long as the target system is static, there is no problem. Emulation resumes normally when the clocks are restarted. Dynamic targets however, may not recover automatically from having their clocks stopped for a few seconds. With dynamic target systems it may be necessary to select a different choice for capturing and/ displaying signal data to avoid this problem — if one is available. Note that some SpeedBridges solve this problem by keeping the target clocks running even when the emulator is not supplying clocks to it. If you expect to have any dynamic target systems, this would be a key issue in selection of an emulator; some emulators do not offer alternate vision choices and hence cannot be used at all with dynamic targets.

19.7.3 Considerations for Software Testing

Most electronic products today have software content — and many of them have extensive software content. The traditional approach to software verification is to wait for (mostly) working silicon to begin software debugging. This makes the hardware and software debugging tasks largely sequential and increases the product's development time. It also means that a serious system problem may not be found until after first silicon, requiring a costly respin and delaying the project for 2 to 3 months. The objective of hardware/software coverification is to make the hardware and software debugging tasks as concurrent as possible. At a minimum, this means starting software debug as soon as the IC is "taped-out," rather than waiting for good silicon. But even greater concurrency is possible. Software debugging with the actual design can begin as soon as the hardware design achieves some level of correct functionality. Starting software debugging early can save from 2 to 6 months of product development time. Early software testing can also uncover a bug in the ASIC early enough to fix it in the ASIC, thereby avoiding a software patch that may degrade usability or performance of product. The increased level of software testing available, with the speed of ICE, also reduces the risk of a software bug going undetected and getting released with the product, and the expensive upgrades, recalls, or lost business that might result.

There are additional benefits to be obtained by starting software verification prior to freezing the hardware design. If system problems or performance issues are found, designers can make intelligent tradeoffs in deciding whether to change the hardware or software, possibly avoiding a degradation in product functionality, reduced performance, or an increase in product cost.

Early software integration on "real" hardware provides enormous value

- Software is developed on "real" hardware before implementation is complete
- Software verification can be done with "live" data (in-circuit)
- Early software development platforms can run faster than with other tools
- There's an opportunity to fix the chip design before tape-out rather than patching the software, allowing an optimal hardware or software solution
- System-level verification allows hardware/software tradeoffs

Whether a microprocessor in-circuit emulator (MP-ICE) or RTOS debugger is used, the software debugging and hardware debugging environments can be synchronized so that hardware/software interface issues can be debugged conveniently. The breakpoint/trigger systems of the emulator and MP-ICE are crossconnected such that the emulator's logic analyzer trigger is one of the MP-ICE breakpoint conditions and the MP-ICE breakpoint trap signal is set as an emulator logic analyzer trigger condition. Therefore, if a software breakpoint is reached, the emulator captures the condition of the ASIC at the same moment. If an ASIC event occurs that triggers the logic analyzer, the software is stopped at that moment. This allows inspection of the hardware events that led to a software breakpoint, or of the ASIC operation resulting from executing a set of software instructions. This kind of coordinated debugging is extremely valuable for understanding subtle problems that occur at the hardware/software interface.

There are a variety of ways to interface to a microprocessor for software testing. JTAG, RS-232, or Ethernet connections can control a resident debug monitor. The OS in the processor may have its own debug environment, or a MP-ICE may be used. The microprocessor itself could be a packaged part or a bonded-out core mounted on the target system, an RTL model which gets mapped into the emulator with the rest of the design, or an instruction-set simulator (ISS) model running on the workstation. Of these, the ISS will suffer a considerable performance penalty compared to the others. Detailed information for these various approaches may be found in [9]. Another approach in which a silicon vendor supports high-performance software development for embedded processors prior to silicon availability is described in [17].

The most common configuration for high-performance software testing is a physical model connected via a JTAG port to the software debugger. The JTAG interface spec does not allow for the JTAG clock to be stopped midstream. This also can cause difficulty with some of the vision choices in emulators. Some emulators do not offer alternate vision choices and, hence cannot be used for software testing with systems using a JTAG port for connection (without suffering huge performance penalties by using an ISS). Other emulators have a single 100% visibility capability which does not allow for operating a JTAG port to the software debugger except by reverting to compiled (and constantly recompiled) probes. As software testing becomes more and more important in achieving time-to-market demands, it is critical to know what software development environments an emulator can support with high performance.

19.7.4 Multi-User Access

Some emulators allow their total capacity to be dynamically partitioned among a number of simultaneous users. This increases the usage considerably, especially early in the design cycle for block-level acceleration and in environments where multiple projects are sharing the emulator. As the design progresses and full-chip and system-level testing is required, a larger percentage of the emulator capacity (perhaps all of it) is used to emulate it.

19.8 Summary

Hardware accelerators and emulators provide much higher verification performance than logic simulators, but require some additional effort to deploy. ICE provides the highest performance — often 10,000 — 1,000,000 times faster than a simulator, but requires an emulation environment to be built around it with speed buffering devices. Accelerators and emulators require the user to be aware of the differences between simulation and silicon (i.e., emulators and chips)

1. Simulation has 12 states or more, silicon has but two states.
2. Simulation generally executes RTL statements sequentially, silicon "executes" RTL concurrently.
3. Simulation is highly interactive, silicon less so.

FPGA-based emulators use commercial FPGAs and are smaller and consume less power, while processor-based emulators require custom silicon designs and consume more power but they compile designs much faster and can deliver higher performance. The key benefit of using emulation for design verification is shorter time-to-market and higher product quality.

References

[1] M.R. Butts et al., U.S. Patent 5,036,473.
[2] N.-P. Chen et al., U.S. Patent 5,475,830.
[3] R.A. Price et al., U.S. Patent 5,259,006.
[4] W. Dai et al., U.S. Patent 5,452,253,
[5] A. Agarwal et al., U.S. Patent 5,596,742.
[6] S.P. Sample et al., U.S. 5,960,191.
[7] W.F. Beausoleil et al., U.S. Patent 5,551,013.
[8] W.F. Beausoleil et al., U.S. Patent 6,035,117.
[9] R. Turner, Accelerated hardware/software co-verification speeds first silicon and first software, *Cadence Des. Sys.*, available on www.cadence.com.
[10] P. Christie and D. Stroobandt, The interpretation and application of Rent's rule, *IEEE Trans.VLSI Syst.*, 8, pp. 639–648, December 2000.
[11] B. Landman and R. Russo, On a pin versus block relationship for partitioning of logic graphs, *IEEE Trans. Comput*, C20, pp. 1469–1479, 1971.
[12] http://www.ecs.umass.edu/ece/tessier/courses/697c/ps3/rents_rule.pdf
[13] J. Babb, R. Tessier, and A. Agarwal. Virtual wires: overcoming pin limitations in FPGA-based logic emulators, Proceedings of IEEE Workshop on FPGA Based Custom Computing Machines, April 1993, pp. 142–151.
[14] K. Kuusilinna, et al., Designing BEE: a hardware emulation engine for signal processing in low-power wireless applications, *EURASIP J. Appl. Signal Process.*, 2003, pp. 502–513, 2003.
[15] J. Varghese, M. Butts, and J. Batcheller, An efficient logic emulation system, *IEEE Trans. VLSI Syst.*, 1, 171–174, 1993.
[16] J. Babb et al., Logic emulation with virtual wires, *IEEE Trans. CAD*, 16, pp. 609–626, 1997.
[17] G. A. Dada, Reducing time to market: parallel software development with emulation and simulation tools for MXC architectures, Motorola, http://www.freescale.com/files/wireless_comm/doc/brochure/MXCSWDEVWP.pdf.

20 Formal Property Verification

Limor Fix
Design Technology, Intel
Pittsburgh, Pennsylvania

Ken McMillan
Cadence Berkeley Laboratories
Berkeley, California

20.1 Introduction

In 1994, Intel spent 500 million dollars to recall the Pentium CPU due to a functional bug in the floating-point division result which was off by 61/10,00,000. The growing usage of hardware and software in our world and their insufficient quality has already cost an enormous price in terms of human life, time, and money. Both hardware and software designs are experiencing a verification crisis since simulation-based logic verification is becoming less and less effective due to the growing complexity of the systems. To fight the low coverage in simulation, large systems are often decomposed into blocks and each of these blocks is simulated separately in its own environment. Only when the block-level simulations are stable, will integration and full-chip/system simulation be carried out. This methodology caused a dramatic increase in the effort invested in simulation-based verification since a complete simulation environment, including tests, checkers, coverage monitors, and environment models, is developed and maintained for each block. Industrial reports show that half of the total design effort is devoted to verification, the number of bugs is growing exponentially, and hundreds of billions of simulation cycles are consumed [1]. Despite the huge investment in verification, only a negligible percentage of all possible executions of the design is actually being tested.

Formal property verification has major advantages over simulation. It offers exhaustive verification technology, which is orders of magnitude more efficient as compared to scalar exhaustive simulation. In formal verification, all possible executions of the design are analysed and no tests need to be developed. The main deficiency of formal verification is its limited capacity compared with simulation. The state-of-the-art formal property verification engines can handle models in the order of 10 K memory elements and with complementary automatic model reduction techniques properties are verified on hardware designs and software programs with 100 to 500 K memory elements.

To give the reader some initial intuition into the formal verification approach, let us consider the simple program:

$$Int\ x;$$

$$read\ (x); \quad if\ x < 0\ then\ x = -x; \quad print(x);$$

Assume that x is a 32-bit integer with a range of 2^{32} different values. To verify fully that for every legal input the program prints a nonnegative number, simulation needs to cover the entire input space, i.e., 2^{32} simulation runs. In a formal verification approach, a single symbolic simulation run provides full coverage of the input space and thus exhaustively verifies the program. In this example, the input to the symbolic simulation is a set of n Boolean variables representing the input bits and the result of the symbolic simulation is a Boolean expression that reflects the program computation over the input variables. The resulting Boolean expression is then checked to be nonnegative.

A *property* is a specification of some aspect of a system's behavior that is considered necessary, but perhaps not sufficient for correct operation. Properties can range from simple "sanity checks" to more complex functional requirements. For example, we might specify that a mutually exclusive pair of control signals is never asserted simultaneously, or we might specify that a packet router chip connected to the PCI bus always conforms to the rules for correct signalling on the bus. In principle, we could even go so far as to specify that packets are delivered correctly, but full functional specifications of this sort are rare, because of the difficulty in verifying them. More often, formal verification is restricted to a collection of properties considered to be of crucial importance, or difficult to check adequately by simulation.

In hardware design, formal property verification is used throughout the design cycle.

As Table 20.1 illustrates, a hardware design project consists of five major simultaneous activities, called here (rows): RTL, Timing, Circuit, Layout, and Post-silicon. Over the duration of the project (columns), each activity has several consecutive subactivities. For example, the RTL activity starts with uArch development, which is followed by RTL development and it ends with RTL validation. At the early stages of the project when the architecture and the microarchitecture are designed, formal models of new ideas and protocols are developed and formally verified against a set of requirements. For example, the formal verification of a new cache coherence protocol may be established at this stage. This verification is usually carried out by the validation team working closely with the architects. An important benefit from this activity is the intimate familiarity of the validation team, early at the design stage, with the new microarchitecture. Later, while the RTL is being developed, the formal property verification

TABLE 20.1 Formal Property Verification in Hardware Design Flow

Project phases	Design	Stabilization	Implementation	Convergence	Debug
RTL	uArch development	RTL development	RTL validation		
Timing	Timing specification at block level	Full timing specification	Converging timing to project goals	Timing converged	
Circuit	Schematic entry	25% schematic	100% schematic	Only bug fix in schematic	
Layout			Layout clean at block level	Layout assembled and clean	
Post-silicon				A0 tape-out	Functional silicon
FPV	Verify micro architectural properties on FV models				
		Verify RTL properties			
			Verify micro architectural properties on the RTL		

activity shifts to RTL verification. The properties to be verified at this stage are either derived from the high-level microarchitectural specification, for example, the IEEE specification of the arithmetic operations, or from the implementation constraints that need to be verified on the RTL, for example verifying that all timing paths are within the required range. Constraints consumed by synthesis, static timing analysis, and power analysis are formally verified to be valid and thus formal verification is offering not only better functional correctness but also debugging of implementation problems. Formal property verification of the RTL has been mostly carried out by validation teams over the last 10 years. However, recently, with the increased capacity of the formal verification technology and increased ease of use, RTL and circuit designers are starting to use formal property verification products. Toward the end of the project and even after tape-out, the focus of formal property verification moves back to the verification of the high-level microarchitecture protocols — but this time, these properties are verified against the RTL and not against artificial formal models. It is still a big effort to verify high-level microarchitecture protocols against the RTL — only a few high-risk areas in the design are selected to be verified and the verification is carried out by the validation teams.

In the hardware electronic industry, tools for formal property verification have been developed since the early 1990s in internal CAD groups of big VLSI companies like Intel, IBM, Motorola, Siemens [2,3], and Bell-labs [4]. In the late 1990s, several start-up companies, like @HDL, 0-in, Real Intent, Verplex, and others, have begun offering similar tools. The large EDA companies are also offering products for formal property verification today.

In advanced electronic systems, software and hardware are becoming more and more interchangeable. One common usage of software is in high-level modeling of complex systems. These systems are designed using high-level software languages like C and SystemC [5] and only later in the design cycle, parts of the system (or all of it) are designated to be implemented in hardware. The other common usage of software is in the design of Embedded Systems. These systems usually consist of embedded software that executes on top of an embedded microprocessor, or increasingly, processors. Modern CPUs may also include an embedded software called *microcode* that is used to simplify the instruction set implemented in hardware.

Simulation is currently the main verification tool for software. However, the increased frequency of fatal and catastrophic software errors is driving the electronic design community to look for better, more exhaustive verification solutions. In the last 5 years, intensive efforts have been directed to finding new and fully automatic ways to apply formal methods to software verification.

Two applications of formal verification are currently under research and development. One is *property verification of software*, which is aimed at establishing the functional correctness of the software with respect to properties developed by the programmer and/or automatically extracted sanity properties that check, for example, array bounds [6–11]. The second application of formal verification to software is *translation validation*, which is aimed at verifying the correct translation of the high-level software model into the corresponding hardware description, e.g., Verilog, or into a lower-level software program [12–15].

Formal property verification of software is just starting to move into the industry — most promising results were reported by the research team in Microsoft, verifying Windows NT drivers. As Bill Gates said in at a Keynote address at WinHec, April 2002: "*Things like even software verification, this has been the Holy Grail of computer science for many decades but now in some very key areas, for example, driver verification we're building tools that can do actual proof about the software and how it works in order to guarantee the reliability.*" [16]

Formal automatic translation validation of software is also at the initial transfer stage in the electronics industry, for example Intel has developed a formal tool for verifying correspondence between microcode programs [15]. A few EDA startups, like Calypto, are offering C to Verilog translation validation. Results reported for academic systems demonstrate ability to verify 10,000 lines of source code in C against the compiler results [17] and being able to verify an academic microprocessor with 550 C lines against its RTL implementation with 1200 latches [14].

20.2 Formal Property Verification Methods and Technologies

20.2.1 Formal Property Specification

Hardware designs and embedded software generally fall into a class known as *reactive systems.* As defined by Pnueli [18], these are systems that interact continuously with their environment, receiving input and producing output. This is in contrast to a computer program, such as a compiler, that receives input once, then executes to termination, producing output once. To specify a reactive system, we need to be able to specify the valid input/output *sequences,* not just the correct function from input to output. For this purpose, Pnueli proposed the use of a formalism known as *temporal logic* that had previously been used to give a logical account of how temporal relationships are expressed in natural language. Temporal logic provides operators that allow us to assert the truth of a proposition at certain times relative to the present time. For example, the formula *Fp* states that p is true at *some* time in the future, while Gp says that p is true at *all* times in the future. These operators allow us to express succinctly a variety of commonly occurring requirements of reactive systems. We can state, for example, that it is never the case that signals `grant1` and `grant2` are asserted at the same time, or that if `req1` is asserted now, that eventually (at some time in the future), `grant1` is asserted. The former is an example of a *safety* property. It says that some bad condition never occurs. The latter property is an example of a *liveness* property. It says that some good condition must eventually occur. To reason about liveness properties, we must consider infinite executions of a system, since the only way to violate the property is to execute infinitely without occurrence of the good condition.

Since Pnueli introduced the use of temporal logic for specification, a number of notations have been developed to make specifications either more expressive or more succinct. For example, we may want to specify that a condition must occur not just eventually, but within a given amount of time. In a real-time temporal logic [19], the formula $F_{(<5)}p$ might be used to specify that p must become true within five time units. Moreover, because hardware systems are typically pipelined, it is common to refer to the conditions occurring in consecutive clock cycles. For this purpose, it is more convenient to use a notation similar to regular expressions. For example, using the PSL language, a standard in the EDA industry, we can say "if `req1` is followed by `grant1`, then eventually `dav` is asserted" like this:

```
(req1; gnt1) |-> F dav
```

The semicolon represents sequencing. Such notations also allow us to express some "counting" properties that have no expression in ordinary temporal logic. For example, we can state that p holds in every other clock cycle like this: (p; true)*.

Modern property specification languages used in EDA provide some additional features that are convenient for hardware specifications. For example, one can specify a particular signal as the "clock" by which time is measured in a given property, and one can specify a "reset" condition that effectively cancels the requirements of the property. The latter is useful for dealing with reset and exception conditions.

20.2.1.1 History of Temporal Specification Languages

Pnueli suggested the use of temporal logic in the late 1970s [20]. Starting in the early 1980s, a variety of logics and notations have been used in research on automatic verification, including linear temporal logic (LTL), computational tree logic (CTL), CCS, dynamic logic, and temporal logic of actions (TLA) [20,21–29]. Some of these notations, notably CCS and TLA, are used for both modelling systems and specifying properties. In the 1990s, property languages specialized to EDA were developed. These include ForSpec, developed at Intel [30], Sugar from IBM [31], temporal "e" from Verisity language, and CBV from Motorola. These languages were donated to the Accellera standards body, and have been the basis for defining a new IEEE standard proposal called PSL [32].

20.2.2 Hardware Formal Verification Technologies/Engines

Having specified properties of a system in a suitable notation, we would then like to be able to verify these properties formally, in as automated a manner as possible. Fully automated verification of temporal properties is generally referred to as *model checking,* in reference to the first such technique developed by

Clarke and Emerson [22]. Model checking can not only verify temporal properties, but also provide *counter-examples* for properties that are false. Counterexamples are traces of incorrect system behavior that are valuable in diagnosing errors.

Since hardware systems are finite state, they can in principle always be verified by model checking. Advances in model checking algorithms have allowed designs of significant size to be checked.

20.2.2.1 The Automata-Theoretic Approach

We can think of a property as specifying a set of acceptable input/output traces of a system. If all possible traces of a given system fall into this set, then the system satisfies the property. Most algorithms for model checking work by translating the property into a *finite automaton* that accepts exactly the set of traces accepted by the property [33]. The automaton is a state graph, whose edges are labelled with input/output pairs. The *language* of the automaton is the set of input/output traces observed along accepting paths in this graph. By defining "accepting path" in various ways, we can define different kinds of languages. For example, if we specify the legal initial and final states of an accepting path, we obtain a set of finite traces. To obtain a language of infinite traces, we can specify the initial states, and the sets of states that may occur "infinitely often" on an accepting infinite path. This allows us to represent liveness properties with automata. Once the property has been translated into an equivalent automaton, we need no longer concern ourselves with the property language.

To check the property on a system, we also represent the system with an equivalent automaton. We now have only to check that every trace of the system automaton is a trace of the property automaton. This can be done by combing the automaton for the complement of the property with the system automaton to produce what is called a "product automaton". If this automaton accepts any traces, then the system does not satisfy the property. This test can be made, in the simplest case, by searching the state graph of the product automaton for an accepting path. For finite traces, this can be accomplished by a depth- or breadth-first search, starting form the initial states. This kind of search-based model checking (referred to as "explicit-state") is exemplified by systems such as COSPAN [34], SPIN [6] and Murphi [7], which are quite effective at verifying software-based protocols, cache coherence protocols, and other systems with relatively small state spaces.

20.2.2.2 In Search of Greater Capacity

The main practical difficulty with the above approach is that the state graphs of the automata can be prohibitively large. The number of states of the system automaton is, in the worst case, the number of possible configurations of the registers (or other state-holding elements) in the system, which is exponential. For most property languages, the number of states in the automaton corresponding to a property is also exponential in the size of the property text. Thus, the brute-force approach described above is hopeless for large hardware designs. Instead, a number of heuristic methods have been developed that avoid explicit construction of the state graph. Here, we discuss a few of these methods that are currently used in EDA.

20.2.2.3 Binary Decision Diagram-Based Symbolic Model Checking

The symbolic model checking (SMC) approach was introduced in the late 1980s [35, 75] and was first implemented in a tool called SMV [36]. In this approach, automata are not explicitly constructed. Instead, a logical formula is used to characterize implicitly the set of possible transitions between states. In this formula, variable x represents the state of register x in the current state while x' represents the state of register x at the next time. The possible transitions of a sequential machine are easily characterized in this way, as a set of Boolean equations (see Figure 20.1).

For most property languages, the property automaton can also be characterized in this way, without an exponential expansion. A breadth-first search can then be accomplished purely by operations on Boolean formulas. To do this, we use a formula to stand for the set of states that satisfies it. The set of successors P' of a set of states P can then be obtained by the following equation, where V is the set of state variables, R the transition formula, and ∃ the existential quantifier over Boolean variables:

$$P'(V') = \exists V.P(V) \wedge R(V,V')$$

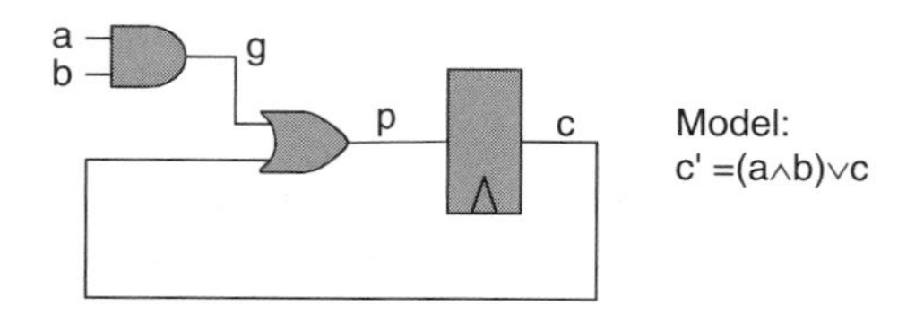

FIGURE 20.1 Characterizing transitions of a circuit.

A symbolic breadth-first search starts with a formula characterizing the initial states, and iterates the above "image" operation until a fixed point occurs (i.e., no new states are obtained).

To use this idea effectively, we need a compact representation for Boolean formulas, on which Boolean operations and existential quantification can be efficiently applied. For this purpose, binary decision diagrams (BDDs) are commonly used. A BDD is a decision graph in which variables occur in the same order along all paths, and common subgraphs are combined. In 1986, Bryant gave an efficient algorithm for Boolean operations on this structure [37]. Symbolic model checking using BDDs provides a means of property checking for circuits of modest size (typically up to a few hundred registers) in cases where the state graph is far too large to be constructed explicitly. This method was used successfully to find errors in the cache coherence protocols of a commercial multiprocessor [38], and has since been used in a number of commercial EDA tools, with many improvements introduced in the 1990s.

20.2.2.4 Satisfiability Problem-Based Bounded and Unbounded Model Checking

The Boolean satisfiability problem (SAT) is to determine whether there exists a truth assignment that makes a given Boolean formula true (or equivalently, whether there exists an input pattern that makes the output of a given Boolean circuit one). The first algorithm to solve this classic NP-complete problem was described in 1960 [76], and since then many improvements have been made. The success of recent SAT solvers such as Chaff [39] and Grasp [40] in solving very large problem instances prompted interest in using SAT solvers for model checking. In 1999, the notion of *bounded model checking* (BMC) using a SAT solver was introduced [41]. As in SMC, the transitions of the system and property automata are characterized by Boolean formulas. By *unfolding* this representation (i.e., making k consecutive copies, corresponding to consecutive time frames) the question of existence of a counter-example of k steps can be posed as a SAT problem. This made it possible to exploit recent advances in SAT solvers to find counter-examples to properties for systems much larger than can be handled by SMC (typically up to a few thousand registers).

20.2.2.5 Unbounded Model Checking Using a SAT Solver

The limitation of BMC is that it can only find counterexamples. It does not provide a practical way of proving that a property holds for all possible behaviors of a system. However, the success of BMC in handling large designs led to new methods for complete model checking that exploit SAT. One method, called k-induction [42], uses a SAT solver to test whether there is a path of k distinct good states leading to a bad state. If not, and if there is no path of k steps from an initial state to a bad state, then by induction the property holds at all times. This method can prove properties using just a SAT solver, though in practice the required k value can be quite large, making the BMC step prohibitive.

Another approach, called *localization*, uses a SAT solver only to decide which components of a system are relevant to proving a particular property. This is done by using the SAT solver to refute either a partial counter-example or all counter-examples of a given length [43,44]. The components of the system used in this refutation are then used to attempt to verify the property using BDD-based SMC or other methods. Since modern SAT solvers are quite effective at ignoring irrelevant facts, this process can produce a significant reduction in the size of the system to be checked. Another approach uses a technique called *interpolation* to derive an approximation of the "successor states" of a given state set from a bounded model-checking run [45].

This allows SMC using just a SAT solver. These techniques allow designs with up to a few thousand latches to be fully verified, at least in cases where only a small part of the system is relevant to the property, the domain in which SAT-based BMC is effective.

20.2.2.6 Symbolic Simulation

In 1990, the technique of *symbolic simulation* was introduced for hardware verification [46]. This technique resembles ordinary simulation in the sense that the user provides an input sequence to drive the design. However, in symbolic simulation, the input values can be symbolic variables as well as numeric zeros and ones. Thus, a single simulation run may in effect represent many runs of the design. The outputs of the simulation are Boolean formulas over the symbolic variables instead of numeric values. The outputs can then be compared to the desired functions. This comparison can be done using BDDs or a SAT solver. The complexity of this check can be reduced by introducing logical "unknown" (X) values at inputs that are considered irrelevant to the case being verified. By itself, symbolic simulation does not prove properties because, like BMC, it simulates only bounded runs of the system. However, complete property verification is possible using the *symbolic trajectory evaluation* (STE) methodology, based on symbolic simulation [47]. This method requires the user to provide part of the proof, but has the advantage that it can be applied to fairly large designs.

20.2.2.7 Theorem Proving Methods

From the1950s into the 1980s, a great deal of progress was made in mechanizing mathematics and logical deduction. Classic papers in the field include [48–55]. Beginning in the 1970s with the Boyer–Moore theorem prover NQTHM [56] and LCF [57], many practical tools have been developed that assist the user in formalizing and proving mathematical theorems. Later systems include ACL2 [68], PVS [59], and HOL [60]. These systems usually provide some form of mechanized *proof search* to aid in the construction of proofs. The proof search mechanism may be fully automatic, as in NQTHM, or customizable by writing *tactics*, as in LCF and its descendants.

To use a theorem prover for formal property verification, we translate both the system model and the desired property into the logic of the prover. This can be done, for example, by automatically translating the program into a logical assertion that characterizes its transition behavior (a *shallow embedding*), or by treating the program itself as an object in the formalism and writing an interpreter for the programming language in the logic (a *deep embedding*). In either case, verifying the property reduces to proving an appropriate theorem in the logic.

To prove properties of complex systems with a theorem prover, a considerable amount of user guidance is required, in the form of lemmas, tactics, or manual guidance of the deduction process. The most common proof approach is to construct an *inductive invariant* of the system (or of the system augmented with auxiliary structures). An inductive invariant is a fact that is true initially, and implies itself at the next time, thus is true always. Once a suitable invariant is obtained, the proof that it is inductive can be aided by the use of a *ground decision procedure* [61]. This approach greatly reduces the required amount of proof guidance. However, manually constructing an inductive invariant is still a time-consuming and intellectually taxing process.

For this reason, a number of tools that combine a theorem prover and a model checker have been developed. In such systems, the overall proof of a property can be reduced to lemmas that can be discharged by a model checker. For example, the HOL-VOSS system [62] and its successor Forte [63] use a theorem prover to reduce the proof to lemmas that can be checked by symbolic simulation. The Cadence SMV system [64] can be used to reduce the proof to verification of temporal logic properties by SMC. Such systems have the potential to shift a substantial amount of the proof effort from the user onto automated tools. To make this possible, however, we require an overall proof strategy that reduces the problem to a scale that can be handled by model checking.

20.2.3 Modularization and Abstraction

The most common proof approaches for dealing with complex systems are *modularization* and *abstraction*. By modularization, we mean breaking a property to be proved (a *goal*) into two or more properties (or *sub-goals*) that are more easily verified automatically, usually because they depend on fewer system components. We must show, in turn, that the subgoals imply the goal. In a strictly modular approach, such as the *assume-guarantee paradigm* [65], we verify a system consisting of two modules A and B, by proving properties of A and B in isolation, then combining these to deduce a property of the system. Strictly modular proofs can be difficult to obtain, however, because of the need to capture in the subgoals all properties of one module that are required by the other. In practice, one can take a more relaxed approach. It is necessary only that the sub-goals be provable in a coarser *abstraction* of the system than the original goal. The abstraction may be obtained by localization (i.e., considering only a subset of system components), or by various methods of abstracting data (including *predicate abstraction*, which is described below).

In fact, the key to all known methods for verifying large, complex systems is *abstraction*. This means replacing a complex system with a simpler system that captures the required properties. An abstract system can be constructed manually, in which case we must prove that the abstract system in fact preserves the properties of interest. Alternatively, an abstraction can be constructed automatically, perhaps according to parameters given by the user. A general framework for this approach is provided by *abstract interpretation* [66]. In this framework, the user chooses a suitable representation for facts about data in the system, and an automatic analysis computes the strongest inductive invariant of the system that can be expressed in this representation. As an example, we might use a representation that can express all linear affine relations between program variables. An important instance of abstract interpretation is *predicate abstraction* [67]. Here, the user provides a set of simple predicates, like $x < y$. The analysis constructs the strongest inductive invariant that can be expressed as a Boolean combination of these predicates.

Of course, the invariant constructed by abstract interpretation may or may not be strong enough to prove the desired property. The key to this approach is to choose an abstraction appropriate to the given property. Thus, in predicate abstraction, the predicates must be chosen carefully, so that they are sufficient to prove the property, without being so numerous as to make the analysis intractable. Recently, considerable attention has been given to automating the selection of predicates [10,68]. This has proved to be especially effective in software model checking as we will see. Abstract interpretation has also been effectively used to prove the subgoals arising in a modularization approach [64]. Here, the key is to choose subgoals that can be proved using a coarse abstraction. By this approach, much of the work of constructing invariants can be done automatically, thus simplifying the manually constructed part of the proof.

20.3 Software Formal Verification

A big challenge in applying formal verification to software is the need to translate the original code into a formal representation that can be reasoned about. In particular difficulties are caused by dynamic memory or process allocations, dynamic loops, dynamic jump-targets, aliases, and pointers. The state explosion problem is another big challenge in formal verification of software — as software programs often have a very large state space. Below is a brief survey of automatic software verification techniques.

20.3.1 Explicit Model Checking of Protocols

In 1991 to 1992, protocols were verified using dedicated modelling languages like Promela [6] and Murphi [7] and explicit model checkers. This approach best fits concurrent asynchronous software systems. The explicit model-checking algorithm performs a search on the state space of the system, and on-the-fly verifies for each newly visited state that the property holds for it. The algorithm stores all the states it encounters in a large hash table. When a state is generated that is already in the hash table, the search does not expand to its successor states. To allow efficient search of the state space, several optimization techniques have been developed. The most dominating one is the partial order reduction. Not all possible successors of a given state are generated and included in the search. However, states and flows of the program which are not visited in the search are proven to be redundant for establishing the correctness of the property. For example, if the program has two processes *P1:: a ; b;* (in process *P1*, statement *a* is performed first and then statement *b*) and *P2:: c ; d ;* then there are six possible execution flows (orders) *abcd, acbd, acdb, cdab, cabd,* or *cadb*. However, if the correctness of the property to be checked is only influenced by the order of execution of the statements *b* and *d*, the search can be limited to check only two flows: *acbd* and *acdb*. This technique has been proven to be successful for protocols that have high degree of concurrency and thus partial reduction can be very beneficial. In particular, industrial cache coherence protocols, cryptographic, and security-related protocols were successfully verified. Known academic and industrial model checkers based on this family of techniques are SPIN [6], Murphi [7] and TLA+ [69]. The main drawback of this approach is that the verification is not applied to the "golden model" of the design, that is, to the actual C and Java programs.

20.3.2 Explicit and Symbolic Model Checking of Java and C Programs

In 1999 to 2001, Java programs were abstracted using abstract interpretation [66] and translated into finite-state models to be verified using explicit and symbolic model checkers [8,9,70,71]. Similar technology has been proposed for translating C programs, using *predicate abstraction* [67] into Boolean programs, and submitting them to a symbolic model checker [10]. The extracted Boolean program has the same control-flow structure as the original program; however, it has only Boolean variables, each representing a Boolean predicate over the variables of the original program. For example, if x and y are complex data structures with the same type, e.g., structures, then a Boolean variable in the extracted program may represent the predicate $x=y$, t, thus resulting in a much smaller state space in the Boolean program.

To give the reader some intuition into predicate abstraction, let x be an integer variable in the original C program and let the Boolean program have only two predicates that refer to x, "the predicate *b1* is defined as $x > 2$" and "the predicate *b2* is defined as $x=5$".

Original program	Boolean program
pc1: x=5;	*pc1: b1=true; b2=true;*
pc2: ..	*pc2: ..*

In this case, it is easy to see that if x is assigned 5 in the original program then both the predicates *b1* and *b2* should become *true* in the corresponding program location, *pc1*, in the Boolean program. A more interesting case is

Original program	Boolean program
pc1: x=x + 1;	*pc1: b1=?; b2=?;*
pc2: ..	*pc2: ..*

In this case, since *b1* represents $x>2$, *b1* should be assigned *true* if $x+1>2$, i.e., if at program location *pc1*, $x>1$ holds. The condition $x>1$ is called the Weakest Precondition that would guarantee that $x>2$ holds after the assignment $x=x+1$. Similarly, since *b2* represents $x=5$, *b2* should be assigned *true* if $x+1=5$, that is, if at program location *pc1*, $x=4$ holds. However, at location *pc1* in the Boolean program, we only have access to the truth value of *b1* and *b2* and not to the values of the predicates $x>1$ and $x=4$ which are not included in the set of predicates used for the generation of the Boolean program. In this case, an automated decision procedure is invoked to strengthen these predicates into an expression over the set of the chosen predicates. In our example, the predicate $x>1$ will be strengthen to $x>2$ and the predicate $x=5$ will strengthen into *false*. Thus, we get:

Original program	Boolean program
pc1: x=x+1;	*pc1: b1=if b1 then true else false;* *b2=false;*
pc2: ..	*pc2: ..*

The predicate abstraction technology has been extended to handle pointers, procedures and procedure calls [10]. For example, let the predicate *b1* be **p>5,*

Original program	Boolean program
pc1: x=3;	*pc1: b1=?;*
pc2: ..	*pc2: ..*

if *x* and **p* are aliases, then *b1* should be assigned *false* since **p* becomes *3*, else *b1* should retain its previous value. Thus, we get:

Original program	Boolean program
pc1: x=3;	*pc1: b1=(if &x=p then false else retain);*
pc2: ..	*pc2: ..*

The SMC algorithm is applied to the Boolean program to check the correctness of the property. If successful, we can conclude that the property holds also for the original program. If failed, the Boolean program is refined to more precisely represent the original program. Additional Boolean predicates are automatically identified from the counter-example and a new Boolean program is constructed. This iterative approach is called iterative abstraction refinement [77]. Using this technique has been proven to be successful for checking safety properties of Windows NT device drivers and for discovering invariants regarding array bounds. Academic and industrial model checkers based on this family of techniques include SLAM by Microsoft [10], Bandera [70], Java Pathfinder [9], TVLA [72], Feaver [73], Java-2-SAL [74], and Blast [68].

20.3.3 Bounded Model Checking of C Programs

Most recently, in 2003 to 2004, C and SpecC programs were translated into propositional formulae and then formally verified using a bounded model checker [14,11]. This approach can handle the entire ANSI-C language consisting of pointers, dynamic memory allocations and dynamic loops, i.e., loops with conditions that cannot be evaluated statically. These programs are verified against user defined safety properties and also against automatically generated properties about pointer safety and array bounds.

The C program is first translated into an equivalent program that uses only *while, if, goto* and *assignment* statements. Then each loop of the form *while (e) inst* is translated into *if (e) inst; if (e) inst; …; if (e) inst; {assertion !e}*, where *if (e) inst* is repeated *n* times. The assertion *!e* is later formally verified and if it does not hold, *n* is increased until the assertion holds. The resulting program that has no loops, and the properties to be checked are then translated into a propositional formula that represents the model after unwinding it *k* times. The resulting formula is submitted to a SAT solver and if a satisfying assignment is found, it represents an error trace. During the *k*-steps unwinding of the program, pointers are handled(punctuation needed?). For example,

```
int a, b, p;
  if (x) p=&a; else p=&b;
*p=5;
```

The above program is first translated into:

```
p1=(x ? &a: p0)∧p2=(x? p1: &b)
*p2=5;
```

That is, three copies of the variable p are created $p0$, $p1$ and $p2$. If x is true then $p1=\&a$ and $p2=p1$, and if x is *false* $p1=p0$ and $p2=\&b$. Then $*p2=5$ is replaced by

$*(x?\ p1;\ \&b)=5$	which is replaced by
$(x\ ?\ *p1;\ *\&b)=5$	which is replaced by
$(x?\ *p1;\ b)=5$	which is replaced by
$(x?\ *(x?\ \&a\ :p0);\ b)=5$	which is replaced by
$(x?\ *\&a\ ;\ b)=5$	which is replaced by
(x? a ; b)=5	

After all these automatic transformation the pointers have been eliminated and the resulting statement is $(x?\ a;\ b)=5$. For details on dynamic memory allocations and dynamic loops see [11].

20.3.4 Translation Validation of Software

Two technological directions are currently pursued for formally verifying the correct translation of software programs. One [12,17], which automatically establishes an abstraction mapping between the source program and the object code, offers an alternative to the verification of synthesizers and compilers. The other direction [13,14], which automatically translates the two programs given in C and Verilog into a BMC formula and submits it to a SAT solver. The value of each Verilog signal at every clock cycle is visible to the C program. Thus, the user can specify and formally verify the desired relation between the C variables and the Verilog signals. Both the C and Verilog programs are unwound for k steps as we have described in the previous section.

20.4 Summary

Formal property verification provides a means of ensuring that a hardware or software system satisfies certain key properties, regardless of the input presented. Automated methods developed over the last two decades have made it possible, in certain cases, to verify properties of large, complex systems with minimal user interaction, and to find errors in these systems. These methods can be used as an adjunct to simulation and testing in the design verification process, to enable the design of more robust and reliable systems.

References

[1] B. Bentley, Validating the Intel Pentium 4 Microprocessor, *Proceedings of the 38th Design Automation Conference (DAC)*, ACM Press, New York, 2001, pp. 244–248.

[2] G. Kamhi, O. Weissberg, L. Fix, Z. Binyamini, and Z. Shtadler, Automatic datapath extraction for efficient usage of HDD, *Proceedings of the 9th International Conference Computer Aided Verification (CAV)*, Lecture Notes in Computer Science, Vol. 1254, Springer, Berlin, 1997, pp. 95–106.

[3] D. Geist and I. Beer, Efficient model checking by automated ordering of transition relation partitions, *Proceedings of Computer-Aided Verification (CAV)*, Springer-Verlag, Berlin,1994.

[4] R.P. Kurshan, Formal verification on a commercial setting, *Proceedings of Design Automation Conference (DAC)*, 1997, pp. 258–262.

[5] T. Grotker, S. Liao, G. Martin, and S. Swan, *System Design with SystemC*, Kluwer Academic Publishers, Dordrecht, 2002.

[6] G.J. Holzmann, *Design and Validation of Computer Protocols*, Prentice-Hall, Englewood Cliffs, NJ, 1991.

[7] D.L. Dill, A.J. Drexler, A.J. Hu, and C. Han Yang, Protocol verification as a hardware design aid, *IEEE International Conference on Computer Design: VLSI in Computers and Processors*, IEEE Computer Society, Washington, DC,, 1992, pp. 522–525.

[8] C. Demartini, R. Iosif, and R. Sisto, A deadlock detection tool for concurrent Java programs. *Software — Practice and Experience*, 29, 577–603, 1999.

[9] K. Havelund and T. Pressburger,. Model checking Java programs using Java PathFinder, *Int. J. Software Tools Technol. Trans.*, 2, 366–381, 2000.
[10] T. Ball, R. Majumdar, T. Millstein, and S.K. Rajamani, Automatic Predicate Abstraction of C Programs, *PLDI 2001, SIGPLAN Notices*, 36, 203–213, 2001.
[11] E. Clarke, D. Kroening, and F. Lerda, A tool for checking ANSI-C program, in *Proceedings of 10th International Conference, on Tools and Algorithms for Construction and Analysis of Systems (TACAS)*, Lecture Notes in Computer Science, Vol. 2988, Springer, Berlin, 2004.
[12] A. Pnueli, M. Siegel, and O. Shtrichman, The code validation tool (CVT)-automatic verification of a compilation process, *Int. J. Software Tools Technol. Trans.* (STTT), 2, 192–201, 1998.
[13] E. Clarke and D. Kroening, Hardware verification using ANSI-C programs as a reference, *Proceedings of ASP-DAC 2003*, IEEE Computer Society Press, Washington, DC, 2003, pp. 308–311.
[14] E. Clarke, D. Kroening, and K. Yorav, Behavioral consistency of C and Verilog programs using bounded model checking, *Proceedings of the 40th Design Automation Conference (DAC)*, ACM Press, New York, 2003, pp. 368–371.
[15] L. Fix, M. Mishaeli, E. Singerman, and A. Tiemeyer, MicroFormal: FV for Microcode. Intel Internal Report, 2004.
[16] Bill Gates, Keynote address at WinHec 2002. www.microsoft.com/billgates/speeches/2002/04-18winhec.asp
[17] L. Zuck, A. Pnueli, Y. Fang, and B. Goldberg, Voc: a translation validator for optimizing compilers, *J. Univ. Comput. Sci.*, 3, 223–247, 2003. Preliminary version in ENTCS, 65, 2002.
[18] A. Pnueli, Specification and development of reactive systems, in *Information Processing 86*, Kugler, H.-J., Ed., IFIP, North-Holland, Amsterdam, 1986, pp. 845–858.
[19] R. Alur and T.A. Henzinger, Logics and models of real time: a survey, in *Real Time: Theory in Practice,* de Bakker, J.W., Huizing, K., de Roever, W.-P., and Rozenberg, G., Eds., *Lecture Notes in Computer Science*, Vol. 600, Springer, Berlin, 1992, pp. 74–106.
[20] A. Pnueli, The temporal logic of programs, *Proceedings of the 18th Symposium on Foundations of Computer Science*, 1977, pp. 46–57.
[21] M. Ben-ari, Z. Manna, and A. Pnueli, The temporal logic of branching time, *Acta Inform.*, 20, 207–226, 1983.
[22] E.M. Clarke and E.A. Emerson, Design and synthesis of synchronization skeletons using branching time temporal logic, *Proceedings of Workshop on Logic of Programs*, Lecture Notes in Computer Science, Vol. 131, Springer, Berlin, 1981, pp. 52–71.
[23] M.J. Fischer and R.E. Lander, Propositional dynamic logic of regular programs, *J. Comput. Syst. Sci.*, 18, 194–211, 1979.
[24] L. Lamport, Sometimes is sometimes "not never" — on the temporal logic of programs, *Proceedings ofthe 7th ACM Symposium on Principles of Programming Languages*, January 1980, pp. 174–185.
[25] B. Misra and K.M. Chandy, Proofs of networks of processes, *IEEE Trans. Software Eng.*, 7, 417–426, 1981.
[26] R. Milner, A calculus of communicating systems, Lecture Notes in Computer Science, Vol. 92. Springer, Berlin, 1980.
[27] A. Pnueli, The temporal semantics of concurrent programs, *Theor. Comput. Sci.*, 13, 45–60, 1981.
[28] J.P. Queille and J. Sifakis, Specification and verification of concurrent systems in Cesar. *Proceedings of 5th International Symposium On Programming*, Lecture Notes in Computer Science, Vol. 137, Springer, Berlin, 1981, pp. 337–351.
[29] P. Wolper, Synthesis of Communicating Processes from Temporal Logic Specifications, Ph.D. thesis, Stanford University, 1982.
[30] R. Armoni, L. Fix, A. Flaisher, R. Gerth, B. Ginsburg, T. Kanza, A. Landver, S. Mador-Haim, E. Singerman, A. Tiemeyer, M. Vardi, and Y. Zbar, The ForSpec temporal logic: a new temporal property specification language, *Proceedings of TACAS*, 2002, pp. 296–311.
[31] I. Beer, S. Ben-David, C. Eisner, D. Fisman, A. Gringauze, and Y. Rodeh, The temporal logic sugar, *Proceedings of Conference on Computer-Aided Verification (CAV'01)*, Lecture Notes in Computer Science, Vol. 2102, Springer, Berlin, 2001, pp. 363–367.
[32] Property Specification Language Reference Manual, Accellera, 2003.
[33] M.Y. Vardi and P. Wolper, An automata-theoretic approach to automatic program verification, *Proceedings of IEEE Symposium on Logic in Computer Science*, Computer Society Press, Cambridge, 1986, pp. 322–331.

[34] R. P. Kurshan, *Computer-Aided Verification of Coordinating Processes*, Princeton Univ. Press, Princeton, 1994.
[35] O. Coudert, C. Berthet, and J.C. Madre, Verification of synchronous sequential machines based on symbolic execution, *Proceedingsof International Workshop on Automatic Verification Methods for Finite State Systems*, Lecture Notes in Computer Science, Vol. 407, Springer, Berlin, 1989.
[36] K.L. McMillan, Symbolic Model Checking: An Approach to the State Explosion Problem, Ph.D. thesis, CMU CS-929131, 1992.
[37] R.E. Bryant, Graph-based algorithms for Boolean function manipulation, *Proceedings of IEEE Transactions on Computers*, Vol. C-35, 1986, pp. 677–691.
[38] K. L. McMillan and J. Schwalbe, Formal verification of the gigamax cache consistency protocol, *International Symposium on Shared Memory Multiprocessing*, 1991, pp. 242–251.
[39] M.W. Moskewicz, C. F. Madigan, Y. Zhao, L. Zhang, and S. Malik, Chaff: engineering an efficient SAT solver, *Proceedings of the 38th Design Automation Conference (DAC'01)*, June 2001.
[40] J.P. Marques-Silva and K.A. Sakallah, GRASP: a new search algorithm for satisfiability, *IEEE Trans. Comput.*, 48, 506–521, 1999.
[41] A. Biere, A. Comatti, E. Clarke, and Y. Zhu, Symbolic model checking without BDDs, *Proceedings of the Workshop on Tools and Algorithms for the Construction and Analysis of Systems (TACAS'99)*, Lecture Notes in Computer Science Springer, Berlin, 1999.
[42] M. Sheeran, S. Singh, and G. Stalmarck, Checking safety properties using induction and a sat-solver, *Proceedings of International Conference on Formal Methods in Computer Aided Design (FMCAD 2000)*, W.A. Hunt and S.D. Johnson, Eds., 2000.
[43] P. Chauhan, E. M. Clarke, J. H. Kukula, S. Sapra, H. Veith, and D. Wang, Automated abstraction refinement for model checking large state spaces using SAT-based conflict analysis, *Proceedings of FMCAD 2002*, Lecture Notes inComputer Science, Vol. 2517, Springer, Berlin, 2002, pp. 33–51.
[44] K.L. McMillan and N. Amla, Automatic abstraction without counterexamples, *Proceedings of Tools and Algorithms for the Construction and Analysis of Systems (TACAS'03)*, Lecture Notes in Computer Science, Vol. 2619, Springer, Berlin, 2003, pp. 2–17.
[45] K.L. McMillan, Interpolation and SAT-based model checking, *Proceedings of CAV 2003*, Lecture Notes in Computer Science, Vol. 2725, Springer, Berlin, 2003, pp. 1–13.
[46] R.E. Bryant and C-J. Seger, Formal verification of digital circuits using symbolic ternary system models, *Workshop on Computer Aided Verification. 2nd International Conference, CAV'90*, Clarke, E.M. and Kurshan, R.P., Eds., Lecture Notes in Computer Science, Vol. 531, Springer, Berlin, 1990.
[47] D.L. Beatty, R.E. Bryant, and C-J. Seger, Formal hardware verification by symbolic ternary trajectory evaluation, *Proceedings of the 28th ACM/IEEE Design Automation Conference.* IEEE Computer Society Press, Washington, DC, 1991.
[48] S.C. Kleene, *Introduction to Metamathematics.* Van Nostrand, Princeton, NJ, 1952.
[49] J. McCarthy, Computer programs for checking mathematical proofs, *Proceedings of the Symposia in Pure Mathematics, Vol. V, Recursive Function Theory.* American Mathematical Society, Providence, RI, 1962.
[50] J.A. Robinson,. A machine-oriented logic based on the resolution principle. *J. ACM*, 12, 23–41, 1965.
[51] D. Scott, Constructive validity, *Symposium on Automatic Demonstration*, Lecture Notes in Mathematics, Vol. 125, Springer, New York, 1970, pp. 237-275.
[52] C.A.R. Hoare, Notes on data structuring, in *Structural Programming*, O.J. Dahl, E.W. Dijkstra, and C.A.R. Hoare, Academic Press, New York, 1972, pp. 83–174.
[53] R. Milner and R. Weyhrauch, Proving computer correctness in mechanized logic, in *Machine Intelligence*, Vol. 7, B. Meltzer and D. Michie, Eds., Edinburgh University Press, Edinburgh, Scotland, 1972, pp. 51–70.
[54] G.P. Huet, A unification algorithm for types lambda-calculus, *Theor. Comput. Sci.*, 1, 27–58, 1975.
[55] D.W. Loveland, *Automated Theorem Proving: A Logical Basis*, North-Holland, Amsterdam, 1978.
[56] R. S. Boyer and J S. Moore, *A Computational Logic*, Academic Press, New York, 1979.
[57] M. Gordon, R. Milner, and C. P. Wadsworth, Edinburgh LCF: a mechanised logic of computation, Lecture Notes in Computer Science, Vol. 78, Springer-Verlag, Berlin, 1979.
[58] M. Kaufmann and J. Moore, An industrial strength theorem prover for a logic based on common lisp, *IEEE Trans. Software Eng.*, 23, April 1997, pp. 203–213.

[59] S. Owre, J. Rushby, and N. Shankar, PVS: a prototype verification system, in *Proceedings of 11th International Conference on Automated Deduction*, D. Kapur, Ed., Lecture Notes in Artificial Intelligence, Vol. 607, Springer, Berlin, 1992, pp. 748–752.
[60] M. J. C. Gordon and T. F. Melham, Eds., *Introduction to HOL: A Theorem Proving Environment for Higher Order Logic*, Cambridge University Press, London, 1993.
[61] G. Nelson and D. C. Oppen, Simplification by cooperating decision procedures, *ACM Trans. Programming Lang. Syst.*, 1, 245—257, 1979.
[62] J. Joyce and C.-J. H. Seger, The HOL-Voss system: model-checking inside a general-purpose theorem-prover, *Proceedings of the 6th International Workshop on Higher Order Logic Theorem Proving and its Applications*, 1993, pp. 185–198.
[63] M. Aagaard, R. B. Jones, C.-Johan H. Seger: Lifted-FL: a pragmatic implementation of combined model checking and theorem proving, in *Theorem Proving in Higher Order Logics (TPHOLs)*, Y. Bertot, G. Dowek, and A. Hirschowitz, Eds., Springer, Berlin, 1999, pp. 323–340.
[64] K.L. McMillan, A methodology for hardware verification using compositional model checking, *Sci. Comput. Programming*, 37, 279–309, 2000.
[65] A. Pnueli, In transition from global to modular temporal reasoning about programs, in *Logics and Models of Concurrent Systems, sub-series F: Computer and System Science*, K.R. Apt, Ed., Springer-Verlag, Berlin, 1985, pp. 123–144.
[66] P. Cousot and R. Cousot, Abstract interpretation: a unified lattice model for static analysis of programs by construction and approximation of fixpoints, in *Proceedings of 4th ACM Symposium on Principles of Programming Languages*, 1977, pp. 238–252.
[67] S. Graf and H. Saidi, Construction of abstract state graphs with PVS, *CAV97: Computer-Aided Verification*, Lecture Notes in Computer Science, Vol. 1254, Springer, Berlin, 1997, pp. 72–83.
[68] T. A. Henzinger, R. Jhala, R. Majumdar, and G. Sutre, Lazy abstraction, *Proceedings of the 29th Annual Symposium on Principles of Programming Languages (POPL)*, ACM Press, 2002, pp. 58–70.
[69] Y. Yu, P. Manolios, and L. Lamport, Model checking TLA+ specifications, *Proceedings of Correct Hardware Design and Verification Methods.*
[70] J. Corbett, M. Dwyer, J. Hatcliff, S. Laubach, C. Pasareanu, Robby, and H. Zheng, Bandera: extraction finite-state models from Java source code, *Proceedings of 22nd International Conference on Software Engineering (ICSE2000)*, June 2000.
[71] M. Dwyer, J. Hatcliff, R. Joehanes, S. Laubach, C. Pasareanu, Robby, W. Visser, and H. Zheng, Tool supported program abstraction for finite-state verification, in *ICSE 01: Software Engineering*, 2001.
[72] T. Lev-Ami and M. Sagiv, TVLA: a framework for Kleene-based static analysis, *Proceedings of the 7th International Static Analysis Symposium*, 2000.
[73] G. Holzmann, Logic verification of ANSI-C code with SPIN, *Proceedings of 7th International SPIN Workshop*, K. Havelund Ed., Lecture Notes in Computer Science, Vol. 1885, Springer, Berlin, 2000, pp. 131–147.
[74] D.Y.W. Park, U. Stern, J.U. Skakkebaek, and D.L. Dill, Java model checking, *Proceedings of the First International Workshop on Automated Program Analysis, Testing and Verification*, June 2000.
[75] J.R. Burch, E.M. Clarke, K.L. McMillan, D.L. Dill, and L.J. Hwang, Symbolic model checking: 10^{20} states and beyond, *Proceedings of the Fifth Annual IEEE Symposium on Logic in Computer Science, IEEE Computer Society Press*, Washington, D.C., 1990. pp. 1–33.
[76] M. Davis and H. Putnam, A computing procedure for quantification theory, *J. ACM*, 7, 201–215, 1960.
[77] E.M. Clarke, O. Grumberg, S. Jha, Y. Lu, and H. Veith, Counterexample-guided abstraction refinement, in *Computer Aided Verification, LNCS 1855, 12th International Conference, CAV 2000*, E.A. Emerson and A.P. Sistla, Eds., Springer, Chicago, IL, USA, July 15–19, 2000, pp. 154–169.

SECTION V

TEST

21

Design-For-Test

Bernd Koenemann
Mentor Graphics, Inc.
San Jose, California

21.1 Introduction

Design-for-test or design-for-testability (DFT) is a name for design techniques that add certain testability features to a microelectronic hardware product design. The premise of the added features is that they make it easier to develop and apply manufacturing tests for the designed hardware. The purpose of manufacturing tests is to validate that the product hardware contains no defects that could adversely affect the product's correct functioning.

Tests are applied at several steps in the hardware manufacturing flow and, for certain products, may also be used for hardware maintenance in the customer's environment. The tests generally are driven by test programs that execute in automatic test equipment (ATE) or, in the case of system maintenance, inside the assembled system itself. In addition to finding and indicating the presence of defects (i.e., the test fails), tests may be able to log diagnostic information about the nature of the encountered test fails. The diagnostic information can be used to locate the source of the failure.

Design-for-test plays an important role in the development of test programs and as an interface for test application and diagnostics.

Historically speaking, DFT techniques have been used at least since the early days of electric/electronic data-processing equipment. Early examples from the 1940s and 1950s are the switches and instruments that allowed an engineer to "scan" (i.e., selectively) probe the voltage/current at some internal nodes in an analog computer (analog scan). Design-for-test is often associated with design modifications that provide improved access to internal circuit elements such that the local internal state can be controlled (controllability) or observed (observability) more easily. The design modifications can be strictly physical in nature (e.g., adding a physical probe point to a net) or add active circuit elements to facilitate controllability/observability (e.g., inserting a multiplexer into a net). While controllability and observability improvements for internal circuit elements definitely are important for the test, they are not the only type of DFT. Other guidelines, for example, deal with the electromechanical characteristics of the interface between the product under test and the test equipment, e.g., guidelines for the size, shape, and spacing of probe points, or the suggestion to add a high-impedance state to drivers attached to probed nets such that the risk of damage from backdriving is mitigated.

Over the years, the industry has developed and used a large variety of more or less detailed and more or less formal guidelines for desired and/or mandatory DFT circuit modifications. The common understanding of DFT in the context of electronic design automation (EDA) for modern microelectronics is shaped to a large extent by the capabilities of commercial DFT software tools as well as by the expertise and experience of a professional community of DFT engineers researching, developing, and using such tools. Much of the related body of DFT knowledge focuses on digital circuits while DFT for analog/mixed-signal circuits takes something of a backseat. The following text follows this scheme by allocating most of the space to digital techniques.

21.2 The Objectives of Design-For-Test for Microelectronics Products

Design-for-test affects and depends on the methods used for test development, test application, and diagnostics. The objectives, hence, can only be formulated in the context of some understanding of these three key test-related activities.

21.2.1 Test Generation

Most tool-supported DFT practiced in the industry today, at least for digital circuits, is predicated on a "structural" test paradigm. "Functional" testing attempts to validate the circuit under test functions according to its functional specification. For example, does the adder really add? Structural testing, by contrast, makes no direct attempt to ascertain the intended functionality of the circuit under test. Instead, it tries to make sure that the circuit has been assembled correctly from some low-level building blocks as specified in a structural netlist. For example, are all logic gates there that are supposed to be there and are they connected correctly? The stipulation is that if the netlist is correct (e.g., somehow it has been fully verified against the functional specification) and structural testing has confirmed the correct assembly of the structural circuit elements, then the circuit should be functioning correctly.

One benefit of the structural paradigm is that the test generation can focus on testing a limited number of relatively simple circuit elements rather than having to deal with an exponentially exploding multiplicity of functional states and state transitions. While the task of testing a single logic gate at a time sounds simple, there is an obstacle to overcome. For today's highly complex designs, most gates are deeply embedded whereas the test equipment is only connected to the primary I/Os and/or some physical test points. The embedded gates, hence, must be manipulated through intervening layers of logic. If the intervening logic contains state elements, then the issue of an exponentially exploding state space and state transition sequencing creates an unsolvable problem for test generation. To simplify test generation, DFT addresses the accessibility problem by removing the need for complicated state transition sequences when trying to control or observe what is happening at some internal circuit element.

Depending on the DFT choices made during circuit design/implementation, the generation of structural tests for complex logic circuits can be more or less automated. One key objective of DFT methodologies, hence, is to allow designers to make trade-offs between the amount and type of DFT and the cost/benefit (time, effort, quality) of the test generation task.

21.2.1.1 Test Application

Complex microelectronic products typically are tested multiple times. Chips, for example may be tested on the wafer before the wafer is diced into individual chips (wafer probe/sort), and again after being packaged (final test). More testing is due after the packaged chips have been assembled into a higher level package such as a printed circuit board (PCB) or a multi-chip module (MCM). For products with special reliability needs, additional intermediate steps such as burn-in may be involved which, depending on the flow details, may include even more testing (e.g., pre- and postburn-in test, or *in situ* test during burn-in). The cost of test in many cases is dominated by the test equipment cost, which in turn depends on the number of I/Os that need to be contacted, the performance characteristics of the tester I/O (i.e., channel) electronics, and the depth/speed of pattern memory behind each tester channel. In addition to the cost of the tester frame itself, interfacing hardware (e.g., wafer probes and prober stations for wafer sort, automated pick-and-place handlers for final test, burn-in-boards (BIBs) for burn-in, etc.) is needed that connects the tester channels to the circuit under test.

One challenge for the industry is keeping up with the rapid advances in chip technology (I/O count/size/placement/spacing, I/O speed, internal circuit count/speed/power, thermal control, etc.) without being forced to upgrade continually the test equipment. Modern DFT techniques, hence, have to offer options that allow next generation chips and assemblies to be tested on existing test equipment and/or reduce the requirements/cost for new test equipment. At the same time, DFT has to make sure that test times stay within certain bounds dictated by the cost target for the products under test.

21.2.2 Diagnostics

Especially for advanced semiconductor technologies, it is expected that some of the chips on each manufactured wafer will contain defects that render them nonfunctional. The primary objective of testing is to find and separate those nonfunctional chips from the fully functional ones, meaning that one or more responses captured by the tester from a nonfunctional chip under test differ from the expected response. The percentage of chips that fail the test, hence, should be closely related to the expected functional yield for that chip type. In reality, however, it is not uncommon that all chips of a new chip type arriving at the test floor for the first time fail (so-called zero-yield situation). In that case, the chips have to go through a debug process that tries to identify the reason for the zero-yield situation. In other cases, the test fall out (percentage of test fails) may be higher than expected/acceptable or fluctuate suddenly. Again, the chips have to be subjected to an analysis process to identify the reason for the excessive test fall out.

In both cases, vital information about the nature of the underlying problem may be hidden in the way the chips fail during test. To facilitate better analysis, additional fail information beyond a simple pass/fail is collected into a fail log. The fail log typically contains information about when (e.g., tester cycle), where (e.g., at what tester channel), and how (e.g., logic value) the test failed. Diagnostics attempt to derive from the fail log the logical/physical location inside the chip at which the problem most likely started. This location provides a starting point for further detailed failure analysis (FA) to determine the actual root cause. Failure analysis, in particular physical FA (PFA), can be very time consuming and costly, since it typically involves a variety of highly specialized equipment and an equally specialized FA engineering team. The throughput of the FA labs is very limited, especially if the initial problem localization from diagnostics is poor. That adversely affects the problem turnaround time and how many problem cases can be analyzed. Additional inefficiency arises if the cases handed over to the FA lab are not relevant for the tester fall out rate.

In some cases (e.g., PCBs, MCMs, embedded, or stand-alone memories), it may be possible to repair a failing circuit under test. For that purpose, diagnostics must quickly find the failing unit and create a work-order for repairing/replacing the failing unit. For PCBs/MCMs, the replaceable/repairable units are the chips and/or the package wiring. Repairable memories offer spare rows/columns and some switching logic that can substitute a spare for a failing row/column. The diagnostic resolution must match the granularity of replacement/repair. Speed of diagnostics for replacement is another issue. For example, cost reasons may dictate that repairable memories must be tested, diagnosed, repaired, and retested in a single test insertion. In that scenario, the fail data collection and diagnostics must be more or less real-time as the test is applied. Even if diagnostics are to be performed offline, failure data collection on expensive production test equipment must be efficient and fast or it will be too expensive.

Design-for-test approaches can be more or less diagnostics-friendly. The related objectives of DFT are to facilitate/simplify fail data collection and diagnostics to an extent that can enable intelligent FA sample selection, as well as improve the cost, accuracy, speed, and throughput of diagnostics and FA.

21.2.3 Product Life-Cycle Considerations

Test requirements from other stages of a chip's product life cycle (e.g., Burn-In, PCB/MCM test, [sub]system test, etc.) can benefit from additional DFT features beyond what is needed for the chip manufacturing test proper. Many of these additional DFT features are best implemented at the chip level and affect the chip design. Hence, it is useful to summarize some of these additional requirements, even though the handbook primarily focuses on EDA for IC design.

21.2.3.1 Burn-In

Burn-in exposes the chips to some period of elevated ambient temperature to accelerate and weed out early life fails prior to shipping the chips. Typically, burn-in is applied to packaged chips. Chips designated for direct chip attach assembly may have to be packaged into a temporary chip carrier for burn-in and subsequently be removed again from the carrier. The packaged chips are put on BIBs, and several BIBs at a time are put into a burn-in oven. In static burn-in the chips are simply exposed to an elevated temperature, then removed from the oven and retested. Burn-in is more effective if the circuit elements on the chips are subjected to local electric fields during burn-in. Consequently, at a minimum, some chip power pads must be connected to a power supply grid on the BiBs. The so-called dynamic burn-in further requires some switching activity, typically meaning that some chip inputs must be connected on the BiBs and be wired out of the burn-in oven to some form of test equipment. The most effective form of burn-in, called *in situ* burn-in, further requires that some chip responses can be monitored for failures while in the oven. For both dynamic and *in situ* burn-in, the number of signals that must be wired out of the oven is of concern because it drives the complexity/cost of the BiBs and test equipment.

Burn-in friendly chip DFT makes it possible to establish a chip burn-in mode with minimal I/O footprint and data bandwidth needs.

21.2.3.2 Printed Circuit Board/Multi-Chip Module Test

The rigors (handling, placing, heating, etc.) of assembling multiple chips into a higher-level package can create new assembly-related defects associated with the chip attach (e.g., poor solder connection) and interchip wiring (e.g., short caused by solder splash). In some cases, the chip internals may also be affected (e.g., bare chips for direct chip attach are more vulnerable than packaged chips). The basic PCB/MCM test approaches concentrate largely on assembly-related defects and at best use very simple tests to validate that the chips are still "alive."

Although functional testing of PCBs/MCMs from the edge connectors is sometimes possible and used, the approach tends to make diagnostics very difficult. In-circuit testing is a widely practiced alternative or complementary method. In-circuit testing historically has used so-called bed-of-nails interfaces to contact physical probe points connected to the interchip wiring nets. If every net connected to the chip is contacted by a nail, then the tester can essentially test the chip as if stand-alone. However, it often is difficult to prevent some other chip driving the same net that the tester needs to control for testing the currently selected chip. To overcome this problem, the in-circuit circuit tester drivers are strong enough to override (backdrive) other chip drivers. Backdriving is considered a possible danger, and reliability problem for some types of chip drivers and some manufacturers may discourage backdriving. Densely packed, double-sided PCBs or other miniaturized packages may not leave room for landing pads on enough nets, and the number and density of nets to be probed may make the bed-of-nail fixtures too unwieldy.

Design-for-test techniques implemented at the chip level can remove the need for backdriving from a physical bed-of-nails fixture or use electronic alternatives to reduce the need for complete physical in-circuit access.

21.2.3.3 (Sub)System Support

Early proto-type bring-up, and in the case of problems, debug, pose a substantial challenge in the development of complex microelectronics systems. It often is very difficult to distinguish between hardware, design, and software problems. Debug is further complicated by the fact that valuable information about the detailed circuit states that could shed light on the problem may be hidden deep inside the chips in the assembly hierarchy. Moreover, the existence of one problem (e.g., a timing problem) can prevent the system from reaching a state needed for other parts of system bring-up, verification, and debug.

System manufacturing, just like PCB/MCM assembly, can introduce new defects and possibly damage the components. The same may apply to hardware maintenance/repair events (e.g., hot-plugging a new memory board).

Operating the system at the final customer's site can create additional requirements, especially if the system must meet stringent availability or safety criteria.

Design-for-test techniques implemented at the chip level can help enable a structural hardware integrity test that quickly and easily validates the physical assembly hierarchy (e.g., chip to board to backplane, etc.), and that the system's components (e.g., chips) are operating. DFT can also increase the observability of internal circuit state information for debug, or the controllability of internal states and certain operating conditions to continue debug in the presence of problems.

21.3 Overview of Chip-Level Design-For-Test Techniques

Design-for-test has a long history with a large supporting body of theoretical work as well as industrial application. Only a relatively small and narrow subset of the full body of DFT technology has found its way into the current EDA industry.

21.3.1 Brief Historical Commentary

Much of the DFT technology available in today's commercial DFT tools has its roots in the electronic data-processing industry. Data-processing systems have been complex composites made up of logic, memory, I/O, analog, human interface, and mechanical components long before the semiconductor industry invented the system-on-chip (SoC) moniker. Traditional DFT as practiced by the large data-processing system companies since at least the 1960s represents highly sophisticated architectures of engineering utilities that simultaneously address the needs of manufacturing, product engineering, maintenance/service, availability, and customer support. The first commercial DFT tools for IC design fielded by the EDA industry, by contrast, were primitive scan insertion tools that only addressed the needs of automatic test pattern generation (ATPG) for random logic. Tools have become more sophisticated and comprehensive, offering more options for logic test (e.g., built-in self-test [BIST]), support for some nonrandom logic design elements (e.g., BIST for embedded memories), and support for higher-level package testing (e.g., boundary scan for PCB/MCM testing). However, with few exceptions, there still is a lack of comprehensive DFT architectures for integrating the bits and pieces, and a lack of consideration for applications besides manufacturing test (e.g., support for nondestructive memory read for debug purposes is not a common offering by the tool vendors).

21.3.2 About Design-For-Test Tools

There are essentially three types of DFT-related tools:

- *Design-for-test synthesis (DFTS).* Design-for-test involves design modification/edit steps (e.g., substituting one flip-flop type with another one) akin to simple logic transformation or synthesis. DFTS performs the circuit modification/edit task.
- *Design rules checking (DRC).* Chip-level DFT is mostly used to prepare the circuit for some ATPG tool or to enable the use of some type of manufacturing test equipment. The ATPG tools and test equipment generally impose constraints on the design under test. DRC checks the augmented design for compliance with those constraints. Note that this DRC should not be confused with physical verification design rule checking (also called DRC) of ICs, as is discussed in the chapter on 'Design Rule Checking' in this handbook.
- *Design-for-test intellectual property (DFT IP) creation, configuration, and assembly.* In addition to relatively simple design modifications, DFT may add test-specific function blocks to the design.

Some of these DFT blocks can be quite sophisticated and may rival some third-party IP in complexity. And, like other IP blocks, the DFT blocks often must be configured for a particular design and then be assembled into the design.

21.3.3 Chip Design Elements and Element-Specific Test Methods

Modern chips can contain different types of circuitry with vastly different type-specific testing needs. Tests for random logic and tests for analog macros are very different, for example. DFT has to address the specific needs of each such circuit type and also facilitate the integration of the resulting type-specific tests into an efficient, high-quality composite test program for all pieces of the chip. System-on-chip is an industry moniker for chips made up of logic, memory, analog/mixed-signal, and I/O components. The main categories of DFT methods needed, and to a reasonable extent commercially available, for today's IC manufacturing test purposes can be introduced in the context of a hypothetical SoC chip.

Systems-on-chip are multiterrain devices consisting of predesigned and custom design elements:

- Digital logic, synthesized (e.g., cell-based) or custom (e.g., transistor-level)
- Embedded digital cores (e.g., processors)
- Embedded memories (SRAM, eDRAM, ROM, CAM, Flash, with or without embedded redundancy)
- Embedded register files (large number, single port, and multiport)
- Embedded Field Programmable Gate Array (eFPGA)
- Embedded Analog/Mixed-Signal (PLL/DLL, DAC, ADC)
- High-Speed I/Os (e.g., SerDes)
- Conventional I/Os (large number, different types, some differential)

The following overview will introduce some key type-specific DFT features for each type of component, and then address chip-level DFT techniques that facilitate the integration of the components into a top-level design.

21.3.3.1 Digital Logic

The most common DFT strategies for digital logic help prepare the design for ATPG tools. ATPG tools typically have difficulties with hard-to-control or hard-to-observe nets/pins, sequential depth, and loops.

21.3.3.1.1 Control/Observe Points.

The job of an ATPG tool is to locally set up suitable input conditions that *excite* a fault (i.e., trigger an incorrect logic response according to the fault definition; for example, to trigger a stuck-at-1 fault at a particular logic gate input, that input must receive a logic 0 from the preceding gates), and that *propagate* the incorrect value to an observable point (i.e., side-inputs of gates along the way must be set to their noncontrolling values). The run-time and success rate of test generation depend not least on the search space the algorithm has to explore to establish the required excitation and propagation conditions.

Control points provide an alternative means for the ATPG tool to more easily achieve a particular logic value. In addition to providing enhanced controllability for test, it must be possible to disable the additional logic such that the original circuit function is retained for normal system operation. In other words, DFT often means the implementation of multiple distinct modes of operation, for example, a *test mode* and a *normal mode.*

The second control point type is mostly used to override unknown/unpredictable signal sources, in particular for signal types that impact the sequential behavior, e.g., clocks. In addition to the two types of control points, there are other types for improved 1-controllability (e.g., using an OR gate) and for "randomization" (e.g., using an XOR gate). The latter type, for example, is useful in conjunction with pseudo-random test methods that will be introduced later. As can be seen from the examples, the implementation of control points tends to add cost due to one or more additional logic levels that affect the path delay and require additional area/power for transistors and wiring. The additional cost for implementing DFT is generally referred to as "overhead," and over the years there have been many, sometimes heated, debates juxtaposing the overhead against the benefits of DFT.

Observe points are somewhat "cheaper" in that they generally do not require additional logic in the system paths. The delay impact, hence, is reduced to the additional load posed by the fan-out and (optional) buffer used to build an observe.

21.3.3.1.2 Scan Design.

Scan design is the most common DFT method associated with synthesized logic. The concept of scan goes back to the very early days of the electronics industry, and it refers to certain means for controlling or observing otherwise hidden internal circuit states. Examples are manual dials to connect measurement instruments to probe points in analog computers, the switches, and lights on the control panel of early digital computers (and futuristic computers in Sci-Fi flicks), and to more automated electronic mechanisms to accomplish the objective, for example the use of machine instructions to write or read internal machine registers. Beginning with the late 1960s or so, scan has been implemented as a dedicated, hardware-based operation that is independent of, and does not rely on, specific intelligence in the intended circuit function.

21.3.3.1.2.1 Implementing the Scan Function. Among the key characteristics of scan architectures are the choice of which circuit states to control/observe, and the choice of an external data interface (I/Os and protocol) for the control/observe information. In basic scan methods, all (full scan) or some (partial scan) internal sequential state elements (latches or flip-flops) are made controllable and observable via a serial interface to minimize the I/O footprint required for the control/observe data. The most common implementation strategy is to replace the functional state elements with dual-purpose state elements (scan cells) that can operate as originally intended for functional purposes and as a serial shift register for scan. The most commonly used type of scan cell consists of an edge-triggered flip-flop with two-way multiplexer (scan mux) for the data input (mux-scan flip-flop).

The scan mux is typically controlled by a single control signal called scan_enable that selects between a scan-data and a system-data input port. The transport of control/observe data from/to the test equipment is achieved by a serial shift operation. To that effect, the scan cells are connected into serial shift register strings called scan chains. The scan-in port of each cell is either connected to an external input (scan-in) for the first cell in the scan chain or to the output of a single predecessor cell in the scan chain. The output from the last scan cell in the scan chain must be connected to an external output (scan-out). The data input port of the scan mux is connected to the functional logic as needed for the intended circuit function.

There are several commercial and proprietary DFTS tools available that largely automate the scan-chain construction process. These tools operate on register transfer level (RTL) and/or gate-level netlists of the design. The tools typically are driven by some rules on how to substitute nonscan storage elements in the prescan design with an appropriate scan cell, and how to connect the scan cells into one or more scan chains. In addition to connecting the scan and data input ports of the scan cells correctly, attention must be given to the clock input ports of the scan cells. To make the shift-registers operable without interference from the functional logic, a particular circuit state (*scan state*), established by asserting designated scan state values at certain primary inputs and/or by executing a designated initialization sequence, must exist that:

1. switches all scan muxes to the scan side (i.e., the local scan_enable signals are forced to the correct value);
2. assures that each scan cell clock port is controlled from one designated external clock input (i.e., any intervening clock gating or other clock manipulation logic is overridden/disabled);
3. all other scan cell control inputs like set/reset are disabled (i.e., the local control inputs at the scan cell are forced to their inactive state);
4. all scan data inputs are sensitized to the output of the respective predecessor scan cell or the respective scan_in port for the first scan cell in the chain (i.e., side-inputs of logic gates along the path are forced to a nondominating value);
5. the output of the last scan_cell in the scan chain is sensitized to its corresponding scan_out port or the side-inputs of logic gates along the path are forced to a nondominating value, such that pulsing the designated external clock (or clocks) once results in shifting the data in the scan chains by exactly one-bit position.

This language may sound pedantic, but the DFTS and DFT DRC tools tend to use even more detailed definitions for what constitutes a valid scan chain and scan state. Only a crisp definition allows the tools to validate the design thoroughly and, if problems are detected, write error messages with enough diagnostic information to help a user find and fix the design error that caused the problem.

21.3.3.1.2.2 About Scan Architectures. Very few modern chips contain just a single scan chain. In fact, it is fairly common to have several selectable scan-chain configurations. The reason is that scan can be used for a number of different purposes. Facilitating manufacturing test for synthesized logic is one purpose. In that case, the scan cells act as serially accessible control and observe points for logic test, and test application essentially follows a protocol such as:

1 Establishing the manufacturing test-scan configuration and associated scan state, and serially load the scan chains with new test input conditions
2 Applying any other primary input conditions required for the test
3 Waiting until the circuit stabilizes, and measure/compare the external test responses at primary outputs;
4 Capturing the internal test responses into the scan cells (typically done by switching scan_enable to the system side of the scan mux and pulsing a clock)
5 Reestablishing the manufacturing test-scan configuration and associated scan state, and serially unload the test responses from the scan chains into the tester for comparison

Steps 1 and 5 in many cases can be overlapped, meaning that while the responses from one test are unloaded through the scan-out pins, new test input data are simultaneously shifted in from the scan-in pins. The serial load/unload operation requires as many clock cycles as there are scan cells in the longest scan chain. Manufacturing test time, and consequently test cost, for scan-based logic tests are typically dominated by the time used for scan load/unload. Hence, to minimize test times and cost, it is preferable to implement as many short, parallel scan chains as possible. The limiting factors are the availability of chip I/Os for scan-in/-out or the availability of test equipment channels suitable for scan. Modern DFT tools can help optimize the number of scan chains and balance their length according to the requirements and constraints of chip-level manufacturing test. Today's scan-insertion flows also tend to include a post-placement *scan reordering* step to reduce the wiring overhead for connecting the scan cells. The currently practiced state of the art generally limits reordering to occur within a scan chain. Research projects have indicated that further improvements are possible by allowing the exchange of scan cells between scan chains. All practical tools tend to give the user some control over partial ordering, keeping subchains untouched, and placing certain scan cells at predetermined offsets in the chains.

In addition to building the scan chains proper, modern DFT tools also can insert and validate *pin-sharing* logic that makes it possible to use functional I/Os as scan-in/-out or scan control pins, thus avoiding the need for additional chip I/Os dedicated to the test. In many practical cases, a single dedicated pin is sufficient to select between normal mode and test mode. All other test control and interface signals are mapped onto the functional I/Os by inserting the appropriate pin-sharing logic.

Besides chip manufacturing test, scan chains often are also used for access to internal circuit states for higher-level assembly (e.g., board-level) testing. In this scenario, it generally is not possible or economically feasible to wire all scan-in/-out pins used for chip testing out to the board connectors. Board-level wiring and connectors are very limited and relatively "expensive." Hence, the I/O footprint dedicated to scan must be kept at a minimum and it is customary to implement another scan configuration in the chips, wherein all scan cells can be loaded/unloaded from a single pair of scan-in/-out pins. This can be done by concatenating the short scan chains used for chip manufacturing test into a single, long scan chain, or by providing some addressing mechanism for selectively connecting one shorter scan chain at a time to the scan-in/-out pair. In either case, a *scan switching network* and associated control signals are required to facilitate the reconfiguration of the scan interface.

In many practical cases, there are more than two scan configurations to support additional engineering applications beyond chip and board-level manufacturing test, for example, debug or system configuration.

The scan architectures originally developed for large data-processing systems in the 1960s and 1970s, for example, were designed to facilitate comprehensive engineering access to all hardware elements for testing, bring-up, maintenance, and diagnostics. The value of comprehensive scan architectures is only now being rediscovered for the complex system-level chips possible with nanometer technologies.

21.3.3.1.3 Timing Considerations and At-Speed Testing.

Timing issues can affect and plague both the scan infrastructure as well as the application of scan-based logic test.

21.3.3.1.3.1 Scan-Chain Timing. The frequently used mux-scan methodology uses edge-triggered flip-flops as storage elements in the scan cells. And the edge-clock is used for both the scan operation and for capturing test responses into the scan cells, making both susceptible to hold-time errors due to clock skew. Clock skew not only exists between multiple *clock domains* but also within each clock domain. The latter tend to be more subtle and easier to overlook. To deal with interdomain issues, the DFT tools have to be aware of the clock domains and clock-domain boundaries. The general rule-of-thumb for scan-chain construction is that each chain should only contain flip-flops from the same clock domain. Also, leading-edge and falling-edge flip-flops should be kept in separate scan chains even if driven from the same clock source. These strict rules of "division" can be relaxed somewhat if the amount of clock skew is small enough to be reliably overcome by inserting a *lock-up latch or flip-flop* between the scan cells. The susceptibility of the scan operation to hold-time problems can further be reduced by increasing the delay between scan cells, for example, by adding buffers to the scan connection between adjacent scan cells.

In practice, it is not at all unusual for the scan operation to fail for newly designed chips. To avoid the likelihood of running into these problems, it is vitally important to perform a very thorough timing verification on the scan mode. In the newer nanometer technologies, signal integrity issues such as static/dynamic IR-drop have to be taken into account in addition to process and circuit variability. A more "radical" approach is to replace the edge-triggered scan clocking with a level-sensitive multiphase clocking approach as, for example, in level-sensitive scan design (LSSD). In this case, the master and slave latches in the scan cells are controlled from two separate clock sources that are pulsed alternately. By increasing the nonoverlap period between the clock phases it is possible to overcome any hold-time problems during scan without the need for lock-up latches or additional intercell delay.

21.3.3.1.3.2 Scan-Based Logic Test Timing Considerations. Clock skew and hold-time issues also affect the reliable data transmission across interdomain boundaries. For example, if the clocks of two interconnected domains are pulsed together for capture, then it may be impossible to predict whether old or new data are captured. If the data-change and clock edge get too close together, the receiving flip-flop could even be forced into metastability. ATPG tools traditionally try to avoid these problems by using a "capture-by-domain" policy in which only one clock domain is allowed to be captured in a test. This is only possible if DFT makes sure that it is indeed possible to issue a capture clock to each clock domain separately (e.g., using a separate test clock input for each domain or by de-gating the clocks of other domains). The "capture-by-domain" policy can adversely affect test time and data volume for designs with many clock domains, by limiting fault detection for each test to a single domain. Some ATPG tools nowadays offer sophisticated multiclock compaction techniques that overcome the "capture-by-domain" limitation (e.g., if clock-domain analysis shows that there is no connection between certain domains, then capture clocks can be sent to all of those domains without creating potential hold-time issues; if two domains are connected, then their capture clocks can be issued sequentially with enough pulse separation to assure predictability of the interface states). Special treatment of the boundary flip-flops between domains in DFT is an alternative method.

21.3.3.1.3.3 Scan-Based At-Speed Testing. In static, fully complementary CMOS logic there is no direct path from power to ground except when switching. If a circuit is allowed to stabilize and settle down from all transitions, a very low power should be seen in the stable (quiescent) state. That expectation is the basis of IDDq (quiescent power) testing. Certain defects, for example shorts, can create power-ground paths and therefore be detectable by an abnormal amount of quiescent power. For many years, low-speed stuck-at testing combined with IDDq testing have been sufficient to achieve reasonable quality levels for many CMOS designs. The normal quiescent background current unfortunately increases with each new technology generation, which reduces the signal-to-noise ratio of IDDq measurements. Furthermore, modern process technologies are increasingly susceptible to interconnect opens and resistive problems that are less easily detectable with IDDq to begin with. Many of these defects cause additional circuit delays and cannot be tested with low-speed stuck-at tests. Consequently, there is an increasing demand for at-speed delay fault testing.

In scan-based delay testing, the circuit is first initialized by a scan operation. Then a rapid sequence of successive input events is applied at tight timings to create transitions at flip-flop outputs in the circuit, have them propagate through the logic, and capture the responses into receiving flip-flops. The responses finally are unloaded by another scan operation for comparison. Signal transitions at flip-flop outputs are obtained by loading the initial value into the flip-flop, placing the opposite final value at the flip-flop's data input, and pulsing the clock. In the case of mux-scan, the final value for the transition can come from the scan side (*release from scan*) or the system side (*release from capture*) of the scan mux, depending on the state of the scan-enable signal. The functional logic typically is connected to the system side of the scan mux. The transitions will, hence, generally arrive at the system side of the receiving flip-flop's scan mux, such that the scan-enable must select the system side to enable capture. The release from scan method, therefore, requires switching the scan-enable from the scan side to the system side early enough to meet setup time at the receiving flip-flop, but late enough to avoid hold-time issues at the releasing flip-flop. In other words, the scan-enable signal is subject to a two-sided timing constraint and accordingly must be treated as a timing-sensitive signal for synthesis, placement, and wiring. Moreover, it must be possible to synchronize the scan-enable appropriately to the clocks of each clock domain. To overcome latency and synchronization issues with high fan-out scan-enables in high-speed logic, the scan-enables sometimes are pipelined. The DFT scan insertion tools must be able to construct viable pipelined or non-pipelined scan-enable trees and generate the appropriate timing constraints and assertions for physical synthesis and timing verification.

Most ATPG tools do not have access to timing information and, in order to generate predictable results, tend to assume that the offset between release and capture clocks is sufficient to avoid completely setup time violations at the receiving flip-flops. That means the minimal offset between the release and capture clock pulses are dominated by the longest signal propagation path between the releasing and receiving flip-flops. If slow maintenance paths, multicycle paths, or paths from other slower clock domains get mixed in with the "normal" paths of a particular target clock domain, then it may be impossible to test the target domain paths at their native speed. To overcome this problem, some design projects disallow multicycle paths and insist that all paths (including maintenance paths) that can be active during the test of a target domain, must fit into the single-cycle timing window of that target domain. Another approach is to add enough timing capabilities to the ATPG software to identify all potential setup and hold-time violations at a desired test timing. The ATPG tool can then avoid sending transitions through problem paths (e.g., holding the path inputs stable) or set the state of all problem flip-flops to "unknown". Yet other approaches use DFT techniques to separate multicycle paths out into what looks like another clock domain running at a lower frequency.

21.3.3.1.3.4 Power and Noise Considerations. Most scan-based at-speed test methods perform the scan-load/-unload operations at a relatively low frequency, not the least to reduce power consumption during shift. ATPG patterns tend to have close to 50% switching probability at the flip-flop outputs during scan, which in some cases can be ten times more than what is expected during normal operation. Such abnormally high switching activity can cause thermal problems, excessive static IR drop, or exceed the tester power-supply capabilities when the scan chains are shifted at full system speed. The reality of pin electronic capabilities in affordable test equipment sets another practical limit to the data rate at which the scan interface can be operated. One advantage of lower scan speeds for design is that it is not necessary to design the scan chains and scan-chain interface logic for full system speed. That can help reduce the placement, wiring, and timing constraints for the scan logic.

Slowing down the scan rate helps reduce average power, but does little for dynamic power (di/dt). In circuits with a large number of flip-flops, in particular when combined with tightly controlled low-skew clocks, simultaneous switching of flip-flop outputs can result in unexpected power/noise spikes and dynamic IR drops, leading to possible scan-chain malfunctions. These effects may need to be considered when allocating hold-time margins during scan-chain construction and verification.

21.3.3.1.3.5 On-Chip Clock Generation. The effectiveness of at-speed tests not only depends on the construction of proper test event sequences but may also critically depend on being able to deliver these sequences at higher speed with higher accuracy than supported by the test equipment. Many modern chips use on-chip clock frequency multiplication (e.g., using phase-locked loops [PLLs]) and phase alignment, and there is an increasing interest in taking advantage of this on-chip clocking infrastructure for testing. To that effect, clock system designers add programmable test waveform generation features to the clock circuitry. Programmable in this context tends to mean the provision of a serially loadable control register that determines the details of which and how many clock phases are generated. The actual sequence generation is triggered by a (possibly asynchronous) start signal that can be issued from the tester (or some other internal controller source) after the scan load operation has completed. The clock generator will then produce a deterministic sequence of edges that are synchronized to the PLL output. Some high-performance designs may include additional features (e.g., programmable delay lines) for manipulating the relative edge positions over and above what is possible by simply changing the frequency of the PLL input clock.

The ATPG tools generally cannot deal with the complex clock generation circuitry. Therefore, the on-product clock generation (OPCG) logic is combined into an OPCG macro and separated from the rest of the chip by cut points. The cut points look like external clock pins to the ATPG tool. It is the responsibility of the user to specify a list of available OPCG programming codes and the resulting test sequences to the ATPG tool, which in turn is constrained to use only the thus specified event sequences. For verification, the OPCG macro is simulated for all specified programming codes, and the simulated sequences appearing at the cut points are compared to the input sequences specified to the ATPG tool.

Some circuit elements may require finer-grained timing that requires a different approach. One example is clock jitter measurement, another memory access time measurement. Test equipment uses analog or digitally controlled, tightly calibrated delay lines (timing verniers), and it is possible to integrate similar features into the chip. A different method for measuring arbitrary delays is to switch the to-be-measured delay path into an inverting recirculating loop and measure the oscillation frequency (e.g., counting oscillations against a timing reference). Small delays that would result in extremely high frequencies are made easier to test by switching them into and out of a longer delay path and comparing the resulting frequencies. Oscillation techniques can also be used for delay calibration to counteract performance variations due to process variability.

21.3.3.1.4 Custom Logic.

Custom transistor-level logic, often used for the most performance-sensitive parts of a design, poses a number of unique challenges to the DFT flow and successful test generation. Cell-based designs tend to use more conservative design practices and the libraries for cell-based designs generally come with premade and preverified gate-level test generation models. For transistor-level custom designs, by contrast, the gate-level test generation models must somehow be generated and verified "after-the-fact" from the transistor-level schematics. Although the commercial ATPG tools may have some limited transistor-level modeling capabilities, their wholesale use generally leads to severe tool run-time problems, and hence is strongly discouraged. The construction of suitable gate-level models is complicated by the fact that custom logic often uses dynamic logic or other performance/area/power-driven unique design styles, and it is not always easy to determine what should be explicitly modeled and what should be implied in the model (e.g., the precharge clocks and precharge circuits for dynamic logic). Another issue for defect coverage as well as diagnostics is to decide which circuit-level nets to explicitly keep in the logic model vs. simplifying the logic model for model size and tool performance.

The often-extreme area, delay, and power sensitivity of custom-designed structures are often met with a partial scan approach, even if the synthesized logic modules have full scan. The custom design team has to make a trade-off between the negative impact on test coverage and tool run-time vs. keeping the circuit overhead small. Typical "rules-of-thumb" will limit the number of nonscannable levels and discourage feed-back loops between nonscannable storage elements (e.g., feed-forward pipeline stages in data paths are often good candidates for partial scan). Area, delay, and power considerations also lead to a more widespread use of pass-gates and other three-state logic (e.g., three-state buses) in custom-design circuitry. The control inputs of pass-gate structures, such as the select lines of multiplexers, and enables of three-state bus drivers tend to require specific decodes for control (e.g., "one-hot") to avoid three-state contention (i.e., establishing a direct path from power to ground) that could result in circuit damage due to burn-out. To avoid potential burn-out, DFT and ATPG must cooperate to assure that safe control states are maintained during scan and during test. If the control flip-flops are included in the scan chains, then DFT hardware may be needed to protect the circuits (e.g., all bus-drivers are disabled during scan).

Other areas of difficulty are complex memory substructures with limited scan and possibly unusual or pipelined decodes that are not easily modeled with the built-in memory primitives available in the DFT/ATPG tools.

21.3.3.1.5 Logic Built-In Self-Test.

Chip manufacturing test (e.g., from ATPG) typically assumes full access to the chip I/Os plus certain test equipment features for successful test application. That makes it virtually impossible to port chip manufacturing tests to higher-level assemblies and into the field. Large data-processing systems historically stored test data specifically generated for in-system testing on disk and applied them to the main processor complex through a serial maintenance interface from a dedicated service processor that was delivered as part of the system. Over the years it became too cumbersome to store and manage vast amounts of test data for all possible system configurations and ECO (engineering change order) levels, and the serial maintenance interface became too slow for efficient data transfer. Hence, alternatives were pursued that avoid the large data volume and data transfer bottleneck associated with traditional ATPG tests.

21.3.3.1.5.1 Pseudo-Random Pattern Testing. The most widely used alternative today is logic built-in self-test (logic BIST) using pseudo-random patterns. It is known from coding theory that pseudo-random patterns can be generated easily and efficiently in hardware, typically using a so-called pseudo-random pattern generator (PRPG) macro utilizing a linear feedback shift register (LFSR). The PRPG is initialized to a starting state called PRPG seed and in response to subsequent clock pulses, produces a state sequence that meets certain tests of randomness. However, the sequence is not truly random. In particular, it is predictable and repeatable if started from the same seed. The resulting pseudo-random logic states are loaded into scan chains for testing in lieu of ATPG data. The PRPGs nowadays are generally built into the chips. That makes it possible to use a large number of relatively short on-chip scan chains because the test data do not need to be brought in from the outside through a narrow maintenance interface or a limited number of chip I/Os. Simple LFSR-based PRPGs connected to multiple parallel scan chains result in undesirable, strong value correlations (*structural dependencies*) between scan cells in adjacent scan chains. The correlations can reduce the achievable test coverage. To overcome this potential problem, some PRPG implementations are based on *cellular automata* rather than LFSRs. LFSR-based implementations use a *phase-shifting network* constructed out of exclusive-OR (XOR) to eliminate the structural dependencies. The phase-shifting network can be extended into a *spreading network* that makes it possible to drive a larger number of scan-chain inputs from a relatively compact LFSR.

More scan chains generally mean shorter scan chains that require fewer clock cycles for load/unload, thus speeding up test application (each scan test requires at least one scan load/unload), assuming that the DFTS tool used for scan insertion succeeds in reducing and balancing the length of the scan chains. Modern DFTS tools are capable of building scan architectures with multiple separately balanced selectable chain configurations to support different test methods on the same chip. It should be noted that the achievable chain length reduction can be limited by the length of preconnected scan-chain segments in hard macros. Hence, for logic BIST it is important to assure that large hard macros are preconfigured with several shorter chain segments rather than a single long segment. As a rule-of-thumb, the maximum chain length for logic BIST should not exceed 500 to 1000 scan cells. That means a large chip with 1M scan cells requires 1K scan chains or more. The overhead for the PRPG hardware is essentially proportional to the number of chains (a relatively constant number of gates per scan chain for the LFSR/CA, phase shifting/spreading network, and scan-switching network).

The flip side of the coin for pseudo-random logic BIST is that pseudo-random patterns are less efficient for fault testing than ATPG-generated, compacted test sets. Hence, ten times as many or more pseudo-random patterns are needed for equivalent nominal fault coverage, offsetting the advantage of shorter chains. Moreover, not all faults are easily tested with pseudo-random patterns. Practical experience with large-scale data-processing systems has indicated that a stuck-at coverage of around up to 95% can be achievable with a reasonable (as dictated by test time) number of pseudo-random test patterns. Going beyond 95% requires too much test application time to be practical. Coverage of 95% can be sufficient for burn-in, higher-level assembly, system, and field testing, but may be unacceptable for chip manufacturing test. Consequently, it is not unusual to see ATPG-based patterns used for chip manufacturing tests and pseudo-random logic BIST for all subsequent tests.

Higher test coverage, approaching that of ATPG, can be achieved with pseudo-random patterns only by making the logic more testable for such patterns. The 50–50 pseudo-random signal probability at the outputs of the scan cells gets modified by the logic gates in the combinational logic between the scan cells. Some internal signal probabilities can be skewed so strongly to 0 or 1 that the effective controllability or observability of downstream logic becomes severely impaired. Moreover, certain faults require many more specific signal values for fault excitation or propagation than achievable with a limited set of pseudo-random patterns (it is like rolling dice and trying to get 100+ sixes in a row). Wide comparators and large counters are typical architectural elements afflicted with that problem. Modern DFT synthesis and analysis tools offer so-called *pseudo-random testability analysis* and *automatic test-point insertion features*. The testability analysis tools use *testability measures* or signal local characteristics captured during good machine fault simulation to identify nets with low pseudo-random controllability or observability for pseudo-random patterns. The test point insertion tools generate a suggested list of control or observe points that should be added to the netlist to improve test coverage. Users generally can control how many and what type of test points are acceptable (control points tend to be more "expensive" in circuit area and delay impact) for manual or automatic insertion. It is not unusual for test points to be "attracted" to timing-critical paths. If the timing-critical nets are known ahead of time, they optionally can be excluded from modification. As a rule-of-thumb, one test point is needed per 1K gates to achieve the 99%+ coverage objective often targeted for chip manufacturing test. Each test point consumes roughly ten gates, meaning that 1% additional logic is required for the test points.

It should be noted that some hard-to-test architectural constructs such as comparators and counters can be identified at the presynthesis RTL, creating a basis for RTL analysis and test point insertion tools.

21.3.3.1.5.2 Test Response Compression. Using pseudo-random patterns to eliminate the need for storing test input data from ATPG solves only part of the data volume problem. The expected test responses for ATPG-based tests must equally be stored in the test equipment for comparison with the actual test responses. As it turns out, LFSR-based hardware macros very similar to a PRPG macro can be used to implement error detecting code (EDC) generators. The most widely used EDC macro implementation for logic BIST is called multiple-input signature register (MISR), which can sample all scan-chain outputs in parallel. As the test responses are unloaded from the scan chains they are simultaneously clocked into the MISR where they are accumulated. The final MISR state after a specified number of scan tests have been applied is called the *signature*, and this signature is compared to an expected signature that has been precalculated by simulation for the corresponding set of test patterns.

The MISR uses only shifting and XOR logic for data accumulation, meaning that each signature bit is the XOR sum of some subset of the accumulated test response bit values. One property of XOR-sums is that even a single unknown or unpredictable summand makes the sum itself unknown or unpredictable. If the signature of a defect-free product under test is unknown or unpredictable, then the signature is useless for testing. Hence, there is a "Golden Rule" for signature-based testing that no unknown or unpredictable circuit state can be allowed to propagate to the MISR. This rule creates additional design requirements over and above what is required for scan-based ATPG. The potential impact is further amplified by the fact that pseudo-random patterns, unlike ATPG patterns, offer little to no ability for intelligently manipulating the test stimulus data. Hence, for logic BIST, the propagation of unknown/unpredictable circuit states (also known as x states) must generally be stopped by hardware means. For example, microprocessor designs tend to contain tens of thousands of three-state nets, and modern ATPG tools have been adapted to that challenge by constructively avoiding three-state contention and floating nets in the generated test patterns. For logic BIST, either test hardware must be added to prevent contention or floating nets, or the outputs of the associated logic must be de-gated (rendering it untestable) so that they cannot affect the signature. Some processor design projects forbid the use of three-state logic, enabling them to use logic BIST. Other design teams find that so radical an approach is entirely unacceptable.

Three-state nets are not the only source of x states. Other sources include unmodeled or incompletely modeled circuit elements, uninitialized storage elements, multiport storage element write conflicts, and set-up/hold-time timing violations. Design-for-test synthesis and DRC tools for logic BIST must analyze the netlist for potential x-state sources and add DFT structures that remove the x-state generation potential (e.g., adding exclusive gating logic to prevent multiport conflicts, or by assuring the proper initialization of storage elements), or add de-gating logic that prevents x-state propagation to the MISR (e.g., at the outputs of uninitialized or insufficiently modeled circuit elements). Likewise, hardware solutions may be required to deal with potential setup/hold-time problems (e.g., clock-domain boundaries, multicycle paths, nonfunctional paths, etc.). High-coverage logic BIST, overall, is considered to be significantly more design-intrusive than ATPG methods where many of the issues can be dealt within pattern generation rather than through design modifications.

21.3.3.1.5.3 At-Speed Testing and Enhanced Defect Coverage with Logic BIST. At-speed testing with logic BIST essentially follows the same scheme as at-speed testing with ATPG patterns. (Historical note: contrary to frequent assertions by logic BIST advocates that pseudo-random pattern logic BIST is needed to enable at-speed testing, ATPG-based at-speed test has been practiced long before logic BIST became popular and is still being used very successfully today.) Most approaches use slow scan (to limit power consumption, among other things) followed by the rapid application of a short burst of at-speed edge events. Just as with ATPG methods, the scan and at-speed edge events can be controlled directly by test equipment or from an OPCG macro. The advantage of using slow scan is that the PRPG/MISR and other scan-switching and interface logic need not be designed for high speed. That simplifies timing closure and gives the placement and wiring tools more flexibility. Placement/wiring consideration may still favor using several smaller, distributed PRPG/MISR macros. Modern DFT synthesis, DRC, fault grading, and signature-simulation tools for logic BIST generally allow for distributed macros.

Certain logic BIST approaches, however, may depend on performing all or some of the scan cycles at full system speed. With this approach, the PRPG/MISR macros and other scan interface logic must be designed to run at full speed. Moreover, scan chains in clock domains with different clock frequencies may be shifted at different frequencies. That affects scan-chain balancing, because scan chains operating at half frequency in this case should only be half as long as chains operating at full frequency. Otherwise they would require twice the time for scan load/unload.

The fact that pseudo-random logic BIST applies ten times (or more) as many tests than ATPG for the same nominal fault coverage can be advantageous for the detection of unmodeled faults and of defects that escape detection by the more compact ATPG test set. This was demonstrated empirically in early pseudo-random test experiments using industrial production chips. However, it is not easy to extrapolate these early results to the much larger chips of today. Today's chips can require 10K or more ATPG patterns. Even ATPG vectors are mostly pseudo-random, meaning that applying the ATPG vectors is essentially equivalent to applying 10K+ logic BIST patterns, which is much more than what was used in the old hardware experiments.

Only recently has the debate over accidental fault/defect coverage been refreshed. The theoretical background for the debate is the different *n-detect* profiles for ATPG and logic BIST. ATPG test sets are optimized to the extent that some faults are only detected by one test (1-detect) in the set. More faults are tested by a few tests and only the remainder of the fault population gets detected ten times (10-detect) or more. For logic BIST tests, by contrast, almost all faults are detected many times. Static bridging fault detection, for example, requires coincidence of a stuck-at fault test for the victim net with the aggressor net being at the fault value. If the stuck-at fault at the victim net is detected only once, then the probability of detecting the bridging fault is determined by the probability of the aggressor net being at the faulty value, e.g., 50% for pseudo-random values. A 2-detect test set would raise the probability to 75% and so on. Hence, multidetection of stuck-at faults increases the likelihood of detecting bridging faults. The trend can be verified by running bridging-fault simulation for stuck-at test sets with different *n*-detect profiles and comparing the results. Recent hardware experiments confirmed the trend for production chips.

It must be noted, however, that modern ATPG tools can and have been adapted to optionally generate test sets with improved *n*-detect profiles. The hardware experiments cited by the BIST advocates in fact were performed with ATPG-generated *n*-detect test sets, not with logic BIST tests. It should also be noted that if the probability of the aggressor net being at the faulty value is low, then even multiple detects may not do enough. ATPG experiments that try to constructively enhance the signal probability distribution have shown some success in that area. Finally, a new generation of tools is emerging that extract realistic bridging faults from the circuit design and layout. ATPG tools can and will generate explicit tests for the extracted faults, and it has been shown that low *n*-detect test sets with explicit clean-up tests for bridging faults can produce very compact test sets with equally high or higher bridging fault coverage than high *n*-detect test sets.

21.3.3.1.5.4 Advanced Logic BIST Techniques. Experience with logic BIST on high-performance designs reveals that test points may be helpful to improve nominal test coverage, but can have some side-effects for characterization and performance screening. One reported example shows a particular defect in dynamic logic implementing a wide comparator that can only be tested with certain patterns. Modifying the counter with test points as required for stuck-at and transition fault coverage creates artificial nonfunctional short paths. The logic BIST patterns use artificial paths for coverage and never sensitize the actual critical path. Knowing that wide comparators were used in the design, some simple weighting logic had been added to the PRPG macro to create the almost-all-1s or almost-all-0s patterns suited for testing the comparators. The defect was indeed only detected by the weighted tests and escaped the normal BIST tests. Overall, the simple weighting scheme measurably increased the achievable test coverage without test points.

If logic BIST is intended to be used for characterization and performance screening, it may also be necessary to enable memory access in logic BIST. It is not uncommon that the performance-limiting paths in high-speed design traverse embedded memories. The general recommendation for logic BIST is to fence embedded memories off with boundary scan during BIST. That again creates artificial paths that may not be truly representative of the actual performance-limiting paths. Hardware experience with high-performance processors shows that enabling memory access for some portion of the logic BIST tests does indeed capture unique fails. Experiments with (ATPG-generated) scan tests for processor's performance binning have similarly shown that testing paths through embedded memories is required for better correlation with functional tests.

As a general rule-of-thumb, if BIST is to be used for characterization, then the BIST logic may have to be designed to operate at higher speeds than the functional logic it is trying to characterize.

21.3.3.1.5.5 Logic BIST Diagnostics. Automated diagnosis of production test fails has received considerable attention recently and is considered a key technology for nanometer semiconductor technologies. In traditional stored pattern ATPG testing, each response bit collected from the device under test is immediately compared to an expected value, and most test equipment in the case of a test fail (i.e., mismatch between actual and expected response values) allows for optionally logging the detailed fail information (e.g., tester cycle, tester channel, fail value) into a so-called fail set for the device under test. The fail sets can then be postprocessed by automated logic diagnostic software tools to determine the most likely root cause locations.

In logic BIST, the test equipment normally does not see the detailed bit-level responses because these are intercepted and accumulated into a signature by the on-chip MISR. The test equipment only sees and compares highly compressed information contained in the accumulated signatures. Any difference between an actual signature and the expected signature indicates that the test response must contain some erroneous bits. It generally is impossible to reconstruct bit-level fail sets from the highly compressed information in the signatures. The automated diagnostic software tools, however, need the bit-level fail sets. The diagnosis of logic BIST fails, hence, require an entirely different analysis approach or some means for extracting a bit-level fail set from the device under test. Despite research efforts aimed at finding alternative methods, practitioners tend to depend on the second approach. To that effect, the logic BIST tests are structured such that signatures are compared after each group of n tests, where n can be 1, 32, 256, or some other number. The tests further must be structured such that each group of n tests is independent. In that case, it can be assumed that a signature mismatch can only be caused by bit-level errors in the associated group of n tests. For fail-set extraction, the n tests in the failing group are repeated and this time the responses are directly scanned out to the test equipment without being intercepted by the MISR (*scan dump operation*). Existing production test equipment generally can only log a limited number of failing bits and not the raw responses. Hence, the test equipment must have stored expect data available for comparison. Conceptually, that could be some "fake" expect vector like all 0s, but then even correct response tests would result in failing bits that would quickly exceed the very limited fail buffers on the testers, meaning that the actual expect data must be used. However, the number of logic BIST tests tends to be so high that it is impractical to store all expect vectors in the production test equipment. Bringing the data in from some offline medium would be too slow. The issue can be overcome if it is only desired to diagnose a small sample of failing devices, for example, prior to sending them to the FA lab. In that case, the failing chips can be sent to a nonproduction tester for retesting and fail-set logging.

Emerging, very powerful, statistical yield analysis and yield management methods require the ability to log large numbers of fail sets during production testing, which means fail-set logging must be possible with minimal impact to the production test throughput. That generally is possible and fairly straightforward for ATPG tests (as long as the test data fit into the test equipment in the first place), but may require some logistical ingenuity for logic BIST and other signature-based test methods.

21.3.3.1.6 Test Data Compression.

Scan-based logic tests consume significant amounts of storage and test time on the ATE used for chip manufacturing test. The data volume in first order is roughly proportional to the number of logic gates on the chip and the same holds for the number of scan cells. Practical considerations and test equipment specifications oftentimes limit the number of pins available for scan-in/-out and the maximum scan frequency. Consequently, the scan chains for more complex chips tend to be longer and it takes commensurately longer to load/unload the scan chains. There is a strong desire to keep existing test equipment and minimize expensive upgrades or replacements. Existing equipment on many manufacturing test floors tend to have insufficient vector memory for the newer chip generations. Memory re-load is very time-consuming and should be avoided if possible. The purpose of test data compression is to reduce the memory footprint of the scan-based logic tests such that they comfortably fit into the vector memory of the test equipment.

21.3.3.1.6.1 Overview. Test equipment tends to have at least two types of memory for the data contained in the test program. One type is the vector memory that essentially holds the logic levels for the test inputs and for the expected responses. The memory allocation for each input bit could include additional space for waveform formats (the actual edge timings are kept in a separate time-set memory space) over and above the logic level. On the output side, there typically are at least two bits of storage for each expected response bit. One bit is a mask value that determines if the response value should be compared or ignored (for example, unknown/unpredictable responses are masked) and the other bit defines the logic level to compare with. The other memory type contains nonvector program information like program op-codes for the real-time processing engine in the test equipment's pin electronics.

Some memory optimization for scan-in data is possible by taking advantage of the fact that the input data for all cycles of the scan load/unload operation use the same format. The format can be defined once upfront and only a single bit is necessary to define the logic level for each scan cycle. A test with a single-scan load/unload operation may thus consume three bits of vector memory in the test equipment, and possibly only two bits of no response masking is needed.

Scan-based logic test programs tend to be simple in structure, with one common loop for the scan/load unload operation and short bursts of other test events in between. Consequently, scan-based test programs tend to consume only very little op-code memory and the memory limitation for large complex chips is only in the vector memory.

It is worth noticing that most production test equipment offers programming features such as branching, looping, logic operations, etc., for functional testing. The scan-based ATPG programs, however, do not typically take advantage of these features for two reasons. First, some of the available features are equipment-specific. Second, much of the data volume is for the expected responses. While the ATPG tools may have control over constructing the test input data, it is the product under test, not the ATPG software that shapes the responses. Hence, taking full advantage of programming features for really significant data reduction would first require some method for removing the dependency on storing the full amount of expected responses.

21.3.3.1.6.2 Input Data Compression. Input data compression, in general, works by replacing the bit-for-bit storage of each logic level for each scan cell with some means for algorithmically constructing multiple input values on-the-fly from some compact source data. The algorithms can be implemented in software or hardware on the tester or in software or hardware inside the chips under test. To understand the nature of the algorithms, it is necessary to review some properties of the tests generated by ATPG tools.

Automatic test pattern generation proceeds by selecting one yet untested fault and generating a test for that one fault. To that effect, the ATPG algorithm will determine sufficient input conditions to excite and propagate the selected fault. That generally requires that specific logic levels must be asserted at some scan cells and primary inputs. The remaining scan cells remain unspecified at this step in the algorithm. The thus constructed, partially specified, vector is called a *test cube.* It has become customary to refer to the specified bits in the test cube as "care bits" and to the unspecified bits as "don't care bits." All ATPG tools used in practice perform vector *compaction*, which means that they try to combine the test cubes for as many faults as possible into a single test vector. The methods for performing vector compaction vary but the result generally is the same. Even after compaction, for almost all tests, there are many more don't care bits than care bits. In other words, scan-based tests generated by ATPG tools are characterized by a low *care bit density* (percentage of specified bits in the compacted multifault test cube). After compaction, the remaining unspecified bits will be filled by some arbitrary fill algorithm. Pseudo-random pattern generation is the most common type of algorithm used for fill. That is noteworthy because it means that the majority of bit values in the tests are generated by the ATPG software algorithmically from a very compact seed value. However, neither the seed nor the algorithm is generally included in the final test data. And, without that knowledge, it tends to be impossible to recompact the data after the fact, with any commonly known compression algorithm (e.g., zip, lzw, etc.). The most commonly practiced test data compression methods in use today focus on using simple software/hardware schemes to regenerate algorithmically fill data in the test equipment or in the chip under test. Ideally, very little memory is needed to store the seed value for the fill data and only the care bits must be stored explicitly.

The "cheapest" method for input fill data compression is to utilize common software features in existing test equipment without modifications of the chip under test. Run length encoding (RLE) is one software method used successfully in practice. In this approach, the pseudo-random fill algorithm in ATPG is replaced with a repeat option that simply repeats the last value until a specified bit with the opposite value is encountered. The test program generation software is modified to search for repeating input patterns in the test data from ATPG. If a repeating pattern of sufficient length is found, it will be translated into a single pattern in vector memory, and a repeat op-code in op-code memory, rather than storing everything explicitly in vector memory. Practical experience with RLE shows that care bit density is so low that repeats are possible and the combined op-code plus vector memory can be 10× less than storing everything in vector memory. As a side-effect, tests with repeat fill create less switching activity during scan and, hence, are less power-hungry than their pseudo-random brethren.

Run length encoding can be effective for reducing the memory footprint of scan-based logic tests in test equipment. However, the fully expanded test vectors comprised of care and don't care bits are created in the test equipment and these are the expanded vectors that are sent to the chip(s) under test. The chips have the same number of scan chains and the same scan-chain length as for normal scan testing without RLE. Test vector sets with repeat fill are slightly less compact than sets with pseudo-random fill. Hence, test time suffers slightly with RLE, meaning that RLE is not the right choice if test time reduction is as important as data volume reduction.

Simultaneous data volume and test time reduction is possible with on-chip decompression techniques. In that scenario, a hardware macro for test input data compression is inserted between the scan-in pins of the chip and the inputs of a larger number of shorter scan chains (i.e., there are more scan chains than scan-in pins). The decompression macro can be combinational or sequential in nature, but in most practical implementations it tends to be linear in nature. Linear in this context means that the value loaded into each scan cell is a predictable linear combination (i.e., XOR sum and optional inversion) of some of the compressed input values supplied to the scan-in pins. In the extreme case, each linear combination contains only a single term, which can, for example, be achieved by replication or shifting.

Broadcast scan, where each scan-in pin simply fans-out to several scan chains without further logic transformation, is a particularly simple decompression macro using replication of input values. In broadcast scan, all scan chains connected to the same scan-in pin receive the same value. That creates strong correlation (replication) between the values loaded into the scan cells of those chains. Most ATPG software implementations have the ability to deal directly with correlated scan cell values and will automatically imply the appropriate values in the correlated scan cells. The only new software needed is for DFT synthesis to create automatically the scan fan-out and DFT DRC for analyzing the scan fan-out network and setting up the appropriate scan cell correlation tables for ATPG. The hard value correlations created by broadcast scan can make some faults hard to test and make test compaction more difficult because the correlated values create care bits even if the values are not required for testing the target fault. It must therefore be expected that ATPG with broadcast scan has longer run-times, creates slightly less compact tests, and achieves slightly lower test coverage than ATPG with normal scan. With a scan fan-out ratio of 1:32 (i.e., each scan-in fans out to 32 scan chains), it is possible to achieve an effective data volume and test time reduction of 20× or so. It is assumed that the scan chains can be reasonably well balanced and that there are no hard macros with preconnected scan segments that are too long.

The more sophisticated decompression macros contain XOR gates and the values loaded into the scan cells are a linear combination of input values with more than one term. Industrial ATPG does understand hard correlations between scan cells but not Boolean relationships like linear combinations with more than one term. Hence, the ATPG flow has to be enhanced to deal with the more sophisticated decompression techniques. Each care bit in a test cube creates a linear equation with the care bit value on the one side and an XOR sum of some input values on the other side. These specific XOR sums for each scan cell can be determined upfront by symbolic simulation of the scan load operation. Since a test cube typically has more than one care bit, a system of linear equations is formed, and a linear equation solver is needed to find a solution for the system of equations. In some cases the system of equations has no solution; for example, if the total number of care bits exceeds the number of input values supplied from the tester, which can happen when too many test cubes are compacted into a single test. In summary, similar to broadcast scan, ATPG for the more sophisticated schemes also adds CPU time, possibly reduces test coverage, and increases the number of tests in the test set. The sophisticated techniques require more hardware per scan chain (e.g., one flip-flop plus some other gates) than the simple fan-out in broadcast scan. However, the more sophisticated methods tend to offer more flexibility and should make more optimal compression results possible.

Although the so-called weighted random pattern (WRP) test data compression approach is proprietary and not generally available, it is worth a brief description. The classical WRP does not exploit the low care bit density that makes the other compression methods possible, but a totally different property of scan-based logic tests generated by ATPG. The property is sometimes called *test cube clustering*, meaning that the specified bit values in the test cubes for groups of multiple faults are mostly identical and only very few care bit values are different (i.e., the test cubes in a cluster have a small Hamming distance from each other). That makes it possible to find more compact data representations for describing all tests in a cluster (e.g., a common base vector and a compact difference vector for each cluster member). Many years of practical experience with WRP confirms that input data volume reductions in excess of 10× are possible by appropriately encoding the cluster information.

It has been suggested that two-level compression should be possible by taking advantage of both the cluster effect and the low care bit density of scan tests. Several combined schemes have been proposed, but they are still in the research stage.

21.3.3.1.6.3 Response Data Compression/Compaction. Since the amount of memory needed for expected responses can be more than for test input data, any efficient data compression architecture must include test response data compression (sometimes also called compaction). WRP, the oldest data compression method with heavy-duty practical production use in chip manufacturing test, for example, borrows the signature approach from logic BIST for response compaction. Both the WRP decompression logic and the signature generation logic were provided in proprietary test equipment. The chips themselves were designed for normal scan and the interface between the chips and the test equipment had to accommodate the expanded test input and test response data. The interface for a given piece of test equipment tends to have a relatively fixed width and bandwidth, meaning that the test time will grow with the gate count of the chip under test. The only way to reduce test time in this scenario is to reduce the amount of data that has to go through the interface bottleneck. That can be achieved by inserting a response compression/compaction macro between the scan chains and the scan-out pins of the chip so that only compressed/compacted response data has to cross the interface.

How that can be done with EDCs using MISR macros was already known from and proven in practice by logic BIST, and the only "new" development required was to make the on-chip MISR approach work with ATPG tests with and without on-chip test input data decompression. From a DFT tool perspective, that means adding DFT synthesis and DFT DRC features for adding and checking the on-chip test response compression macro, and adding a fast signature simulation feature to the ATPG tool. The introduction of signatures instead of response data also necessitated the introduction of new data types in the test data from ATPG and corresponding new capabilities in the test program generation software, for example, to quickly reorder tests on the test equipment without having to resimulate the signatures.

One unique feature of the MISR-based response compression method is that it is not necessary to monitor the scan-out pins during the scan load/unload operation (the MISR accumulates the responses on-chip and the signature can be compared after one or more scan load/unload operations are completed). The test equipment channels normally used to monitor the scan-outs can be reallocated for scan-in, meaning that the number of scan-ins and scan chains can be doubled, which reduces test time by a factor of 2 if the scan chains can be rebalanced to be half as long as before. Furthermore, no expected values are needed in the test equipment vector memory for scan, thus reducing the data volume by a factor of 2 or more (the latter comes into play if the test equipment uses a two-bit representation for the expected data and the input data can be reformatted to a one-bit representation).

Mathematically speaking, the data manipulations in an MISR are very similar to those in a linear input data decompression macro. Each bit of the resulting signature is a linear combination (XOR sum) of a subset of test response bit values that were accumulated into the signature. Instead of using a sequential state machine like an MISR to generate the linear combinations, it is also possible to use a combinational XOR network to map the responses from a large number of scan-chain outputs to a smaller number of scan-out pins. Without memory, the combinational network cannot accumulate responses on-chip. Hence, the scan-out pins must be monitored and compared to expected responses for each scan cycle. The data reduction factor is given by the number of internal scan-chain outputs per scan-out pin.

Selective compare uses multiplexing to connect one out of several scan-chain outputs to a single scan-out pin. The selection can be controlled from the test equipment directly (i.e., the select lines are connected to chip input pins) or through some intermediate decoding scheme. Selective compare is unique in that the response value appearing at the currently selected scan-chain output is directly sent to the test equipment without being combined with other response values, meaning that any mismatches between actual and expected responses can be directly logged for analysis. The flip side of the coin is that responses in the currently de-selected scan chains are ignored, which reduces the overall observability of the logic feeding into the de-selected scan cells and could impair the detection of unforeseen defects. It also should be noted that in addition to monitoring the scan-out pins, some input bandwidth is consumed for controlling the selection.

Mapping a larger number of scan-chain outputs onto a smaller number of scan-out pins through a combinational network is sometimes referred to as *space compaction.* A MISR or similar sequential state machine that can accumulate responses over many clock-cycles performs *time compaction.* Of course, it is possible to combine space and time compaction into a single architecture.

21.3.3.1.6.4 X-State Handling. Signatures are known for not being able to handle unknown/unpredictable responses (x states). In signature-based test methods, e.g., logic BIST, it is therefore customary to insist that x-state sources should be avoided or disabled, or if that is not possible then the associated x states must under no circumstances propagate into the signature generation macro.

x-state avoidance can be achieved by DFT circuit modifications (e.g., using logic gates instead of pass-gate structures) or by adjusting the test data such that the responses become predictable (e.g., asserting enable signals such that no three-state conflicts are created and no unterminated nets are left floating). The latter is in many cases possible with ATPG patterns; however, that typically increases ATPG run-time and can adversely affect test coverage. Design modifications may be considered too intrusive, especially for high-performance designs, which has hampered the widespread acceptance of logic BIST.

Disabling x-state propagation can be achieved by local circuit modifications near the x-state sources or by implementing a general-purpose *response-masking* scheme. Response masking typically consists of a small amount of logic that is added to the inputs of the signature generation macro (e.g., MISR). This logic makes it possible to selectively de-gate (i.e., force a known value) onto the MISR input(s) that could carry x states. A relatively simple implementation, for example, consists of a serially loadable *mask vector* register with one mask bit for each MISR input. The value loaded into the mask bit determines whether the response data from the associated scan-chain output are passed through to the MISR or are de-gated. The mask vector is serially preloaded prior to scan load/unload and could potentially be changed by reloading during scan load/unload. A single control signal directly controlled from the test equipment can optionally activate or de-activate the effect of the masking on a scan cycle by scan cycle basis. More sophisticated implementations could offer more than one dynamically selectable mask bit per scan-chain or decoding schemes to dynamically update the mask vector. The dynamic control signals as well as the need to preload or modify mask vectors do consume input data bandwidth from the tester and add to the input data volume. This impact generally is very minimal if a single mask vector can be preloaded and used for one or more than one full scan load/unload without modification. The flip side of the coin in this scenario is some loss of observability due to the fact that a number of predictable responses that could carry defect information may be masked in addition to the x states.

If a combinational space compaction network is used and the test equipment monitors the outputs, then it becomes possible to let x states pass through and mask them in the test equipment. Flexibility could, for example, be utilized to reduce the amount of control data needed for a selective compare approach (e.g., if a "must-see" response is followed by an x state in a scan chain, then it is possible to leave that chain selected and to mask the x state in the tester rather than expending input bandwidth to change the selection).

If an XOR network is used for space compaction, then an x state in a response bit will render all XOR-sums containing that particular bit value equally unknown/unpredictable, masking any potential fault detection in the other associated response bits. The "art" of constructing x-tolerant XOR networks for space compaction is to minimize the danger of masking "must-see" response bit values. It has been shown that suitable and practical networks can indeed be constructed if the number of potential x states appearing at the scan-chain outputs in any scan cycle is limited.

Newer research adds some memory to the XOR-network approach. Unlike in MISR, there is no feedback loop in the memory. The feedback in the MISR amplifies and "perpetuates" x states (each x state, once captured, is continually fed back into the MISR and will eventually contaminate all signature bits). The x-tolerant structures with memory, by contrast, are designed such that x states are flushed out of the memory fairly quickly (e.g., using a shift-register arrangement without feedback).

It should be noted that many of the x-tolerant schemes are intended for designs with a relatively limited number of x states. However, certain test methodologies can introduce a large and variable number of x states. One example is delay-test with aggressive strobe timings that target short paths but cause setup time violations on longer paths.

21.3.3.1.6.5 Logic Diagnostics. Automated logic diagnostics software generally is set up to work from a bit-level fail set collected during test. The fail set identifies which tests failed and were data-logged, and within each data-logged failing test, which scan cells or output pins encountered a mismatch between the expected and the actual response value. Assuming binary values, the failing test can be expressed as the bitwise XOR sum of the correct response vector and an *error vector*. The diagnosis software can use a *precalculated fault dictionary or posttest simulation*. In both cases, fault simulation is used to associate individual faults from a list of model faults with error-simulated vectors. Model faults like stuck-at or transition faults are generally attached to gate-level pins in the netlist of the design under test. Hence the fault model carries gate-level locality information with it (the pin the fault is attached to). The diagnosis algorithms try to find a match between the error vectors from the data-logged fail sets and simulated error vectors that are in the dictionary or are created on the fly by fault simulation. If a match is found, then the associated fault will be added to the so-called *call-out* list of faults that partially or completely match the observed fail behavior.

Response compression/compaction reduces the bitwise response information to a much smaller amount of data. In general, the reduction is lossy in nature, meaning that it may be difficult or impossible to unambiguously re-construct the bit-level error vectors for diagnosis. That leaves essentially two options for diagnosing fails with compressed/compacted responses. The first option is to first detect the presence of fails in a test, then reapply the same test and extract the bit-level responses without compression/compaction. This approach has already been discussed in the section on logic BIST. Quick identification of the failing tests in signature-based methods can be facilitated by adding a reset capability to the MISR macro and comparing signatures at least once for each test. The reset is applied in between tests and returns the MISR to a fixed starting state even if the previous test failed and produced an incorrect signature. Without the reset, the errors would remain in the MISR, causing incorrect signatures in subsequent tests even if there are no further mismatches. With the reset and signature compared at least once per test, it is easier to determine the number of failing tests and schedule all or some of them for retest and data logging.

As already explained in the section on logic BIST, bit-level data logging from retest may make it necessary to have the expected responses in the test equipment's vector memory. To what extent that is feasible during production test is a matter of careful data management and sampling logistics. Hence, there is considerable interest in enabling meaningful *direct diagnostics* that use the compacted responses without the need for retest and bit-level data logging.

Most response compression/compaction approaches used in practice are linear in nature, meaning that the *difference vector* between a nonfailing compacted response and the failing compacted response is only a function of the bit-level error vector. In other words, the difference vector contains reduced information about the error vector. The key question for direct diagnostics is to what extent it is possible to associate this reduced information with a small enough number of "matching" faults to produce a meaningful callout list. The answer depends on what type of and how much fault-distinguishing information is preserved in the mapping from the un-compacted error vector to the compacted difference vector, and how that information can be accessed in the compacted data. Also important is what assumptions can be realistically made about the error distributions and what type of encoding is used for data reduction. For example, if it is assumed that for some grouping of scan chains, at most one error is most likely to occur per scan cycle in a group, then using an error-correcting code (ECC) for compaction would permit complete reconstruction of the bit-level error vector. Another issue is what matching criteria are used for deciding whether to include a fault in the call-out. A complete match between simulated and actual error vector generally also entails a complete match between simulated and observed compacted difference vectors. That is, the mapping preserves the matching criterion (albeit the fact that the information reduction leads to higher ambiguity). On the other hand, partial matching success based on proximity in terms of Hamming distance may not be preserved. For example, MISR-based signatures essentially are hash codes that do not preserve proximity. Nor should we forget the performance considerations. The bit-level error information is used in posttest simulation methods not only to determine matching but also to greatly reduce the search space by only including faults from the back-trace cones feeding into the failing scan cells or output pins.

Selective compare at first blush appears to be a good choice for diagnostics, because it allows for easy mapping of differences in the compacted responses back to bit-level errors. However, only a small subset of responses is visible in the compacted data and it is very conceivable that the ignored data contains important fault-distinguishing information.

To make a long story short, direct diagnostics from compacted responses are an area for potentially fruitful research and development. Some increasingly encouraging successes of direct diagnosis for certain linear compression/compaction schemes have been reported recently.

21.3.3.1.6.6 Scan-Chain Diagnostics. Especially for complex chips in new technologies it must be expected that defects or design issues affect the correct functioning of the scan load/unload operation. For many design projects, scan is not only important for test but also for debug. Not having fully working scan chains can be a serious problem.

Normal logic diagnostics assume fully working scan chains and are not immediately useful for localizing scan chain problems. Having a scan-chain problem means that scan cells downstream from the problem location cannot be controlled and scan cells upstream from the problem location cannot be observed by scan. The presence of such problems can mostly be detected by running scan-chain integrity tests, but it generally is difficult or impossible to derive the problem location from the results. For example, a hold-time problem that results in race-through makes the scan chain look too short and it may be possible to deduce that from the integrity test results. However, knowing that one or more scan cells were skipped does not necessarily indicate which cells were skipped.

Given that the scan load/unload cannot be reliably used for control and observation, scan-chain diagnostics tend to rely on alternative means of controlling/observing the scan cells in the broken scan chains. If mux-scan is used, for example, most scan cells have a second data input (namely the system data input) other than the scan data input and it may be possible to control the scan cell from that input (e.g., utilizing scan cells from working scan chains). Forcing known values into scan cells through nonscan inputs for the purpose of scan-chain diagnostics is sometimes referred to as *lateral insertion*. And most scan cell outputs not only feed to the scan data input of the next cell in the chain, but also feed functional logic that in turn may be observable. It also should be noted that the scan cells downstream from the problem location are observable by scan.

Design-for-test can help with diagnostics using lateral insertion techniques. A drastic approach is to insist that all scan cells have a directly controllable set and clear to force a known state into the cells. Other scan architectures use only the clear in combination with inversion between all scan cells to help localize scan-chain problems. Having many short scan chains can also help as long as the problem is localized and affects only a few (ideally one) chain. In that scenario the vast majority of scan chains are still working and can be used for lateral insertion and observation. In this context, it is very useful to design the scan architecture such that the short scan chains can be individually scanned out to quickly determine which chains are working and which not.

21.3.3.2 Embedded Memories

The typical DFT Synthesis tools convert only flip-flops or latches into scan cells. The storage cells in embedded memories generally are not automatically converted. Instead, special memory-specific DFT is used to test the memories themselves as well as the logic surrounding the memories.

21.3.3.2.1 Types of Embedded Memories.

Embedded memories come in many different flavors, varying in functionality, usage, and in the way they are implemented physically. From a test perspective, it is useful to grossly distinguish between register files and dense custom memories.

Register files are often implemented using design rules and cells similar to logic cells. The sensitivity to defects and failure modes is similar to that of logic, making them suitable for testing with typical logic tests. They tend to be relatively shallow (small address space) but can be wide (many bits per word), and have multiple ports (to enable read or write from/to several addresses simultaneously). Complex chips can contain tens or even many hundreds of embedded register files.

Dense custom memories, by contrast, tend to be hand-optimized and may use special design rules to improve the density of the storage cell array. Some memory types, for example embedded DRAM (eDRAM), use additional processing steps. Because of these special properties, dense custom memories are considered to be subject to special and unique failure modes that require special testing (e.g., retention time testing, and pattern sensitivities). On the other hand, the regular repetitive structure of the memory cell arrays, unlike "random" logic, lends itself to algorithmic testing. Overall, memory testing for stand-alone as well as embedded memories has evolved in a different direction than logic testing.

In addition to "normally" addressed random access memories (RAMs) including register files, static RAMs (SRAMs), and dynamic RAMs (DRAMs), there are read-only memories (ROMs), content addressable memories (CAMs), and other special memories to be considered.

21.3.3.2.2 Embedded Memories and Logic Test.

Logic testing with scan design benefits from the internal controllability and observability that comes from converting internal storage elements into scan cells. With embedded memories, the question arises as to whether and to what extent the storage cells inside the memories should be likewise converted into scan cells. The answer depends on the type of memory and how the scan function is implemented. Turning a memory element into a scan cell typically entails adding a data port for scan data, creating master–slave latch pairs or flip-flops, and connecting the scan cells into scan chains. The master–slave latch pair and scan-chain configuration can be fixed or dynamic in nature.

In the fixed configuration, each memory storage cell is (part of) one scan cell with a fixed, dedicated scan interconnection between the scan cells. This approach requires modification of the storage cell array in the register file, meaning that it only can be done by the designer of the register file. The overhead, among other things, depends on what type of cell is used for normal operation of the register file. If the cells are already flip-flops, then the implementation is relatively straightforward. If the cells are latches, then data port and a single-port scan-only latch can be added to each cell to create a master–slave latch pair for shifting. An alternative is to only add a data port to each latch and combine cell pairs into the master–slave configuration for shifting. The latter implementation tends to consume less area and power overhead, but only half of the words can be controlled or observed simultaneously (pulsing the master or slave clock overwrites the data in the corresponding latch types). Hence, this type of scan implementation is not entirely useful for debug where a nondestructive read is preferred, and such a function may have to be added externally if desired.

In the dynamic approach, a shared set of master or slave latches is temporarily associated with the latches making up one word in the register file cell array, to establish a master–slave configuration for shifting. Thus, a serial shift register through one word of the memory at a time can be formed using the normal read/write to first read the word and then write it back shifted one bit position. By changing the address, all register file bits can be serially controlled and observed. Because no modification of the register cell array is needed, and normal read/write operations are used, the dynamic scan approach could be implemented by the memory user. The shared latch approach and reuse of the normal read/write access for scan keep the overhead limited and the word-level access mechanism is very suitable for debug (e.g., reading/writing one particular word). The disadvantage of the dynamic approach is that the address-driven scan operation is not supported by the currently available DFT and ATPG tools.

Regardless of the implementation details, scannable register files are modeled at the gate level and tested as part of the chip logic. However, they increase the number of scan cells as well as the size of the

ATPG netlist (the address decoding and read/write logic are modeled explicitly at the gate level), and thereby can possibly increase test time. Moreover, ATPG may have to be enhanced to recognize the register files and create intelligently structured tests for better compaction. Otherwise, longer than normal test sets can result.

Modern ATPG tools tend to have some sequential test generation capabilities and can handle nonscannable embedded memories to some extent. It should be noted in this context that the tools use highly abstract built-in memory models. Neither the memory cell array nor the decoding and access logic are explicitly modeled, meaning that no faults can be assigned to them for ATPG. Hence, some other means must be provided for testing the memory proper. Sequential test generation can take substantially longer and result in lower test coverage. It is recommended to make sure that the memory inputs can be controlled from and the memory outputs can be observed at scan cells or chip pins through combinational logic only. For logic BIST and other test methods that use signatures without masking, it must also be considered that nonscannable embedded memories are potential x-state sources until they are initialized. If memory access is desired as part of such a test (e.g., it is not unusual for performance-limiting paths to include embedded memories such that memory access may be required for accurate performance screening/characterization), then it may be necessary to provide some mechanism for initializing the memories. Multiple write ports can also be a source of x states if the result cannot be predicted when trying to write different data to the same word. To avoid multiport write conflicts it may be necessary to add some form of port-priority, for example, by adding logic that detects the address coincidence and degates the write-clock(s) for the port(s) without priority, or by delaying the write-clock for the port with enough priority to assure that its data will be written last.

For best predictability in terms of ATPG run-times and achievable logic test coverage, it may be desirable to remove entirely the burden of having to consider the memories for ATPG. That can be accomplished, for example, by providing a memory bypass (e.g., combinational connection between data inputs and data outputs) in conjunction with observe points for the address and control inputs. Rather than a combinational bypass, boundary scan can be used. With bypass or boundary scan, it may not be necessary to model the memory behavior for ATPG and a simple black-box model can be used instead. However, the bypass or boundary scan can introduce artificial timing paths or boundaries that potentially mask performance-limiting paths.

21.3.3.2.3 Testing Embedded Memories.

It is possible to model smaller nonscannable embedded memories at the gate level. In that case, each memory bit cell is represented by a latch or flip-flop, and all decoding and access logic is represented with logic gates. Such a gate-level model makes it possible to attach faults to the logic elements representing the memory internals and to use ATPG to generate tests for those faults. For large, dense memories this approach is not very practical because the typical logic fault models may not be sufficient to represent the memory failure modes and because the ATPG fault selection and cube compaction algorithms may not be suited for generating compact tests for regular structures like memories. As a consequence, it is customary to use special memory tests for the large, dense memories and, in many cases, also for smaller memories. These memory tests are typically specified to be applied to the memory interface pins. The complicating factor for embedded memories is that the memory interface is buried inside the chip design and some mechanism is needed to transport the memory tests to/from the embedded interface through the intervening logic.

21.3.3.2.3.1 Direct Access Testing. *Direct access testing* requires that all memory interface pins are individually accessible from chip pins through combinational access paths. For today's complex chips it is rare to have natural combinational paths to/from chip pins in the functional design. It is, hence, up to DFT to provide the access paths and also provide some chip-level controls to selectively enable the access paths for testing purposes and disable them for normal functional chip operation. With direct access testing, the embedded memory essentially can be tested as if it were a stand-alone memory. It requires that the memory test program is stored in the external test equipment. Memory tests for larger memories are many cycles long and could quickly exceed the test equipment limits if the patterns for each cycle have to be stored in vector memory. This memory problem does not arise if the test equipment contains dedicated algorithmic pattern generator (APG) hardware that can be programmed to generate a wide range of memory test sequences from a very compact code. In that case, direct access testing has the benefit of being able to take full advantage of the flexibility and high degree of programmability of the APG and other memory-specific hardware/software features of the equipment.

If the chip contains multiple embedded memories or, memories with a too many pins, and there are not enough chip pins available to accommodate access for all memory interface pins, then a multiplexing scheme with appropriate selection control has to be implemented. It should be noted that the need to connect the access paths to chip pins can possibly create considerable wiring overhead if too many paths have to be routed over long distances. It should also be noted that particularly for high-performance memories, it could be difficult to control the timing characteristics of the access paths accurately enough to meet stringent test timing requirements. Variants of direct access testing may permit the use of tightly timed pipeline flip-flops or latches in the data and nonclock access paths to overcome the effect of inaccurate access path timings. The pipelining and latency of such sequential access paths must be taken into account when translating the memory test program from the memory interface to chip interface.

Another potential method for reducing the chip pin footprint required for direct access testing is to serialize the access to some memory interface pins, e.g., the data pins. To that effect, scan chains are built to shift serially wide bit patterns for the data words from/to a small number of chip pins. It may take many scan clock cycles to serially load/unload an arbitrary new bit pattern and the memory may have to "wait" until that is done. The serial access, hence, would not be compatible with test sequences that depend on back-to-back memory accesses with arbitrary data changes. Test time also is a concern with serial access for larger memories.

21.3.3.2.3.2 Memory BIST. Although direct access testing is a viable approach in many cases, it is not always easy to implement, can lead to long test times if not enough bandwidth is available between the chip under test and the test equipment, and may consume excessive vector memory if no APG hardware is available. Moreover, it may not be easy or possible to design the access paths with sufficient timing accuracy for a thorough performance test. *Memory BIST* has become a widely used alternative. For memory BIST, one or more simplified small APG hardware macros, also known as *memory BIST controllers*, are added to the design and connected to the embedded macros using a multiplexing interface. The interface selectively connects the BIST hardware resources or the normal functional logic to the memory interface pins.

A variety of more or less sophisticated controller types are available. Some implementations use pseudo-random data patterns and signature analysis similar to logic BIST in conjunction with simple address stepping logic. Signature analysis has the advantage that the BIST controller does not have to generate expected responses for compare after read. This simplification comes at the expense of limited diagnostic resolution in case the test fails.

The memory test sequences used with APG hardware in the test equipment are constructed from heavily looped algorithms that repeatedly traverse the address space and write/read regular bit patterns into/from the memory cell array. Most memory BIST implementations used in practice follow the same scheme. *Hardwired controllers* use customized finite state machines to generate the data-in pattern sequences, expected data-out pattern sequences, as well address traversal sequences for some common memory test algorithms (e.g., march-type algorithms). The so-called *programmable controllers* generally contain several hardwired test programs that can be selected at run-time by loading programming register in the controller. *Microcoded controllers* offer additional flexibility by using a dedicated microengine with a memory-test-specific instruction set and an associated code memory that must be loaded at run-time to realize a range of user customizable algorithms.

The complexity of the controller not only depends on the level of programmability but also on the types of algorithms it can support. Many controllers are designed for so-called linear algorithms in which the address is counted up/down so that the full address space is traversed a certain number of times. In terms of hardware, the linear traversal of the address space can be accomplished by one address register with increment/decrement logic. Some more complex nonlinear algorithms jump back and forth between a test address and a disturb address, meaning that two address registers with increment/decrement and more complex control logic are required. For data-in generation, a data register with some logic manipulation features (invert, shift, rotate, mask, etc.) is needed. The register may not necessarily have the full data word width as long as the data patterns are regular and the missing bits can be created by replication. If bit-level compare of the data-outs is used for the BIST algorithm, then similar logic plus compare logic is required to generate the expected data-out patterns and perform the comparison. Clock and control signals for the embedded memory are generated by timing circuitry that, for example, generates memory clock and control waveforms from a single free-running reference clock. The complexity of the timing circuitry depends on how flexible it is in terms of generating different event sequences to accommodate different memory access modes and how much programmability it offers in terms of varying the relative edge offsets.

Certain memory tests, like those for retention and pattern-sensitive problems, require that specific bit values are set up in the memory cell array according to physical adjacency. The physical column/row arrangement is not always identical to the logical bit/word arrangement. Address/bit scrambling as well as cell layout details may have to be known and taken into account to generate physically meaningful patterns.

The impact of memory BIST on design complexity and design effort depends, among other things, on methodology and flow. In some ASIC flows, for example, the memory compiler returns the memories requested by the user complete with fully configured, hardwired, and preverified memory BIST hardware already connected to the memory. In other flows it is entirely up to the user to select, add, and connect memory BIST after the memories are instantiated in the design. Automation tools for configuring, inserting, and verifying memory BIST hardware according to memory type and configuration are available. The tools generate customized BIST RTL or gate-level controllers and memory interfaces from input information about memory size/configuration, number/type of ports, read/write timings, BIST/bypass interface specification, address/bit scrambling, and physical layout characteristics. Some flows may offer some relief through optional *shared controllers* where a single controller can drive multiple embedded memories. For shared controllers, the users have to decide and specify the sharing method (e.g., testing multiple memories in parallel or one after the other).

In any scenario, memory BIST can add quite a bit of logic to the design and this additional logic should be planned and taken into account early enough in the physical design planning stage to avoid unpleasant surprises late in the design cycle. In addition to the additional transistor count, wiring issues and BIST timing closure can cause conflicts with the demands of the normal function mode. Shared controllers make placement and wiring between the controller and the memory interface more complicated and require that the trade-off between wiring complexity and additional transistor count is well understood and the sharing strategy is planned appropriately (e.g., it may not make sense to share a controller for memories that are too far away from each other).

21.3.3.2.3.3 Complex Memories and Memory Substructures. In addition to embedded single-/multiport embedded SRAMs, some chips contain more complex memory architectures. Embedded DRAMs can have more complex addressing and timing than "simple" SRAMs. eDRAMs are derived from stand-alone DRAM architectures, and, for example, they may have time multiplexed addressing (i.e., the address is supplied in two cycles and assembled in the integrated memory controller) and programmable access modes (e.g., different latency, fast page/column, etc.). The BIST controller and algorithm design also has to accommodate the need for periodic refresh, all of which can make BIST controllers for eDRAM more complex.

Other fairly widely used memory types include logic capabilities. Content addressable memories, for example, contain logic to compare the memory contents with a supplied data word. BIST controllers for CAMs use enhanced algorithms to thoroughly test the compare logic in addition to the memory array. The algorithms depend on the compare capabilities and the details of the circuit design and layout.

High-performance processors tend to utilize tightly integrated custom memory subsystems where, for example, one dense memory is used to generate addresses for another dense memory. Performance considerations do not allow for separating the memories with scan. Likewise, compare logic and other high-performance logic may be included as well. Memories subsystems of this nature generally cannot be tested with standard memory test algorithms, and hence require detailed analysis of possible failure modes and custom development of suitable algorithms.

Some micro-controllers and many chips for Smartcards, for example, contain embedded nonvolatile flash memory. Flash memory is fairly slow and poses no challenge for designing a BIST controller with sufficient performance, but flash memory access and operation are different from SRAM/DRAM access. Embedded flash memory is commonly accessible via an embedded microprocessor that can, in many cases, be used for testing the embedded flash memory.

21.3.3.2.3.4 Performance Characterization. In addition to use in production test, a BIST approach may be desired for characterizing the embedded memory performance. The BIST controller, memory interface, and timing circuitry in this scenario have to be designed for higher speed and accuracy than what is required for production testing alone. It may also be necessary to add more algorithm variations or programmability to enable stressing particular areas of the embedded memory structure. The BIST designers in this case also may want to accommodate the use of special lab equipment used for characterization.

If BIST is to be used for characterization and speed binning, it should be designed with enough performance headroom to make sure that the BIST circuitry itself is not the performance-limiter that dominates the measurements. Designing the BIST circuitry to operate at "system cycle time," as advertised by some BIST tools, may not be good enough.

The quality of speed binning and performance characterization not only depends on the performance of BIST controller and memory interface, but also on the accuracy and programmability of the timing edges used for the test. The relative offset between timing edges supplied from external test equipment generally can be programmed individually and in very small increments. Simple timing circuitry for memory BIST, by contrast, tends to be aligned to the edges of a reference clock running at or near system speed. The relative edge offset can only be changed by changing the frequency, and affects all edge offsets similarly.

Some memory timing parameters can be measured with reasonable accuracy without implementing commensurate on-chip timing circuitry. For example, data-out bit values can be fed back to the address. The memory is loaded with appropriate values such that the feedback results in oscillation and the oscillation frequency is an indicator of read access time. By triggering a clock pulse from a data-out change, it may be similarly possible to obtain an indication of write access time.

21.3.3.2.3.5 Diagnosis, Redundancy, and Repair. Embedded memories can be yield limiters and useful sources of information for yield improvement. To that effect, the memory BIST implementation should permit the collection of memory fail-bit maps. The fail-bit maps are created by data logging the compare status or compare data and, optionally, BIST controller status register contents.

A simple data logging interface with limited diagnostic resolution can be implemented by issuing a BIST-start edge when the controller begins with the test and a cycle-by-cycle pass/fail signal that indicates whether a mismatch occurred during compare or not. Assuming the relationship between cycle count and BIST progress is known, the memory addresses associated with cycles indicating mis-compare can be derived. This simple method allows for creating an address-level fail log. However, the information is generally not sufficient for FA.

To create detailed fail-bit maps suitable for FA, the mis-compares must be logged at the bit level. This can, for example, be done by unloading the bit-level compare vectors to the test equipment through a fully parallel or a serial interface. The nature of the interface determines how long it takes to log a complete fail-bit map. It also has an impact on the design of the controller. If a fully parallel interface is used that is fast enough to keep up with the test, then the test can progress uninterrupted while the data are logged. If a serial or slower interface is used, then the test algorithm must be interrupted for data logging. This approach sometimes is referred to as *stop-on-nth-error*. Depending on the timing sensitivity of the tests, the test algorithm may have to be restarted from the beginning or from some suitable checkpoint each time the compare data have been logged for one fail. In other cases, the algorithm can be paused and can resume after logging.

Stop-on-*n*th-error with restart and serial data logging is relatively simple to implement, but it can be quite time consuming if mis-compares happen at many addresses (e.g., defective column). Although that is considered tolerable for FA applications in many cases, there have been efforts to reduce the data-logging time without needing a high-speed parallel interface by using on-chip data reduction techniques. For example, if several addresses fail with the same compare vector (e.g., defective column), then it would be sufficient to log the compare data once and only log the addresses for subsequent fails. To that effect, the on-chip data reduction hardware would have to remember one or more already encountered compare vectors and for each new compare vector, check whether it is different or not.

Large and dense embedded memories, for example, those used for processor cache memory, offer *redundancy* and *repair* features, meaning that the memory blocks include spare rows/columns and address manipulation features to replace the address/bit corresponding to the failing rows/columns with those corresponding to the spares. The amount and type of redundancy depend on the size of the memory blocks, their physical design, and the expected failure modes, trying to optimize postrepair memory density and yield. Memories that have either spare rows or spare columns, but not both, are said to have *one-dimensional* redundancy, and memories that have both spare rows and columns are said to have *two-dimensional* redundancy. It should be noted that large memories may be composed of smaller blocks and it is also possible to have block-level redundancy/repair.

Hard repair means that the repair information consisting of the failing row/column address(es) is stored in nonvolatile form. Common programming mechanisms for hard repair include laser-programmable fuses, electrically programmable fuses, or flash-type memory. The advantage of hard repair is that the failing rows/columns only need to be identified once upfront. However, hard repair often requires special equipment for programming, and one-time programmable fuses are not suitable for updating the repair information later in the chip's life cycle. *Soft repair*, by contrast, uses volatile memory, for example flip-flops, to store the repair information. This eliminates the need for special equipment and the repair information can be updated later. However, the repair information is not persistent and must be redetermined after each power-on.

All embedded memory blocks with redundancy/repair must be tested prior to repair, during manufacturing production test, with sufficient data logging to determine whether they can be repaired and if so, which rows/columns must be replaced. Chips with hard repair fuses may have to be brought to a special programming station for fuse-blow and then be returned to a tester for postrepair retest. Stop-on-*n*th-error data logging with a serial interface is much too slow for practical repair during production test. Some test equipment has special hardware/software features to determine memory repair information from raw bit-level fail data. In that scenario, no special features for repair may be required in the memory BIST engine, except for providing a data-logging interface that is fast enough for real-time logging.

If real-time data logging from memory BIST is not possible or not desired, then on-chip redundancy analysis (ORA) is required. Designing a compact ORA macro is relatively simple for one-dimensional redundancy. In essence all that is needed is memory to hold row/column addresses corresponding to the number of spares, and logic that determines whether a new failing row/column address already is in memory; if not, the memory uses the new memory if there still is room, and, if there is no more room, sets a flag indicating that the memory cannot be repaired. Dealing with two-dimensional redundancy is much more complicated, and ORA engines for two-dimensional redundancy are not commonly available. Embedded DRAM, for example, may use two-dimensional redundancy and come with custom ORA engines from the eDRAM provider. The ORA engines for two-dimensional redundancy are relatively large, and typically one such engine is shared by several eDRAM blocks, meaning that the blocks must be tested one after the other. To minimize test time, the eDRAM BIST controller should stagger the tests for the blocks such that ORA for one block is performed when the other blocks are idle for retention testing.

If hard repair is used, the repair data are serially logged out. Even if hard repair is used, the repair information may also be written into a repair data register. Then repair is turned on, and the memory is retested to verify its full postrepair functionality.

Large complex chips may contain many memory blocks with redundancy, creating a large amount of repair data. In that scenario, it can be useful to employ data compression techniques to reduce the amount of storage needed for repair or to keep the size of the fuse bay for repair small enough.

21.3.3.3 Embedded Digital Cores

The notion of embedded cores has become popular with the advent of SoC products. Embedded cores are predesigned and preverified blocks that are assembled with other cores and user-defined blocks into a complex chip.

21.3.3.3.1 Types of Embedded Digital Cores.

Embedded cores are generally classified into *hard cores*, which are already physically completed and delivered with layout data, and *soft cores*, which in many cases are delivered as synthesizable RTL code. In some cases the cores may already be synthesized to a gate-level netlist that still needs to be placed and wired (*firm cores*).

For hard cores, it is important whether or not a "test-ready" detailed gate-level netlist is made available (*white box*), or the block internals are not disclosed (*black box*). In some cases, partial information about some block features may be made available (*gray box*). Also, the details of what DFT features have been implemented in the core can be very important.

21.3.3.3.2 Merging Embedded Cores.

Soft cores and some white cores can be *merged* with other compatible cores and user-defined blocks into a single netlist for test generation.

For hard cores, the success of merging depends on whether the DFT features in the core are compatible with the test methodology chosen for the combined netlist. For example, if the chosen methodology is full-scan ATPG, then the core should be designed with full scan and the scan architecture as well as the netlist should be compatible with the ATPG tool. If logic BIST or test data compression is used, then the scan-chain segments in the core should be short enough and balanced to enable scan-chain balancing in the combined netlist.

21.3.3.3.3 Direct Access Test.

Direct access test generally requires the inputs of the embedded core to be individually and simultaneously controlled from, and the core outputs to be individually and simultaneously observed at, chip pins, through combinational access paths. This is only possible if enough chip pins are available and the test equipment has enough appropriately configured digital tester channels. Some direct access test guidelines may allow for multiplexing subsets of outputs onto common chip pins. In that case, the test must be repeated with a different output subset selected until all outputs have been observed. That is, the required chip output pin footprint for direct access can be somewhat reduced at the expense of test time. All inputs must however remain controllable at all times. For multiple identical cores, it may be possible to broadcast the input signals to like core inputs from a common set of chip input pins and test the cores in parallel, as long as enough chip output pins are available for observing the core outputs and testing the cores together fits does not exceed the power budget. Concurrent testing of nonidentical cores seems conceptually possible if all core inputs and outputs can be directly accessed simultaneously. However, even with parallel access it may not be possible to align the test waveforms for nonidentical cores well enough to fit within the capabilities of typical production test equipment (the waveforms for most equipment must fit into a common tester cycle length and limited set time set memory; the tester channels of some equipment, on the other hand, can be partitioned into multiple groups that can be programmed independently). If the chip contains multiple cores that cannot be tested in parallel, then some chip-level control scheme must be implemented to select one core (or a small enough group of compatible cores) at a time for direct access test.

In addition to implementing access paths for direct access testing, some additional DFT may be required to make sure that a complete chip test is possible and does not cause unwanted side effects. For example, extra care may be required to avoid potential three-state burn-out conditions resulting from a core with three-state outputs and some other driver on the same net trying to drive opposite values. In general, it must be expected that currently de-selected cores and other logic could be exposed to unusual input conditions during the test of the currently selected core(s). The integration of cores is easier if there is a simple control state to put each core into safe state that protects the core internals from being affected by unpredictable input conditions, asserts known values at the core outputs, and keeps power/noise for/from the core minimal. For cores that can be integrated into chips in which IDDq testing is possible, the safe state or another state should prevent static current paths in the core. The core test selection mechanism should activate the access paths for the currently selected core(s), while asserting the appropriate safe/IDDq state for nonselected cores.

The inputs of black-box cores cannot be observed and the black-box outputs cannot be fully controlled. Other observe/control means must be added to the logic feeding the core inputs and fed by the core outputs (*shadow logic*) to make that logic fully testable. Isolating and diagnosing core-internal defects can be very difficult for black-box cores or cores using proprietary test approaches that are not supported by the tools available and used for chip-level test generation.

There are some DFT synthesis and DRC tools that help with the generation, connection, and verification of direct access paths. Also, there are some test generation tools that help translate core test program pin references from the core pins to the respective chip pins, and integrate the tests into an overall chip-level test program.

21.3.3.3.4 Serializing the Embedded Core Test Interface.

There may not be enough chip pins available or the overhead for full parallel access to all core pins may be unacceptable. In that case, it may be possible and useful to serialize the access to some core pins. One common serialization method is to control core input pins and observe core output pins with scan cells. Updating the input pattern or comparing the output pattern in this scenario entails a scan load/unload operation. The core test program must be able to allow insertion of multiple tester cycles worth of wait time for the scan load/unload to complete. Moreover, the output of scan cells change state during scan. If the core input pin and core test program cannot tolerate multiple arbitrary state changes during scan load/unload, then a hold latch or flip-flop may have to be provided between the scan cell output and core input to retain the previous state required by the core test program until scan load/unload is done and the new value is available. For digital logic cores, data pins may be suitable for serialized access, assuming there is no asynchronous feedback and the internal memory cells are made immune to state changes (e.g., by turning off the clocks) during the core-external scan load/unload. Although it is conceptually possible to synthesize clock pulses and arbitrary control sequences via serialized access with hold, it generally is recommended or required to provide direct combinational control from chip pins for core clock and control pins.

Functional test programs for digital logic cores can be much shorter than memory test programs for large embedded memories. Serialized access for digital logic cores can, hence, be more practical and is more widely practiced for digital logic cores. In many cases, it is possible to find existing scan cells and to sensitize logic paths between those scan cells and the core pins with little or no need for new DFT logic. This minimizes the design impact and permits testing through functional paths rather than test-only paths.

However, it should be noted that with serialized access, it generally is not possible to create multiple arbitrary back-to-back input state changes and to observe multiple back-to-back responses. Hence, serialized access may not be compatible with all types of at-speed testing.

There are some DFT synthesis/DRC and test program translation tools that work with serialized access. Test program translation in this case is not as simple as changing pin references (and possibly polarity) and adjusting time sets as in the case of combinational direct access testing. Input value changes and output measures on core pins with serialized access entail inserting a core-external scan load/unload procedure. Additional scan load operations may be required to configure the access paths.

The capabilities of the pattern translation tool may have a significant impact on the efficiency of the translated test program. For example, if the embedded macro is or contains an embedded memory with BIST, then at least a portion of the test may consist of a loop. If the translation tool cannot preserve the loop, then the loop must be unrolled into parallel vectors prior to translation. That can result in excessive vector memory demand from the test equipment. Black-box cores that come with internal scan and with a scan-based test program are another special case. If the core scan-in/-out pins are identified and the associated scan load/unload procedures are appropriately defined and referenced in the core test program, then the pattern translation software may (or may not) be able to retain the scan load/unload procedure information in the translated test program, resulting in a more (or less) efficient test program.

21.3.3.3.5 Standardized Embedded Core Access and Isolation Architectures.

There have been several proprietary and industry-wide attempts to create an interoperable architecture for embedded core access and isolation. These architectures generally contain at least two elements. One element is the core-level test interface, often referred to as a core test *wrapper*, and the other one is a chip-level test access mechanism (TAM) to connect the interfaces among themselves and to chip pins, as well as some control infrastructure to selectively enable/disable the core-level test interfaces.

When selecting/designing a core-level test interface and TAM, it is important to have a strategy for both, i.e., for testing the core itself and for testing the rest of the chip. For testing the core itself, the core test interface should have an internal test mode in which the core input signals are controlled from the TAM and the core output signals are observed by the TAM without needing participation from other surrounding logic outside of the core. The issue for testing the surrounding logic is how to observe the signals connected to the core inputs and how to control the signals driven by the core outputs. If the core is a black-box core for example, the core inputs cannot be observed and the core outputs cannot be controlled. Even if the core is a white box, there may be no internal DFT that would make it easy. To simplify test generation for the surrounding logic, the core test interface should have an *external test mode* that provides alternate means for observing the core input signals and for controlling the core output signals, without needing participation from the core internals. Even if the core internals do not have to participate in the external test mode, it may be important to assure that the core does not go into some illegal state (e.g., three-state burn-out) or create undesired interference (e.g., noise, power, and drain) with the rest of the chip. It therefore may be recommended or required that the core test interface should have controls for a *safe mode* that protects the core itself from arbitrary external stimulation and vice versa. Last but not the least, there has to be a *normal mode* in which the core pins can be accessed for normal system function, without interference from the TAM and core test interface. Layout considerations may make it desirable to allow for routing access paths for one core "through" another core. The core test interface may offer a dedicated *TAM bypass mode* (combinational or registered) to facilitate the daisy chaining of access connections.

Overall, the TAM, so to speak, establishes switched connections between the core pins and the chip pins for the purpose of transporting test data between test equipment and embedded cores. It should be noted that some portions of the test equipment could be integrated on the chip (e.g., a BIST macro) such that some data sources/sinks for the TAM are internal macro pins and others are chip I/Os. The transport mechanism in general can be parallel or serialized. It is up to the core provider to inform the core user about any restrictions and constraints regarding the type and characteristics of the TAM connections as well as what the required/permissible data sources/sinks are for each core pin. Different core pins, e.g., clock pins and data pins, tend to have different restrictions and constraints. Depending on the restrictions and constraints, there probably will be some number of pins that must have parallel connections (with or without permitted pipe-lining), some number of pins that could be serialized (with some of those possibly requiring a hold latch or flip-flop). If certain core input pins need to be set to a specific state to initialize the core to some mode of operation, then it may be sufficient to assure that the respective state is asserted (e.g., decoded from some test mode control signals) without needing full blown independent access from chip pins. It further should be noted that certain cores may have different modes of operations, including different test modes (e.g., logic test and memory BIST), such that there could be different mode-specific access restrictions and constraints on any given pin. The TAM will have to accommodate that and if necessary, be dynamically adjustable.

Serialization of the core pin access interface reduces the number of chip pins and the bit-width (e.g., affecting the wiring overhead) of the TAM required for access, at the expense of test time. In addition to serializing some portion of the TAM for an individual core, there often is a choice to be made about testing multiple cores in parallel vs. serially. If the chip contains many cores, the trade-off becomes increasingly complex. For example, the TAM for all or subsets of cores could be arranged in a daisy-chain or star configuration, and decisions need to be made about the bit-width of each TAM. More decisions need to be made about testing groups of cores in parallel or in sequence. The decisions are influenced by the availability of pin, wiring and test equipment resources for parallel access, by the power/noise implications of testing several cores in parallel, by test time, and more.

Core test wrappers can be predesigned into a core by the core designer/provider (so-called *wrapped core*) or not (*unwrapped core*). In the latter case, the core user could wrap the core prior to, during, or after assembling it into a chip. In some cases, the wrapper overhead can be kept smaller by taking advantage of already existing core-internal design elements (e.g., flip-flops on the boundary of the core to implement a serial shift register for serialized access). If performance is critical, then it could be better to integrate the multiplexing function required for test access on the core input side with the first level of core-logic, or take advantage of hand-optimized custom design optimization that is possible in a custom core but not in synthesized logic. In any case, building a wrapper that requires modification of the core logic proper in hard cores in general can only be done by the core designer. Prewrapping a core can have the disadvantage of not being able to optimize the wrapper for a particular usage instance and of not being able to share wrapper components between cores. The once popular expectation that each third-party core assembled into a chip would be individually wrapped and tested using core-based testing has given way to a more pragmatic approach where un-wrapped cores are merged into larger partitions based on other design flow and design (team, location, schedule, etc.) management decisions, and only the larger partitions are wrapped and tested using core-based testing.

A chip-level control infrastructure is needed in addition to the wrapper and TAM components to create an architecture for core-based test. The control infrastructure is responsible for distributing and (locally) decoding instructions that configure the core/wrapper modes and the TAM according to the intended test objective (e.g., testing a particular core vs. testing the logic in between cores). The distribution mechanism of the control infrastructure is, in some architectures. combined with a mechanism for retrieving local core test status/result information. For example, local core-attached instruction and status/result registers can be configured into serial scan chains (dedicated chains for core test or part of "normal" chains) for core test instruction load and status/result unload.

21.3.3.4 Embedded Field Programmable Gate Arrays

Although eFPGAs are not very prevalent yet, they can have some unique properties that affect chip-level testing.

21.3.3.4.1 Embedded Field Programmable Gate Array Characteristics.

The term "field programmable" means that the eFPGA function can be programmed in the field, that is long after manufacturing test. At manufacturing test time, the final function generally is not known. FPGAs contain both functional resources and programming resources. Both types of resources tend to have their own I/O interface. The functional resources are configured by the programming resources prior to actual "use" to realize the intended function, and the rest of the chip logic communicates with the functional resources of an eFPGA through the functional I/O interface of the eFPGA. The normal chip functional logic in general does not see or interact directly with the programming resources or the programming I/O interface of the eFPGAs.

The customers of chips with eFPGAs expect that all functional and programming resources of the eFPGAs are working. Hence, it is necessary at chip manufacturing test time, to fully test all functional and programming resources (even if only a few of them will eventually be used).

21.3.3.4.2 Embedded Field Programmable Gate Array Test and Test Integration Issues.

The chip-level DFT for chips with eFPGAs will have to deal with the duality of the programming and functional resources. Many eFPGAs, for example, use an SRAM-like array of storage elements to hold the programming information. However, the outputs of the memory cells are not connected to data selection logic for the purpose of reading back a data word from some memory address. Instead, the individual memory cell outputs are intended to control the configuration of logic function blocks and of interconnect switches that constitute the functional resources of the eFPGA. The data-in side of the eFPGA programming memory array may also be different than in a normal SRAM that is designed for random access read/write. The design is optimized for loading a full eFPGA configuration program from some standardized programming interface. The programming interfaces tend to be relatively narrow even if the internal eFPGA programming memory is relatively wide. Hence, the address/data information may have to be brought in sequentially in several chunks.

Another idiosyncrasy of FPGAs is that the use of BIST techniques for testing does not necessarily mean that the test is (nearly) as autonomous as logic BIST, for example. Several BIST techniques have been proposed and are being used for testing the functional resources of the FPGA. To that effect, the reprogrammability of the FPGA is used to temporarily configure pattern generation and response compression logic from some functional resources and configure other functional resources as test target. Once configured, the BIST is indeed largely autonomous and needs only a little support from external test equipment. The difference is that potentially large amounts of programming data need to be downloaded from the external test equipment to create a suite of BIST configurations with sufficient test coverage. The programming data can consume large amounts of vector memory and downloading them costs test time. Non-BIST tests still need programming data and have the additional problem of needing test equipment support at the functional I/O interface during test application.

The chip-level DFT approach for chips with eFPGAs has to deal with the vector memory and test time issue as well as implement a suitable test interface for the I/O interface of the functional resources (even for eFPGAs with BIST there is a need to test the interface between the chip logic and the functional resources of the eFPGA). Also to be considered is that an unconfigured eFPGA is of little help in testing the surrounding logic, and a decision has to be made whether to exploit (and depend on) the programmability of the eFPGA's functional resources or to hardwire the DFT circuitry.

21.3.3.4.3 Embedded Field Programmable Gate Array Test Access Techniques.
To test fully an eFPGA, it is necessary to test both the programming resources and the functional resources, meaning that test access to the interfaces of the two different types of resources is needed. Of course, it is possible to treat the eFPGA like other embedded cores and implement an interface wrapper connected to the chip-level TAM. This approach however, does not take advantage of the fact that the chip already provides access to the eFPGA programming interface for functional configuration. The chip-level programming interface may be a good starting point for test access, especially if signature-based BIST is used to test the functional resources such that there is limited return traffic from the eFPGAs during test.

Regardless of whether the programming interface is adapted or another TAM is used, it may be desirable to test multiple sFPGA cores in parallel to reduce the demand on vector memory and test time. For example, if there are multiple identical core instances, then it makes sense to broadcast the same programming data to all instances simultaneously. The normal programming interface may have no need for and, therefore, not offer a broadcast load option for multiple eFPGA macros in parallel. Hence, such an option may have to be added to make the normal programming interface more useable for testing. Likewise, if another TAM is used, then the TAM should be configurable for the optional broadcast of programming data.

21.4 Conclusion

There are many other topics that could be covered in this chapter. These include the issues of embedded analog/mixed-signal DFT (which is covered in the chapter on "Analog Test" by Bozena Kaminska, in this handbook); and DFT and I/Os, both normal and high speed. In addition, there are issues of top-level DFT, including the integration of DFT elements, boundary scan for high-level assembly test, and test interface considerations (including IEEE 1149.1 protocol, BSDL, HSDL, COP/ESP, interrupts, burn-in, etc.). However, considerations of space and time do not let us go any further into these many interesting details at this point; perhaps in a future edition of this handbook, we will be able to cover them in some detail.

References

[1] K.J. Lee et al., Using a single input to support multiple scan chains, *Proceedings of ICCAD*, 1998, pp. 74–78 (broadcast scan).
[2] I.M. Ratiu and H.B. Bakoglu, Pseudorandom built-in self-test and methodology and implementation for the IBM RISC System/6000 processor, *IBM J. R&D*, Vol. 34, 78–84, 1990.
[3] B.L. Keller and D.A. Haynes, Design automation for the ES/9000 series processor, *Proceedings of ICCD*, 1991, pp. 550–553.
[4] A. Samad and M. Bell, Automating ASIC design-for-testability — the VLSI test assistant, *Proceedings of ITC*, 1989, pp. 819–828 (DFT automation including direct access macro test).

[5] V. Immaneni and S. Raman, Direct access test scheme — design of block and core cells for embedded ASICs, *Proceedings of ITC*, 1990, pp. 488–492.
[6] B.H. Seiss et al., Test point insertion for scan-based BIST, *Proceedings of the European Test Conference*, 1991, pp. 253–262.
[7] D. Kay and S. Mourad, Controllable LFSR for BIST, *Proceedings of the IEEE Instrument and Measurement Technology Conference*, 2000, pp. 223–229 (streaming LFSR-based decompressor).
[8] G.A. Sarrica and B.R. Kessler, Theory and implementation of LSSD scan ring & STUMPS channel test and diagnosis, *Proceedings of the International Electronics Manufacturing Technology Symposium*, 1992, pp. 195–201 (chain diagnosis for LBIST, lateral insertion concept).
[9] R. Rajski and J. Tyszer, Parallel Decompressor and Related Methods and Apparatuses, US Patent, US 5,991,909, 1999.
[10] I. Bayraktaroglu and A. Orailoglu, Test volume and application time reduction through scan chain concealment, *Proceedings of DAC*, 2001, pp. 151–155 (combinational linear decompressor).
[11] D.L. Fett, Current Mode Simultaneous Dual-Read/Write Memory Device, US Patent, US 4,070,657, 1978 (scannable register file).
[12] N.N. Tendolkar, Diagnosis of TCM failures in the IBM 3081 processor complex, *Proceedings of DAC*, 1983, pp. 196–200 (system-level ED/FI).
[13] J. Reilly et al., Processor controller for the IBM 3081, *IBM J. R&D*, Vol. 26, 22–29, January 1982 (system-level scan architecture).
[14] H.W. Miller, Design for test via standardized design and display techniques, *Elect. Test*, 108–116, 1983 (system-level scan architecture; scannable register arrays, control, data, and shadow chains; reset and inversion for chain diagnosis).
[15] A.M. Rincon et al., Core design and system-on-a-chip integration, *IEEE Des. Test*, Vol. 14, 26–35, 1997 (test access mechanism for core testing).
[16] E.K. Vida-Torku et al., Bipolar, CMOS and BiCMOS circuit technologies examined for testability, *Proceedings of the 34th Midwest Symposium on Circuits and Systems*, 1992, pp. 1015–1020 (circuit-level inductive fault analysis).
[17] J. Dreibelbis et al. Processor-based built-in self-test for embedded DRAM, *IEEE J. Solid-State Circ.*, Vol. 33, 1731–1740, November 1998 (microcoded BIST engine with 1-dimensional on-chip redundancy allocation).
[18] U. Diebold et al., Method and Apparatus for Testing a VLSI Device, European Patent EP 0 481 097 B1, 1995 (LFSR re-seeding with equation solving).
[19] D. Westcott, The self-assist test approach to embedded arrays, *Proceedings of ITC*, 1981, pp. 203–207 (hybrid memory BIST with external control interface).
[20] E.B. Eichelberger and T.W. Williams, A logic design structure for LSI testability, *Proceedings of DAC*, 1977, pp. 462–468.
[21] D.R. Resnick, Testability and maintainability with a new 6k gate array, *VLSI Design*, Vol. IV, 34–38, March/April 1983 (BILBO implementation).
[22] Zasio J.J., Shifting away from probes for wafer test, *COMPCON S'83*, 1983, pp. 317–320 (boundary scan and RPCT).
[23] P. Goel, PODEM-X: an automatic test generation system for VLSI logic structures, *Proceedings of DAC*, 1981, pp. 260–268.
[24] N. Benowitz et al., An advanced fault isolation system for digital logic, *IEEE Trans. Comput.*, Vol. C-24, 489–497, May 1975 (pseudo-random logic BIST).
[25] P. Goel and M.T. McMahon, Electronic chip-in-place test, *Proceedings of ITC*, 1982, pp. 83–90 (sort-of boundary scan).
[26] Y. Arzoumanian and J. Waicukauski, Fault diagnosis in an LSSD environment, *Proceedings of ITC*, 1981, pp. 86–88 (post-test simulation/dynamic fault dictionary; single location paradigm with exact matching).
[27] K.D. Wagner, Design for testability in the Amdahl 580, *COMPCON 83*, 1983, pp. 383–388 (random access scan).
[28] H-J. Wunderlich, PROTEST: a tool for probabilistic testability analysis, *Proceedings of DAC*, 1985, pp. 204–211 (testability analysis).
[29] H.C. Godoy et al., Automatic checking of logic design structures for compliance with testability ground rules, *Proceedings of DAC*, 1977, pp. 469–478 (early DFT DRC tool).

[30] B. Koenemann, J. Mucha, and G. Zwiehoff, Built-in test for complex digital integrated circuits, *IEEE J. Solid State Circ.*, Vol. 15, 315–319, June 1980 (hierarchical TM-Bus concept).
[31] B. Koenemann et al., Built-in logic block observation techniques, *Proceedings of ITC*, 1979, pp. 37–41.
[32] R.W. Berry et al., Method and Apparatus for Memory Dynamic Burn-in, US Patent, US 5,375,091, 1994 (design for burn-in).
[33] J.A. Waicukauski et al., Fault simulation for structured VLSI, *VLSI Syst. Des.*, 20–32, December 1985 (PPSFP).
[34] P.H. Bardell and W.H. McAnney, Self-testing of multi-chip logic modules, *Proceedings of ITC*, 1982, pp. 200–204 (STUMPS).
[35] K. Maling and E.L. Allen, A computer organization and programming system for automated maintenance, *IEEE Trans. Electr. Comput.*, 887–895, 1963 (early system-level scan methodology).
[36] R.D. Eldred, Test routines based on symbolic logic statements, *ACM J.*, Vol. 6, 33–36, January 1959.
[37] M. Nagamine, An automated method for designing logic circuit diagnostics programs, *Design Automation Workshop*, 1971, pp. 236–241 (parallel pattern compiled code fault simulation).
[38] P. Agrawal and V.D. Agrawal, On improving the efficiency of Monte Carlo test generation, *Proceedings of FTCS*, 1975, pp. 205–209 (weighted random patterns).
[39] V.S. Iyengar et al., On computing the sizes of detected delay faults, *IEEE Trans. CAD*, Vol. 9, 299–312, March 1990 (small delay fault simulator).
[40] Y. Aizenbud et al., AC test quality: beyond transition fault grading, *Proceedings of ITC*, 1992, pp. 568–577 (small delay fault simulator).
[41] R.C. Wong, An AC test structure for fast memory arrays, *IBM J. R&D*, Vol. 34, 314–324, March/May 1990 (SCAT-like memory test timing).
[42] M.H. McLeod, Test Circuitry for Delay Measurements on a LSI chip, US Patent No. 4,392,105, 1983 (on chip delay measurement method).
[43] W.S. Klara et al., Self-Contained Array Timing, US Patent, US 4,608,669, 1986.
[44] J.A. Monzel et al., AC BIST for a compilable ASIC embedded memory library, Digital Papers North Atlantic Test Workshop, 1996 (SCAT-like approach).
[45] L. Ternullo et al., Deterministic self-test of a high-speed embedded memory and logic processor subsystem, *Proceedings of ITC*, 1995, pp. 33–44 (BIST for complex memory subsystem).
[46] W.V. Huott et al., Advanced microprocessor test strategy and methodology, *IBM J. R&D*, Vol. 41, 611–627, 1997 (microcoded memory BIST).
[47] H. Koike et al., A BIST scheme using microprogram ROM for large capacity memories, *Proceedings of ITC*, 1990, pp. 815–822.
[48] S.B. Akers, The use of linear sums in exhaustive testing, *Comput. Math Appl.*, 13, 475–483, 1987 (combinational linear decompressor for locally exhaustive testing).
[49] D. Komonytsky, LSI self-test using level sensitive design and signature analysis, *Proceedings of ITC*, 1982, pp. 414–424 (scan BIST).
[50] S.K. Jain and V.D. Agrawal, STAFAN: an alternative to fault simulation, *Proceedings of DAC*, 1984, pp. 18–23.
[51] B. Nadeau-Dostie et al., A serial interfacing technique for built-in and external testing, *Proceedings of CICC*, 1989, pp. 22.2.1–22.2.5.
[52] T.M. Storey and J.W. Barry, Delay test simulation, *Proceedings of DAC*, 1977, pp. 492–494 (simulation with calculation of test slack).
[53] E.P. Hsieh et al., Delay test generation, *Proceedings of DAC*, 1977, pp. 486–491.
[54] K. Kishida et al., A delay test system for high speed logic LSI's, *Proceedings of DAC*, 1986, pp. 786–790 (Hitachi delay test generation system).
[55] A. Toth and C. Holt, Automated database-driven digital testing, *IEEE Comput.*, 13–19, January 1974 (scan design).
[56] A. Kobayashi et al., Flip-flop circuit with FLT capability, *Proceedings of IECEO Conference*, 1968, p. 962.
[57] M.J.Y. Williams and J.B. Angell, Enhancing testability of large scale circuits via test points and additional logic, *IEEE Trans. Comput.*, Vol. C-22, 46–60, 1973.
[58] R.R. Ramseyer et al., Strategy for testing VHSIC chips, *Proceedings of ITC*, 1982, pp. 515–518.

22

Automatic Test Pattern Generation

Kwang-Ting (Tim) Cheng
University of California
Santa Barbara, California

Li-C. Wang
University of California
Santa Barbara, California

22.1 Introduction

Test development for complex designs can be time-consuming, sometimes stretching over several months of tedious work. In the past three decades, various test development automation tools have attempted to address this problem and eliminate bottlenecks that hinder the product's time to market. These tools, which automate dozens of tasks essential for developing adequate tests, generally fall into four categories: design-for-testability (DFT), test pattern generation, pattern grading, and test program development and debugging. The focus of this chapter is on automatic test pattern generation (ATPG).

Because ATPG is one of the most difficult problems for electronic design automation, it has been researched for more than 30 years. Researchers, both theoreticians and industrial tool developers, have focused on issues such as scalability, ability to handle various fault models, and methods for extending the algorithms beyond Boolean domains to handle various abstraction levels.

Historically, ATPG has focused on a set of *faults* derived from a gate-level fault model. For a given target fault ATPG consists of two phases: *fault activation* and *fault propagation. Fault activation* establishes a signal value at the fault site opposite that produced by the fault. *Fault propagation* propagates the fault effect forward by sensitizing a path from the fault site to a primary output. The objective of ATPG is to find an input (or test) sequence that, when applied to the circuit, enables testers to distinguish between the correct circuit behavior and the faulty circuit behavior caused by a particular fault. Effectiveness of ATPG is measured by the fault coverage achieved for the fault model and the number of generated vectors, which should be directly proportional to test application time.

ATPG efficiency is another important consideration. It is influenced by the fault model under consideration, the type of circuit under test (full scan, synchronous sequential, or asynchronous sequential), the level of abstraction used to represent the circuit under test (gate, register transistor, switch), and the required test quality.

As design trends move toward nanometer technology, new ATPG problems are emerging. During design validation, engineers can no longer ignore the effects of crosstalk and power supply noise on reliability and performance. Current modeling and vector-generation techniques must give way to new techniques that consider timing information during test generation, that are scalable to larger designs, and that can capture extreme design conditions. For nanometer technology, many current design validation problems are becoming manufacturing test problems as well, so new fault-modeling and ATPG techniques will be needed.

This chapter is divided into five sections. Section 22.2 introduces gate-level fault models and concepts in traditional combinational ATPG. Section 22.3 discusses ATPG on gate-level sequential circuits. Section 22.4 describes circuit-based Boolean satistifiability (SAT) techniques for solving circuit-oriented problems. Section 22.5 illustrates ATPG for faults such as crosstalk and power supply noise, which involve timing and applications other than manufacturing testing. Section 22.6 presents sequential ATPG approaches that go beyond the traditional gate-level model.

22.2 Combinational ATPG

A fault model is a hypothesis of how the circuit may go wrong in the manufacturing process. In the past several decades, the most popular fault model used in practice is the *single stuck-at fault* model. In this model, one of the signal lines in a circuit is assumed to be stuck at a fixed logic value, regardless of what inputs are supplied to the circuit. Hence, if a circuit has n signal lines, there are potentially $2n$ stuck-at faults defined on the circuit, of which some can be viewed as being equivalent to others [1].

The stuck-at fault model is a *logical fault* model because no delay information is associated with the fault definition. It is also called a *permanent fault* model because the faulty effect is assumed to be permanent, in contrast to *intermittent* and *transient* faults that can appear randomly through time. The fault model is *structural* because it is defined based on a structural gate-level circuit model.

A stuck-at fault is said to be detected by a test pattern if, when applying the pattern to the circuit, different logic values can be observed, in at least one of the circuit's primary outputs, between the original circuit and the faulty circuit. A pattern set with 100% stuck-at fault coverage consists of tests to detect every possible stuck-at fault in a circuit.

Stuck-at fault coverage of 100% does not necessarily guarantee high quality. Earlier studies demonstrate that not all fault coverages are created equal [2,3] with respect to the quality levels they achieve. As fault coverage approaches 100%, additional stuck-at fault tests have diminishing chances to detect nontarget defects [4]. Experimental results have shown that in order to capture all nontarget defects, generating multiple tests for a fault may be required [5]. Generating tests to observe faulty sites multiple times may help to achieve higher quality [6,7]. Correlating fault coverages to test quality is a fruitful research area beyond the scope of this chapter. We use stuck-at fault as an example to illustrate the ATPG techniques.

A test pattern that detects a stuck-at fault satisfies two criteria simultaneously: *fault activation* and *fault propagation*. Consider Figure 22.1 as an example. In this example, input line a of the AND gate is assumed to be stuck-at 0. In order to activate this fault, a test pattern must produce logic value 1 at line a. Then, under the good-circuit assumption, line a has logic value 1 when the test pattern is applied. Under the faulty-circuit assumption, line a has logic value 0. The symbol $D = 1/0$ is used to denote the situation. D needs to be *propagated* through a sensitized path to one of the primary outputs. In order for D to be propagated from line a to line c, input line b has to be set at logic value 1. The logic value 1 is called the *non controlling value* for an AND gate. Once b is set at the noncontrolling value, line c will have whatever logic value that line a has.

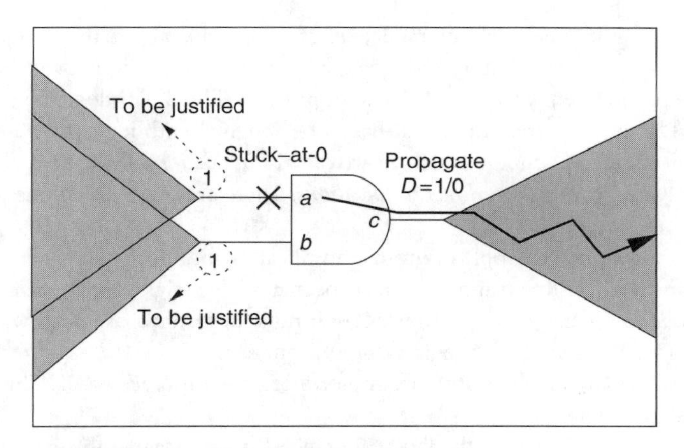

FIGURE 22.1 Fault activation and fault propagation for a stuck-at.

The ATPG process involves simultaneous justification of the logic value 1 at lines *a* and *b*, and propagation of the fault difference *D* to a primary output. In a typical circuit with reconvergent fanouts, the process involves a search for the right decisions to assign logic values at primary inputs and at internal signal lines in order to accomplish both justification and propagation. The ATPG problem is an NP-complete problem [8]. Hence, all known algorithms have an exponential worst-case run time.

Algorithm 22.1: BRANCH-AND-BOUND ATPG*(circuit, a fault)*

```
Solve()
if (bound_by_implication() = FAILURE)
  then return (FAILURE)
if (error difference at PO) and all lines are justified
  then return (SUCCESS)
while (there is an untried way to solve the problem)
     { make a decision to select an untried way to propagate or to justify
  do { if (Solve() = SUCCESS)
     { then return (SUCCESS)
return (FAILURE)
```

Algorithm 22.1 illustrates a typical branch-and-bound approach to implement ATPG. The efficiency of this algorithm is affected by two things:

- The *bound_by_implication()* procedure determines whether there is a conflict in the current value assignments. This procedure helps the search to avoid making decisions in subspaces that contain no solution.
- The decision-making step inside the while loop determines how branches should be ordered in the search tree. This determines how quickly a solution can be reached. For example, one can make decisions to propagate the fault difference to a primary output (PO) before making decisions to justify value assignments.

22.2.1 Implication and Necessary Assignments

The ATPG process operates on at least a five-value logic defined over $\{0,1,D,\overline{D},X\}$; *X* denotes unassigned value [9]. *D* denotes that in the good circuit, the value should be logic 1, and in the faulty circuit the value should be logic 0. $\overline{D}$ is the complement of *D*. Logical AND, OR, and NOT can be defined based on these five values [1,9]. When a signal line is assigned with a value, it can be one of the four values $\{0,1,D,\overline{D}\}$.

After certain value assignments have been made, *necessary assignments* are those implied by the current assignments. For example, for an n-input AND gate, its output being assigned with logic 1 implies that all its inputs have to be assigned with logic 1. If its output is assigned with logic 0 and $n - 1$ inputs are assigned with logic 1, then the remaining input has to be assigned with logic 0. These necessary assignments derive from an analysis of circuit structure are called *implications*. If the analysis is done individually for each gate, the implications are *direct implications*. There are other situations where implications can be *indirect*.

Figure 22.2(a) shows a simple example of indirect implication. Suppose that a value 0 is being justified backward through line d where line b and line c have been already assigned with logic 1. To justify $d = 0$, there are three choices to set the values of the AND's inputs. Regardless of which way is used to justify $d = 0$, the line a must be assigned with logic value 0. Hence, in this case $d = 0$, $b = 1$, $c = 1$ together implies $a = 0$. Figure 22.2 (b) shows another example where $f = 1$ implies $e = 1$. This implication holds regardless of other assignments.

Figure 22.3 shows an example where the fault difference D is propagated through line a. Suppose that there are two possible paths to propagate D, one through line b and the other through line c. Suppose that both paths eventually converge at a 3-input AND gate. In this case, regardless of which path is chosen as the propagation path, line d must be assigned with logic value 1. Therefore, in this case $a = D$ implies $d = 1$.

The *bound_by_implication()* procedure in Algorithm 22.1 performs implications to derive all necessary assignments. A *conflict* occurs if a line is assigned with two different values after all the necessary assignments have been derived. In this case the procedure returns failure. It can be seen that the greater the number of necessary assignments that can be derived by implications, the more likely that a conflict can be detected. Because of this, efficiently finding necessary assignments through implications has been an important research topic for improving the performance of ATPG since the introduction of the first complete ATPG algorithm in Ref. [9].

In addition to the direct implications that can easily be derived based on the definitions of logic gates, indirect implications can be obtained by analyzing circuit structure. This analysis is called *learning* where

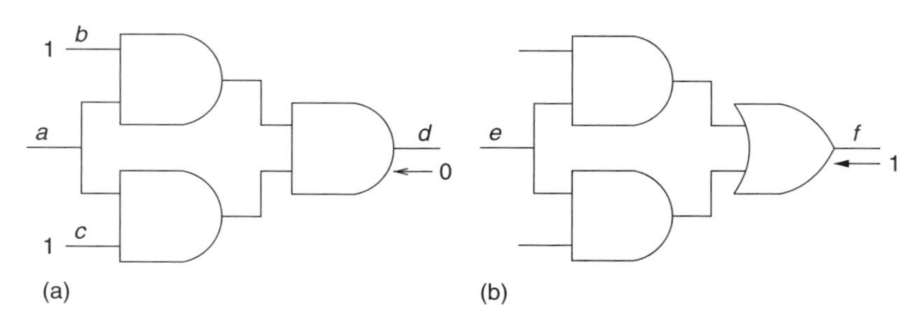

FIGURE 22.2 Implication examples [10]. (a) $d = 0$ implies $a = 0$ when $b = 1$, $c = 1$; (b) $f = 1$ implies $e = 1$.

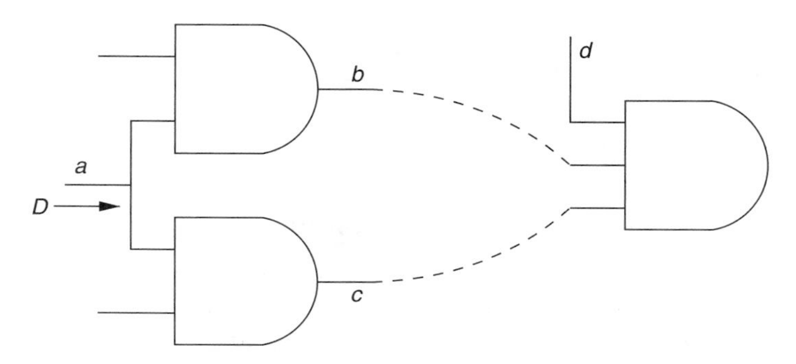

FIGURE 22.3 Implication $d = 1$ because of the unique propagation path [10].

correlations among signals have been established by simple and efficient methods. Learning methods should be simple and efficient enough so that the speedup in search by exploring the resulting implications should outweigh the cost of establishing the signal correlations. There are two types of learning approaches proposed in the literature:

Static learning Signal correlations are established before the search. For example, in Figure 22.2(b), forward logic simulation with $e = 0$ obtains $f = 0$. Because $e = 0$ implies $f = 0$, we obtain that $f = 1$ implies $e = 1$. For example, in Figure 22.3, the implication $a = D \Rightarrow d = 1$ is universal. This can be obtained by analyzing the *unique propagation path* from signal line a. These implications can be applied at any time during the search process.

Dynamic learning Signal correlations are established during the search. If the learned implications are conditioned on a set of value assignments, these implications can only be used to prune the search subspace based on those assignments. For example, in Figure 22.2(a), $d = 0$ implies $a = 0$ only when b and c have been assigned with 1. However, unconditional implications of the form $(x = v_1) \Rightarrow (y = v_2)$ can also be learned during dynamic learning.

The concepts of static and dynamic learning were suggested in [10]. A more sophisticated learning approach called *recursive learning* was presented in [11]. Recursive learning can be applied statically or dynamically. Because of its high cost, it is more efficient to apply recursive learning dynamically. Conflict-driven recursive learning and conflict-learning for ATPG were recently proposed in [12].

Knowing when to apply a particular learning technique is crucial in dynamic learning to ensure that the gain from learning outweighs the cost of learning. For example, it is more efficient to apply recursive learning on hard-to-detect faults where most of the subspaces during search contain no solution [11]. Because of this, search in these subspaces is inefficient. On the other hand, recursive learning can quickly prove that no solution exists in these subspaces. From this perspective, it appears that recursive learning implements a complementary strategy with respect to the decision-tree-based search strategy, since one is more efficient for proving the absence of a solution but the other is more efficient for finding a solution. Conflict-learning is another example in which learning is triggered by a conflict [12]. The conflict is analyzed and the cause of the conflict is recorded. The assumption is that during a search in a neighboring region, the same conflict might recur. By recording the cause of the conflict, the search subspace can be pruned more efficiently in the neighboring region. Conflict learning was first proposed in [13] with application in SAT. The authors in [12] implement the idea in their ATPG with circuit-based techniques.

22.2.2 ATPG Algorithms and Decision Ordering

One of the first complete ATPG algorithms is the D-algorithm [9]. Subsequently, other algorithms were proposed, including PODEM [15], FAN [16], and SOCRATES [10].

D-algorithm

The D-algorithm is based on the five-value logic defined on $\{0,1,D,\overline{D},X\}$. The search process makes decisions at primary inputs as well at internal signal lines. The D-algorithm is able to find a test even though a fault difference may necessitate propagation through multiple paths. Figure 22.4 illustrates such an example.

Suppose that fault difference D is propagated to line b. If the decision is made to propagate D through path d,g,j,l, we will require setting $a = 1$ and $k = 1$. Since $a = 1$ implies $i = 0$, $k = 1$ implies $h = 1$ which further implies $e = 1$. A conflict occurs between $e = 1$ and $b = D$. If the decision is made to propagate D through path e,h,k,l, we will require setting $i = 0$ and $j = 1$. $j = 1$ implies $g = 1$, which further implies $d = 1$. Again, $d = 1$ and $b = D$ cause a conflict. In this case, D has to be propagated through both paths. The required assignments are setting $a = 1$, $c = 1$, $f = 1$. This example illustrates a case when *multiple path sensitization* is required to detect a fault.

The D-algorithm is the first ATPG algorithm that can produce a test for a fault even though it requires multiple path sensitization. However, because the decisions are based on a five-value logic system, the search

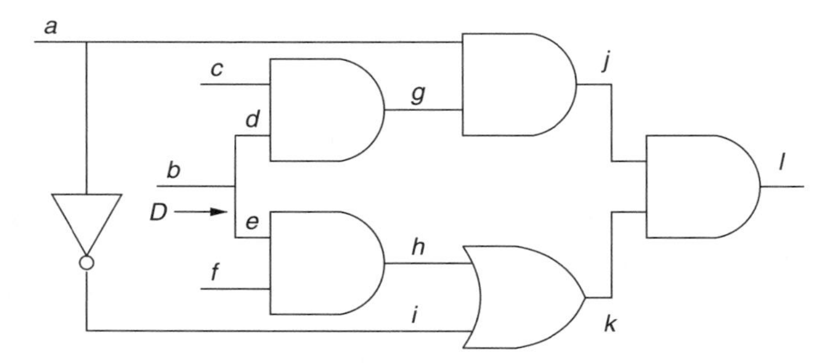

FIGURE 22.4 Fault propagation through multiple paths.

can be time-consuming. In practice, most faults may require only single-path sensitization and hence, explicit consideration of multiple path sensitization in search may become an overhead for ATPG [14].

PODEM

The D-algorithm can be characterized as an *indirect search* approach because the goal of an ATPG is to find a test at primary inputs, while the search decisions in D-algorithm are made on primary inputs and internal signal lines. PODEM implements a *direct search* approach where value assignments are made only on primary inputs, so that potentially the search tree is smaller. PODEM was proposed in [15]. A recent implementation of PODEM, called ATOM, is presented in [17].

FAN

FAN [16] adds two new concepts to PODEM. First, decisions can be made on internal *head lines* that are the end points of a tree logic cone. Therefore, a value assigned to a head line is guaranteed to be justifiable because of the tree circuit structure. Second, FAN uses a *multiple-backtrace* procedure so that a set of objectives can be satisfied simultaneously. In contrast, the original PODEM tries to satisfy one objective at a time.

SOCRATES

SOCRATES [10] is a FAN-based implementation with improvements in the implication and multiple-backtrace procedures. It also offers an improved procedure to identify a *unique sensitization path* [16].

The efficiency of an ATPG implementation depends primarily on the decision ordering it takes. There can be two approaches to influence the decision ordering: one by analyzing the fault-difference propagation paths and the other by measuring the testability of signal lines.

Analyzing the potential propagation paths can help to make decisions more likely to reach a solution. For example, a simple X-path check [15] can determine whether there exists a path from the point of a *D* line to a primary output, where no signal lines on the path have been assigned any value. *Unique sensitization* [16] can identify signal lines necessary for *D* propagation regardless of which path it takes. The *Dominator* approach [18] can identify necessary assignments for *D* propagation. Henftling et al. in [14] propose a single-path-oriented ATPG where fault propagation is explicitly made to have a higher priority than value justification.

Decision ordering can be guided through *testability measures* [1,19]. There are two types of testability measures: *controllability measures* and *observability measures.* Controllability measures indicate the relative difficulty of justifying a value assignment to a signal line. Observability measures indicate the relative difficulty of propagating the fault difference *D* from a line to a primary output. One popular strategy is to select a more difficult problem to solve before selecting the easier ones [1]. However, there can be two difficulties with a testability-guided search approach. First, the testability measures may not be sufficiently accurate. Second, always solving the hard problems first may bias the decisions too much in some cases. Wang et al. in [12] suggest a *dynamic decision ordering* approach in which failures in the justification process will trigger changes in decision ordering.

22.2.3 Boolean Satisfiability-Based ATPG

ATPG can also be viewed as solving a SAT problem. SAT-based ATPG was originally proposed in [20]. This approach duplicates part of the circuit which is influenced by the fault and constructs a satisfiability circuit instance by combining the good circuit with the faulty part. An input assignment that differentiates the faulty circuit from the good circuit is a test to detect the fault. Several other SAT-based ATPG approaches were developed later [21] [22]. Recently, a SAT-based ATPG called SPIRIT [23] was proposed, which includes almost all known ATPG techniques with improved heuristics for learning and search. ATPG and SAT will be discussed further in Section 22.4.

Practical implementation of an ATPG tool often involves a mixture of learning heuristics and search strategies. Popular commerical ATPG tools support full-scan designs where ATPG is mostly combinational. Although ATPG efficiency is important, other considerations such as test compression rate and diagnosability are also crucial for the success of an ATPG tool.

22.3 Sequential ATPG

The first ATPG algorithm for sequential circuits was reported in 1962 by Seshu and Freeman [24]. Since then, tremendous progress has been made in the development of algorithms and tools. One of the earliest commercial tools, LASAR [25], was reported in the early 1970s.

Due to the high complexity of the sequential ATPG, it remains a challenging task for large, highly sequential circuits that do not incorporate any design for testability (DfT) scheme. However, these test generators, combined with low-overhead DfT techniques such as partial scan, have shown a certain degree of success in testing large designs. For designs that are sensitive to area and performance overhead, the solution of using sequential-circuit ATPG and partial scan offers an attractive alternative to the popular full-scan solution, which is based on combinational-circuit ATPG.

It requires a sequence of vectors to detect a single stuck-at fault in a sequential circuit. Also, due to the presence of memory elements, the controllability and observability of the internal signals in a sequential circuit are in general much more difficult than those in a combinational circuit. These factors make the complexity of sequential ATPG much higher than that of combinational ATPG.

Sequential-circuit ATPG searches for *a sequence of vectors* to detect a particular fault through the space of all possible vector sequences. Various search strategies and heuristics have been devised to find a shorter sequence and to find a sequence faster. However, according to reported results, no single strategy/heuristic outperforms others for all applications/circuits. This observation implies that a test generator should include a comprehensive set of heuristics.

In this section, we will discuss the basics and give a survey of methods and techniques for sequential ATPG. We focus on the methods that are based on gate-level circuit models. Examples will be given to illustrate the basics of representative methods. The problem of sequential justification, sometimes referred to as *sequential SAT*, will be discussed in more detail in Section 22.4.

Figure 22.5 shows the taxonomy for sequential test generation approaches. Few approaches can directly deal with the timing issues present in highly asynchronous circuits. Most sequential-circuit test-generation approaches neglect the circuit delays during test generation. Such approaches primarily target synchronous or almost synchronous (i.e., with some asynchronous reset/clear and/or few asynchronous loops) sequential circuits, but they cannot properly handle highly asynchronous circuits whose functions are strongly related to the circuit delays and are sensitive to races and hazards. One engineering solution to using such approaches for asynchronous circuits is to divide the test-generation process into two phases. A potential test is first generated by ignoring the circuit delays. The potential test is then simulated using proper delay models in the second phase to check its validity. If the potential test is invalid due to race conditions, hazards, or oscillations, test generation is called again to produce a new potential test.

The approaches for (almost) synchronous circuits can be classified according to the level of abstraction at which the circuit is described. A class of approaches uses the state transition graph (STG) for test generation [26–29]. This class is suitable for pure controllers for which the STGS are either readily available or

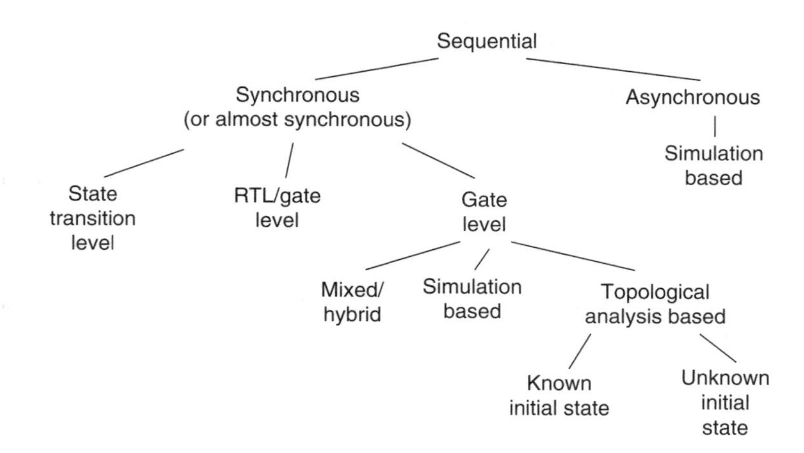

FIGURE 22.5 Sequential test generation: Taxonomy.

easily extractable from a lower level description. For data-dominated circuits, if both register transfer level (RTL) and gate-level descriptions are provided, several approaches can effectively use the RTL description for state justification and fault propagation [30–32].

Most of the commercial test generators are based on the gate-level description. Some of them employ the iterative array model [33,34] and use topological analysis algorithms [35–38] or they might be enhanced from a fault simulator [24,40–43]. Some use the mixed/hybrid methods that combine the topological-analysis-based methods and the simulation-based methods [44–46]. Most of these gate-level approaches assume an unknown initial state in the flip-flops, whereas some approaches assume a known initial state to avoid initialization of the state-holding elements [47–49]. The highlighted models and approaches in Figure 22.5 are those commonly adopted in most of today's sequential ATPG approaches.

22.3.1 Topological-Analysis-Based Approaches

Many sequential circuit test generators have been devised on the basis of fundamental combinational algorithms. Figure 22.6(a) shows the Huffman model of a sequential circuit. Figure 22.6(b) shows an array of combinational logic through *time-frame expansion*. In any time frame, logic values can be assigned only to the primary inputs (PIs). The values on the next state lines (NSs) depend on the values of the current state lines (PSs) at the end of the previous time frame. The iterative combinational model is used to approximate the timing behavior of the circuit. Topological analysis algorithms that activate faults and sensitize paths through these multiple copies of the combinational circuit are used to generate input assignments at the primary inputs. Note that a single stuck-at fault in a sequential circuit will correspond to a multiple stuck-at fault in the iterative array model where each time frame contains the stuck-at fault at the corresponding fault site.

The earliest algorithms extended the D-algorithm [9] based on the iterative array model [33,34]. It starts with one copy of the combinational logic and sets it to time frame 0. The D-algorithm is used for time frame 0 to generate a combinational test. When the fault effect is propagated to the next state lines, a new copy of the combinational logic is created as the next time frame, and the fault propagation continues. When there are values required at the present state lines, a new copy of the combinational logic is created as the previous time frame. The state justification is then performed backwards in the previous time frame. The process continues until there is no value requirement at the present state lines, and a fault effect appears at a primary output.

Muth [50] pointed out that the five-value logic based on $\{0,1,D,D,X\}$ used in the D-algorithm is not sufficient for sequential ATPG. A nine-value logic is suggested to take into account the possible repeated effects of the fault in the iterative array model. Each of the nine values is defined by an ordered pair of

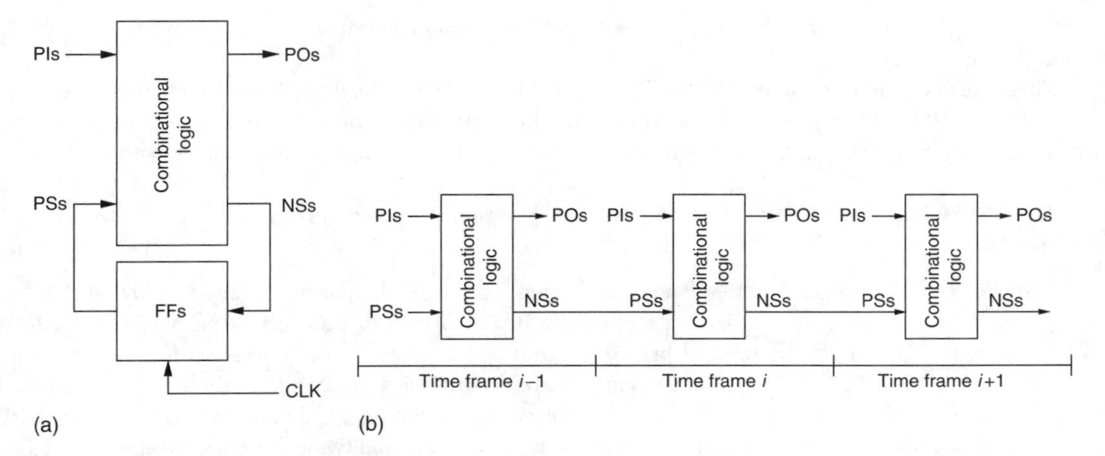

FIGURE 22.6 Synchronous sequential circuit model and time-frame expansion. (a) Huffman model; (b) Iterative array model.

binary values – the first value of the pair represents the ternary value (0,1,or X) of a single line in the fault-free circuit, and the second value represents the ternary value of the signal line in the faulty circuit. Hence, for a signal, there are possibly nine distinct ordered pairs (0/0, 0/1, 0/X, 1/0, 1/1, 1/X, X/0, X/1 and X/X).

The extended D-algorithm and the nine-value-based algorithm use *mixed forward and reverse time processing* techniques during test generation. The requirements created during the forward process (fault propagation) have to be justified by the backward process later. The mixed time processing techniques have some disadvantages. The test generator may need to maintain a large number of time frames during test generation because all time frames are partially processed, and the implementation is somewhat complicated.

The reverse time processing (RTP) technique used in the extended backtrace algorithm (EBT) [35] overcomes the problems caused by the mixed time processing technique. RTP works backwards in time from the last time frame to the first time frame. For a given fault, it preselects a path from the fault site to a primary output. This path may involve several time frames. The selected path is then sensitized backwards starting from the primary output. If the path is successfully sensitized, backward justification is performed for the required value at the fault site. If the sensitization process fails, another path is selected.

RTP has two main advantages: (1) At any time during the test-generation process, only two time frames need to be maintained: the current time frame and the previous one. For such a unidirectional algorithm, the backward justification process is done in a breadth-first manner. The value requirements in time frame n are completely justified before the justification of the requirements in time frame $n-1$. Therefore, the justified values at internal nodes of time-frame n can be discarded when the justification of time frame $n-1$ starts. As a result, the memory usage is low and the implementation is easier. Note that the decision points and their corresponding circuit status still need to be stacked for the purpose of backtracking. (2) It is easier to identify repetition of state requirements. A state requirement is defined as the state specified at the present state lines of a time frame during the backward-justification process. If a state requirement has been visited earlier during the current backward-justification process, the test generator has found a loop in the state-transition diagram. This situation is called *state repetition.* The backward-justification process should not continue to circle that loop, so backtracking should take place immediately. Since justification in time frame n is completed before the justification in time frame n - 1, state repetition can be easily identified by simply recording the state requirement after the completion of backward justification of each time frame, and then comparing each newly visited state requirement with the list of previously visited state requirements. Therefore, the search can be conducted more effectively. Similarly, the test generator can maintain a list of illegal states, i.e., the states that have been previously determined as unjustifiable. Each newly visited state requirement should also be compared against this list to determine whether the state requirement is an identified illegal state, in order to avoid repetitive and unnecessary searches.

There are two major problems with the EBT algorithm: (1) Only a single path is selected for sensitization. Faults that require multiple-path sensitization for detection may not be covered. (2) The number of possible paths from the fault site to the primary outputs can be very large; trying path-by-path may not be efficient.

After the EBT approach, several other sequential ATPGs were proposed, including the BACK algorithm [36], HITEC [37], and FASTEST [38].

BACK : The BACK algorithm [36] is an improvement over the EBT algorithm. It also employs the RTP technique. Instead of preselecting a path, the BACK algorithm preselects a primary output. It assigns a D or $\overline{D}$ to the selected primary output and justifies the value backwards. A testability measure (called drivability) is used to guide the backward D-justification from the selected primary output to the fault site. Drivability is a measure associated with a signal that estimates the effort of propagating a D or $\overline{D}$ from the fault site to the signal. The drivability measurement is derived based on the SCOAP [51] controllability measurement of both fault-free and faulty circuits. For a given fault, the drivability measure of each signal is computed before test generation starts.

HITEC : HITEC [37] employs several new techniques to improve the performance of test generation. Even though it uses both forward and reverse time processing, it clearly divides the test-generation process into two phases. The first is the forward time processing phase in which the fault is activated and propagated to a primary output. The second phase is the justification of the initial state determined in the first phase using the reverse time processing. Due to the use of the forward time processing for fault propagation, several efficient techniques (such as the use of dominators, unique sensitization, and mandatory assignments [10,16,18,39] used in combinational ATPG) can be extended and applied in phase one. In the reverse time processing algorithms, such techniques are of no use. Also, no drivability is needed for the fault-propagation phase which further saves some computing time.

FASTEST : FASTEST [38] uses only forward time processing and uses PODEM [15] as the underlying test-generation algorithm. For a given fault, FASTEST first attempts to estimate the total number of time frames required for detecting the fault and also to estimate at which time frame the fault is activated. The estimation is based on SCOAP[51]-like controllability and observability measures. An iterative array model with the estimated number of time frames is then constructed. The present state lines of the very first time frame have an unknown value and cannot be assigned to either binary value. A PODEM-like algorithm is employed where the initial objective is to activate the target fault at the estimated time frame. After an initial objective has been determined, it backtraces starting from the line of the initial objective until it reaches an unassigned primary input or a present state line in the first time frame. For the later case, backtracking is performed immediately. This process is very similar to the PODEM algorithm except that the process now works on a circuit model with multiple time frames. If the algorithm fails to find a test within the number of time frames currently in the iterative array, the number of time frames is increased, and test generation is attempted again based on the new iterative array.

Compared to the reverse time processing algorithms, the main advantage of the forward time processing algorithm is that it will not waste time in justifying unreachable states and will usually generate a shorter justification sequence for bringing the circuit to a hard-to-reach state. For circuits with a large number of unreachable states or hard-to-reach states, the reverse time processing algorithms may spend too much time in proving that unreachable states are unreachable or generating an unduly long sequence to bring the circuit to a hard-to-reach state. However, the forward time processing algorithm requires a good estimate of the total number of time frames and the time frame for activating each target fault. If that estimation is not accurate, the test generator may waste much effort in the smaller-than-necessary iterative array model.

22.3.2 Undetectability and Redundancy

For combinational circuits or full-scan sequential circuits, a fault is called *undetectable* if no input sequence can produce a fault effect at any primary output. A fault is called *redundant* if the presence of the fault does not change the input/output behavior of the circuit. The detectability is associated with a test-generation procedure whereas the redundancy is associated with the functional specification of a design.

A fault is combinationally redundant if it is reported as undetectable by a complete combinational test generator [52]. The definitions of detectability and redundancy for (nonscan) sequential circuits are much more complicated [1,53,54] and these two properties (the redundancy and undetectability of stuck-at faults) are no longer equivalent [1,53,54].

It is pointed out in [54] that undetectability could be precisely defined only if a test strategy is specified, and redundancy cannot be defined unless the operational mode of the circuit is known. The authors give formal and precise definitions of undetectability with respect to four different test strategies — namely fullscan, reset, multiple observation time, and single observation time. They also explain redundancies with respect to three different circuit operational modes — namely reset, synchronization, and nonsynchronization [54].

A fault is called undetectable under full scan if it is combinationally undetectable [55a]. In the case where hardware reset is available, a fault is said to be undetectable under the reset strategy if no input sequence exists such that the output response of the fault-free circuit is different from the response of the faulty circuit, both starting from their reset states. In the case where hardware reset is not available, there are two different test strategies: the multiple observation time (MOT) strategy and the single observation time (SOT) strategy.

Under the SOT strategy, a sequence detects a fault only if a fault effect appears at the same primary output *Oi* and at the same vector *vj* for all power-up initial state-pairs of the fault-free and faulty circuits (O_i could be any primary output, and vj could be any vector in the sequence). Most gate-level test generators and the sequential ATPG algorithms mentioned above assume the SOT test strategy.

Under the MOT strategy, a fault can be detected by multiple input sequences — each input sequence produces a fault effect at some primary output for a subset of power-up initial state-pairs and the union of the subsets covers all possible power-up initial state-pairs (for an n-flip-flop circuit, there are 2^{2n} power-up initial state-pairs). Under the MOT strategy, it is also possible to detect a fault using a single test sequence for which fault effects appear at different primary outputs or different vectors for different power-up initial state-pairs.

22.3.3 Approaches Assuming a Known Reset State

To avoid generating an initialization sequence, a class of ATPG approaches assumes the existence of a known initial state. For example, this assumption is valid for circuits like controllers that usually have a hardware reset (i.e., there is an external reset signal, and the memory elements are implemented by resettable flip-flops). Approaches like STALLION [47], STEED [48], and VERITAS [49] belong to this category.

STALLION

STALLION first extracts the STG for the fault-free circuit. For a given fault, it finds an activation state *S* and a fault-propagation sequence *T* that will propagate the fault effect to a primary output. This process is based on PODEM and the iterative array model. There is no backward state justification in this step. Using the STG, it then finds a state transfer sequence T0 from the initial state S0 to the activation state *S*. Because the derivation of the state transfer sequence is based on the state graph of the fault-free circuit, the sequence may be corrupted by the fault and hence, may not bring the faulty circuit into the required state *S*. Therefore, fault simulation for the concatenated sequence T0 T is required. If the concatenated sequence is not a valid test, an alternative transfer sequence or propagation sequence will be generated.

STALLION performs well for controllers for which the STG can be extracted easily. However, the extraction of STG is not feasible for large circuits. To overcome this deficiency, STALLION constructs a partial STG only. If the required transfer sequence cannot be derived from the partial STG, the partial STG is then dynamically augmented.

STEED

STEED is an improvement upon STALLION. Instead of extracting the complete or partial STG, it generates ON-set and OFF-set for each primary output and each next-state line for the fault-free circuit during the preprocessing phase. The ON-set (OFF-set) of a signal is the complete set of cubes (in terms of the primary inputs and the present state lines) that produces a logic 1 (logic 0) at a signal. The ON-sets and OFF-sets of the primary outputs and next state lines can be generated using a modified PODEM algorithm.

For a given fault, PODEM is used to generate one combinational test. The state transfer sequence and fault propagation sequence are constructed by intersecting the proper ON/ OFF-sets. In general, ON/OFF-set is a more compact representation than the STG. Therefore, STEED can handle larger circuits than STALLION. STEED shows good performance for circuits that have relatively small ON/OFF-sets. However, generating, storing, and intersecting the ON/ OFF-sets can be very expensive (in terms of both CPU time and memory) for certain functions such as parity trees. Therefore, STEED may have difficulties generating tests for circuits containing such function blocks. Also, like STALLION, the transfer and fault-propagation sequences derived from the ON/OFF-sets of the fault-free circuit may not be valid for the faulty circuit and therefore need to be verified by a fault simulator.

VERITAS

VERITAS is a BDD-based test generator which uses the binary decision diagram (BDD) [55b] to represent the state transition relations as well as sets of states. In the preprocessing phase, a state enumeration algorithm based on such BDD representations is used to find the set of states that are reachable from the reset state and the corresponding shortest transfer sequence for each of the reachable states. In the test-generation phase, as with STEED, a combinational test is first generated. The state transfer sequence to drive the machine into the activation state is readily available from the data derived from reachability analysis done in the preprocessing phase. Due to the advances in BDD representation, construction and manipulation, VERITAS in general achieves better performance than STEED.

In addition to the assumption of a known reset state, another common principle used by the above three approaches is to incorporate a preprocessing phase to (explicitly or implicitly) compute the state transition information. Such information could be used during test generation to save some repeated and unnecessary state justification effort. However, for large designs with huge state space, such preprocessing could be excessive. For example, complete reachability analysis used in the preprocessing phase of VERITAS typically fails (due to memory explosion) for designs with several hundreds of flip-flops. Either using partial reachability analysis, or simply performing state justification on demand during test generation, is a necessary modification to these approaches for large designs.

22.3.4 Summary

The presence of flip-flops and feedback loops substantially increases the complexity of the ATPG. Due to the inherent intractability of the problem, it remains infeasible to automatically derive high quality tests for large, nonscan sequential designs. However, because considerable progress has been made during the past few decades, and since robust commercial ATPG tools are now available, the partial-scan design methodology that relies on such tools for test generation might become a reasonable alternative to the full-scan design methodology [56].

As is the case with most other CAD tools, there are many engineering issues involved in building a test generator to handle large industrial designs. Industrial designs may contain tristate logic, bidirectional elements, gated clocks, I/O terminals, etc. Proper modeling is required for such elements, and the test-generation process would also benefit from some modifications. Many of these issues are similar to those present in the combinational ATPG problem that have been addressed (See e.g. [57–59]).

Developing special versions of ATPG algorithms/tools for circuits with special circuit structures and properties could be a good way to further improve the ATPG performance. For example, if the target circuit has a pipeline structure and is feedback-free, the algorithm described in [60pp. 98–101] is much more efficient than any algorithm surveyed in this section, which focuses on circuits with a more general structure. Many partial-scan circuits have unique circuit structures. A more detailed comparison of various sequential ATPG algorithms, practical implementation issues, and applications with partial-scan designs can be found in the survey [56].

22.4 ATPG and SAT

SAT has attracted tremendous research effort in recent years, resulting in the development of various efficient SAT solver packages. Popular SAT solvers [13,61–64] are designed based upon the conjunctive normal form (CNF).

Given a finite set of variables, V, over the set of Boolean values $\mathbf{B} \in \{0,1\}$, a *literal*, l or $\bar{l}$ is an instance of a variable v or its complement $\neg v$, where $v \in V$. A *clause* c_i, is a disjunction of literals $(l_1 \vee l_2 \vee \ldots \vee l_n)$. A formula f, is a conjunction of clauses $c_1 \wedge c_2 \wedge \ldots \wedge c_m$. Hence, a clause is considered as a set of literals, and a formula as a set of clauses. An assignment A *satisfies* a formula f if $f(A) = 1$. In a SAT problem, a formula f is given and the problem is to find an assignment A to satisfy f or prove that no such assignment exists.

22.4.1 Search in SAT

Modern SAT solvers are based on the search paradigm proposed in GRASP [13], which is an extension from the original DPLL [65] search algorithm. Algorithm 22.2 [13] describes the basic GRASP search procedure.

Algorithm 22.2: SAT()

comment: B is the backtracking decision level

comment: d is the current decision level

```
Search(d,B)
if (decided(d) = SUCCESS)
  then return (SUCCESS)
while (true)
  do { if deduce(d) ≠ (CONFLICT)
         then { if (Search(d + 1 B) = SUCCESS)
                  then return (SUCCESS)
                if (B ≠ d)
                  then erase(); return (CONFLICT)
       if (diagnose(d,B) = CONFLICT)
         then erase(); return (CONFLICT)
       erase();
```

In the algorithm, function *decide()* selects an unassigned variable and assigns it with a logic value. This variable assignment is referred to as a *decision*. If no unassigned variable exists, *decide()* will return *SUCCESS* which means that a solution has been found. Otherwise *decide()* will return *CONFLICT* to invoke the *deduce()* procedure to check for conflict. A decision level d is associated with each decision. The first decision has decision level 1, and the decision level increases by one for each new decision.

The purpose of *deduce()* is to check for conflict by finding all necessary assignments induced by the current decisions. This step is similar to performing *implications* in ATPG. For example, in order to satisfy f, every clause of it must be satisfied. Therefore, if a clause has only one unassigned literal, and all the other literals are assigned with 0, then the unassigned literal must be assigned with value 1. A conflict occurs when a variable is assigned with both 1 and 0 or a clause becomes unsatisfiable.

The purpose of *diagnose()* is to analyze the reason that causes the conflict. The reason can be recorded as a *conflict clause.* The procedure can also determine a backtracking level other than backtracking to the previous decision level, a feature which can be used to implement *nonchronological backtracking* [66]. The *erase()* procedure deletes the value assignments at the current decision level.

In a modern SAT solver, one of the key concepts is *conflict-driven learning. Conflict-driven learning* is a method to analyze the causes of a conflict and then record the reason as a conflict clause to prevent the search from reentering the same search subspace. Since the introduction of conflict-driven learning, a hot research topic has been to find ways to derive conflict clauses that could efficiently prune the search space.

22.4.2 Comparison of ATPG and Circuit SAT

From all appearances, the problem formulation of ATPG is more complicated. ATPG involves fault activation and fault propagation, whereas circuit SAT concerns only justifying the value 1 at the single primary output of a circuit. However, as we have mentioned in Section 22.2.3, the ATPG problem can also be converted into a SAT problem [20].

Conflict-driven learning was originally proposed for SAT. One nice property of conflict-driven learning is that the reason for a conflict can be recorded as a *conflict clause* whose representation is consistent with that of the original problem. This simplifies the SAT solver implementation, in contrast to ATPGs where various learning heuristics are used, each of which may require a different data structure for efficient implementation. Silva and Sakallah in [22] argued that this simplification in implementation might provide benefits for runtime efficiency.

For circuit SAT, the conflict clauses can also be stored as gates. Figure 22.7 illustrates this. During the application of SAT search, constraints on the signal lines can be accumulated and added onto the circuit. These constraints are represented as OR gates where the outputs are set with logic value 1. These constraints encode the signal correlations that have to hold, due to the given circuit structure.

The idea of conflict-driven learning was implemented in the ATPG in [12]. For many other applications in computer-aided design (CAD) of integrated circuits, applying SAT to solve a circuit-oriented problem often requires transformation of the circuit gate-level netlist into its corresponding CNF format [67]. In a typical circuit-to-CNF transformation, the topological ordering among the internal signals is obscured in the CNF formula. In CNF format, all signals become (input) *variables.*

For solving circuit-oriented problems, circuit structural information has proved to be very useful. Tafertshofer et al. [68] developed a structural graph model called an *implication graph* for efficient implication and learning in SAT. Methods were also provided in [69,70] to utilize structural information in

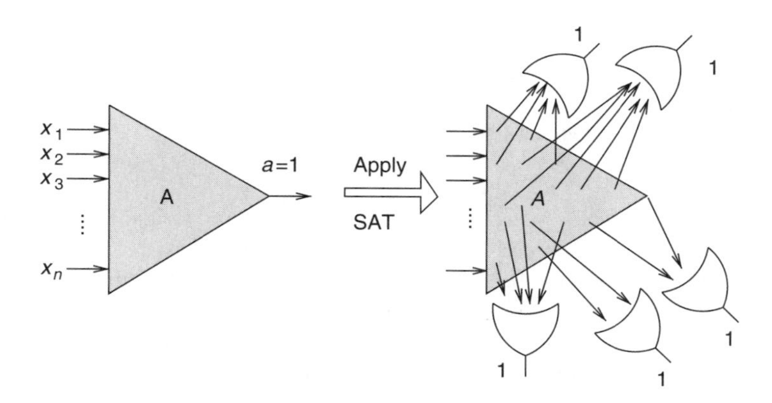

FIGURE 22.7 Learned gates accumulated by solving $a = 1$ in a circuit SAT problem.

SAT algorithms, which required minor modifications to the existing SAT algorithms. Gupta et al. [71] implemented a circuit-based SAT-solver that used structural information to identify unobservable gates and to remove the clauses for those gates. The work in [72] represented Boolean circuits in terms of 2-input AND gates and inverters. Based on this circuit model, a circuit SAT solver could be integrated with BDD sweeping [73].

Ganai et al. [74] developed a circuit-based SAT solver that adopted the techniques used in CNF-based SAT solver zChaff [62], e.g., the watched literal technique for efficient implication. Ostrowski et al. [75] tried to recover the structural information from CNF formulas, utilizing the structural information to eliminate clauses and variables. Theoretical results regarding circuit-based SAT algorithms were presented in[76]. Lu et al. [77] developed a SAT solver employing circuit-based implicit and explicit learning, applying the solver on industrial hard cases [78]. Lu et al. [79] developed a sequential circuit SAT solver for the sequential justification problem. Below, we summarize the ideas in [77,79] to illustrate how circuit information can be used in circuit SAT.

22.4.3 Combinational Circuit SAT

Consider the circuit in Figure 22.8 where shaded area B contains shaded area A, and shaded area C contains shaded area B. Suppose we want to solve a circuit SAT problem with the output objective $c = 1$. When we apply a circuit SAT solver to prove that $c = 0$ or to find an input assignment to make $c = 1$, potentially the search space for the solver is the entire circuit. Now suppose we identify, in advance, two internal signals a and b, such that $a = 1$ and $b = 0$ will be very unlikely outcomes when random inputs are supplied to the circuit. Then, we can divide the original problem into three subproblems: (1) solving $a = 1$, (2) solving $b = 0$, and then (3) solving $c = 1$.

Since $a = 1$ is unlikely to happen, when a circuit SAT solver makes decisions trying to satisfy $a = 1$, it is likely to encounter conflicts. As a result, much conflict-driven information can be learned and stored as *conflict gates* (as illustrated in Figure 22.7). If we assume that solving $a = 1$ is done only based upon the *cone of influence* headed by the signal a (the shaded area A in Figure 22.8), then the conflict gates will be based upon the signals contained in the area A only.

As the solver finishes solving $a = 1$ and starts solving $b = 0$, all the learned information regarding the circuit area A can be used to help in solving $b = 0$. In addition, if $a = 1$ is indeed unsatisfiable, then signal a can be assigned with 0 when the solver is solving $b = 0$. Similarly, learned information from solving $a = 1$ and $b = 0$ can be reused to help in solving $c = 1$.

Intuitively, we believe that solving the three subproblems following their topological order could be accomplished much faster than directly solving the original problem. This is because when solving $b = 0$, hopefully fewer (or no) decisions are required to go into area A. Hence, the search space is more restricted within the portion of area B that is not part of area A. Similarly, solving c = 1 requires most decisions to be made only within the portion of area C which is not part of area B. Moreover, the conflict gates accumulated by solving $a = 1$ could be smaller because they are based upon the signals in area A

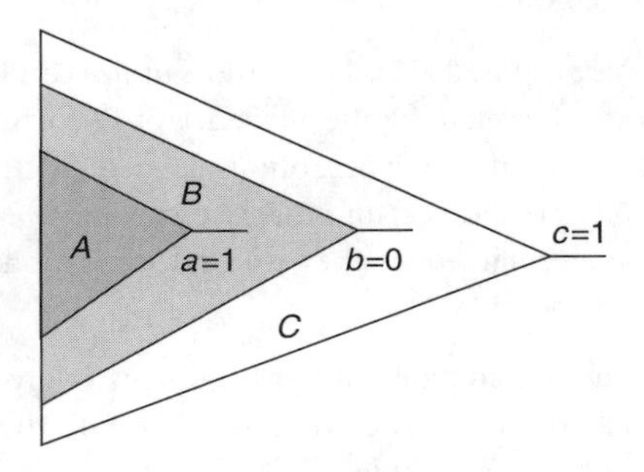

FIGURE 22.8 An example for incremental SAT solving.

only. Similarly, the conflict gates accumulated during the solving of $b = 0$ could be smaller. Conceptually, this strategy allows solving a complex problem incrementally.

Two observations can be made: (1) the incremental process suggests that we can guide the search process by solving a sequence of preselected subproblems following their topological order. (2) The selection of the sub-problems such as $a = 1$ and $b = 0$ should be those most likely to be unsatisfiable. Intuitively, we might expect that the search for a solution to a probably unsatisfiable subproblem will encounter more conflicts and can accumulate conflict-driven learning information more quickly.

If few or no conflicts arise in solving $a = 1$ and solving $b = 0$, then there may be not much information to be learned from solving these two problems. In this case, the time to spend on solving $a = 1$ may seem to be unnecessary overhead. Usually, this observation would suggest the ineffectiveness of the method used to determine that, both $a = 1$ and $b = 0$ are unlikely to be true. Moreover, if solving $c = 1$ does not depend much on signals a and b, then the above incremental strategy cannot be effective either. For example, if c is the output signal of a 2-input AND gate with two inputs as a function g and its complement g, then c = 1 can be directly proved to be unsatisfiable whether or not the actual function g is known.

Intuitively, we think the above incremental strategy would not be effective for solving a circuit SAT problem whose input is given in CNF form. This is because by treating the CNF form as a two-level OR–AND circuit structure, the topological ordering among the signals is lost. With a two-level structure, the incremental strategy has little room to proceed. In the example discussed above, both a and b become primary inputs in the 2-level OR–AND CNF circuit. Then, the ordering for solving the subproblems may become solving $b = 0$ followed by solving $a = 1$.

The authors in Ref. [77] implement a circuit SAT solver based on the idea just described. The decision ordering in the SAT solver is guided through *signal correlations* identified before the search process. A group of signals $s_1, s_2, \ldots, s_i$ (where $i > 1$) are said to be correlated if their values satisfy a certain Boolean function $f(s_1, s_2, \ldots, s_i)$ during random simulation, e.g., the values of s_1 and s_2 satisfy $s_1 = s_2$ during random simulation. Examples of signal correlations are *equivalence correlation, inverted equivalence correlation,* and *constant correlation.*

Two signals s_1 and s_2 have an equivalence correlation (inverted equivalence correlation) if and only if the values of the two signals satisfy $s_1 = s_2$ ($s_1 = s_2$) during random simulation. If s_2 is replaced by constant 0 or 1 in the notation, then that is a constant correlation. Note that the correlations are defined based on the results of random simulation. Signals may be correlated in one run of random simulation, yet not be correlated in another run of random simulation.

In [77], signal correlations can be used in two types of learning strategies: *explicit learning* and *implicit learning.* In explicit learning, the search is constrained by solving a sequence of subproblems constructed from the signal correlations to implement the incremental search as described above. In implicit learning, the decision-making in solving the original SAT objective is influenced by the signal correlations. No explicit solving of subproblems is performed. When the circuit SAT solver was applied to some industrial hard cases, it obtained promising performance results [78].

22.4.4 Sequential Circuit SAT

Recently, a sequential SAT solver was proposed [79]. The authors utilize combined ATPG and SAT techniques to implement a sequential SAT solver by retaining the efficiency of Boolean SAT and being complete in the search. Given a circuit following the Huffman synchronous sequential-circuit model, *Sequential SAT* (or *sequential justification)* is the problem of finding an *ordered input assignment sequence* such that a desired objective is satisfied, or proving that no such sequence exists. Under this model, a sequential SAT problem can fit into one of the following two categories:

- In a weak SAT problem, an initial-state value assignment is given. The problem is to find an ordered sequence of input assignments such that together with the initial state, the desired objective is satisfied or proved to be unsatisfiable.

- In a strong SAT problem, no initial state is given. Hence, it is necessary to identify an input sequence to satisfy the objective starting from the unknown state. To prove unsatisfiability, a SAT solver needs to prove that no input sequence can satisfy the given objective for *all* reachable initial states.

A strong SAT problem can be translated to a weak SAT problem by encoding technique [80]. In sequential SAT, a sequential circuit is conceptually unfolded into multiple copies of the combinational circuit through time frame expansion. In each time frame, the circuit becomes combinational and hence, a combinational SAT solver can be applied. In each time frame, a state element such as a flip-flop is translated into two corresponding signals: a pseudo primary input (PPI) and a pseudo-primary output (PPO). The initial state is specified with the PPIs in time frame 0. The objective is specified with the signals in time frame n (the last time frame, where n is unknown before solving the problem). During search in the intermediate time frames, intermediate solutions are produced at the PPIs, and they become the intermediate PPO objectives to be justified in the previous time frames.

Given a sequential circuit, a *state clause* is a clause consisting only of state variables. A state clause encodes a state combination where no solution can be found. Due to the usage of state clauses, the time frame expansion can be implemented implicitly by keeping only one copy of the combinational circuit.

22.4.4.1 Sequential SAT and State Clauses

To illustrate the usage of state clauses in sequential SAT, Figure 22.9 depicts a simple example circuit with three primary inputs a b c, one primary output f, and three state-holding elements (i.e., flip-flops) x,y,z. The initial state is $x = 1$, $y = 0$, $z = 1$. Suppose the SAT objective is to satisfy $f = 1$.

Starting from time frame n where n is unknown, the circuit is treated as a combinational circuit with state variables duplicated as PPOs and PPIs. This is illustrated as (1) in the figure. Since this represents a combinational SAT problem, a combinational circuit SAT solver can be applied.

Suppose after the combinational SAT solving, we can identify a solution $a = 1$, $b = 0$, $c = 0$, $PPI_x = 0$, $PPI_y = 1$ $PPI_z = 0$ to satisfy $f = 1$ (step (2)). The PPI assignment implies a state assignment $x = 0$, $y = 1$, $z = 0$. Since it is not the initial state, at this point, we may choose to continue the search by expanding into time frame $n - 1$ (this follows a depth-first search strategy).

Before solving in time frame $n - 1$, we need to examine the solution state $x = 0$, $y = 1$, $z = 0$ more closely. This is because this solution may not represent the minimal assignment to satisfy the objective $f = 1$.

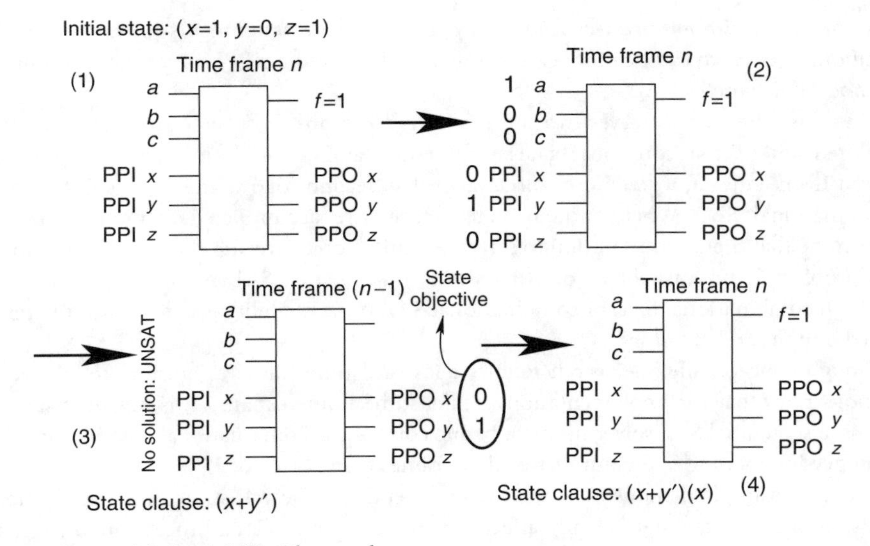

FIGURE 22.9 Sequential circuit SAT with state clauses.

Suppose that after analysis, we can determine that $z = 0$ is unnecessary. In other words, by keeping the assignment "$x = 0$, $y = 1$," we may discover that $f = 1$ can still be satisfied. This step is called *state minimization.*

After state minimization, backward time-frame expansion is achieved by adding a *state objective* $PPO_x=0$, $PPO_y = 1$ to the combinational copy of the circuit. Also, a state clause "$(x + y')$" is generated to prevent reaching those same state solutions defined by the state assignment "$x = 0$, $y = 1$." The new combinational SAT instance is then passed to the combinational circuit SAT solver for solving.

Suppose in time frame $n - 1$, no solution can be found. Then, we need to backtrack to time frame n to find another solution other than the state assignment "$x = 0$, $y = 1$." In a way, we have proved that from state "$x = 0$, $y = 1$," there exists no solution. This implies that there is no need to remove the state clause "$(x+y)$." However, at this point, we may want to perform further analysis to determine whether both $PPO_x = 0$ and $PPOy = 1$ are necessary to cause the unsatisfiability. Suppose that after *conflict analysis*, we discover that PPO_x 0 alone is sufficient to cause the conflict, then, for the sake of efficiency, we want to add another state clause "(x)." This is illustrated in (4) of the Figure 22.9. This step is called *state conflict analysis.*

The new combinational SAT instance now has the state clauses "$(x+y')(x)$" that record the non solution state subspaces previously identified. The solving continues until either one of the following two conditions is reached:

1. After state minimization, a solution is found with a state assignment *containing* the initial state. For example, a solution with the state assignment "$x=1$, $z = 1$" contains the initial state "$x=1$, $y = 0$, $z = 1$." In this case, a solution for the sequential SAT problem is found. We note that a solution without any state assignment and with only PI assignments is considered as one containing any initial state.
2. If in time frame n, the initial objective $f = 1$ and state clauses together cannot be satisfied, then the original problem is unsatisfiable. This is equivalent to saying that, without adding any PPO objective, if the initial objective $f = 1$ and state clauses together cannot be satisfied, then the original problem is unsatisfiable. Here, with the state clauses, the time-frame expansion is conducted implicitly rather than explicitly, i.e., only one copy of the combinational part of the circuit is required to be kept in the search process.

The above example illustrates several important concepts in the design of a sequential-circuit SAT solver:

- The *state minimization* involves finding the minimal state assignments for an intermediate PPI solution. The search can be more efficient with an intermediate PPI solution containing a smaller number of assigned states.
- The use of state clauses serves two purposes: (1) to record those state subspaces that have been explored, and (2) to record those state subspaces containing no solution. The first is to prevent the search from entering a *state justification loop.* The second follows the same spirit as that in the combinational SAT. When enough state clauses are accumulated, combinational SAT can determine that there is no solution to satisfy the initial objective and the state clauses. In this case, the problem is unsatisfiable. Note that while in sequential SAT, unsatisfiability is determined through combinational SAT, in combinational SAT, unsatisfiability is determined by implications based on the *conflict clauses.*
- Although conceptually, the search follows backward time frame expansion, the above example demonstrates that for implementation, explicit time frame expansion is not necessary. In other words, a sequential SAT solver needs only one copy of the combinational circuit. Moreover, there is no need to memorize the number of time frames being expanded.
- The above example demonstrates the use of a depth-first search strategy. As mentioned before, decision ordering can significantly affect the efficiency of a search process. In sequential circuit SAT, the issue is when to proceed with time-frame expansion. In depth-first fashion, time-frame

expansion is triggered by trying to solve the most recent state objective generated. In breadth-first fashion, the newly generated state objectives are resolved after solving all state objectives in the current time frame. A hybrid search strategy can be implemented by managing the state objective queue in various orders.

A *frame objective* is an objective to be satisfied which is passed to the combinational SAT solver. A frame objective can be either the initial objective or a state objective. *A frame solution* is an assignment on the PIs and PPIs, which satisfies a given frame objective without conflicting with the current state clauses.

When a frame objective is proved by the combinational SAT to be unsatisfiable, conflict analysis is applied to derive additional state clauses based on only the PPOs. In other words, conflict analysis traces back to PPOs to determine which of of them actually contribute to the conflict. Here, the conflict analysis can be similar to that in the combinational SAT [81], but the goal is to analyze the conflict sources up to the PPOs only.

As in the combinational SAT, where conflict clauses are accumulated through the solving process, in sequential SAT design, state clauses are accumulated through the solving process. The sequential solving process consists of a sequence of combinational solving tasks based on given frame objectives. At the beginning, the frame objective is the initial objective. As the solving proceeds, many state objectives become frame objectives. Hence, frame objectives also accumulate through the solving process.

A frame objective can be removed from the *objective array* only if it is proved to be unsatisfiable by the combinational SAT. If it is satisfiable, the frame objective stays in the *objective array.* The sequential SAT solver stops only when the *objective array* becomes empty. This means that it has exhausted all the state objectives and has also proved that the initial objective is unsatisfiable based on the accumulated state clauses.

During each step of combinational SAT, conflict clauses also accumulate through the combinational SAT solving process. When the sequential solving switches from one frame objective to another, these conflict clauses stay. Hence, in the sequential solving process, the conflict clauses generated by the combinational SAT are also accumulated. Although these conflict clauses can help to speed up the combinational SAT solving, experiences show that for sequential SAT, managing the state clauses dominates the overall sequential search efficiency [79].

Algorithm 22.3: SEQUENTIAL SAT(C,obj,s_0)

```
comment: C is the circuit with PPIs and PPOs expanded
comment: obj is the initial objective
comment: s_0 is the initial state
comment: FO is the objective array
FO ← {obj};
while (FO ≠ 0)
do {
    f_obj ← selecta_frame_objective(FO);
    f_sol ← combinational_solve_a_frame_objective(C,f_obj);
    if (f_sol = NULL)
        { clause ← PPO_state_conflict_analysis(C,f_obj);
        { add_state_clause(C,clause);
        { FO ← FO − {f_obj};
    else {
        stateassignment ← state_minimization(C,f_obj,f_sol);
        if (s_0 ∈ stateassignment)
        {return (SAT);
        else { clause ← convert_to_clause(stateassignment);
             { add_state_clause(C,clause);
             { FO ← FO + (stateassignment);
return (UNSAT)
```

The overall algorithm of the sequential circuit SAT solver is described in Algorithm 22.3. Note that in this algorithm, solving each frame objective produces only one solution. However, it is easy to extend this algorithm so that solving each frame objective produces many solutions at once. The search strategy is based on the selection of one frame objective at a time from the objective array *FO*, where different heuristics can be implemented. The efficiency of the sequential circuit SAT solver highly depends on the selection heuristic [79].

22.5 Applications of ATPG

In this section, we show that ATPG technology, in addition to generating high-quality tests for various fault models, also offers efficient techniques for analyzing designs during design verification and optimization. Already, ATPG has been used to generate tests not only to screen out chips with manufacturing defects but also to identify design errors and timing problems during design verification. It has also been used as a powerful logic-analysis engine for applications such as logic optimization, timing analysis, and design-property checking.

22.5.1 ATPG for Delay Faults and Noise Faults

The move toward nanometer technology is introducing new failure modes and a new set of design and test problems [82]. Device features continue to shrink as the number of interconnect layers and gate density increases. The result is increased current density and a higher voltage drop along the power nets as well as increased signal interference from coupling capacitance. All this gives rise to noise-induced failures, such as power supply noise or crosstalk. These faults may cause logic errors or excessive propagation delays which degrade circuit performance.

Demands for higher circuit operating frequencies, lower cost, and higher quality mean that testing must ascertain that the circuit's timing is correct. Timing defects can stay undetected after logic-fault testing such as testing of stuck-at faults, but they can be detected using delay tests. Unlike ATPG for stuck-at faults, ATPG for delay faults is closely tied to the test application strategy [83]. Before tests for delay faults are derived, the test application strategy has to be decided. The strategy depends on the circuit type as well as on the test equipment's speed.

In structural delay testing, detecting delay faults requires applying 2-vector patterns to the combinational part of the circuit at the circuit's intended operating speed. However, because high-speed testers require huge investments, most testers could be slower than the designs being tested. Testing high-speed designs on slower testers requires special test application and test-generation strategies — a topic that has been investigated for many years [83,84]. Because an arbitrary vector pair cannot be applied to the combinational part of a sequential circuit, ATPG for delay faults may be significantly more difficult for these than for full-scan circuits. Various testing strategies for sequential circuits have been proposed, but breakthroughs are needed [84]. Most researchers believe that some form of DFT is required to achieve high-quality delay testing for these circuits.

Noise faults must be detected during both design verification and manufacturing testing. When coupled with process variations, noise effects can exercise worst-case design corners that exceed operating conditions. These corners must be identified and checked as part of design validation. This task is extremely difficult, however, because noise effects are highly sensitive to the input pattern and to timing. Timing analysis that cannot consider how noise effects influence propagation delays will not provide an accurate estimation of performance, nor will it reliably identify problem areas in the design.

An efficient ATPG method must be able to generate validation vectors that can exercise worst-case design corners. To do this, it must integrate accurate timing information when the test vectors are derived. For manufacturing testing, ATPG techniques must be augmented and adapted to new failure conditions introduced by nanometer technology. Tests for conventional fault models, such as stuck-at and transition faults, obviously cannot detect these conditions. Thus, to check worst-case design corners, test vectors must sensitize the faults and propagate their effects to the primary outputs, as well as activate the conditions of worst-case noise effects. They must also scale to increasingly larger designs.

Power supply noise. For a highly integrated system-on-a-chip, more devices are switching simultaneously, which increases power supply noise. One component of this noise, inductive noise, results from sudden current changes on either the package lead or wire/substrate inductance. The other component, net IR voltage drop, is caused by current flowing through the resistive power and ground lines. The noise can cause a voltage glitch on these lines, resulting in timing or logic errors. Large voltage drops through the power supply lines can cause electromigration, which in turn can cause short or open circuits. To activate these defects and propagate them to the primary outputs, ATPG must carefully select test vectors.

Power supply noise can affect both reliability and performance. It reduces the actual voltage level that reaches a device, which in turn can increase cell and interconnection propagation delays. One way to detect these effects is to apply delay tests. Unfortunately, most existing delay techniques are based on simplified, logic-level models that cannot be directly used to model, and test timing defects in high-speed designs that use deep sub-micron technologies. New delay testing strategies are needed to close the gap between the logic-level delay fault models and physical defects. The tests must produce the worst-case power supply noise along the sensitized paths, and thus cause the worst-case propagation delays on these paths [85,86].

Crosstalk effects. The increased design density in deep-submicron designs leads to more significant interference between the signals because of capacitive coupling, or crosstalk. Crosstalk can induce both Boolean errors and delay faults. Therefore, ATPG for worst-case crosstalk effects must produce vectors that can create and propagate crosstalk pulses as well as crosstalk-induced delays [87–91].

Crosstalk-induced pulses are likely to cause errors on hazard-sensitive lines such as inputs to dynamic gates, clock, set/reset, and data inputs to flip-flops. Crosstalk pulses can result in logic errors or degraded voltage levels, which increase propagation delays. ATPG for worst-case crosstalk pulse aims to generate a pulse of maximum amplitude and width at the fault site and propagate its effects to primary outputs with minimal attenuation [92].

Studies show that increased coupling effects between signals can cause signal delay to increase (slowdown) or decrease (speedup) significantly. Both conditions can cause errors. Signal slowdown can cause delay faults if a transition is propagated along paths with small slacks. Signal speedup can cause race conditions if transitions are propagated along short paths. To guarantee design performance, ATPG techniques must consider how worst-case crosstalk affects propagation delays [88,89].

22.5.2 Design Applications

ATPG technology has been applied successfully in several areas of IC design automation, including logic optimization, logic equivalence checking, design property checking, and timing analysis.

22.5.2.1 Logic Optimization

To optimize logic, design aids can either remove redundancy or restructure the logic by adding and removing redundancy.

Redundancy Removal Redundancy is the main link between test and logic optimization. If there are untestable stuck-at faults, there is likely to be redundant logic. The reasoning is that, if a stuck-at fault does not have any test (the fault is untestable), the output responses of the faulty circuit (with this untestable fault) will be identical to the responses of the fault-free circuit for all possible input patterns applied to these two circuits. Thus, the faulty circuit (with an untestable stuck-at fault) is indeed a valid implementation of the fault-free circuit. Therefore, when ATPG identifies a stuck-at-1 (stuck-at-0) fault as untestable, one can simplify the circuit by setting the faulty net to logic

1(0) and thus effectively removing the faulty net from the circuit. This operation, called *redundancy removal*, also removes all the logic driving the faulty net.

Figure 22.10 illustrates an example. However, note that output Z in Figure 22.10(a) is hazard-free, but output Z in Figure 22.10(b) may have glitches. Testers must ensure that redundancy is removed only if having glitches is not a concern (as in synchronous design, for example).

Because this method only removes logic from the circuits, the circuit is smaller when the process ends; the topological delay of the longest paths will be shorter than or at most equal to that of the original circuit. The power dissipation of the optimized circuit will also be lower.

Logic Restructuring Removing a redundant fault can change the status of other faults. Those that were redundant might no longer be redundant, and *vice versa.* Although these changes complicate redundancy removal, they also pave the way for more rigorous optimization methods. Even for a circuit with no redundancies, designers can add redundancies to create new redundancies elsewhere in the circuit. By removing the created new redundancies, they may obtain an optimized circuit. This technique is called logic restructuring. For example, Figure 22.11 shows a circuit example that has no redundant logic. In Figure 22.12(a), a signal line is artificially added that does not change the function of the circuit but does create redundant logic. Figure 22.12(b) shows the resulting circuit after redundancy removal. This circuit is simpler than the one in Figure 22.11.

Efficient algorithms for finding effective logic restructuring [93] have been proposed in the past few years. By properly orienting the search for redundancy, these techniques can be adapted to target several optimizing goals

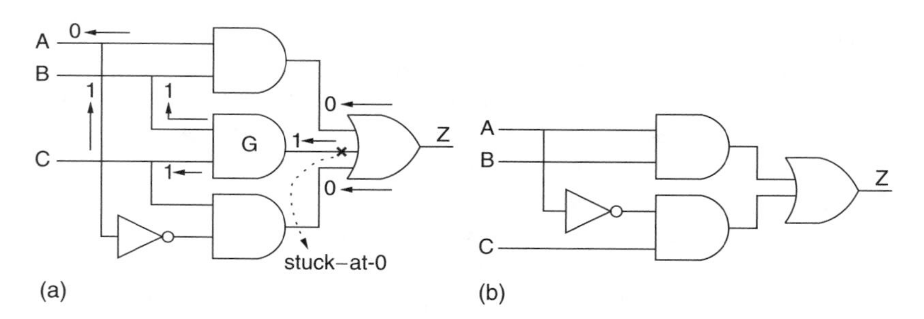

FIGURE 22.10 How ATPG works for redundancy removal: (a) the stuck-at 0 fault is untestable; (b) remove gate G and simply the logic.

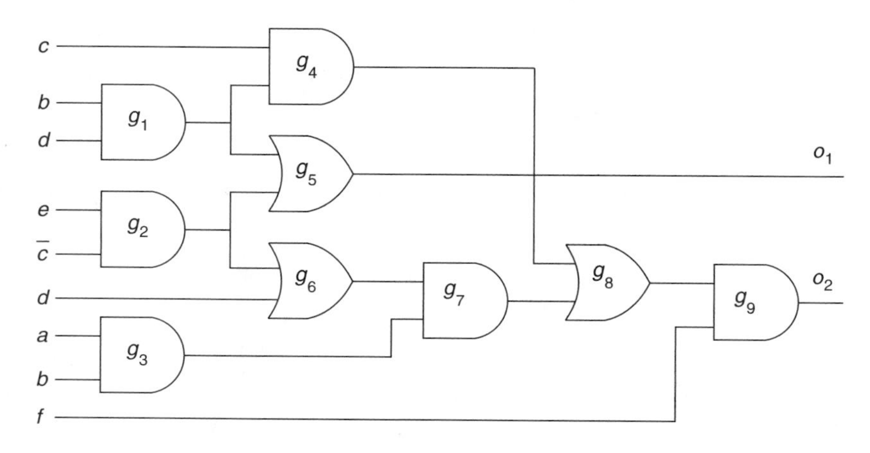

FIGURE 22.11 A circuit that is not redundant.

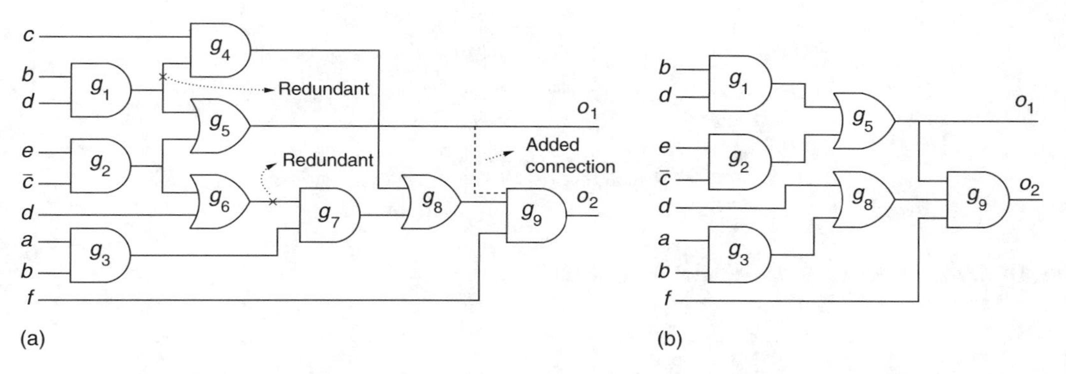

FIGURE 22.12 An example of logic restructuring by adding and removing redundancy: (a) adding a redundant connection; (b) optimized circuit after redundancy removal.

22.5.2.2 Design Verification

Techniques used to verify designs include checking logic equivalence and determining that a circuit does or does not violate certain properties.

Logic Equivalence Checking. It is important to check the equivalence of two designs described at the same or different levels of abstraction. Checking the functional equivalence of the optimized implementation against the RTL specification, for example, guarantees that no error is introduced during logic synthesis and optimization, especially if part of the process is manual. Checking the equivalence of the gate-level implementation and the gate-level model extracted from the layout assures that no error is made during physical design.

Traditionally, designers check the functional equivalence of two Boolean functions by constructing their canonical representations, as truth tables or BDDs, for example. Two circuits are equivalent if and only if their canonical representations are isomorphic.

Consider the comparison of two Boolean networks in Figure 22.13. A joint network can be formed by connecting the corresponding primary input pairs of the two networks and by connecting the corresponding primary output pairs to XOR gates. The outputs of these XOR gates become the new primary outputs of the joint network. The two networks are functionally equivalent if the primary output response of the joint network is 0 for any input vector. Therefore, to prove that two circuits are equivalent, designers must merely prove that no input vector produces 1 at this model's output signal *g*.

Another way to do equivalence checking is to formulate it as a problem that searches for a distinguishing vector, for which the two circuits under verification produce different output responses. If no distinguishing vector can be found after the entire space is searched, the two circuits are equivalent. Otherwise, a counterexample is generated to disprove equivalence. Because a distinguishing vector is also a test vector for the stuck-at-0 fault on the joint network's output *g*, equivalence checking becomes a test-generation process for *g's* stuck-at-0 fault. However, directly applying ATPG to check the output equivalence (finding a test for stuck-at-0 fault) could be CPU-intensive for large designs.

Figure 22.14 shows how complexity can be reduced substantially by finding an internal functional similarity between the two circuits being compared [94]. Designers first use naming information or structure analysis to identify a set of potentially equivalent internal signal pairs. They then build a model, as in Figure 22.14(a), where signals *a1* and *a2* are candidate internal equivalent signals. To check the equivalence between these signals, we run ATPG for a stuck-at-0 fault at signal line f. If ATPG concludes that no test exists for that fault, the joint network can be simplified to the one in Figure 22.14(b), where signal *a1* has been replaced with signal *a2*

With the simplified model, the complexity of ATPG for the output *g* stuck-at-0 fault will be reduced. The process identifies internal equivalent pairs sequentially from primary inputs to primary outputs. By the time it gets to the output of the joint network, the joint network could be substantially smaller, and

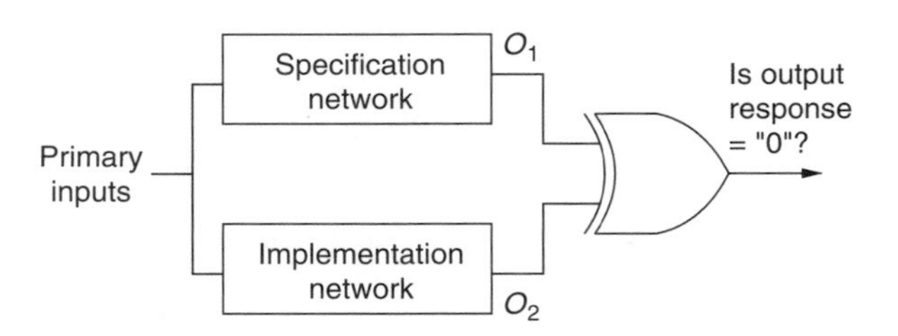

FIGURE 22.13 Circuit model for equivalence checking of two networks.

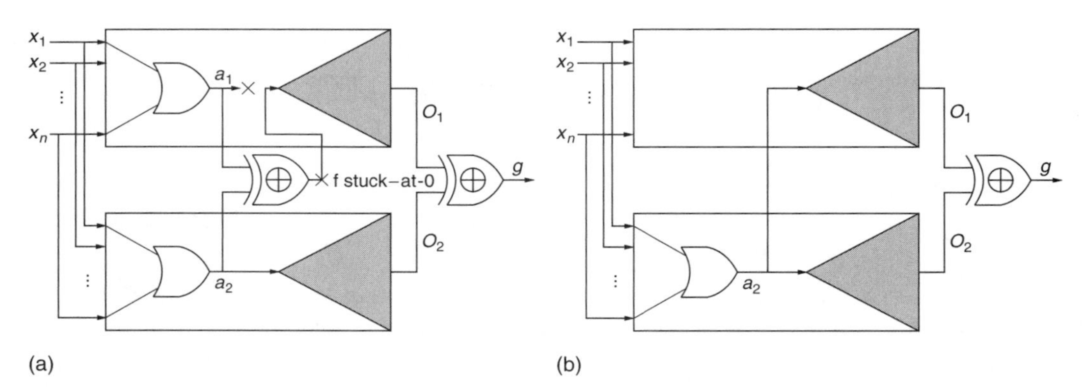

FIGURE 22.14 Pruning a joint network by finding internal equivalent pair. (a) Model for checking if signals *a1* and *a2* are equivalent; (b) Reducing complexity by replacing *a1* with *a2*.

ATPG for the *g* stuck-at-0 fault will be quite trivial. Various heuristics for enhancing this idea and combining it with BDD techniques have been developed in the past few years [95]. Commercial equivalence checking tools can now handle circuit modules of more than a million gates within tens of CPU minutes.

Property Checking An ATPG engine can find an example for proving that the circuit violates certain properties or, after exhausting the search space, can prove that no such example exists and thus that the circuit meets certain properties [96,97]. One example of this is checking for tristate bus contention, which occurs when multiple tristate bus drivers are enabled and their data is not consistent. Figure 22.15 shows a sample application. If the ATPG engine finds a test for the output stuck-at-0 fault, the test found will be the vector that causes bus contention. If no test exists, the bus can never have contention. Similarly, ATPG can check to see if a bus is floating — all tristate bus drivers are disabled — simply by checking for a vector that sets all enable lines to an inactive state.

ATPG can also identify races, which occur when data travels through two levels of latches in one clock cycle. Finally, an ATPG engine can check for effects (memory effect or an oscillation) from asynchronous feedback loops that might be in a pure synchronous circuit [96]. For each asynchronous loop starting and ending at a signal S, the ATPG engine simply checks to see whether there is a test to sensitize this loop. If such a test exists, the loop will cause either a memory effect (the parity from S to S is even) or an oscillation (the parity is odd).

Timing Verification and Analysis Test vectors that sensitize selected long paths are often used in simulations to verify circuit timing. In determining the circuit's clock period, designers look for the slowest true path. Various sensitization criteria have been developed for determining a true path. Such criteria set requirements (in terms of logic values and the arrival times of the logic values) at side inputs of the gates along the path. These requirements are somewhat similar to those for deriving tests for path delay faults. Thus, an ATPG engine can be used directly for this application.

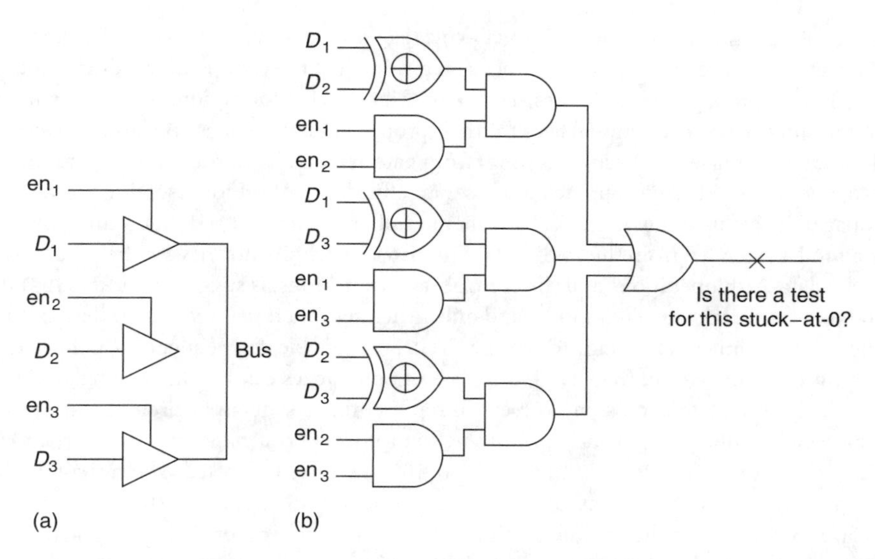

FIGURE 22.15 Application to bus contention checking. (a) A bus example; (b) An ATPG model for checking bus contention.

22.5.3 Summary

ATPG remains an active research area in both the computer-aided design and the test communities, but the new emphasis is on moving ATPG operations toward higher levels of abstraction and on targeting new types of faults in deep submicron devices. Test tools have evolved beyond merely gate-level test generation and fault simulation. Most design work now takes place at RTL and above, and test tools must support RTL handoff. New noise faults, including those from power supply noise, and crosstalk-induced noise, as well as substrate and thermal noise, will need models for manufacturing testing. The behaviors of these noise faults need to be modeled at levels of abstraction higher than the electrical, circuit, and transistor levels. Finding test vectors that can cover these faults is a challenge for ATPG.

22.6 High-Level ATPG

Test generation could be significantly speeded up if a circuit model at a higher level of abstraction is used. In this section, we discuss briefly the principles of approaches using RTL models and STGs. Here, we do not intend to give a detailed survey for such approaches, only a brief description of representative methods.

Approaches using RTL models have the potential to handle larger circuits because the number of primitives in an RTL description is much smaller than the gate count. Some methods in this class of approaches use only RTL description of the circuit [98–103], while others assume that both gate-level and RTL models are available [30–32]. Note that automatic extraction of the RTL description from a lower level of description is still not possible and therefore the RTL descriptions must be given by the designers. It is also generally assumed that data and control can be separated in the RTL model.

For approaches using both RTL and gate-level models [30–32], typically a combinational test is first generated using the gate-level model. The fault-free justification sequence and the fault propagation sequence are generated using the (fault-free) RTL description. Justification and fault propagation sequences generated in such a manner may not be valid and therefore need to be verified by a fault simulator. These approaches, in general, are suitable for data-dominated circuits but are not appropriate for control-dominated circuits.

For approaches using only RTL models [98–103], functional fault models at RTL are targeted, instead of the single-stuck-at fault model at the gate-level. The approaches in [98,99] target microprocessors and functional fault models are defined for various functions at the control-sequencing level. Because tests are generated for the functional fault models, a high coverage for gate-level stuck-at faults cannot be guaranteed. The methods suggested in [100,101] focus on minimizing the value conflicts during the value justification and fault propagation processes, using the high-level information. The technique in [103] intends to guarantee that the functional tests for their proposed functional faults achieve a complete gate-level stuck-at fault coverage. To do so, mappings from gate-level faults to functional operations of modules need to be established. This approach also uses an efficient method for resolving the value conflicts during propagation/justification at the RTL . A method of characterizing a design's functional information using a model extended from the traditional finite-state machine model, with the capability of modeling both the data-path operations and the control state transitions, is suggested in Ref.[102]. However, this method does not target any fault model and only generates functional vectors for design verification.

For finite-state machines (FSMs) for which the STGS are available, test sequences can be derived using the state transition information. In general, this class of approaches can handle only relatively small circuits due to the known state-explosion problem in representing a sequential circuit using its state table. However, successful applications of such approaches to protocol performance testing [104] and to testing the boundary-scan Test Access Port (TAP) controller [105] have been reported. The earliest method is the checking experiment [26] that is based on distinguishing sequences. The distinguishing sequence is defined as an input sequence that produces different output responses for each initial state of the FSM. This approach is concerned with the problem of determining whether or not a given state machine is distinguishable from all other possible machines with the same number of states or fewer. No explicit fault model is used. The distinguishing sequence may not exist, and the bound on length, if it exists, is proportional to the factorial of the number of states. This method is impractical because of the long test sequence. Improved checking experiments, based on either the simple I/O sequence [27] or the unique input/output (UIO) sequence [104] of the FSM, significantly reduce the test length.

A functional fault model in the state transition level has recently been used in a test generator FTG for FSMs [29]. In the single-state-transition (SST) fault model, a fault causes the destination state of a single state transition to be faulty. It has been shown [29] that the test sequence generated for the SST faults in the given STG achieves high fault coverages for the single stuck-at faults as well as the transistor faults in its multi-level logic implementation. As an approximation, FTG uses the fault-free STG to generate the fault-propagation sequence. AccuraTest [28] further improves the technique by using both fault-free and faulty circuits' STGs to generate accurate test sequences for the SST faults as well as for some multiple-state-transition faults.

More recent works in high-level ATPG can be found in [106–109]. The work in [106] utilizes *program slicing*, which was originally proposed as a static program analysis technique, for hierarchical test generation. Program slicing extracts environmental constraints for a given module, after which the module with the constraints can be synthesized into a gate-level model for test generation using a commercial sequential ATPG tool. Since program slicing extracts only relevant constraints for a module and ignores the rest of the design, it can significantly reduce the complexity of ATPG for each individual module. Huang and Cheng [107] propose a *word-level* ATPG combined with an arithmetic constraint solver in an ATPG application for property checking. The word-level ATPG involves a *word-level* implication engine. The arithmetic constraint solver is used as a powerful implication engine on the arithmetic datapath. Zhang et al. [109] propose a sequential ATPG guided by RTL information represented as an *assignment decision diagram* (ADD). STGs extracted from ADDs are used to guide the ATPG process. Iyer [108] developed a constraint solver for application in functional verification based on *random test program generation* (RTPG) methodology. The constraint solver utilizes word-level ATPG techniques to derive functional test sequences under user-supplied functional constraints.

References

[1] Abramovici, M., Breuer, M.A., and Friedman, A.D., *Digital Systems Testing and Testable Design*, Rev. Print Edition, Wiley-IEEE Press, 1994.

[2] Maxwell, P.C., Aitken, R.C., Johansen, V., and Chiang, I., The effect of different test sets on quality level prediction: When is 80% better than 90%? *Proceedings of the International Test Conference*, Washington, DC, 1991, pp. 26–30, 358–364.

[3] Maxwell, P.C. and Aitken, R.C., Test sets and reject rates: All fault coverages are not created equal, *IEEE Des. Test Comput.*, 10, 42–51, 1993.
[4] Wang, L.-C., Mercer, M.R., Kao, S.W., and Williams, T.W., On the decline of testing efficiency as fault coverage approaches 100%, *13th IEEE VLSI Test Symposium*, Princeton, NJ, 30 April–3 May 1995, pp. 74–83.
[5] Ma, S.C., Franco, P., and McCluskey, E.J., An experimental chip to evaluate test techniques experiment results, *Proceeding of the International Test Conference*, Washington, DC, 21–25 October 1995, pp. 663–672.
[6] Grimaila, M.R., Sooryong Lee., Dworak, J., Butler, K. M., Stewart, B., Balachandran, H., Houchins, B., Mathur, V., Jaehong Park, Wang, L.-C., and Mercer, M. R., REDO-random excitation and deterministic observation – first commercial experiment, *17th IEEE VLSI Test Symposium*, Dana Point, California, 25–29 April 1999, pp. 268–274.
[7] Dworak, J., Wicker, J.D., Lee, S., Grimaila, M.R., Mercer, M.R., Butler, K.M., Stewart, B., and Wang, L.-C., Defect-oriented testing and defective-part-level prediction, *IEEE Des. Test Comput.*, 18, 31–41, 2001.
[8] Fujiwara, H. and Toida, S., The complexity of fault detection problems for combinational logic circuits, *IEEE Trans. Comput.*, 31, 555–560, 1982.
[9] Roth, J.P., Diagnosis of automata failure: a calculus & a method, *IBM J. Res. Develop.*, 10, 278–291, 1966.
[10] Schulz, M.H., Trischler, E., and Sarfert, T.M., SOCRATES: a highly efficient automatic test pattern generation system, *IEEE Trans. Comput.-Aided Des. ICs*,. 7, 126–137, 1988.
[11] W. Kunz, and Pradhan, D.K., Recursive learning: an attractive alternative to the decision tree for test generation in digital circuits, *International Test Conference*, Washington, DC, 1992, pp. 816–825.
[12] Wang, C., Reddy, S.M., Pomeranz, I., Lin, X., and Rajski, J., Conflict driven techniques for improving deterministic test pattern generation, *International Conference on Computer-Aided Design*, San Jose, CA, 2002, pp. 87–93.
[13] Marques-Silva, J.P., and Sakallah, K.A., GRASP: a search algorithm for propositional satisfiability, *IEEE Trans. Comput.*, 48, 506–521, 1999.
[14] Henftling, M., Wittmann, H.C., and Antreich, K.J., A Single-Path-Oriented Fault-Effect Propagation in Digital Circuits Considering Multiple-Path Sensitization, *International Conference on Computer-Aided Design*, 1995, pp. 304–309.
[15] Goel, P., An implicit enumeration algorithm to generate tests for combinational logic circuits, *IEEE Trans. Comput.*, C–30, 215–222, 1981.
[16] Fujiwara, H., and Shimono, T., On the acceleration of test generation algorithms. *IEEE Trans. Comput.*, C-32, 1137–1144, 1983.
[17] Hamzaoglu, I., and Patel, J.H., New techniques for deterministic test pattern generation, *IEEE VLSI Test Symposium*, Monterey, CA, 1998, pp. 446–451.
[18] Kirkland, T., and Mercer, M.R., A topological search algorithm for ATPG, *24th ACM/IEEE Design Automation Conference*, Miami Beach, FL, 1987, pp. 502–508.
[19] Ivanov, A., and Agarwal, V.K., Dynamic testability measures for ATPG, *IEEE Trans. Comput.-Aided Des. ICs*, 7, 1988, pp. 598–608.
[20] Larrabee, T., Test pattern generation using Boolean Satisfiability. *IEEE Trans. Comput.-Aided Des. ICs*, 11, 4–15, 1992.
[21] Stephan, P., Brayton, R.K., and Sagiovanni-Vincentelli. A.L., Combinational test generation using satisfiability, *IEEE Trans. Comput.-Aided Des. ICs*, 15, 1167–1176, 1996.
[22] Marques-Silva, J.P., and Sakallah, K.A., Robust search algorithms for test pattern generation, *IEEE Fault-Tolerance Computing Symposium*, Seattle, WA, 1997, pp. 152–161.
[23] Gizdarski, E., and Fujiwara, H., SPIRIT: A highly robust combinational test generation algorithm. *IEEE VLSI Test Symposium*, Marina Del, Rey, CA, 2001, pp. 346–351.
[24] Seshu, S., and Freeman, D.N., The diagnosis of asynchronous sequential switching systems, *IRE Trans. Electron. Computing*, EC-11, 459–465, 1962.
[25] Thomas, J.J., Automated diagnostic test program for digital networks, *Comput. Des.*, 63–67, 1971.
[26] Hennie, F.C., Fault-detecting experiments for sequential circuits, *5th Annual Symposium on Switching Circuit Theory and Logical Design*, 1964, pp. 95–110.
[27] Hsieh, E.P., Checking experiments for sequential machines, *IEEE Trans. Comput.*, C-20, 1152–1166, 1971.

[28] Pomeranz, I., and Reddy, S.M., On achieving a complete fault coverage for sequential machines using the transition fault model, *28th Design Automation Conference*, San Francisco, CA, 1991, pp. 341–346.
[29] Cheng, K.-T., and Jou, J.-Y., A functional fault model for sequential circuits, *IEEE Trans. Comput.-Aided Des. ICs*, 11, 1992, pp. 1065–1073.
[30] Hill, F.J., and Huey, B., SCIRTSS: A search system for sequential circuit test sequences, *Transactions on Computers*, C-26, 490–502, 1977.
[31] Breuer, M.A., and Friedman, A.D., Functional level primitives in test generation, *IEEE Trans. Comput.*, C-29, 223–235, 1980.
[32] Ghosh, A., Devadas, S., and Newton, A.R., Sequential test generation at the register-transfer and logic levels, *27th ACM/IEEE Design Automation Conference*, Orlando, FL, June 1990, pp. 580–586.
[33] Kubo, H., A procedure for generating test sequences to detect sequential circuit failures, NEC *Res. Dev.*, 12, 69–78, 1968.
[34] Putzolu, G.R, and Roth, J.P., A heuristic algorithm for the testing of asynchronous circuits, *IEEE Trans. Comput.* C-20, 639–647, 1971.
[35] Marlett, R., EBT, A comprehensive test generation technique for highly sequential circuits, *15th Design Automation Conference*, Las Vegas, NV, 1978, pp. 332–339.
[36] Cheng, W.T., The BACK algorithm for sequential test generation, *International Conference on Computer Design*, Rye Brook, NY, 1988, pp. 66–69.
[37] Niermann, T., and Patel, J.H., HITEC: A test generation package for sequential circuits, *European Design Automation Conference*, Amsterdam, The Netherlands, 1991, pp. 214–218.
[38] Kelsey, T.P., Saluja, K.K., and Lee, S.Y., An efficient algorithm for sequential circuit test generation, *IEEE Trans. Comput.* 42–11, 1361–1371, 1993.
[39] Schulz, M.H., and Auth, E., Advanced automatic test pattern generation and redundancy identification techniques, *18th International Symposium on Fault-Tolerant Computing*, Tokyo, Japan, 1988, pp. 30–35.
[40] Cheng, K.-T., and Agrawal, V.D., *Unified methods for VLSI simulation and test generation*, Kluwer, Dordrecht, 1989.
[41] Saab, D.G., Saab, Y.G., and Abraham, J.A., CRIS: A test cultivation program for sequential VLSI circuits, *International Conference on Computer-Aided Design*, San Jose, CA, 1992, pp. 216–219.
[42] Rudnick, E.M., Patel, J.H., Greenstein, G.S., and Niermann, T.M., Sequential circuit test generation in a genetic algorithm framework, *ACM/IEEE Design Automation Conference*, San Diego, CA, 1994, pp. 698–704.
[43] Prinetto, P. Rebaudengo, M., and Sonza Reorda, M., An automatic test pattern generator for large sequential circuits based on genetic algorithm, *International Test Conference*, Baltimore, MD, 1994, pp. 240–249.
[44] Saab, D.G., Saab, Y.G., and Abraham, J.A., Iterative [simulation-based genetics + deterministic techniques] = complete ATPG, *International Conference on Computer-Aided Design*, San Jose, CA, 1994, pp. 40–43.
[45] Rudnick, E.M., and Patel, J.H., Combining deterministic and genetic approaches for sequential circuit test generation, *32nd Design Automation Conference*, San Francisco, CA, 1995, pp. 183–188.
[46] Hsiao, M.S., Rudnick, E.M., and Patel, J.H., Alternating strategy for sequential circuit ATPG, *European Design and Test Conference*, Paris, France, 1996, pp. 368–374.
[47] Ma, H-K. T., Devadas, S., Newton, A.R., and Sangiovanni-Vincentelli, A., Test generation for sequential circuit, *IEEE Trans. Comput.-Aided Design*, 7, 1081–1093, 1988.
[48] Ghosh, A., Devadas, S., and Newton, A.R., Test generation and verification for highly sequential circuits, *IEEE Trans. Comput.-Aided Design*, 10, 652–667, 1991.
[49] Cho, H., Hachtel, G. D., Somenzi, F., Redundancy identification/removal and test generation for sequential circuits using implicit state enumeration, *IEEE Trans. CAD*, 12, 935–945, 1993.
[50] Muth, P., A Nine-valued circuit model for test generation, *IEEE Trans. Comput.* C-25, 630–636, 1976.
[51] Goldstein, L.H., Controllability/observability analysis for digital circuits, *IEEE Trans. Circuits and Syst.* CAS-26, 685–693, 1979.
[52] Agrawal, V.D., and Chakradhar, S.T., Combinational ATPG theorems for identifying untestable faults in sequential circuits, *IEEE Trans. Comput.-Aided Des.* 14, 1155–1160, 1995.

[53] Cheng, K.-T., Redundancy removal for sequential circuits without reset states, *IEEE Trans.-Aided Des.*, 12, 13–24, 1993.
[54] Pomeranz, I., and Reddy, S.M., Classification of faults in sequential circuits, *IEEE Trans. Comput.* 42, 1066–1077, 1993.
[55a] Devadas, S., Ma, H.-K. T., and Newton, A.R., Redundancies and don't-cares in sequential logic synthesis, *J. Electron. Test. (JETTA)*, 1, 15–30, 1990.
[55b] Bryant, R.E., Graph-based algorithms for boolean function manipulation, *IEEE Trans. Comput.*, C-35, 677–691, 1986.
[56] Cheng, K.-T., Gate-level test generation for sequential circuits, *ACM Trans Des. Autom. Electron. Syst.*, 1, 405–442, 1996.
[57] Breuer, M.A., Test generation models for busses and tri-state drivers, *IEEE ATPG Workshop*, March 1983, pp. 53–58.
[58] Ogihara, T., Murai, S., Takamatsu, Y., Kinoshita, K., and Fujiwara, H., Test generation for scan design circuits with tri-state modules and bidirectional terminals, *Design Automation Conference*, Miami Beach, FL, June 1983, pp. 71–78.
[59] Chakradhar, S.T., Rothweiler, S., and Agrawal, V D., Redundancy removal and test generation for circuits with non-boolean primitives, *IEEE VLSI Test Symposium*, Princeton, NJ, 1995, pp. 12–19.
[60] Miczo, A., Digital logic testing and simulation, Harper & Row, New York, 1986.
[61] Zhang, H., SATO: an efficient propositional prover, *Proc. Int. Conf. Automated Deduction*, 1249, 272–275, 1997.
[62] Moskewicz, M., Madigan, C., Zhao, Y., Zhang, L., and Malik, S., Chaff: engineering an efficient SAT solver, *Proceedings of, Design Automation Conference*, Las Vegas, NV, 2001, pp. 530–535.
[63] Goldberg, E., and Novikov, Y., BerkMin: a fast and robust Sat-Solver, *Proceeding of, Design, Automation and Test in Europe*, Paris, France, 2002, pp. 142–149.
[64] Ryan, L., the siege satisfiability solver. http://www.cs.sfu.ca/ loryan/personal/.
[65] Davis, M., Longeman, G., and Loveland, D., A machine program for theorem proving, *Commn. ACM*, 5, 394–397, 1962.
[66] McAllester, D.A., An outlook on truth maintenance, AIMemo 551, MIT AI Laboratory, 1980.
[67] Tseitin, G.S., On the complexity of derivation in propositional calculus, in *studies in Constructive Mathematics and Mathematical Logic*, Part 2, 1968, pp. 115–125. Reprinted in Siekmann, J., and Wrightson, G., Eds., *Automation of Reasoning*, Vol. 2, Springer-Heiduberg., 1983, pp. 466–483.
[68] Tafertshofer, P., Ganz, A., and Henftling, M., A SAT-based implication engine for efficient ATPG, equivalence checking, and optimization of netlists, *Proceeding of, International Conference on Computer-Aided Design*, San Jose, CA, 1997, pp. 648–657.
[69] Silva, L., Silveira, L., and Marques-Silva, J.P., Algorithms for solving boolean satisfiability in combinational circuits, *Proceedinf of, Design, Automation and Test in Europe*, Munich, Germany, 1999, pp. 526–530.
[70] Silva, L., and Silva, J.M., Solving satisfiability in combinational circuits, *IEEE Design and Test of Computers*, 2003, pp. 16–21.
[71] Gupta, A., Yang, Z., and Ashar, P., Dynamic detection and removal of inactive clauses in SAT with application in image computation, *Proceeding of the, ACM/IEEE Design Automation Conference*, Las Vegas, NV, 2001, pp. 536–541.
[72] Kuehlmann, A., Ganai, M., and Paruthi, V., Circuit-based boolean reasoning, *Proceeding of, the ACM/IEEE Design Automation Conference*, Las Vegas, NV, 2001, pp. 232–237.
[73] Kuehlmann, A., and Krohm, F., Equivalence checking using cuts and heaps, *Proceedings Design Automation Conference*, Anahein, California, *1997*, pp. 263–268.
[74] Ganai, M.K., Zhang, L., Ashar, P., Gupta, A., and Malik, S., Combining strengths of circuit-based and CNF-based algorithms for a high-performance SAT solver, *Proceeding of the, ACM/IEEE Design Automation Conference*, New Orleans, Louisiana, 2002, pp. 747–750.
[75] Ostrowski, R., Grgoire, E., Mazure, B., and Sas, L., Recovering and exploiting structural knowledge from CNF formulas, *Principles and Practice of Constraint Programming (CP '02)*, Van Henten-ryck, P., Ed., LNCS 2470, Springer-Heidelberg, 2002, pp. 185–199.
[76] Broering, E., and Lokam, S.V., Width-based algorithms for SAT and CIRCUIT-SAT, *Proceeding of Theory and Applications of Satisfiability Testing*, Santa Margherita Ligure, Italy, 2003, pp. 162–171.
[77] Lu, F., Wang, L.C., Cheng, K.T., and Huang, R., A circuit SAT solver with signal correlation guided learning, *Proceeding of the Design, Automation and Test in Europe*, Munich, Germany, 2003, pp. 92–97.

[78] Lu, F., Wang, L.C., Cheng, K.-T., Moondanos, J., and Hanna. Z., A signal correlation guided ATPG solver and its applications for solving difficult industrial cases. *Proceeding of the IEEE/ACM Design Automation Conference*, Anaheim, CA, 2003, pp. 436–441.
[79] Lu, F., Iyer, M.K., Parthasarathy, G., Wang, L.-C., Cheng, K.-T., and Chen, K.C., An efficient sequential SAT solver with improved search strategies, *Proceeding of the European Design Automation and Test Conference*, Munich, Germany, 2005.
[80] Tafertshofer, P., Ganz, A., and Henftling, M., A SAT-based implication engine for efficient ATPG, equivalence checking, and optimization of netlists. *Proceeding of the IEEE/ACM International Conference on Computer-Aided Design*, San Jose, CA, 1997, pp. 648–655.
[81] Zhang, L., Madigan, C., Moskewicz, M., and Malik, S., Efficient conflict driven learning in a boolean satisfiability solver, *Proceeding of the International Conference on Computer-Aided Design*, San Jose, CA, 2001, pp. 279–285.
[82] Cheng, K.-T., Dey, S., Rodgers, M., and Roy, K., Test challenges for deep sub-micron technologies, *Design Automation Conference*, Los Angeles, CA, 2000, pp. 142–149.
[83] Special issue on speed test and speed binning for complex ICs, *IEEE D& T*, 20, 2003.
[84] Krstic, A., and Cheng, K.-T., *Delay Fault Testing for VLSI Circuits*, Kluwer, Boston, 1998.
[85] Jiang, Y.-M., Cheng, K.-T., and Deng, A.-C., Estimation of maximum power supply noise for deep submicron designs, *International Symposium on Low Power Electronics and Design*, ACM Order Dept., New York, NY, Monterey, CA, 1998, pp. 233–238.
[86] Chang, Y.-S., Gupta, S. K., and Breuer. M.A., Test generation for maximizing ground bounce considering circuit delay, *IEEE VLSI Test Symposium*, Napa Valley, CA, 2003, pp. 151–157.
[87] Lee, K.T., Nordquist, C., and Abraham, J., Automatic test pattern generation for crosstalk glitches in digital circuits, *IEEE VSLI Test Symposium*, Monterey, CA, 1998, pp. 34–39.
[88] Chen, L.H., and Marek-Sadowska, M., Aggressor alignment for worst-case coupling noise, *International Symposium Physical Design*, San Diego, CA, 2000, pp. 48–54.
[89] Chen, W.Y., Gupta, S.K., and Breuer, M.A., Test generation for crosstalk-induced delay in integrated circuits, *International Test Conference*, Atlantic City, NJ, 1999, pp. 191–200.
[90] Krstic, A., Liou, J.-J., Jiang, Y.-M., and Cheng, K.-T., Delay testing considering crosstalk-induced effects, *International Test Conference*, Baltimore, MD, 2001, pp. 558–567.
[91] Chen, L.-C., Mak, T.M., Breuer, M.A., and Gupta, S.A., Crosstalk test generation on pseudo industrial circuits: a case study, *International Test Conference*, Baltimore, MD, 2001, pp. 548–557.
[92] Chen, W., Gupta, S.K., and Breuer, M.A., Analytic models for crosstalk delay and pulse analysis under non-ideal inputs, *International Test Conference*, Washington, DC, 1997, pp. 809–818.
[93] Entrena, L.A., and Cheng, K.-T., Combinational and sequential logic optimization by redundancy addition and removal, *IEEE Trans. Comput.-Aided Des.* 14, 909–916, 1995.
[94] Brand, D., Verification of large synthesized designs, *International Conference on Computer-Aided Design*, 1993, pp. 534–537.
[95] Huang, S.-Y., and Cheng, K.-T., *Formal Equivalence Checking and Design Debugging*, Kluwer, Boston, 1998.
[96] Keller, B., McCauley, K., Swenton, J., and Youngs, J., ATPG in practical and non-traditional applications, *IEEE International Test Conference*, Washington, DC, 1998, pp. 632–640.
[97] Wohl, P., and Waicukauski, J., Using ATPG for clock rules checking in complex scan designs, *IEEE VLSI Test Symposium*, Monterey, CA, 1997, pp. 130–136.
[98] Thatte, S.M., and Abraham, J.A., Test generation for microprocessors, *IEEE Trans. Comput.* C-29, 429–441, 1980.
[99] Brahme, D., and Abraham, J.A., Functional testing of microprocessors, *IEEE Trans. Comput.* C-33, pp. 475–485, 1984.
[100] Lee, J., and Patel, J.H., A signal-driven discrete relaxation technique for architectural level test generation, *International Conference on Computer-Aided Design*, 1991, pp. 458–461.
[101] Lee, V., and Patel, J.H., Architectural level test generation for microprocessors, *IEEE Trans. Compu.-Aided Des.ICs*, 1310, 1288–1300, 1994.
[102] Cheng, K.-T., and Krishnakumar, A.S., Automatic generation of functional vectors using the extended finite state machine model, *ACM Trans. Des. Autom. Electron. Syst.*, 1, 57–79, 1996.
[103] Hansen, M.C., and Hayes, J.P., High-level test generation using symbolic scheduling, *International Test Conference*, Washington, DC, 1995, pp. 586–595.

[104] Sabnani, K., and Dahbura, A., A protocol test generation procedure, *Comput. Networks*, 15, 285–297, 1988.
[105] Dahbura, A.T., Uyar, M.U., and Yau, C.sW., An optimal test sequence for the JTAG/IEEE P1 149.1 test access port controller, *International Test Conference*, Washington, DC, 1989, pp. 55–62.
[106] Vedula, V. M., Abraham, A., and Bhadra., Program slicing for hierarchical test generation, *IEEE VLSI Test Symposium*, Monterey, CA, 2002, pp. 237-243.
[107] Huang, C., and Cheng, K.-T., Using word-level ATPG and modular arithmetic constraint-solving techniques for assertion property checking, *IEEE Trans. CAD*, 20, 381–391, 2001.
[108] Iyer, M.A., RACE: a word-level ATPG-based constraints solver system for smart random simulation, *International Test Conference*, Charlotte, NC, 2003, pp. 299–308.
[109] Zhang, L., Ghosh, I., and Hsiao, M., Efficient sequential ATPG for functional RTL Circuits, *International Test Conference*, Charlotte, NC, 2003, pp. 290–298.

23

Analog and Mixed Signal Test

Bozena Kaminska
Simon Fraser University and Pultronics Incorporated
Burnaby, British Columbia
Canada

23.1 Introduction

With the ever-increasing levels of integration of system-on-chip (SoC) designs, more and more of which include analog and mixed-signal (A/M-S) elements, test equipment, test development and test execution times, and costs are being increasingly impacted. The convergence of computer, telecommunication, and data communications technologies is driving the trend toward integrating more A/M-S circuit elements into larger deep submicron chips. The forces of "smaller, cheaper, faster" will only increase as we move into the future. But increased integrated functionality comes at a price when test is included in the equation.

This is especially true when the whole chip-synthesis model becomes untenable, as is the case when processor and other cores are designed by, or acquired from multiple sources, either in-house or from a third party. Adding the A/M-S elements, for which no truly effective synthesis and automatic test pattern generation (ATPG) techniques exist, aggravates the test issues considerably. Capital equipment cost increases to millions of dollars. Test development time can exceed functional circuit development time. It is thus necessary for design and test engineers to work together, early in the SoC architecture design phase, in order to keep the testing costs under control. Trade-offs between traditional automatic test equipment (ATE) and new integrated techniques need to be considered, and the test resources partitioned between internal and external test methods.

There is a trend [1–3] toward increasing use of design-for-test (DfT) methodologies that focus on testing the structure of a design rather than its macrofunctionality. Design for test with the purpose of testing the structure of the device is called structural DfT. This trend is being driven by several factors. The traditional driving forces for the use of DfT have been, and remain, observability into the design for diagnostics and system check-out, and achieving acceptable fault coverage levels, typically between 95 and 98%, in a predictable timeframe. As the designs have become larger with the advent of the SoC methodologies, DfT is being used to provide test portability for reusable intellectual-property (IP) blocks or cores. Additionally, DfT tools have advanced to permit a more comprehensive device test that, in some cases, has been proven to eliminate the need for a traditional functional test. As a consequence of these advances, DfT methods are seen as an enabling technology to break the cost trend. The reduced cost trend will cause a shift in the market acceptance of these nontraditional methods over the traditional ones. This rationale stems from the fact that much of the "tester" is now on-chip or in the form of partitioned test.

This also becomes a motivation for the focus and the detailed overview of the available test techniques for A/M-S circuits and, in particular, when they are part of an SoC. The fault modeling and test specification is reviewed, followed by practical DfT with emphasis on some specific mixed-signal blocks.

23.2 Analog Circuits and Analog Specifications

The following terms are used in discussing A/M-S testing:

- A *circuit* is a system containing a set of components (elements) connected together.
- *Parameters* are circuit characteristics obtained by measuring the output signals. A mathematical function in terms of some components describes each parameter. For example, the mathematical expression for the cut-off frequency is $f_c = 1/(2RC)$.
- A *nominal value* is the value of the component or the parameter of a good circuit.
- A *relative deviation* indicates the deviation of the element or the parameter from its nominal value, divided by its nominal value.

In designing electronic circuits, the designer should know the circuit's performance deviations due to changes in the value of its elements. The deviation of the elements' values from their nominal values depends on the manufacturing process and on the temperature of the element [4].

Various measurement techniques prove the diverse output parameters that characterize analog circuits and analog specifications. Four categories of commonly used measurements can be distinguished:

1. *Harmonic* techniques measure the frequency response of the circuit under test. From this measurement, we can extract diverse parameters — for example, gain at a known frequency, cut-off frequency, Q factor, and phase as a function of frequency. The input stimulus used in harmonic measurement is usually a sinusoidal waveform with a variable frequency.
2. *Time-domain* measurements use pulse signals, including a square wave, step, and pulse train, as the input stimuli of a circuit. We can then observe the circuit's transient response at the output. We can derive many parameters from this measurement and use them to predict the defective components in the analog circuit. Some of these parameters are rise, delay, and fall times.
3. *Static* measurements attempt to determine the parameters of the stable states of the circuit under test. This measurement includes the determination of the DC operating point, leakage currents, output resistance, transfer characteristics, and offset.
4. *Noise* measurements determine the variations in signal that appear at the circuit's output when the input is set to zero.

Tests involving sinusoidal signals for excitation are the most common among linear circuits, such as amplifiers, data converters, and filters. Among all waveforms, the sinusoid is unique in that its shape is not altered by its transmission through a linear circuit; only its magnitude and phase are changed. In contrast, a nonlinear circuit will alter the shape of a sinusoidal input. The more nonlinear the circuit is, the greater the change in the shape of the sinusoid. One means of quantifying the extent of the nonlinearity present in a circuit is by observing the power distributed in the frequency components contained in the output signal using a Fourier analysis. For example, a circuit can be excited by a sinusoid signal. At the output of a band pass filter that is connected to the output of a circuit under test, a power spectral density plot can be observed. In general, the fundamental component of the output signal is clearly visible, followed by several harmonics. The noise floor is also visible. By comparing the power contained in the harmonics to that in the fundamental signal, a measure of total harmonic distortion (TDH) is obtained. By comparing the fundamental power to the noise power over a specified bandwidth, one obtains the signal-to-noise ratio (SNR). By altering the frequency and amplitude of the input sinusoid signal, or by adding an additional tone with the input signal, other transmission parameters can be derived from the power spectral density plot [5].

Faults in analog circuits can be categorized as *catastrophic* (*hard*) *and parametric* (*soft*) (see Figure 23.1). Catastrophic faults are open and short circuits, caused by sudden and large variations of components. These usually induce a complete loss of correct functionality. Parametric faults are caused by an abnormal deviation of parameter values and result in altered performance [6–8]. In analog circuits, the concern is the selection of parameters to be tested, and the accuracy to which they should be tested, to detect the deviation of faulty components.

The most accepted test strategy for analog circuits depends on *functional testing* that is based on the verification of a circuit's functionality by applying stimuli signals at the input and verifying its outputs [5,9]. This type of test is convenient for complex analog circuits. Its major drawbacks, however, are (1) the difficulty of detecting and identifying the defective elements, (2) the complexity in writing and executing the test program, and (3) the access to the primary inputs and outputs of a circuit. The access problem is becoming especially critical as complexity and SoC integration increases. The test bus standards 1149.4 and 1149.1 (see below) are increasingly used to assure the access mechanism to the embedded A/M-S blocks [10,11]. This technique is known as design-for-functional-testing (DfFT).

The above-mentioned drawbacks have been a subject of intense research in recent years. Two main directions can be observed: one that deals with an extensive effort in fault modeling and test specification, and the second, which in contrast researches DfT techniques. Design for test includes built-in-self-test (BIST), and test access mechanisms such as test bus standard 1149.4.

The following observations can be made based on the current industry practices. A successful test strategy is domain-specific in digital circuits. For example, there are dedicated memory BIST solutions, random logic is well suited for full scan, and boundary scan is useful to tie the whole chip together. For mixed-signal testing, success is associated with the functional definition of a module and its related test specifications; for example, analog-to-digital converter (ADC), digital-to-analog converter (DAC), phase lack loop (PLL), filter, and transceiver. DfT and BIST are largely developed for each specific functional module and very often even for a particular design style or implementation. The trend of maintaining functional testing despite the recognized benefits of fault-based testing is much stronger for A/M-S modules. The standalone circuits as well as embedded modules (for example, in the case of SoC) are tested for the specification

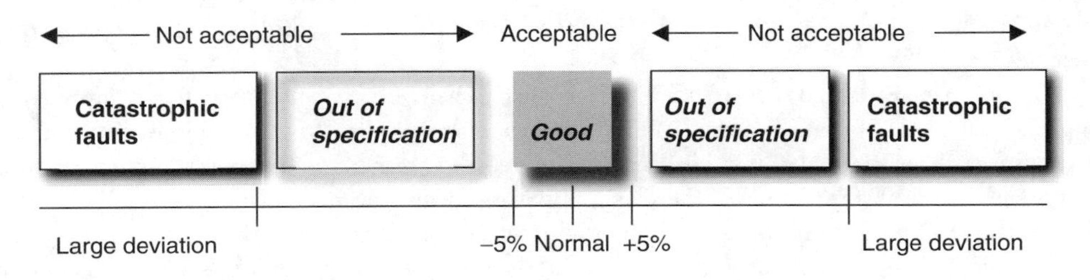

FIGURE 23.1 Fault classification: parametric and catastrophic faults.

verification and their related tolerances. A notion of DfFT becomes an SoC practice to deal with the access problem and specification verification.

In practice, functional testing can include the following tests:

1. *parametric*, which verifies the analog characteristics within a specified tolerance (for example, voltages, currents, impedances, and load conditions);
2. *dynamic*, which verifies the dynamic characteristics of the system under test — in particular, a transient analysis in the time domain;
3. *static*, which verifies the stable states of the system.

There is a common understanding that mixed signal represents the interface between the real world and digital processing. ADC or DAC are some examples. But it is not the only meaning of A/M-S. Otherwise, why are SPICE-like simulators developed for fully analog simulation used for digital-cell design? Each time a critical performance is considered, such as delay, jitter, or rise and fall times, an analog simulation takes place. The same is true with testing. Performance testing of digital cells is primarily an analog function. So, together, digital and analog design and test form the mixed-signal environment. Fault modeling and tolerance analysis relate to both digital and analog circuits. Performance testing requires analog techniques and instruments.

23.3 Testability Analysis

Testability analysis in analog circuits is an important task and a desirable approach for producing testable complex systems [12,70]. The densities of analog circuits continue to increase and the detection and isolation of faults in these circuits becomes more difficult owing to the nature of analog faults, the densities of today's analog circuits, and very often their embedded placement in SoC. Testability information is useful to designers who must know which nodes to make accessible for testing, and to test engineers who must plan test strategies. By analyzing testability, we can predict what components can be isolated by a given set of tests, and what kind of defects can be observed at a given test node. Analyzing testability in the frequency domain is an approach introduced for analog circuits to choose adequate test frequencies for increasing fault diagnosis [6]. It has also been observed that not all test points and frequencies are equally useful, and hence some selection criteria must be applied for a robust test selection.

The first published testability evaluation methods are based on calculating the rank-test algorithm and determining the solvability of a set of fault diagnosis equations describing the relation between measurements and parameters. Recently, a more practical approach [1,6,7] based on a test selection that allows maximization of the deviation between the output voltage of the good and faulty circuits, has been developed. The input stimulus that magnifies the difference between the response of the good circuit and that of the faulty one is derived. A more comprehensive study is presented in [4,13,14] and is based on the analog-fault observability concept and multifrequency testability analysis.

In [7], test nodes and test frequencies are chosen that increase the testability of a circuit in terms of searching for an effective way to observe defective components. A fault is judged to be easy to test if its effect can be observed at one of the outputs. The proposed multifrequency analysis consists of applying a sinusoidal input signal with a variable frequency and observing the amplitude of the output signals at the various test nodes. The influence of a defective component's deviation on the output is different from one interval of frequencies to another. This means that the observability of a fault depends on the frequencies.

In [6], it has been observed that for a parameter (such as gain) that reacts poorly to a component (for example, a resistance), deviation makes it impossible to observe a defect in this component. The defect *observability* has been defined as the sensitivity of the output parameter with respect to the variations of a component. As the absolute value of the sensitivity of an output parameter's variation with respect to a component's variation is high, the observability of a defect in a circuit is also high. We can see that the sensitivity gives information about component deviation observability and the observability depends also on the selected output parameters. A component may be highly observable with respect to one parameter, but may not be observable at all with respect to another parameter. A definition of sensitivity is provided in [15].

To analyze various symptoms of a fault, we can compute or measure the different observabilities in the frequency domain either from the sensitivity of the output signals' amplitudes with respect to each component of the circuit, or from the sensitivity of transfer functions relating the output test nodes to the input. Thus, it is possible to optimize the number of parameters necessary for testing and, at the same time, to achieve high fault coverage with high fault observability while reducing the test time.

To increase observability of the defective components in a circuit, a parameter that has a high sensitivity with respect to this component needs to be tested. Concepts such as fault masking, fault dominance, fault equivalence, and fault isolation for analog circuits have been defined in [6,16,17] and are based on a sensitivity concept. It can be noted that the testability analysis not only plays an important role in test set specification, but also in defining test access points for functional testing of embedded devices in SoC and for implementation of test bus standard 1149.4 for DfFT. Finally, testability analysis constitutes an important source of information about a circuit of interest for design, test, and application purposes.

23.4 Fault Modeling and Test Specification

Test generators have become an important part of functional verification of digital circuits. In the advent of the ever-increasing complexity of designs, decreasing design cycles, and cost-constrained projects resulting in an increased burden on verification engineers, design teams are becoming increasingly dependent on automatic test generators. Analog design teams do not have the same level of tool support. Automatic test specification is still an open research area.

The introduction of the *stuck-at* fault model for digital circuits enabled digital testing to cope with the exponential growth in the digital circuit size and complexity. Indeed, the stuck-at fault model enabled functional testing to be replaced by structural testing and acted as a measure to quantify the quality of the test plan, permitting test requirements and benchmarking of DfT strategies.

Hard and soft fault modeling and simulation, which address A/M-S circuits, have been the subject of many publications [2,8,16,18–23]. Millor et al. [24] reported on a test generation algorithm for detecting catastrophic faults under normal parameter variations. The later extension resulted in an approach based on a statistical process fluctuation model that was derived to select a subset of circuit specifications that detect parametric faults and minimize the test time. Test generation is formulated in [25] as a quadratic programming problem. This approach was developed for parametric faults and it determines an input stimulus $x(t)$ that maximizes the quadratic difference of response from the good and the faulty circuits, with all other parameters at their nominal values. The test generation approach for hard and parametric faults based on sensitivity analysis and tolerance computation was proposed in [26]. In this approach, the worst-case performance was expressed in terms of sensitivity and parameter tolerance; however, frequency analysis was not considered, and the model was a linearization obtained from first-order partial derivatives. Based on [26], Ben Hamida et al. [20] developed an automated sensitivity tool that allows adjunct network-based sensitivity analysis for designing fault-resistant circuits and for generating test vectors for parametric and catastrophic faults under normal parameter variations.

A method presented in [18] is founded on a fault model and sensitivity. For a given fault list, perturbation of sensitivity with respect to frequency is used to find the direction toward the best test frequency. In [17], Huynh et al. derived a multifrequency test generation technique based upon testability analysis and fault observability concepts. The test frequencies selected are those where the output performance sensitivity is a maximum with respect to the faulty component deviation. In these approaches [17,18], the masking effect due to variations of the fault-free components in their tolerance box are not considered and the test frequencies may be not optimal. A DC test generation technique for catastrophic faults was developed in [27]. It is formulated as an optimization problem and includes the effects of normal parameter variations.

23.5 Catastrophic Fault Modeling and Simulation

The majority of approaches presented for A/M-S fault simulation are based on *cause–effect* analysis and do not allow parallel fault simulation. Indeed, cause–effect analysis enumerates all the possible faults (causes) existing in a fault model and determines all their corresponding responses (effects) to a given applied test in a serial manner. The required simulation time can become impractically long, especially for large analog designs.

Fault simulation is used to construct a fault dictionary. Conceptually, a fault dictionary stores the signatures (effects) of the faults for all stimuli T. This approach requires computing the response of every possible fault before testing, which is impractical. In [22], a new type of fault dictionary that does not store the signatures (effects) of the faults f, but computes and stores the fault value R_{fault} (cause) that if added to the circuit, will drive the output parameter out of its tolerance box. This approach allows parallel fault simulation. The following steps are required for fault dictionary construction:

1. Generate the fault list; all possible shorts and opens in a circuit. Two fault-list extractors can be used: a layout-based fault list (standard inductive fault analysis) and a schematic-based fault-list extractor.
2. Compute the output sensitivities with respect to all hard faults in the fault list in parallel, for example, using the adjoint network method [28]. For example, the initial value for R_{fault} can be defined as zero value resistance for the opens and as zero value conductance for the shorts. The computed R_{fault} value (cause) is defined as the smallest resistance value that if added to the circuit, will only deviate the output parameter to the edge of the tolerance box.
3. From the fault-free circuit output tolerance $\eth_{out}$ (effect) and the fault-free output sensitivities S with respect to all hard faults in the fault list (obtained in step 2), the following equation is used to compute the fault value R_{fault} (cause) for all the defects in the fault list:

$$R_{fault} = \eth_{out}/S_{out}\ (R_{fault})$$

The adjoint network method for sensitivity computation in AC, DC, and transient domains has been described in [21–23]. Other parallel sensitivity computations are possible as well.

After creating the fault dictionary, the best stimulus (test vector) for detecting each fault needs to be found. The stimulus should be selected to maximize the fault observability. First, knowing the component tolerance, the output parameter distribution of the fault-free and faulty circuit can be estimated. Then the resistor value R_{fault} that will cause the output parameter to go out of the tolerance range is computed. Then, the dominance of faults is analyzed to minimize the number of test vectors. The details of the method, an example of implementation and practical results are in [22]. This is a structural approach and a parallel fault simulation and test generation.

23.6 Parametric Faults, Worst-Case Tolerance Analysis, and Test Generation

A robust test set should detect parametric faults under maximum masking effect due to the tolerances of the circuit parameters. Indeed, the worst masking effect of fault-free parameters may create a region around the tolerance box of a parameter where its faulty deviations cannot be detected. The size of the region usually depends on the applied test. A test set is robust if it minimizes the area of the regions associated with the parameters and if the boundaries of these regions are accurately determined. Only in this case, the quality of the test set may be considered and guaranteed. If a test set is not robust, then some faulty circuits may be classified as good. In [4,13,14], a series of optimization problems is formulated and two solutions proposed to guarantee the detectability of parametric variations under the worst-case tolerance analysis: by the nonlinear programming method SQP (sequential quadratic programming) available in MATLAB, and by constraint logic programming (CLP) using relational interval arithmetic. The multifrequency testing, which is much better suited for subtle parameter variations than DC testing, is adopted.

23.7 Design for Test — An Overview

The impetus for the DfT and BIST techniques is to facilitate the use of low-performance test instrumentation or perhaps to eliminate any need whatsoever by adding self-test capabilities and strategic control and observation points within the silicon. DfT is very often *ad hoc* and largely a matter of early engagement with the design community to specify design block architectures, signal routing and external control, and access points to ensure maximal visibility and provide bypass control to critical functional blocks.

Great strides have been made in BIST techniques [1–3,5,9] for A/M-S and RF devices, but robust, production deployment of these techniques is not yet widespread. This is due in part to the challenges of soft (parametric) fault behaviors and blocks being difficult to model or simulate accurately enough to generate alternative structural test stimuli. In addition, many of the proposed BIST techniques consume an unattractive amount of silicon die area and are intrusive to sensitive circuits and design methodology, or impede the post-silicon debug process.

In a traditional production test approach for testing of the A/M-S circuits, the functional specifications are measured using the appropriate tester resources and using the same kind of test stimuli and configuration with respect to which the specification is defined, e.g., multitone signal generator for measuring distortion, gain for codec, ramp generator for measuring integral nonlinearity (INL), differential nonlinearity (DNL) of ADCs and DACs. The measurement procedures are in agreement with the general intuition of how the module behaves and, hence, the results of the measurement are easy to interpret, in contrast to the fault-based testing.

In the BIST approaches, the external ATE functionality is designed inside the device for applying appropriate test stimuli and measuring the test response corresponding to the specification. In [29], adjustable delay generators and counters are implemented next to the feedback path of the PLL to measure the RMS jitter. Since the additional circuitry does not modify the operation of the PLL, the same BIST circuitry can be employed online. Reference [29] also discusses different ways of measuring properties like loop gain, capture range, and lock-in time by modifying the feedback path to implement dedicated phase delay circuitry. All these components are automatically synthesized using the digital libraries available in the manufacturing process. This kind of automation provides scalability and easy migration to different technologies. The approach of Seongwon and Soma [30] is similar to this approach in the sense that the extra tester circuitry is all-digital and can be easily integrated into an IEEE 1149.1 interface. In this work, the BIST approaches reuse the charge pump and the divide-by-N counter of the PLL in order to generate a defect-oriented test approach, which can structurally verify the PLL. While [29] can also be implemented on a tester board, Ref. [30] is limited to the BIST since a multiplexer must be inserted into the delay-sensitive path between the phase detector and the charge pump. Since both examples employ all-digital test circuitry, their application is limited to a few analog components like PLLs, where digital control is possible.

In [5], an attempt is made to implement simple on-chip signal generators and on-chip test response data capture techniques for testing the performance of high-frequency analog circuits. The communication between the BIST hardware and the external world takes place through a low-frequency digital channel.

It can be observed that there are the following groups of approaches: (1) a device under test is modified to improve testability (DfT), (2) a signal generator and signal analysis devices are incorporated on chip (BIST) or on a tester board, (3) test access is improved by implementing test bus standards 1144.1 and 1149.4, and (4) a device under test is reconfigured into an oscillator for test purposes. The typical solutions in these categories will be reviewed, followed by a dedicated discussion of the most popular mixed-signal devices and their respective test techniques.

23.8 Analog Test Bus Standard

The significant advancement [10,11,31] in mixed-signal DfT area is the IEEE P1149.4 bus standard. It is a new IEEE standard that aims at providing a complete solution for testing analog and digital I/O pins and the interconnection between mixed-signal ICs. The secondary objective is to provide access to internal cores based on the Test Access Bus concept. It includes the IEEE 1149.1 boundary scan Test Access Port (TAP) controller and therefore provides a support infrastructure for the BIST and test setup.

Figure 23.2 shows the IEEE P1149.4 architecture that includes the following elements:

- Test Access Port (TAP) comprising a set of four dedicated test pins: test data in (TDI), test data out (TDO), test mode select (TMS), test clock (TCLK), and one optional test pin: test reset (TRSTn).
- Analog test access port (ATAP) comprising two dedicated pins: Analog test stimulus (AT1) and analog test output (AT2), and two optional pins: inverse AT1 (AT1n) and inverse AT2 (AT2n) for differential signals.
- Test bus interface circuit (TBIC).
- An analog boundary module (ABM) on each analog I/O.
- A digital boundary module (DBM) on each digital I/O.
- A standard TAP controller and its associated registers.

23.9 Oscillation-Based DFT/BIST

Currently, oscillation-based approaches seem to be the most applicable in practical implementations for all types of A/M-S devices. From the time the oscillation methodology was introduced as OBIST [32 to 46], many researchers have successfully explored more areas of application, better performance, and other innovations [1,47,48]. The commercial products have been developed and industrial implementations have been

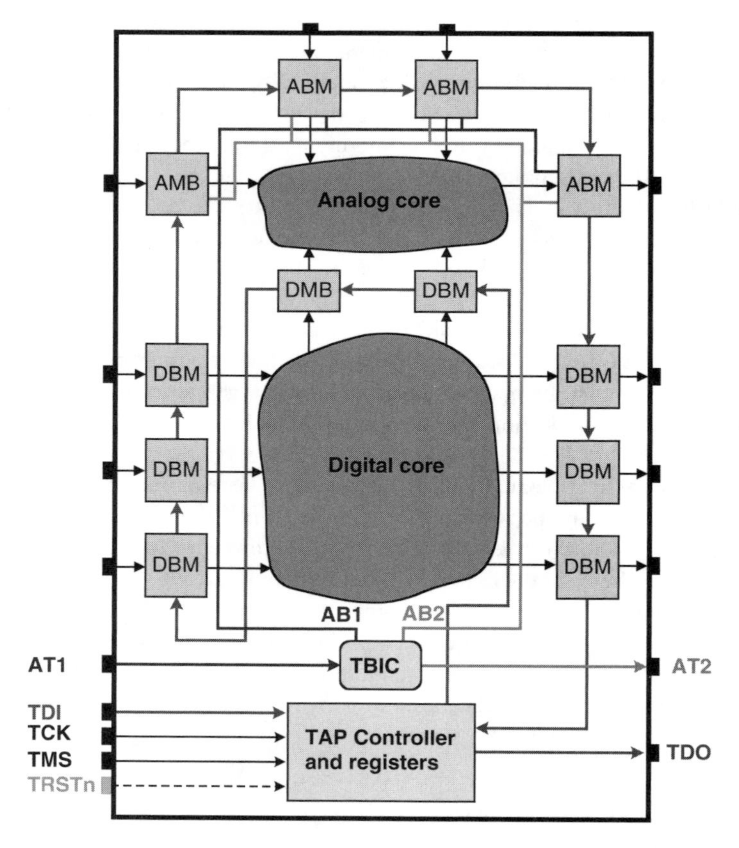

FIGURE 23.2 IEEE 1149.4 architecture.

successful as well [49]. The popularity of oscillation-based DfT/BIST techniques results from the following main characteristics:

- Adaptable for various functional modules: A/M-S, digital, and MEMS devices.
- Signal generator is not required.
- Immune to noise and technology-independent.
- Delivers very good fault coverage for parametric, hard, and functional errors.
- Easy to understand.

Oscillation DfT/BIST is a testing technique based upon converting a functional circuit to an oscillator and inferring something about the functionality of the circuit from the oscillation frequency (see Figure 23.3). To a first order, in any practical circuit, the oscillation frequency will be a function of several parameters of the circuit and also the elements that were added to convert the circuit to an oscillator. How then can oscillation BIST be used to definitively generate a go/no go decision in production test? To measure the parameters of the circuit under test? To characterize that circuit?

In a mathematical sense, the answer to these questions is very simple. If

<Frequency of oscillation> = Known_function [$p1, p2, \ldots, p4, \ldots, pn$]

and pi are all unknown circuit performance parameters, then n separate oscillation frequency measurements made under varying conditions can be used to calculate pi. So, by changing the feedback circuit topology, its frequency selection, its attenuation, etc., varying conditions may be imposed upon the circuit under test and varying frequencies of oscillation thereby determined. These n different equations with n unknowns can be solved for pi.

In a practical sense, not all of the pi may be of interest. Furthermore, the exact value of pi may be irrelevant. In most cases, what is needed is an assurance that each pi lies within a specific range of values, i.e., the criterion that was assumed when the circuit was designed and simulated — and if properly designed, the circuit will be functional over a known range of values for each of the pi. These are the values that can be used in a simulation of the circuit, with the BIST in place, to determine an acceptable range of frequencies of oscillation. Thus, the combination of a simulation and a single OBIST frequency measurement can be used to determine, with some level of certainty that the circuit under test is functioning properly. The level of certainty (including the feedback) relates to the parameters of its elements (the higher the sensitivity, the more certain that a single OBIST measurement will suffice). It is also a function of how tight the performance specifications of the circuit are. The tighter the specifications are, the more likely that additional measurements will be required to guarantee all the performance specifications.

The most extensive practical and commercial applications have been developed for PLL [49,50], general timing test, and characterization (including delay and 'stuck-at' fault testing in digital circuits [29,32,45,46]),

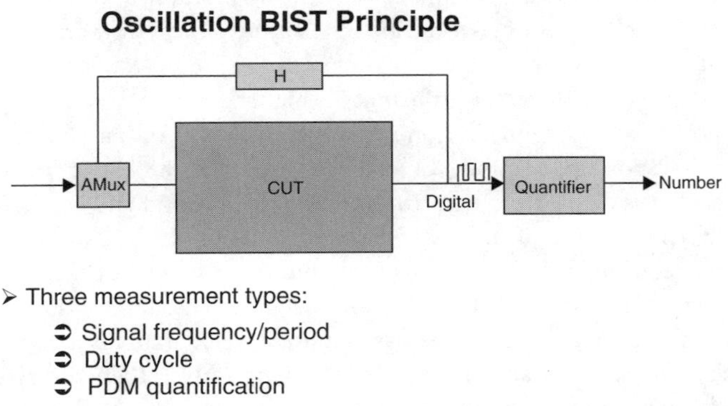

FIGURE 23.3 The principle of the oscillation technique- transforming a device into an oscillator.

operational amplifiers, and for a multitude of data converters. The differences in the implementations are mostly related to: (1) the configuration method of a device into an oscillator (choice of a transfer function, number of oscillators, control technique), (2) result processing techniques (from a custom signal processing to classical DSP to integrated method of histograms (HBIST): see below). In particular, each OBIST measurement can be targeted at a single parameter (or a small group of parameters) of a device under test by judicious selection of the feedback circuit. For example, a single application of OBIST can guarantee that the low-frequency gain of an operational amplifier is higher than some specified value. This is done by using a high attenuation factor in the feedback circuit, combined with reactive elements (one or more capacitors) which cause a low-frequency oscillation, provided the gain of the operational amplifier is greater than the specified value. Another example would be to guarantee operation at a certain input common mode voltage by feedback of the desired voltage, along with the feedback signal designed to make it oscillate into the input of the circuit to be tested. If the circuit oscillates, then the input range parameter requirement is fulfilled. A final example involves the measurement of the gain-bandwidth product of an operational amplifier. The operational amplifier is required to have a certain minimum gain in order to overcome the attenuation of the feedback loop — at a specific high frequency determined by frequency selective feedback.

In general, reconfiguration of the application circuit is accomplished by using digital circuits to control analog switches or multiplexers. The switches and multiplexers are designed to have a minimal impact on the measurement.

23.10 PLL, VCO, and Jitter Testing

PLL voltage-controlled oscillator (VCO) testing has generated significant interest recently due to the widespread integration of embedded, high-performance PLLs in mixed-signal communications, and data processing devices. In popular serial data communication standards such as Fiber Channel, FireWire, and GigaBit Ethernet, the data and clock are embedded within the signal codes. Because of these characteristics the receiver must have special clock recovery circuitry implemented with the PLL that extracts the data and clock received through the media. These clock recovery circuits are sensitive to input jitter (time distortion) and level distortion. To guarantee proper data reception in any network, the transmitter and receiver must meet a certain jitter budget.

Typical PLL applications include frequency synthesis, phase demodulation, clock distribution, and timing recovery — essential operations for systems like wireless phones, optical fiber links, and microcomputers, along with multimedia, space, and automotive applications. Serial data techniques and parameters also apply to hard disk drive read/write channels. Although the serial data and embedded clock information are stored on and retrieved from a magnetic medium rather than being transmitted over cable, the sensitivity to jitter is similar to datacomm and telecomm applications.

In recent years, the use of PLLs for clock generation has been widely popular for modern microprocessors. This is because PLL has the advantage of allowing multiplication of the reference clock frequency and allowing phase alignment between chips. It is impossible, however, to have these advantages if PLL has excessive jitter or variation in phase alignment. If the jitter is too large, then the cycle time for logic propagation is reduced and the error probability is higher.

Before discussing how to measure jitter, it is important to distinguish what jitter means for two application areas: noncommunication applications, for example, processors and serial communication applications. In the first case, jitter is defined as the variation of the clock period and is referred to as period jitter or cycle-to-cycle jitter. The measure of the phase variation between the PLL's input and output clocks is known as a long-term or tracking jitter. In serial data communication, on the other hand, jitter is defined as the short-term variation of a digital signal's significant instants,for example, rising edges with respect to their ideal position in time. Such a jitter is known as accumulative jitter and is described as a phase modulation of a clock signal. Jitter sources can come from power supply noise, thermal noise from the PLL components, and limited bandwidth of the transmitting media. There are some applications where the absolute jitter is important, for example, in clock synthesis circuits; there is a need for the use of a jitter-free or low-jitter reference signal. The difference between the position of corresponding edges of the signal of interest and the reference signal indicates the jitter. Although in a production environment the focus is on overall jitter, for characterization purposes, the following jitter components are distinguished: data-dependent, random, and duty-cycle distortion. Data-dependent or periodic jitter is caused by one or more sine waves and their harmonics. Random jitter has a probability distribution function — usually assumed to be Gaussian, but often it is not — and has a power spectral density that is a function of frequency. Duty-cycle distortion is caused by differing propagation delays for positive and negative data transitions. Based on the above definitions, a number of samples are collected to determine the jitter characteristics of interest. The most common characteristics include RMS, peak-to-peak, and frequency.

23.10.1 High-Speed Serial Links

In serial data communication, jitter plays a key role in clock extraction and network timing. The recovery of the clock from the data signal poses more stringent requirements on the jitter of the data signal than would exist when shipping a synchronous clock along with the data signal. The latter is typically done in very short links, where the effort of using a full-fledged clock-data-recovery receiver does not pay off in terms of silicon area and power consumption. Bit-error rate (BER) measurements or eye-diagram plots characterize the signal quality of serial links. BER measurements are subject to statistical uncertainties because of the tradeoff between test time and measurement accuracy.

The easiest way to measure jitter is to plot eye diagrams on an oscilloscope and apply histogram functions to the zero crossings, as Figure 23.10 shows. This procedure yields values for the horizontal or vertical eye opening as well for the total peak-to-peak jitter. However, this type of jitter measurement does not provide much insight into jitter properties. Some improvements come with the sampling scopes that add jitter measurement functionality based on the clever bookkeeping of sampling-point positions. Due to the sub-rate function of a sampling scope, it does not analyze every bit, which leads to accuracy problems at low error rates. A more adequate method of measuring jitter is based on the scan technique which measures the BER of every single bit within the data eye, and fits the resulting BER curve to a mathematical jitter model to obtain the required jitter properties. For data stems with several gigabits-per-second data rates, a BER scan tests many billions of bits each second and thus maintains accuracy for low BERs. In eye-diagram plots, distributed transitions of the threshold as data toggle between the logic states indicate jitter. The histograms measured at the zero crossing represent probability density function of the jitter, and statistically describe the data transitions' temporal locations. To derive a mathematical jitter model, jitter must be subdivided into different categories. In general, it is possible to split jitter into random and deterministic components. Random jitter has a Gaussian distribution and stems, for instance, from the phase noise of a VCO or power supply noise. Deterministic jitter can be divided into different subcategories of origin. Predominant types are sinusoidal and data-dependent jitter as well as jitter arising from duty-cycle distortion. Sinusoidal jitter can stem from slow variations of the supply voltage, the temperature, or the clock reference. The jitter components are not necessarily sinusoidal waveforms in reality, but we can model this jitter contribution with a sinusoidal representation. Data-dependent jitter of deterministic jitter comes from circuit asymmetries or duty-cycle distortion.

23.11 Review of Jitter Measurement Techniques

From the user perspective, there are three established ways of making jitter measurements: using ATE equipment, a real-time sampling oscilloscope, or a dedicated jitter measurement instrument. Because jitter is becoming such an important factor in the overall system performance, it is worthwhile to examine the pros and cons of using each of these techniques. A fourth technique based on BIST or just DFT has recently emerged (see Table 23.1).

23.11.1 Spectrum Analyzer Measurement

An analog spectrum analyzer can be used to measure the jitter of a signal in the frequency domain in terms of phase noise. For this measurement, the jitter is modeled as phase modulation. For example, a

TABLE 23.1 A Comparison of the Four Different Ways to Measure Jitter

	ATE Testers	Real-Time Sampling Oscilloscope	Dedicated Instrumentation	BIST/DFT
Performance	Limited accuracy, resolution, and throughput	Very high accuracy, resolution, and throughput	High accuracy, resolution, and moderate throughput	High accuracy, resolution, and throughput
Additional hardware	None or dedicated instrumentation	High-speed digitizer system per channel	Digital jitter measurement system per ATE	Modular, flexibility with different levels of integration
Additional software	Minimal	Significant signal processing and application-specific code	Insignificant signal processing and application-specific code	Insignificant signal processing and application-specific code
Cost	Nil or ATE-related	About $10,000 per channel for large channel counts, more for small channel counts	About $100,000 per ATE	Implementation dependence,low
Application	>30 psec jitter in CMOS production test	1 to 5 psec jitter on a few critical signals/clocks precise characterization of communications channels, optical fiber clocks, prediction of jitter	>1 psec jitter lock jitter, in ICs. serial data communications, carrier jitter in integrated RF subsystems	1 or more psec jitter Clock , data jitter, serial data communication, measurements in a noisy environment
Manufacturers	ATE companies	Tektronix, LeCroy, Agilent	Wavecrest, Guidetech	Emerging companies

sine wave signal is represented as a perfect sine wave with amplitude and phase modulation. With a spectrum analyzer, one can measure the power of the signal as a frequency with wide dynamic range; however, one cannot distinguish between the amplitude and phase-modulation components. A common assumption made when using a spectrum analyzer to measure jitter is that the amplitude modulation component of the signal is negligible. This assumption may be valid for signals internal to a pure digital system where undistorted square waves or pulses are the norm. This assumption is typically not true for a serial communication or data channel. Isolating the noise from the actual signal frequency components and translating that into jitter is nontrivial.

23.11.2 Real-Time Time Interval Analyzer Measurements

This technique measures the time interval between a reference voltage crossing of the transmitted signal. There is no need for an abstract model of the signal when using this technique because the time intervals are measured directly. The real-time interval analyzer gives complete knowledge of the nature of the jitter and the components of the jitter. With this measurement technique the position and time of every edge is measured, thus allowing statistical space and frequency-based models of the signal, as well as absolute peak-to-peak measurements. The clear advantages of this technique are that there are no skipped edges and measurement acquisition time is limited only by the signal itself. In practice, instrumentation that has the necessary acquisition rate and resolution to test gigabit data rates does not exist.

23.11.3 Repetitive Start/Stop Measurements

This is a common technique that gives high resolution and accuracy using a direct time measurement. Time measurements are made by starting a counter on the first occurrence of an edge, and stopping the counter on the next edge. Enhancements to this technique include skipping multiple edges for cumulative measurements, comparing two different signals, and time interpolation for achieving resolution greater than the counter clock period. Re-triggering of time interval measurement normally requires a significant dead time, particularly when time interpolation is used. After collecting many of these time interval measurements, postprocessing is applied to extract statistical parameters and jitter components. This technique has been used to good effect in ATE equipment to measure jitter of low-frequency clock (<100 MHz) or in some bench-top instrument implementations, such as Wavecrest, that include special software features for jitter component characterization.

23.11.4 ATE-Based Equipment

Using automatic tester equipment, a signal may be repeatedly acquired at slightly different time settings for the capture clock. The distribution of signal timing can be determined from the average number of successful acquisitions at each clock setting. Although using ATE equipment lowers the cost for those designers who have the test equipment, it does take some time to make multiple acquisitions of a given transition. What is more of a concern is that the accuracy and resolution are limited by the tester resolution, accuracy, and jitter. One plus point is that the processing load imposed on the tester's computer is minimal.

23.11.5 Real-Time Digital Sampling Oscilloscope

When using a real-time digital sampling oscilloscope to measure jitter, a signal is acquired by the oscilloscope's oversampling clock and transitions through a fixed voltage threshold that is determined by filtering and interpolation. The variation of the transitions with respect to a fixed clock is interpreted with special jitter software. Advanced real-time oscilloscopes are typically used to measure 1 to 5 psec jitter on a few critical signals/clocks for precise signal characterization in communications channels, optical fiber clocks, and predicting the BER. Making jitter measurements with an oscilloscope requires a high-speed digitizer on each of the instrument's channels, along with a high-speed memory system and possibly DSP hardware. The oscilloscope's system has the potential for very high throughput and accuracy because the signal is being continuously observed and the threshold transitions can be interpolated to very high precision, in part, because of the oscilloscope's multibit acquisition.

The drawback of this approach is that the processing load can be quite high and increases with the channel count, and so it may become impractical to realize a tolerable throughput when there are multiple channels involved. Moreover, the power dissipation in practical systems is of the order of 15 W per channel, so large channel counts may become problematic for this reason, too. Also with most oscilloscopes, when the full complement of channels is in use, the sampling acquisition is no longer continuous, and so the probability of capturing infrequent jitter faults drops off quickly.

23.11.6 Dedicated Jitter Instrumentation

Dedicated jitter instrument hardware is used to measure jitter directly and report the result as a number. A dedicated jitter measurement instrument is used, offloading the measurement burden from the ATE or oscilloscope. However, since jitter measurements are now made independent of general-purpose equipment, the overall test time is increased to accommodate these special measurements. Moreover, a method must be provided to switch the measurement system to the required channels. There are two main providers of this instrumentation: Wavecrest and Guidetech. Guidetech's hardware is based on time-interval analyzer (TIA) technology, whereas Wavecrest's hardware is based on counter–timer technology. TIAs create precise time steps of trigger events, and within limits (Guidetech's limit is 2 million triggers/sec) can measure the timing of every trigger event. At higher frequencies, however, TIAs can make precise measurements only on some fraction of all of the cycles received. Counter-based analyzers, on the other hand, make timing measurements by "stretching" cycles, for example, by ramping an analog integrator rapidly until a trigger event is detected, and then ramping the integrator slowly back down to zero, so that a relatively low-speed counter can measure the ramp-down time. This technique is allegedly slower than that used in TIAs, and, in the case of the Wavecrest boxes, limits the maximum frequency of signals that can be completely sampled (sampled in every cycle) to about 30,000 waveforms/sec.

23.11.7 BIST and DFT

All the methods discussed above rely on external measurement equipment to examine jitter. In contrast, the new BIST/DFT approaches can be used for SoC/ICs that exhibit jitter as low as 1 to 5 psec. Applications can include clock and data jitter, differential jitter, serial data communications, and measurements in a noisy environment.

23.11.7.1 DFT/ BIST Based on Oscillation Method

The PLL test circuit, DFT/BIST, which follows the oscillation methodology, is based on the Vernier principle transposed to the time domain. Due to a very high measurement resolution, it is practical to measure the frequency and the jitter of PLL circuits. As a Vernier caliper allows for a precise measurement of a linear distance between two points, the same principle applied in a time space allows for a precise measurement of time interval between two events. Instead of two linear scales having a very small difference in linear step, two oscillators with a very small difference in oscillation frequency are used for the PLL test circuit. The measurement resolution is that of the time difference between the periods of two oscillators. These oscillators have two essential characteristics:

1. When triggered, the oscillation will start virtually instantaneously and with a fixed phase relationship to the external trigger pulse.
2. When oscillating, the stability of the frequency is preserved.

The system uses phase-startable oscillators in both the start ($T1$) and stop ($T2$) channels. The stop oscillator has a slightly shorter period of oscillation than the reference (start) oscillator, such that once started, they will reach coincidence some number of cycles later, depending on the measured time interval. If the reference (start) oscillator period is $T0$, then the stop oscillator has period equal to $T0 - \Delta T$, where ΔT is the difference between the periods of both the oscillators. The measured time interval T is expressed as follows:

$$T = N1\,T1 - N2\;T2 = n(T1 - T2) = n\Delta T$$

where $N1$ is the number of start oscillator pulses to coincidence and equals n if the frequency difference is sufficiently small, and $N2$ the number of stop oscillator pulses to coincidence and equals n if the frequency difference is sufficiently small.

A practical limitation on the interpolation factor n is imposed by the inherent noise, and there is little advantage to be gained by having an interpolation factor capable of giving resolutions substantially better than this. The recently implemented [51] interpolation scheme provides better than 1 psec single-shot resolution. Other reported implementations (like that by Fluence [52]) give tens of picoseconds single-shot resolution.

The high-single-shot resolution allows the collection of meaningful statistical information. Important characteristics concerning time intervals like maximum, minimum, and standard deviation values are collected in addition to the mean value. With low-speed data collection, full histogram-building information is obtained. In many time interval measurement situations, these statistical properties are of prime importance. For example, in PLL jitter testing, the standard deviation gives an excellent indication of jitter.

In all practical implementations [48,53], the test circuit contains two matched phase-startable oscillators (see Figure 23.4). The coincidence is detected by a coincidence detector. The test circuit consists of two main parts: the functional circuit and the result processing circuit or software. The functional circuit is equivalent to about 2000 to 2500 gates, depending on the implementation. For result processing, different techniques can be adopted depending on the application requirements. It can be a hardware-implemented DSP technique (such as histogram building [54–56], used by Fluence) or custom software module [51]. The result processing is, in general, very fast, and can easily be executed in real time. The dedicated algorithms for fast-performance computation include: RMS and peak-to-peak jitter, instantaneous period, frequency, phase jitter, delay, and similar characteristics.

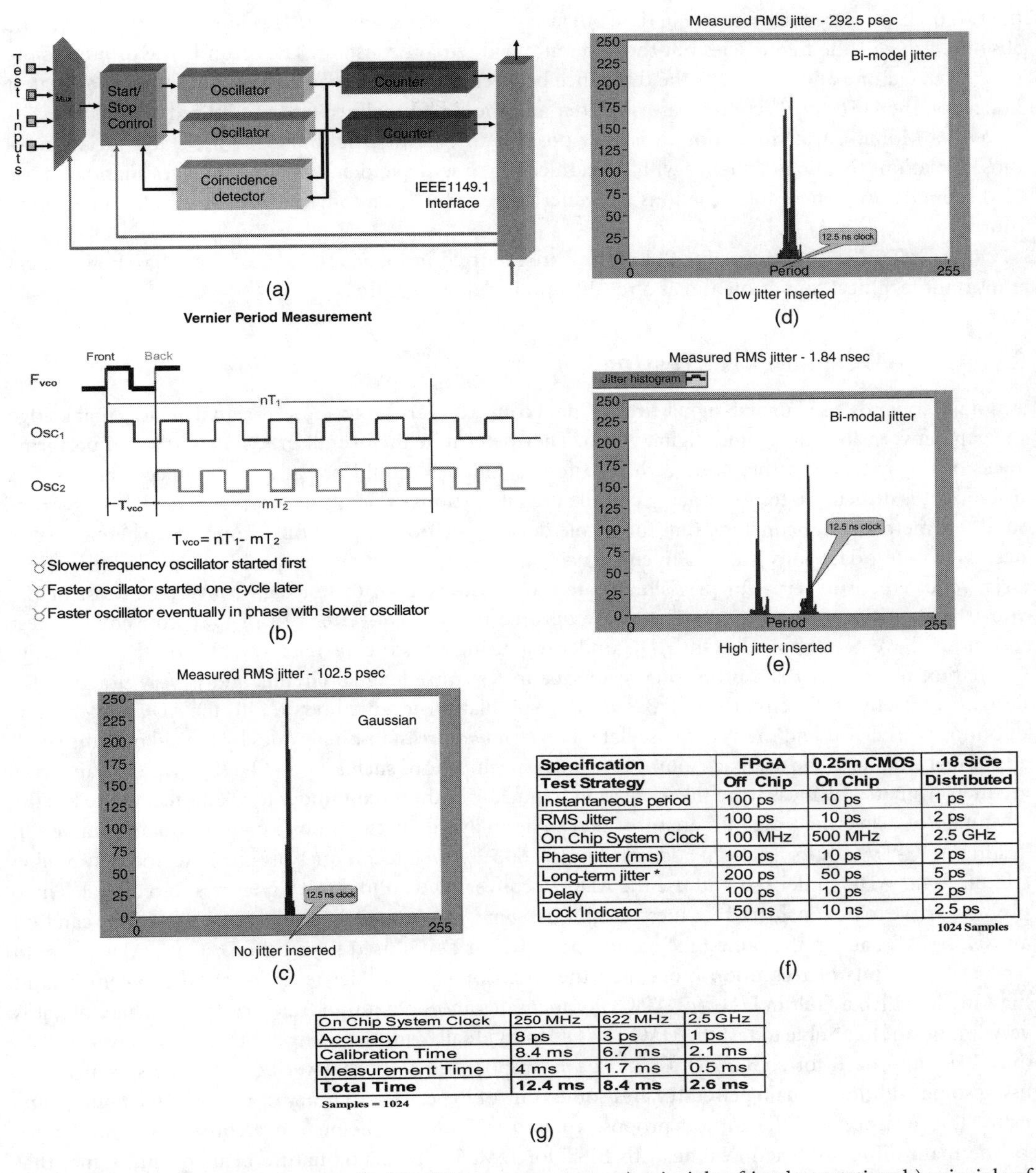

Specification	FPGA	0.25m CMOS	.18 SiGe
Test Strategy	Off Chip	On Chip	Distributed
Instantaneous period	100 ps	10 ps	1 ps
RMS Jitter	100 ps	10 ps	2 ps
On Chip System Clock	100 MHz	500 MHz	2.5 GHz
Phase jitter (rms)	100 ps	10 ps	2 ps
Long-term jitter *	200 ps	50 ps	5 ps
Delay	100 ps	10 ps	2 ps
Lock Indicator	50 ns	10 ns	2.5 ps

1024 Samples

(f)

On Chip System Clock	250 MHz	622 MHz	2.5 GHz
Accuracy	8 ps	3 ps	1 ps
Calibration Time	8.4 ms	6.7 ms	2.1 ms
Measurement Time	4 ms	1.7 ms	0.5 ms
Total Time	**12.4 ms**	**8.4 ms**	**2.6 ms**

Samples = 1024

(g)

FIGURE 23.4 Jitter measurements principle, Jitter measurements: a) principle of implementation; b) principle of operation; c, d, e) measurement examples, f) performance example for different implementations; g) example of final possible results as a function of implementation technology.

Besides its impressive jitter measurement capability, there are a multitude of benefits that flow from this DFT approach to PLL measurement. First, the dedicated test circuit based on the DFT approach requires only minor tester capabilities, which in turn lowers test cost development and equipment requirements. The test circuit can be placed directly on a low-cost tester [51] or can follow the integrated implementation. In the case of a fully integrated version, time-dependent process variations incurred during the fabrication of the SoC/IC can have an impact on the jitter of the oscillators used in the test circuitry. What should be noted is that the oscillator itself can inject jitter depending on process variations. Basically, there are two types of jitter that can occur in these particular situations. The first is the correlated jitter, where the two oscillation frequencies vary in time but always by the same amount. This jitter will only impact the absolute value of the frequencies but the difference will remain constant. The second type of jitter is the noncorrelated one and it impacts the difference between the two frequencies. The latest has the greatest impact on the DFT approach to measuring jitter and should be analyzed and minimized.

With a Monte-Carlo simulation, it is only possible to quantify the impact of the static variations—those variations that do not change with time. Injecting time-dependent variation involves an indirect use of the simulation, where the variations are generated externally and injected in the simulation as noise sources.

Other DFT/BIST techniques for PLL testing are reported in [5,29,53]. Each of them has, however, an important technical limitation and as a result, a limited applicability.

23.11.8 ADC and DAC Testing

Among frequently used mixed-signal circuits, data converters are typical mixed-signal devices that bridge the gap between the analog and digital world. They determine the overall precision and speed performances of the system and therefore, dedicated test techniques should not affect their specifications. For instance, it is difficult to test the analog and the digital portions of data converters separately using structural test methods and conclude that the whole device specifications are fully respected. Therefore, it is necessary to test data converters as an entity using at least some functional specifications. There is also a strict requirement in terms of precision related to hardware used to test data converters. Most of the efforts in on-chip testing data converters are devoted to ADC converters. In [57,58], conventional test techniques have been applied to the ADC under test, using a microcontroller available on the same chip. Oscillation-test strategy is a promising technique to test mixed-signal circuits and is very practical for designing effective BIST circuits [40]. Based on the oscillation-test method, in [40] the ADC under test is put into an oscillator and the system oscillates between two preestablished codes by the aid of some small additional circuits in the feedback loop. Functional specifications such as offset, DNL, INL, and gain error are then evaluated by measuring the oscillation frequency of the circuit under test. This technique has the advantage of delivering a digital signature that can be analyzed on-chip or by a conventional digital tester. Testing ADC–DAC pairs has been addressed in [59,60,72]. Such techniques use the DAC to apply analog stimuli to the ADC under test and use the ADC to convert DAC under test signatures into digital. Three problems have to be considered for these techniques. First, it is limited to applications, where one can find an ADC–DAC pair on the same IC. Second, the ADC (or DAC) used to test the DAC (or ADC) should have at least 2 bits of resolution more than the DAC (or ADC) under test. The third problem is fault masking in which a fault in DAC (or ADC) compensates another fault in ADC (or DAC). Therefore, it is very important to be able to test the DAC or ADC individually without using another data converter. The only BIST approach for solitary DACs has been proposed in [60], that verifies all static specifications using some additional analog circuitry and some control logic. The accuracy of the analog circuitry limits the test precision and the authors propose an auto-calibration scheme to overcome this limitation.

The main difficulty when dealing with BIST for DAC is the analog nature of its output signal that requires one to design high-resolution but still area-efficient analog signature analyzers. Oscillation-test strategy deals with this problem by establishing a closed-loop oscillation including the circuit under test, in which one does not have to apply analog input stimuli, and the test output is a pure digital signal.

Figure 23.5 shows the implementation of oscillation-test method to apply BIST to embedded data converters. Other interesting references include [47,61–64].

23.11.8.1 ADC Testing by Available ATE Systems

Testing A/M-S circuits often involves much more data and much more computation than is necessary for testing digital-only circuits. Traditional methods for A/M-S testing usually involve adding large suites of expensive analog instrumentation to the highest speed, highest pin count digital ATE system available. For testing an ADC, the most expensive instrument is usually the very precise, very low-noise floor stimulus source. In addition, there is the challenge of getting that stimulus signal to the input pin(s) of the

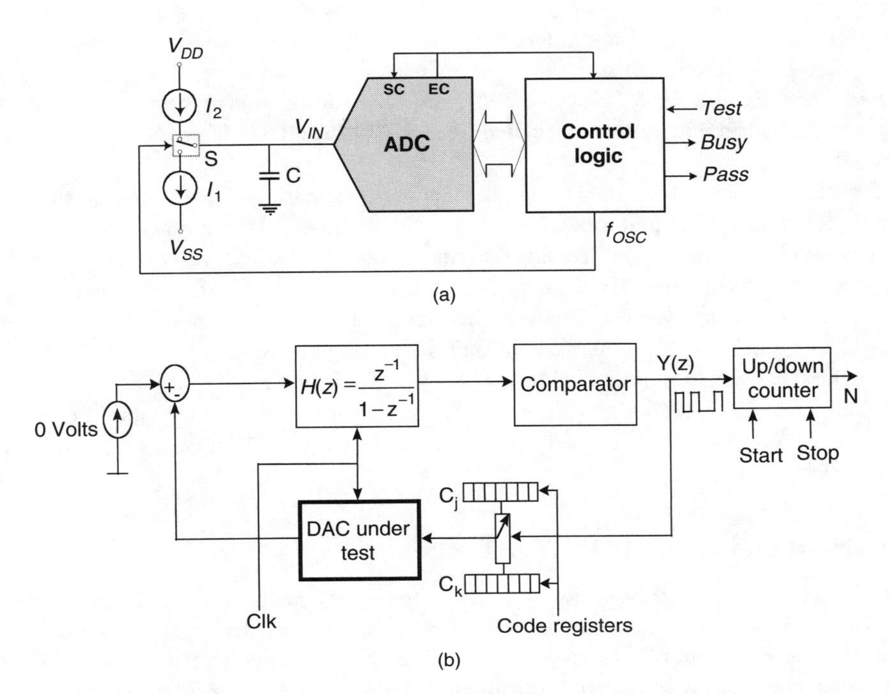

FIGURE 23.5 (a) Oscillation-test method used to apply BIST to ADCs; (b) Oscillation-test used to apply BIST to embedded DACs.

device under test, and the most time-consuming process is applying the requisite precision of input values to the ADC, while simultaneously capturing the digital output codes that must then be analyzed to determine the quality of the ADC portion of the design.

The input stimulus for ADC testing is most often either a ramp signal or a sine wave. The ramp signal is usually the preferred stimulus since it is easier to generate, using either a DAC or a specially designed counter driving analog voltage or current sources, rather than the sine wave. And if histogram-based techniques are used, the ramp signal usually results in a more even distribution of information on results. This has the advantage of reducing the data storage size and the resulting analysis time. The ramp signal stimulus is, however, unusable for AC-coupled designs. These designs require the constant movement typified by the sine wave signal, typically at a frequency higher than the ADC sampling rate, and need the use of either over-sampling or under-sampling techniques. Coherency between the analog and digital signals associated with the ADC is also required with a sine wave stimulus — a requirement that is not necessary when using a ramp signal.

Another issue associated with ADC testing is the type of testing that should be done. ADC test types are usually divided into two categories — static and dynamic. The so-called static tests include mainly INL and DNL, from which other test results such as gain, offset, and missing codes can be determined. The so-called dynamic tests include, for example, directly measured (as compared to computed) SNR, signal including noise and distortion (SINAD), and total harmonic distortion (THD). However, these parameters are only obtainable by computing results using fast Fourier transform (FFT) techniques.

It is also interesting to note that there is currently much debate over the advisability and fault coverage of "structural" testing techniques as compared to traditional functional performance of the testing methods. The static ADC tests — INL, DNL, offset, gain, and missing codes — will identify structural defects. Testing for "performance" defects requires using actual SNR, SINAD, and THD measurement techniques, or the use of sophisticated software routines to calculate these values from the static test results. So, in addition to the philosophical differences in the two testing strategies, there are also instrument and computation time trade-offs to be taken into consideration.

Linearity— INL, DNL: Integral and differential nonlinearity are the two basic building blocks for ADC testing. They can be used to detect structural defects in a chip.

Noise and distortion — SNR, SINAD, and THD: Noise and distortion can be measured directly using sophisticated ATE instrumentation or calculated from the INL and DNL figures while taking clock jitter into account.

Sophisticated instrumentation is used to supply the stimulus, and the ATE digital pin electronics are used to capture the resulting ADC output codes. These codes are then analyzed using either direct bit-by-bit, clock cycle-by-clock cycle comparisons, or by reading the acquired digital states and analyzing them with digital signal processing (DSP) techniques. The DSP approach is often much faster than the discrete comparison approach and can be used to yield qualitative information using tolerance values. The direct comparison approach will only provide go/no go information and is very unforgiving.

The histogram-based testing described below can also be applied to ADC testing using BIST. A specially designed linear feedback shift register (LFSR) is used to create the histogram generator whose results are stored in memory and accessed via the IEEE 1149.1 (JTAG) port that is often already present in the SoC design.

23.11.9 Histogram-Based DfT/BIST

Histogram-based methods have been used for analog testing for many years. In Tektronix PTS101 test system, a histogram is used to extract information about the code density in the output of A to D converter using a sine wave input. Reference [55] describes the use of the code density histogram-based method for distortion and gain tracking measurements on an A to D converter. Histograms have also been used to make quantitative measurements in digital systems, e.g., for performance analysis, a histogram built in real time gives an accurate picture of the way a digital computer spends its time [55].

In contrast to these informal test methods that somehow use histograms, an efficient integrated version based on histograms has been developed and commercialized [50,52,54],, as HBIST. The HBIST permits use of the low-impact applications including ADCs, DACs, and PLLs. Employing HBIST results in digital test results which can be analyzed for such parameters as INL and DNL, gain and offset errors, effective least-significant bit (LSB), and clipping and modulation distortion.

During an HBIST implementation, essential information about the signal under test is converted to a histogram, from which characteristics can be studied to gain valuable information about circuit performance. The sample-and-hold circuit and ADC perform the conversion from analog domain to the digital domain (Figure 23.6). Once the data from the ADC are fed to the histogram generator, the results can be downloaded and read by a digital ATE system. The technique uses under-sampling of the analog signal(s) under test to quantify how long the signal remains at each amplitude level, placing the values in various bins in a histogram (Figure 23.7). The histogram characterizes the waveform of the signal under test, capturing its essential elements.

Using software simulation tools, an ideal histogram for each signal under test can be created, as can histograms for signals due to certain defects, like stuck bits and various types of nonlinearity and distortion. These signatures for various types of faulty circuit behavior can be stored for use in determining the pass/fail status of analog circuits under test during production testing. Should the signal under test vary from the expected signal, the histogram normally undergoes significant changes (Figure 23.8). The clipped sine wave shown does not spend nearly as much time at the high and low boundaries. Therefore, the resulting histogram has fewer entries in the outside bins and many more entries in the bins adjacent to them. Subtracting the acquired histogram from the ideal histogram creates a difference histogram that can be analyzed to determine which defects are present in the circuit under test.

In addition, the histogram-based method can be deployed to test the ADC that is part of the circuit itself when a proper stimulus signal, usually a ramp, is applied to its input. The ramp signal can be supplied either by an external signal generator or by a DAC that might already be present in the design. Multiplexers at the DAC inputs and output allow it to be employed not only for on-chip functional purposes, but also

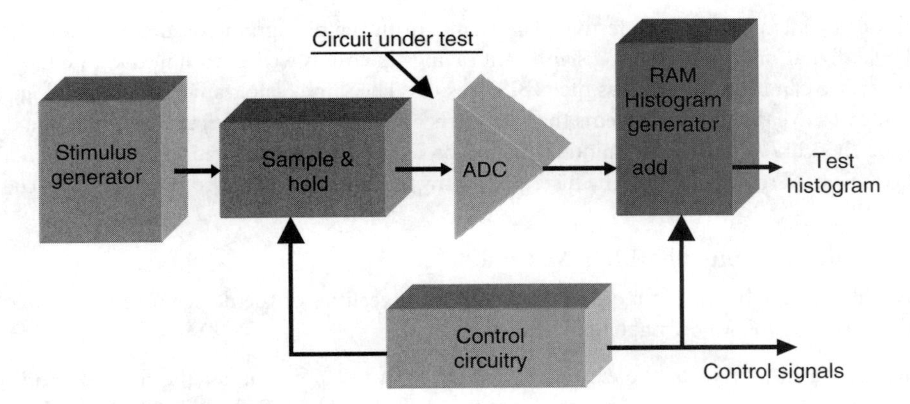

FIGURE 23.6 A typical HBIST configuration for analog circuit testing.

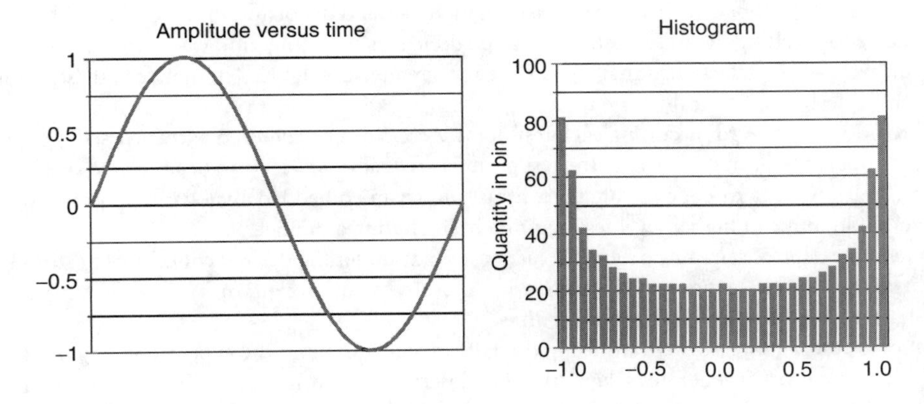

FIGURE 23.7 An example of histogram building from a sine wave.

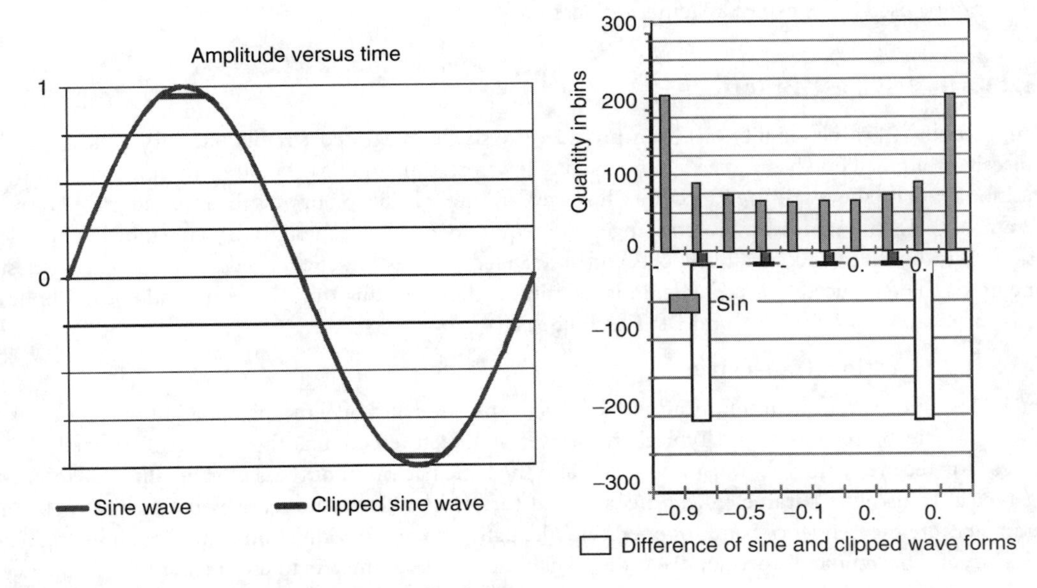

FIGURE 23.8 A difference histogram when clipping occurs.

as the stimulus generator for ADC testing. The results from the histogram-based BIST circuitry can be accessed by a digital-only ATE, offering significant savings in cost as well as complexity. The IEEE 1149.1 test bus interface can be used to access the HBIST results. The same histogram-based technology applies to DACs. It can solve the many problems that designers encounter when DAC is embedded in a complex SoC design. The PLL/VCO BIST technique based on oscillators, as described above, can also benefit from the histogram-based result analysis. The histogram infrastructure can be shared on an SoC for converters and PLL testing.

23.11.9.1 The Elements of HBIST Method

HBIST is highly adaptable to test requirements and to the realities of the device under test. A complete summary is given by the following items:

1. Generate a histogram of the expected signal at each test point under the desired conditions of stimulus. The histograms can be generated theoretically — e.g., from a SPICE simulation, or they can be experimentally derived, as from testing a "golden device." Stimulus can be supplied from external or built-in sources.
2. Determine the range of variance of offset, gain, noise, and distortion that are acceptable in the signature for each signal under test. This can be done by simulating limit cases with the limits of circuit parameters selected so that signals are marginally acceptable. Alternatively, these parameters can be derived empirically.
3. Provide a method for accessing each test point. Access can be achieved with: an oscilloscope probe or input cable — in cases where the test point is accessible; an electron beam probe such as a scanning electron microscope in voltage-contrast mode; an embedded analog buffer, probe, or sample and hold; built-in higher level test equipment, including a digitizer.
4. Generate a histogram of the signal at each test point under the desired conditions of stimulus. The histogram can be generated by built-in high-level test equipment, or by sending a replica of the signal at the test point to external facilities.
5. Process the acquired histogram using the HBIST algorithm and the expected histogram as a template. This process generates a signature that describes the difference between the histogram of the signal at the test point and its template.
6. Use the HBIST signature as both a go/no go test criterion and as a basis for the diagnosis of the cause of test failure. The part of the signature that results in test failure contains information that can be used by an expert system for diagnosis.

23.11.10 RF Test Practices

Until recently, RF functionality has been provided by single integrated circuits like mixers, PLLs, and transceivers. In fact, one has succeeded in reaching low cost, time to market, and low ppm, owing to the traditional test methods. Future RF circuits, however, are either in silicon or on substrate-integrated (SoC, SiP, and MCM), and represent new challenges and requirements. Since these RF IPs are embedded in the system, it is difficult to access all RF ports and as such, current test practices need to be revised. The test time needs to be reduced to acceptable limits within the digital testing time domains, and it also implies the incorporation of DfT, BIST, and DfFT techniques.

23.11.10.1 Testing Transceiver

The inclusion of wireless communication facilities in the SoCs and SiPs means the addition of a transceiver that can be seen as an IP. Typical transceiver architecture contains three predefined partitions, namely, the receiver path, the transmitter path, and the logic part. Without loss of generality, the following typical transceiver parameters can be specified for testing: (1) at the receiver path some of these parameters are the frequency bands of operation, the path gain, intermodulation distortion, noise figure, blocking, and harmonic distortion; (2) at the transmitter path parameters to consider are the frequency bands, transmitted power, automatic gain control, and the frequency spectrum; (3) the logic part deals with issues like VCO tuning, sensitivity, and leakage. The RF architecture is similar for different products, but small variations that are product-specific, imply a dedicated set of test parameters, which need to be included into test program. Very often, a tester is unable to measure some critical parameters.

The typical test strategies include block-by-block test and system-level test. Characterizing an isolated block in a whole transmitter is rather laborious since it involves proper loading conditions from one building block to another, and the requirement of special DfT to place isolation switches. In a system test strategy the transceiver is tested as used by end user, and also the product is specified in terms of communication standards. The preferred system test is based on BER measurements. A BER test compares a string of bits coming out of the device under test, against its corresponding golden sequence. BER is actually the ratio of the total number of errors to the total number of bits checked. Typically, a standard limit is one error for every 1000 demodulated bits. BER tests are applied as modulated and control signals to the device under test, such that it is possible to test parameters such as sensitivity, blocking signals, and other sorts of channel interference.

During the test development process, the important phase comprises the definition of test modes and failure conditions, followed by the required test parameter set. The appropriate DfT and the possible BIST can be decided based on the test requirements and test resource availability. External equipment often imposes the limitations on achievable performance of test. Reference [65] describes a number of the best RF test practices that are currently in use.

23.11.10.2 Transceiver Loop—Back Technique

Few methods in the literature address DfT and BIST for RF circuits. These techniques are based on the basic idea of loop-back in order to reuse the transmitter or receiver section [49,66–68,71,73] in a transceiver. The output of a system is routed back to its input directly, without using the wireless link.

In the case of integrated transceivers in the SoC/SiP environment, the loop-back test technique can reuse the DSP and memory resources already available in the system for the test response analyzer and signal generation. This approach may be able to reach the lowest test cost possible, if only switches and attenuators are needed. This may be a very important quality, mainly for systems operating at frequencies in the order of gigahertz. Another advantage is the lower effort needed in order to implement the test, and high flexibility, as the test is implemented in software and does not depend on the technology of the transceiver.

There are some disadvantages of the loop-back technique, mainly related to the fact that the complete transceiver is tested as a whole. This way, faults in the transmitter could be masked by the receiver, reducing fault coverage and ruling out fault diagnosis. There are some attempts to increase observability and controllability of the signal path [69]. The transmitter portion of the transceiver is used to create modulated signals, testing sensitivity, and several other RF tests. Further improvement can be found in [3] as shown in Figure 23.9. The masking effects are eliminated and high-test performance is achieved. A sample test result is shown in Figure 23.10.

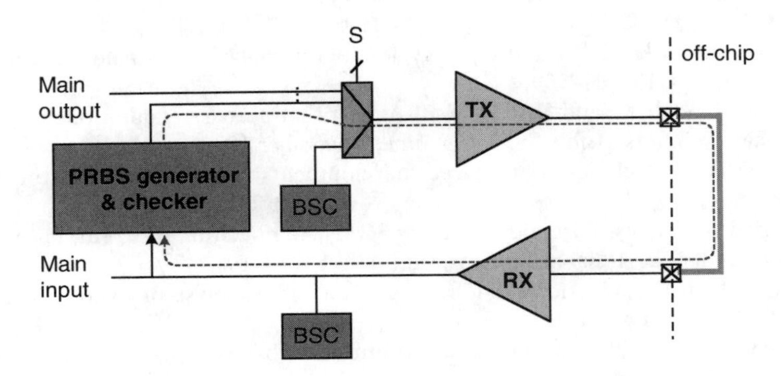

FIGURE 23.9 Enhanced loop-back technique (BSC - Boundary Scan Cell).

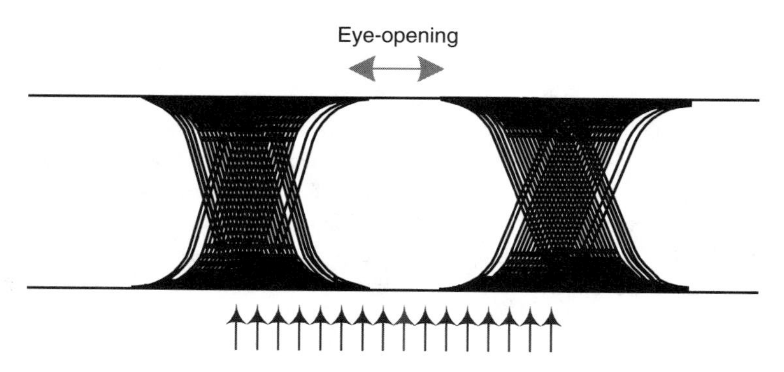

FIGURE 23.10 Sampling points in test mode to measure eye-opening.

It can be noted that the enhanced loop-back technique represents a functional BIST solution aimed at enabling current tester platforms to test high-speed interfaces at-speed, with increased functional test coverage. As presented in [3], the tests include: (1) supports external and internal loop-back from transmitter to receiver; (2) pseudo random bit sequence (PRBS) generator to generate test pattern; (3) ability to inject jitter and phase hit at the transmitter; (4) ability to check for correct incoming data at the receiver and count BER; (5) ability to measure eye-opening at the receiver; (6) ability to measure receiver sensitivity; (7) ability to uncover the fault masking effect between the receiver and the transmitter; and (8) offer embedded boundary scan cell (BSC) to reduce their impact on performance.

23.12 Summary

A short overview of A/M-S testing and practical DFT techniques has been presented. The practical aspects of such typical blocks as PLL, ADC, and DAC have been highlighted by reviewing testing by traditional and DfT approaches.

References

[1] *Proceedings IEEE Mixed Signal Testing Workshops*, 1994–2005.

[2] B. Vinnakota, Ed., *Analog and Mixed Signal Test*, Prentice-Hall, Englewood Cliffs, NJ, 1998.

[3] Kaminska, B. and Arabi, K., Mixed-signal testing — a concise overview, tutorial, *IEEE ICCAD 2003*, Santa Clara, November 2004.

[4] Abderrahman, A., Cerny, E., and Kaminska, B., Worst-case tolerance analysis and CLP-based multifrequency test generation for analog circuits, *IEEE Trans. CAD*, 18, 332–345, 1999.

[5] Burns, M. and Roberts, G., *An Introduction to Mixed Signal IC Test and Measurement*, Oxford University Press, Oxford, UK, 2001.

[6] Slamani, M. and Kaminska, B., Soft large deviation and hard fault multifrequency analysis in analog circuits, *IEEE Des. Test Comput.*, 12, 70–80, 1995.

[7] Slamani, M., Kaminska, B., Multifrequency testability analysis for analog circuits, *IEEE Trans. Circuits Syst., Part II*, 13, 134–139, 1996.

[8] Ben-Hamida, N., Saab, K., and Kaminska, B., A perturbation-based fault modeling and simulation for mixed-signal circuits, *Asian Test Symposium*, November 1997.

[9] Roberts, G., Metrics, techniques and recent developments in mixed-signal testing, *IEEE ICCAD*, November 1996.

[10] IEEE Standard Test Access Port and Boundary SCAN Architecture, IEEE Std. 1149.1-2001.

[11] IEEE Standard for Mixed-Signal Test Bus, IEEE Std. 1149.4-1999.

[12] Hemnek, G. J., Meijer, B. W., and Kerkhoff, H., G., Testability analysis of analog systems, *IEEE Trans. CAD*, 9, 573–583, 1990.

[13] Abderrahman, A., Cerny, E., Kaminska, B., Optimization-based multifrequency test generation for analog circuits, *J. Electron. Testing: Theory Appl.*, 13, 59–73, 1996.

[14] Abderrahman A., Cerny, E., and Kaminska, B., CLP-based multifrequency test generation for analog circuits, *IEEE VLSI Test Symposium*, April 1997.
[15] Slamani, M. and Kaminska, B., Fault diagnosis of analog circuits based on sensitivity computation and functional testing, *IEEE Des. Test Comput.*, 9, 30–39, 1992.
[16] Slamani, M., Kaminska, B., and Quesnel, G., An integrated approach for analog circuit testing with a minimum number of detected parameters, *IEEE International Test Conference*, Washington, DC, October 1994, pp. 631–640.
[17] Huynh, S. Dd. et al., Automatic analog test signal generation using multifrequency analysis, *IEEE Trans. Circuits Syst. II*, 46, 565–576, 1999.
[18] Balivada, A. Zheng, H., Nagi, N., Chatterjee, A., and Abraham, J., A unified approach for fault simulation of linear mixed-signal circuits, *J. Electron. Testing: Theory Appl.*, 9, 29–41, 1996.
[19] Harvey, J. A. et al., Analog fault simulation based on layout-dependent fault models, *IEEE ITC*, 1995, pp. 641–649.
[20] Ben Hamida, N., Saab, K., Marche, D., and Kaminska, B., LimSoft: automated tool for design and test integration of analog circuits, *IEEE International Test Conference*, Washington, DC, October 1996, pp. 571–580.
[21] Saab, K., Ben Hamida, N., Marche, D., and Kaminska, B., LIMSoft: automated tool for sensitivity analysis and test vector generation, *IEEE Proc. Circuits, Devices Syst.*, 143, 1998, pp. 118–124.
[22] Saab, K., Benhamida, N., Kaminska, B., Closing the gap between analog and digital testing, *IEEE Trans. CAD*, 20, 307–314, 2001.
[23] Saab, K. and Kaminska, B., Method for Parallel Analog and Digital Circuit Fault Simulation and Test Set Specification, U.S. Patent No. 09/380,386, June 2002.
[24] Millor, L. et al., Detection of catastrophic faults in analog IC, *IEEE Trans. CAD*, 8, 114–130, 1989.
[25] Tsai, S.J., Test vector generation for linear analog devices, *IEEE ITC*, 1990, pp. 592–597.
[26] Ben Hamida, N. and Kaminska, B., Analog circuits testing based on sensitivity computation and new circuit modeling, *IEEE International Test Conference*, 1993, pp. 652–661.
[27] Devarayandurg, G. and Soma, M., Analytical fault modeling and static test generation for analog IC, *IEEE ICCAD*, 1994, pp. 44–47.
[28] Director, S.W. and Rohrer, A., The generalized adjoint network and network sensitivities, *IEEE Trans. Circuit Theory*, CT-16, 318–323, 1969.
[29] Sunter, S. and Roy, A., BIST for PLL SIN digital applications, *IEEE International Test Conference*, September 1999, pp. 532–540.
[30] Seongwon, K., and Soma, M., An all-digital BIST for high-speed PLL, *IEEE Trans. Circuits Syst. II*, 48, 141–150, 2001.
[31] Sunter, S., The P1149.4 mixed-signal test bus: cost and benefits, *Proceedings of the IEEE ITC*, 1995, pp. 444–45.
[32] Arabi, K., and Kaminska, B., Method of Dynamic On-Chip Digital Integrated Circuit Testing, U.S. Patent No. 6,223,314, B1, April 2001.
[33] Arabi, K., Kaminska, B., Oscillation–Based Test Method for Testing an at least Partially Analog Circuit, U.S. Patent No. 6,005,407, December 1999.
[34] Arabi, K., and Kaminska, B., A new BIST scheme dedicated to digital-to-analog and analog-to-digital converters, *IEEE Des. Test Comput.*, 13, 37–42, 1996.
[35] Arabi, K., and Kaminska, B., Oscillation-test strategy for analog and mixed-signal circuits, *IEEE VLSI Test Symposium*, Princeton, 1996, pp. 476–482.
[36] Arabi, K., and Kaminska, B., Design for testability of integrated operational amplifiers using oscillation-test strategy, *IEEE International Conference on Computer Design (ICCD)*, Austin, October 1996, pp. 40–45.
[37] Arabi, K., and Kaminska, B., Testing analog and mixed-signal integrated circuits using oscillation-test method, *IEEE Trans. CAD IC&S*, 16, 745–753, 1997.
[38] Arabi, K. and Kaminska, B., Oscillation built-in self-test scheme for functional and structural testing of analog and mixed-signal integrated circuits, *IEEE International Test Conference*, November 1997.
[39] Arabi, K., and Kaminska, B., Parametric and catastrophic fault coverage of oscillation built-in self-test, *IEEE VLSI Test Symposium*, April 1997.
[40] Arabi, K., and Kaminska, B., Efficient and accurate testing of analog-to-digital converters using oscillation-test method, *European Design & Test Conference*, Paris, France, March 1997.

[41] Arabi, K., and Kaminska, B., Design for testability of embedded integrated operational amplifiers, *IEEE J. Solid-State Circuits*, , 33, 573–581, 1998.
[42] Arabi, K., and Kaminska, B., Integrated temperature sensors for on-line thermal monitoring of microelectronic structure, *J. Electron. Testing, JETTA*, 12,81–92, 1998.
[43] Arabi, K. and Kaminska, B., Oscillation built-in self-test of mixed-signal IC with temperature and current monitoring, *JETTA Special Issue on On-Line Testing*, 12, 93–100, 1998.
[44] Arabi, K. and Kaminska, B., Oscillation test methodology for low-cost testing of active analog filters, *IEEE Trans. Instrum. Meas.*, 48, 798–806, 1999.
[45] Arabi, B., His, H., and Kaminska, B., Dynamic digital integrated circuit testing using oscillation test method, *IEE Electron. Lett.*, 34, 762–764, 1998.
[46] Arabi, K., Hs, H., Dufaza, C., and Kaminska, B., Digital oscillation test method for delay and structural testing of digital circuits, *IEEE International Test Conference*, October 1998, pp. 91–100.
[47] Norworthy, R., Schreier, R., and Temes, G.C., *Delta–Sigma Data Converters: Theory, Design, and Simulation*, IEEE Press, Washington, DC, 1997.
[48] Tabatabaei, S. and Ivanov, A., An embedded core for subpicoseconds timing measurements, *IEEE ITC*, , October 2002, pp. 129–137.
[49] Kossel, M. and Schmatz, M.L., Jitter measurements of high-speed serial links, *IEEE Des. Test Comput.*, 536–543, 2004..
[50] Frisch, A. and Rinderknecht, T., Jitter Measurement System and Method, U.S. Patent No. 6,295,315, Fluence Technologies, September 2001.
[51] Kaminska, B. and Sokolowska, E., Floating infrastructure IP, *IEEE ITC Test Week*, 2003.
[52] OPMAXX, Fluence, Credence Corporation — Technical Notes, 1997–2002.
[53] Ong, C. et al., A scalable on-chip jitter extraction technique, *IEEE VTS*, , 2004, pp. 267–272.
[54] Frisch, A. and Almy, T., HBIST: histogram-based analog BIST, *IEEE ITC*, November 1997.
[55] Kerman et al., Hardware histogram processor, *COMPCON'1979*, p. 79.
[56] Crosby, P., Data Occurrence Frequency Analyzer, U.S. Patent No. 5,097,428, 1989.
[57] Browing, C., Testing A/D converter on microprocessors, *IEEE ITC*, 1985, pp. 818–824.
[58] Bobba, R. et al., Fast embedded A/D converter testing using the microcontrollers' resources, *IEEE ITC*, 1990, pp. 598–604.
[59] Arabi, K., Kaminska, B., and Rzeszut, J., BIST for DAC and ADC, *IEEE Des.Test Comput.s*, 13, 40–49, 1996.
[60] Arabi, K., Kaminska, B., and Rzeszut, J., A new BIST approach for medium to high-resolution DAC, *The 3rd Asian Test Symposium*, Nara, Japan, November 1994.
[61] Teraoca, E. et al., A built-in self test for ADC and DAC in a single-chip speech Codec, *IEEE ITC*, 1997, pp. 791–796.
[62] Aziz, P.M. et al., An overview of sigma–delta converters, *IEEE Signal Processing Magazine*, January 1996, pp. 61–84.
[63] Kenney, J.G. and Carley, L.R., Design of multi-bit noise-shaping data converters, *J. Analog Integrated Circuits Signal Process.*, 3, 259–272, 1993.
[64] Vazquez, D. et al., On-chip evaluation of oscillation-based test output signals for switched capacitor circuits, *Analog IC Signal Process.*, 33, 201–211, 2002.
[65] Pineda, J. et al., RF test best practices, *RF Test Workshop, DATE 2004*, Paris, 2004, pp. 43–49.
[66] Akbay, S. and Chatterjee, A., Feature extraction-based built-in alternate test of RF components using a noise reference, *IEEE VTS*, April 2004, pp. 273–278.
[67] Jarwala, M. et al., End-to-end strategy for wireless systems, *IEEE ITC*, 1995, pp. 940-946,
[68] Ferrario, J. et al., Architecting millisecond test solution for wireless phone RFIC's, *IEEE ITC*, 2003, pp.1325–1332.
[69] Force, C., Reducing the cost of high-frequency test, *RF Test Workshop, DATE 2004*, Paris, 2004, pp. 36–39.
[70] Kaminska, B. et al., Analog and mixed-signal benchmark circuits — First release, *IEEE International Test Conference*, November 1997.
[71] Kaminska, B., Multi-gigahertz electronic components: practicality of CAD tools, *DATE 2004 —— W3*, Paris, March 2004.
[72] Arabi, K., Kaminska, B., and Rzeszut, J., A new built-in self-test approach for digital-to-analog and analog-to-digital converters, *IEEE/ACM International Conference on CAD*, San Jose, CA, November 1994, pp. 491–494.
[73] www.logicvision.com.

Index

A

D

E

I

J

K

L

M

N

O

R

S

T

X

Z